M000288690

Modern Perspectives in Lattice QCD

Quantum Field Theory and High Performance Computing

École de Physique des Houches
Session XCIII, 3–28 August 2009

Modern Perspectives in Lattice QCD

Quantum Field Theory and High Performance Computing

Edited by

Laurent Lellouch, Rainer Sommer, Benjamin Svetitsky,
Anastassios Vladikas, Leticia F. Cugliandolo

OXFORD
UNIVERSITY PRESS

OXFORD
UNIVERSITY PRESS

Great Clarendon Street, Oxford OX2 6DP

Oxford University Press is a department of the University of Oxford.
It furthers the University's objective of excellence in research, scholarship,
and education by publishing worldwide in

Oxford New York

Auckland Cape Town Dar es Salaam Hong Kong Karachi
Kuala Lumpur Madrid Melbourne Mexico City Nairobi
New Delhi Shanghai Taipei Toronto

With offices in

Argentina Austria Brazil Chile Czech Republic France Greece
Guatemala Hungary Italy Japan Poland Portugal Singapore
South Korea Switzerland Thailand Turkey Ukraine Vietnam

Oxford is a registered trade mark of Oxford University Press
in the UK and in certain other countries

Published in the United States
by Oxford University Press Inc., New York

© Oxford University Press 2011

The moral rights of the authors have been asserted
Database right Oxford University Press (maker)

First published 2011

All rights reserved. No part of this publication may be reproduced,
stored in a retrieval system, or transmitted, in any form or by any means,
without the prior permission in writing of Oxford University Press,
or as expressly permitted by law, or under terms agreed with the appropriate
reprographics rights organization. Enquiries concerning reproduction
outside the scope of the above should be sent to the Rights Department,
Oxford University Press, at the address above

You must not circulate this book in any other binding or cover
and you must impose the same condition on any acquirer

British Library Cataloguing in Publication Data

Data available

Library of Congress Cataloging in Publication Data

Data available

Typeset by SPI Publisher Services, Pondicherry, India
Printed in Great Britain
on acid-free paper by
CPI Antony Rowe, Chippenham, Wiltshire

ISBN 978–0–19–969160–9

1 3 5 7 9 10 8 6 4 2

École de Physique des Houches
Service inter-universitaire commun à l'Université Joseph Fourier de Grenoble et à
l'Institut National Polytechnique de Grenoble

Subventionné par l'Université Joseph Fourier de Grenoble,
le Centre National de la Recherche Scientifique, le
Commissariat à l'Énergie Atomique

Directeur:
Leticia Cugliandolo, Université Pierre et Marie Curie – Paris VI, France

Directeurs scientifiques de la session XCIII:
Laurent Lellouch, CPT-Marseille, CNRS, Marseille, France
Rainer Sommer, NIC, DESY, Zeuthen, Germany
Benjamin Svetitsky, Tel Aviv University, Tel Aviv, Israel
Anastassios Vladikas, INFN-Tor Vergata, Rome, Italy
Leticia Cugliandolo, LPTHE, Université Paris VI, Paris, France

Previous sessions

Publishers

- Session VIII: Dunod, Wiley, Methuen
- Sessions IX and X: Herman, Wiley
- Session XI: Gordon and Breach, Presses Universitaires
- Sessions XII–XXV: Gordon and Breach
- Sessions XXVI–LXVIII: North Holland
- Session LXIX–LXXVIII: EDP Sciences, Springer
- Session LXXIX–LXXXVIII: Elsevier
- Session LXXXIX– : Oxford University Press

Preface

The next generation of lattice field theorists will finally resolve fundamental, non-perturbative questions about QCD without uncontrolled approximations. The aim of the XCIII session of the École de Physique des Houches, held in August 2009, was to familiarize this generation with the principles and methods of modern lattice field theory.

The emphasis of the school was on the theoretical developments that have shaped the field in the last two decades and that have turned lattice gauge theory into a robust approach to the determination of low-energy hadronic quantities and of fundamental parameters of the Standard Model. Nowadays, we master light sea-quark masses that are ever closer to their physical values. Algorithmic developments and new computational strategies enabling these calculations were covered at review level. Finally, in the "extreme" mass regimes—close to the chiral limit, and out to the bottom quark mass—lattice calculations must be combined with the techniques of effective field theory. This, too, was part of our curriculum.

Many younger researchers (and not only they) are not properly trained in the fundamentals, as they are overwhelmed by the time-consuming work of running heavy-duty computations. The many lattice conferences and workshops are understandably centered on the presentation of ongoing numerical work. While several excellent textbooks have been published recently, we feel it right to offer this compendium of extensive lecture notes and reviews to serve both as an introduction and as an up-to-date review for the advanced student.

By way of introduction, the courses of the school began by covering lattice theory basics (P. Hernández), lattice renormalization and improvement (P. Weisz and A. Vladikas), and the many faces of chirality (D. B. Kaplan). A later course introduced QCD at finite temperature and density (O. Philipsen). The aim was to organize the students' knowledge of the general notions of the field before going on to more specialized subjects.

Computational techniques have always developed in parallel with theoretical approaches, driving the successes of the field and being driven by them. A broad view of lattice computation from the basics to recent developments was offered in the corresponding course (M. Lüscher). The students learned the basics of lattice computation in a hands-on tutorial (S. Schaefer)—a first at Les Houches! The tutorial course received enthusiastic reviews from the students. We hope that readers of this book will try the tutorial out for themselves—the software is available at the web site of the school, http://nic.desy.de/leshouches.

One of the most important recent advances in the field is the possibility of approaching the chiral limit. This is a result of better theoretical control of chiral symmetry in discretized QCD as well as of better computational methods. Chiral perturbation theory is used to extrapolate to physical quark masses; at the same time, it offers a framework for parameterizing the low-energy physics by means of effective coupling constants. These themes were covered in the course on chiral perturbation theory (M. Golterman).

The remainder of the courses highlighted specific fields where the impact of QCD is growing quickly. The ongoing successes of B factories and the prospect of the coming LHC experiments have focused attention on quark-flavor changing weak transitions. Their interpretation requires non-perturbative matrix elements in QCD with heavy flavors. A course in heavy-quark effective theories (R. Sommer), an essential tool for performing the relevant lattice calculations, covered HQET from its basics to recent advances.

A number of shorter courses rounded out the school and broadened its purview. These included recent applications to the nucleon–nucleon interaction (S. Aoki) and flavor phenomenology and CP violation (L. Lellouch). The course on physics beyond the Standard Model (T. Appelquist and E.T. Neil) showed how the methods of lattice QCD are being applied to other gauge theories with very different physics.

The planning of the school's scientific programme was meticulous. In the early stages of its preparation, we were privileged to be advised and encouraged by the International Advisory Committee, whose members were

- Norman Christ (Columbia University, New York, USA);
- Martin Lüscher (CERN, Geneva, Switzerland);
- Tetsuya Onogi (Kyoto University, Kyoto, Japan);
- Giancarlo Rossi (University of Rome Tor Vergata, Rome, Italy).

The importance of their help cannot be overemphasized.

The organizers of this Les Houches session happily acknowledge the generous financial support of CNRS, DESY, INFN, Forschungszentrum Jülich, and the Lawrence Livermore National Laboratory. We also acknowledge the significant financial contribution of the following nodes of "Flavianet," an FP7 European Network: Spain (Valencia), Switzerland (Bern), Germany-North, UK, Italy, and France.

The organizers wish to express their gratitude to the Les Houches administrative staff—M. Gardette, I. Lelievre, B. Rousset, and T. Rousset—for their invaluable assistance throughout the organization of this session. Our thanks also go to the staff members of Les Houches for solving all kinds of practical problems and, especially, for transforming our respites from work into relaxing and edifying culinary experiences. We thank K. Cichy, S. Meinel, and E. Yurkovsky, as well as O. Philipsen, for providing photographs of the school. Of course we thank the lecturers for their great effort and enthusiasm in preparing and giving the long and detailed courses. Finally, we appreciate that we were very fortunate in having an excellent group of students, whose lively curiosity and healthy skepticism were essential to making the school a success.

Marseilles, Paris, Rome, Tel Aviv, and Zeuthen, August 2010,

L. Lellouch
R. Sommer
B. Svetitsky
A. Vladikas
L. F. Cugliandolo

Contents

List of participants

ORGANIZERS

LELLOUCH LAURENT
CPT Marseille, Case 907, CNRS Luminy, F-13288 Marseille, Cedex 9, France

SOMMER RAINER
NIC, DESY, Platanenallee 6, 15738 Zeuthen, Germany

SVETITSKY BENJAMIN
School of Physics and Astronomy, Tel Aviv University, 69978 Tel Aviv, Israël

VLADIKAS ANASTASSIOS
INFN, c/o Department of Physics, Universtiy of Rome "Tor Vergata", via della Ricerca Scientifica 1, I-00133 Rome, Italy

CUGLIANDOLO LETICIA
LPTHE, Tour 24-5ème étage, 4 place Jussieu, F-75232 Paris Cedex 05, France

LECTURERS

AOKI SINYA
University of Tsukuba, Graduate School of Pure and Applied Sciences, Ten'nodai 1-1-1, Tsukuba, Ibaraki 305-8571, Japan

APPELQUIST THOMAS AND NEIL ETHAN
Yale University, SPL 44, 217 Prospect Street, New Haven, CT 06511-3712, USA

GOLTERMAN MAARTEN
San Francisco State University, Department of Physics and Astronomy, 1600 Holloway Avenue, San Francisco, CA 94132, USA

HERNANDEZ GAMAZO PILAR
IFIC, Edificio Institutos de Investigacion, Universidad de Valencia, Apartment 22085, E-46071 Valencia, Spain

KAPLAN DAVID
Institute for Nuclear Theory Box 351550 Seattle, WA 98195-1550 USA

LELLOUCH LAURENT
CPT Marseille, Case 907, CNRS Luminy, F-13288 Marseille, Cedex 9, France

LÜSCHER MARTIN
Physics Department, CERN, Route de Meyrin, 1211 Genève 23, Switzerland

PHILIPSEN Owe
University of Münster, ITP, Wilhelm-Klemm-Str. 9, D-48149 Münster, Germany

SCHÄFER Stefan
Humboldt Universität, Institut Für Physik (AG COM), Newtonstrasse 15, 12489 Berlin, Germany

VLADIKAS Anastassios
NFN, c/o Department of Physics, Via della Ricerca Scientifica 1, I-0013 Rome, Italy

WEISZ Peter
Max-Planck-Institut für Physik, Werner-Heisenberg-Institut, Föhringer Ring 6, D-80805, Munich, Germany

Participants

ALLAIS Andrea
Center for Theoretical Physics, 6-308, Massachusetts Institute of Technology, 77 Massachusetts Avenue, 02139 Cambridge, MA, USA

BANERJEE Debasish
Tata Institute of Fundamental Research, A 301, Department of Theoretical Physics, 1, Homi Bhabha Road, Colaba, Mumbai, 400 005, India

BAUMGARTNER David
Institute for theoretical physics, University of Bern, Sidlerstrasse 5, 3012 Bern, Switzerland

BERNARDONI Fabio
IFIC, Edificio Institutos de Ivestigaciòn, Apartado de Correos 22085, E-46071 Valencia-Spain

BRAMBILLA Michele
Universita degli Studi di Parma, Viale Degli Usberti, 7/A I-43100, Parma, Italy

BUIVIDOVICH Pavel
JIPNR, National Academy of Science, Acad. Krasin Street 99, 220 109 Minsk, Belarus

CHENG Michael
Lawrence Livermore National Laboratory, 7000 East Avenue L-415, Livermore, CA, 94550, USA

CICHY Krzysztof
Adam Mickiewicz University, Faculty of Physics, Quantum Physics Division, Umultowska 85, 61-614 Poznan, Poland

DINTER Simon
NIC, DESY, Platanenallee 6, 15738 Zeuthen, Germany

DI VITA Stefano
Universita degli Studi Roma Tre, Dipartimento di fisica "E. Amaldi", Via della Vasca Navale 84, 00146 Rome, Italy

DRACH VINCENT
LPSC, 53 Avenue des Martyrs, F-38026 Grenoble, France

ENGEL GEORG
Institut für Physik, Fachbereich Theoretische Physik, Karl-Franzens-Universität Graz, Universitätsplatz 5, 8010 Graz, Austria

FENG XU
NIC, DESY, Platanenallee 6, 15738 Zeuthen, Germany

FRISON JULIEN
Centre de Physique Théorique, Bureau 424, Campus Luminy, Case 907, F-13288 Marseille, France

FRITZSCH PATRICK
School of Physics and Astronomy, University of Southampton, SO17 1BJ, UK

GREEN JEREMY
Center for Theoretical Physics, Massachusetts Institute of Technology, 77 Massachusetts Avenue, Cambridge, MA 02139, USA

HARRAUD PIERRE-ANTOINE
LPSC, 53 rue des Martyrs, F-38026 Grenoble, France

HESSE DIRK
DESY Zeuthen, 15738 Zeuthen, Platanenallee 6, Germany

HEYBROCK SIMON
University of Regensburg, Fakultät für Physik, Universitäts Str. 31, 93040, Regensburg, Germany

KEEGAN LIAM
University of Edinburgh, JCMB, King's Buildings, Mayfield Road, Edinburgh EH9 3JZ, Scotland

KERRANE EOIN
School of Physics and Astronomy, University of Edinburgh, James Clerk Maxwell Building, Kingõs Buildings, Mayfield Road, Edinburgh EH9 3JZ, Scotland

KORCYL PIOTR
Jagiellonian University, Institute of Physics, Ul. Reymonta 4, 30-059 Krakow, Poland

KUJAWA AGNIESZKA
Adam Mickiewicz University, Faculty of Physics, Solid State Theory Division, Umultowska 85, 61-614, Poznan, Poland

LANGELAGE JENS
Institute for Theoretical Physics, Westfälische Wilhelms-Universität Münster, Wilhelm-Klemm-Str. 9, 48149 Münster, Germany

LIN MEIFENG
Center for Theoretical Physics Massachusetts Institute of Technology 77 Massachusetts Avenue Cambridge, MA 02139, USA

LIU YUZHI
Department of Physics and Astronomy, The University of Iowa, 203 Van Allen Hall, Iowa City, IA 52242-1479 USA

SCHIEL RAINER
University of Regensburg, Department of Physics, Universitätsstr. 31, 93040 Regensburg, Germany

UNO SHUMPEI
Department of Physics, Graduate School of Science, Nagoya University, Furo-cho, Chikusa-ka, Nagoya, 464-8602, Japan

USUI KOUTA
Department of Physics, Faculty of Science, University of Tokyo, Hongo 7-3-1, Bunkyo-ku, Tokyo, 113-0033 Japan

VAQUERO AVILES-CASCO ALEJANDRO MARINO
Facultad de Ciencias, Departamento de Fisica Teorica, C/Pedro Cerbuna 12, Cp 50009, Zaragoza, Spain

VIROTTA FRANCESCO
NIC, DESY, Platanenallee 6, 15738 Zeuthen, Germany

VULVERT GREGORY
Centre de Physique Théorique, CNRS, Campus de Luminy, Case 907, F-13288 Marseille Cedex 09, France

WEIR DAVID
Theoretical Physics Group, Blackett Laboratory, Imperial College, Prince Consort Road, London SW7 2AZ, UK

YURKOVSKY EVGENY
Tel-Aviv University, School of Physics and Astronomy, IL-69978 Tel Aviv, Israël

ZHOU RAN
University of Connecticut, U 3046, 2152 Hillside Road, Storrs, CT, 06269, USA

1
Lattice field theory fundamentals

M. Pilar HERNÁNDEZ

Departamento de Física Teórica and IFIC
Universidad de Valencia and CSIC
Apart. 22085, 46071 Valencia

In memory of my collaborator and friend Jan Wennekers

Overview

This is the first time a Les Houches summer school is fully devoted to lattice field theory (LFT). This is timely as the progress in the field has been spectacular in the last years.

In these lectures I will concentrate on the basics and I will deal mostly with the analytical formulation of LFT. The remaining lectures will discuss in detail the essential algorithmic aspects, as well as the modern perspectives.

There are many good introductory books on the subject (Creutz, 1983; Lüscher, 1989; Montvay and Münster, 1994; Smit, 2002; Rothe, 2005; DeGrand and DeTar, 2006; Gattringer and Lang, 2010). The goal of my lectures will be to provide a short summary of the more basic contents of those books.

1.1 On the need for a lattice formulation of quantum field theory

There is firm experimental evidence that the laws of particle physics are accurately described by a quantum field theory (QFT).

The experiments at LEP and at flavor factories of the last decades (Amsler *et al.*, 2008) have established the validity of the Standard Model (SM) up to a level of precision of 1% or better. The Standard Model is a renormalizable QFT with a simple Lagrangian that fits in a t-shirt (Fig. 1.1).

The pure gauge interactions depend on just three free parameters (the three coupling constants associated to the three gauge groups), and preserve the three discrete symmetries: parity (P), charge conjugation (C) and time-reversal (T). The matter–gauge interactions do not introduce any further free parameter in the model, but they violate P and C maximally, and preserve T. On the other hand, the interactions of

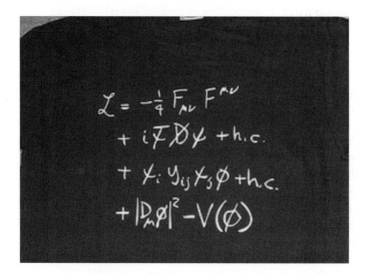

Fig. 1.1 t-shirt Standard Model.

the scalar Higgs field, that will be tested soon at the LHC, are poorly understood theoretically, as they bring real havoc to the theory. This sector contains many new free parameters that are required to fit data: 22–24 (depending on whether neutrinos are Majorana particles or not). It is responsible for the spontaneous breaking of the gauge symmetry, which is the fundamental pillar of the SM, and a mechanism we do not really understand. This sector also has the key to the subtle violation of CP and T symmetries in the SM.

Perturbation theory has given a great deal of information about the SM, but fails in various situations:

- In processes involving particles with $SU(3)$ interactions: quarks and gluons described by the beautifully simple QCD Lagrangian

$$\mathcal{L}_{\text{QCD}} = -\frac{1}{4g^2} F^a_{\mu\nu} F^a_{\mu\nu} + \sum_q \bar{\psi}_q (i\slashed{D} + m_q)\psi_q. \tag{1.1}$$

QCD is a strongly coupled theory at low energies, resulting in several phenomena that cannot be understood in perturbation theory: *confinement*, a *mass gap* and *spontaneous chiral-symmetry breaking*. A precise quantitative understanding of QCD interactions is, furthermore, needed to test the quark–flavor sector of the SM, that is expected to be quite sensitive to new physics.

- The Higgs self-interactions are completely untested. The SM version of the Higgs potential:

$$V(\Phi) = -\frac{\mu^2}{2}\Phi^\dagger\Phi + \frac{\lambda}{4!}(\Phi^\dagger\Phi)^2, \tag{1.2}$$

is probably too naive. It suffers from the so-called *triviality problem*: the fact that the only renormalized value of the coupling is zero

$$\lim_{\Lambda\to\infty} \lambda_R = 0, \tag{1.3}$$

and is therefore a trivial theory, i.e. not phenomenologically viable. Only if there is a physical cutoff, can the renormalized coupling be non-vanishing, which would imply that the SM is an effective theory, valid only below some energy scale. If this is the case, however, the Higgs mass is expected to receive large quadratic corrections from the higher-energy scales. This is the so-called hierarchy problem. Establishing the triviality of the SM Higgs potential requires to go beyond perturbation theory.

- Beyond the SM interactions (BSM): there are many alternatives to address the various open questions in the SM. Even though there is presently no compelling proposal to extend the SM at high energies (all alternatives involve more free parameters that the SM), it is quite likely that the SM is not the whole story. In many of the most popular theories BSM non-perturbative effects come into play: in SUSY non-perturbative effects are often invoked to break SUSY at low energies, technicolor theories are up-scaled versions of QCD and the fashionable nearly conformal gauge theories also require a non-perturbative approach.

- Origin of chirality. The breaking of parity by the weak interactions is probably the most intriguing feature of the SM. It is notoriously difficult to ensure chiral gauge symmetry non-perturbatively. Finding such a formulation is likely to shed some light on the symmetry principles of the SM.

For all these reasons, having a non-perturbative tool to solve QFTs is essential. The only first-principles method is the regularization on a space-time grid that provides a non-perturbative definition of a regularized QFT (at least those that are asymptotically free), and can be treated in principle by numerical methods. Clearly this is not an easy task and the efforts of the lattice community in the past few decades were concentrated on Yang Mills theories (YM) and QCD.

Solving YM is not only the original goal of the lattice formulation, but it is still one of the famous Millennium Prize problems.[1] It will require the proof of the existence of the theory and the presence of a mass gap.[2] It would be great if any of the students in this school would solve this problem, becoming rich in more than one way...

In these lectures, I will review the foundations of the lattice formulation of scalar, fermion and gauge field theories, as well as QCD.

1.2 Basics of quantum field theory

Quantum field theory is the synthesis of quantum mechanics and special relativity, which can be reached following two very different routes, as the relativistic limit of a system of many identical quantum particles and from the canonical quantization of classical fields.

1.2.1 Relativistic quantum mechanics and Fock space

The Hilbert space of a *fixed* number of quantum particles is not sufficient to describe the dynamics of a quantum system in the relativistic domain, because particles can be created/destroyed in collisions. The appropriate space to describe a relativistic quantum system is *Fock space*, the sum of all Hilbert spaces with any fixed number of particles:

$$\mathcal{F} = |0\rangle \oplus \mathcal{H}_1 \oplus \mathcal{H}_2 \ldots \oplus \mathcal{H}_\infty, \tag{1.4}$$

where $|0\rangle$ is the vacuum state, which we assume normalized.

Owing to the (anti-)symmetrization properties of the physical states under permutation of identical particles, the states can be characterized by the occupation numbers, N_i, i.e. the number of particles in the energy level E_i. It is possible to define creation and annihilation operators $\hat{a}_i^\dagger$ and $\hat{a}_i$ in Fock space that create/destroy one particle in the ith level. If the particles are bosons, these operators have the following commutation relations:

$$\left[\hat{a}_i^\dagger, \hat{a}_j^\dagger\right] = 0, \quad \left[\hat{a}_i, \hat{a}_j\right] = 0, \quad \left[\hat{a}_i, \hat{a}_j^\dagger\right] = \delta_{ij}. \tag{1.5}$$

We can use them to construct all the states of Fock space from the vacuum state $|0\rangle$:

$$|N_1, \ldots, N_n\rangle = \left(\hat{a}_1^\dagger\right)^{N_1} \ldots \left(\hat{a}_n^\dagger\right)^{N_n} |0\rangle. \tag{1.6}$$

An arbitrary observable is an operator in Fock space that can always be written in terms of creation/annihilation operators as:

$$\hat{O} = \sum_{n_1 \ldots n_n; m_1, \ldots m_n} O_{n_1 \ldots n_n m_1 \ldots m_n} (\hat{a}_1)^{n_1} \ldots (\hat{a}_n)^{n_n} (\hat{a}_1^\dagger)^{m_1} \ldots (\hat{a}_n^\dagger)^{m_n}, \tag{1.7}$$

where $O_{n_1 \ldots m_n}$ are numbers such that $\hat{O}$ is Hermitian.

In the case of free particles, the one-particle states can be chosen to be momentum eigenstates and the spectrum is continuous:

$$|\mathbf{p}\rangle = \hat{a}_\mathbf{p}^\dagger |0\rangle, \quad E_\mathbf{p} = \sqrt{\mathbf{p}^2 + m^2}. \tag{1.8}$$

The Lorentz-invariant normalization of these states is

$$\langle \mathbf{p} | \mathbf{p}' \rangle = (2\pi)^3 2 E_\mathbf{p} \, \delta(\mathbf{p} - \mathbf{p}'). \tag{1.9}$$

We can define the so-called field operator:

$$\hat{\phi}^\dagger(\mathbf{x}) \equiv \int \frac{d^3 p}{(2\pi)^3 2 E_\mathbf{p}} e^{-i p \cdot x} \hat{a}_\mathbf{p}^\dagger, \tag{1.10}$$

which acts on the vacuum as

$$\hat{\phi}^\dagger(\mathbf{x})|0\rangle = \int \frac{d^3 p}{(2\pi)^3 2 E_\mathbf{p}} e^{-i p \cdot x} |\mathbf{p}\rangle \simeq |\mathbf{x}\rangle, \tag{1.11}$$

and can therefore be interpreted as creating a particle at point $\mathbf{x}$.

1.2.2 Canonical field quantization

We can also start with a real classical field $\phi(\mathbf{x}, t)$ with classical Lagrangian and Hamiltonian given by:

$$\mathcal{L}(\phi) = \int d^3 x \frac{1}{2} \left(\dot{\phi}^2 - (\nabla \phi)^2 - m_0^2 \phi^2 \right), \tag{1.12}$$

$$\mathcal{H}(\phi, \pi) = \int d^3 x \frac{1}{2} \left(\pi^2 + (\nabla \phi)^2 + m_0^2 \phi^2 \right), \tag{1.13}$$

where

$$\pi = \frac{\partial \mathcal{L}}{\partial \dot{\phi}}. \tag{1.14}$$

This classical system can be quantized canonically by identifying the pair of canonical variables, $\{\phi, \pi\}$, with quantum operators $\{\hat{\phi}, \hat{\pi}\}$.

In momentum space, it is easy to see that the Hamiltonian describes an infinite number of harmonic oscillators, one for each momentum, with one quantum of energy being

$$E_{\mathbf{p}} = \sqrt{\mathbf{p}^2 + m_0^2}. \tag{1.15}$$

The well-known ladder operators $\hat{a}_{\mathbf{p}}^{\dagger}$ and $\hat{a}_{\mathbf{p}}$ of the harmonic oscillator are related to the quantum field operator by

$$\hat{\phi}(\mathbf{x}, t) = \int \frac{d^3 p}{(2\pi)^3} \frac{1}{2E_{\mathbf{p}}} \left\{ \hat{a}_{\mathbf{p}} e^{-i(E_{\mathbf{p}} t - \mathbf{p}\mathbf{x})} + h.c. \right\}. \tag{1.16}$$

This operator resembles the time-evolved field operator in Fock space of Eq. (1.10), if we identify the ladder operators of the harmonic oscillators with creation/annihilation operators in Fock space, providing therefore a particle interpretation to the quantized field.

Summarizing, the most important physical intuition is that a quantum field is a bunch of harmonic oscillators, whose ladder operators correspond to creation/annihilation operators in a Fock space. Particles are therefore interpreted as excitations of a quantum harmonic oscillator. This connection goes also in the opposite direction. Indeed, Weinberg has shown (Weinberg, 1995) that operators (observables) in Fock space, Eq. (1.7), that satisfy the following conditions

$$\text{Cluster decomposition} \leftrightarrow \text{locality}$$
$$\text{Hermiticity} \leftrightarrow \text{unitarity}$$
$$\text{Lorentz invariance} \leftrightarrow \text{causality}$$

are necessarily functions of quantum field operators, such as Eq. (1.16), resulting therefore in a quantum field theory.

Continuous symmetries such as Lorentz invariance act in Fock space, according to Wigner's theorem, as unitary operators.[3] It is easy to see that the operator of space translations by the vector $\mathbf{x}$ is

$$\hat{U}_{\mathbf{x}} \equiv e^{-i\hat{\mathbf{P}}\mathbf{x}}, \tag{1.17}$$

where $\hat{\mathbf{P}}$ is the momentum operator.

The operator that implements time translations, by x_0, i.e. the quantum evolution operator is given in terms of the Hamiltonian by

$$\hat{U}_{x_0} \equiv e^{-i\hat{H}x_0}. \tag{1.18}$$

1.2.3   Field correlation functions and physical observables

The essential assumption that goes into the definition of cross-sections and decay widths, is the existence of asymptotic states, which correspond to a bunch of non-interacting 1-particle states in the infinite past $t \to -\infty$ (in-states) as well as in the infinite future $t \to \infty$ (out-states). For this to happen two conditions are required:

- localization of one-particle states or wave packets;
- localization of the interaction: only when particles get sufficiently close are their interactions significant.

The relation between the scattering matrix elements and time-ordered field correlation functions is given by the LSZ (Lehmann, Symanzik and Zimmermann, 1955) reduction formula:

$$\prod_{i=1}^{n} \int d^4 x_i e^{ip_i \cdot x_i} \prod_{j=1}^{k} \int d^4 y_j e^{-iq_j \cdot y_j} \langle 0| T \left(\hat{\phi}(x_1) \ldots \hat{\phi}(x_n) \hat{\phi}(y_1) \ldots \hat{\phi}(y_k) \right) |0\rangle$$

$$\simeq_{p_i^0 \to E_{\mathbf{p}_i}, q_j^0 \to E_{\mathbf{q}_j}} \prod_{i=1}^{n} \left(\frac{i\sqrt{Z}}{p_i^2 - m^2 + i\epsilon} \right) \prod_{j=1}^{k} \left(\frac{i\sqrt{Z}}{q_j^2 - m^2 + i\epsilon} \right) \langle \mathbf{p}_1, \ldots, \mathbf{p}_n, out | \mathbf{q}_1, \ldots, \mathbf{q}_k; in \rangle,$$

$$(1.19)$$

where the scattering amplitudes, $\langle \mathbf{p}_1, \ldots, \mathbf{p}_n, out | \mathbf{q}_1, \ldots, \mathbf{q}_k; in \rangle$, correspond to k asymptotic *in* states (one-particles states at $t = -\infty$) that end up as n *out* states (one-particle states at $t = \infty$), and T() stands for time-ordering. Z and m are the field renormalization constant and mass that characterize the asymptotic one-particle states. They can be extracted from the spectral representation of the two-point correlation functions, or Källen–Lehmann representation (Lehmann, 1954; Källen, 1972), that we derive in the next section. For a derivation of the LSZ formula see for example (Itzykson and Zuber, 1980; Peskin and Schroeder, 1995).

1.2.3.1   *Källen–Lehmann spectral representation of the propagator*

The two-point correlation function or propagator is a fundamental quantity that allows the identification of the asymptotic states, their masses and field renormalization factors. We review why this is so.

The invariance of the Hamiltonian under translations implies $\left[\hat{H}, \hat{\mathbf{P}} \right] = 0$, i.e. the eigenstates of the Hamiltonian are also momentum eigenstates;

$$\hat{\mathbf{P}} | \alpha(\mathbf{p}) \rangle = \mathbf{p} | \alpha(\mathbf{p}) \rangle,$$

$$\hat{H} | \alpha(\mathbf{p}) \rangle = E_{\mathbf{p}}(\alpha) | \alpha(\mathbf{p}) \rangle,$$

$$(1.20)$$

with

$$E_{\mathbf{p}}^2(\alpha) = m(\alpha)^2 + \mathbf{p}^2.$$

$$(1.21)$$

Here, $m(\alpha)$ is not necessarily a one-particle mass, since $|\alpha(\mathbf{p})\rangle$ could represent a multiparticle state with total momentum $\mathbf{p}$, in which case it is simply the energy of the system in the rest frame, i.e. where the total momentum vanishes. Therefore, α labels all the energy eigenstates of zero momentum.

Using the completeness relation for the full Hilbert space:

$$\hat{1} = |0\rangle\langle 0| + \sum_\alpha \int \frac{d^3p}{(2\pi)^3 2E_{\mathbf{p}}(\alpha)} |\alpha(\mathbf{p})\rangle\langle\alpha(\mathbf{p})| \tag{1.22}$$

we can write the propagator as

$$\langle 0| T\left(\hat{\phi}(x)\hat{\phi}(0)\right)|0\rangle\Big|_{x_0>0} = \langle 0|\hat{\phi}(x)\hat{1}\hat{\phi}(0)|0\rangle = \langle 0|\hat{\phi}(0)e^{-i\hat{P}\cdot x}\hat{1}\hat{\phi}(0)|0\rangle$$

$$= \sum_\alpha \int \frac{d^3p}{(2\pi)^3 2E_{\mathbf{p}}(\alpha)} \, e^{-ip\cdot x}\Big|_{p_0=E_{\mathbf{p}}(\alpha)} |\langle 0|\hat{\phi}(0)|\alpha(\mathbf{p})\rangle|^2, \tag{1.23}$$

where we have used that the operator that implements space-time translations by x is $\hat{U}_x = e^{-i\hat{P}x}$, where $\hat{P}_0 = \hat{H}$, according to Eqs. (1.17) and (1.18). We have furthermore assumed that the vacuum is invariant under temporal and spatial translations ($\hat{U}_x|0\rangle = |0\rangle$), and that the vacuum expectation value of the field vanishes ($\langle 0|\hat{\phi}(0)|0\rangle = 0$).

We can now relate the state with momentum $\mathbf{p}$ and that with zero momentum by the unitarity transformation that implements the corresponding boost, $\hat{U}_{\mathbf{p}}$:

$$|\alpha(\mathbf{p})\rangle = \hat{U}_{\mathbf{p}}|\alpha(0)\rangle. \tag{1.24}$$

Since the operator $\hat{\phi}(0)$, being a scalar, and the vacuum state are invariant under this boost

$$\langle 0|\hat{\phi}(0)|\alpha(\mathbf{p})\rangle = \langle 0|\hat{\phi}(0)|\alpha(0)\rangle, \tag{1.25}$$

we finally obtain the famous Källen–Lehmann (KL) formula

$$\langle 0|T\hat{\phi}(x)\hat{\phi}(0)|0\rangle = \sum_\alpha \int \frac{d^3p}{(2\pi)^3 2E_{\mathbf{p}}(\alpha)} \, e^{-ip\cdot x}\Big|_{p_0=E_{\mathbf{p}}(\alpha)} |\langle 0|\hat{\phi}(0)|\alpha(\mathbf{p})\rangle|^2$$

$$= i\sum_\alpha \int \frac{d^4p}{(2\pi)^4} e^{-ip\cdot x} \frac{|\langle 0|\hat{\phi}(0)|\alpha(0)\rangle|^2}{p^2 - m(\alpha)^2 + i\epsilon}, \tag{1.26}$$

where the last equality is easy to show by performing a contour integration over p_0.

Two observations are in order

- For each state labelled by α there is a field renormalization constant

$$Z_\alpha \equiv |\langle 0|\hat{\phi}(0)|\alpha(0)\rangle|^2, \tag{1.27}$$

which is the same quantity that characterizes the asymptotic states in the LSZ relation of Eq. (1.19).

- The states do not have to be discrete, therefore the sum over α is really an integral. It is common to write the KL relation in terms of a spectral density:

$$\langle 0|T\hat{\phi}(x)\hat{\phi}(0)|0\rangle = \int_0^\infty \frac{dM^2}{2\pi} \rho(M^2)\Delta(x; M^2),\qquad(1.28)$$

with

$$\Delta(x; M^2) \equiv i \int \frac{d^4 p}{(2\pi)^4} \frac{e^{-ip\cdot x}}{p^2 - M^2 + i\epsilon},\qquad(1.29)$$

and

$$\rho(M^2) = \sum_{\alpha\in 1\mathrm{particle}} (2\pi)\, Z_\alpha\, \delta(M^2 - m(\alpha)^2) + \text{continuum}.\qquad(1.30)$$

1.2.3.2   Wick rotation

The LSZ reduction formula demonstrates that correlation functions of time-ordered products of fields:

$$W_n(t_1, \mathbf{x_1}; \dots, t_n, \mathbf{x}_n) = \langle 0|\hat{\phi}(t_1, \mathbf{x}_1)\dots\hat{\phi}(t_n, \mathbf{x}_n)|0\rangle,\quad t_1 \geq t_2 \dots \geq t_n,\ (1.31)$$

contain *all* the physical information of the theory. These objects are therefore the primary quantities to be computed on the lattice. However, this is not done in Minkowski but in Euclidean space, after an analytic continuation.

It is possible to show under general conditions that these functions can be continuously extended to analytic functions in the complex domain of the variables $t_1, \dots t_n$ so that

$$\text{Im } t_1 \leq \text{Im } t_2 \leq \dots .\text{Im } t_n.\qquad(1.32)$$

The Euclidean correlation functions or Schwinger functions are defined as:

$$S_n(x_1, \dots, x_n) = W_n(-ix_1^0, \mathbf{x}_1; \dots - ix_n^0, \mathbf{x}_n),\qquad(1.33)$$

where the Euclidean times are $x_i^0 = it_i$ and

$$x_1^0 \geq x_2^0 \dots \geq x_n^0.\qquad(1.34)$$

The computation of these functions is sufficient to solve the theory. This Euclidean approach shows all its power in the functional integral representation that we now describe.

1.2.4 Functional formulation of a scalar field theory

Feynman reformulated quantum mechanics via the so-called path integral (Feynman, 1948), that represents of the basic time-evolution operator of the quantum theory as an integral over classical paths.

1.2.4.1 Path integral in quantum mechanics

As stated before, the quantum operator that evolves states from time t_i to t_f is

$$\hat{U}(t_f, t_i) = e^{-i\hat{H}(t_f - t_i)}, \tag{1.35}$$

where $\hat{H}$ is the quantum Hamiltonian.

Let us consider a system of one particle with a Hamiltonian $\hat{H} = \frac{\hat{p}^2}{2m} + V(\hat{x})$. Let us divide the time interval in a large number, N, of infinitesimal intervals of width τ:

$$t_n = t_i + n\tau, \quad n = 0, \ldots, N, \quad \tau \equiv \frac{t_f - t_i}{N}. \tag{1.36}$$

We can therefore write the evolution as the composition of infinitesimal evolutions

$$\hat{U}(t_f, t_i) = \hat{U}(t_f, t_{N-1})\hat{U}(t_{N-1}, t_{N-2})\ldots\hat{U}(t_1, t_i). \tag{1.37}$$

At each time slice t_n we can include the identity operator as the projector on a complete basis, such as the position basis

$$\hat{1} = \int d^3x_n \, |\mathbf{x}_n\rangle\langle\mathbf{x}_n|, \tag{1.38}$$

therefore

$$\hat{U}(t_f, t_i) = \left(\prod_{n=1}^{N-1} \int d^3x_n\right) \hat{U}(t_f, t_{N-1})|\mathbf{x}_{N-1}\rangle\langle\mathbf{x}_{N-1}|\ldots.|\mathbf{x}_1\rangle\langle\mathbf{x}_1|\hat{U}(t_1, t_i)$$

$$= \left(\prod_{n=1}^{N-1} \int d^3x_n\right) \hat{T}|\mathbf{x}_{N-1}\rangle \left(\prod_{n=2}^{N-1} \langle\mathbf{x}_n|\hat{T}|\mathbf{x}_{n-1}\rangle\right) \langle\mathbf{x}_1|\hat{T}, \tag{1.39}$$

where we have denoted the evolution operator in each interval by the *transfer operator*, $\hat{T}$:

$$\hat{U}(t_{n+1}, t_n) = e^{-i\hat{H}\tau} \equiv \hat{T}. \tag{1.40}$$

The next step is to *define* a new transfer operator $\hat{T}_F$ that coincides with $\hat{T}$ in the limit $\tau \to 0$, and that makes it easy to evaluate the matrix elements $\langle\mathbf{x}_n|\hat{T}_F|\mathbf{x}_{n-1}\rangle$. A possible definition is

$$\hat{T}_F \equiv e^{-i\frac{\tau}{2}V(\hat{x})} \, e^{-i\tau\frac{\hat{p}^2}{2m}} e^{-i\frac{\tau}{2}V(\hat{x})}, \tag{1.41}$$

which implies

$$\langle \mathbf{x_{n+1}}|\hat{T}_F|\mathbf{x_n}\rangle = \sqrt{\frac{m}{2\pi i \tau}} \exp\left[i\tau\left(\frac{m}{2}\left(\frac{\mathbf{x}_{n+1}-\mathbf{x}_n}{\tau}\right)^2 - \frac{V(\mathbf{x}_{n+1})+V(\mathbf{x}_n)}{2}\right)\right]$$

$$= \sqrt{\frac{m}{2\pi i \tau}}\, e^{i\tau \mathcal{L}(t_n)}, \tag{1.42}$$

where the function $\mathcal{L}$ is the time-discretized version of the classical Lagrangian

$$\mathcal{L}(t) \equiv \frac{1}{2}m\left(\frac{d\mathbf{x}}{dt}\right)^2 - V(\mathbf{x}), \tag{1.43}$$

and $\mathbf{x}(t_n) = \mathbf{x}_n$.

Finally, the evolution operator is given by

$$\langle \mathbf{x_f}|\hat{U}(t_f,t_i)|\mathbf{x_i}\rangle = \lim_{N\to\infty}\left(\sqrt{\frac{m}{2\pi i \tau}}\right)^N \prod_{n=1}^{N-1}\int d^3x_n\, e^{i\tau \sum_{n=0}^{N-1}\mathcal{L}(t_n)}\Bigg|_{\mathbf{x}(t_f)\equiv \mathbf{x_f};\mathbf{x}(t_i)\equiv \mathbf{x_i}}$$

$$\equiv c\int \mathcal{D}x(t)\, e^{i\int_{t_i}^{t_f} dt \mathcal{L}(t)}, \tag{1.44}$$

where c is a constant. This amplitude is the *path integral* over all paths that pass by the space-time points $(t_i,\mathbf{x}_i)$ and $(t_f,\mathbf{x}_f)$.

Obviously we have not proven here the equivalence between the two representations (canonical and functional), since the two definitions of the transfer operator in Eqs. (1.40) and (1.41) agree only for small τ.

The path-integral representation is therefore an *alternative formulation* of quantum mechanics. Clearly the link between the world of quantum operators and that of functional integrals is the transfer operator $\hat{T}_F$, the Hamiltonian being a derived quantity:

$$\hat{H}_F \equiv \frac{i}{\tau}\ln \hat{T}_F. \tag{1.45}$$

$\hat{H}_F$ and $\hat{H}$ do not coincide, although they are expected to lead to the same physics.

As we have explained above, the quantum time-evolution operator can be analytically continued to imaginary time $t \to -ix_0$, and so does the path-integral representation we have just introduced. The transfer operator in Euclidean space is the positive operator:

$$\hat{T}_F^E = \exp\left(-\frac{\tau}{2}V(\hat{x})\right)\exp\left(-\tau\frac{\hat{P}^2}{2m}\right)\exp\left(-\frac{\tau}{2}V(\hat{x})\right), \tag{1.46}$$

and the relation with the Euclidean Hamilton operator is therefore

$$\hat{H}_F^E = -\frac{1}{\tau}\hat{T}_F^E. \tag{1.47}$$

From here onwards we will eliminate the F and E indices for simplicity, and denote the Euclidean transfer operator by $\hat{T}$.

An important role is played in the following by the *partition function*, which can be defined as

$$\mathcal{Z} \equiv \text{Tr}\left[\hat{U}(T/2, -T/2)\right] \equiv \lim_{N \to \infty} \text{Tr}\left[\hat{T}^N\right] = \int_{\text{PBC}} \mathcal{D}x(t)e^{-\int_{-T/2}^{T/2} \mathcal{L}dt}, \qquad (1.48)$$

where PBC stands for periodic boundary conditions, since now the integration is over all classical paths that are periodic, i.e $\mathbf{x}_i = \mathbf{x}_f$ in Eq (1.44) and we sum over $\mathbf{x}_i$.

1.2.4.2 *Path integral in quantum field theory*

We have reviewed the canonical quantization of a scalar field in Section 1.2.2, which amounts to considering not one but an infinite number of quantum operators $\hat{\mathbf{x}}$ and $\hat{\mathbf{P}}$, one pair for each point in space, satisfying the canonical commutation relations:

$$\left\{\hat{x}_i, \hat{P}_i\right\} \to \left\{\hat{\phi}(\mathbf{x}), \hat{\pi}(\mathbf{x})\right\}, \quad \left[\hat{\phi}(\mathbf{x}), \hat{\pi}(\mathbf{y})\right] = i\delta(\mathbf{x} - \mathbf{y}), \qquad (1.49)$$

The quantum Hamiltonian is given, for a generic potential, by:

$$\hat{H} \equiv \int d^3\mathbf{x} \left[\frac{1}{2}\hat{\pi}^2 + \frac{1}{2}(\nabla\hat{\phi})^2 + V(\hat{\phi})\right]. \qquad (1.50)$$

The equivalent of the complete position basis is now[4]

$$\hat{1} = \int \prod_{\mathbf{x}} d\phi(\mathbf{x})|\phi\rangle\langle\phi|, \qquad (1.51)$$

where the states $|\phi\rangle$ are the eigenstates of the field operator

$$\hat{\phi}(\mathbf{x})|\phi\rangle = \phi(\mathbf{x})|\phi\rangle. \qquad (1.52)$$

Following the same steps as in the case of one degree of freedom, we can represent the time evolution operator in Euclidean time by discretizing time as before in terms of a transfer operator:

$$\hat{U}(t_f, t_i) = \hat{U}(t_f, t_{N-1})\hat{U}(t_{N-1}, t_{N-2})....\hat{U}(t_1, t_i) \equiv \hat{T}^N, N\tau = t_f - t_i. \quad (1.53)$$

$$\hat{U}(t_f, t_i) = \int \prod_{n=1}^{N-1} d\phi_n(\mathbf{x}_n) \, \hat{T}|\phi_{N-1}\rangle\langle\phi_{N-1}|\hat{T} \ldots |\phi_1\rangle\langle\phi_1|\hat{T}. \qquad (1.54)$$

The transfer operator is the analogous of Eq. (1.41) and can be defined as:

$$\hat{T} = \exp\left(-\frac{\tau}{2}\hat{H}_V\right)\exp\left(-\tau\hat{H}_K\right)\exp\left(-\frac{\tau}{2}\hat{H}_V\right), \qquad (1.55)$$

where

$$\hat{H}_V \equiv \int d^3\mathbf{x} \left[\frac{1}{2}(\nabla\hat{\phi})^2 + V(\hat{\phi})\right], \tag{1.56}$$

$$\hat{H}_K \equiv \int d^3\mathbf{x} \frac{1}{2}(\hat{\pi})^2. \tag{1.57}$$

We can compute the matrix elements of this transfer operator easily (the operator $\hat{H}_V$ is diagonal in the ϕ basis, while $\hat{H}_K$ is diagonal in the momentum basis) and the result is

$$\langle\phi_{n+1}|\hat{T}|\phi_n\rangle = \exp\left[-\frac{\tau}{2}\int d^3x \left(\frac{\phi_{n+1}-\phi_n}{\tau}\right)^2 + (\nabla\phi_n)^2 + V(\phi_n) + V(\phi_{n+1})\right]$$

$$= \exp\left(-\tau\mathcal{L}(\phi_n)\right), \tag{1.58}$$

where $\mathcal{L}$ is the time-discretized version of the classical Euclidean Lagrangian. Finally,

$$\langle\phi_f|\hat{U}(t_f, t_i)|\phi_i\rangle = \lim_{N\to\infty}\int\left[\prod_{n=0}^{N}d\phi_n(\mathbf{x_n})\right]\exp\left(-\tau\sum_{n=0}^{N}\mathcal{L}(\phi_n)\right)$$

$$\equiv \int_{\substack{\phi(\mathbf{x},t_i)=\phi_i(\mathbf{x})\\\phi(\mathbf{x},t_n)=\phi_f(\mathbf{x})}}\mathcal{D}\phi\exp\left(-\int dt\mathcal{L}(\phi)\right). \tag{1.59}$$

We can also define the partition function as

$$\mathcal{Z} = \lim_{N\to\infty}\mathrm{Tr}\left[\hat{T}^N\right] = \int_{PBC}\mathcal{D}\phi\,e^{-\int dt\mathcal{L}(\phi)} = \int_{PBC}\mathcal{D}\phi\,e^{-\mathcal{S}[\phi]}, \tag{1.60}$$

where

$$S[\phi] = \int dt\mathcal{L}(\phi) = \int d^4x\left\{\frac{1}{2}(\partial_\mu\phi(x))^2 + V(\phi(x))\right\}, \tag{1.61}$$

is the classical Euclidean action.

1.2.4.3　*Correlation functions in the functional formalism*

We are interested in correlation functions. We can easily derive their functional representation similarly by noting that for any operator

$$\langle 0|\hat{O}(\mathbf{x},t)|0\rangle = \lim_{T\to\infty}\frac{\mathrm{Tr}\left[\hat{O}e^{-\hat{H}T}\right]}{\mathrm{Tr}\left[e^{-\hat{H}T}\right]} = \lim_{T\to\infty}\frac{\mathrm{Tr}\left[\hat{O}e^{-\hat{H}T}\right]}{\mathcal{Z}}, \tag{1.62}$$

provided $|0\rangle$ is the lowest-energy state, since the contribution to the trace of the excited states is exponentially suppressed.

Then, we can write the time-ordered correlation function

$$S_n = \langle 0|\hat{\phi}(\mathbf{x}_1, t_1)\dots\hat{\phi}(\mathbf{x}_n, t_n)|0\rangle = \lim_{T\to\infty}\mathrm{Tr}\left[\hat{\phi}(\mathbf{x}_1, t_1)\dots\hat{\phi}(\mathbf{x}_n, t_n)e^{-\hat{H}T}\right]/\mathcal{Z} \quad (1.63)$$

and applying the same procedure of discretizing time we find the functional representation of the n-point function

$$S_n = \frac{\int_{\mathrm{PBC}}\mathcal{D}\phi\, e^{-S[\phi]}\phi(\mathbf{x}_1, t_1)\dots\phi(\mathbf{x}_n, t_n)}{\int_{\mathrm{PBC}}\mathcal{D}\phi\, e^{-S[\phi]}} \equiv \langle\phi(x_1)\dots\phi(x_n)\rangle, \quad (1.64)$$

where the integrals are over periodic classical fields, as defined above.

Some useful definitions in the context of perturbation theory are:

- The generating functional of correlation functions is

$$Z[J] = \langle e^{\int d^4x J(x)\phi(x)}\rangle, \quad (1.65)$$

where we have introduced an external source density $J(x)$ so that

$$\frac{\delta}{\delta J(x_1)}\dots\frac{\delta}{\delta J(x_n)}Z[J]\bigg|_{J=0} = \langle\phi(x_1)\dots\phi(x_n)\rangle. \quad (1.66)$$

We define a functional derivative as

$$\frac{\delta}{\delta J(x)}J(y) = \delta(x-y), \qquad \frac{\delta}{\delta J(x)}\int d^4y J(y)\phi(y) = \phi(x). \quad (1.67)$$

It is easy to compute $Z[J]$ in the scalar field theory we are considering for the free case, i.e. for $V(\phi) = \frac{m_0^2}{2}\phi^2$. The path integral is Gaussian and the result is

$$Z[J] = \exp\left(\frac{1}{2}\int d^4x d^4y J(x)K^{-1}(x,y)J(y)\right), \quad (1.68)$$

where

$$K \equiv -\partial^\mu\partial_\mu + m_0^2, \quad (1.69)$$

is a linear operator acting in the space of real scalar fields. The free propagator is

$$\langle\phi(x)\phi(y)\rangle = \frac{\delta^2 Z[J]}{\delta J(x)\delta J(y)}\bigg|_{J=0} = K^{-1}(x,y) = \int d^4p\,\frac{e^{i(x-y)}}{p^2 + m_0^2}. \quad (1.70)$$

- The generating functional of connected correlation functions $W[J] \equiv \ln Z[J]$ satisfies:

$$\frac{\delta}{\delta J(x_1)}\dots\frac{\delta}{\delta J(x_n)}W[J] = \langle\phi(x_1)\dots\phi(x_n)\rangle_{\mathrm{conn}}. \quad (1.71)$$

- The generating functional of vertex functions, which are connected and one-particle amputated correlation functions, also called one-particle irreducible or

1PI[5], can be obtained from the Legendre transform of $W[J]$:

$$\Gamma[\Phi] = W[J] - \int d^4x J(x)\Phi(x)\Big|_{J[\Phi]}, \qquad (1.72)$$

where $J[\Phi]$ is defined from the solution of equation

$$\frac{\delta W[J]}{\delta J(x)} = \Phi(x). \qquad (1.73)$$

The functional derivatives of $\Gamma[\Phi]$ generate the 1PI correlation functions

$$\Gamma^{(n)}(x_1, \ldots, x_n) = \frac{\delta}{\delta\Phi(x_1)} \cdots \cdot \frac{\delta}{\delta\Phi(x_n)}\Gamma[\Phi] = \langle \phi(x_1)\phi(x_2)\ldots\phi(x_n)\rangle_{\mathrm{conn,1PI}},$$

$$(1.74)$$

or vertex functions that represent the interaction vertices in the Lagrangian and are therefore the basic objects in the renormalization procedure. More details can be found in standard books (Peskin and Schroeder, 1995).

All these generating functionals are easy to find in the free case, but not in the interacting case. At this point one can follow two approaches:

- Perturbation theory.
- A non-perturbative evaluation of the correlation functions, which can be achieved via a discretization of space-time, known as the lattice formulation. This Euclidean functional formulation of QFT provides a link between QFT and statistical mechanics. After the discretization of space-time, the functional integrals of Eqs. (1.60) and (1.64) become finite-dimensional ones, and in many cases can be treated by statistical importance sampling methods. I refer to the lectures of M. Lüscher (Lüscher, 2009) for a general discussion of these methods.

1.2.5 Symmetries and Ward identities

Noether's theorem establishes the connection between continuous symmetries of the Lagrangian and conserved currents. In the functional formulation, symmetries of the Lagrangian imply relations between correlation functions that are usually referred to as *Ward–Takahashi identities* (Ward, 1950; Takahashi, 1957). These identities are easy to derive at tree level and can be shown to hold also at the quantum level (Peskin and Schroeder, 1995).

Let us consider an infinitesimal local field transformation of the form:

$$\phi(x) \to \phi(x) + \epsilon_a(x)\delta_a\phi(x), \qquad (1.75)$$

which will usually correspond to a unitary transformation. The Lagrangian changes at first order by

$$\delta\mathcal{L}[\phi] = \frac{\delta\mathcal{L}}{\delta\epsilon_a(x)}\epsilon_a(x) + \frac{\delta\mathcal{L}}{\delta\partial_\mu\epsilon_a(x)}\partial_\mu\epsilon_a(x) + \mathcal{O}(\epsilon^2). \qquad (1.76)$$

Now let's consider the generating functional

$$Z[J] = \int D\phi e^{-S[\phi]+\int d^4x J(x)\phi(x)}, \tag{1.77}$$

on which we can perform the change of variables of Eq. (1.75)

$$Z[J] = \int D\phi' e^{-S[\phi']+\int d^4x J(x)\phi'(x)} = Z[J] + \delta Z[J]. \tag{1.78}$$

Since this should be true for arbitrary $\epsilon_a(x)$, and assuming the measure does not change $(D\phi' = D\phi)$ we have

$$\frac{\delta Z[J]}{\delta\epsilon^a(x)} = \int D\phi e^{-S[\phi]+\int d^4x J(x)\phi(x)} \left(\partial_\mu \mathcal{J}_\mu^a - \frac{\delta \mathcal{L}}{\delta \epsilon_a(x)}\bigg|_{\epsilon=0} + J(x)\delta_a\phi(x) \right) = 0, \tag{1.79}$$

where $\mathcal{J}_\mu^a$ coincides with the classically conserved Noether current,

$$\mathcal{J}_\mu^a(x) \equiv \frac{\delta \mathcal{L}(\phi + \epsilon^a \delta_a \phi)}{\delta \partial_\mu \epsilon_a(x)}\bigg|_{\epsilon=0}. \tag{1.80}$$

The nth functional derivatives with respect to the external sources, J, of the functional in Eq. (1.79) give relations between the correlation functions of the following type:

$$\frac{\partial}{\partial x_\mu} \langle \phi(x_1)\phi(x_2) \dots \phi(x_n)\mathcal{J}_\mu^a(x) \rangle = \langle \phi(x_1)\phi(x_2) \dots \phi(x_n) \frac{\delta \mathcal{L}}{\delta \epsilon_a(x)}\bigg|_{\epsilon=0} \rangle$$

$$- \sum_i \delta(x_i - x)\langle \phi(x_1)..\delta_a\phi(x_i)..\phi(x_n)\rangle, \tag{1.81}$$

where the last term is the sum of contact terms that vanish if $x \neq x_1, \dots, x_n$. For more details on the derivation of these identities see for example (Collins, 1984).

1.2.6   Perturbation theory in the functional formalism

Correlation functions in the interacting case, i.e. for

$$V(\phi) = \frac{1}{2}m_0^2\phi^2 + \frac{\lambda}{4!}\phi^4 \tag{1.82}$$

can be obtained by perturbing in λ. We just need to separate the free and interacting parts of the classical action:

$$S[\phi] = S^{(0)}[\phi] + S^{(1)}[\phi], \tag{1.83}$$

with

$$S^{(0)}[\phi] \equiv \int d^4x \left\{ \frac{1}{2}\left[(\partial_\mu\phi(x))^2 + m_0^2\phi^2\right] \right\}, S^{(1)}[\phi] = \int d^4x \frac{\lambda}{4!}\phi^4. \tag{1.84}$$

The generating functional can therefore be Taylor-expanded in the coupling constant, λ:

$$Z[J] = \frac{\langle e^{\int d^4x J(x)\phi(x)} e^{-S^{(1)}[\phi]}\rangle_0}{\langle e^{-S^{(1)}[\phi]}\rangle_0}$$

$$\equiv \frac{\int \mathcal{D}\phi\, e^{-S^{(0)}[\phi]+\int d^4x J(x)\phi(x)} \sum_n \frac{1}{n!}\left(-S^{(1)}[\phi]\right)^n}{\int \mathcal{D}\phi\, e^{-S^{(0)}[\phi]} \sum_n \frac{1}{n!}\left(-S^{(1)}[\phi]\right)^n}, \tag{1.85}$$

where $\langle\rangle_0$ is the average with respect to the unperturbed theory and therefore can be evaluated in terms of the free generating functional of Eq. (1.68). The nth Schwinger function is given by

$$S_n = \frac{\langle \phi(x_1)\phi(x_2)\ldots\phi(x_n)e^{-S^{(1)}[\phi]}\rangle_0}{\langle e^{-S^{(1)}[\phi]}\rangle_0}, \tag{1.86}$$

and a similar Taylor expansion in λ allows to compute S_n in terms of free correlation functions. Three observations are in order:

- Wick's theorem holds. All contributions can be obtained from functional derivatives of the free generating functional, $Z^{(0)}[J]$, evaluated at $J = 0$. Therefore, Wick's theorem is reproduced because $Z^{(0)}[J]$ is quadratic in the currents and therefore the fields have to be paired up in propagators to give a non-vanishing contribution:

$$\langle \phi(x_1)\phi(x_2)\ldots.\phi(x_{2n})\rangle_0 = \sum_{\text{perm}} \langle \phi(x_1)\phi(x_2)\rangle_0 \ldots.\langle \phi(x_{2n-1})\phi(x_{2n})\rangle_0. \tag{1.87}$$

  Correlation functions are therefore obtained from products of propagators.
- The denominator in Eq. (1.86) ensures that all contributions with disconnected parts that do not contain any external leg cancel (i.e. vacuum polarization diagrams).
- For each insertion of $S^{(1)}[\phi]$ there is an integration over space-time that can give rise to ultraviolet divergences (UV).

A similar perturbative expansion can be trivially defined for the generating functionals of connected and 1PI diagrams.

1.2.7   Perturbative renormalizability

In order to ensure the UV finiteness of the perturbative contributions, it is sufficient to consider the 1PI diagrams, where the propagators attached to the external legs are amputated. Let us consider a general diagram of an Nth 1PI correlation function in momentum space for the scalar theory, Eq. (1.82). The contribution of a diagram with I internal lines (i.e. propagators linking two vertices) and L loops is generically of the form:

$$\Gamma^{(N)}(p_1, \ldots, p_N) \sim \int \prod_{l=1}^{L} d^4 q_l \prod_{i=1}^{I} \frac{1}{k_i(q_l, p_j)^2 + m^2}, \tag{1.88}$$

where the q_l stand for the L loop momenta, p_j for the N external momenta and k_i are the momenta of the I internal lines, that can in general be written as linear combinations of the external and loop momenta. The loop momentum integrals give rise to UV divergences. If these integrals are cutoff at some scale Λ, the diagram behaves as $\sim \Lambda^\omega$ when Λ is scaled to ∞, where ω is the power of the leading divergence, also called the *superficial degree of divergence*. Scaling the loop momenta with Λ in Eq. (1.88), the following relation follows:

$$\omega \equiv 4L - 2I. \tag{1.89}$$

ω must therefore be negative for the diagram to be finite, although this condition is in general not sufficient.

There is a topological relation between I, the number of vertices V and external legs N of the diagram:

$$2I + N = 4V, \tag{1.90}$$

since each vertex involves four fields and each leg is either external or linked to another internal line.

Finally, the number of loops, L, is related to V and N. Each propagator involves an integral over momentum, Eq. (1.70). Each vertex involves an integration over space-time, giving rise to $\delta(\sum_i p_i)$, where the sum is over all momenta attached to the vertex. One of these deltas corresponds to the conservation of the external momenta, while the others allow us to reduce $V - 1$ of the loop integrations. Therefore, the number of loops of the diagram satisfies, using Eq. (1.90),

$$L = I - V + 1 = V - N/2 + 1, \tag{1.91}$$

and substituting Eqs. (1.90) and (1.91) in Eq. (1.89) we find

$$\omega = 4 - N. \tag{1.92}$$

ω does not depend on the number of loops or vertices. It is fixed by the number of external legs. Only 1PI diagrams with $N = 2, 4$ might have a non-negative degree of divergence. It can be shown that the UV divergences in these diagrams give contributions to the vertex functions of the form

$$\delta\Gamma^{(2)}[\Phi] = A\partial_\mu \Phi \partial_\mu \Phi + B\Phi^2 \tag{1.93}$$

$$\delta\Gamma^{(4)}[\Phi] = C\Phi^4, \tag{1.94}$$

where A, B, C are divergent, but since they have the same structure as the terms already present in the Lagrangian, they can be reabsorbed in a redefinition of m_0^2, λ

and the normalization of the field itself. For this reason, we say that this theory is *perturbatively renormalizable*.

More generically, we can consider a theory where $S^{(1)}$ has other interactions such as

$$S^{(1)}[\phi] = \frac{\lambda}{4!}\phi^4 + \frac{\lambda'}{6!}\phi^6 + \dots, \tag{1.95}$$

while λ has no mass dimension, the additional couplings in general do, e.g. $[\lambda'] = -2$.

Let us consider more generally a vertex with N_∂ derivatives and N_ϕ fields. The corresponding coupling, g_V, must have mass dimension

$$[g_V] = 4 - N_\phi - N_\partial. \tag{1.96}$$

We can repeat the power-counting exercise above to evaluate the superficial degree of divergence of a vertex function that contains V vertices of this type and we find that the relations of Eqs. (1.89) and (1.90) are modified to

$$\omega = 4L - 2I + N_\partial V, \quad 2I + N = N_\phi V \tag{1.97}$$

and therefore

$$\omega = 4 - N - [g_V]V. \tag{1.98}$$

We find a very different behavior as the order of the perturbative expansion grows depending on the sign of $[g_V]$:

$[g_V] > 0$ diagrams become less divergent with V: *superrenormalizable theory*
$[g_V] = 0$ the divergence does not depend on V: *renormalizable theory*
$[g_V] < 0$ divergences for larger N as V grows: *non-renormalizable theory*

Even if one considers only renormalizable theories, the proof of perturbative renormalizability is rather involved, because a diagram with $\omega < 0$ does not have to be finite. In general, there are subdivergences (that is divergences that show up when a subset of all the internal momenta are scaled with Λ). The proof of renormalizability in the continuum therefore takes the following steps:

- Prove a power-counting theorem to characterize divergent and finite diagrams.
- Recursive procedure to subtract subdivergences: e.g. in the BPHZ (Bogoliubov and Parasiuk, 1957; Hepp, 1966; Zimmermann, 1969) subtraction scheme, the superficial degree of divergence of a diagram is reduced by subtracting the Taylor expansion of the diagram in the external momenta up to order equal to the degree of divergence. A forest formula establishes the recursive procedure to subtract subdivergences.
- All-orders proof.

The conclusions to all orders in perturbation theory are the same as those based on the superficial degree of divergence. For more details about perturbative renormalizability we refer to P. Weisz's lectures (Weisz, 2009).

1.2.8 Wilsonian renormalization group

The old concept of renormalizability that looked like a sacred requirement of any sensible quantum field theory is now updated. Thanks to Wilson and others we know now that there is nothing special about a bare Lagrangian that is renormalizable. In fact the consequence of the point of view of assuming the existence of a fundamental cutoff (such as the one existing in a theory defined on the lattice) is that renormalizability is an emergent effective phenomenon. If such a theory induces correlation lengths that tend to infinity in units of the cutoff, it can be accurately represented by a renormalizable theory, as long as we are interested in describing physics at scales of the order of this long correlation length. For a classical reference see (Wilson and Kogut, 1974) and references therein.

1.2.8.1 Renormalization group transformations

K. Wilson studied the connection of renormalizability and critical phenomena via his celebrated *renormalization group transformations*. Let us assume that we have a real cutoff, such as a space-time lattice spacing $a = \Lambda^{-1}$, as we will see later. Taking the continuum limit $a \to 0$ is therefore like taking the cutoff to infinity, and the hope is that a finite limit exists, in which physical scales stay finite and therefore

$$m_{\text{phys}}a \to 0. \tag{1.99}$$

Seen as a statistical system this implies that the correlation length (rate of the exponential decay of the two-point correlator), $\xi \sim m_{\text{phys}}^{-1}$, goes to infinity in units of the lattice spacing

$$\xi/a \to \infty, \tag{1.100}$$

and this is what we call in statistical mechanics a *critical point*. The continuum limit of a QFT must therefore be a critical point.

It is an empirical fact that many systems near critical points behave in similar ways, a property called *universality* (the long-range properties of many systems do not depend on the details of the microscopic interactions). It was the contribution of Wilson and others that established the link

Universality in critical statistical systems $\leftrightarrow$ Renormalizability in QFT

Both phenomena can be understood in terms of *fixed points* of the renormalization group.

Let us suppose that we have a lattice scalar theory on a lattice of spacing a that describes physics scales $m \ll a^{-1}$. The most general theory that is local can be written as

$$S(a) = \sum_{\alpha} g_{\alpha}(a) \sum_{x} O_{\alpha}(\phi(x)), \tag{1.101}$$

where O_α are local operators (of the field and its derivatives) with arbitrary dimension that respect the lattice symmetries. This is a very complicated system with many coupled degrees of freedom, however, if we are interested only in the long-distance properties, many of the degrees of freedom (those at short distance or large momenta) induce effects that can be absorbed in a change in the couplings g_α, as we will see.

In order to understand what happens when we take the limit $a \to 0$ keeping the physical scale fixed, we can follow Wilson's recipe and do it in little steps. We consider a series of lattice spacings that decrease by a factor $1 - \epsilon$ at a time:

$$a \geq a_1 \geq a_2 \ldots \geq a_n = (1 - \epsilon)^n a, \quad \epsilon \ll 1. \tag{1.102}$$

We want to compare the actions defined in the series of lattices and we do this by defining, at each step n, an effective action at the original scale a, $S^{(n)}(a)$. This action is obtained from the nth action at the scale a_n, after integrating out recursively the extra degrees of freedom that appear at each step. These are short-ranged (momentum scales between a_{n-1}^{-1} and a_n^{-1}), and therefore result in a local action, which must then have the same generic form of Eq. (1.101), but with different couplings in general:

$$S^{(n)}(a) = \sum_\alpha g_\alpha^{(n)}(a) \sum_x O_\alpha(\phi(x)). \tag{1.103}$$

We call a *renormalization group (RG) transformation*, the function that defines the change in the couplings:

$$R_\alpha : g_\alpha^{(n)} \to g_\alpha^{(n+1)} \quad g_\alpha^{(n+1)} = R_\alpha(g^{(n)}). \tag{1.104}$$

Obviously we can make this transformation a continuous one and then we talk about the RG flow of the coupling constants. While the couplings change we are changing the physics obviously, but if we perform sufficiently many transformations we can hit a fixed point if it exists. A *fixed point* corresponds to some point in coupling space g_α^* such that

$$R_\alpha(g^*) = g_\alpha^*. \tag{1.105}$$

It is at these points that physics would no longer change as we move towards the continuum limit, since the action remains unchanged. The fixed points are therefore critical points:

$$\lim_{n \to \infty} m_{phys}(g^*) a_n \to 0, \tag{1.106}$$

unless the physical critical mass diverges, which would be uninteresting for a QFT.

Now it turns out that such fixed points, if they exist, are rather universal, because they can be approached by tuning just a few parameters , called relevant couplings. A *priori*, one could imagine having to tune all the couplings $\alpha = 1, \ldots, \infty$ to reach a given fixed point, but this is usually not the case and this is the essence of renormalizability and universality. Near a fixed point the evolution of the couplings reads at linear order

$$g_\alpha^{(n+1)} - g_\alpha^* = \sum_\beta \left.\frac{\partial R_\alpha}{\partial g_\beta}\right|_{g^*} (g_\beta^{(n)} - g_\beta^*), \tag{1.107}$$

so the distance to the fixed point $\Delta g^{(n)}$ changes according to the following equation:

$$\Delta g_\alpha^{(n+1)} = \sum_\beta M_{\alpha\beta} \Delta g_\beta^{(n)}, \quad M_{\alpha\beta} \equiv \left.\frac{\partial R_\alpha}{\partial g_\beta}\right|_{g^*}. \tag{1.108}$$

We can find different situations depending on the eigenvalues, λ, of the matrix M:

$\lambda > 1$ $\Delta g_\alpha^{(n)}$ increases as $n \to \infty$ α is a relevant direction

$\lambda = 1$ $\Delta g_\alpha^{(n)}$ stays the same as $n \to \infty$ α is a marginal direction

$\lambda < 1$ $\Delta g_\alpha^{(n)}$ decreases as $n \to \infty$ α is an irrelevant direction

   In the first case, the distance to the fixed point grows in the corresponding direction, these are *relevant couplings* that would need to be tuned. In the third case, the distance to the fixed point decreases and these are *irrelevant couplings*. In the second case, the couplings are called *marginal* and might need tuning or not depending on subtle quantum effects that always make λ slightly different from one. The fact that the number of relevant directions is finite and usually small is behind the two related properties: universality of the fixed point and the renormalizability of QFT.

Gaussian fixed point.   We will make this discussion a bit more explicit by considering the Gaussian fixed point of scalar theories. First, we note that the free massless point of a scalar theory is a fixed point. Consider the action

$$S(a) = \int_{BZ(a)} \frac{d^4p}{(2\pi)^4} \frac{1}{2}\phi(-p)p^2\phi(p), \tag{1.109}$$

where $BZ(a)$ is the Brillouin zone $[-\pi/a, \pi/a]$ in each momentum direction.

   When we do the first RG transformation we start with the same action but in a lattice of spacing $a_1 = (1 - \epsilon)a$. Since the fields at different momenta are independent variables, we can integrate over those at momenta $\pi/a \leq |p_\mu| \leq \pi/a_1$ so that the partition function:

$$\mathcal{Z}^{(1)} = \int \prod_{p \in BZ(a_1)} d\phi(p) e^{-\int_{BZ(a_1)} \frac{d^4p}{(2\pi)^4} \frac{1}{2}\phi(-p)p^2\phi(p)} \tag{1.110}$$

$$= C \int \prod_{p \in BZ(a)} d\phi(p) e^{-\int_{BZ(a)} \frac{d^4p}{(2\pi)^4} \frac{1}{2}\phi(-p)p^2\phi(p)}, \tag{1.111}$$

where C is some constant that comes from the integration of the momentum modes of $BZ(a_1)$ that lay out of $BZ(a)$. The effective action after integrating the high-frequency modes up to scale a^{-1} is therefore $S^{(1)}(a) = S(a)$. The original action is a fixed point of the renormalization group. Note that since there is no mass term, it is also a critical point, as expected.

Now we can see why the Gaussian fixed point is the one responsible for the renormalizability of $\lambda\phi^4$. We start with an arbitrary lattice action that is quadratic in the fields, but including all terms that have the lattice symmetries.

$$S(a) = \int_{BZ(a)} \frac{d^4 p}{(2\pi)^4} \frac{1}{2} \phi(-p) \left(p^2 + \frac{1}{a^2} m_0^2 + g_1 a^2 p^4 + \dots \right) \phi(p), \qquad (1.112)$$

where we have expressed all the couplings in units of the lattice spacing to make them dimensionless:

$$[m_0] = [\alpha] = \dots = 0. \qquad (1.113)$$

This action is also diagonal in momentum space and therefore the integration over the momentum modes in a slice of momenta in $BZ(a_1)$ and out of $BZ(a)$ can be done as before so the action for the modes up to a^{-1} is $S(a_1)$, but erasing the high-momentum modes, i.e.:

$$S^{(1)}(a) = \int_{BZ(a)} \frac{d^4 p}{(2\pi)^4} \frac{1}{2} \phi(-p) \left(p^2 + \frac{1}{a^2} \left(\frac{a}{a_1} \right)^2 m_0^2 + g_1 a^2 \left(\frac{a_1}{a} \right)^2 p^4 + \dots \right) \phi(p), \qquad (1.114)$$

therefore the action is no longer a fixed point, because all the couplings except the kinetic term have changed:

$$\begin{pmatrix} m_0^{(1)\,2} \\ g_1^{(1)} \\ \dots \end{pmatrix} = M \begin{pmatrix} m_0^2 \\ g_1 \\ \dots \end{pmatrix}, \qquad M = diag\left((1-\epsilon)^{-2}, (1-\epsilon)^2, \dots \right). \qquad (1.115)$$

The only eigenvalue of M that is above one is the first one, therefore there is one relevant direction, that of m_0^2 and all the rest are irrelevant. After a large number of RG transformations (as we approach the continuum limit) these directions disappear. On the other hand, m_0^2, which fixes the physical mass gap grows, therefore it needs to be tuned to remain finite in the continuum limit. The continuum limit of this theory, even if it has non-renormalizable terms should correspond to a free massive renormalizable scalar QFT.

Finally, in the fully interacting case, the situation is more complicated, but still near the Gaussian fixed point (sufficiently small couplings) the continuum limit corresponds to a renormalizable scalar field theory. In this case the action contains all terms, including interactions

$$S(a) = \sum_x \frac{1}{2} \partial_\mu \phi \partial_\mu \phi + \frac{1}{2a^2} m_0^2 \phi^2 + \frac{\lambda}{4!} \phi^4 + \frac{\lambda'}{6!} \phi^6 + g_{\frac{1}{2}} a^2 \phi \partial^4 \phi + \dots \qquad (1.116)$$

Now, the integration over the momentum shell $\pi/a \leq |p_\mu| \leq \pi/a_1$ cannot be done analytically. But for sufficiently small couplings it can be done in perturbation theory, see for example (Peskin and Schroeder, 1995). It gives

$$S^{(1)}(a) = \sum_x \frac{Z^{(1)}}{2} \partial_\mu \phi \partial_\mu \phi + \frac{1}{2a^2} m_0^{(1)^2} \phi^2 + \frac{\lambda^{(1)}}{4!} \phi^4 + a^2 \frac{\lambda'^{(1)}}{6!} \phi^6 + \frac{g_1^{(1)}}{2} a^2 \phi \partial^4 \phi + \cdots,$$

$$(1.117)$$

where

$$Z^{(1)} = 1 + \mathcal{O}(\lambda^2), \tag{1.118}$$

and

$$m_0^{(1)^2} = (m_0^2 + \delta m_0^2)(1 - \epsilon)^{-2}, \tag{1.119}$$

$$\lambda^{(1)} = \lambda + \delta\lambda, \tag{1.120}$$

$$\lambda'^{(1)} = (\lambda' + \delta\lambda')(1 - \epsilon)^2, \tag{1.121}$$

$$g_1^{(1)} = (g_1 + \delta g_1)(1 - \epsilon)^2. \tag{1.122}$$

All δ terms depend on the couplings $\lambda, \lambda', \ldots$, but vanish for small enough couplings. Therefore, for small enough couplings, the matrix M in this case has one relevant direction, many irrelevant ones and just one marginal. It is for this marginal direction that the value of $\delta\lambda$, even if small, is important since it determines the fate of this direction. At lowest order of perturbation theory it is

$$\delta\lambda = \frac{3\lambda^2}{16\pi^2} \ln(1 - \epsilon) < 0, \tag{1.123}$$

therefore $\lambda^{(1)} < \lambda$ and the direction is marginally irrelevant. The change is much slower than for an irrelevant direction since it is only logarithmic. The continuum theory is therefore again a massive free scalar theory, at least within this perturbative analysis.

Summarizing Wilson's approach to renormalization shows the following intuitive physical picture:

QFT with a cutoff $\leftrightarrow$ Statistical system near criticality

Renormalized QFT $\leftrightarrow$ Statistical system at a fixed point

This picture is of course an essential ingredient for the definition of QFT on a lattice, because it implies that we do not have to worry about the precise definition of $S(a)$, the continuum limit will correspond to the fixed point of the statistical system nevertheless. We need to ensure, however, that the fixed point corresponds to the QFT we want to describe. For this we need to make sure that

- the action has the right degrees of freedom;
- it is local;

- has the right symmetries to flow to the desired fixed point (for example if we break some symmetry we might artificially increase the number of relevant directions).

Under these very general assumptions we are otherwise free to make our choice.

Exercise 1.1 Consider the 1D Ising model with an action:

$$S = -\beta \sum_x \sigma_x \sigma_{x+1} \quad \beta > 0, \tag{1.124}$$

where the spin variables $\sigma_x = \pm 1$. Identify the quantum operator and the transfer operator for this model. Diagonalize the transfer operator. Compute the correlator from this result, i.e.

$$\langle \sigma_x \sigma_y \rangle = \lim_{N \to \infty} \mathrm{Tr}[\hat{T}^{N-(x-y)} \hat{\sigma} \hat{T}^{(x-y)} \hat{\sigma}] / \mathrm{Tr}[\hat{T}^N], \tag{1.125}$$

show that the correlation length is

$$\xi^{-1} = -\ln \tanh \beta, \tag{1.126}$$

and therefore only diverges at $\beta = \infty$ (zero temperature).

1.3   Lattice scalar field theory

The definition of a scalar quantum field theory on the lattice assumes that the field lives in a discretized space-time. The simplest choice is to consider the lattice spacing a to be the same in all space-time directions, that is a cubic lattice:

$$\phi(x), \quad x = na, \quad n = (n_0, n_1, n_2, n_3), \quad n_i \in Z^4. \tag{1.127}$$

Therefore,

$$\int dx_i \to a \sum_{n_i \in Z}, \quad \int d^4x \to a^4 \sum_x \equiv a^4 \sum_{n \in Z^4}. \tag{1.128}$$

Canonical quantization goes through identically, the only change is that the labelling of degrees of freedom is discrete and not continuous.

The Fourier transform therefore becomes a Fourier series. Any function defined on a cubic lattice, $F(na)$, has a Fourier transform that is periodic in the Brillouin zone (BZ):

$$\tilde{F}(p) = a^4 \sum_n e^{-ipna} F(na), \quad \tilde{F}(p) = \tilde{F}\left(p + \frac{2\pi}{a} m\right), \quad m \in Z^4. \tag{1.129}$$

It is easy to invert the relation of Eq. (1.129):

$$\int_{-\pi/a}^{\pi/a} \frac{d^4p}{(2\pi)^4} e^{ipna} \tilde{F}(p) = F(na). \tag{1.130}$$

Therefore, lattice four-momenta are cutoff at scale $|p_i| \leq \pi/a$ and therefore the inverse lattice spacing, a^{-1}, is also an energy cutoff, i.e. the theory is regularized.

A very useful formula is Poisson's summation formula:

$$\sum_{n \in Z^4} e^{inz} = (2\pi)^4 \sum_{n \in Z^4} \delta(z - 2\pi n) \equiv (2\pi)^4 \delta_P(z). \tag{1.131}$$

The functional approach to quantization in Euclidean space-time involves the partition function.

$$\mathcal{Z} = \int \mathcal{D}\phi \, e^{-S[\phi]}, \quad \mathcal{D}\phi \to \prod_x d\phi(x), \tag{1.132}$$

and $S[\phi]$ is some discretized version of the action of Eq. (1.61), which is not unique. According to Wilson's RG all actions should be equivalent in the continuum limit provided they satisfy the same symmetries (in this case $\phi \leftrightarrow -\phi$). The simplest choice is:

$$S[\phi] \to a^4 \sum_x \left\{ \frac{1}{2} \hat{\partial}_\mu \phi(x) \hat{\partial}_\mu \phi(x) + \frac{1}{2} m_0^2 \phi(x)^2 + \frac{\lambda}{4!} \phi(x)^4 \right\}, \tag{1.133}$$

where we have defined the *forward lattice derivative*

$$\hat{\partial}_\mu \phi(x) \equiv \frac{1}{a} \left(\phi(x + \hat{\mu}a) - \phi(x) \right). \tag{1.134}$$

We can also define a *backward derivative*

$$\hat{\partial}_\mu^* \phi(x) \equiv \frac{1}{a} \left(\phi(x) - \phi(x - \hat{\mu}a) \right). \tag{1.135}$$

As in the continuum we can obtain the correlation functions from the generating functional

$$Z[J] \equiv \int \prod_x d\phi(x) e^{-S[\phi] + a^4 \sum_x J(x)\phi(x)} / \mathcal{Z}. \tag{1.136}$$

1.3.1 Free lattice scalar theory

As in the continuum, it is easy to solve the lattice theory in the free case, that is for $\lambda = 0$. We can rewrite the action as

$$S^{(0)}[\phi] = a^4 \sum_x \left\{ \frac{1}{2} \hat{\partial}_\mu \phi \hat{\partial}_\mu \phi + \frac{m_0^2}{2} \phi^2 \right\} = \frac{a^4}{2} \sum_{x,y} \phi(x) K_{xy} \phi(y), \tag{1.137}$$

with

$$K_{xy} \equiv -\frac{1}{a^2} \sum_{\hat{\mu}=0}^{3} \left(\delta_{x+a\hat{\mu}y} + \delta_{x-a\hat{\mu}y} - 2\delta_{xy} \right) + m_0^2 \delta_{xy}. \tag{1.138}$$

The corresponding generating functional is

$$Z^{(0)}[J] = e^{\frac{a^4}{2} \sum_{x,y} J_x (K^{-1})_{xy} J_y} \det\left(a^4 K\right)^{-1}, \tag{1.139}$$

where we have used $a^4 \sum_y K_{xy} K_{yz}^{-1} = \delta_{xz}$.

We can then compute the propagator:

$$\langle \phi(x)\phi(y)\rangle_0 = \frac{1}{a^8} \frac{\partial Z^{(0)}[J]}{\partial J_x \partial J_y}\bigg|_{J=0} = \frac{1}{a^4} K_{xy}^{-1}. \tag{1.140}$$

To get a more familiar expression we go to Fourier space. Using Poisson's formula Eq. (1.131), after some easy manipulations we find

$$\tilde{K}_{pq} = a^8 \sum_{xy} e^{-ipx} e^{-iqy} K_{xy} = a^4 (2\pi)^4 \delta_P(p+q) \left\{ m_0^2 + \frac{2}{a^2} \sum_\mu (1 - \cos p_\mu a) \right\}$$

$$= a^4 (2\pi)^4 \delta_P(p+q) \left\{ m_0^2 + \sum_\mu \hat{p}_\mu^2 \right\}, \tag{1.141}$$

where

$$\hat{p}_\mu \equiv \frac{2}{a} \sin\left(\frac{p_\mu a}{2}\right) \qquad \hat{p}^2 \equiv \sum_\mu \hat{p}_\mu^2. \tag{1.142}$$

Therefore,

$$K_{xy} = a^4 \int \frac{d^4 p}{(2\pi)^4} e^{ip\cdot(x-y)} \left(\hat{p}^2 + m_0^2\right). \tag{1.143}$$

It is easy to see that the inverse is

$$\langle \phi(x)\phi(y)\rangle = a^{-4} K_{xy}^{-1} = \int \frac{d^4 p}{(2\pi)^4} \frac{e^{ip\cdot(x-y)}}{\hat{p}^2 + m_0^2}. \tag{1.144}$$

Since in the free theory all correlation functions are products of propagators, this is enough to construct all correlation functions.

It is instructive to understand in this very simple context two important questions:

- What is the particle interpretation?
- What happens in the continuum limit?

According to the discussion in Section 1.2.3.1, the spectral representation of the propagator at large times provides a direct link between the Euclidean formulation and the particle interpretation. Indeed, we can identify the one-particle asymptotic states from the Källen–Lehmann spectral representation of the propagator, Eq.(1.26):

$$\langle \phi(x)\phi(0)\rangle|_{x_0>0} = \sum_\alpha \int \frac{d^3 p}{(2\pi)^3 2E_\mathbf{p}(\alpha)} |\langle 0|\hat{\phi}(0)|\alpha(0)\rangle|^2 e^{-E_\mathbf{p}(\alpha)x_0} e^{i\mathbf{p}\cdot\mathbf{x}}, \tag{1.145}$$

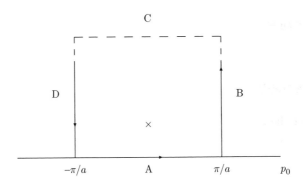

Fig. 1.2

with $E_{\mathbf{p}}(\alpha) = \sqrt{m_\alpha^2 + \mathbf{p}^2}$.

Starting with the free propagator, Eq. (1.144), we can perform the integral over $p_0 \in \left[-\frac{\pi}{a}, \frac{\pi}{a}\right]$ (contour A) using the residuum theorem (see Fig. 1.2): We consider the closed contour including the interval A , the contour B $\left[\frac{\pi}{a}, \frac{\pi}{a} + i\infty\right]$, the contour C $\left[\frac{\pi}{a} + i\infty, -\frac{\pi}{a} + i\infty\right]$ and the contour D $\left[-\frac{\pi}{a} + i\infty, -\frac{\pi}{a}\right]$. We have then

$$\int_A (\ldots) + \int_B (\ldots) + \int_C (\ldots) + \int_D (\ldots) = 2\pi i \sum_{\text{poles}} \text{Residues}. \tag{1.146}$$

By periodicity of the function in the BZ, we have

$$\int_B (\ldots) + \int_D (\ldots) = 0, \tag{1.147}$$

while for $x_0 > 0$, the integral over C vanishes, $\int_C (\ldots) = 0$. Therefore, we end up with the relation

$$\int_A (\ldots) = 2\pi i \sum_{\text{poles}} \text{Residues}. \tag{1.148}$$

Single poles occur at the solutions of the equation:

$$\hat{p}^2 + m^2 = 0 \Rightarrow p_0 = \pm i\omega(\mathbf{p}) \left(\text{mod } \frac{2\pi}{a}\right), \tag{1.149}$$

which are purely complex in the BZ. $\omega(\mathbf{p})$ is a real number satisfying:

$$\cosh \omega(\mathbf{p})a = 1 + \frac{a^2}{2}\left(m_0^2 + \frac{4}{a^2}\sum_{i=1}^{3}\sin^2\frac{p_i a}{2}\right). \tag{1.150}$$

There is only one pole within the closed contour, $p_0 = +i\omega(\mathbf{p})$ with residue

$$\text{Residue}[p_0 = +i\omega(\mathbf{p})] = \frac{1}{2\bar{\omega}(\mathbf{p})}, \qquad \bar{\omega}(\mathbf{p}) \equiv \frac{1}{a}\sinh\left(\omega(\mathbf{p})a\right), \tag{1.151}$$

and therefore

$$\langle \phi(x)\phi(0)\rangle = \int_A (\ldots) = \int d^3 \frac{p}{(2\pi)^3} \frac{1}{(2\bar{\omega}(\mathbf{p}))} e^{-\omega(\mathbf{p})x_0} e^{i\mathbf{p}\cdot\mathbf{x}}. \tag{1.152}$$

We indeed recover the expected behavior of Eq. (1.145) if we identify the one-particle energies $E_{\mathbf{p}}(\alpha) \to \omega(\mathbf{p})$, while the matrix elements

$$|\langle 0|\hat{\phi}(0)|\alpha\rangle| \to \sqrt{\frac{\omega(\mathbf{p})}{\bar{\omega}(\mathbf{p})}}. \tag{1.153}$$

Had we started with the canonical quantization of the free lattice scalar field we would have arrived at the same result.

The continuum limit $a \to 0$ can be readily obtained:

$$\lim_{a\to 0} \omega(\mathbf{p}) = \lim_{a\to 0} \bar{\omega}(\mathbf{p}) = \sqrt{m_0^2 + \mathbf{p}^2} + \mathcal{O}(a^2), \tag{1.154}$$

as expected.

Exercise 1.2 Show that the free scalar Euclidean propagator in a periodic box of extent T and L is given by

$$\langle \phi(x)\phi(0)\rangle = L^{-3} \sum_{\mathbf{p}} \frac{\cosh\left[E_{\mathbf{p}}(\frac{T}{2} - x_0)\right]}{2E_{\mathbf{p}} \sinh\left(\frac{T}{2}E_{\mathbf{p}}\right)}, \qquad E_{\mathbf{p}} = \sqrt{\mathbf{p}^2 + m^2}. \tag{1.155}$$

Use:

$$\sum_{n=1}^{\infty} \frac{\cos nx}{n^2 + \alpha^2} = -\frac{1}{2\alpha^2} + \frac{\pi}{2\alpha}\frac{\cosh\left(\alpha(\pi - x)\right)}{\sinh(\alpha\pi)}, \qquad 0 \leq x \leq 2\pi. \tag{1.156}$$

Show that in the infinite-volume limit, the correct KL representation is obtained.

1.3.2 Interacting lattice scalar theory

When $\lambda \neq 0$ the theory cannot be solved analytically, however, one can rigorously prove the fundamental property of unitarity or the existence and uniqueness of the Hilbert-space representation. This can be done by the following steps:

- Identification of a transfer operator $\hat{T}$ and field operator $\hat{\phi}$ such that

$$\langle \phi(x_1)\ldots\phi(x_n)\rangle = \lim_{T\to\infty} \frac{\text{Tr}\left[\hat{T}^{(T/2-x_1^0)/a}\hat{\phi}(0,\mathbf{x}_1)\hat{T}^{(x_1^0-x_2^0)/a}\hat{\phi}(0,\mathbf{x}_2)\ldots\hat{T}^{(T/2+x_n^0)/a}\right]}{\text{Tr}[\hat{T}^{T/a}]}. \tag{1.157}$$

In general, it takes some guesswork to identify the transfer operator. In this case, it is easy to see that it may be chosen as that of Eq. (1.55) by simply substituting the continuum derivatives by the discrete ones, and the integrals over space by sums.

- Prove that $\hat{T}$ is strictly positive (see exercise). For any $|\Psi\rangle$:

$$\langle\Psi|\hat{T}|\Psi\rangle > 0, \qquad \langle\Psi|\Psi\rangle = 1. \tag{1.158}$$

- Prove that $\hat{T}$ and $\hat{\phi}$ are unique (up to unitary transformations). This is the content of the *reconstruction theorem* (Streater and Wightman (1964)).

All these conditions imply that the quantum Hamiltonian $\hat{H} \equiv -\frac{1}{a}\ln\hat{T}$ is self-adjoint and unique.

Alternatively, one can invoke the *Osterwalder–Schrader reflection positivity condition* which ensures unitarity as a result of a property of Euclidean correlation functions (i.e. without the need to identify the Hilbert-space transfer operator). The *time-reflection positivity* condition is the following. Let O be any product of the classical fields at positive times:

$$O(x_1^0,\ldots,x_n^0) = \phi(x_1)\ldots.\phi(x_n), \qquad x_i^0 > 0. \tag{1.159}$$

We define the operation, $\theta[\ldots]$ of time reflection as

$$\theta\left[O(x_1^0,\ldots,x_n^0)\right] = O(-x_1^0,\ldots,-x_n^0). \tag{1.160}$$

If for any such polynomial it is true that

$$\langle\theta\left[O^\dagger\right]O\rangle \geq 0, \tag{1.161}$$

we say that the theory has reflection positivity, which ensures (Osterwalder and Schrader, 1973; 1975)

- positivity of the scalar product in Hilbert space;
- positivity of $\hat{T}^2$, which is the operator that generates times translations by $2a$ and therefore a Hermitian Hamiltonian $\hat{H} = -\frac{1}{2a}\ln\hat{T}^2$.

Exercise 1.3 Prove the unitarity of the lattice scalar model by

a) showing that the transfer matrix is a positive operator;
b) showing that the lattice formulation has the property of reflection positivity.
   To show this, show that the action can be written as

$$S = S_+ + S_0 + S_-, \tag{1.162}$$

where S_0 depends only on the fields at $x_0 = 0$, S_+ on the fields at $x_0 > 0$ and S_- on the fields at $x_0 < 0$. Show that

$$\theta(S_+) = S_-. \tag{1.163}$$

Rewrite the correlation function of Eq. (1.161) in a manifestly positive way.

Fig. 1.3 One-loop contributions to $\Gamma^{(2)}$ and $\Gamma^{(4)}$.

1.3.3 Lattice perturbation theory

Deriving the perturbative expansion and Feynman rules from the lattice theory is completely analogous to the continuum. We treat

$$S^{(1)} = a^4 \sum_x \frac{\lambda}{4!} \phi(x)^4, \tag{1.164}$$

as a perturbation in the path integral, Eq. (1.85).

The Feynman rules for this theory are just like those in the continuum with the propagator substituted by the lattice one of Eq. (1.144), while the vertex is the same: it connects four scalar lines with strength $-\lambda$. The combinatorial factors coming from Wick contractions are also just like in the continuum.

Let's consider the one-loop corrections to the two- and four-vertex functions (Fig. 1.3):

$$\Gamma^{(2)}(p, -p) = -(\hat{p}^2 + m_0^2) - \frac{\lambda}{2} \int_{BZ} \frac{d^4 k}{(2\pi)^4} \frac{1}{\hat{k}^2 + m_0^2} \equiv -(\hat{p}^2 + m_0^2) - \frac{\lambda}{2} I_1(a, m_0)$$

$$\Gamma^{(4)}(p_1, p_2, p_3, p_4) = -\lambda + \left(\frac{\lambda^2}{2} \int_{BZ} \frac{d^4 k}{(2\pi)^4} \frac{1}{(\hat{k}^2 + m_0^2)(\widehat{k + p_1 + p_2})^2 + m_0^2} + \text{perm} \right)$$

$$\equiv -\lambda + \frac{\lambda^2}{2} \left(I_2(a, m_0, p_1 + p_2) + \text{perm.} \right). \tag{1.165}$$

All Feynman graphs satisfy the following properties in momentum space:

- periodic functions of momenta with periodicity $2\pi/a$ in each momentum direction;
- loop momenta are integrated only in the BZ and are therefore finite.

On the lattice, divergences are expected when we try to approach the continuum limit $a \to 0$. The expectation from perturbative renormalizability is that a continuum limit can be taken provided a tuning of m, λ and the field normalization are performed. It is easy to check that this is indeed the case at the one-loop order.

The $\Gamma^{(2)}$ above does not have a finite continuum limit since

$$I_1(a, m_0) = \int_{BZ} \frac{d^4 k}{(2\pi)^4} \frac{1}{\hat{k}^2 + m_0^2} = \frac{1}{a^2} F(m_0 a), \tag{1.166}$$

and the function $F(x)$ does not vanish for small x:

$$F(0) = \int_{-\pi}^{\pi} \frac{d^4 k}{(2\pi)^4} \frac{1}{\sum_\mu (\sin k_\mu / 2)^2} = 0.154933 \ldots . \tag{1.167}$$

The first derivative is, however, not defined at $m_0 a = 0$, because it has a logarithmic divergence. Isolating this divergence, we find:

$$I_1(a, m_0) = \frac{1}{a^2} F(m_0 a) = \frac{F(0)}{a^2} - m_0^2 \left(-\frac{1}{16\pi^2} \ln(m_0 a)^2 + C + \mathcal{O}(m_0 a)^2 \right), \tag{1.168}$$

where $C = 0.030345755 \ldots$.

In this simple example, it is easy to show that the divergent constant of Eq. (1.166) can be reabsorbed by a redefinition of m_0^2

$$\Gamma^{(2)}(p, -p) = -(\hat{p}^2 + m_0^2) - \frac{\lambda}{2} I_1(a, m_0) \equiv -(\hat{p}^2 + m_R^2). \tag{1.169}$$

Similarly, if we consider the $\Gamma^{(4)}$ vertex function we find that the integral I_2 is divergent. If we consider the Taylor expansion with respect to external momenta, we find that the divergence is present only in the leading term (i.e. at zero external momenta):

$$I_2(a, m_0, 0) = \int_{BZ} \frac{d^4 k}{(2\pi)^4} \frac{1}{(\hat{k}^2 + m_0^2)^2} = -\frac{d}{dm_0^2} I_1(a, m_0)$$

$$= C - \frac{1}{16\pi^2} (\ln(m_0 a)^2 - 1) + \mathcal{O}(a^2), \tag{1.170}$$

therefore the corresponding divergence can be reabsorbed in λ:

$$\Gamma^{(4)}(0, 0, 0, 0) = -\lambda + \frac{3\lambda^2}{2} I_2(a, m_0, 0) \equiv -\lambda_R. \tag{1.171}$$

The renormalized quantities are therefore

$$m_R^2 = m_0^2 + \frac{\lambda}{2} \left(\frac{F(0)}{a^2} + \frac{m_0^2}{16\pi^2} \ln(m_0 a)^2 - C m_0^2 \right),$$

$$\lambda_R = \lambda + \frac{3\lambda^2}{2} \left(-C + \frac{1}{16\pi^2} (\ln(m_0 a)^2 + 1) \right). \tag{1.172}$$

This way of redefining the renormalized couplings corresponds to the usual mass-shell scheme:

$$\Gamma^{(2)}(0,0) = -m_R^2, \quad \left.\frac{d\Gamma^{(2)}(p,-p)}{dp^2}\right|_{p=0} = 1, \quad \Gamma^{(4)}(0,0,0,0) = -\lambda_R. \tag{1.173}$$

That this must hold to all orders of perturbation theory requires a non-trivial theorem known as the *Reisz power-counting theorem* (Reisz, 1988). It is the analog of the continuum one, and permits to carry the BPHZ recursive renormalization procedure over to the lattice regularization. This has been discussed in P. Weisz's lectures (Weisz, 2009).

1.3.4   Callan–Symanzik equations. Beta functions

We have already discussed the renormalization group and why approaching the continuum limit can be seen as a flow in the space of couplings. As we have seen above, the continuum scalar theory that we are trying to describe has one relevant direction, m_0, and one marginal one λ. As we approach the continuum limit, the quantities of Eq. (1.172) must be tuned.

   In the Wilsonian RG we have seen that as we approach the continuum limit, the effective couplings change smoothly in a way that is locally determined by the effective couplings themselves. We can therefore derive a differential equation to describe this change. These are the famous *Callan–Symanzik equations*. Let us consider a fixed λ and let us see how λ_R changes with a. We tune m so that m_R is fixed to the physical mass as we approach the continuum limit. Differenciating the second Eq. (1.172) we find at leading order in the perturbative expansion:

$$\beta(\lambda_R) \equiv a\left.\frac{d\lambda_R}{da}\right|_\lambda = \frac{3}{(16\pi^2)}\lambda^2 + \mathcal{O}(\lambda^3) = \frac{3}{(16\pi^2)}\lambda_R^2 + \mathcal{O}(\lambda_R^3). \tag{1.174}$$

This is the Callan–Symanzik beta function.

   This function can be computed to higher orders, for instance the two-loop result is

$$\beta(\lambda) = \beta_0\lambda^2 + \beta_1\lambda^3 + \dots \tag{1.175}$$

and the coefficients β_0 and β_1 can be shown to be universal (do not depend on the regularization scheme):

$$\beta_0 = \frac{3}{16\pi^2}, \quad \beta_1 = -\frac{17}{3(16\pi^2)^2}. \tag{1.176}$$

The equation can be integrated to give

$$a = Ce^{-1/(\beta_0\lambda_R)}\lambda_R^{-\beta_1/\beta_0^2}(1 + \mathcal{O}(\lambda_R)), \tag{1.177}$$

where C is some integration constant that must be determined from initial conditions. This equation shows that as we approach the continuum limit

$$\lim_{a\to 0} \lambda_R(a)\Big|_{\lambda} \sim \lim_{a\to 0} \frac{1}{\ln a} = 0, \tag{1.178}$$

so the continuum theory has a vanishing renormalized coupling, i.e. it is *trivial*. Unfortunately, this argument is not a sufficient proof of triviality, because it is based on perturbation theory. The question is of course if one could use the lattice formulation to go beyond.

1.3.5 Triviality in lattice $\lambda\phi^4$ (and in the SM)

The Higgs sector of the Standard Model is a multicomponent $\lambda\phi^4$ theory, with a continuous global symmetry that is spontaneously broken. The β function of the bare coupling, λ, has the same properties as in the single-scalar case: the renormalized coupling decreases as we approach the continuum limit at fixed bare coupling. Therefore, one should worry that actually this theory cannot be defined without a cutoff, or if one does then it is a trivial theory $\lambda_R = 0$, which would not be in agreement with phenomenology. In particular, the Higgs mass is related to the renormalized coupling in the following way:

$$\frac{m_H^2}{v^2} = \frac{\lambda_R}{3}. \tag{1.179}$$

Therefore, taking the cutoff to ∞ would imply in particular a massless Higgs.

If we do not remove the cutoff, we can try to maximize the value of λ_R modifying λ in all its possible range: $\lambda \in [0, \infty)$. For example, we could lower the cutoff as much as possible:

$$\frac{\Lambda}{m_H} \geq 2, \tag{1.180}$$

so that the cutoff is higher than two times the Higgs mass (otherwise the SM would not make sense, not even as an effective theory). Such a condition implies an upper bound on λ_R:

$$\lambda_R \leq \lambda_R^{\max}, \tag{1.181}$$

and therefore an upper bound to the Higgs mass, according to Eq. (1.179).

This problem has a very definite answer in the lattice regularization and it was studied extensively in the late 1980s. The picture that emerged from numerical studies as well as analytically is that indeed the only IR fixed point in the discretized scalar theories is the trivial one and the theory is trivial in the continuum limit.

The method followed by Lüscher–Weisz (Lüscher and Weisz, 1988; 1989) can be summarized as follows. The (m_0, λ) space can be mapped to the $(\kappa, \bar{\lambda})$ space, where the original lattice action is written as

$$S = a^4 \sum_x \left[\phi(x)^2 + \bar{\lambda}(\phi(x)^2 - 1)^2 - \kappa \sum_\mu (\phi(x)\phi(x + \hat{\mu}) + \phi(x)\phi(x - \hat{\mu})) \right],$$

$$(1.182)$$

after the change of variables

$$\phi(x) \to \sqrt{2\kappa}\phi(x), \quad a^2 m_0^2 \to \frac{1 - 2\bar{\lambda}}{\kappa} - 8, \quad \lambda \to \frac{6\bar{\lambda}}{\kappa^2}. \qquad (1.183)$$

There is a critical line $\kappa_c(\bar{\lambda})$, where the mass vanishes, where the continuum limit should lie. For values of κ sufficiently far from this line, the so-called hopping parameter expansion (or high-temperature expansion), a Taylor series in κ, is convergent. The strategy to study the triviality of the theory follows the following steps:

- Use the hopping parameter expansion or high-temperature expansion to compute m_R and λ_R (as defined by some renormalization prescription such as the on-shell one, defined above) in a region of κ not too close to κ_c. The fact that the series has been computed to very high order, allows us to control very well the truncation error for values of λ_R that are already in the perturbative domain

$$m_R a = \frac{1}{\sqrt{\kappa}} \sum_n \alpha_n(\bar{\lambda})\kappa^n,$$

$$\lambda_R = \sum_n \beta_n(\bar{\lambda})\kappa^n. \qquad (1.184)$$

For $m_R a \sim 0.5$, we are sufficiently far from the critical line to have an accurate description, while λ_R is rather small.

- Solve the perturbative Callan–Symanzik equations for the renormalized coupling in order to approach the critical line with initial conditions given by the results of the hopping expansion. Since the initial λ_R is small enough and it gets smaller as we approach the continuum limit, the procedure is under control.

In this way, Lüscher–Weisz could map the lines of constant (m_R, λ_R) as the cutoff changes. As $m_R a$ decreases along these lines, we get closer to $\bar{\lambda} = \infty$, which is the furthest we can get, so one can read the bound on λ_R by considering this value of the bare coupling. The result can be plotted in the renormalized plane $(m_R a)^{-1}$ vs. m_R/v_R at $\bar{\lambda} = \infty$ as shown in Fig. 1.4. At $m_R a \sim 0.5$ we can read the value of m_R/v_R, resulting in the limit (Lüscher and Weisz, 1988)

$$m_H \leq 630 \text{ GeV}, \qquad (1.185)$$

for the $O(4)$ model. These results agree with the numerical studies e.g. (Montvay, Münster and Wolff, 1988; Hasenfratz *et al.*, 1987), therefore the issue is settled, to the extent that neglecting fermion and gauge field effects in the SM is a good approximation. For a review of the triviality problem see (Callaway, 1988).

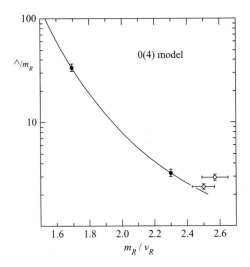

Fig. 1.4 Value of $m_R a$ as a function of m_R/v in the $O(4)$ scalar model as obtained by Lüscher and Weisz (Lüscher and Weisz, 1988).

1.4   Free fermions on the lattice

The Fock space of fermions can be reconstructed from the vacuum acting with creation and annihilation operators $\hat{a}_k$ and $\hat{a}_k^\dagger$, satisfying the following canonical anticommutation relations

$$\{\hat{a}_k, \hat{a}_l\} = \{\hat{a}_k^\dagger, \hat{a}_l^\dagger\} = 0, \quad \{\hat{a}_k, \hat{a}_l^\dagger\} = \delta_{kl}. \tag{1.186}$$

An arbitrary normalized state $|\psi\rangle$ can be written

$$|\psi\rangle = \sum_p \frac{1}{p!} \psi_{k_1,\ldots k_p} \hat{a}_{k_1}^\dagger \ldots \hat{a}_{k_p}^\dagger |0\rangle. \tag{1.187}$$

In the functional formalism (Berezin, 1966), the fermion classical fields are elements of a Grassmann algebra. The generators are a set of anticommuting variables $c_1, \ldots c_n$ and $\bar{c}_1, \ldots \bar{c}_n$, with the following anticommutation properties:

$$\{c_i, c_j\} = \{c_i, \bar{c}_j\} = \{\bar{c}_i, \bar{c}_j\} = 0, \tag{1.188}$$

which imply that $c_i^n = 0, n \geq 2$. The elements of the algebra are elements of the form

$$X_{n_1,\ldots,n_n,m_1,\ldots,m_n} = c_1^{n_1} \ldots \bar{c}_n^{m_n}, \quad n_i, m_i \in \{0,1\}. \tag{1.189}$$

Any function of the Grassmann variables can be represented by a series expansion:

$$f(c, \bar{c}) = \sum_{n_i, m_i} f_{n_1 \ldots m_n} X_{n_1, \ldots m_n}. \tag{1.190}$$

We can define the integral over all Grassmann variables as:

$$\int d\bar{c}\, dc\, f(c,\bar{c}) = f_{111\ldots1}. \tag{1.191}$$

Note that this implies in particular

$$\int dc_i = 0, \quad \int dc_i c_i = 1. \tag{1.192}$$

In defining the partition function for fermions we will find integrals of the form

$$\mathcal{Z}_F \equiv \int d\bar{c}\, dc\, \exp\left\{-\sum_{i,j}\bar{c}_i M_{ij} c_j\right\} = \frac{(-1)^n}{n!}\int d\bar{c}\, dc\left(\sum_{i;j}\bar{c}_i M_{ij} c_j\right)^n$$

$$= \frac{(-1)^{\frac{n(n-1)}{2}}}{n!}\sum_{i_1,\ldots i_n;j_1,\ldots j_n}\epsilon_{i_1,\ldots i_n}\epsilon_{j_1,\ldots j_n}M_{i_1 j_1}\ldots\ldots M_{i_n j_n} = (-1)^{\frac{n(n-1)}{2}}\det(M).$$

$$\tag{1.193}$$

And for correlation functions involving fermions we need integrals of the form:

$$\langle c_{k_1}\bar{c}_{l_1}\ldots\ldots c_{k_m}\bar{c}_{l_m}\rangle_F \equiv \mathcal{Z}_F^{-1}\int d\bar{c}\, dc\, c_{k_1}\bar{c}_{l_1}\ldots\ldots c_{k_m}\bar{c}_{l_m}\exp\left\{-\sum_{i,j}\bar{c}_i M_{ij} c_j\right\}$$

$$= \sum_{\text{perm}}(-1)^{\sigma(\text{perm})}\langle c_{k_1}\bar{c}_{l_1}\rangle_F\ldots\langle c_{k_m}\bar{c}_{l_m}\rangle_F, \tag{1.194}$$

where each contraction is between a c and a $\bar{c}$ variable

$$\langle c_{k_m}\bar{c}_{l_m}\rangle_F = (M^{-1})_{k_m l_m}. \tag{1.195}$$

The Euclidean action for free Dirac fermions of mass m is given by

$$S[\psi,\bar{\psi}] = \int d^4x\, \frac{1}{2}\left[\bar{\psi}(x)\gamma_\mu\partial_\mu\psi(x) - \partial_\mu\bar{\psi}(x)\gamma_\mu\psi(x)\right] + m\bar{\psi}(x)\psi(x), \tag{1.196}$$

where we can choose the *chiral representation* of the γ matrices:

$$\gamma_\mu = \begin{pmatrix} 0 & e_\mu \\ e_\mu^\dagger & 0 \end{pmatrix}, \tag{1.197}$$

and the 2×2 matrices are taken to be:

$$e_0 \equiv -I, \quad e_k \equiv -i\sigma_k, \tag{1.198}$$

where σ_k are the Pauli matrices. It is easy to check the following properties

$$\gamma_\mu^\dagger = \gamma_\mu, \quad \{\gamma_\mu,\gamma_\nu\} = 2\delta_{\mu\nu}. \tag{1.199}$$

We also define

$$\gamma_5 = \gamma_0\gamma_1\gamma_2\gamma_3, \tag{1.200}$$

satisfying

$$\gamma_5^\dagger = \gamma_5, \qquad \gamma_5^2 = 1. \tag{1.201}$$

The mapping of a single Dirac fermion on the Grassmann algebra is

$$\{c_1, \ldots, c_n; \bar{c}_1, \ldots \bar{c}_n\} \rightarrow \{\psi_\alpha(x); \bar{\psi}_\alpha(x)\}_x^{\alpha=1,..4}. \tag{1.202}$$

The number of c and $\bar{c}$ Grassmann variables to represent a general fermion is therefore $4 \times N_{\text{flavor}} \times N_{\text{color}} \times$ space-time points. The partition function is

$$\mathcal{Z}_F = \int d\bar{\psi}d\psi e^{-S[\psi, \bar{\psi}]}. \tag{1.203}$$

As in the scalar theory, the propagator of the theory gives us information on the one-particle asymptotic states of the theory via the Källen–Lehmann representation of the propagator. At large Euclidean time, the fermion propagator should behave as

$$\langle 0|\psi(x)\bar{\psi}(0)|0\rangle_F\big|_{x_0>0} = \sum_\alpha \int \frac{d^3p}{(2\pi)^3} Z_\alpha^2 \left.\frac{i\gamma_\mu p_\mu - m}{2ip_0}\right|_{p_0=i\sqrt{m_\alpha^2+\mathbf{p}^2}} e^{-E_p(\alpha)x_0} e^{i\mathbf{p}\mathbf{x}},$$

$$\tag{1.204}$$

with $E_p(\alpha) = \sqrt{m_\alpha^2 + \mathbf{p}^2}$. This is the KL representation for fermions that can be derived analogously to the scalar case.

1.4.1 Naive fermions

Let us now try to discretize the Euclidean action in the same way we did for the scalar fields. The fields are now defined at the lattice points only and the derivatives are substituted by their discrete versions. We find therefore the so-called *naive fermion action*:

$$S[\psi, \bar{\psi}] = a^4 \sum_{x,\alpha,\mu} \bar{\psi}_\alpha(x)(\gamma_\mu)_{\alpha\beta} \left[\frac{1}{2}(\hat{\partial}_\mu + \hat{\partial}_\mu^*) + m\right]\psi_\beta(x) = a^4 \sum_{x,y} \bar{\psi}_\alpha(x)K_{xy}^{\alpha\beta}\psi_\beta(y), \tag{1.205}$$

where

$$K_{xy}^{\alpha\beta} \equiv \sum_\mu \frac{1}{2a}(\gamma_\mu)_{\alpha\beta}\left(\delta_{yx+a\hat{\mu}} - \delta_{yx-a\hat{\mu}}\right) + m\delta_{\alpha\beta}\delta_{xy}. \tag{1.206}$$

We can understand the particle interpretation of this theory by studying the Källen–Lehmann representation of the propagator. According to the Grassmann integration rules, Eq. (1.195), it is given by

$$\langle\psi_\alpha(x)\bar{\psi}_\beta(y)\rangle_F = \frac{1}{a^4}\left(K^{-1}\right)_{xy}^{\alpha\beta}. \tag{1.207}$$

The propagator can be easily computed in momentum space:

$$K_{pq}^{\alpha\beta} = a^4 \left[\sum_\mu \frac{i}{a} \gamma_\mu \sin(q_\mu a) + m \right]_{\alpha\beta} (2\pi)^4 \delta_P(p+q), \qquad (1.208)$$

so that

$$\langle \psi_\alpha(x) \bar{\psi}_\beta(y) \rangle_F = \int_{BZ} \frac{d^4 p}{(2\pi)^4} \frac{e^{ip(x-y)}}{\sum_\mu i\gamma_\mu \frac{\sin(p_\mu a)}{a} + m}. \qquad (1.209)$$

As we did in the case of the scalar field, we first perform the integration over p_0. We deform the integration into the complex plane depicted in Fig. 1.2. The integral can then be written as a sum of residues of single poles in the band $|\mathrm{Re}\, p_0| \le \pi/a$ and $\mathrm{Im}\, p_0 \ge 0$. Contrary to the scalar case, we find two poles in this region, satisfying:

$$e^{ip_0 a} = \pm e^{-\omega_{\mathbf{p}} a} \equiv \pm \left(\sqrt{1 + M_{\mathbf{p}}^2} - M_{\mathbf{p}} \right), \qquad (1.210)$$

with

$$M_{\mathbf{p}}^2 \equiv m^2 a^2 + \sum_{k=1}^{3} \sin(p_k a)^2. \qquad (1.211)$$

The integral can be easily performed and gives:

$$\langle \psi_\alpha(x) \bar{\psi}_\beta(0) \rangle_F = \int \frac{d^3 p}{(2\pi)^3} \frac{e^{i\mathbf{p}\mathbf{x}} e^{-\omega_{\mathbf{p}} x_0}}{\sinh(2\,\omega_{\mathbf{p}} a)} \left[\left(\gamma_0 \sinh \omega_{\mathbf{p}} a - i \sum_k \gamma_k \sin p_k a + ma \right) \right.$$
$$\left. + (-1)^{x_0/a} \left(-\gamma_0 \sinh \omega_{\mathbf{p}} a - i \sum_k \gamma_k \sin p_k a + ma \right) \right]_{\alpha\beta}. \qquad (1.212)$$

Two new features appear with respect to the scalar case:

- There are two terms in the sum with the same energy, $\omega_{\mathbf{p}}$, but different residue.
- The energy, $\omega_{\mathbf{p}}$, as a function of the spatial momenta in one direction (the others are set to zero) is shown in Fig. 1.5. There are two different minima in the BZ (the one at $-\pi$ is the same as that at π by periodicity). More generically, we find 2^3 minima at

$$p_k = \bar{p}_k \equiv n_k \frac{\pi}{a} \qquad n_k = 0, 1. \qquad (1.213)$$

As we approach the continuum limit:

$$\lim_{a \to 0} \omega_{\mathbf{p}} \big|_{p_k = n_k \pi/a} = m. \qquad (1.214)$$

Therefore, the minima correspond to the same energy.

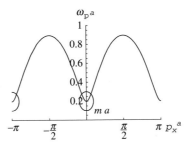

Fig. 1.5 ω_{p} as a function of $p_x a$ for $p_y = p_z = 0$.

Near the continuum limit, it is justified to consider the contribution near these momenta, so let us consider the expansion around them:

$$p_j = \bar{p}_j^{(i)} + k_j, \quad k_j a \ll 1, \tag{1.215}$$

where $i = 1, \ldots, 2^3$. It is easy to see that

$$\sinh 2\omega_{\mathrm{p}} a \simeq 2 a k_0 + O(a^2), \quad \sinh \omega_{\mathrm{p}} a \simeq a k_0 + O(a^2),$$
$$\sin p_j a \simeq \cos(\bar{p}^{(i)} a) k_j a + O(a^2), \quad k_0 \equiv \sqrt{m^2 + \mathbf{k}^2}. \tag{1.216}$$

Putting it all together

$$\sum_{i=1}^{8} e^{i\mathbf{p}^{(i)}\mathbf{x}} \int \frac{d^3 k}{(2\pi)^3} \frac{e^{i\mathbf{k}\mathbf{x}} e^{-\omega_{\mathrm{p}} t}}{2k_0} \left[\left(\gamma_0 k_0 - i \sum_j \gamma_j \cos(\bar{p}_j^{(i)} a) k_j + m \right) \right.$$
$$\left. + (-1)^{t/a} \left(-\gamma_0 k_0 - i \sum_j \cos(\bar{p}_j^{(i)} a) \gamma_j k_j + m \right) \right]$$
$$= \sum_{\alpha=1}^{16} e^{i\bar{p}^{(\alpha)}x} \int \frac{d^3 k}{(2\pi)^3} \frac{e^{i\mathbf{k}\mathbf{x}} e^{-\omega_{\mathrm{p}} X_0}}{2k_0} \left[\gamma_0 \cos(\bar{p}_0^{(\alpha)} a) k_0 - i \sum_j \gamma_j \cos(\bar{p}_j^{(\alpha)} a) k_j + m \right], \tag{1.217}$$

where we have used the fact that the second term can be written in the same form as the first, corresponding to a different temporal momenta $\bar{p}_0 = \pi/a$. The 16 terms now correspond to

$$\bar{p}_\mu = (n_0, n_1, n_2, n_3) \frac{\pi}{a}, \quad n_\mu = 0, 1. \tag{1.218}$$

We can find unitarity operators S_a such that

$$S_\alpha \gamma_\mu S_\alpha^\dagger = \gamma_\mu \cos(\bar{p}_\mu^{(\alpha)} a). \tag{1.219}$$

For example

$$S_\alpha = \prod_\mu (i\gamma_\mu\gamma_5)^{n_\mu^{(\alpha)}}, \tag{1.220}$$

satisfies this property. Therefore, we can write the continuum limit as

$$\sum_{\alpha=1}^{16} e^{i\bar{p}^{(\alpha)}x} \int \frac{d^3k}{(2\pi)^3} \frac{e^{i\mathbf{k}\mathbf{x}}e^{-\omega_\mathbf{p}t}}{2k_0} S_\alpha \left[\left(\gamma_0 k_0 - i\sum_k \gamma_k k_k + m\right)\right] S_\alpha^{-1}. \tag{1.221}$$

We can now recognize in each term the contribution of a relativistic fermion in the continuum, Eqs. (1.204), since S_α is just a similarity transformation: an equivalent representation of the γ matrices.

Summarizing, we have found that the continuum limit contains 16 relativistic free fermions instead of 1. This is the famous *doubling problem* (Wilson, 1975; Susskind, 1977).

1.4.2 Doubling and chiral symmetry

There is a deep connection between the doubling problem and the difficulty to regularize chirality (Nielsen and Ninomiya, 1981).

It is well known that in the absence of a mass term, the free-fermion action has a global symmetry under chiral rotations:

$$\psi(x) \to e^{i\alpha\gamma_5}\psi(x). \tag{1.222}$$

The naive discretization we just considered also has an exact global symmetry of this form. The invariance under chiral rotations implies that the Dirac spinor representation is actually reducible to its chiral components. Therefore, in the continuum we can consider a free Weyl fermion as the left or right chiral component that we can define by applying a projector on the Dirac field

$$\psi_L \equiv \frac{1-\gamma_5}{2}\psi, \quad \psi_R \equiv \frac{1+\gamma_5}{2}\psi. \tag{1.223}$$

Let us see what happens with the doublers when we naively discretize the action for a Weyl fermion. The naive propagator, Eq. (1.209), is (for $m = 0$):

$$\langle \psi_L(x)\bar{\psi}_L(0)\rangle_F = \int_{BZ} \frac{d^4p}{(2\pi)^4} \frac{e^{ipx}}{ia^{-1}\gamma_\mu \sin p_\mu a} \left(\frac{1-\gamma_5}{2}\right)$$

$$\approx \sum_{\alpha=1}^{16} e^{i\bar{p}^{(\alpha)}x} \int \frac{d^3k}{(2\pi)^3} \frac{e^{i\mathbf{k}\mathbf{x}}e^{-\omega_\mathbf{p}t}}{2k_0} S_\alpha \left[\left(\gamma_0 k_0 - i\sum_k \gamma_k k_k\right) S_\alpha^{-1}\left(\frac{1-\gamma_5}{2}\right)\right]$$

$$= \sum_{\alpha=1}^{16} e^{i\bar{p}^{(\alpha)}x} \int \frac{d^3k}{(2\pi)^3} \frac{e^{i\mathbf{k}\mathbf{x}}e^{-\omega_\mathbf{p}t}}{2k_0} S_\alpha \left[\left(\gamma_0 k_0 - i\sum_k \gamma_k k_k\right)\left(\frac{1-\epsilon^a\gamma_5}{2}\right)\right] S_\alpha^{-1},$$

$$\tag{1.224}$$

where $\epsilon^\alpha = (-1)^{\sum_\mu n_\mu^{(\alpha)}}$. Therefore, each of the doublers contributes either a left-handed relativistic Weyl fermion for $\epsilon^\alpha = 1$ or a right-handed one $\epsilon^\alpha = -1$ in the continuum. It turns out that the number of right and left movers is the same!

$$\text{Left}: 1 + 6 + 1 = 8, \tag{1.225}$$

$$\text{Right}: 4 + 4 = 8. \tag{1.226}$$

This result can be generalized to rather arbitrary forms of the fermionic action. It is the content of the famous *Nielsen–Ninomiya theorem* (Nielsen and Ninomiya, 1981). In its Euclidean version, the theorem considers actions of the form

$$S_F = a^4 \sum_{x,y} \bar{\psi}(x)\gamma_\mu F_\mu(x-y)(1-\gamma_5)\psi(y), \tag{1.227}$$

satisfying the following properties:

- action quadratic in the fermion fields;
- invariant under lattice translations (i.e. diagonal in momentum space);
- local (smooth Fourier transform);
- hermitian action: $F_\mu(x)^* = -F_\mu(x)$ (implies a real Fourier transform of F_μ field).

We also assume that the function F_μ has some isolated zeros (in order to have a continuum limit). Let us call $\bar{p}^\alpha$ the zeros of $F_\mu(p)$. Sufficiently close we can approximate

$$F_\mu(p) \simeq M_{\mu\nu}^{(\alpha)}(p - \bar{p}^\alpha)_\nu + \ldots = M_{\mu\nu}^{(\alpha)} k_\nu^{(\alpha)} + \ldots, \tag{1.228}$$

where $M_{\mu\nu}^{(\alpha)}$ is a real matrix that can be decomposed in general as

$$M_{\mu\nu}^{(\alpha)} = O_{\mu\rho}^{(\alpha)} S_{\rho\nu}^{(\alpha)}, \tag{1.229}$$

where O^α is an orthogonal matrix and S^α is a positive real symmetric matrix. The orthogonal matrix can be reabsorbed in a unitarity rotation of the fields for the following reason. Consider a rotation in $d+1$ (with d even) Euclidean space that acts in the first d coordinates as O and in the last coordinate it multiplies by $\det^{-1} O = \pm 1$. Such a rotation therefore belongs to $SO(d+1)$. The spinor representation of such rotations are the $d+1$ γ matrices: (γ_μ, γ_5). There must exist therefore a unitary matrix that implements the rotation in the spinor representation such that

$$\Lambda^{(\alpha)}\gamma_\nu\Lambda^{(\alpha)-1} = O_{\mu\nu}^{(\alpha)}\gamma_\mu \qquad \Lambda^{(\alpha)}\gamma_5\Lambda^{(\alpha)-1} = \det^{-1} O^{(\alpha)}\gamma_5. \tag{1.230}$$

Therefore, we can rewrite the action as

$$\sum_\alpha \int \frac{d^4 k^{(\alpha)}}{(2\pi)^4} \bar{\psi}(-k^{(\alpha)})\Lambda^{(\alpha)}\gamma_\rho S_{\rho\nu}^{(\alpha)} k_\nu^{(\alpha)}(1 - \det O^{(\alpha)}\gamma_5)\Lambda^{(\alpha)-1}\psi(k^{(\alpha)}). \tag{1.231}$$

The real positive matrix $S^{(\alpha)}$ is harmless and can be reabsorbed in a rescaling of the momentum. However, we see that there are left-movers and right-movers depending on the sign of $\det O^{(\alpha)}$. A theorem by Poincaré–Hopf states that $\sum_\alpha \det O^{(\alpha)}$ is the Euler characteristic of the manifold on which the vector $F_\mu(p)$ is defined. It is zero for the Brillouin zone (which is topologically a four-torus). Therefore, there must be as many zeros with $\det O^{(\alpha)} = 1$ as those with $\det O^{(\alpha)} = -1$. In particular, this implies the number of zeros cannot be one!

Intuitively, this is a generalization of a simpler version of the theorem for one-dimensional functions: a smooth and periodic function that crosses zero must do it an even number of times with opposite signs of the derivatives at the zeros.

Not surprisingly the easiest way to get rid of doublers is to break chiral symmetry. This is Wilson's solution to the doubling problem (Wilson, 1975).

1.4.3 Wilson fermions

K. Wilson proposed to add to the naive action the following term

$$\Delta_W S = -a^4 \sum_x \bar{\psi}(x) \frac{ra}{2} \hat{\partial}_\mu^* \hat{\partial}_\mu \psi(x), \tag{1.232}$$

where r is some arbitrary constant of $O(1)$. Note that this term does break explicitly chiral symmetry since it is like a momentum-dependent mass term. It is easy to see that the propagator in momentum space is modified to

$$\langle \psi_\alpha(x) \bar{\psi}_\beta(y) \rangle_F = \int_{BZ} \frac{d^4p}{(2\pi)^4} \frac{e^{ip(x-y)}}{\sum_\mu i\gamma_\mu \frac{\sin(p_\mu a)}{a} + m + \frac{r}{a} \sum_\mu (1 - \cos p_\mu a)}. \tag{1.233}$$

As before, the integration over p_0 can be performed as a sum of residues of the solutions, in the region Im $p_0 > 0$, $-\pi < \text{Re } p_0 < \pi$, of

$$\sum_\mu \sin^2 p_\mu + \left(m + \frac{r}{a} \sum_\mu (1 - \cos p_\mu a) \right)^2 = 0. \tag{1.234}$$

For $r = 1$ (Wilson's choice) the only solution is at $p_0 = i\omega_{\mathbf{p}}$ satisfying

$$\cosh \omega_{\mathbf{p}} = \frac{1 + \sum_k \sin^2 p_k a + (ma + 1 + \sum_k (1 - \cos p_k a))^2}{2(ma + 1 + \sum_k (1 - \cos p_k a))}. \tag{1.235}$$

The pole corresponding to the temporal doubler is absent. Also, the spatial momenta of Eq. (1.213), have an energy

$$\omega_{\mathbf{p}}^{(\alpha)} = \frac{1}{a} \ln \left(1 + ma + 2 \sum_k n_k^{(\alpha)} \right), \tag{1.236}$$

therefore the only pole that survives in the continuum limit (i.e. $\lim_{a \to 0} a\omega_{\mathbf{p}} = 0$) corresponds to $n_k^{(\alpha)} = 0$ for all k. The others have energies of the order of the cutoff.

Wilson's solution therefore removes doublers at the cost of breaking chiral symmetry.

Exercise 1.4 Symmetries of Wilson fermions. Show that the Wilson Dirac operator satisfies γ_5-hermiticity:

$$D^\dagger = \gamma_5 D \gamma_5$$

and is invariant under the discrete symmetries: C, P and T:

$$P : \psi(x) \to \gamma_0 \psi(x_P) \tag{1.237}$$

$$\bar\psi(x) \to \bar\psi(x_P)\gamma_0 \tag{1.238}$$

$$T : \psi(x) \to \gamma_0\gamma_5\psi(x_T) \tag{1.239}$$

$$\bar\psi(x) \to \bar\psi(x_T)\gamma_5\gamma_0 \tag{1.240}$$

$$C : \psi(x) \to C\bar\psi^T(x) \tag{1.241}$$

$$\bar\psi(x) \to -\psi^T(x)C^{-1}, \tag{1.242}$$

$$\tag{1.243}$$

where $x_P = (x_0, -\mathbf{x})$, $x_T = (-x_0, \mathbf{x})$ and $C = \gamma_0\gamma_2$, which satisfies $C\gamma_\mu C = -\gamma_\mu^* = -\gamma_\mu^T$.

Exercise 1.5 Show that the γ_5-hermiticity implies that for the complex eigenvalues of D, the corresponding eigenvectors satisfy

$$v_\lambda^\dagger \gamma_5 v_\lambda = 0, \quad \lambda^* \neq \lambda. \tag{1.244}$$

Real eigenvalues on the other hand can have non-zero chirality.

1.4.3.1 *Transfer matrix of Wilson fermions and unitarity*

Actually, Wilson fermions with $r = 1$ are the only fermion regularization for which the transfer matrix has been proven to be positive (Wilson, 1975; Lüscher, 1977; Smit, 1991).

As in the scalar case, we will proceed by finding a transfer operator $\hat{T}$ acting on Fock space such that

$$\mathcal{Z}_F = \lim_{N\to\infty} \text{Tr}[\hat{T}^N], \tag{1.245}$$

and proving that it is positive in such a way that the Hamiltonian $\hat{H} = -\frac{1}{a}\ln\hat{T}$ is well defined.

We need the equivalent of the Schrödinger representation of states. For the scalar field we defined the basis $|\phi\rangle$ (the analog of the position basis in ordinary QM), such that

$$\hat\phi(x)|\phi\rangle = \phi(x)|\phi\rangle. \tag{1.246}$$

In the fermion case, similarly, we define a basis $|a\rangle$ (Smit, 2002), such that

$$\hat{a}_k |a\rangle = a_k |a\rangle, \tag{1.247}$$

where $\hat{a}_k$ are the annihilation operators in Fock space and a_k are Grassmann variables that represent the classical fermion field, which can be shown to anticommute with the operators.

One can show that the state $|a\rangle$ can be constructed from the vacuum as:

$$|a\rangle = \prod_k e^{-a_k \hat{a}_k^\dagger} |0\rangle. \tag{1.248}$$

Using the properties of the Grassmann integrals, one can also show that the basis $|a\rangle$ satisfies the completeness relation

$$\int da^\dagger da \, \frac{|a\rangle\langle a|}{\langle a|a\rangle} = 1, \tag{1.249}$$

where

$$\langle a|a\rangle = \prod_k e^{a_k^\dagger a_k} \equiv e^{a^\dagger a}, \quad a^\dagger a = \sum_k a_k^\dagger a_k. \tag{1.250}$$

Any arbitrary state in Fock space can be written as

$$|\psi\rangle = \sum_p \frac{1}{p!} \psi_{k_1,\ldots k_p} \hat{a}_{k_1}^\dagger \ldots \hat{a}_{k_p}^\dagger |0\rangle. \tag{1.251}$$

It has a wave function in the $|a\rangle$ basis:

$$\langle a|\psi\rangle \equiv \psi(a^\dagger) = \sum_p \frac{1}{p!} \psi_{k_1,\ldots k_p} a_{k_1}^\dagger \ldots a_{k_p}^\dagger. \tag{1.252}$$

Let us consider any normal-ordered operator $\hat{A}$

$$\hat{A} = \sum_{p,q} \frac{1}{p!q!} A_{k_1 \ldots k_p} \hat{a}_{k_1}^\dagger \ldots \hat{a}_{k_p}^\dagger \hat{a}_{l_q} \ldots \hat{a}_{l_1}. \tag{1.253}$$

The matrix elements of the operators in this basis can be shown to be:

$$A(a^\dagger, a) \equiv \langle a|\hat{A}|a\rangle = \langle a|a\rangle \sum_{p,q} \frac{1}{p!q!} A_{k_1 \ldots k_p} a_{k_1}^\dagger \ldots a_{k_p}^\dagger a_{l_q} \ldots a_{l_1}. \tag{1.254}$$

Finally, the following relations can also be derived (Smit, 2002):

- Trace:

$$\text{Tr}\hat{A} \equiv \sum_p \frac{1}{p!} \sum_{k_1,\ldots,k_p} \langle k_1,\ldots,k_p|\hat{A}|k_1,\ldots k_p\rangle = \int da^\dagger da \, e^{-a^\dagger a} A(a^\dagger, -a). \tag{1.255}$$

- The product of three operators, $\hat{A}, \hat{B}$ and $\hat{C}$, where $\hat{B}/\hat{C}$ only depend on creation/destruction operators, respectively, while $\hat{A}$ depends on both, satisfies:

$$\langle a|\hat{B}\hat{A}\hat{C}|a\rangle = B(a^\dagger)A(a^\dagger, a)C(a).\tag{1.256}$$

- Operators of the exponential form

$$\hat{A} = \exp\left(\sum_{kl} \hat{a}_k^\dagger M_{kl}\hat{a}_l\right),\tag{1.257}$$

satisfy

$$A(a^\dagger, a) = \exp\left(a^\dagger e^M a\right).\tag{1.258}$$

Let us see now how we can identify the transfer operator

$$\mathrm{Tr}[\hat{T}^N] = \int da_N^\dagger da_N e^{-a_N^\dagger a_N}\langle a_N|\hat{T}^N| - a_N\rangle$$

$$= \int \prod_n \left(da_n^\dagger da_n\right) e^{-a_N^\dagger a_N}\langle a_N|\hat{T}|a_{N-1}\rangle e^{-a_{N-1}^\dagger a_{N-1}}$$

$$\langle a_{N-1}|\hat{T}|\ldots.|a_1\rangle e^{-a_1^\dagger a_1}\langle a_1|\hat{T}| - a_N\rangle,\tag{1.259}$$

that should be compared with Eq. (1.203). As in the scalar case, we should somehow identify the a_n with the ψ at fixed times.

The Wilson fermion action (for $r = 1$) can be written as

$$S_W[\psi, \bar{\psi}] = a^3 \sum_{x_0}\sum_{\mathbf{x,y}} \psi^\dagger(\mathbf{x}, x_0)(\gamma_0 A_{\mathbf{xy}} + B_{\mathbf{xy}})\psi(\mathbf{y}, x_0)$$

$$+ a^4 \sum_{x_0, y_0}\sum_{\mathbf{x}} \psi^\dagger(\mathbf{x}, x_0)\frac{1}{2a}\left(P_-\delta_{y_0 x_0 + a} - P_+\delta_{y_0 x_0 - a}\right)\psi(\mathbf{x}, y_0),\tag{1.260}$$

where

$$A_{\mathbf{xy}} \equiv (ma + 4)\,\delta_{\mathbf{xy}} - \frac{1}{2}\sum_k(\delta_{\mathbf{yx}+\hat{k}a} + \delta_{\mathbf{yx}-\hat{k}a})\tag{1.261}$$

$$B_{\mathbf{xy}} \equiv \frac{1}{2}\sum_k \gamma_0\gamma_k(\delta_{\mathbf{yx}+\hat{k}a} - \delta_{\mathbf{yx}-\hat{k}a})\tag{1.262}$$

and $P_\pm = (1 \pm \gamma_0)/2$ are projectors in spinor space, with $P_+ + P_- = 1$.

Let us now decompose the fermions into their $\pm$ components and let us define a basis of the Grassmann variables a_{x_0} (we omit for simplicity the index that runs over $\mathbf{x}$ and the spinor indices) in the following way:

$$(a_{x_0}^\dagger P_+)^T \equiv P_+ \psi(x_0) a^{3/2}, \quad P_- a_{x_0} \equiv P_- \psi(x_0 + a) a^{3/2}, \tag{1.263}$$

$$a_{x_0}^\dagger P_- \equiv \psi^\dagger(x_0) P_- a^{3/2}, \quad (P_+ a_{x_0})^T \equiv \psi^\dagger(x_0 + a) P_+ a^{3/2}, \tag{1.264}$$

so that the $\pm$ components of ψ correspond to those of the a variables at different timeslices. With these identifications, we can rewrite the action as

$$S_W[\psi, \bar\psi] = \sum_{x_0} \left(a_{x_0}^\dagger a_{x_0} - a_{x_0}^\dagger A a_{x_0 - a} + a_{x_0 - a} P_+ B P_- a_{x_0 - a} + a_{x_0}^\dagger P_- B P_+ a_{x_0}^\dagger \right). \tag{1.265}$$

Therefore, we find an exact matching if we identify $|a_n\rangle \to |a_{x_0}\rangle$ so that

$$e^{-\sum_{x_0} a_{x_0}^\dagger a_{x_0}} \to e^{-\sum_n a_n^\dagger a_n} \tag{1.266}$$

$$\langle a_n | \hat{T} | a_{n-1} \rangle \to \langle a_{x_0} | \hat{T} | a_{x_0 - a} \rangle$$
$$= \exp(-a_{x_0}^\dagger P_+ B P_- a_{x_0}^\dagger) \exp(a_{x_0}^\dagger A a_{x_0 - a}) \exp(-a_{x_0 - a} P_+ B P_- a_{x_0 - a}), \tag{1.267}$$

which implies, according to Eq. (1.258),

$$\hat{T} = \exp(-\hat{a}^\dagger P_+ B P_- \hat{a}^\dagger) \exp(\hat{a}^\dagger \ln(A) \hat{a}) \exp(-\hat{a} P_- B P_+ \hat{a}). \tag{1.268}$$

Probing the positivity is now straightforward. The operator in the middle is positive if A is positive. In momentum space the operator is

$$A(p) = ma + 4 + \sum_k \cos p_k a > 0. \tag{1.269}$$

Since the transfer matrix has the structure

$$\hat{T} = \hat{T}_1^\dagger \hat{T}_2 \hat{T}_1, \tag{1.270}$$

for any state $|\psi\rangle$

$$\langle \psi | \hat{T}_1^\dagger \hat{T}_2 \hat{T}_1 | \psi \rangle = \langle \xi | \hat{T}_2 | \xi \rangle > 0, \quad |\xi\rangle = \hat{T}_1 | \psi \rangle, \tag{1.271}$$

and $\hat{T}$ is a positive Hermitian operator, from which a Hermitian Hamiltonian can be defined. Therefore, the lattice formulation of Wilson fermions with $r = 1$ has a direct Hilbert-space interpretation, just as the scalar theory.

The case $r \neq 1$ cannot be treated in the same way, and in fact positivity has not been proven. Reflection positivity on the other hand can be proved for $r \leq 1$ (Osterwalder and Seiler, 1978; Menotti and Pelissetto, 1987).

1.4.4 Kogut–Susskind or staggered fermions

Given that naive lattice fermions correspond to 2^d Dirac fermions in the continuum, one idea would be to use some of the doublers to represent the $4 = 2^{d/2}$ spinor components of a Dirac fermion. Kogut and Susskind (Kogut and Susskind, 1975) proved that this can be done, therefore reducing the doubling problem to that of $2^d/2^{d/2}$ replicas instead of 2^d. The advantage is that the lattice action can be shown to have an extra exact $U(1)$ symmetry compared to the Wilson action.

Let us briefly review the construction of the Kogut–Susskind action. There are two steps

- Perform a local unitary rotation of the fermion fields that diagonalizes the action in spinor space. That is, find a unitary S_x such that

$$S_x^\dagger \gamma_\mu S_{x+\hat\mu a} = \rho_{x\mu}\mathbf{I}. \tag{1.272}$$

It is easy to prove that the choice

$$S_x \equiv \gamma_0^{n_0} \cdots \gamma_3^{n_3} = \prod_\mu \gamma_\mu^{n_\mu} \quad x = a\,(n_0, n_1, n_2, n_3) \tag{1.273}$$

satisfies Eq. (1.272) with $\rho_{x\mu} = (-1)^{\sum_{\rho<\mu} n_\rho}$. We can therefore perform the transformation of the spinors

$$\psi(x)_\alpha \to (S_x)_{\alpha\beta}\psi_\beta(x) \equiv \chi_\alpha(x), \tag{1.274}$$

and the action factorizes in the four spinor components

$$S_{KS} = a^4 \sum_{x,\alpha} \left[\sum_\mu \rho_{x\mu}\bar\chi^\alpha(x)\frac{1}{2}\left(\chi^\alpha(x+a\hat\mu) - \chi^\alpha(x-a\hat\mu)\right) + m\bar\chi^\alpha(x)\chi^\alpha(x) \right]. \tag{1.275}$$

We can therefore consider just *one* of these replicas that we call χ.

- Reconstruction of the Dirac field

In order to reconstruct the Dirac field using the doublers associated to the variable χ, one needs to consider a lattice with a doubled lattice spacing $2a$. The Dirac fields will be Grassmann variables defined on this coarser lattice. We can define therefore coordinates in the new lattice as

$$y_\mu = 2aN_\mu, \tag{1.276}$$

while the coordinates of the points in the original lattice can be labelled as

$$x = an_\mu = 2aN_\mu + az_\mu, \quad z_\mu = 0, 1. \tag{1.277}$$

Therefore, we define new fields on the coarser lattice that will have space-time coordinates $y_\mu = 2aN_\mu$ and also other 2^d internal components labelled by z_μ:

$$\chi(na) \equiv \psi_z(2Na). \tag{1.278}$$

It is easy to show that

$$\chi(n + \hat{\mu}) = \sum_{z'} \delta_{z+\hat{\mu}z'}\psi_{z'}(N) + \delta_{z-\hat{\mu}z'}\psi(N + \hat{\mu})_{z'} \tag{1.279}$$

$$\chi(n - \hat{\mu}) = \sum_{z'} \delta_{z-\hat{\mu}z'}\psi_{z'}(N) + \delta_{z+\hat{\mu}z'}\psi(N - \hat{\mu})_{z'}. \tag{1.280}$$

Defining

$$\Gamma^{\mu}_{zz'} \equiv \rho_{z\mu}(\delta_{z+\hat{\mu}z'} + \delta_{z-\hat{\mu}z'}), \quad \Gamma^{5\mu}_{zz'} \equiv \rho_{z\mu}(\delta_{z-\hat{\mu}z'} - \delta_{z+\hat{\mu}z'}), \tag{1.281}$$

where

$$\rho_{z\mu} = (-1)^{\sum_{\nu \leq \mu} z_{\nu}} \tag{1.282}$$

it is easy to show that in terms of the new fields the action is

$$S_{KS} = a^4 \sum_{N,z} \sum_{\mu} \bar{\psi}_z(N) \frac{1}{4} \left[\Gamma^{\mu}_{zz'}(\hat{\partial}_{\mu} + \hat{\partial}^*_{\mu}) + \Gamma^{5\mu}_{zz'} a \hat{\partial}^*_{\mu} \hat{\partial}_{\mu} \right] \psi_{z'}(N), \tag{1.283}$$

where $\hat{\partial}_{\mu}, \hat{\partial}^*_{\mu}$ are the forward and backward derivatives in the coarser lattice.

Furthermore, we can show that

$$\Gamma^{\mu}_{zz'} = \text{Tr}\left[S^{\dagger}_z \gamma_{\mu} S_{z'} \right], \tag{1.284}$$

$$\Gamma^{5\mu}_{zz'} = \text{Tr}\left[S^{\dagger}_z \gamma_5 S_{z'} \gamma_5 \gamma_{\mu} \right], \tag{1.285}$$

with

$$S_z \equiv \prod_{\nu} \gamma^{z_{\nu}}_{\nu}. \tag{1.286}$$

Finally, defining

$$\Psi^{\alpha i}(N) \equiv \sum_z (S_z)_{\alpha i} \psi_z(N), \quad \bar{\Psi}^{\alpha i}(N) \equiv \sum_z \bar{\psi}_z(N) \left(S^{\dagger}_z \right)_{i\alpha}, \tag{1.287}$$

we can get back four Dirac spinors with spinor index α and flavor index i. After a simple normalization we get the Kogut–Susskind action in terms of the new variables:

$$S_{\text{KS}} = (2a)^4 \sum_{\mu,N} \left[\bar{\Psi}(N)(\gamma_{\mu} \otimes 1)\frac{1}{2}(\hat{\partial}_{\mu} + \hat{\partial}^*_{\mu})\Psi(N) + a\bar{\Psi}(N)(\gamma_5 \otimes \gamma^T_{\mu}\gamma^T_5)\frac{1}{2}a \, \hat{\partial}_{\mu}\hat{\partial}^*_{\mu}\Psi(N) \right]$$

$$+ (2a)^4 m \sum_N \bar{\Psi}(N)\Psi(N), \tag{1.288}$$

where the first γ matrices act on spinor variables, and the second acts on flavor ones.

A few comments are in order:

- In the naive continuum limit, the "Wilson"-type term vanishes and the action goes to the continuum action of four free massive Dirac spinors.
- The action looks quite similar to the Wilson action. The difference is the Dirac/flavor structure of the Wilson term.
- The action has an exact $U(1)$ chiral symmetry for $m = 0$ under spin-flavor rotations of the form

$$\Psi_N \to e^{i\alpha(\gamma_5 \otimes \gamma_5^T)} \Psi_N, \quad \bar{\Psi}_N \to \bar{\Psi}_N e^{i\alpha(\gamma_5 \otimes \gamma_5^T)}. \tag{1.289}$$

This symmetry can be preserved in the interacting case, and ensures a chiral symmetry in the continuum limit without extra fine tunings.

The transfer matrix operator has been constructed also for staggered fermions (Smit, 2002), but it is not positive, therefore there is no guarantee that the formulation has a Hilbert-space formulation at finite lattice spacing. The hope is therefore that unitarity is recovered in the continuum limit, which seems to be the case in the lowest orders of perturbation theory. Further subtleties of staggered fermions in the interacting case can be found in the lectures of M. Golterman (Golterman, 2009).

Recently, very significant progress has been achieved in the constructions of fermion actions that preserve a lattice chiral symmetry. These new developments are covered in the lectures of D. Kaplan (Kaplan, 2009).

Exercise 1.6 Two flavors of twisted-mass Wilson fermions are defined by the Wilson action with a mass term that has the form

$$m + i\mu\gamma_5\tau_3, \quad \tau_3 = \begin{pmatrix} 1 & 0 \\ 0 & -1 \end{pmatrix}. \tag{1.290}$$

Show that in the naive continuum limit the action is equivalent to the standard Dirac action by performing a chiral rotation of the form

$$\psi \to e^{i\frac{\alpha}{2}\gamma_5\tau_3}\psi, \quad \bar{\psi} \to \bar{\psi}e^{i\frac{\alpha}{2}\gamma_5\tau_3}\psi, \quad \tan\alpha = \frac{\mu}{m}. \tag{1.291}$$

1.5   Lattice gauge fields

1.5.1   Lattice gauge field theories: abelian case

K. Wilson figured out how to formulate a quantum field theory of gauge fields on the lattice preserving an exact gauge invariance (Wilson, 1974).

We will first derive the lattice formulation of an abelian gauge theory and then we will generalize the construction to other gauge theories such as $SU(3)$ describing the color interactions.

An easy way to understand how this is done is to consider the case of charged particles (scalar ones to make it simple). A scalar charged particle is described by a complex scalar field. The results of Section 1.3 can be readily applied to a complex scalar field.

It is well known that gauge invariance in quantum mechanics corresponds to a symmetry under local rephasing of the wave functions describing the charged particles, and a shift of the gauge potentials. Maxwell's equations are invariant under

$$A_\mu(x) \rightarrow A_\mu(x) + \partial_\mu \Lambda(x), \tag{1.292}$$

while the Schrödinger equation describing a particle with charge q in this field is also invariant if there is a simultaneous rephasing of the charged field wave function by

$$\phi(x) \rightarrow e^{iq\Lambda(x)}\phi(x) \equiv \Omega(x)\phi(x). \tag{1.293}$$

Let us consider the case in which the electromagnetic field strength vanishes in all space, that is we consider a *pure gauge* configuration, i.e. $A_\mu = \partial_\mu F(x)$ for arbitrary $F(x)$. We can then choose a gauge in which the gauge potential vanishes. In this gauge it is easy to discretize the scalar action, it is just the one corresponding to a free scalar field:

$$S = \frac{a^4}{2} \sum_{x,y} \phi^\dagger(x) K_{xy} \phi(y), \tag{1.294}$$

with

$$K_{xy} = -\frac{1}{a^2} \sum_{\hat\mu} \left(\delta_{x+a\hat\mu y} + \delta_{x-a\hat\mu y} - 2\delta_{xy}\right) + m^2 \delta_{xy}. \tag{1.295}$$

Now, let us change the gauge, which implies a rephasing of the charged fields,

$$\phi(x) \rightarrow e^{iq\Lambda(x)}\phi(x) = \phi'(x) \quad \phi(x)^\dagger \rightarrow \phi(x)^\dagger e^{-iq\Lambda(x)} = \phi'(x)^\dagger, \tag{1.296}$$

and a change of the gauge field to $A'_\mu = \partial_\mu \Lambda(x)$.

The action in terms of the new fields, $\phi'(x)$, should therefore correspond to the action of a scalar field coupled to the gauge field $A'_\mu = \partial_\mu \Lambda$. Substituting $\phi(x)$ in terms of $\phi'(x)$ in Eq. (1.294), we find:

$$S = \frac{a^4}{2} \sum_{x,y} \phi'^\dagger(x) K_{xy}^\Lambda \phi'(y), \tag{1.297}$$

where

$$K_{xy}^\Lambda = -\frac{1}{a^2} \sum_{\hat\mu} \left(\delta_{x+a\hat\mu y} U_\mu(x) + \delta_{x-a\hat\mu y} U_\mu^\dagger(x - a\hat\mu) - 2\delta_{xy}\right) + m^2 \delta_{xy}. \tag{1.298}$$

We have introduced the so-called *link variables*, defined as

$$U_\mu(x) \equiv e^{iq\Lambda(x)} e^{-iq\Lambda(x+a\hat\mu)} = e^{-iq \int_x^{x+a\hat\mu} \partial_\mu \Lambda(x) dx_\mu} \equiv e^{-iq \int_x^{x+a\hat\mu} dx_\mu A'_\mu(x)}. \tag{1.299}$$

The link variable is simply a *parallel transporter* between two adjacent points on the lattice, $x + a\hat{\mu}$ and x:

$$P(x, x + a\hat{\mu}) \equiv \exp\left(iq \int_{x+a\hat{\mu}}^{x} A'_\mu(x) dx_\mu\right). \tag{1.300}$$

The integral can be done along the straight line:

$$x_\mu(t) = x + ta\hat{\mu}, \quad t = [0, 1], \tag{1.301}$$

and this is why we associate it to a link. In the following, we will absorb the charge q in the gauge potential.

We can now check that the action in Eq. (1.297) in terms of A'_μ is gauge invariant for any gauge field (not just the pure gauge configurations we started with). Consider a general continuum gauge field $A_\mu(x)$, not necessarily with vanishing field strength. The gauge transformation of a parallel transporter between points x and y is

$$P^\Lambda(y, x) = \exp\left(i \int_x^y (A_\mu + \partial_\mu \Lambda) dx_\mu\right) = \exp\left(i \int_x^y A_\mu dx_\mu + i\Lambda(y) - i\Lambda(x)\right)$$

$$= \exp(i\Lambda(y)) P(y, x) \exp(-i\Lambda(x)) = \Omega(y) P(y, x) \Omega^\dagger(x). \tag{1.302}$$

The lattice action of Eq. (1.297) is indeed invariant under the gauge transformation

$$\phi'(x) \to \Omega(x)\phi'(x) \quad U_\mu(x) \to \Omega(x)U_\mu(x)\Omega^\dagger(x + a\hat{\mu}). \tag{1.303}$$

It is easy to generalize this procedure to fermions or any other charged fields. Starting with the free action we can couple the field to a gauge field by substituting the partial derivatives by covariant ones:

$$\hat{\partial}_\mu \psi(x) = \frac{1}{a}(\psi(x + a\hat{\mu}) - \psi(x)) \to \nabla_\mu \psi(x) = \frac{1}{a}(U_\mu(x)\psi(x + a\hat{\mu}) - \psi(x)),$$

$$\hat{\partial}_\mu^* \psi(x) = \frac{1}{a}(\psi(x) - \psi(x - a\hat{\mu})) \to \nabla_\mu^* \psi(x) = \frac{1}{a}(\psi(x) - U_\mu^\dagger(x - a\hat{\mu})\psi(x - a\hat{\mu})). \tag{1.304}$$

1.5.1.1 *Path integral*

Now that we have identified the parallel transporters as the basic gauge variables on the lattice, we still need to construct the Euclidean lattice path integral to represent the continuum one

$$\mathcal{Z} = \int dA_\mu e^{-S[A_\mu]} \quad S[A_\mu] \equiv \frac{1}{4} \int d^4x F_{\mu\nu} F_{\mu\nu}, \tag{1.305}$$

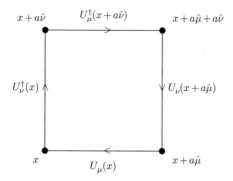

Fig. 1.6 Plaquette.

where $F_{\mu\nu} = \partial_\mu A_\nu - \partial_\nu A_\mu$ is the field strength. Obviously we must do this ensuring that Eq. (1.303) remains a symmetry.

Let us consider any ordered loop of parallel transporters, a so-called *Wilson loop*. A Wilson loop starting and ending in the point x transforms as

$$W(x) \equiv P(x, y_1)P(y_1, y_2).\ldots.P(y_n, x) \to \Omega(x)W(x)\Omega(x)^\dagger. \tag{1.306}$$

In the case of an abelian group, $W(x)$ is therefore invariant.

Since we want our action to be local, we can try with the smallest Wilson loop, which is a loop around the basic plaquette, see Fig. 1.6:

$$U_{\mu\nu}(x) \equiv U_\mu(x)U_\nu(x + a\hat{\mu})U_\mu^\dagger(x + a\hat{\nu})U_\nu^\dagger(x), \tag{1.307}$$

indeed under a gauge transformation

$$U_{\mu\nu}(x) \to \Omega(x)U_{\mu\nu}(x)\Omega(x)^\dagger. \tag{1.308}$$

It is an easy exercise to check that if we define a lattice gauge field $\hat{A}_\mu$ by

$$U_\mu \equiv e^{iqa\hat{A}_\mu(x)}, \tag{1.309}$$

then

$$U_{\mu\nu}(x) = e^{-iqa^2 \hat{F}_{\mu\nu} + \mathcal{O}(a^3)}, \tag{1.310}$$

where

$$\hat{F}_{\mu\nu}(x) \equiv \hat{\partial}_\mu \hat{A}_\nu(x) - \hat{\partial}_\nu \hat{A}_\mu(x), \tag{1.311}$$

and the derivatives are discrete ones, Eq. (1.134).

From this result it is easy to guess a good lattice action for the link variables:

$$S[U] = \frac{1}{q^2}\sum_x \sum_{\mu \leq \nu}\left[1 - \frac{1}{2}\left(U_{\mu\nu}(x) + U_{\mu\nu}^\dagger(x)\right)\right], \tag{1.312}$$

which satisfies the basic properties:

- it is local;
- it is real;
- it is gauge invariant;
- it has the right classical continuum limit:[6]

$$\lim_{a \to 0} S[U] = \int d^4x \frac{1}{4} F_{\mu\nu}^2 + \mathcal{O}(a^2). \tag{1.313}$$

We still need to define the measure over the link variables. Since the link variables are elements of $U(1)$ we can define a gauge invariant measure as

$$dU \equiv \prod_{\mu,x} d\phi_\mu(x) \quad U_\mu(x) = e^{i\phi_\mu(x)}, \quad 0 \le \phi_\mu(x) \le 2\pi. \tag{1.314}$$

Since the variables at different points are independent, and a gauge transformation induces a constant shift of the phase of each link variable,

$$\phi_\mu(x) \to \phi'_\mu(x) = \Lambda(x) + \phi_\mu(x) - \Lambda(x + a\hat{\mu}), \quad d\phi_\mu(x) = d\phi'_\mu(x), \tag{1.315}$$

this measure is gauge invariant. We will see that the measure is less trivial in the non-abelian case.

1.5.2 Lattice gauge field theories: non-abelian case

The color interactions in QCD are based on the non-abelian gauge symmetry $SU(3)$. In the continuum, the Yang–Mills theory based on a group $SU(N)$ is a quantum field theory of the vector gauge potential $A_\mu(x)$ that takes values in the Lie algebra of the gauge group:

$$A_\mu(x) = A_\mu^a(x) T^a, \tag{1.316}$$

where the coefficients $A_\mu^a(x)$ are real and $T^a = (T^a)^\dagger$ are the Hermitian generators of the algebra. The Yang–Mills field tensor is defined by

$$F_{\mu\nu}(x) = \partial_\mu A_\nu(x) - \partial_\nu A_\mu(x) - i[A_\mu(x), A_\nu(x)], \tag{1.317}$$

which is also an element of the algebra.

A gauge transformation is:

$$A_\mu(x) \to \Omega(x) A_\mu(x) \Omega(x)^{-1} + i\Omega(x) \partial_\mu \Omega(x)^{-1}, \tag{1.318}$$

where $\Omega(x) \in SU(N)$. It implies the following transformation of the field tensor

$$F_{\mu\nu}(x) \to \Omega(x) F_{\mu\nu}(x) \Omega(x)^{-1}. \tag{1.319}$$

The Euclidean Yang–Mills action is given by

$$S[A_\mu] = \frac{1}{2g_0^2} \int d^4x \, \text{Tr} \, [F_{\mu\nu} F_{\mu\nu}], \tag{1.320}$$

and is therefore gauge invariant.

A colored scalar field in the fundamental representation of this symmetry group transforms as

$$\phi(x) \rightarrow \phi'(x) = \Omega(x)\phi(x) \quad \Omega \in SU(N). \tag{1.321}$$

The main difference with the $U(1)$ case is that now the Ω are $N \times N$ matrices that do not commute.

We can proceed as for the $U(1)$ case above and start with the free action of colored scalar fields Eq. (1.294). We can then identify the way gauge fields appear in the lattice action by performing a gauge transformation of the colored fields, $\phi(x) \rightarrow \phi'(x)$. The field $\phi'(x)$ will then be coupled to a gauge field, according to Eq. (1.318),

$$A'_\mu(x) = i\Omega(x)\partial_\mu\Omega(x)^{-1}. \tag{1.322}$$

When we do this, we find the same result as in the $U(1)$ case, Eq. (1.297), provided we define the link variables as

$$U_\mu(x) \equiv \Omega(x)\Omega(x + a\hat{\mu})^\dagger. \tag{1.323}$$

This corresponds to the parallel transporter of the non-abelian gauge field Eq. (1.322) from $x + a\hat{\mu}$ to x.

To see this, let us recall the definition of a parallel transporter for $SU(N)$. Consider an N-component vector $\mathbf{v}$ of unit length and a curve in R^4 that can be parametrized by $z_\mu(t)$. A parallel transport of $\mathbf{v}$, along such a curve from points t_0 to t, in the presence of the field A_μ is the solution of the equation

$$\left[\frac{d}{dt} - i\frac{dz_\mu(t)}{dt} A_\mu(z_\mu(t)) \right] \mathbf{v}(t) = 0. \tag{1.324}$$

The parallel transporter from $z_\mu(t_0) = x$ to $z_\mu(t) = y$, $P(y, x)$, is the matrix that satisfies

$$\mathbf{v}(t) = P(y, x)\mathbf{v}(t_0). \tag{1.325}$$

In the abelian case, the solution to this equation is Eq. (1.300). For the non-abelian case, the solution can be written as a series in A_μ:

$$
\mathbf{v}(t) = \left(I + i \int_{t_0}^t dt_1 \dot{z}_\mu(t_1) A_\mu(z(t_1)) \right.
$$
$$
\left. - \int_{t_0}^t dt_1 \dot{z}_\mu(t_1) A_\mu(z(t_1)) \int_{t_0}^{t_1} dt_2 \dot{z}_\nu(t_2) A_\nu(z(t_2)) + \dots \right) \mathbf{v}(t_0)
$$
$$
\equiv P \exp \left(i \int_x^y A_\mu(z) dz_\mu \right) \mathbf{v}(t_0). \tag{1.326}
$$

Now it is easy to check (see exercise), using the definition, Eqs. (1.324) and (1.325), that the parallel transporter from two adjacent points on the lattice $x + a\hat{\mu}$ and x is the vector potential of Eq. (1.322) is given by Eq. (1.323).

Similarly, from the definition it is easy to show that the gauge transformation of a parallel transporter is

$$
P(y, x) \to \Omega(y) P(y, x) \Omega^\dagger(x). \tag{1.327}
$$

Exercise 1.7 Prove, using the definition of the parallel transporter of Eq. (1.324), that

- the link variable

$$
U_\mu(x) \equiv \Omega(x) \Omega(x + a\hat{\mu})^\dagger, \tag{1.328}
$$

  is a parallel transporter from $x + a\hat{\mu}$ to x;
- the gauge transformation of a parallel transporter is

$$
U(x, y) \to \Omega(x) U(x, y) \Omega^\dagger(y). \tag{1.329}
$$

These properties are sufficient to ensure the gauge invariance of the plaquette action also for $SU(N)$:

$$
S[U] \equiv C \sum_x \sum_{\mu < \nu} \mathrm{Tr} \left[1 - \frac{1}{2} \left(U_{\mu\nu}(x) + U_{\mu\nu}^\dagger(x) \right) \right]. \tag{1.330}
$$

The coefficient C can be chosen to recover the conventional normalization in the classical continuum limit, Eq. (1.320):

$$
C \equiv \frac{2}{g_0^2}, \tag{1.331}
$$

where g_0 is the gauge coupling.

1.5.2.1   *Gauge measure and path integral*

In order to define the path integral we still need to define the measure over the link variables in a gauge-invariant way. Since the link variables are elements of a compact group $SU(N)$, the measure is simply the Haar measure on the group, which can be proven to be the unique measure that obeys two essential properties

- it is gauge invariant

$$\int_{SU(N)} dU\, f(U) = \int_{SU(N)} f(VU)dU = \int_{SU(N)} f(UV)dU, \qquad (1.332)$$

  for any $V \in SU(N)$
- it is normalized

$$\int_{SU(N)} dU = 1. \qquad (1.333)$$

Let us consider any parametrization of the group in terms of n coordinates, z_i, then

$$dU = w(z)dz_1 dz_2 \ldots dz_n, \qquad (1.334)$$

where n must be the number of generators of the algebra. The invariance of the measure requires

$$dU(z') = d(VU(z)W^\dagger) = dU(z), \qquad (1.335)$$

for arbitrary $V, W \in SU(N)$. Therefore,

$$w(z')dz_1' \ldots dz_n' = w(z')|\det(\partial z_a'/\partial z_b)|dz_1 \ldots .dz_n = w(z)dz_1 \ldots dz_n \qquad (1.336)$$

or

$$w(z') = w(z)/J(z, z'), \qquad (1.337)$$

where $J(z, z')$ is the Jacobian of the transformation $z \to z'$. If there were two different measures satisfying this property, the only possibility is that both functions are proportional up to a constant. The constant is then fixed by the normalization condition and the measure is unique, given a set of coordinates.

Once we are sure that the measure is unique, we can find it by explicit construction. We can define a metric tensor in the group by

$$g_{kl} \equiv -2\mathrm{Tr}[(U\partial_k U^{-1})(U\partial_l U^{-1})], \qquad (1.338)$$

which can be shown to be positive definite and gauge invariant. The measure in these coordinates can then be defined as

$$w(z) = c\sqrt{\det g(z)}, \qquad (1.339)$$

where c is obtained from the normalization condition.

Exercise 1.8 Using the Haar measure, Eqs. (1.338) and (1.339), show that

$$\int dU\, f(U) = \int dU\, f(U^*) = \int dU\, f(U^{-1}).\qquad(1.340)$$

We will now show a few of the most commonly used coordinates in the simplest case of $SU(2)$.

1.5.2.2 *Examples of coordinate systems for* $SU(2)$

1) For $SU(2)$ a useful parametrization maps the group elements to a three-dimensional sphere S^3:

$$U = x_0 + ix_a\sigma_a, \qquad x^2 = x_0^2 + \sum_{a=1}^{3} x_a^2 = 1,\qquad(1.341)$$

where σ_a are the Pauli matrices. The Haar measure is simply

$$dU = \frac{1}{\pi^2}\delta(x^2 - 1)d^4x.\qquad(1.342)$$

2) From a sphere S^3 we can easily go to R^3 via a stereographic projection, leaving undefined only the element at the north pole $U = -1$. The stereographic coordinates, $\mathbf{z} = (z_1, z_2, z_3)$, can be related to those on the sphere by

$$x_0 = \frac{(1 - \mathbf{z}^2)}{(1 + \mathbf{z}^2)}, \qquad x_a = \frac{2z_a}{(1 + \mathbf{z}^2)}.\qquad(1.343)$$

The Haar measure is

$$dU = d^3z\, \frac{4}{\pi^2(1 + \mathbf{z}^2)^3}.\qquad(1.344)$$

3) Finally, the exponential mapping that is useful in perturbation theory

$$U = \exp(i\phi n_a\sigma^a/2) = \cos\frac{\phi}{2} + i\sin\frac{\phi}{2}\mathbf{n}\cdot\boldsymbol{\sigma},\ 0 \le \phi \le 2\pi,\qquad(1.345)$$

where $\mathbf{n}$ are unit vectors in R^3. The invariant measure is

$$dU = \frac{1}{4\pi^2}d\phi\, d\Omega(\mathbf{n})\sin(\phi/2)^2,\qquad(1.346)$$

where $\Omega(\mathbf{n})$ is the uniform measure in S^2.

Defining a global coordinate system for $SU(N)$ is more complicated. Very often, however, explicit expressions of the measure are not needed, because integrals can be solved by invariant tensor methods.

Exercise 1.9 Work out the Haar measure in $SU(2)$ in terms of the variables α_k:

$$U = \exp(i\alpha_k \sigma_k), \tag{1.347}$$

where σ_k are the Pauli matrices. Compute the constant c so that the measure is properly normalized.

Two observations are in order:

- the integrals over the link variables are finite, there is no need to fix the gauge;
- the integrals can be done via importance sampling methods, because the action is real and positive definite

$$S[U] \sim \sum_P \text{Tr}[2 - U_P - U_P^\dagger] = \sum_P \text{Tr}[(1 - U_P)(1 - U_P^\dagger)] \geq 0, \tag{1.348}$$

the equality being obtained only when all plaquettes are unity: $U_P = 1$.

Before including the sources we are interested in, we should find out what are the operators that should represent the particle excitations in this theory. In order to understand this we should make contact with the operator formulation via the transfer matrix, which defines the Hamiltonian (Wilson, 1974, Lüscher, 1989, Smit, 2002).

1.5.3   Transfer matrix and unitarity of the plaquette action

As for the scalar and fermion lattice field theories, we want to make sure that there is a Hilbert-space representation of the lattice gauge theory. We need therefore to identify the field operators at $t = 0$ that represent creation and annihilation of particles. We also need to identify the transfer operator that evolves the operators at $t = 0$ in time.

In the Schrödinger picture, the physical states are described by wave functions that depend on the basic field variables at time $t = 0$. The time evolution of these fields is related to the Hamiltonian. At a fixed time $t = 0$, we can identify the spatial links:

$$U_k(\mathbf{x}, 0) \quad k = 1, 2, 3 \tag{1.349}$$

with the wave function coordinates, so the Schröndiger wave function of an arbitrary state $|\psi\rangle$ in this basis depends only on these links:

$$\psi[U_k(\mathbf{x}, 0)] = \langle U|\psi\rangle, \tag{1.350}$$

where $|U\rangle$ are the eigenbasis of the spatial link operators (i.e. analogous to the position basis in ordinary QM):

$$\hat{U}_k(\mathbf{x})|U\rangle = U_k(\mathbf{x}, 0)|U\rangle. \tag{1.351}$$

The states $|U\rangle$ form an orthonormal and complete basis

$$\langle U|U'\rangle = \prod_{\mathbf{x},k} \delta(U'_k(\mathbf{x},0) - U_k(\mathbf{x},0)) \tag{1.352}$$

with

$$\int dU\, \delta(U,U') = 1. \tag{1.353}$$

The scalar product of two such wave functions is therefore

$$\langle\psi|\phi\rangle \equiv \int \prod_{\mathbf{x},k} dU_k(\mathbf{x})\ \psi[U]^\dagger \phi[U]. \tag{1.354}$$

A gauge transformation leaves the scalar product invariant thanks to the invariance of the Haar measure and therefore the symmetry transformation must correspond to a unitarity operator,

$$\psi[U^\Omega] = \omega\psi[U], \tag{1.355}$$

where ω is the unitary operator that implements the gauge transformation in the space of wave functions.

In contrast with the ϕ^4 model previously discussed, however, the Hilbert space of physical states includes only those wave functions that are gauge invariant:

$$\psi[U^\Omega] = \psi[U]. \tag{1.356}$$

We can define a projector on gauge-invariant wave functions in the following way

$$\psi_{\text{phys}}[U] = \mathcal{P}_{\text{phys}}\psi[U] = \int \prod_{\mathbf{x}} d\Omega(\mathbf{x})\psi[U^\Omega]. \tag{1.357}$$

It is trivial to check, using the invariance of the measure, that the wave function $\psi_{\text{phys}}[U]$ is gauge invariant, also that the projector acts trivially on gauge-invariant wave functions.

The transfer matrix, $\hat{T}$, is an operator in the Hilbert space of wave functions that must satisfy that

$$\mathcal{Z} = \lim_{N\to\infty} \text{Tr}[\hat{T}^N], \quad N = T/a. \tag{1.358}$$

We can rewrite, inserting the completeness relation for the $|U\rangle$, basis:

$$\int dU^{(m)}|U^{(m)}\rangle\langle U^{(m)}| = 1, \tag{1.359}$$

$$\text{Tr}[\hat{T}^N] = \int dU^{(0)} \int dU^{(1)} \dots \langle U^{(0)}|\hat{T}|U^{(1)}\rangle \langle U^{(1)}|\hat{T}| \dots \dots |U^{(N-1)}|\hat{T}|U^{(0)}\rangle$$

$$= \prod_{m=0}^{N-1} \int dU^{(m)} \langle U^{(m)}|\hat{T}|U^{(m+1)}\rangle, \tag{1.360}$$

where $|U^{(m)}\rangle$ are basis states at the timeslice $x_0 = mT/N = ma$ and $|U^{(N)}\rangle = |U^{(0)}\rangle$. We can therefore identify the coordinates $U_k^{(m)}(\mathbf{x})$ with the spatial links at time ma:

$$U_k^{(m)}(\mathbf{x}) \to U_k(\mathbf{x}, ma), \tag{1.361}$$

with periodic boundary conditions in time.

Let us rewrite the plaquette action in the following way

$$S[U] = \frac{1}{g_0^2} \sum_m \sum_{\mathbf{x}} \left(\sum_{k<l} \text{Tr}[2 - U_{kl}(\mathbf{x}, ma) - U_{kl}^{\dagger}(\mathbf{x}, ma)] \right.$$

$$\left. + \sum_k \text{Tr}[2 - U_{k0}(\mathbf{x}, ma) - U_{k0}^{\dagger}(\mathbf{x}, ma)] \right)$$

$$= \sum_m \left\{ V[U^{(m)}] + K[U^{(m)}, U^{(m+1)}] + V[U^{(m+1)}] \right\}, \tag{1.362}$$

where

$$V[U^{(m)}] \equiv \frac{1}{2g_0^2} \sum_{\mathbf{x}} \sum_{k<l} \text{Tr}[2 - U_{kl}(\mathbf{x}, ma) - U_{kl}^{\dagger}(\mathbf{x}, ma)], \tag{1.363}$$

$$K[U^{(m)}, U^{(m+1)}] \equiv \frac{1}{g_0^2} \text{Tr}[2 - (U_k(\mathbf{x}, ma)U_0(\mathbf{x} + a\hat{k}, ma)U_k^{\dagger}(\mathbf{x}, ma + a)U_0^{\dagger}(\mathbf{x}, ma) + h.c.)]. \tag{1.364}$$

Now we can rewrite the path integral separating the integration over spatial and temporal links:

$$\mathcal{Z} = \prod_m \int \prod_{\mathbf{x},k} dU_k(\mathbf{x}, ma)dU_0(\mathbf{x}, ma) \exp\left[-(V[U^m] + K[U^{(m)}, U^{(m+1)}] + V[U^{(m+1)}])\right], \tag{1.365}$$

which can be written in the form of Eq. (1.360) if we identify

$$\langle U^{(m)}|\hat{T}|U^{(m+1)}\rangle = \int \prod_{\mathbf{x}} dU_0(\mathbf{x}, ma) \exp\left[-(V[U^m] + K[U^{(m)}, U^{(m+1)}] + V[U^{m+1}])\right]. \tag{1.366}$$

Therefore, the operator $\hat{T}$ has the form

$$\hat{T} = e^{-\hat{V}}\hat{T}_K e^{-\hat{V}}, \tag{1.367}$$

where $\hat{V}$ is an Hermitian operator diagonal in the $|U\rangle$ basis

$$\langle U'|\hat{V}|U\rangle = V[U]\delta(U', U), \tag{1.368}$$

and $V[U]$ is defined in Eq. (1.363). The kinetic operator satisfies

$$\langle U'|\hat{T}_K|U\rangle = \int \prod_{k,\mathbf{x}} d\Omega(\mathbf{x}) \exp\left(-\frac{1}{g_0^2}\text{Tr}[2 - (U_k'(\mathbf{x})\Omega(\mathbf{x} + a\hat{k})U_k^\dagger(\mathbf{x})\Omega^\dagger(\mathbf{x}) + h.c.)]\right).$$

$$\tag{1.369}$$

We define the operator $\hat{T}_K^0$:

$$\langle U'|\hat{T}_K^0|U\rangle \equiv \exp\left(-\frac{1}{g_0^2}\text{Tr}[2 - (U_k'(\mathbf{x})U_k^\dagger(\mathbf{x}) + h.c.)]\right). \tag{1.370}$$

It is easy to show that

$$\langle U'|\hat{T}_K|U\rangle = \langle U'|\hat{T}_K^0 \mathcal{P}_{\text{phys}}|U\rangle = \langle U'|\mathcal{P}_{\text{phys}}\hat{T}_K^0|U\rangle, \tag{1.371}$$

for all $|U\rangle, |U'\rangle$, where $\mathcal{P}_{\text{phys}}$ is the projector we have defined in Eq. (1.357). From this we can easily show two important properties:

- $\hat{T}_K$ commutes with the projector on physical states, and therefore only transforms physical states (gauge invariant under time-independent gauge transformations) to physical states:

$$\hat{T}_K \mathcal{P}_{\text{phys}} = \hat{T}_K^0 \mathcal{P}_{\text{phys}}^2 = \hat{T}_K^0 \mathcal{P}_{\text{phys}} = \hat{T}_K = \mathcal{P}_{\text{phys}}\hat{T}_K^0 = \mathcal{P}_{\text{phys}}\hat{T}_K. \tag{1.372}$$

The same is true for $\hat{T}_V$. It is easy to see this by realizing that it is diagonal in the U basis and that the eigenvalues are gauge invariant.

- The transfer matrix is positive definite. We need to show that

$$\langle\psi|\hat{T}|\psi\rangle > 0 \tag{1.373}$$

for all physical states $|\psi\rangle$ ($\langle\psi|\psi\rangle = 1$ and $\mathcal{P}_{\text{phys}}|\psi\rangle = |\psi\rangle$). Since

$$\langle\psi|e^{-\hat{V}}\hat{T}_K e^{-\hat{V}}|\psi\rangle = \langle\phi|\hat{T}_K|\phi\rangle \tag{1.374}$$

where $|\phi\rangle \equiv e^{-\hat{V}}|\psi\rangle$, it is sufficient to prove the positivity of $\hat{T}_K^0$

$$\langle\psi|\hat{T}_K^0|\psi\rangle = e^{-\frac{2N}{g_0^2}} \int dU\, dU' \psi[U']^* \psi[U] \, \exp\left(\frac{1}{g_0^2}\sum_{k,\mathbf{x}}\text{Tr}[U_k'(\mathbf{x})U_k^\dagger(\mathbf{x}) + h.c.]\right).$$

$$\tag{1.375}$$

We can now expand the exponential in powers of $1/g_0^2$. Each term in the series can be shown to be positive by noticing that each term is a positive constant (coefficients of the Taylor expansion of the exponential) times an integral of the form

$$\int dU \ dU' \psi[U']^* \psi[U] \ (U'_{\alpha\beta}U^*_{\alpha\beta})^n (U_{\gamma\delta}U'^*_{\gamma\delta})^m$$

$$= r^*_{\alpha_1...\alpha_n,\beta_1,...,\beta_n,\gamma_1,...\gamma_m,\delta_1,...\delta_m} r_{\alpha_1...\alpha_n,\beta_1,...,\beta_n,\gamma_1,...\gamma_m,\delta_1,...\delta_m} \geq 0, \quad (1.376)$$

where

$$r_{\alpha_1...\alpha_n,\beta_1,...,\beta_n,\gamma_1,...\gamma_m,\delta_1,...\delta_m} \equiv \int dU \, \psi[U] U^*_{\alpha_1\beta_1} \cdots U^*_{\alpha_n\beta_n} U_{\gamma_1\delta_1} \cdots U_{\gamma_m\delta_m}.$$

$$(1.377)$$

All terms must vanish for it to be zero. If the integral of a function $\psi[U]$ with any power of U and U^* is zero, the function must vanish. Therefore, for all normalizable wave functions, the positivity condition must hold.

Summarizing, we have identified the field operators, $\hat{U}_k(\mathbf{x})$, which represent the spatial links at $x_0 = 0$, and a positive transfer operator that determines their time evolution. Euclidean correlation functions of gauge-invariant combinations of such operators at arbitrary times can be represented by the corresponding functional integrals in the Wilson formulation of lattice gauge theories. The simplest gauge-invariant field operator is the spatial plaquette.

Having a unitary theory is reassuring, but the infrared behavior of this theory is highly non-trivial. We believe two fundamental phenomena take place:

- generation of a mass gap (in spite of the absence of dimensionful couplings);
- confinement or the property that asymptotic states are gauge singlets.

A very useful intuition can be obtained from the strong-coupling expansion of the lattice theory, as first realized by Wilson (Wilson, 1974), where both phenomena can be shown to take place.

1.5.4 Strong-coupling expansion: confinement, mass gap

The strong-coupling expansion is an expansion in inverse powers of the coupling g_0, which by the structure of the path integral is equivalent to a high-temperature expansion of the statistical system:

$$\mathcal{Z} = C \int \prod_l dU_l e^{+\frac{\beta}{2N} \sum_p [\chi(U_p) + \chi(U_p^\dagger)]}, \quad (1.378)$$

where l, p is a short-hand notation for the links and plaquettes, respectively,

$$\chi(U_p) \equiv \text{Tr}\,[U_p], \quad (1.379)$$

is the character of U_p and

$$\beta \equiv \frac{2N}{g_0^2}. \quad (1.380)$$

Therefore, a series expansion for large g_0 corresponds to an expansion for small β or high temperature:

$$\mathcal{Z} = \int \prod_l dU_l \prod_p \sum_n \frac{1}{n!} \left(\frac{\beta}{2N} \right)^n (\chi(U_p) + \chi(U_p)^*)^n. \tag{1.381}$$

Since $\chi(U_p)$ is bounded, the series has a finite radius of convergence, in contrast with the small g_0 expansion. The strong-coupling expansion has been worked out to very high orders. A more detailed discussion can be found in the literature (Montvay and Münster, 1994).

Working out the leading contribution to a given observable is quite simple noticing two facts. The leading order-contribution has the lowest number of plaquettes. All link variables must be shared by at least two plaquettes, since any unpaired link results in a zero contribution by

$$\int dU \ U_{\alpha\beta} = 0. \tag{1.382}$$

For the following two examples, the only non-trivial integral needed is that of two links

$$\int dU \ U_{\alpha\beta} U^\dagger_{\gamma\delta} = \frac{1}{N} \delta_{\alpha\delta} \delta_{\beta\gamma}. \tag{1.383}$$

1.5.4.1   Plaquette–plaquette correlator and mass gap

We have seen that correlation functions of spatial plaquettes should be able to describe the propagation and scattering of physical particles. Since these objects are gauge invariant, they cannot be gluons and they are called generically *glueballs*. According to the Källen–Lehmann representation, we should be able to find out the presence of a mass gap in the theory by studying the correlator of two spatial plaquettes at large time separation.

Let us consider a plaquette $U_{kl}(\mathbf{x}, x_0)$ in any two spatial directions, $\hat{k}$ and $\hat{l}$, fixed at a position $(\mathbf{x}, x_0)$ and another one parallel and with opposite orientation to the first at the position $(\mathbf{x}, x_0 + T)$. The leading diagram with paired links and the minimum tiling of plaquettes is given by a rectangle linking the two external plaquettes, Fig. 1.7. The β and N dependence is given by

$$\left(\frac{\beta}{2N} \right)^{N_p} \left(\frac{1}{N} \right)^{N_i} N^{N_v}, \tag{1.384}$$

since each internal plaquette brings a factor $\beta/2N$, each integral over two paired links brings in a factor $1/N$, Eq. (1.383), and each vertex gives a factor of N. In this case we have

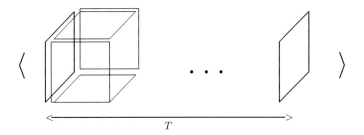

Fig. 1.7 Minimum tiling of the plaquette two-point correlator.

$$N_p = \#\text{plaquettes} = 4T/a \tag{1.385}$$

$$N_i = \#\text{integrals} = \#\text{links}/2 = 2(N_p + 2) \tag{1.386}$$

$$N_v = \#\text{vertices} = N_v = 4(T/a + 1) \tag{1.387}$$

Putting all this together we find

$$C_{pp}(T) \sim \left(\frac{\beta}{2N^2}\right)^{4T/a} = \exp\left(-\frac{4}{a}\ln\left(\frac{2N^2}{\beta}\right)T\right), \tag{1.388}$$

therefore the correlator decays exponentially in time as expected in a theory with a finite mass gap, where correlators decay as $\exp(-mT)$. In this case, the mass gap is

$$m \sim \frac{4}{a}\ln\left(\frac{2N^2}{\beta}\right). \tag{1.389}$$

Unfortunately, no continuum limit can be reached in the strong-coupling expansion since $\lim_{a\to 0} ma = $ finite. It is only in the continuum limit where we expect to find the universal behavior of Yang–Mills field theory and therefore this result is not enough to prove the existence of a mass gap. One would need to ensure by other means that this behavior survives in the continuum limit.

1.5.4.2 *Wilson loop and the static potential*

Let us consider a rectangular loop with two spatial sides and two temporal ones, W_{RT}, Fig. 1.8. The spatial side length is R and the temporal one is T. It is easy to work out the leading-order strong-coupling behavior of such an observable. It corresponds to the diagram where the loop is tiled up with plaquettes parallel to the loop. The behavior is

$$\langle W_{RT}\rangle = \left(\frac{\beta}{2N}\right)^{N_p}\left(\frac{1}{N}\right)^{N_i} N^{N_v}, \quad N > 2, \tag{1.390}$$

where it is easy to count plaquettes, paired links and vertices:

$$N_p = (R/a)(T/a) \quad N_i = 2N_p + (R/a + T/a) \quad N_v = (R/a + 1)(T/a + 1), \tag{1.391}$$

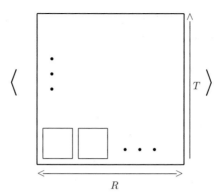

Fig. 1.8 Minimum tiling of the Wilson loop, W_{RT}.

so the final result is

$$\langle W_{RT} \rangle \sim N \left(\frac{\beta}{2N^2} \right)^{RT/a^2} \sim \exp\left(-\ln\left(\frac{2N^2}{\beta}\right) \frac{RT}{a^2} \right) \sim \exp\left(-\sigma \text{Area} \right). \qquad (1.392)$$

Therefore, the rate of the exponential decay as the temporal extent increases goes with the area encircled by the Wilson loop. This behavior is called *area-law* and is a criterion for confinement. We now discuss why this is so.

The Wilson loop is related to the *static potential*, that is the potential of two point sources infinitely heavy and separated by a distance R. Let us consider for simplicity the case of scalar particles (the result will not depend on the spin). The static limit corresponds to an action where the spatial derivatives (spatial momenta) are neglected:

$$S_{\text{stat}}[\phi] = a^4 \sum_x \frac{1}{2} \left[\left(\hat{\partial}_0 \phi\right)^* \hat{\partial}_0 \phi + m^2 |\phi|^2 \right], \qquad |\hat{\partial}_k \phi| \ll m\phi, \qquad (1.393)$$

the field values at different space points $\mathbf{x}$ are independent variables.

One can show (see exercise) that the correlator in the static approximation and in the presence of a background gauge field is

$$\langle \phi(\mathbf{x}, x_0) \phi^\dagger(\mathbf{y}, y_0) \rangle_\phi = \frac{a}{2 \sinh(a\omega)} e^{-(x_0 - y_0)\omega} \delta(\mathbf{x} - \mathbf{y}) U(\mathbf{x}, x_0; \mathbf{y}, y_0), \qquad (1.394)$$

where $U(\mathbf{x}, x_0; \mathbf{y}, y_0)$ is the parallel transporter and

$$\cosh(a\omega) = 1 + \frac{1}{2} a^2 m^2. \qquad (1.395)$$

Exercise 1.10 Prove Eq. (1.394) in the absence of gauge fields, i.e. with the parallel transporter set to the identity. Show that in the presence of gauge fields the static propagator

Eq. (1.394) satisfies

$$(-\nabla_0^* \nabla_0 + m^2)\langle \phi(x)\phi^\dagger(y)\rangle_\phi = \delta(x - y). \tag{1.396}$$

The simplest gauge-invariant operator representing a quark and antiquark separated by some spatial distance $|\mathbf{y} - \mathbf{x}| = R$ at time t is

$$\mathcal{O}(t) = \phi^\dagger(\mathbf{y}, t)U(\mathbf{y}, t; \mathbf{x}, t)\phi(\mathbf{x}, t). \tag{1.397}$$

The correlator at large times $T \to \infty$,

$$C_{q\bar{q}}(T) \equiv \langle \mathcal{O}^\dagger(T)\mathcal{O}(0)\rangle_{\phi,U} \tag{1.398}$$

represents a quark–antiquark pair separated by a distance R that is created at time $x_0 = 0$ and evolves until time T. Integrating over the scalar fields, using Eq. (1.394), and neglecting factors that do not depend on R (e.g. $\exp(-T\omega)$) we get

$$C_{q\bar{q}}(T) \sim \langle \text{Tr}[U(\mathbf{y}, T; \mathbf{y}, 0)U(\mathbf{y}, 0; \mathbf{x}, 0)U(\mathbf{x}, 0; \mathbf{x}, T)U^\dagger(\mathbf{y}, T; \mathbf{x}, T)]\rangle_U = \langle W_{RT}\rangle, \tag{1.399}$$

that is, the R dependence of this correlator is the same as that of a Wilson loop of area RT. We expect therefore that the exponential decay in time of such a correlator gives us information about the energy of this system. The energy will contain an R-independent contribution, but it will also depend on the distance due to the potential energy between the quark and antiquark. We therefore expect

$$C_{q\bar{q}}(T) \sim \exp(-E(R)T), \tag{1.400}$$

where

$$E(R) = E_0 + V(R). \tag{1.401}$$

Relating Eqs. (1.392) and Eq. (1.400), we have

$$\lim_{\beta \to 0} V(R) = \frac{R}{a^2} \ln\left(\frac{2N^2}{\beta}\right) + \ldots = \sigma R + \ldots, \tag{1.402}$$

where σ is called *the string tension*:

$$\lim_{\beta \to 0} \sigma = \frac{1}{a^2} \ln\left(\frac{2N^2}{\beta}\right). \tag{1.403}$$

The linear behavior of the potential as a function of R is a criterion for confinement, because the potential energy grows without bound when the quark and the antiquark are pulled apart.

Unfortunately, once more, the finite string tension that we find in the strong-coupling limit does not imply that there is one in the continuum limit because

$$\lim_{a \to 0} a^2 \sigma = \text{finite}, \tag{1.404}$$

and therefore σ diverges in the continuum limit.

These two simple examples show that the strong-coupling analysis gets all the qualitative behavior right, but there is no continuum limit in this approximation. We will see that a continuum limit can be shown to exist in the opposite extreme of small coupling, as expected from perturbative renormalizability.

1.5.5 Weak-coupling expansion

Perturbatively we know that Yang–Mills theories are renormalizable and this, according to Wilson's renormalization group, implies that a continuum limit can be defined in lattice perturbation theory.

On the lattice, the weak-coupling expansion corresponds to a saddle-point expansion around the configurations with vanishing action. We have seen that these correspond to all plaquettes being the identity:

$$U_p = 1. \tag{1.405}$$

These in turn are pure gauge configurations that are gauge equivalent to the configuration with all links set to the identity:

$$U_\mu(x) = 1. \tag{1.406}$$

Near this configuration, a convenient parametrization of the link variables is the exponential mapping

$$U_\mu(x) = \exp\left(-ig_0 a T^a A_\mu^a(x)\right), \tag{1.407}$$

but it is necessary to fix the gauge if we are going to integrate over unbounded gauge fields A_μ^a, just as in the continuum.

1.5.5.1 *Gauge fixing*

The gauge-fixing procedure on the lattice follows closely that in the continuum.

1) Choose a gauge-fixing condition such as

$$G[U] = 0. \tag{1.408}$$

The gauge-fixing functional $G[U]$ is a function of the link variables and it is well defined for U near the identity. It must also satisfy that for any U near the identity, there is one and only one gauge transformation g such that

$$G[U^g] = 0. \tag{1.409}$$

2) Include the gauge fixing in the path integral by

$$\int dU e^{-S[U]} = \int d\Omega \int dU e^{-S[U]} \delta(G[U^\Omega]) \Delta[U], \tag{1.410}$$

where

$$\Delta[U]^{-1} \equiv \int d\Omega \delta(G[U^\Omega]), \tag{1.411}$$

which is gauge invariant. Using the invariance of the measure, dU, it is easy to see that the integrand of Eq. (1.410) does not depend on Ω and since the integral $\int d\Omega = 1$, we have

$$\int dU e^{-S[U]} = \int dU e^{-S[U]} \delta(G[U]) \Delta[U]. \tag{1.412}$$

3) Rewrite the operator $\Delta[U]$ as a local ghost contribution using the Faddeev–Popov trick (Peskin and Schroeder, 1995). For any configuration U, let $\Omega_0(U)$ be the gauge transformation that satisfies

$$G[U^{\Omega_0(U)}] = 0. \tag{1.413}$$

Consider an infinitesimal gauge transformation, $\Omega_\epsilon = \exp(i\epsilon_a T^a)$:

$$G[U^{\Omega_0(U)\Omega_\epsilon}] = G[U^{\Omega_0(U)}] + \left.\frac{\partial G^a[U^{\Omega_0(U)\Omega_\epsilon}]}{\partial \epsilon_b}\right|_{\epsilon=0} \epsilon_b + \ldots \equiv M_{ab}[U]\epsilon_b + \ldots \tag{1.414}$$

If $\det(M[U]) \neq 0$, we can restrict the Ω integration in Eq. (1.411) to the neighborhood of Ω_0

$$\Delta[U]^{-1} = \int d\Omega \delta(G[U^\Omega]) = \int d\epsilon \delta(M[U]\epsilon) = \frac{1}{\det(M[U])}. \tag{1.415}$$

The determinant can now be included as an integral over Grassmann variables, or ghost fields

$$\Delta[U] = \int d\bar{c}dc \, e^{-S_{FP}[c,\bar{c},U]}, \quad S_{FP}[c,\bar{c},U] \equiv \bar{c}^a M_{ab}[U]c^c. \tag{1.416}$$

This the Faddeev–Popov term.

4) Rewrite the delta function as a Gaussian integral.
Consider a different gauge fixing functional $G'[U] = G[U] + k^a T^a$, with k^a some constants. Since $M'[U] = M[U]$, $\Delta[U]$ is the same and the partition function does not depend on k^a. We can therefore integrate over them with a Gaussian weight

$$\mathcal{Z} \sim \int \prod_a dk_a e^{-\frac{1}{2\alpha}\sum_a k_a^2} \int d\bar{c}dcdU e^{-S[U]-S_{FP}[c,\bar{c},U]} \, \delta(G(U)+k)$$

$$= \int d\bar{c}dcdU \, e^{-S[U]-S_{FP}[c,\bar{c},U]-\frac{1}{2\alpha}\sum_a G^a(U)G^a(U)}. \tag{1.417}$$

This is the starting point of lattice perturbation theory.

A commonly used gauge is the *Lorentz gauge*:

$$G[U] = \sum_\mu \hat{\partial}^*_\mu A_\mu(x), \tag{1.418}$$

where the field A_μ is defined via the exponential mapping Eq. (1.407).

The last ingredient that we should specify is the measure for the exponential mapping. It can be shown that the Haar measure for each of the link variables can be written as

$$dU = \prod_{x,\mu} dU_\mu(x) = \prod_{x,\mu} \exp\left(-\mathrm{Tr}\left[\ln\left(\frac{2}{\omega}\sinh\left(\frac{\omega}{2}\right)\right)\right]\right) dA_\mu \tag{1.419}$$

with

$$\omega(U)_{ab} \equiv g_0 f_{abc} A^c_\mu(x), \tag{1.420}$$

and f_{abc} are the structure constants of the group, satisfying

$$[T^a, T^b] = i f_{abc} T^c. \tag{1.421}$$

Exercise 1.11 Show the following properties for $SU(N)$.

- Let $U(\alpha) \equiv \exp(i\alpha^a T^a)$. For λ a real number, define

$$R(\lambda)_{ab} \equiv 2\mathrm{Tr}\left[U(\lambda\alpha)T^a U^\dagger(\lambda\alpha)T^b\right]. \tag{1.422}$$

Show that $R(\lambda)$ satisfies the following differential equation

$$\frac{\partial R}{\partial \lambda} = -i\hat{\alpha}R \quad \hat{\alpha} \equiv \alpha_a t^a \quad (t^a)_{bc} = -i f_{abc}, \tag{1.423}$$

and therefore that $R(1) = \exp(-i\hat{\alpha})$.

- Next, define

$$M(\lambda) \equiv U(\lambda\alpha)U^\dagger(\lambda(\alpha + \epsilon)). \tag{1.424}$$

Neglecting terms of $\mathcal{O}(\epsilon^2)$ show that

$$\frac{\partial M}{\partial \lambda} = -i\epsilon^a R_{ab}(\lambda)T^b + \mathcal{O}(\epsilon^2), \tag{1.425}$$

and that this implies

$$M(1) = 1 - i\epsilon^a \left(\frac{1 - \epsilon^{-i\hat{\alpha}}}{i\hat{\alpha}}\right)_{ab} T^b + \dots \tag{1.426}$$

- Show that the previous results imply ($X, Y \in su(n)$):

$$(e^{iX} Y e^{-iX})_a = Y^b (e^{-i\hat{X}})_{ba} \tag{1.427}$$

$$e^{iX} \partial_a e^{-iX} = -i \left(\frac{1 - e^{-i\hat{X}}}{i\hat{X}} \right)_{ab} T^b. \tag{1.428}$$

- Use these properties to show that the Haar measure for the exponential coordinates, α_a, can be written as

$$dU = \exp \left(\text{Tr} \left[\ln \left(\frac{2}{\hat{\alpha}} \sin(\hat{\alpha}/2) \right) \right] \right) \prod_a d\alpha_a. \tag{1.429}$$

- Show that in $SU(N)$, for $U_\mu = \exp(-ig_0 A_\mu)$ and for the Lorentz gauge

$$G[U] = g_0 \sum_\mu \hat{\partial}_\mu^* A_\mu(x) \tag{1.430}$$

the operator M that enters in the ghost action is

$$M(U) = \hat{\partial}_\mu^* \left\{ \frac{ig_0 \hat{A}_\mu}{(1 - \exp(+ig_0 \hat{A}_\mu))} \hat{\partial}_\mu - ig_0 \hat{A}_\mu \right\}, \tag{1.431}$$

where $\hat{A}_\mu \equiv A_\mu^a t^a$.

1.5.5.2 Feynman rules

The derivation of the Feynman rules for the gauge-fixed lattice action

$$S_{GF}[c, \bar{c}, U] = S[U] + S_{FP}[c, \bar{c}, U] + \frac{1}{2\alpha} \sum_a G^a(U) G^a(U) \tag{1.432}$$

is conceptually straightforward but quite complicated! As usual we go over to momentum space and assign the gauge potential, defined from the exponential map, Eq. (1.407) to the points in the middle of the link $x + \hat{\mu}a/2$:

$$A_\mu(p) = a^4 \sum_x e^{ip(x + a\frac{\hat{\mu}}{2})} A_\mu(x). \tag{1.433}$$

The leading contribution $\mathcal{O}(g_0^0)$ is only quadratic in the fields. The gauge part is:

$$S^{(0)}[U] = \frac{1}{2} \int_{BZ} \frac{d^4 k}{(2\pi)^4} A_\mu^a(-k) e^{\frac{ik_\mu a}{2}} \left(\delta_{\mu\nu} \hat{k}^2 - (1 - \alpha) \hat{k}_\mu \hat{k}_\nu \right) e^{\frac{-ik_\nu a}{2}} A_\nu(k), \tag{1.434}$$

with

$$\hat{k}_\mu = \frac{2}{a} \sin \left(\frac{k_\mu a}{2} \right) \qquad \hat{k}^2 = \sum_\mu \hat{k}_\mu^2. \tag{1.435}$$

The corresponding Feynman rules for the gauge and ghost propagators are:

$$-\frac{\delta_{ab}}{\hat{k}^2}\left[\delta_{\mu\nu} - (1-\alpha)\frac{\hat{k}_\mu \hat{k}_\nu}{\hat{k}^2}\right],$$

$$\frac{\delta_{ab}}{\hat{k}^2}.$$

At higher order in g_0 there are diagrams that have continuum analogs, such as:

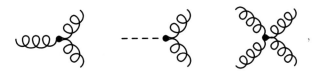

and some more that do not appear in the continuum, such as a gluon mass term

$$-(2\pi)^4 \frac{Ng_0^2}{12a^2}\,\delta_{ab}\delta_{\mu\nu},$$

or a two-gluon–two-ghost vertex:

For more details see (Rothe, 2005).

At one loop one can explicitly check renormalizability. By power counting we expect that only the 2, 3 and 4 vertex functions are divergent in the continuum limit and one can show that the divergence in $\Gamma^{(2)}_{\mu\nu;ab}$ can be reabsorbed in a field redefinition

$$Z^{1/2}A_{\mu R} = A_\mu, \tag{1.436}$$

while the divergences in 3- and 4-point vertex functions are related by gauge invariance and can be reabsorbed in a redefinition of the coupling

$$Z_g g_R = g_0. \tag{1.437}$$

A large number of seemingly miraculous cancellations take place, ensuring the absence of disastrous contributions, such as a gluon mass term or Lorentz-non-invariant counterterms. The miracle can be shown to occur to all orders by realizing that there is an exact BRST invariance satisfied also on the lattice. I refer to the lectures of P. Weisz for further details (Weisz, 2009).

There seems to be a continuum limit in perturbation theory, but we still do not know if this perturbative analysis provides a solid proof of renormalizability.

We can look at the Callan–Symanzik equations to understand the gauge-coupling flow as we approach the continuum limit. Let us consider the momentum subtraction scheme:

$$\Gamma^{(2)}(k)|_{k^2=\mu^2} = \text{tree} - \text{level}$$

$$\Gamma^{(4)}(k_1, k_2, k_3)|_{k_i \cdot k_j = \frac{1}{2}(3\delta_{ij}-1)\mu^2} = \text{tree} - \text{level}, \tag{1.438}$$

where $\mu a \ll 1$. From these two conditions the renormalized coupling is found to be:

$$g_R^2(\mu) = g_0^2 \left(1 - \frac{g_0^2}{16\pi^2} \frac{11N_c}{3} (\ln(a^2\mu^2) + c') \right). \tag{1.439}$$

As we approach the continuum limit we must tune the coupling $g_0(a)$ so as to keep the physical coupling fixed. Neglecting scaling violations $\mathcal{O}(\mu a)$ we have the following RG equation:

$$\beta(g_0) \equiv -a \left. \frac{\partial g_0}{\partial a} \right|_{g_R \text{ fixed}} = -\beta_0 g_0^3 - \beta_1 g_0^5 + \dots, \tag{1.440}$$

where β_0, β_1 are universal (i.e. do not depend on the regularization scheme). For $SU(3)$:

$$\beta_0 = \frac{11}{16\pi^2} > 0, \quad \beta_1 = \frac{102}{(16\pi^2)^2}. \tag{1.441}$$

This equation shows that g_0 decreases as we approach the continuum limit, so perturbation theory becomes more accurate as we approach this limit. In fact $g_0 = 0$ is a zero of the β function, i.e. an *UV fixed point*, therefore our target continuum limit corresponds to $g_0 = 0$. Therefore, the perturbative analysis of renormalizability is perfectly justified. Note that this would not be the case if the fixed point occurs at a different value of g_0.

We can integrate the RG equation to get

$$a = c \exp\left(\frac{-1}{2\beta_0 g_0^2} \right) (g_0^2)^{-\frac{\beta_1}{2\beta_0^2}}, \tag{1.442}$$

where c is a constant of integration and does not depend on a, even though it has the same dimensions. It is common practice to define a Λ parameter in terms of this constant

$$a\Lambda \equiv \exp\left(\frac{-1}{2\beta_0 g_0^2} \right) (\beta_0 g_0^2)^{-\frac{\beta_1}{2\beta_0^2}}, \tag{1.443}$$

which remains constant in the continuum limit. All scales should be proportional to Λ as we approach the continuum limit. It can therefore be taken as a reference scale, although it should always be remembered that this scale depends on the regularization scheme.

1.5.6 Topological charge

$SU(3)$ gauge fields in the continuum fall into distinct topological sectors labelled by the integer

$$Q = -\frac{1}{32\pi^2} \int d^4x \; \epsilon_{\mu\nu\rho\sigma} \; \text{Tr}\,[F_{\mu\nu} F_{\rho\sigma}], \tag{1.444}$$

called the *topological charge or instanton number*.

That configurations with $Q \neq 0$ exist and are important for physics was understood when classical (i.e. finite action) solutions were found with $Q = 1$ (instantons) (Belavin *et al.*, 1975). They are believed to play an important role ('t Hooft, 1976; Witten, 1979; Veneziano, 1979) in several fundamental problems such as the

- $U_A(1)$ problem in QCD;
- violation of $B + L$ in the Standard Model (where B is baryon number and L is lepton number).

Consider a continuum gauge field in a box of size $T \times L^3$, with the following boundary conditions (periodicity up to a gauge transformation):

$$A_\mu(x)|_{x_i=L} = A_\mu(x)|_{x_i=0}, \quad \hat{k} = 1, 2, 3$$

$$A_\mu(x)|_{x_0=T} = \Omega(\mathbf{x}) A_\mu(x)|_{x_0=0} \Omega^\dagger(\mathbf{x}) + i\Omega(\mathbf{x}) \partial_\mu \Omega(\mathbf{x})^\dagger, \quad (1.445)$$

where $\Omega(\mathbf{x})$ is periodic in the spatial directions, that is, it is the map of the 3-torus on $SU(3)$. Such functions fall in homotopy classes characterized by the winding number of the map that coincides with Q.

If $Q \neq 0$, no smooth gauge transformation $g(x)$ exists such that

$$g(0, \mathbf{x}) = I, \quad g(T, \mathbf{x}) = \Omega(\mathbf{x}). \quad (1.446)$$

Otherwise, the winding could be gauged away, which must not be possible, since Q is gauge invariant.

Example: Let us consider the simpler case of $U(1)$ in 2D. Let us consider the function Ω that maps the circle, T^1 into $U(1)$:

$$\Omega : T_1 \to U(1) \quad (1.447)$$

$$x \to e^{i2\pi \frac{x}{L} q}. \quad (1.448)$$

The topological charge in 2D is given by

$$Q = -\frac{1}{2\pi} \int d^2 x \sum_{\mu<\nu} \epsilon_{\mu\nu} F_{\mu\nu} = \frac{i}{2\pi} \int_0^L dx \Omega \partial_x \Omega^\dagger = q, \quad (1.449)$$

where we have used Eq. (1.445).

Exercise 1.12 A geometrical definition of topological charge in compact $U(1)$ in two dimensions. Consider the following local quantity

$$q_n = \frac{-i}{2\pi} \sum_{\mu<\nu} \epsilon_{\mu\nu} \ln U_{\mu\nu}(n), \quad (1.450)$$

and its global sum:

$$Q = \sum_n q_n, \tag{1.451}$$

in a periodic lattice. Check that q_n is gauge invariant. Show that Q is an integer. Show that its naive continuum limit is the 2D topological charge.

Is there topology in the lattice formulation? If for example we have the lattice boundary conditions

$$U_k(x)|_{x_0=T} = \Omega(\mathbf{x}) \, U_k(x)|_{x_0=0} \, \Omega(\mathbf{x} + a\hat{k})^\dagger, \tag{1.452}$$

it is obvious that Ω can be gauged away by changing the temporal link variables at the border as

$$U_0'(\mathbf{x}, T - a) = U_0(\mathbf{x}, T - a)\Omega^\dagger(\mathbf{x}), \tag{1.453}$$

which does not change the measure $dU_0' = dU_0$. Does this mean that we are missing $Q \neq 0$ configurations on the lattice?

No, it means that all topological sectors correspond to periodic boundary conditions, and that all are connected by lattice gauge transformations. However, the configurations at the boundaries between the different topological sectors do not have a continuum limit, i.e. they become singular in the continuum limit, so there is no contradiction.

For some purposes it might be useful to define an integer topological charge also at finite a. This can be done in various ways, for example:

- Geometrical definition (Lüscher, 1982). A local density $q(x)$ can be defined with the following properties

$$\sum_x q(x) \in Z, \quad \lim_{a \to 0} q(x) = -\frac{1}{32\pi^2} \int d^4x \, \epsilon_{\mu\nu\rho\sigma} \, \mathrm{Tr} \, [F_{\mu\nu} F_{\rho\sigma}]. \tag{1.454}$$

One can also show that Q does not change if a gauge-invariant constraint is set on the plaquettes $\mathrm{Tr}[U_p] \geq 1 - \epsilon$.

- Fermionic zero modes. The index theorem (Atiyah and Singer, 1971) establishes the following relation:

$$Q = n_R - n_L, \tag{1.455}$$

where $n_{R/L}$ are the righ-handed/left-handed zero modes of the Dirac operator on the background gauge configuration with charge Q. Lattice fermions that satisfy a chiral symmetry at finite a such as Ginsparg–Wilson fermions allow definition of a topological charge from the number of zero modes (Hasenfratz *et al.*, 1998).

1.6 Lattice QCD

The original investigation on lattice field theory was motivated by the need to make predictions in QCD. There is by now little doubt that QCD is the right theory of the strong interactions. It is an $SU(3)$ gauge theory, with six flavors of quarks in the fundamental representation. The Euclidean action in the continuum is

$$S_{\text{QCD}} = \int d^4x \sum_q \bar{\psi}_q(\gamma_\mu D_\mu + m_q)\psi_q - \frac{1}{2g_0^2}\text{Tr}[F_{\mu\nu}F_{\mu\nu}], \qquad (1.456)$$

which has therefore seven free parameters: the gauge coupling and six quark masses.

Let us review briefly the main properties of QCD.

- Symmetries. At the classical level the symmetries of this action are
  - Lorentz invariance;
  - $SU(3)$ gauge invariance;
  - discrete symmetries: C, P and T;
  - quark number: $\psi_q \to e^{i\alpha_q}\psi_q$.

  In the absence of quark masses, there is a much larger global symmetry group that is a chiral $U(6)_L \times U(6)_R$:

  $$P_R\psi \to U_R P_R\psi \quad P_L\psi \to U_L P_L\psi \quad U_R, U_L \in U(6), \qquad (1.457)$$

  where $\psi = (\psi_u, \psi_d, \psi_d, ..)$.

- Spontaneous chiral-symmetry breaking
  The chiral flavor group is believed to be broken to $U(6)_V$ spontaneously by a quark condensate

  $$-\langle \bar{\psi}_i\psi_j \rangle \neq \Sigma\delta_{ij}, \qquad (1.458)$$

  which is only invariant under $U_R = U_L = U_V$.

- Anomalous breaking of $U_A(1)$
  The $U(1)_A$ is broken by a different mechanism: via an anomaly. Indeed, even if there is no symmetry breaking by the vacuum or by the explicit mass terms, the current associated with this symmetry is not conserved. According to the Noether theorem, the axial current

  $$J_\mu^5 = \sum_q \bar{\psi}_q\gamma_\mu\gamma_5\psi_q \qquad (1.459)$$

  should be conserved. However, a one-loop computation (Adler and Bardeen, 1969) shows that

  $$\partial_\mu J_\mu^5 = \frac{g_0^2}{16\pi^2}\epsilon_{\alpha\beta\gamma\delta}\text{Tr}[F_{\alpha\beta}F_{\gamma\delta}], \qquad (1.460)$$

  while the vector current is of course conserved. We identify the topological charge on the right-hand side!

The spontaneous breaking of a global symmetry implies the presence of as many Nambu–Goldstone (Nambu, 1960; Goldstone, 1961) massless particles, as broken generators: the generators of the $SU(6)$ axial rotations (since $U_A(1)$ is not broken spontaneously).

In reality quark masses are not zero. They are plotted in a logarithmic scale in Fig. 1.9, where we see that they encompass five orders of magnitude. Compared to the mass gap of the pure gauge theory, $\sim 1\text{GeV}$, there are three quarks: u, d, s that can be considered light, while another three c, b, t are heavy. Therefore, the approximate flavor symmetry in the presence of quark masses is at most $SU(3)$ and not $SU(6)$. We expect therefore that the spontaneous symmetry breaking results in eight lighter states corresponding to the Nambu–Goldstone bosons, which have the quantum numbers of the pseudoscalar mesons. Indeed the lightest excitations in QCD are the octet of pseudoscalar mesons: $\pi^{\pm}, \pi_0, K^{\pm}, K_0, \bar{K}_0, \eta$. The η', being the mode associated to $U(1)_A$, is significantly more massive, because it is not a Nambu-Goldstone boson.

Witten and Veneziano (Witten, 1979; Veneziano, 1979) obtained a prediction for the mass of this special meson in the large-N_c limit:

$$\frac{F_\pi^2 m_{\eta'}^2}{2N_f} = \chi_{top} \equiv \int d^4x \langle Q(x)Q(0)\rangle, Q \equiv \frac{g_0^2}{32\pi^2}\epsilon_{\alpha\beta\gamma\delta}\text{Tr}[F_{\alpha\beta}F_{\gamma\delta}]. \tag{1.461}$$

The η' mass can then be determined from a purely gauge observable, such as the topological susceptibility, which in the large N_c limit can be determined in the pure gauge theory!

QCD is a renormalizable theory, but perturbation theory does not provide a good description of its phenomenology at large distances or low energies, because the theory is strongly interacting. Indeed, the main features of QCD that determine to a large extent its phenomenology are intrinsically non-perturbative: mass gap, confinement, spontaneous chiral-symmetry breaking, anomalous currents, etc. Obviously the goal of a non-perturbative approach to QCD would be to understand from first principles all these phenomena, and to provide an accurate description of QCD phenomenology, such as the hadron spectrum and other properties.

Even though in most cases these would be postdictions, it is nevertheless extremely important to finalize with success this long-term project for several reasons:

- The flavor sector of the SM is poorly understood, and it is rather generic that models beyond the SM induce non-standard effects in flavor violating processes in the quark sector. Having precise predictions in the SM is therefore indispensable to search for such non-standard effects.

- It would allow study of QCD at high density and temperature, conditions of the early Universe and of very dense systems such as neutron stars, that are not easy to reproduce in the laboratory. We refer to the lectures by U. Philipsen (Philipsen, 2009).

- QCD is in some sense a model field theory for many extensions of the SM, as well as for the lattice approach. In QCD we know where the UV fixed point lies so we

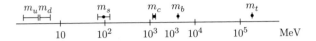

Fig. 1.9   Quark masses.

know where the continuum limit is and how to approach it. The lattice method might be necessary to study other field theories, such as technicolor models or theories with dynamical gauge symmetry breaking, where things might not be as easy. Clearly having solved QCD is a benchmark to guide future investigations.

Giving the spread of quark masses that span six orders of magnitude, dealing with all quarks in a lattice simulation is very difficult since approaching the continuum limit in controlled conditions requires

$$am_q \ll 1, \tag{1.462}$$

and therefore extremely fine lattices. This brute force approach is not practical. Fortunately, when we try to describe the low-energy regime, the effect of the heavy quarks can be accurately described by an effective theory that results from integrating them out. A consequence of the decoupling theorem (Appelquist and Carazzone, 1975) (which is another realization of Wilsonian renormalization group) is that the effects of the heavy quarks in the low-energy dynamics are well represented by local operators of the light fields only (gluons and the lighter quarks). The effect of the heavy scales is reabsorbed in the couplings. This implies that in order to study hadron processes at energies much lower than the heavy-quark mass scales, we can simply ignore the heavy quarks.

We are also interested, however, in processes involving heavy hadrons. A way to do this is to consider them as static sources, as is done in the heavy-quark effective theory. I refer to R. Sommer's lectures (Sommer, 2009) for a detailed discussion of this effective theory as an efficient tool to study heavy flavors on the lattice.

1.6.1   Wilson formulation of lattice QCD

By now, it should be clear how to discretize this action following for example the Wilson approach

$$S_{\text{QCD}}[U, \bar{\psi}, \psi] = S[U] + S_W[U, \bar{\psi}, \psi], \tag{1.463}$$

where $S[U]$ is the plaquette action of Eq. (1.330) and $S_W[U, \bar{\psi}, \psi]$ is the Wilson action for each of the quark fields:

$$S_W[U, \bar{\psi}, \psi] = a^4 \sum_{q,x} \bar{\psi}_q [D_W + m_q] \psi_q, \tag{1.464}$$

where the Wilson operator is

$$D_W \equiv \frac{1}{2} \left(\gamma_\mu (\nabla_\mu + \nabla_\mu^*) - ar\nabla_\mu^* \nabla_\mu \right) \tag{1.465}$$

and

$$\nabla_\mu \psi(x) = \frac{1}{a}\left[U_\mu(x)\psi(x+a\hat{\mu}) - \psi(x)\right],$$

$$\nabla^*_\mu \psi(x) = \frac{1}{a}\left[\psi(x) - U_\mu(x-a\hat{\mu})^\dagger \psi(x-\hat{\mu})\right]. \tag{1.466}$$

The action can be rewritten as

$$S_W = a^4 \left\{ \sum_{q,x} \bar{\psi}_q(x) \left[m_q + \frac{4r}{a}\right] \psi_q(x) + \frac{1}{2a} \sum_{q,x,\mu} \bar{\psi}_q(x)\left(\gamma_\mu - r\right) U_\mu(x)\psi_q(x+a\hat{\mu}) \right.$$

$$\left. - \bar{\psi}_q(x)\left(\gamma_\mu + r\right) U^\dagger_\mu(x-a\hat{\mu})\psi_q(x-a\hat{\mu}) \right\}. \tag{1.467}$$

It is common practice to rewrite the fermionic action in terms of the parameter κ:

$$S_W = a^4 \sum_{q,x} \bar{\psi}_q(x)\psi_q(x) - \kappa_q \sum_{q,x,\mu} \left(\bar{\psi}_q(x)(\gamma_\mu - r)U_\mu(x)\psi_q(x+a\hat{\mu})\right.$$

$$\left. + \bar{\psi}_q(x)(\gamma_\mu + r)U^\dagger_\mu(x-a\hat{\mu})\psi_q(x-a\hat{\mu})\right), \tag{1.468}$$

where:

$$\kappa_q \equiv \frac{1}{2am_q + 8r}. \tag{1.469}$$

In the free case, the massless limit corresponds to the critical value $\kappa_c = \frac{1}{8r}$.

The measures over the gauge links and the Grassmann variables are the same as defined before and therefore the partition function is

$$\mathcal{Z} = \int dU\, d\bar{\psi}d\psi\, e^{-S_{QCD}[U,\bar{\psi},\psi]} = \int dU\, \mathcal{Z}_F[U]e^{-S_g[U]}, \tag{1.470}$$

where

$$\mathcal{Z}_F[U] \equiv \int d\bar{\psi}d\psi\, e^{-S_W[U,\bar{\psi},\psi]}. \tag{1.471}$$

Since the action is quadratic in the fermion fields, the integration over the Grassmann fields can be performed analytically giving

$$\mathcal{Z}_F[U] = \prod_q \det\left(D_W + m_q\right). \tag{1.472}$$

For sufficiently large m_q, the det() factors are positive, so they can be exponentiated to a real contribution to the gauge action. The integral over the gauge degrees of freedom can still be solved by importance-sampling methods. I refer to the lectures of M. Lüscher for more details (Lüscher, 2009).

The integration over Grassmann variables can always be done analytically for any correlation function involving fermion fields. For the quark propagator we have

$$\langle \psi_{\alpha,i}(x)\bar{\psi}_{\beta,j}(y)\rangle = \mathcal{Z}^{-1} \int DU \langle \psi_{\alpha,i}(x)\bar{\psi}_{\beta,j}(y)\rangle_F \prod_q \det\left(D_W + m_q\right)\ e^{-S_g[U]},$$

(1.473)

where α,β and i,j are spin and flavor indices, respectively, and

$$\langle \psi(x)_{\alpha i}\bar{\psi}(y)_{\beta j}\rangle_F = \mathcal{Z}_F^{-1}\int D[\bar{\psi}]D[\psi]\psi(x)_{\alpha i}\bar{\psi}(y)_{\beta j}e^{-S_F[U,\bar{\psi},\psi]} = \delta_{ij}\left[(D_W + m_i)^{-1}\right]_{xy}^{\alpha\beta}.$$

(1.474)

All fermion integrals result in products of propagators, as expected from Eq. (1.194).

1.6.1.1   *Positivity of the transfer matrix and Hilbert-space interpretation*

The positivity of the transfer matrix, $\hat{T}$, can be proved from the results obtained for the gauge fields and the free fermions. Indeed, the transfer matrix can be written as

$$\hat{T} = \hat{T}_F^{1/2}\hat{T}_g\hat{T}_F^{1/2}\hat{\mathcal{P}}_{\text{phys}},$$

(1.475)

where $\hat{T}_F$ is the transfer matrix for fermions, Eq. (1.268), coupled to the gauge fields in the temporal gauge. The positivity of $\hat{T}_F$ that can be proved in completely analogy with the free fermion case for $r = 1$. $\hat{T}_g$ is the transfer operator for gauge fields, Eq. (1.367). The positivity of $\hat{T}$ follows from that of $\hat{T}_g$ and $\hat{T}_F$ (Lüscher, 1977).

1.6.1.2   *Perturbative expansion, renormalization and continuum limit*

The perturbative expansion can be worked out like in the pure gauge theory. The Feynman rules are supplemented by the fermion vertices with one, two and an arbitrary number of gluons. In the presence of fermions besides the 1PI divergent graphs we considered in the pure gauge case, there is also the fermion two-point vertex graph. A one-loop computation shows that this divergence can be reabsorbed in a redefinition of the fermion mass and wave function, in agreement with the expectation of renormalizability.

On general grounds, we know that in order to warrant that we approach the continuum limit we should make sure that the symmetries are those of the QCD action. It is easy to check that the lattice action is invariant under C, P and T. It is also invariant under the flavor vector symmetries in the limit $m_q = 0$. However, all axial symmetries are broken by the Wilson term. The question is then: what are the additional relevant or marginal operators that can appear in the continuum limit? The only renormalizable operator that can be induced as a result of chiral symmetry breaking is of the form $\bar{\psi}\psi$. It is indeed a relevant operator that is generated with a coefficient $\frac{1}{a}$ and needs to be tuned, just as the mass of the scalar needed to be tuned to reach the critical line. Therefore, the continuum limit of this action even for massless fermions requires a more complicated tuning:

$$g_0 \to 0, \quad \kappa^q \to \kappa_c^q.$$

(1.476)

The theory is asymptotically free, just like the pure gauge theory.

A very useful procedure to define the massless point, beyond perturbation theory, is to impose the PCAC relation.

1.6.1.3 Lattice symmetries and scaling violations

We have seen that the Wilson term breaks chiral symmetry, and in QCD the full chiral flavor symmetry group. This is in principle a disaster, because the low-energy properties of QCD depend in a strong way on the fact that this symmetry is broken only spontaneously as we discussed. It is therefore essential to make sure that the continuum limit is taken in such a way that QCD is recovered. The symmetries in the functional formalism result in a series of Ward identities (WI), as we discussed in section 1.2.5. Therefore, a way to ensure that the symmetry is recovered in the continuum limit is to ensure that renormalized Ward identities are satisfied up to terms that vanish in the continuum limit.

Bochicchio *et al.* studied for the first time how the chiral WI is recovered in the continuum limit of Wilson fermions (Bochicchio *et al.*, 1985). To derive the WIs we consider the following non-singlet transformation ($\text{Tr}[T^a] = 0$)

$$\delta\psi(x) \rightarrow i\epsilon_a(x)T^a\gamma_5\psi(x),$$
$$\delta\bar\psi(x) \rightarrow i\epsilon_a(x)\bar\psi(x)T^a\gamma_5. \qquad (1.477)$$

Performing such a change of variables in the expectation value of the operator O we get:

$$\langle\delta_\epsilon S_W \, O\rangle = \langle\delta_\epsilon O\rangle, \qquad (1.478)$$

where

$$\delta_\epsilon S_W = a^4 \sum_x \epsilon^a(x) \left\{ i\bar\psi(x)\gamma_5\{m, T^a\}\psi(x) - i\sum_\mu \hat\partial_\mu^* A_\mu^a(x) + iX^a(x) \right\}, \qquad (1.479)$$

and

$$X^a(x) \equiv -\frac{r}{2a}\sum_\mu \left[\bar\psi(x)T^a\gamma_5 U_\mu(x)\psi(x+a\hat\mu) + \bar\psi(x)T^a\gamma_5 U_\mu^\dagger(x-a\hat\mu)\psi(x-a\hat\mu) \right.$$

$$+ \bar\psi(x-a\hat\mu)U_\mu(x-a\hat\mu)T^a\gamma_5\psi(x) + \bar\psi(x+a\hat\mu)U_\mu^\dagger(x)T^a\gamma_5\psi(x)$$

$$\left. - 4\bar\psi(x)T^a\gamma_5\psi(x) \right] = \delta_\epsilon(\text{Wilson term}) \qquad (1.480)$$

$$A_\mu^a(x) \equiv \frac{1}{2}\left[\bar\psi(x)\gamma_\mu\gamma_5 T^a U_\mu(x)\psi(x+a\hat\mu) + \bar\psi(x+a\hat\mu)\gamma_\mu\gamma_5 T^a U_\mu^\dagger(x)\psi(x) \right]. \qquad (1.481)$$

m is the quark mass matrix. In the naive continuum limit, we find that $X^a \rightarrow 0$, while $A_\mu(x)$ goes to the continuum axial current.

So, the WI on the lattice reads:

$$\langle O(y)\hat{\partial}^*_\mu A^a_\mu(x)\rangle = \langle O(y)\bar{\psi}(x)\gamma_5\{m, T^a\}\psi(x)\rangle + \langle O(y)X^a(x)\rangle - i\left\langle \frac{\delta O(y)}{\delta\epsilon^a(x)}\right\rangle.$$

(1.482)

The anomalous term, X^a, even though vanishing in the naive continuum limit will generate divergences that need to be renormalized. Being a local operator of $d = 5$ will generically mix with the operators $\hat{\partial}^*_\mu A_\mu$ and with the pseudoscalar density $P^a = \bar{\psi}(x)T^a\gamma_5\psi(x)$, so in general

$$X^a = -2\bar{m}P^a - (Z_A - 1)\hat{\partial}^*_\mu A_\mu + X^a_R,$$

(1.483)

where the last term is a renormalized operator that vanishes in the continuum limit and $\bar{m}$ and $Z_A - 1$ are the mixing coefficients of X^a with the lower-dimensional operators. $\bar{m} \sim a^{-1}$, while Z_A can be shown to be finite. Therefore,

$$\lim_{a\to 0} \langle O(y)Z_A\hat{\partial}^*_\mu A^a_\mu\rangle = \lim_{a\to 0} \langle O(y)\bar{\psi}(x)\gamma_5\{m - \bar{m}, T^a\}\psi(x)\rangle - i\left\langle \frac{\delta O(y)}{\delta\epsilon^a(x)}\right\rangle.$$

(1.484)

In the continuum limit we recover the standard chiral WI, with the lattice current normalized by Z_A and the quark mass is proportional to $m - \bar{m}$. In general, the scaling violations are $O(a)$, however, the improvement program described in (Weisz, 2009) allows us, to reach $O(a^2)$.

In summary, the consequence of the explicit chiral-symmetry breaking by the Wilson term is twofold:

- The bare mass m needs to be tuned non-perturbatively to fix the quark mass, for example, the so-called PCAC quark mass can be obtained from the ratio (up to a multiplicative renormalization)

$$\frac{\langle\hat{\partial}^*_\mu A^a_\mu(x)P^a(0)\rangle}{\langle P^a(x)P^a(0)\rangle} \sim m_{PCAC}.$$

(1.485)

- The axial current is renormalized. For a method to determine Z_A non-perturbatively and further details on the uses of WIs see the lectures of P. Weisz (Weisz, 2009) and A. Vladikas (Vladikas, 2009).

Exercise 1.13 Show that the Wilson action for QCD has the following discrete symmetries

$$P : \psi(x) \to \gamma_0\psi(x_P)$$

(1.486)

$$\bar{\psi}(x) \to \bar{\psi}(x_P)\gamma_0$$

(1.487)

$$U_0 \to U_0(x_P)$$

(1.488)

$$U_k \to U^\dagger_k(x_P - a\hat{k})$$

(1.489)

$$T : \psi(x) \rightarrow \gamma_0 \gamma_5 \psi(x_T) \tag{1.490}$$

$$\bar{\psi}(x) \rightarrow \bar{\psi}(x_T) \gamma_5 \gamma_0 \tag{1.491}$$

$$U_0 \rightarrow U_0^\dagger (x_T - a\hat{0}) \tag{1.492}$$

$$U_k \rightarrow U_k(x_T) \tag{1.493}$$

$$C : \psi(x) \rightarrow C\bar{\psi}^T(x) \tag{1.494}$$

$$\bar{\psi}(x) \rightarrow -\psi^T(x) C^{-1} \tag{1.495}$$

$$U_\mu \rightarrow U_\mu^*, \tag{1.496}$$

where $x_P = (x_0, -\mathbf{x})$, $x_T = (-x_0, \mathbf{x})$ and $C = \gamma_0 \gamma_2$, satisfying $C\gamma_\mu C = -\gamma_\mu^* = -\gamma_\mu^T$.

Exercise 1.14 Show that the Wilson action for QCD is invariant under global $U_V(N_f)$ in the quark-mass degenerate limit

$$q \rightarrow Uq \quad \bar{q}_f \rightarrow \bar{q} U^\dagger \quad U \in U(N_f). \tag{1.497}$$

Derive the lattice WI for the $U_V(N_f)$ symmetry and identify the conserved vector current.

1.6.2 Observables

We will briefly discuss a few of the observables that are routinely measured in lattice QCD. The first important question is of course the low-lying spectrum. Computing the meson and baryon masses requires the computation of two-point correlators of appropriate operators. The Källen–Lehmann representation implies that the large-time behavior of these two point functions is dominated by the lightest one-particle states with the same quantum numbers.

How do we choose the operator? We have seen, from the transfer matrix construction that operators with a Hilbert interpretations are gauge invariant products of the fundamental fields $\psi, \bar{\psi}$ and the spatial plaquettes at fixed times. In principle, any operator in the Hilbert space can be represented by creation and annihilation operators that create the one-particle asymptotic states in the interacting theory. Ensuring that the quantum numbers are the right ones (spin, color, isospin, parity, etc.) the operator will generically have an overlap with the one-particle states. Obviously we do not know *a priori* which operator maximizes this overlap and there are several techniques to improve it (variational techniques, smearing, etc.), which we will not discuss here.

1.6.2.1 Mesons

The simplest operators that are used to compute meson correlation functions are of the form:

$$M^a(x) \equiv \bar{\psi}_{\alpha i c}(x) \Gamma_{\alpha\beta} T_{ij}^a \psi_{\beta j c}(x), \tag{1.498}$$

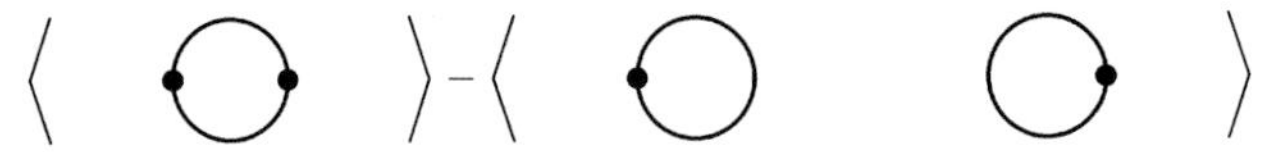

$$\left\langle \; \bigcirc \; \right\rangle - \left\langle \; \bigcirc \qquad \bigcirc \; \right\rangle \qquad (49.5)$$

Fig. 1.10 Connected and disconnected contribution to a meson correlator.

where

$$\Gamma = \{1, \gamma_5, \gamma_\mu, \gamma_\mu \gamma_5, \dots\} \tag{1.499}$$

for the scalar, axial, vector and axial vector... T^a is a matrix in flavor space and the color indices are summed over since a meson is a singlet of color. The proper choice of the matrix T^a ensures the right flavor composition or isospin (or $SU(3)$) flavor quantum numbers. In order to improve the signal it is common practice to project on the zero spatial momentum states by computing the correlator

$$C_M(x_0) = \sum_{\mathbf{x}} \langle M^a(x_0, \mathbf{x}) M^a(0, \mathbf{0}) \rangle. \tag{1.500}$$

As usual, the Grassmann integrations can be readily performed and the result is

$$C_m(x_0) = \frac{1}{Z} \int DU e^{-S_g[U]} \det(D_W + m) \sum_{\mathbf{x}}$$

$$\times \left\{ -\mathrm{Tr}[(D_W + m)_{0,x}^{-1}(\Gamma \otimes T^a)(D_W + m)_{x,0}^{-1}(\Gamma \otimes T^b)] \right.$$

$$\left. + \mathrm{Tr}[(D_W + m)_{0,0}^{-1}(\Gamma \otimes T^a)] \mathrm{Tr}[(D_W + m)_{x,x}^{-1}(\Gamma \otimes T^b)] \right\}. \tag{1.501}$$

The two terms correspond to the connected and disconnected contributions, shown in Fig. 1.10. The latter are much harder to compute numerically because the sum over $\mathbf{x}$ would require the inversion of the Dirac operator as many times as there are spatial points, while the connected contribution can be obtained with a single inversion per spin and color.

1.6.2.2   *Baryons*

Baryons are qqq color singlets. We can take the following operators:

$$B^{abc}_{\alpha\beta\gamma} = \psi(x)_\alpha \equiv \epsilon_{c_1 c_2 c_3} \psi_{\alpha a c_1} \psi_{\beta b c_2} \psi_{\gamma c c_3}, \tag{1.503}$$

where a, b, c are the flavor indices and α, β, γ the spinor ones. The contraction of this three-quark object with appropriate tensors of both set of indices will ensure the right flavor and spin, respectively.

For example, consider the proton, which is a $J = 1/2$, $P = +1$ and $I = 1/2$ state made up of two u quarks and one d quark. In order to combine these three, we can first combine the d and one u in a $J = 0$, $I = 0$ diquark state and then add the third one. We need therefore to combine the u and d antisymmetrically both in flavor and spin, obtaining

$$(u_\alpha d_\beta - d_\alpha u_\beta)(C\gamma_5)_{\alpha\beta}, \tag{1.504}$$

where $C\gamma_5$ ensures that the quark states with up and down spin are combined anti-symmetrically and are therefore a singlet under rotations. Since $C\gamma_5$ is antisymmetric the two terms are the same and the possible proton operator is given by

$$p_\gamma = u_\gamma u_\alpha d_\beta (C\gamma_5)_{\alpha\beta} = u^T C\gamma_5 d u_\gamma, \tag{1.505}$$

where the color indices are not shown but are contracted with the ϵ tensor.

The corresponding antiproton is

$$\bar{p}_\gamma = \bar{d} C\gamma_5 \bar{u}^T \bar{u}_\gamma. \tag{1.506}$$

The two-point correlation functions of those operators at large x_0 separation, are dominated by the lightest one-particle state in the corresponding channel:

$$\lim_{x_0 \to 0} \sum_x \langle B(x)B(0) \rangle = \lim_{x_0 \to 0} \sum_x \langle 0|T(\hat{B}(x)\hat{B}(0))|0 \rangle_E = \frac{Z_L}{2}(1+\gamma_0)e^{-m_L x_0}, \tag{1.507}$$

where m_L is the mass of the lightest state in this channel, $|L\rangle$, and $Z_L = |\langle 0|\hat{B}(0)|L\rangle|^2$, the vacuum-to-this-state matrix element.

Exercise 1.15 Write down an interpolating operator for the Ω baryon ($J^P = 3/2^+$) made of three strange quarks and an interpolating operator for the ρ^+ meson.

1.6.3 Decay constants: pion to vacuum matrix elements

A consequence of the chiral Ward indentity is the PCAC relation, i.e. the coupling of the axial current to the single pseudoscalar meson states, the lightest of them being the pion $|\pi\rangle$. The corresponding matrix element is the decay constant. This is the matrix element needed for determining the leptonic decays widths of pseudoscalar mesons, from which several of the elements of the CKM matrix are best determined.

$$\langle 0|A^a_\mu(x)|\pi(p)\rangle = iF_\pi p_\mu e^{-ipx}. \tag{1.508}$$

Therefore, F_π can be obtained from the normalization of the axial-current two-point correlator, provided it is appropriately renormalized:

$$-\lim_{x_0 \to \infty} Z_A^2 \sum_x \langle A_0(x)A_0(0) \rangle = \frac{F_\pi^2 M_\pi}{2} \exp(-M_\pi x_0). \tag{1.509}$$

An essential requirement is therefore to obtain Z_A. I refer to P. Weisz's (Weisz, 2009) and A. Vladikas's lectures (Vladikas, 2009).

1.6.4 Form factors: single state matrix elements of current operators

In order to describe other processes, such as meson semileptonic decays in which a meson decays into a lighter one emitting two leptons, e.g. $B \to \pi l \nu_l$ (important in the determination of V_{ub}) we need to know the matrix element of the weak current between the two initial and final meson states.

$$\langle M|\bar{q}T^a\gamma_\mu(1-\gamma_5)q|M'\rangle, \tag{1.510}$$

where the flavor quantum numbers of M, M' and T^a should be appropriately fixed for the given process. According to the LSZ reduction formulae, this matrix element can be obtained from the expectation value of the time-ordered product of three operators: the vector current and the two operators that have an overlap with the initial and final meson states,

$$\lim_{x_0,y_0\to+\infty,-\infty}\sum_{\mathbf{x},\mathbf{y}}\langle M^a(x)J_\mu^b(0)M^c(y)\rangle. \tag{1.511}$$

In contrast with two point functions that depend on a single momentum, the three-point functions depend on two and therefore the matrix element has a non-trivial momentum dependence dictated by Lorentz invariance such as:

$$\langle\pi(p)|J_\mu(q)|B(p')\rangle = f^+(q^2)\left[p'+p-\frac{m_B^2-m_\pi^2}{q^2}q\right]_\mu + f^0(q^2)\frac{m_B^2-m_\pi^2}{q^2}q_\mu. \tag{1.512}$$

The coefficients $f^+(q^2)$, $f^0(q^2)$ are called form factors and in principle they must be determined in the whole kinematical range of q^2.

1.6.5 Two-body decays

Other processes such as $K\to\pi\pi, \rho\to\pi\pi$, etc. involve also three-point functions. However, their large-time behavior does not contain sufficient information to reconstruct the corresponding S-matrix element (Maiani and Testa, 1990). A similar problem affects other scattering processes.

It is important to point out that there is nothing wrong with LSZ reduction formula on the Euclidean infinite lattice (Lüscher, 1988). Any S-matrix element can be computed by:

- Computing the connected Euclidean correlation functions in momentum space

$$\sum_{x_n}\cdots\sum_{x_1}e^{-iq_1x_1}\ldots e^{-iq_nx_n}\langle O(x_1)\ldots O(x_n)\rangle = S_n(q_1,\ldots,q_n). \tag{1.513}$$

- Wick rotating them back to Minkowski:

$$W_n(E_1,\ldots E_n) = S_n(q_1,\ldots,q_n)\big|_{q_i^0=(i-\epsilon)E_i}. \tag{1.514}$$

- The S-matrix element is then given by

$$\langle\mathbf{p}_3,\ldots,\mathbf{p}_n; out|\mathbf{p}_1\mathbf{p}_2; in\rangle = \prod_k\frac{(E_k^2-\omega(\mathbf{p}_k))}{\sqrt{Z_k}}W_n\bigg|_{E_i=\pm\omega(\mathbf{p}_i)}. \tag{1.515}$$

This method is, however, numerically hopeless. There are smarter ways to proceed, by using finite-size scaling techniques. QCD in a box is a wonderful laboratory from which physical information can be extracted. A few examples of the uses of a finite volume are

- Finite-size dependence of one-particle masses is related to the forward elastic scattering amplitude (Lüscher, 1983; 1986*a*).
- Two-particle spectra in a box is related to the scattering phase shifts and unstable particle widths (Lüscher, 1986*b*). See the lectures of S. Aoki (Aoki, 2009) where some applications are discussed.
- The Nambu–Goldstone bosons in a box behave in a way that can be predicted by chiral perturbation theory and provides a different regime to match QCD with the chiral Lagrangian: the so-called ϵ-regime (Gasser and Leutwyler, 1987).
- Non-perturbative renormalization: the renormalization scale is set by the box size (Jansen *et al.*, 1996). See the lectures of (Weisz, 2009).

this list is probably not exhausted...

An important message is that in lattice QCD simulations the optimal conditions to extract physical parameters are not necessarily the same conditions as in real experiments. We surely need to prove the universality of our results by taking the limit $a \to 0$, but we should also exploit as much as possible the possibilities that the lattice offers of probing QCD in new conditions (unphysical quark masses, finite volume, etc...)

Acknowledgments

I wish to thank the organizers of this wonderful school: L. Lellouch, R. Sommer, B. Svetitsky and especially A. Vladikas for their support and their invaluable help in editing these proceedings. I profited greatly from the critical reading of these lectures by F. Bernardoni and A. Donini. I would also like to thank my co-lecturers S. Aoki, D. Kaplan, M. Golterman and P. Weisz for their advice, their patient sitting through my lectures, and even answering some hard questions for me. Last but not least I wish to warmly thank the students for the stimulating enviroment of the school, and the director of the school Leticia Cugliandolo for her support in academic as well as family matters.

Notes

1. http://www.claymath.org/millennium/Yang-Mills_Theory/yangmills.pdf
2. *Prove that for any compact simple gauge group G, a non-trivial quantum YM theory exists in R^4 and has a non-vanishing mass gap (existence includes establishing axiomatic properties such as Osterwalder and Schrader).*
3. Some discrete symmetries such as time reversal are implemented by anti-unitary operators, but we will not consider this case here.

4. In the Schrödinger picture, the wave function is no longer a function of **x** but of $\phi(\mathbf{x})$.
5. These are the correlation functions that cannot be made disconnected by cutting out one particle propagator.
6. Whether a continuum limit of this discretized theory exists is of course not warrantied from this property.

References

Adler, S. L. and Bardeen, W. A. (1969). Absence of higher order corrections in the anomalous axial vector divergence equation. *Phys. Rev.*, **182**, 1517–1536.

Amsler, C. et al. (2008). Review of particle physics. *Phys. Lett.*, **B667**, 1.

Aoki, S. (2009). lectures on lattice QCD and nuclear physics, Les Houches École d'Été de Physique Théorique.

Appelquist, T. and Carazzone, J. (1975). Infrared singularities and massive fields. *Phys. Rev.*, **D11**, 2856.

Atiyah, M. F. and Singer, I. M. (1971). The index of elliptic operators. 5. *Annals Math.*, **93**, 139–149.

Belavin, A. A., Polyakov, A. M., Schwartz, A. S., and Tyupkin, Y. S. (1975). Pseudoparticle solutions of the Yang-Mills equations. *Phys. Lett.*, **B59**, 85–87.

Berezin, F. A. (1966). *The method of second quantization.* Academic Press, New York.

Bochicchio et al. (1985). Chiral symmetry on the lattice with Wilson fermions. *Nucl. Phys.*, **B262**, 331.

Bogoliubov, N. and Parasiuk, O. (1957). On the multiplication of the causal function in the quantum theory of fields. *Acta Math.*, **97**, 227–266.

Callaway, David J. E. (1988). Triviality pursuit: Can elementary scalar particles exist? *Phys. Rept.*, **167**, 241.

Collins, J. C. (1984). *Renormalization.* Cambridge University Press, Cambridge.

Creutz, M. (1983). *Quarks, gluons and lattices.* Cambridge Monographs on Mathematical Physics, Cambridge University Press, Cambridge.

DeGrand, T. and DeTar, C. (2006). *Lattice methods for quantum chromodynamics.* World Scientific Publishing New Jersay, USA.

Feynman, R. (1948). The space-time formulation of nonrelativistic quantum mechanics. *Rev. Mod. Phys.*, **20**, 367.

Gasser, J. and Leutwyler, H. (1987). Thermodynamics of chiral symmetry. *Phys. Lett.*, **B188**, 477.

Gattringer, C. and Lang, C.B. (2010). *Quantum chromodynamics on the lattice: An introductory presentation.* Lect. Notes Phys. 788, Springer, Berlin Heidelberg.

Goldstone, J. (1961). Field theories with superconductor solutions. *Nuovo Cim.*, **19**, 154–164.

Golterman, M. (2009). Lectures on applications of chiral perturbation theory to lattice QCD, Les Houches École d'Été de Physique Théorique.

Hasenfratz, A. et al. (1987). The triviality bound of the four component phi**4 model. *Phys. Lett.*, **B199**, 531.

Hasenfratz, P., Laliena, V., and Niedermayer, F. (1998). The index theorem in QCD with a finite cut-off. *Phys. Lett.*, **B427**, 125–131.

Hepp, K. (1966). Proof of the Bogolyubov-Parasiuk theorem on renormalization. *Commun. Math. Phys.*, **2**, 301–326.

Itzykson, C. and Zuber, J. B. (1980). *Quantum field theory*. McGraw-Hill Inc New york.

Jansen, K. et al. (1996). Non-perturbative renormalization of lattice QCD at all scales. *Phys. Lett.*, **B372**, 275–282.

Källen, G. (1972). *Quantum electrodynamics*. Springer-verl./New York, 233p.

Kaplan, D. B. (2009). Lectures on chiral symmetry and lattice fermions, Les Houches École de Physique.

Kogut, J. B. and Susskind, L. (1975). Hamiltonian formulation of Wilson's lattice gauge theories. *Phys. Rev.*, **D11**, 395.

Lehmann, H (1954). On the properties of propagation functions and renormalization contants of quantized fields. *Nuovo Cim.*, **11**, 342–357.

Lehmann, H., Symanzik, K., and Zimmermann, W. (1955). On the formulation of quantized field theories. *Nuovo Cim.*, **1**, 205–225.

Lüscher, M. (1977). Construction of a selfadjoint, strictly positive transfer matrix for Euclidean lattice gauge theories. *Commun. Math. Phys.*, **54**, 283.

Lüscher, Martin (1982). Topology of lattice gauge fields. *Commun. Math. Phys.*, **85**, 39.

Lüscher, M. (1983). On a relation between finite size effects and elastic scattering processes. Lecture given at Cargese Summer Inst., Cargese, France, Sep 1–15, 1983.

Lüscher, M. (1986*a*). Volume dependence of the energy spectrum in massive quantum field theories. 1. Stable particle states. *Commun. Math. Phys.*, **104**, 177.

Lüscher, M. (1986*b*). Volume dependence of the energy spectrum in massive quantum field theories. 2. Scattering states. *Commun. Math. Phys.*, **105**, 153–188.

Lüscher, M. (1988). Selected topics in lattice field theory. Lectures given at Summer School 'Fields, Strings and Critical Phenomena', Les Houches, France, Jun 28 – Aug 5, 1988.

Lüscher, M. (1989), Lattice Field Theory, unpublished notes.

Lüscher, M. (2009). Lectures on computational strategies, Les Houches École d'Été de Physique Théorique.

Lüscher, M. and Weisz, P. (1988). Is there a strong interaction sector in the standard lattice Higgs model? *Phys. Lett.*, **B212**, 472.

Lüscher, M. and Weisz, P. (1989). Scaling laws and triviality bounds in the lattice phi**4 theory. 3. N component model. *Nucl. Phys.*, **B318**, 705.

Maiani, L. and Testa, M. (1990). Final state interactions from Euclidean correlation functions. *Phys. Lett.*, **B245**, 585–590.

Menotti, P. and Pelissetto, A. (1987). General proof of Osterwalder-Schrader positivity for the Wilson Action. *Commun. Math. Phys.*, **113**, 369.

Montvay, I. and Münster, G. (1994). *Quantum fields on a lattice*. Cambridge Monographs on Mathematical Physics, Cambridge University Press Cambridge.

Montvay, I., Münster, G., and Wolff, U. (1988). Percolation cluster algorithm and scaling behaviour in the four-dimensional Ising model. *Nucl. Phys.*, **B305**, 143.

Nambu, Y. (1960). Quasi-particles and gauge invariance in the theory of superconductivity. *Phys. Rev.*, **117**, 648–663.

Nielsen, H. B. and Ninomiya, M. (1981). No go theorem for regularizing chiral fermions. *Phys. Lett.*, **B105**, 219.

Osterwalder, K. and Schrader, R. (1973). Axioms for Euclidean Green's functions. *Commun. Math. Phys.*, **31**, 83–112.

Osterwalder, K. and Schrader, R. (1975). Axioms for Euclidean Green's functions. 2. *Commun. Math. Phys.*, **42**, 281.

Osterwalder, K. and Seiler, E. (1978). Gauge field theories on the lattice. *Ann. Phys.*, **110**, 440.

Peskin, M. E. and Schroeder, D. V. (1995). *An introduction to quantum field theory.* Addison-Wesley Publishing Company Reading USA.

Philipsen, U. (2009). Lectures on QCD at finite temperature and quark density, Les Houches École d'Été de Physique Théorique.

Reisz, T. (1988). A power counting theorem for Feynman integrals on the lattice. *Commun. Math. Phys.*, **116**, 81.

Rothe, H.J. (2005). *Lattice gauge theories: An introduction.* World Scientific Lecture Notes in Physics Vol. 74.

Smit, J. (1991). Transfer operator and lattice fermions. *Nucl. Phys. Proc. Suppl.*, **20**, 542–545.

Smit, J. (2002). *Introduction to quantum fields on a lattice.* Cambridge Lecture Notes in Physics, Cambridge University Press, Cambridge.

Sommer, R. (2009). Lectures on heavy quark effective theory, Les Houches École d'Été de Physique Théorique.

Streater, R. and Wightman, A. S. (1964). *PCT, spin and statistics and all that.* Mathematical Physics Monograph Series, Benjamin, New York.

Susskind, L. (1977). Lattice fermions. *Phys. Rev.*, **D16**, 3031–3039.

't Hooft, G. (1976). Symmetry breaking through Bell-Jackiw anomalies. *Phys. Rev. Lett.*, **37**, 8–11.

Takahashi, Y. (1957). On the generalized Ward identity. *Nuovo Cim.*, **6**, 371.

Veneziano, G. (1979). U(1) without instantons. *Nucl. Phys.*, **B159**, 213–224.

Vladikas, A. (2009). Lectures on non-perturbative renormalization, Les Houches École d'Été de Physique Théorique.

Ward, J. C. (1950). An identity in quantum electrodynamics. *Phys. Rev.*, **78**, 182.

Weinberg, S. (1995). *The quantum theory of fields, Vol. 1: Foundations.* Cambridge University Press Cambridge.

Weisz, P. (2009). Lectures on renormalization and improvement, Les Houches École d'Été de Physique Théorique.

Wilson, K.G. (1974). Confinement of quarks. *Phys. Rev.*, **D10**, 2445–2459.

Wilson, K. G. (1975). Quarks and strings on a lattice. New phenomena In subnuclear physics. Part A. Proceedings of the First Half of the 1975 International School of Subnuclear Physics, Erice, Sicily, July 11 – August 1, 1975, ed. A. Zichichi, Plenum Press, New York, 1977, p. 69, CLNS-321.

Wilson, K. G. and Kogut, J. B. (1974). The renormalization group and the epsilon expansion. *Phys. Rept.*, **12**, 75–200.

Witten, E. (1979). Current algebra theorems for the U(1) Goldstone boson. *Nucl. Phys.*, **B156**, 269.

Zimmermann, W. (1969). Convergence of Bogolyubov's method of renormalization in momentum space. *Commun. Math. Phys.*, **15**, 208–234.

2

Renormalization and lattice artifacts

Peter WEISZ

Max-Planck-Institut für Physik, Föhringer Ring 6,
D-80805 München, Germany

2.1 Perturbative renormalization

2.1.1 Introduction

Most of our present knowledge on the structure of renormalization of quantum field theories comes from perturbation theory (PT), a small coupling expansion around a free field theory. In this framework Feynman rules, derived formally in the continuum from the Gell-Mann–Low formula (Gell-Mann and Low, 1951) or from the path-integral approach, express amplitudes at a given order as sums of expressions associated with Feynman diagrams (see e.g. Nakanishi). Tree diagrams are associated with well-defined amplitudes, but for diagrams involving loops and associated integration over internal momenta one soon encounters divergent expressions, e.g. for massive $g_0\phi^4$ theory in four euclidian dimensions the diagram in Fig. 2.1 is associated with the expression

$$I = g_0^2 \int^\Lambda \mathrm{d}^4k \frac{1}{(k^2 + m^2)([(k + p)^2 + m^2])}, \tag{2.1}$$

which is logarithmically divergent as the ultraviolet (UV) cutoff $\Lambda \to \infty$. There are many ways to introduce an UV cutoff Λ – a specification defines a particular *regularization*. The process of *renormalization* is then a well-defined prescription how to map regularized expressions to amplitudes that are finite when $\Lambda \to \infty$ while maintaining desired properties that are summarized by the Osterwalder–Schrader axioms[1] (Osterwalder and Schrader, 1973, 1975) order by order in PT.

For lattice QCD, renormalization is not needed if one is only interested in obtaining the low-energy (LE) spectrum and scattering data. For such purposes we need knowledge of the phase diagram, the location of critical points where the continuum limit is reached, and the nature of the approach, e.g. the question whether ratios of masses tend to their continuum limit as powers in the lattice spacing a:

$$\frac{m_1(a)}{m_2(a)} = \frac{m_1(0)}{m_2(0)} + C_{12}(am_1)^p, \quad p > 0\,(?). \tag{2.2}$$

Lattice artifacts will be the topic of Sections 2.4–2.6. But perturbative and non-perturbative renormalization will be needed for (1) computing matrix elements of composite operators (describing probes of other interactions, finite-temperature transport coefficients, ...), (2) relating LE to high-energy (HE) scales (e.g. computing

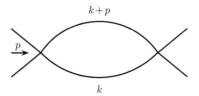

Fig. 2.1 Simple 1–loop diagram.

running couplings and running masses), and (3) giving hints on the nature of lattice artifacts.

In this Section we shall mainly consider perturbative renormalization; in Section 2.2 we shall discuss renormalization group equations that follow from multiplicative renormalization, and aspects of non-perturbative renormalization will be the subject of Section 2.3.

2.1.2 History and basic concepts

In a series of papers Reisz (1988*b*–1989) has proven perturbative renormalizability of lattice Yang–Mills theory and QCD (for a large class of actions). His proof is based mainly on methods developed with continuum regularization, (with some important modifications that we will mention later), so we will start the discussion with these.

Renormalization theory has a long and interesting history; here I reproduce a part of Wightman's delightful discussion (Wightman, 1975) since the reference is not always easily available.

"One of the first steps in natural history is to establish a classifactory nomenclature. I will do this for perturbative renormalization theory, but in so doing, I want to tell stories with a moral for an earnest student: Renormalization theory has a history of egregious errors by distinguished savants. It has a justified reputation for perversity; a method that works up to 13th order in the perturbation series fails in the 14th order. Arguments that sound plausible often dissolve into mush when examined closely. The worst that can happen often happens. The prudent student would do well **to distinguish sharply between what has been proved and what has been made plausible,** *and in general he should watch out!*

My first cautionary tale has to do with the early days of renormalization theory. When F.J. Dyson analyzed the renormalization theory of the S-matrix for quantum electrodynamics of spin one-half particles in his two great papers of 1948-9, (Dyson, 1949a, 1949b) he laid the foundations for most later work on the subject, but his treatment of one phenomenon, overlapping divergences was incomplete. Among the methods offered to clarify the situation, that of J. Ward (1951) seemed outstandingly simple, so much so that it was adopted in Jauch and Rohrlich's standard textbook (Jauch and Rohrlich, 1955). Several years later Mills and Yang noticed that unless further refinements are introduced the method does not work for the photon self energy (Wu, 1962). The lowest order for which the trouble manifests itself is the fourteenth e.g. in the (7–loop) graph Fig. 2.2. Mills and Yang repaired the method and sketched some of the steps in a proof that it would yield a finite renormalized amplitude (Mills and Yang, 1966). An innocent reading of the textbook of Jauch and Rohrlich, would never suspect such refinements are necessary.

Another attempt to cope with the overlapping divergences was made by Salam (1951b, 1951a). I will not describe it, if for no other reason than that I never have succeeded in understanding it. Salam and Matthews commenting on this and related work somewhat later (1951) remarked "... The difficulty, as in all this work is to find a notation which is both concise and intelligible to at least two people of whom

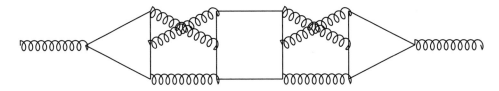

Fig. 2.2 Troublesome 14th-oder QED diagram

one may be the author". The belief is widespread that when Salam's work is combined with later significant work by S. Weinberg (1960), the result should be a mathematically consistent version of renormalization theory. At least that is what one reads in the text book of Bjorken and Drell for quantum electrodynamics (Bjorken and Drell, 1965), and in the work of R. Johnson (1970) and the lectures of K. Symanzik for meson theories (Symanzik, 1961). So apparently the Matthews–Salam criterion has been satisfied. I only wish they had spelled it out a little for the peasants.

Another foundation of renormalization theory with a rather different starting point was put forward by Stueckelberg and Green (1951). It was refounded and brought to a certain stage of completion in the standard text book of Bogoliubov and Shirkov (1959). The mathematical nut that had to be cracked is in the paper of Bogoliubov and Parasiuk (1957), (amazingly, not quoted in the English translation of Bogoliubov and Shirkov).[2] This paper introduces a systematic combinatorial and analytic scheme for overcoming the overlapping divergence problem. This paper is very important for later developments. Unfortunately it was found by K. Hepp (1966) that Theorem 4 of the paper is false, and that consequently the proof of the main result is incomplete as it stands. However Hepp found that Theorem 4 is not essential to derive the main result and he could fill all the gaps. Thus it is appropriate to introduce the initials BPH to stand for the renormalization method described in (Bogoliubov and Parasiuk, 1957) and (Hepp, 1966). So far as I know it was the first version of renormalization theory on a mathematically sound basis."

The rest of the article is also highly recommendable. Reading his article one can understand why many workers in the field who had gone through these historical developments were wary about renormalization theory.[3] A real breakthrough in the ease of understanding was supplied by Zimmermann (1968, 1973a, 1973b), (in particular also concerning renormalization of composite operators). I strongly recommend Lowenstein's article (1972) for a clear exposition of Zimmermann's methods.

It is the purpose of this chapter to give an overview of the important steps and concepts of perturbative renormalization, stating the main results without proofs; for the latter the interested reader must consult further literature. Moreover, I will only discuss the standard approach using expansions in Feynman diagrams, but I would like to mention a powerful alternative approach using flow equations based on the renormalization group developed by Wilson and co-workers (1971a, 1971b, 1974) and improved by Polchinski (1984), a framework that is well suited for proving structural results, see e.g. the works of Keller, Kopper and Salmhofer (1992), (Salmhofer, 1999).

2.1.3   Perturbative classification of theories

Concerning the nature of a Feynman diagram the first important concept is its *superficial degree of divergence*. Consider a diagram with E external lines, V internal vertices, P internal lines and ℓ independent loops. There are some relations between these the first of which is the topological relation[4]

$$V + \ell - P = \text{number of connected components.} \tag{2.3}$$

Other relations depend on the precise nature of the vertices in the theory, e.g. with only a 4-point vertex:

$$4V = 2P + E. \tag{2.4}$$

Then, for a connected diagram Γ with ℓ loops the superficial degree of divergence $\delta(\Gamma)$ estimates the behavior of the associated integral when all internal momenta are large. For ϕ_d^4 theory (in d space-time dimensions):

$$\delta(\Gamma) = d\ell - 2P \tag{2.5}$$

$$= d - \frac{1}{2}(d-2)E + (d-4)V. \tag{2.6}$$

For $d = 4$ this simplifies to $\delta(\Gamma) = 4 - E$, which is negative for $E > 4$ indicating possible overall convergence of the integral in this case.

Exercise 2.1 Show $\delta(\Gamma) = 6 - 2E$ for ϕ_6^3 theory.

This theory is often discussed in the literature because of the comparative simplicity of the diagrams involved. In this case $\delta(\Gamma)$ is negative for $E > 3$.

It is clear that considerations of $\delta(\Gamma)$ alone are not sufficient to establish convergence of the integral because Γ could have divergent subdiagrams as in Fig. 2.3. However, it does lead to the following classification of perturbative QFT:

Fig. 2.3   2–loop diagram in ϕ_6^3 theory with a nested divergence

- *super-renormalizable*: theories where only a finite number of diagrams have $\delta \geq 0$;
- *renormalizable*: there exists $E_0 < \infty$ such that all diagrams with $E > E_0$ have $\delta < 0$;
- *non-renormalizable*: all E-point diagrams have $\delta \geq 0$ for sufficiently high number of loops ℓ.

Remarks: i). Non-renormalizable theories are not necessarily unphysical or devoid of predictive power (an interesting example concerns gravity (Donoghue, 1995)). They can be good effective theories, such as chiral Lagrangians discussed in Goltermann's lectures.

ii) Non-perturbative formulations of perturbatively renormalizable theories may be *trivial* in the limit that the ultraviolet cutoff is sent to infinity.[5] Examples of such theories are thought to be ϕ^4 and QED in 4 dimensions, albeit there is at present no rigorous proof of this conventional wisdom.

iii) There are some rigorous non-perturbative constructions of super-renormalizable theories, e.g. ϕ_2^4 (Glimm and Jaffe, 1968), Yukawa theory in 2 dimensions (Seiler, 1975), ϕ_3^4 (Feldman, 1975). The Schwinger functions in the topologically trivial sector of SU(2) Yang–Mills theory in 4 dimensions in a sufficiently small volume have also been rigorously constructed (Magnen *et al.*, 1992, 1993). For all of these cases the structural information on renormalization found in perturbation theory carry over to the non-perturbative framework.

In the following, we restrict attention to renormalizable theories. A generic procedure to put perturbative renormalization on a firm mathematical basis consists of three steps:

1. Introduce an UV regularization.
2. Construct a mapping of a Feynman integral to an absolutely convergent integral when the cutoff is removed.
3. Show that the map in step 2 is equivalent to a renormalization of the bare parameters and fields in the original Lagrangian, which formally ensures the desired axiomatic properties of the resulting amplitudes.

There are many perturbative regularization procedures,[6] each with their own advantages and disadvantages. In all cases known they give equivalent physical results.

2.1.4 Zimmermann's forest formula

Let us first restrict attention to theories with massive bare propagators. Returning to the simple 1-loop example (2.1), the integral can obviously be made convergent if the integrand is subtracted at external momentum $p = 0$:

$$I \to R = g_0^2 \int^\Lambda \mathrm{d}^4 k \left[\frac{1}{(k^2 + m^2)([(k+p)^2 + m^2)} - \frac{1}{(k^2 + m^2)^2} \right]. \tag{2.7}$$

We note that since the subtraction is a constant (independent of the external momenta) it can be absorbed in a renormalization of the bare coupling g_0.

The problem now is to find the generalization of the subtraction in the simple case above for an arbitrary diagram. For this purpose we can restrict attention to *one particle irreducible (1PI) diagrams*, diagrams that remain connected after cutting any internal line, since an arbitrary diagram can be constructed from 1PI parts joined by propagators.

Here, we consider mappings where subtractions are made directly on the integrands, as adopted by Reisz for the lattice regularization.[7] Let $I_\Gamma(p,k)$ be the original integrand of a diagram with external momenta $p = \{p_1, p_2, \dots\}$ and internal momenta $k = \{k_1, k_2, \dots\}$. The regularized integral will then be given by

$$R_\Gamma(p) = \int \prod_j d^d k_j \, R_\Gamma(p,k), \tag{2.8}$$

with

$$R_\Gamma(p,k) = I_\Gamma(p,k) - \text{subtractions} \tag{2.9}$$

$$= \left(1 - t_p^{\delta(\Gamma)}\right) \overline{R}_\Gamma(p,k), \tag{2.10}$$

where t_p^δ is the Taylor operator of degree δ:

$$t_p^\delta F(p) = F(0) + p_i^\mu \frac{\partial}{\partial p_i'^\mu} F(p')|_{p'=0} + \dots + \frac{1}{\delta!} p_{i_1}^{\mu_1} \dots p_{i_\delta}^{\mu_\delta} \frac{\partial}{\partial p_{i_1}'^{\mu_1}} \dots \frac{\partial}{\partial p_{i_\delta}'^{\mu_\delta}} F(p')|_{p'=0}. \tag{2.11}$$

Dyson's original proposal (1949*b*) was

$$\overline{R}_\Gamma(p,k) = \prod_{\gamma, 1PI \subset \Gamma} \left(1 - t_{p^\gamma}^{\delta(\gamma)}\right) I_\Gamma(p,k),$$

where p^γ are the external momenta of γ. On the rhs of this equation the term involving γ_1 is to the right of that for γ_2 if γ_1 is *nested* in γ_2, i.e. $\gamma_1 \subset \gamma_2$. If sets of lines of γ_1 and γ_2 do not intersect then the order is irrelevant. The proposal cures the problem of nested divergences, however, no prescription is given how to order *overlapping divergences*, that is divergent subdiagrams that have non-trivial intersection but are not nested such as in Fig. 2.4.[8]

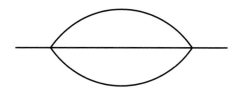

Fig. 2.4 2–loop diagram with overlapping divergences

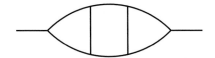

Fig. 2.5   3–loop diagram in ϕ_3^3 theory

The correct prescription presented by Zimmermann (1968) is obtained by expanding Dyson's product and dropping all terms containing products of terms involving overlapping subdiagrams. The result is called *Zimmermann's forest formula*:

$$\overline{R}_\Gamma(p,k) = \sum_{\mathcal{F}} \prod_{\gamma \in \mathcal{F}} \left(-t_{p^\gamma}^{\delta(\gamma)}\right) I_\Gamma(p,k), \tag{2.12}$$

where the sum is over all forests:

$$\mathcal{F} = \{\gamma | \gamma \text{ 1PI}, \delta(\gamma) \geq 0;$$

$$\gamma_1, \gamma_2 \in \mathcal{F} \Rightarrow \gamma_1 \subset \gamma_2 \text{ or } \gamma_2 \subset \gamma_1 \text{ or } \gamma_1, \gamma_2 \text{ non-overlapping}\}.$$

The empty set ϕ is included as a special forest.

Exercise 2.2   Give the set of forests for the diagram in Fig. 2.5 (it has 8 elements).

The following rules are used to specify the momenta flow in Γ and its subgraphs. The momenta flowing in the σth internal line $L_{ab\sigma}$ joining vertices a, b is specified as $l_{ab\sigma} = P_{ab\sigma}(p) + K_{ab\sigma}(k)$ with $K_{ab\sigma} = \sum_i \epsilon_{ab\sigma i} k_i$; $\epsilon_{ab\sigma i} = 1$ if $l_{ab\sigma} \in$ loop $\mathcal{C}_i$; $\epsilon_{ab\sigma i} = -1$ if $l_{ba\sigma} \in$ loop $\mathcal{C}_i$; $\epsilon_{ab\sigma i} = 0$ otherwise. The momentum $P_{ab\sigma}$ is solved by Kirchoff's laws: $\sum_{b\sigma} P_{ab\sigma} + p_a = 0$. $\sum_{L_{ab\sigma} \in \mathcal{C}_i} r_{ab\sigma} P_{ab\sigma} = 0$. Here, "resistances" $r_{ab\sigma}$ can be chosen for convenience (including zero) but no closed loop of zero resistances.

Similarly for subdiagrams γ: $l_{ab\sigma}^\gamma = P_{ab\sigma}^\gamma(p^\gamma) + K_{ab\sigma}^\gamma(k)$. First, solve for $P_{ab\sigma}^\gamma$ keeping the same resistances and momentum conservation at the vertex. Then determine the $K_{ab\sigma}^\gamma(k)$ from $l_{ab\sigma}^\gamma = l_{ab\sigma}$.

After applying the forest formula the resulting integrand has a form

$$F(p,k) = \frac{V(k,p,m)}{\prod_{i=1}^n (l_i^2 + m^2)}, \tag{2.13}$$

where

$$l_i = K_i(k) + P_i(p). \tag{2.14}$$

To proceed, it is first necessary to give conditions on such an integrand required to guarantee the convergence of the corresponding integral

$$F(p) = \int d^d k_1 \ldots d^d k_\ell \, F(p, k). \tag{2.15}$$

Here and in the following, we will assume that the internal momenta have been selected in a "natural way" so that $K_i = k_i$ for $i = 1, \ldots, \ell$ (this is always possible). Consider the set

$$\mathcal{L} = \{K_1, \ldots, K_n\}, \tag{2.16}$$

and let $u_1, \ldots, u_r, v_1, \ldots, v_{\ell-r}$ $r \geq 1$, be ℓ linearly independent elements of $\mathcal{L}$. Let H be the (Zimmermann) subspace spanned by the u_i. Then, the *upper degree* of a function $f(u, v)$ with respect to H is given by

$$\overline{\deg}_{u|v} f(u, v) = \bar{\nu} \quad \text{if} \quad \lim_{\lambda \to \infty} \lambda^{-\bar{\nu}} f(\lambda u, v) \neq 0, \infty, \tag{2.17}$$

and for the integral (2.15) the upper degree wrt H is defined by

$$\overline{\deg}_H F(p) = dr + \overline{\deg}_{u|v} F(p, k(u, v, p)). \tag{2.18}$$

The important statement is *Dyson's power-counting theorem*: $F(p)$ in Eq. (2.15) is absolutely convergent if $\overline{\deg}_H F(p) < 0$ for all subspaces H.

The theorem was first proven by Weinberg (1960), and later a simpler proof was given by Hahn and Zimmermann (1968).

It can be shown that the forest formula yields integrals $F(p) = R_\Gamma(p)$ whose integrands satisfy the conditions of the theorem. Once this is done it remains to prove step 3 of the renormalization procedure (see Section 2.1.3). This is not at all obvious from the forest formula as it stands. For this purpose it is advantageous to show that the forest formula is a solution to the *Bogoliubov–Parasiuk–Hepp recursion relations* (Bogoliubov and Shirkov, 1959, Bogoliubov and Parasiuk, 1957, Hepp, 1966):[9]

$$\overline{R}_\Gamma = \sum_\psi \prod_{\gamma \in \psi} (-t^{\delta(\gamma)}) \overline{R}_\gamma, \tag{2.19}$$

where ψ are sets of disjoint 1PI subgraphs. This formula serves as a basis for an inductive proof of step 3; again the proof is too long to present here;[10] a tactic to prove multiplicative renormalizability of gauge theories will be given in Section 2.1.6.

2.1.5 Renormalization of theories with massless propagators

For theories with massless bare propagators one must control possible infrared (IR) divergences. To this end for a Zimmermann subspace one defines a *lower degree*:

$$\underline{\deg}_{v|u} f(u, v) = \underline{\nu} \quad \text{if} \quad \lim_{\lambda \to 0} \lambda^{-\underline{\nu}} f(\lambda u, v) \neq 0, \infty, \tag{2.20}$$

and for the integral (2.15):

$$\underline{\deg}_{H'} F(p) = d(\ell - r) + \underline{\deg}_{v|u} F(p, k(u, v, p)). \qquad (2.21)$$

The appropriately modified power-counting theorem then states that $F(p)$ is absolutely convergent if $\overline{\deg}_H F(p) < 0$ and $\underline{\deg}_{H'} F(p) > 0$ for all H and for all *non-exceptional external momenta* p (that is a set for which no partial sum of external momenta vanish, except for the complete sum expressing total momentum conservation).

    To achieve this one has to modify the R operation described above to avoid IR divergences introduced by Taylor subtractions at $p = 0$. One way, introduced by Lowenstein and Zimmermann (1975) and adopted by Reisz (1988a, 1988c) is to introduce an auxiliary mass in the massless propagators $l^2 \to l^2 + (1 - s)\mu^2$. UV subtractions are then made at $p = 0, s = 0$, and afterwards s is set to 1 in the remaining parts. In order to avoid further IR singularities it is usually necessary to make extra finite subtractions to ensure 2- and 3-point functions vanish at exceptional external momenta singularities. In gauge theories this is guaranteed by the Slavnov–Taylor identities, which we consider in the next section.

2.1.6   Renormalization of 4D Yang–Mills theory

In this section we will outline the proof of multiplicative renormalizability of Yang–Mills (YM) theory in 4 dimensions. Perturbation theory is an expansion around a saddle point. For gauge theories, such as pure YM theory, gauge invariance of the action implies severe degeneracy of the saddle point. To apply perturbation theory one needs to lift the degeneracy by *gauge fixing* (see Lüscher's elegant discussion 1988).[11] It is convenient to employ linear gauge-fixing functions, e.g. the Lorentz gauge, which transform under the adjoint representation of the global gauge group so that the gauge-fixed action is invariant under such transformations. Because of the non-linearity of the YM theory this way of fixing the gauge introduces extra terms in the functional integral, which can be expressed as a local Lagrangian involving *Faddeev–Popov ghost fields* $c, \bar{c}$ as introduced in Hernandez' lectures. The action in this case is

$$S_0 = -2 \int \mathrm{tr} \left\{ \frac{1}{4} F^2 + \frac{\lambda_0}{2} (\partial A)^2 - \bar{c}\partial_\mu [D_\mu, c] \right\} \qquad (2.22)$$

$$= S_A + S_{\mathrm{gf}} + S_{\mathrm{FP}}, \qquad (2.23)$$

where $D_\mu = \partial_\mu + g_0 A_\mu$ and $F_{\mu\nu} = \partial_\mu A_\nu - \partial_\nu A_\mu + g_0 [A_\mu, A_\nu]$. Here, $A_\mu = T_d A_\mu^d$, where T_d denote a basis of the Lie algebra of SU(N) with $\mathrm{tr}(T_d T_e) = -\frac{1}{2}\delta_{de}$.

    The action is no longer gauge invariant but an important property is that S_A and $S_{\mathrm{gf}} + S_{\mathrm{FP}}$ are separately invariant under *BRST transformations* (Becchi *et al.*, 1976):

$$\delta A_\mu = \epsilon s A_\mu = \epsilon [D_\mu, c], \qquad (2.24)$$

$$\delta c = \epsilon s c = -\epsilon g_0 c^2, \qquad (2.25)$$

$$\delta \bar{c} = \epsilon s \bar{c} = \epsilon \lambda_0 \partial_\mu A_\mu, \tag{2.26}$$

where ϵ is an infinitesimal Grassmann parameter.

Exercise 2.3 Show the BRST invariance of $S_{\text{gf}} + S_{\text{FP}}$.

Note that s is nilpotent: $s^2 H[A, \bar{c}, c] = 0$ for all functionals H. This property and the BRST invariance of the action and measure imply the powerful *Slavnov–Taylor identities* (Slavnov, 1972), (Taylor, 1971);[12] for any functional $H[A, \bar{c}, c]$:

$$\langle sH \rangle = 0. \tag{2.27}$$

For example, if we take $H = \bar{c}^a(x) \partial_\nu A_\nu^b(y)$ we obtain

$$\langle \lambda_0 \partial_\mu A_\mu^a(x) \partial_\nu A_\nu^b(y) \rangle = \langle \bar{c}^a(x) \partial_\nu [D_\nu, c]^b (y) \rangle \tag{2.28}$$

$$= \delta^{ab} \delta(x - y), \tag{2.29}$$

where in the last line the ghost-field equation of motion has been used. It follows that the full (bare) gluon propagator has the form

$$\widetilde{D'}_{\mu\nu}^{ab}(k) = \frac{\delta_{ab}}{k^2} \left[\frac{\delta_{\mu\nu} - k_\mu k_\nu / k^2}{1 + \Pi(k^2)} + \frac{1}{\lambda_0} \frac{k_\mu k_\nu}{k^2} \right]. \tag{2.30}$$

More generally, BRST invariance gives relations between bare Green functions and those containing composite operator insertions that describe the non-linear symmetry transformations of the fields. Many features of the theory derive from these Ward identities. In particular, they are *sufficient to restrict the structure of the counterterms and show multiplicative renormalizability*, i.e. the theory is renormalized by adjusting the bare parameters and fields of the bare action. Also, the renormalized theory satisfies similar symmetry properties as the bare one,[13] which is also crucial to prove the IR finiteness of correlation functions with non-exceptional kinematics.

To proceed systematically the WIs are summarized in equations for generating functionals (Zinn-Justin, 1974), (Becchi *et al.*, 1976), (Zinn-Justin, 2002) such as

$$Z_0(J, \bar{\xi}, \xi, K, L; g_0, \lambda_0) = N_0 \int D[A] D[c] D[\bar{c}] \exp\Big\{ \mathcal{E}_0(A, c, \bar{c}, K, L; g_0, \lambda_0)$$

$$+ \frac{1}{\hbar} S_c(A, c, \bar{c}; J, \bar{\xi}, \xi) \Big\}, \tag{2.31}$$

where N_0 is chosen such that $Z_0(\text{all sources} = 0; g_0, \lambda_0) = 1$. The term S_c contains source terms for the basic fields

$$S_c(A, c, \bar{c}; J, \bar{\xi}, \xi) = \int_x \sum_b \big\{ J_\mu^b(x) A_\mu^b(x) + \bar{\xi}^b(x) c_b(x) + \bar{c}^b(x) \xi_b(x) \big\}, \tag{2.32}$$

and

$$\mathcal{E}_0(A, c, \bar{c}, K, L; g_0, \lambda_0) = \frac{1}{\hbar} \int_x \sum_b \left\{ K_\mu^b(x) s A_\mu^b(x) + L^b(x) s c^b(x) \right\}$$

$$- \frac{1}{\hbar} S_0(A, c, \bar{c}; g_0, \lambda_0), \tag{2.33}$$

where the new sources K, L must be introduced for the composite fields appearing in the BRS transformations. Note that since s is nilpotent further sources in Z_0 are not needed. The factors of $\hbar$ in diagrammatic contributions to correlation functions count the number of associated loops.

Now, the BRS transformations imply that Z_0 satisfies

$$\int_x \sum_b \left\{ J_\mu^b(x) \frac{\partial}{\partial K_\mu^b(x)} - \bar{\xi}_b \frac{\partial}{\partial L^b(x)} - \lambda_0 \xi^b(x) \partial_\mu \frac{\partial}{\partial J_\mu^b(x)} \right\} Z_0 = 0, \tag{2.34}$$

and the same equation holds for the generating functional of the connected Green functions $W_0 = \hbar \ln Z_0$. For considerations of renormalizability we have already noted that it is more convenient to consider 1PI diagrams. The generating functional Γ_0 for these is called the *vertex functional* (see Hernandez' lectures) that is obtained from W_0 via a Legendre transformation,

$$W_0(J, \bar{\xi}, \xi, K, L; g_0, \lambda_0) = \Gamma_0(A, c, \bar{c}, K, L; g_0, \lambda_0)$$

$$+ \int_x \left\{ J_\mu^b(x) A_\mu^b(x) + \bar{\xi}^b(x) c^b(x) + \bar{c}^b(x) \xi^b(x) \right\}, \tag{2.35}$$

where

$$A_\mu^b(x) = \frac{\partial W_0}{\partial J_\mu^b(x)}, \quad c^b(x) = \frac{\partial W_0}{\partial \bar{\xi}^b(x)}, \quad \bar{c}^b(x) = -\frac{\partial W_0}{\partial \xi^b(x)}, \tag{2.36}$$

with inverse relations

$$J_\mu^b(x) = -\frac{\partial \Gamma_0}{\partial A_\mu^b(x)}, \quad \bar{\xi}^b(x) = \frac{\partial \Gamma_0}{\partial c^b(x)}, \quad \xi^b(x) = -\frac{\partial \Gamma_0}{\partial \bar{c}^b(x)}. \tag{2.37}$$

The vertex functional satisfies the equation

$$\int_x \left\{ \frac{\partial \Gamma_0}{\partial A_\mu^b(x)} \frac{\partial \Gamma_0}{\partial K_\mu^b(x)} + \frac{\partial \Gamma_0}{\partial c^b(x)} \frac{\partial \Gamma_0}{\partial L^b(x)} - \lambda_0 \partial_\mu A_\mu^b(x)) \frac{\partial \Gamma_0}{\partial \bar{c}^b(x)} \right\} = 0. \tag{2.38}$$

Setting

$$\bar{\Gamma}_0 = \Gamma_0 + \int_x (\lambda_0/2) \left(\partial_\mu A_\mu^b(x) \right)^2, \tag{2.39}$$

and using the FP ghost field equation

$$\partial_\mu \frac{\partial \overline{\Gamma}_0}{\partial K_\mu^b(x)} + \frac{\partial \overline{\Gamma}_0}{\partial \bar{c}^b(x)} = 0, \tag{2.40}$$

we can write the WI in a slightly simpler form

$$\int_x \left\{ \frac{\partial \overline{\Gamma}_0}{\partial A_\mu^b(x)} \frac{\partial \overline{\Gamma}_0}{\partial K_\mu^b(x)} + \frac{\partial \overline{\Gamma}_0}{\partial c^b(x)} \frac{\partial \overline{\Gamma}_0}{\partial L^b(x)} \right\} = 0. \tag{2.41}$$

We now seek a functional $\mathcal{E}_\mathrm{R} = \mathcal{E}_0 + \mathrm{O}(\hbar)$ such that Green functions generated from

$$Z_\mathrm{R}(J, \bar{\xi}, \xi, K, L; g_0, \lambda_0) = N_\mathrm{R} \int D[A] D[c] D[\bar{c}] \exp\Big\{ \mathcal{E}_\mathrm{R}(A, c, \bar{c}, K, L; g, \lambda)$$

$$+ (1/\hbar) S_c(A, c, \bar{c}; J, \bar{\xi}, \xi) \Big\} \tag{2.42}$$

are finite order by order in PT.[14] Such a functional can be found and is simply characterized by the following

Renormalization theorem: There exist renormalization constants $Z_1, Z_3, \tilde{Z}_3$ that are formal power series in $\hbar$, relating bare fields and renormalized fields

$$A^0 = Z_3^{1/2} A, \ L^0 = Z_3^{1/2} L, \ \lambda_0 = Z_3^{-1} \lambda, \ g_0 = Z_1 g, \tag{2.43}$$

$$c^0 = \tilde{Z}_3^{1/2} c, \ \bar{c}^0 = \tilde{Z}_3^{1/2} \bar{c}, \ K^0 = \tilde{Z}_3^{1/2} K, \tag{2.44}$$

such that

$$\mathcal{E}_\mathrm{R}(A, c, \bar{c}, K, L; g_0, \lambda_0) = \mathcal{E}_0(A^0, c^0, \bar{c}^0, K^0, L^0; g_0, \lambda_0). \tag{2.45}$$

Furthermore, the Zs have only (multi-) poles in ϵ with dimensional regularization (DR) ('t Hooft and Veltman, 1972), (Breitenlohner and Maison, 1977a, 1977b), (Collins, 1984), or depend only logarithmically on a momentum cutoff with other regularizations.

The renormalization theorem implies $\overline{\Gamma}_\mathrm{R}(A, c, \bar{c}, K, L; g, \lambda) = \overline{\Gamma}_0(A^0, c^0, \bar{c}^0, K^0, L^0; g_0, \lambda_0)$ and the renormalized vertex functional satisfies the same equation as Γ_0. The form of the WI indicates universality of the renormalized theory, (the correlation functions being fixed by a finite number of specified normalization conditions). The measure and renormalized action $S_\mathrm{R}(A, c, \bar{c}, g, \lambda) = S_0(A^0, c^0, \bar{c}^0, g_0, \lambda_0)$ are invariant under $\delta_\epsilon A_\mu = Z_A^{-1/2} \delta_\epsilon A_\mu^0, \ldots$ The same holds true for the contributions to $\mathcal{E}_\mathrm{R}$ that are linear in K, L.

The proof of the theorem is by induction on the number of loops. Assume the theorem holds to order $n - 1$ in the loop expansion and hence the functional Γ_R satisfies the identities. The key point is that the equations for Γ are non-linear, and one obtains simple equations (involving functional derivatives of the action) for the divergences in the limit of large cutoff that occur at order n. Since all subdivergences have been removed, only overall divergences remain that by the power-counting theorem are polynomials in momentum space of the order of the divergence degrees of the corresponding diagrams.[15] The equations from the WI suffice to restrict all the coefficients (i.e. the renormalization constants) such that the theorem can be proven at order n.

From the renormalized BRS WI for the functionals follow various relations for correlation functions. Starting from the identities (2.38, 2.41) for Γ_0 or Γ_R and differentiations wrt the FP fields and the gauge fields. As mentioned before one gets, e.g., *Slavnov–Taylor identities* for 2-,3-point vertices, which imply absence of mass terms in the gluon self-energy and small-momenta behavior of 3-point vertex function.

2.1.7 Perturbative renormalization of lattice gauge theory

For an introduction to perturbation theory with the lattice regularization please see the lectures of Hernandez. Here, I just summarize some of the salient points. First, the lattice action is chosen to be gauge invariant and such that it has the desired classical continuum limit. The Feynman rules are algebraically more complicated than in the continuum, e.g. one also encounters vertices involving more than 4 gauge fields, but they are straightforwardly derived and now usually computer generated. There are also extra terms coming from the measure (the Jacobian of the change of variables U to A). The integral over the gauge field A is extended to ∞, which formally modifies the integral by only comparably negligible non-perturbative terms. For performing the perturbative expansion in this way, gauge fixing is needed, as in the continuum. Finally, there is an exact BRS symmetry on the lattice since this symmetry doesn't originate from the continuum aspects but is a general consequence of gauge-fixing in the presence of a gauge-invariant cutoff (see Lüscher 1988).

So, we have practically all the ingredients to prove perturbative renormalizability following the continuum procedure apart from *a power-counting theorem for the lattice regularization*. The latter has been provided by Reisz (1988*b*); very nice accounts can be found in (Luscher, 1988) and (Reisz, 1988*d*). An influential earlier paper considering renormalization of lattice gauge theories is by Sharatchandra (1978). The difficulty comes from the fact that in momentum space the domain of integration is compact, internal momenta k are restricted to the Brillouin zone $\mathcal{B}$, $|k_\mu| \leq \pi/a$, and the integrand is a periodic function wrt the loop momenta. Also, as mentioned above, often the vertices are quite complex and there are additional "irrelevant" vertices, i.e. those that vanish in the naive continuum limit. In this situation the usual convergence theorems in the continuum do not apply.

Let us first consider the case with massive propagators. A general lattice Feynman diagram corresponds to an integral of the form[16]

$$I = \int_{\mathcal{B}} \mathrm{d}^4 k_1 \ldots \mathrm{d}^4 k_\ell \, \frac{V(k, p, m, a)}{C(k, p, m, a)}, \tag{2.46}$$

where the numerator V of the integrand contains all the vertices and numerators of propagators, and the denominator C is the product of the denominators of the propagators,

In order to make rigorous statements on the convergence Reisz specified the following restrictions on V, C that, however, hold for a large class of lattice actions:

For V: 1) $\exists\ \omega$ st $V(k, p, m, a) = a^{-\omega} F(ak, ap, am)$,

where F is 2π periodic in ak_i and a polynomial in am.

    2) $\lim_{a \to 0} V(k, p, m, a) = P(k, p, m)$ exists.

For C: 1) $C = \prod_{i=1}^{n} C_i(l_i, m, a)$, with

$$C_i(l_i, m, a) = a^{-2} G_i(al_i, am), \tag{2.47}$$

$$l_i(k, p) = \sum_j a_{ij} k_j + \sum_l b_{il} p_l, \quad a_{ij} \in \mathbb{Z}, \quad (\hat{l}_\mu = \frac{2}{a} \sin \frac{al_\mu}{2}), \tag{2.48}$$

the latter (integer) condition maintaining the 2π-periodicity of G_i.

    2) $\lim_{a \to 0} C_i(l_i, m, a) = l_i^2 + m^2$.

    3) $\exists\ a_0, A$ st $|C_i(l_i, m, a)| \geq A(\hat{l}_i^2 + m^2)\ \forall\ a \leq a_0, \forall\ i$,

the latter condition ensuring that $1/|C_i| \sim O(a^2)$ at the boundary of $\mathcal{B}$.

With these specifications, for $a > 0$ and $m^2 > 0$ the Feynman integrals are absolutely convergent and the dependence on external momenta is smooth. However, the lattice integral does not necessarily converge in the continuum limit if the continuum limit of the integrand is absolutely integrable, e.g.

$$\int_{\mathcal{B}} \frac{d^4 k}{(2\pi)^4} \frac{1 - \cos(ak_\mu)}{\hat{k}^2 + m^2} = \frac{1}{8a^2} + O(a^0), \tag{2.49}$$

i.e. the integral is quadratically divergent although the continuum limit of the integrand vanishes.

A lattice degree of divergence involves the behavior of the integrand for large internal momenta k_i and simultaneously for small a. For a Zimmermann subspace (defined as before) the (upper) degree of divergence is defined by

$$\overline{\deg}_{u|v} F = \bar{\nu} \text{ if } F(\lambda u, v, m, a/\lambda) \sim_{\lambda \to \infty} k\lambda^{\bar{\nu}}, \tag{2.50}$$

$$\overline{\deg}_{u|v} I = 4r + \overline{\deg}_{u|v} V - \underline{\deg}_{U|V} C. \tag{2.51}$$

With this definition of the degree of divergence Reisz proved the *Theorem*: If $\overline{\deg}_H I < 0$ for all Zimmermann subspaces, then the continuum limit of I exists and is given by the corresponding continuum expression.

To prove the renormalizability one must also specify a lattice R-operation corresponding to subtractions of *local* lattice counterterms, e.g.

$$t^\delta \to \hat{t}^\delta, \quad \hat{t}_p^\delta f(p) = f(0) + \mathring{p}_\mu \frac{\partial}{\partial p'_\mu} f(p')|_{p'=0} + \dots, \tag{2.52}$$

where $\mathring{k}_\mu = a^{-1}\sin(ak_\mu)$ is 2π periodic in ak.

The proof of the theorem is lengthy and cannot be presented here. Let me just mention that the subtractions are organized by the Zimmermann forest formula as in the continuum, and zero mass propagators are treated as by Lowenstein and Zimmermann mentioned in Section 2.1.5. Some fine points involve the treatment of the measure terms. The use of the BRS Ward identities also proceeds similarly to that for continuum regularizations (Reisz, 1989).

Finally, matter fields are included straightforwardly. Note that criterion 3) is satisfied for Wilson fermions (which require an additional additive mass renormalization) for which $C_i(p) = (1 + am)\hat{p}^2 + m^2 + \frac{1}{2}a^2 \sum_{\mu<\nu} \hat{p}_\mu^2 \hat{p}_\nu^2$. It is not satisfied by staggered fermions, however, an extension of the power-counting theorem in this case has been supplied by Giedt (2007). This paves the way for a proof of perturbative renormalizability of staggered fermions.[17]

2.2   Renormalization group equations

In the last section we learned that correlation functions of the basic QCD fields are multiplicatively renormalizable provided bare parameters are also multiplicatively renormalized. For perturbation theory in the continuum we usually employ dimensional regularization where Feynman rules are developed in $D = 4 - 2\epsilon$ dimensions (Collins, 1984). Bare amplitudes have (multi-) poles at $\epsilon = 0$ and in the *minimal subtraction (MS) renormalization scheme* one subtracts just these. Renormalization constants then have the form

$$Z = 1 + g^2 \frac{z_1}{\epsilon} + g^4 \left(\frac{z_2}{\epsilon^2} + \frac{z_3}{\epsilon}\right) + \cdots, \tag{2.53}$$

where g is the (dimensionless) renormalized coupling related to the bare coupling by

$$g^2 = \mu^{-2\epsilon} g_0^2 Z_g, \tag{2.54}$$

with μ the renormalization scale. Note the MS scheme is a *mass-independent renormalization scheme*, i.e. a scheme for which the renormalization constants do not depend on the quark masses.

Renormalized correlation functions involving r gauge fields, n quark–antiquark pairs and l ghost–antighost pairs are given by

$$G_R^{r,n,l}(\mu, p; g, \lambda, m_j) = \left(Z_3^{-1/2}\right)^r \left(Z_2^{-1/2}\right)^{2n} \left(\widetilde{Z}_3^{-1/2}\right)^{2l} G_0^{r,n,l}(p; g_0, \lambda_0, m_{0j}), \tag{2.55}$$

where the renormalized gauge parameter and renormalized quark masses are given by

$$\lambda = Z_3 \lambda_0, \quad m_j = Z_m m_{0j}. \tag{2.56}$$

The *renormalization group equations* (Callan, 1970), (Symanzik, 1970, 1971), follow immediately from the simple observation that bare correlation functions are independent of the renormalization scale μ:

$$\left[\mu\frac{\partial}{\partial\mu} + \beta\frac{\partial}{\partial g} + \tau m_j\frac{\partial}{\partial m_j} + \delta\frac{\partial}{\partial\lambda} + r\gamma_3 + 2n\gamma_2 + 2l\widetilde{\gamma}_3\right] G_{\mathrm{R}}^{r,n,l} = 0, \tag{2.57}$$

where the coefficient functions are defined through:

$$\widetilde{\beta}(\epsilon, g) = \mu\frac{\partial g}{\partial\mu}\Big|_{g_0,\lambda_0,m_{0j}} = -\epsilon g\left\{1 - \frac{1}{2}g\frac{\partial}{\partial g}\ln Z_g\right\}^{-1} \tag{2.58}$$

$$= -\epsilon g + \beta(g), \tag{2.59}$$

$$m_i\tau(g) = \mu\frac{\partial m_i}{\partial\mu}\Big|_{g_0,\lambda_0,m_{0j}}, \tag{2.60}$$

$$\delta(g) = \mu\frac{\partial\lambda}{\partial\mu}\Big|_{g_0,\lambda_0,m_{0j}}, \tag{2.61}$$

$$\gamma_3(g) = \frac{1}{2}\mu\frac{\partial}{\partial\mu}\ln Z_3\Big|_{g_0,\lambda_0,m_{0j}}, \tag{2.62}$$

and similarly for $\gamma_2, \widetilde{\gamma}_3$. Note that for any function f depending only on g, ϵ,

$$\mu\frac{\partial}{\partial\mu}f(g, \epsilon)\Big|_{g_0,\lambda_0,m_{0j}} = \widetilde{\beta}(\epsilon, g)\frac{\partial}{\partial g}f(g, \epsilon). \tag{2.63}$$

The functions β and τ have the following perturbative expansions:

$$\beta(g) = -g^3\sum_{k=0}^{\infty} b_k g^{2k}, \tag{2.64}$$

$$\tau(g) = -g^2\sum_{k=0}^{\infty} d_k g^{2k}, \tag{2.65}$$

with leading coefficients given by

$$b_0 = (4\pi)^{-2}\left[\frac{11}{3}N - \frac{2}{3}N_{\mathrm{f}}\right], \tag{2.66}$$

$$b_1 = (4\pi)^{-4}\left[\frac{34}{3}N^2 - \left(\frac{13}{3}N - \frac{1}{N}\right)N_{\mathrm{f}}\right], \tag{2.67}$$

$$d_0 = (4\pi)^{-2}\frac{3(N^2 - 1)}{N}. \tag{2.68}$$

Note that in the MS scheme β and τ are independent of the gauge parameter λ. To show this start from the general relations between renormalized and bare quantities in the form:

$$g_0 = \mu^\epsilon \left\{ g + \sum_{r=1}^{\infty} a_r(g, m_k/\mu, \lambda)\epsilon^{-r} \right\}, \tag{2.69}$$

$$m_{0j} = m_j + \mu \sum_{r=1}^{\infty} b_{jr}(g, m_k/\mu, \lambda)\epsilon^{-r}, \tag{2.70}$$

$$\lambda_0 = \lambda + \sum_{r=1}^{\infty} c_r(g, m_k/\mu, \lambda)\epsilon^{-r}, \tag{2.71}$$

where the coefficients a_r, b_{jr}, c_r are independent of ϵ. Now define

$$\rho \equiv \frac{\partial g}{\partial \lambda_0} \Big/ \frac{\partial \lambda}{\partial \lambda_0}, \quad \nu_j \equiv \frac{\partial m_j}{\partial \lambda_0} \Big/ \frac{\partial \lambda}{\partial \lambda_0}, \tag{2.72}$$

where the differentiation wrt λ_0 is at fixed g_0, m_{0j}, μ. For any renormalized Green function G of gauge-invariant operators we must have an equation of the form:

$$\left\{ \frac{\partial}{\partial \lambda} + \rho \frac{\partial}{\partial g} + \sum_j \nu_j \frac{\partial}{\partial m_j} + \sigma_G \right\} G = 0, \tag{2.73}$$

i.e. a change in the gauge parameter must be compensated by a change of the renormalized parameters and a multiplicative renormalization of the operators appearing in the definition of G. Now, differentiate Eqs. (2.69), (2.70) wrt λ_0 to obtain

$$A\left(\begin{array}{c} \rho \\ \mu^{-1}\nu_j \end{array}\right) + v = 0, \quad v = \left(\begin{array}{c} \sum_{r=1} \frac{\partial a_r}{\partial \lambda}\epsilon^{-r} \\ \sum_{r=1} \frac{\partial b_{jr}}{\partial \lambda}\epsilon^{-r} \end{array}\right), \tag{2.74}$$

where

$$A = \left(\begin{array}{cc} 1 + \sum_{r=1} \frac{\partial a_r}{\partial g}\epsilon^{-r} & \sum_{r=1} \frac{\partial a_r}{\partial(m_j/\mu)}\epsilon^{-r} \\ \sum_{r=1} \frac{\partial b_{jr}}{\partial g}\epsilon^{-r} & \delta_{jk} + \sum_{r=1} \frac{\partial b_{jr}}{\partial(m_k/\mu)}\epsilon^{-r} \end{array}\right). \tag{2.75}$$

Since ρ and ν_j must be finite we conclude that $A^{-1}v$ must also be finite. It follows that since a_r, b_{jr} do not depend on ϵ we must have

$$\rho = 0, \quad \nu_j = 0 \ \forall \ j, \tag{2.76}$$

and hence

$$\frac{\partial a_r}{\partial \lambda} = 0, \quad \frac{\partial b_{jr}}{\partial \lambda} = 0 \ \forall \ j. \tag{2.77}$$

The functions β, τ_j are determined by a_1, b_{j1} and are hence also independent of λ.

2.2.1 Physical quantities, Λ parameter and RGI masses

A physical quantity P is independent of wave-function renormalization and independent of λ and hence satisfies the simplified RG equation:

$$\left[\mu \frac{\partial}{\partial \mu} + \beta \frac{\partial}{\partial g} + \tau m_j \frac{\partial}{\partial m_j} \right] P = 0. \tag{2.78}$$

Now, every solution of this equation can be expressed in terms of special solutions. First, the Λ *parameter* that doesn't involve quark masses:

$$\Lambda = \mu \ell(g), \tag{2.79}$$

where $\ell(g)$ satisfies the equation

$$\left[1 + \beta \frac{\partial}{\partial g} \right] \ell(g) = 0, \tag{2.80}$$

and is completely fixed by its behavior for small g:

$$\ell(g) = (b_0 g^2)^{-b_1/(2b_0^2)} e^{-1/(2b_0 g^2)} \exp \left\{ - \int_0^g dx \left[\frac{1}{\beta(x)} + \frac{1}{b_0 x^3} - \frac{b_1}{b_0^2 x} \right] \right\}. \tag{2.81}$$

The other parameters are the *RG-invariant masses*

$$M_i = m_i \theta(g), \tag{2.82}$$

where $\theta(g)$ satisfies

$$\left[\beta \frac{\partial}{\partial g} + \tau \right] \theta(g) = 0, \tag{2.83}$$

which is also fixed (i.e. also its normalization) by its behavior as $g \to 0$:

$$\theta(g) = (2 b_0 g^2)^{-d_0/(2b_0)} \exp \left\{ - \int_0^g dx \left[\frac{\tau(x)}{\beta(x)} - \frac{d_0}{b_0 x} \right] \right\}. \tag{2.84}$$

Consider as an example the case when $P(Q^2, \mu, g, m_j)$ is a dimensionless physical quantity depending on a Euclidean momentum Q, then:

$$P = \tilde{P} \left(Q^2/\mu^2, g, m_j/\sqrt{Q^2} \right). \tag{2.85}$$

We can now appreciate the power of the RG equation; it enables us to deduce the behavior of P for large Q^2, since it implies

$$P = \tilde{P} \left(1, \bar{g}(t), \bar{m}_j(t)/\sqrt{Q^2} \right), \quad t \equiv \ln(Q^2/\Lambda^2), \tag{2.86}$$

where $\bar{g}(t)$ is a *running coupling* and $\bar{m}_j(t)$ is the *running mass*. The running coupling satisfies the equation

$$Q\frac{\partial \bar{g}}{\partial Q} = \beta(\bar{g}), \tag{2.87}$$

and is implicitly defined by

$$t = \frac{1}{b_0 \bar{g}^2} + \frac{b_1}{b_0^2} \ln(b_0 \bar{g}^2) + 2 \int_0^{\bar{g}} dx \left[\frac{1}{\beta(x)} + \frac{1}{b_0 x^3} - \frac{b_1}{b_0^2 x} \right]. \tag{2.88}$$

As $Q^2 \to \infty$ the running coupling tends to zero (logarithmically)

$$\bar{g}^2(t) = \frac{1}{b_0 t} \left\{ 1 - \frac{b_1}{b_0^2 t} \ln(t) + O(t^{-2}) \right\} \tag{2.89}$$

a property known as *asymptotic freedom* (Gross and Wilczek, 1973), (Politzer, 1973). The running mass is given by

$$\bar{m}_j(t) = M_j \theta^{-1}\left(\bar{g}(t)\right), \tag{2.90}$$

and satisfies the equation

$$Q\frac{\partial \bar{m}_j}{\partial Q} = \bar{m}_j \tau(\bar{g}). \tag{2.91}$$

For $Q^2 \to \infty$ it decreases according to

$$\bar{m}_j(t) = M_j \left(\frac{2}{t}\right)^{d_0/(2b_0)} \left\{ 1 - \frac{d_0 b_1}{2 b_0^3 t}(1 + \ln(t)) + \frac{d_1}{2 b_0^2 t} + O(t^{-2}) \right\}. \tag{2.92}$$

Any two mass-independent schemes can be related by finite parameter renormalizations;

$$g' = g\sqrt{\chi_g(g)}, \quad \chi_g(g) = 1 + \chi_g^{(1)} g^2 + \ldots, \tag{2.93}$$

$$m' = m\chi_{\mathrm{m}}(g). \tag{2.94}$$

Physical quantities are scheme independent and so (for the case where the renormalization scales are equal)

$$P(\mu, g, m_j) = P'(\mu, g', m_j'), \tag{2.95}$$

satisfies Eq. (2.78) in the first scheme and a corresponding equation in the second scheme with new coefficients β', τ' that are related to β, τ by:

$$\beta'(g') = \left\{ \beta(g)\frac{\partial g'}{\partial g} \right\}\big|_{g=g(g')}, \tag{2.96}$$

$$\tau'(g') = \left\{ \tau(g) + \beta(g)\frac{\partial}{\partial g} \ln \chi_{\mathrm{m}}(g) \right\}\big|_{g=g(g')}. \tag{2.97}$$

Exercise 2.4 Show that it follows that $b'_0 = b_0, b'_1 = b_1, d'_0 = d_0$, i.e. these coefficients are universal but the higher loop coefficients are not, e.g.

$$d'_1 = d_1 + 2b_0\chi_\mathrm{m}^{(1)} - d_0\chi_g^{(1)}. \tag{2.98}$$

Also,

$$\Lambda' = \Lambda \exp\left\{\frac{\chi_g^{(1)}}{2b_0}\right\}, \tag{2.99}$$

$$M' = M, \tag{2.100}$$

i.e. Λ parameters are scheme dependent (albeit their relations require just 1-loop computations), but the *RG-invariant masses are scheme independent*.

2.2.2 The MS lattice coupling

With the lattice regularization of QCD we can also define a MS scheme by just subtracting powers of logarithms in the lattice cutoff a. The MS perturbatively renormalized lattice coupling then has the form

$$g_\mathrm{latt} = g_0 - b_0 g_0^3 \ln(a\mu) + \dots. \tag{2.101}$$

The 2-loop relation between g_latt and $g_{\overline{\mathrm{MS}}}$ has been computed for a large class of actions, also including fermions using sophisticated algebraic programs for automatic generation of the Feynman rules (see e.g. (Luscher and Weisz, 1995), (Hart *et al.*, 2009), (Constantinou and Panagopoulos, 2008)). As discussed in the last section, the 1-loop relation gives the ratio of Λ parameters. This was first computed for the standard Wilson action by Hasenfratz and Hasenfratz (1980); in that case for $N = 3, N_\mathrm{f} = 0$ one finds a large number $\Lambda_{\overline{\mathrm{MS}}}/\Lambda_\mathrm{latt} = 28.8$. Close to the continuum limit, assuming that $g_0 = 0$ is the critical point that should be approached when taking the continuum limit, we should find for a physical mass m in the limit of zero-mass quarks

$$m/\Lambda_\mathrm{latt} = c_m + \mathrm{O}(a^p), \tag{2.102}$$

$$\Rightarrow ma \sim_{g_0 \to 0} c(b_0 g_0^2)^{-b_1/2b_0^2} e^{-1/(2b_0 g_0^2)} R(g_0), \quad R(g_0) = 1 + \mathrm{O}(g_0^2). \tag{2.103}$$

An observation of the leading behavior above is called "asymptotic scaling"; present spectral measurements are considered consistent with these expectations although the correction factor $R(g_0)$ is a power series in g_0 and hence usually not slowly varying in the regions where the simulations are performed.

We remark here again that for lattice regularizations that break chiral symmetry, such as Wilson fermions (see Hernandez' lectures), the quark mass also needs an additive renormalization

$$m = Z_\mathrm{m} m_\mathrm{q}, \quad m_\mathrm{q} = m_0 - m_\mathrm{c}. \tag{2.104}$$

Computation of the quark self-energy to 1-loop gives am_c, Z_2, Z_m to $O(g_0^2)$, and subsequent comparison to the result in the $\overline{\text{MS}}$ scheme yields $\chi_m^{(1)}$.

2.2.3   Renormalization of composite operators

Consider a perturbative computation of a correlation function involving a bare composite operator and a product of basic fields at physically separated space-time points. After performing the required renormalization of the bare parameters and basic fields the resulting expression either 1) stays finite as the UV cutoff is removed or 2) diverges. If case 1) holds to all orders of PT then this is (usually) due to the fact that the composite operator is a conserved current or one satisfying a current algebra. In case 2) the simplest situation is that the correlation function becomes finite when the composite operator is multiplicatively renormalized,

$$\phi_R(\mu) = Z_\phi(\mu)\phi^{\text{bare}}.$$

But in general, operators mix with other operators having the same conserved quantum numbers and the same canonical dimension (or less)

$$\mathcal{O}_{R\sigma} = \sum_\tau Z_{\sigma\tau}\mathcal{O}_\tau^{\text{bare}} + \text{"}Z \times \text{ lower dimension ops"}. \tag{2.105}$$

A correlation function involving an insertion of a purely multiplicatively renormalizable operator with a product of diagonally multiplicatively (gauge-invariant) renormalized operators, all located at physically separated points, satisfies the RG equation (for the case of massless quarks):

$$\left[\left(\mu\frac{\partial}{\partial\mu} + \beta\frac{\partial}{\partial g} - \sum_i \gamma_{\phi_i}\right)\delta_{\tau\sigma} - \gamma_{\tau\sigma}\right]G^\sigma_{R;1,\ldots,n} = 0, \tag{2.106}$$

where

$$\gamma_{\phi_i} = \mu\frac{\partial}{\partial\mu}\ln Z_{\phi_i}, \quad \gamma_{\tau\sigma} = \mu\frac{\partial Z_{\tau\rho}}{\partial\mu}\left(Z^{-1}\right)_{\rho\sigma}. \tag{2.107}$$

For the physical interpretation it is often advantageous to define *RG-invariant operators* by (in the simpler cases)

$$\phi_{\text{RGI}\,i} = C_i(\mu/\Lambda)\phi_{R\,i}, \tag{2.108}$$

where C_i is a solution to the equation

$$\left[\mu\frac{\partial}{\partial\mu} + \beta\frac{\partial}{\partial g} + \gamma_{\phi_i}\right]C_i = 0. \tag{2.109}$$

It is given (with conventional normalization) by

$$C_i(\mu/\Lambda) = \left(2b_0\bar{g}^2(\mu)\right)^{\gamma_{\phi_i}^{(0)}/(2b_0)}\exp\left\{-\int_0^{\bar{g}}dx\left[\frac{\gamma_{\phi_i}(x)}{\beta(x)} + \frac{\gamma_{\phi_i}^{(0)}}{b_0 x}\right]\right\}, \tag{2.110}$$

where we have assumed that $\gamma_{\phi_i}(g) = \gamma^{(0)}_{\phi_i} g^2 + O(g^4)$.

As an example of a lattice regularization, let us consider the case of Wilson's fermions. It has an isovector current

$$V^a_\mu(x) = \frac{1}{2}\left\{\overline{\psi}(x)\frac{\tau^a}{2}(\gamma_\mu - 1)U(x,\mu)\psi(x + a\hat{\mu})\right.$$

$$\left. +\overline{\psi}(x + a\hat{\mu})\frac{\tau^a}{2}(\gamma_\mu + 1)U(x,\mu)^\dagger\psi(x)\right\}, \tag{2.111}$$

where τ^a are the Pauli matrices acting on the flavor indices, which is exactly conserved (in the case of degenerate quark masses) i.e. on-shell, $\partial^*_\mu V^a_\mu(x) = 0$, where ∂^*_μ is the lattice backward derivative. It follows that this bare operator doesn't need any renormalization. No analogous conserved axial vector current exists for Wilson fermions even for $m_q = 0$. Often, in practical numerical computations simpler lattice currents are employed:

$$V^a_\mu(x) = \overline{\psi}(x)\frac{\tau^a}{2}\gamma_\mu\psi(x), \quad A^a_\mu(x) = \overline{\psi}(x)\frac{\tau^a}{2}\gamma_\mu\gamma_5\psi(x), \tag{2.112}$$

which are expected to be conserved up to lattice artifacts. In PT this is indeed the case, but in order that the currents obey the correct current algebra (up to O(a) artifacts) they require a finite renormalization

$$V_R = Z_V V, \quad A_R = Z_A A, \quad Z_V/Z_A = 1 + O(g_0^2). \tag{2.113}$$

Digression: It is probably not so well known among students specializing in lattice theory that there is a problem with "naive γ_5" in dimensional regularization. Namely, the algebraic rules

$$\{\gamma_\mu, \gamma_\nu\} = 2\delta_{\mu\nu}, \quad \delta_{\mu\mu} = D, \quad \{\gamma_\mu, \gamma_5\} = 0, \quad \text{tr}\gamma_5 = 0, \quad \text{and cyclicity of trace}$$

imply the unwanted relation[18] $\text{tr}\,(\gamma_5\gamma_\mu\gamma_\nu\gamma_\rho\gamma_\lambda) = 0$ unless $D = 2$ or $D = 4$. The modified algebra proposed by 't Hooft and Veltman (1972) is to define $\gamma_5 = \gamma_0\gamma_1\gamma_2\gamma_3$ so that

$$\{\gamma_\mu, \gamma_5\} = 0 \text{ for } \mu \leq 3, \quad [\gamma_\mu, \gamma_5] = 0 \text{ for } \mu > 3.$$

The algebra is then consistent but results in the necessity of having to introduce an infinite renormalization for the bare axial current

$$(A^a_\mu)^{MS} = Z^{MS}_A(g)\overline{\psi}\frac{\tau^a}{2}\frac{1}{2}[\gamma_\mu, \gamma_5]\,\psi, \tag{2.114}$$

$$Z^{MS}_A(g) = 1 + g^4\frac{1}{(4\pi)^2}2b_0C_F\frac{1}{\epsilon} + \dots, \quad C_F = \frac{N^2 - 1}{2N}, \tag{2.115}$$

and we need a further renormalization $\chi_A(g) = 1 - g^2\frac{4}{(4\pi)^2}C_F + \dots$ to obtain a correctly normalized current $(\mu\frac{\partial}{\partial\mu}(\chi_A Z^{MS}_A) = 0)$.

2.2.4   Ward identities

A way to renormalize currents, especially also non-perturbatively, is to enforce the continuum Ward identities (Ward, 1950), (Takahashi, 1957) (in some cases only up to lattice artifacts), which for the case of flavor $SU(N_f) \times SU(N_f)$ are equivalent to current algebra in Minkowski space. General Ward identities are obtained by making infinitesimal transformations in the functional integral. For transformations that leave the functional measure invariant we obtain relations of the form

$$\langle \delta S \, \mathcal{O} \rangle = \langle \delta \mathcal{O} \rangle. \tag{2.116}$$

For axial transformations (e.g. for $N_f = 2$) (Luscher *et al.*, 1996)

$$\delta_A \psi(x) = \omega^a(x) \frac{1}{2} \tau^a \gamma_5 \psi(x), \quad \delta_A \overline{\psi}(x) = \omega^a(x) \overline{\psi}(x) \frac{1}{2} \tau^a \gamma_5, \tag{2.117}$$

and working formally in the continuum (with an assumed chiral invariant regularization) the action transforms as

$$\delta_A S = \int_{\mathcal{R}} \mathrm{d}^4 x \, \omega^a(x) \left[-\partial_\mu A_\mu^a(x) + 2m P^a(x) \right], \tag{2.118}$$

where $P^a = \overline{\psi} \frac{\tau^a}{2} \gamma_5 \psi$ is the pseudoscalar density and we have assumed that $\omega^a(x) = 0$ for x outside a bounded region $\mathcal{R}$. For example, if the observable $\mathcal{O} = \mathcal{O}_{\text{ext}}$ has no support in $\mathcal{R}$ then the WI becomes

$$\langle \left[-\partial_\mu A_\mu^a(x) + 2m P^a(x) \right] \mathcal{O}_{\text{ext}} \rangle = 0, \tag{2.119}$$

the famous *PCAC relation* that has many applications (see Section 2.5.3).

If $\mathcal{O} = \mathcal{O}_{\text{int}} \mathcal{O}_{\text{ext}}$, having support inside and outside $\mathcal{R}$, we obtain

$$\int_{\mathcal{R}} \mathrm{d}^4 x \, \omega^a(x) \langle \left[-\partial_\mu A_\mu^a(x) + 2m P^a(x) \right] \mathcal{O}_{\text{int}} \mathcal{O}_{\text{ext}} \rangle = \langle \delta_A \mathcal{O}_{\text{int}} \mathcal{O}_{\text{ext}} \rangle, \tag{2.120}$$

which in the limit $\omega^a(x) \rightarrow$ constant inside $\mathcal{R}$, and $m = 0$ simplifies to

$$\omega^a \int_{\partial \mathcal{R}} \mathrm{d}\sigma_\mu(x) \, \langle A_\mu^a(x) \mathcal{O}_{\text{int}} \mathcal{O}_{\text{ext}} \rangle = -\langle \delta_A \mathcal{O}_{\text{int}} \mathcal{O}_{\text{ext}} \rangle, \tag{2.121}$$

where $\mathrm{d}\sigma_\mu$ is the outward normal to the boundary of $\mathcal{R}$. For the case $\mathcal{O}_{\text{int}} = A_\nu^b(y)$ and $\mathcal{R}$ the region between two fixed-time hyperplanes at t_2, t_1, (and using $\delta_A A_\nu^b(x) = -i\omega^a(x) \epsilon^{abc} V_\nu^c(x)$) the WI reads ($t_2 > t_1$),

$$\int \mathrm{d}^3 x \, \langle [A_0^a(\mathbf{x}, t_2) - A_0^a(\mathbf{x}, t_1)] A_\nu^b(y) \mathcal{O}_{\text{ext}} \rangle = i\epsilon^{abc} \langle V_\nu^c(y) \mathcal{O}_{\text{ext}} \rangle, \tag{2.122}$$

which is equivalent to the current algebra relation $\left[A_0^a(\mathbf{x}, t), A_0^b(\mathbf{y}, t) \right] = i\delta^{(3)}(\mathbf{x} - \mathbf{y}) \epsilon^{abc} V_0^c(\mathbf{x}, t)$ in Minkowski space. All CA relations can be obtained analogously.

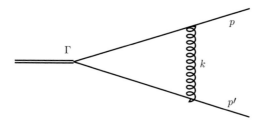

Fig. 2.6 1-loop diagram contributing to a $\overline{\psi}\Gamma\psi$ vertex function

2.2.5 Scale-dependent renormalization

Most composite operators require scale-dependent renormalization. The non-perturbative renormalization of such operators will be postponed to later sections. Here, we just outline the simple perturbative 1-loop computation of the anomalous dimensions of bilinear quark operators $\overline{\psi}\Gamma\psi$ (e.g. the pseudoscalar density $\Gamma = \gamma_5$ or $\Gamma = \tau^a/2$), in the continuum MS scheme. At 1-loop the bare vertex function of the density with a quark–antiquark pair is given by the diagram with a gluon exchange in Fig. 2.6: using dimensional regularization

$$\Gamma(p,p') = -T^a T^a g^2 \int \frac{\mathrm{d}^D k}{(2\pi)^D} \frac{\gamma_\nu \gamma(k+p)\Gamma\gamma(k-p')\gamma_\mu}{(k+p)^2(k+p')^2} \frac{1}{k^2} \left[\delta_{\mu\nu} - (1-\lambda^{-1})\frac{k_\mu k_\nu}{k^2} \right].$$

$$(2.123)$$

Noting that the integral is only logarithmically divergent for $D = 4$, for the computation of the UV divergent parts we can set $p = p' = 0$ in the numerator. Without any previous experience of dimensional regularization one can accept that the integral involved is singular as $\epsilon \to 0$,

$$\int \frac{\mathrm{d}^D k}{(2\pi)^D} F(k,p,p') \sim \frac{C}{\epsilon}, \quad F = \frac{1}{(k+p)^2(k+p')^2}, \quad C = \frac{1}{(4\pi)^2},$$

$$(2.124)$$

and it immediately follows for the integrals appearing in Eq. (2.123),

$$\int \frac{\mathrm{d}^D k}{(2\pi)^D} \frac{k_\rho k_\tau}{k^2} F(k,p,p') \sim \frac{C}{D\epsilon} \delta_{\rho\tau},$$

$$(2.125)$$

$$\int \frac{\mathrm{d}^D k}{(2\pi)^D} \frac{k_\mu k_\nu k_\rho k_\tau}{(k^2)^2} F(k,p,p') \sim \frac{C}{D(D+2)\epsilon} (\delta_{\mu\nu}\delta_{\rho\tau} + 2 \text{ perms}).$$

$$(2.126)$$

Then, we easily deduce

$$\Gamma(p,p')_{\mathrm{div}} \sim -T^a T^a g^2 \frac{C}{\epsilon} \left[\frac{1}{D}\gamma_\mu \gamma_\rho \Gamma \gamma_\rho \gamma_\mu - (1-\lambda^{-1})\Gamma \right].$$

$$(2.127)$$

For example,[19] noting for our normalization $T^a T^a = -C_F$,

$$\text{for } \Gamma = \gamma_\mu \gamma_5 : \quad \Gamma(p, p')_{\text{div}} = g^2 \frac{CC_F}{\epsilon} \gamma_\mu \gamma_5 \left[\frac{(2 - D)^2}{D} - (1 - \lambda^{-1}) \right]$$

$$\simeq g^2 \frac{C_F}{(4\pi)^2 \epsilon} \lambda^{-1} \gamma_\mu \gamma_5, \tag{2.128}$$

$$\text{for } \Gamma = \gamma_5 : \quad \Gamma(p, p')_{\text{div}} = g^2 \frac{CC_F}{\epsilon} \gamma_5 \left[D - (1 - \lambda^{-1}) \right]$$

$$\simeq g^2 \frac{C_F}{(4\pi)^2 \epsilon} (3 + \lambda^{-1}) \gamma_5. \tag{2.129}$$

To complete the computation of the renormalization factor we have to compute the quark-field renormalization factor Z_2 to 1-loop. I leave this as an exercise; but if we accept that the axial current doesn't need any divergent renormalization to one loop this contribution must just cancel the contribution above $Z_2^{(1)} = -\frac{C_F \lambda^{-1}}{(4\pi)^2} \frac{1}{\epsilon}$, and we deduce that the divergent part of the pseudoscalar density is given by

$$Z_P = 1 - \frac{3C_F}{(4\pi)^2} \frac{1}{\epsilon} g^2 + \dots . \tag{2.130}$$

Note $Z_P^{(1)} = -d_0/2$ as should be the case for $Z_P Z_m = 1$. For lattice regularization we would obtain $Z_P = 1 + g_0^2 d_0 \ln(a\mu) + \dots$ (in the MS scheme).

2.2.6 Anomalies

Any account on renormalization would not be complete without mentioning anomalies. These involve symmetries present in the classical theory that are violated in the process of regularization and that cannot be regained in the limit that the UV cutoff is removed. This vast and important subject will be covered in the lectures by Kaplan and thus no details will be given here. Let me just mention that we have already encountered one example, which is that massless QCD breaks scale invariance at the quantum level. There is a mass parameter in the renormalized theory and the trace of the energy momentum tensor is non-zero

$$\theta_{\mu\mu} = \frac{\beta(g)}{2g} N[F^2]. \tag{2.131}$$

For a discussion of the energy-momentum tensor in lattice gauge theories see (Caracciolo *et al.*, 1992).

Another famous example (see Kaplan's lectures) is the U(1) axial anomaly (Adler, 1969) expressing the fact that the U(1) axial current is not conserved in the limit of massless quarks:

$$\partial_\mu A^0_{R\mu} = 2m N[\bar{\psi} \gamma_5 \psi] + \frac{g^2}{32\pi^2} N[\epsilon_{\mu\nu\rho\lambda} F_{\mu\nu} F_{\rho\lambda}]. \tag{2.132}$$

This anomaly must be reproduced by a given formulation of lattice fermions in order that it can be considered an acceptable regularization of QCD. For the many formulations available the way that this achieved varies quite considerably. For example, with Wilson fermions the measure is invariant under infinitesimal axial U(1) transformations, but the Wilson term in the action breaks the symmetry and produces the correct anomaly in the continuum limit as was first shown by Karsten and Smit (1981). In the Ginsparg–Wilson formulation it is the measure that is not invariant under the (modified) chiral U(1) symmetry transformations (whereas the action is); see Section 2.6.

2.2.7 Operator product expansions

Consider correlation functions involving renormalized local gauge-invariant operators $\langle A(x)B(0)\phi(y_1)\ldots\phi(y_r)\rangle$ with the y_i physically separated from $x, 0$ and from each other. In the limit $x \to 0$ singularities appear that are described by local operators $\mathcal{O}_n$ having the same global symmetries as the formally combined operator AB:

$$A(x)B(0) \sim_{x\to 0} \sum_n C_{AB}^{(n)}(x)\mathcal{O}_n(0), \qquad (2.133)$$

where the coefficients $C_{AB}^{(n)}(x)$ are c-numbers. This so-called *Wilson's operator product expansion* has been shown to hold in some generality in the framework of perturbation theory by Wilson and Zimmermann (1972). The relation is structural and thought to hold also at the non-perturbative level.[20] If the fields $A, B, \mathcal{O}_n$ are multiplicatively renormalizable then the coefficients obey the RG equation

$$\left[\mu\frac{\partial}{\partial\mu} + \beta(g)\frac{\partial}{\partial g} - \gamma_A - \gamma_B + \gamma_n\right] C_{AB}^{(n)}(x) = 0. \qquad (2.134)$$

The engineering dimension of the coefficients are, in AF theories given by those of the operators involved. A famous example is the OPE of vector currents

$$V_\mu(x)V_\nu^\dagger(0) \sim_{x=0} C_{\mu\nu}^{(0)}(x) + C_{\mu\nu}^{(1)}(x)N[F^2](0) + \ldots. \qquad (2.135)$$

The leading coefficient $C_{\mu\nu}^{(0)}(x)$ multiplying the identity operator behaves like $1/(x^2)^3$ (for $x \to 0$) up to logs, and (in the case of electromagnetic currents) describes the leading high-energy behavior in e^+e^- annihilation. It has been computed to high order in PT (Baikov *et al.*, 2009). Because of gauge invariance the next operators occurring here (in the massless theory) have dimension 4 (only one term has been exhibited above) and so the corresponding coefficients, e.g. $C_{\mu\nu}^{(1)}(x) \sim 1/x^2$. As stressed by many authors long ago, e.g. by David (1986), this does not mean that there are no terms in the associated physical amplitudes behaving like $1/(x^2)^2$. Unfortunately, one still encounters the contrary statement in the literature; there is reference to "the gluon condensate" as a non-perturbative effect, while forgetting that there can be non-perturbative effects in the coefficients $C_{\mu\nu}^{(n)}(x)$ and the fact that to my knowledge there is at present no regularization independent definition of the gluon condensate! These considerations do of course not negate the strength of the OPE;

PT gives the leading short-distance behavior of the coefficients and the subleading terms in the OPE give the leading effects in processes with non-vacuum external states.

Another useful example of an application of the OPE is in non-leptonic decays in the framework of the Standard Model (e.g. $K \to 2\pi$ (Gaillard and Lee, 1974a), (Altarelli and Maiani, 1974), which will be discussed in detail in the lectures of Lellouch). I would just like to emphasize a few points below and in this discussion neglect quark masses. The typical Minkowski-space amplitude for initial and final hadronic states I, F has the form

$$T_{FI} \propto \int \mathrm{d}^4x \, D(x, m_W) \langle F | T J_{1\mu}^L(x) J_{2\mu}^L(0) | I \rangle, \tag{2.136}$$

involving left-handed currents J_1^L, J_2^L, where $D(x, m_W)$ is the scalar function occurring in the W-meson propagator. Since the physical W-meson mass m_W is much larger than typical strong interaction scales involved, short distances dominate the integral. The simplest case to consider is that where the currents involve different flavored quarks; in that case the OPE implies

$$\int \mathrm{d}^4x \, D(x, m_W) T J_{ru\mu}^L(x) J_{sv\mu}^L(0) \sim \sum_{\sigma=\pm} h^\sigma(\mu/m_W, g) O_{rsuv}^\sigma(0), \tag{2.137}$$

with composite operators

$$O_{rsuv}^\sigma = O_{rsuv} \pm O_{rsvu}, \tag{2.138}$$

$$O_{rsuv} = N\left[J_{ru\mu}^L J_{sv\mu}^L \right], \quad J_{ru\mu}^L = \bar{\psi}_r \gamma_\mu \frac{1}{2}(1 - \gamma_5)\psi_u. \tag{2.139}$$

The operators $O^\pm$ renormalize diagonally if one has a regularization preserving chiral symmetry.[21] Restricting to that case we write the rhs in terms of the RGI operators introduced in Eq. (2.108) as

$$\sum_{\sigma=\pm} k^\sigma(m_W/\Lambda) O_{rsuv}^{\mathrm{RGI}\sigma}(0), \tag{2.140}$$

with

$$k^\sigma(m_W/\Lambda) = h^\sigma(1, \bar{g}^2(m_W))/C_\sigma(m_W/\Lambda) \tag{2.141}$$

$$= (2b_0 \bar{g}^2(m_W))^{-\gamma_\sigma^{(0)}/2b_0} \left[1 + k_{\sigma 1} \bar{g}^2(m_W) + \ldots \right]. \tag{2.142}$$

The coefficients can be computed in PT, e.g. $k_{\sigma 1} = h_{\sigma 1} - (\gamma_\sigma^{(1)} - \gamma_\sigma^{(0)} b_1/b_0)/(2b_0)$, where $h_{\sigma 1}$ is determined by 1-loop matching of the full amplitude with that of the OPE, and $\gamma_\sigma^{(1)}$ is a two-loop anomalous dimension. There remains the important job of the lattice community to determine the non-perturbative amplitudes $\langle F | O^{\mathrm{RGI}\sigma}(0) | I \rangle$.

Wilson fermions break chiral symmetry and this has the effect, as first pointed out by Martinelli (1984), that with this regularization the parity-even part of $O^\pm$ mixes with other operators e.g. $O_{\pm}^{VA} \propto (VV - AA) \pm (u \leftrightarrow v)$. Although the number of such operators is restricted by CPS symmetry (charge conjugation, parity, and $SU(N_f)$ in the limit $m_q = 0$), there are still 3 such operators, which makes Wilson fermions a rather awkward regularization for computing the desired physical amplitudes in this case.

2.3   Non-perturbative renormalization

So far, we have mainly considered renormalization in the framework of perturbation theory, which for QCD is only applicable to a class of high-energy processes. But QCD is a candidate theory for the hadronic interactions at all energies. In particular, in most numerical QCD computations we are attempting to determine low-energy observables. In such studies we fix the bare quark masses by fixing a sufficient number of scales, e.g. $m_\pi/f_\pi, m_K/f_\pi$ .. to their physical values, and then $af_\pi(g_0)$ gives the lattice spacing $a(g_0)$ for the pion decay constant f_π fixed. This is called a *hadronic renormalization scheme*.

In order to connect a hadronic scheme to a perturbative scheme one could in principle proceed by computing a non-perturbatively defined running coupling, e.g. using a 2-current vacuum correlation function, over a large range of energies. Eventually at high energies (after taking the continuum limit) we can compare to perturbation theory and estimate the scales, e.g. $f_\pi/\Lambda_{\overline{MS}}$. We can proceed similarly for running masses and scale-dependent renormalization constants. However, despite the huge increase in available computational resources and advances in algorithmic development, to measure physical high-energy E observables with small lattice artifacts and negligible finite-volume effects still raises the old practical problem that the lattices required need too many points ($a \ll 1/E \ll 1/f_\pi < L$).

Many procedures have been applied in order to attempt to overcome this difficulty. The most naive way is to try to use the perturbative relation between the $\overline{MS}$ coupling and the bare lattice coupling $\alpha_0 = g_0^2/(4\pi)$ mentioned in Section 2.2.2:

$$\alpha_{\overline{MS}}(\mu) = \alpha_0 + \alpha_0^2 d_1(a\mu) + \dots \tag{2.143}$$

$$d_1(a\mu) = -8\pi b_0 \ln(a\mu) + k, \tag{2.144}$$

where the constant k depends on the lattice action. As a non-perturbative input one computes a mass scale, e.g. a charmonium mass splitting, to give the lattice spacing $a(\alpha_0)$ in physical units. Now, one can use the relation above to obtain an estimate for $\alpha_{\overline{MS}}(\mu = s/a)$ (for some chosen factor s). This procedure encounters many basic problems; first, it is difficult to separate the lattice (and finite-volume) artifacts and estimate the systematic errors, and secondly for many actions (e.g. the standard plaquette action) one encounters large perturbative coefficients, in fact we have already seen this in the large ratio between the lattice and $\overline{MS}$ Λ parameters. It was first observed by Parisi (1985)[22] that large contributions to this ratio come from tadpole diagrams, and that similar diagrams appear in the computation of the average plaquette

$$P = \frac{1}{N} \langle \operatorname{tr} U(p) \rangle = 1 + \mathrm{O}(g_0^2). \tag{2.145}$$

Mean-field improved bare PT is an expansion in an alternative bare coupling

$$\alpha_P \equiv \alpha_0/P. \tag{2.146}$$

If one now re-expresses $\alpha_{\overline{\mathrm{MS}}}$ in terms of α_P one usually finds that the perturbative coefficients are reduced; e.g. choosing the scale μ such that the 1-loop term is absent; for the standard action $N = 3, N_f = 0$ one obtains

$$\alpha_{\overline{\mathrm{MS}}} = \alpha_P + 2.185\alpha_P^3 + \dots, \quad \text{for } \mu a = 2.6, \tag{2.147}$$

with a reasonably small 2-loop coefficient. Computation of a mass scale and the plaquette expectation value gives an estimate of the $\overline{\mathrm{MS}}$ coupling at $\mu = 2.6/a$. There are obviously many variants and the technique has been perfected by Lepage and Mackenzie (1993). But I think it is still true to say that systematic errors in these determinations are difficult to estimate and the scales achieved are not so high that one can be confident to use the result as initial conditions for running the RG equation with perturbative beta-functions to higher energies.

2.3.1   Intermediate regularization-independent momentum scheme

Another approach is to use an *intermediate regularization independent momentum scheme*, which will be mentioned again in Section 2.5.3. In this approach one usually considers correlation functions involving some basic fields and thus one has to fix a gauge (and tackle the Gribov ambiguity problem if a covariant gauge is used). Using similar ideas non-perturbative running couplings can be defined as they are in PT, e.g. from the 3-point gluon vertex function (Alles *et al.*, 1997). For a covariant gauge the full gluon propagator in the continuum has the form (2.30), and the 3-gluon vertex function has the structure ($p_1 + p_2 + p_3 = 0$)

$$\Gamma_{\mu_1\mu_2\mu_3}^{a_1a_2a_3}(p_1,p_2,p_3) = -if^{a_1a_2a_3} F_{\mu_1\mu_2\mu_3}(p_1,p_2,p_3) + id^{a_1a_2a_3} D_{\mu_1\mu_2\mu_3}(p_1,p_2,p_3), \tag{2.148}$$

where f, d are the SU(N) invariant tensors. At the "symmetric point" (SP) $p_i^2 = M^2, p_i p_j = -\frac{1}{2}M^2, i \neq j$ we have for the first amplitude

$$F_{\mu_1\mu_2\mu_3}|_{\mathrm{SP}} = F(M^2)\left\{\delta_{\mu_1\mu_2}(p_1 - p_2)_{\mu_3} + 2\,\mathrm{perms}\right\} + R_{\mu_1\mu_2\mu_3}, \tag{2.149}$$

where $n_{\mu_1} R_{\mu_1\mu_2\mu_3} = 0$ if $np_i = 0$ for all p_i. The symmetric point was also introduced by Lee and Zinn-Justin (1972) as a set of symmetric non-exceptional momenta in order to avoid potential IR problems (at least in the framework of PT). One can now define a renormalized running coupling by (Celmaster and Gonsalves, 1979b, 1979a)

$$g_{\mathrm{MOM}}(M^2) = F(M^2)\left[1 + \Pi(M^2)\right]^{-3/2}, \tag{2.150}$$

where $\Pi(k^2)$ is the dynamical function appearing in the full gluon propagator (2.30).

With an analogous construction for the lattice regularization, one then measures g_{MOM} at various values of allowed $M^2 = k \left(\frac{2\pi}{L}\right)^2$ [23] at a given value of the bare coupling g_0 at which one also measures a mass scale in order to specify M^2 in physical units. One then, if possible, repeats the procedure at other values of g_0 in order to attempt a continuum-limit extrapolation. Then, at the largest values of M^2 one has to resort to perturbative evolution in order to reach high energies. The method avoids the use of bare PT, but for presently feasible lattice sizes one cannot really reach sufficiently large M^2, and in the most optimistic case one only has a small window of M^2 without too severe lattice artifacts.

2.3.2 Recursive finite-size technique

Lattice simulations are necessarily performed at finite volumes and these effects are a source of systematic error if the measured physical quantities are desired in infinite volume. On the other hand, we can also make use of the finite volume as a probe of the system, as developed in QFT to a high degree by Luscher (1991*a,b*). For example, in Aoki's lectures he described how one can extract infinite-volume scattering data from measurements of finite-volume effects on spectra at large volumes.

Our interest here is to overcome the renormalization problem mentioned above and for this purpose it is useful to define a renormalized coupling depending on the volume α_{FV}, e.g. one determined in terms of the force between two static quarks

$$\alpha_{q\bar{q}}(L) \propto \left\{ r^2 F_{q\bar{q}}(r, L) \right\}_{r=L/2}. \tag{2.151}$$

There are infinitely many acceptable choices and at large L their behaviors can be completely different. The important feature that characterizes them is that at small L (where the spectral properties are vastly different from familiar infinite-volume spectra) we can use PT to compute them as a power series in $\alpha_{\overline{\text{MS}}}$ (starting linearly):

$$\alpha_{\text{FV}}(L) = \alpha_{\overline{\text{MS}}}(\mu) + \alpha_{\overline{\text{MS}}}(\mu)^2 \left[8\pi b_0 \ln(\mu L) + c_{\text{FV}} \right] + \dots. \tag{2.152}$$

Let us assume that we have decided on the definition of the finite-size coupling. The tactic to connect a hadronic scheme to a perturbative one, which goes under the name of *the recursive finite-size technique*,[24] consists of the following steps.

1) Set the scale on the lattice with largest physical extent $L = L_{\text{max}}$ so that L_{max} is known in physical units say ~ 0.5 fm, and compute $\alpha_{\text{FV}}(L_{\text{max}})$.
2) Now perform non-perturbative evolution until $\alpha_{\text{FV}}(L)$ is known on a lattice of much smaller size say ~ 0.005 fm.
3) Assuming perturbative evolution has apparently set in at the scale reached in step 2, one continues with perturbative evolution and eventually obtains the Λ parameter in the FV scheme Λ_{FV} in physical units.
4) Relate the coupling to the $\overline{\text{MS}}$-scheme to 1-loop to obtain the ratio of Λ parameters, and hence $\Lambda_{\overline{\text{MS}}}$ in physical units. One can then use perturbative running to compute $\alpha_{\overline{\text{MS}}}$ at any HE scale say $\mu = m_W$.

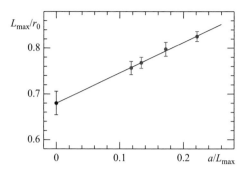

Fig. 2.7 Extrapolation of $L_{\max}/r_0$ to the continuum limit

In the following, we will outline some of the steps in more detail. The order that some of the steps are carried out in practice is not fixed, but let us start with step 1. Let us define

$$L_{\max} : \text{volume where } \bar{g}_{\mathrm{FV}}^2(L_{\max}) = \text{fixed value } 3.48, \tag{2.153}$$

the precise latter value is of course not essential but chosen after initial test runs to correspond to lattice sizes ~ 1 fm. We would like to determine $L_{\max}$ in terms of some physical unit. Considering first the pure Yang–Mills theory, a convenient quantity is Sommer's scale (Sommer, 1994) r_0 defined by $r_0^2 F_{q\bar{q}}(r_0) = 1.65$, which in phenomenological heavy-quark (charmonium) potential models corresponds to a distance ~ 0.5 fm. To compute $L_{\max}/r_0$ we first select a value of the bare coupling g_0 on a large lattice (say $L/a \sim 48$) where one can measure r_0/a accurately with negligible finite-volume effects. At the same g_0 one measures $\bar{g}_{\mathrm{FV}}^2$ on smaller lattices $L/a = 6, 8, \ldots 16$ and obtains $L_{\max}/a$ by interpolation, and hence $(L_{\max}/r_0)(g_0)$. Now, the procedure is repeated at other values and subsequently the data is extrapolated to the continuum limit using the theoretically expected form of the artifacts (discussed in the next section) as illustrated in Fig. 2.7.

Step 2, measuring the evolution was historically done in the reverse order from the description above and the standard notation in the following is suited for this. In the continuum limit there is a well-defined function σ, *the step scaling function*, relating the coupling at one volume to that at double the volume:

$$\bar{g}_{\mathrm{FV}}^2(2L) = \sigma(\bar{g}_{\mathrm{FV}}^2(L)). \tag{2.154}$$

With a lattice regularization this is modified to

$$\bar{g}_{\mathrm{FV}}^2(2L) = \Sigma(\bar{g}_{\mathrm{FV}}^2(L), a/L). \tag{2.155}$$

One then starts with a convenient small lattice, say $L/a = 8$, and tunes g_0 such that $\bar{g}_{\mathrm{FV}}^2(L)$ equals some small value u. At the same g_0 one computes $\bar{g}_{\mathrm{FV}}^2(2L) = \Sigma(u, a/L)$. One then repeats this for a manageable range of a/L as illustrated in Fig. 2.8, and extrapolates the result to the continuum limit thus obtaining a value for $\sigma(u) = u'$.

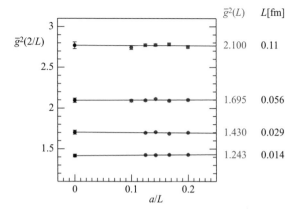

Fig. 2.8   Extrapolation of Σ to the continuum limit à la Symanzik

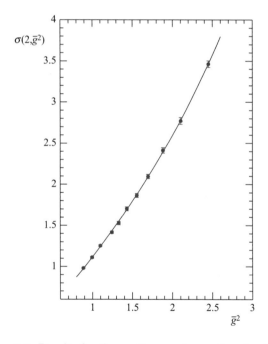

Fig. 2.9   Results for the continuum step scaling function

The whole procedure is then repeated but this time starting with an initial value $\bar{g}^2_{\mathrm{FV}}(L) = u'$ (or a value close to u'). After this has been done many (O(10)) times, one ends up with a sequence of points for the continuum step-scaling function as illustrated in Fig. 2.9:

$$u_k = \sigma(u_{k+1}), \quad u_0 = 3.48, \tag{2.156}$$

giving $\bar{g}_{\mathrm{FV}}^2(L)$ at $L = 2^{-k} L_{\mathrm{max}}$. At the small values of L one can check whether $\sigma(u)$ is well described by the perturbative expectation

$$\sigma(u) = u + (2 \ln 2) b_0 u^2 + \dots \tag{2.157}$$

and if this is the case one can use the beta function with perturbative coefficients to compute Λ_{FV} in physical units. After that the steps are straightforward; a 1-loop computation relating α_{FV} to $\alpha_{\overline{\mathrm{MS}}}$ yields the ratio of Λ parameters and hence the desired value of the product $\Lambda_{\overline{\mathrm{MS}}} r_0$ relating the scales of the hadronic (LE) and perturbative (HE) renormalization schemes.

It is clear that for the success of the RFS method described above, we need a definition of the coupling that satisfies the following criteria: a) it should be accurately measurable, b) it has preferably small lattice artifacts, and c) it should be relatively easily computable in PT.

2.3.3 The Schrödinger functional

After some extensive R&D members of the Alpha Collaboration found that couplings based on the Schrödinger functional (SF)[25] satisfy the above requirements (Luscher *et al.*, 1992). It was further realized that the setup is also well suited for the computation of renormalization constants in general, and that it is easily extended to include fermions and to compute their running masses.

In this framework, one studies the system in a cylindrical volume Λ with Dirichlet boundary conditions in one (the temporal) direction and periodic bc in the other (spatial) directions (illustrated in Fig. 2.10):

$$A_k(\mathbf{x}, 0) = C_k, \quad A_k(\mathbf{x}, T) = C_k', \quad A_k(\mathbf{x} + L, t) = A_k(\mathbf{x}, t). \tag{2.158}$$

Formally, in the continuum the SF is given by the functional integral

$$Z(C, C') = \int_{SF \ \mathrm{bc}} D[A] \, e^{-S}, \tag{2.159}$$

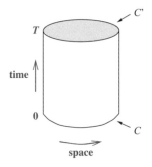

Fig. 2.10 Geometry of the Schrödinger functional

which is properly defined so that it is equal to the transition amplitude $\langle C'|e^{-\mathbb{H}T}\mathbb{P}|C\rangle$ ($\mathbb{H}$ the Hamiltonian, $\mathbb{P}$ the projector on gauge invariant states) in the Hamiltonian formulation.

The renormalization of the SF in scalar field theories was first studied by Symanzik (1981a). He found that apart from the usual renormalization of the bare parameters and fields in the bulk one just requires some extra terms on the boundaries, specifically spatial integrals over local fields of dimension ≤ 3. Lüscher's paper (1985) gives a clear introduction to the subject. There are no such local gauge-invariant operators in pure Yang–Mills theory and so the (bare) SF should in this case not need any renormalization besides the usual coupling-constant renormalization. We remark that one of the first papers considering the related topic of the structure of Yang–Mills theories in the temporal gauge were by Rossi and Testa (1980a,b).

For small bare coupling g_0 the functional integral is dominated by fields around the absolute minimum of the action described by some background field B. The SF then has a perturbative expansion

$$-\ln Z(C, C') \sim \Gamma(B) = g_0^{-2}\Gamma_0(B) + \Gamma_1(B) + \dots. \tag{2.160}$$

If the boundary fields depend on a parameter η then one can define a renormalized running coupling as

$$\bar{g}_{\mathrm{SF}}^2(L) = \left(\frac{\partial\Gamma_0(B)/\partial\eta}{\partial\Gamma(B)/\partial\eta}\right)_{\eta=0,T=L}. \tag{2.161}$$

Regularizing the gauge theory on the lattice the Scrödinger functional is an integral over all configurations of link matrices in SU(N):

$$Z(C, C') = \int D[U]e^{-S[U]}, \tag{2.162}$$

with the Haar measure and the U satisfying periodic boundary conditions in the spatial directions and Dirichlet bc in the time direction:

$$U(x, k)|_{x_0=0} = W(x, k), \quad U(x, k)|_{x_0=T} = W'(x, k), \quad k = 1, 2, 3, \tag{2.163}$$

where

$$W(x, k) = \mathcal{P}\exp\left\{a\int_0^1 dt\, C_k\left(x + (1-t)a\hat{k}\right)\right\}, \tag{2.164}$$

i.e. the SF is considered as a functional of the continuum fields C, C' and the continuum limit $a \to 0$ is taken with C, C' fixed.

One can work in principle with any acceptable lattice action, the simplest being Wilson's plaquette action

$$S = \frac{1}{g_0^2}\sum_p w(p)\Re\,\mathrm{tr}\,(1 - U(p)), \tag{2.165}$$

where the sum is over all plaquettes p and the weight $w(p) = 1$ except for those lying on the boundary that is chosen $w(p) = 1/2$ to avoid a classical $O(a)$ effect.

Note that the derivative entering the definition of the coupling is

$$\frac{\partial \Gamma}{\partial \eta} = \left\langle \frac{\partial S}{\partial \eta} \right\rangle. \tag{2.166}$$

The expectation value appearing on the rhs involves only "plaquettes" localized on the boundary. These are accurately measurable, hence satisfying criteria (a) above.

As for the particular choice of the boundary fields C, C' to make the perturbation expansion well defined we need the following *stability condition*: if $V(x, \mu) = \exp[aB_\mu(x)]$ is a configuration of least action (with bc C, C') then any other gauge field with the same action is gauge equivalent to V. Secondly, we would like to have criterion (b). How to make optimal choices satisfying these demands is not at all obvious. Again, after some experimentation, the Alpha Collaboration made the choice of abelian bc, e.g. for SU(3):

$$C_k(x) = \frac{i}{L} \begin{pmatrix} \phi_1 & 0 & 0 \\ 0 & \phi_2 & 0 \\ 0 & 0 & \phi_3 \end{pmatrix}, \quad \sum_{\alpha=1}^{3} \phi_\alpha = 0, \quad \phi_\alpha \text{ indep. of } x, k, \tag{2.167}$$

and similarly for C' involving elements ϕ'_i. The induced background field is abelian and given by $(T = L)$

$$B_0 = 0, \quad B_k(x) = C + (C' - C)x_0/L, \quad C = C(\eta). \tag{2.168}$$

Stability has been proven (Luscher *et al.*, 1992) provided the ϕ_αs satisfy

$$\phi_1 < \phi_2 < \phi_3, \quad |\phi_\alpha - \phi_\beta| < 2\pi,$$

(and similarly for ϕ'_α), and provided TL/a^2 is large enough, albeit the bound not being very restrictive, e.g. $TL/a^2 > 2\pi^2$ for $N = 3$.

With this setup the Alpha Collaboration produced measurements of a running coupling (in the continuum limit as far as it could be controlled) over a large range of energies,[26] as depicted in Fig. 2.11 (Capitani *et al.*, 1999b). At high energies the running is consistent with perturbative expectations, giving convincing numerical support to the (yet unproven) conventional wisdom that the critical coupling is $g_c = 0$ and that the continuum limit of the lattice theory is asymptotically free.[27] Contrary to widespread opinion the latter property is non-trivial (so far lacking rigorous proof) and some authors have questioned its validity (Patrascioiu and Seiler, 2000).

2.3.4 Inclusion of fermions

The inclusion of fermions in the SF framework was first considered by Sint (1994, 1995). In the continuum it has been argued by Lüscher (2006) that "natural boundary conditions" involve linear conditions for the fields of lowest dimension. For Dirac fermions these take the form

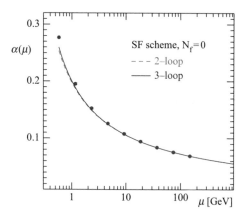

Fig. 2.11 SF running coupling $\alpha(\mu) = \bar{g}_{\mathrm{SF}}^2(L)/4\pi$, $\mu = 1/L$ ($N_{\mathrm{f}} = 0$)

$$\mathcal{B}\psi|_{\mathrm{bdy}} = 0, \tag{2.169}$$

where, in order to obtain a non-trivial propagator, the matrix $\mathcal{B}$ must not have maximal rank. If one demands invariance under space, parity, time reflections ($x_0 \to T - x_0$) and charge conjugation, one is left with the possibility

$$P_+\psi(x) = \overline{\psi}(x)P_- = 0, \quad x_0 = 0; \quad P_-\psi(x) = \overline{\psi}(x)P_+ = 0, \quad x_0 = T, \tag{2.170}$$

(or $P_+ \leftrightarrow P_-$), where $P_\pm = \frac{1}{2}(1 \pm \gamma_0)$. Sint showed that with these homogeneous boundary conditions the SF is renormalizable without the necessity of including any extra boundary terms.

With the lattice regularization where continuity considerations are *a priori* missing, boundary conditions are implicit in the specification of the dynamical fields (those to be integrated in the functional integral) and the precise form of the action close to the boundary. For Wilson fermions the terms in the action coupling close to the boundary, e.g. near $x_0 = 0$ have the form $\propto \overline{\psi}(a)P_+\psi(0) + \overline{\psi}(0)P_-\psi(a)$. It is thus natural in the corresponding lattice SF to declare fields $\psi(x), \overline{\psi}(x)$ away from the boundary, i.e. $0 < x_0 < T$ as the dynamical variables and expect that the bcs (2.170) are recovered dynamically in the continuum limit. Often in the SF literature one sees the equations

$$P_+\psi(0, \mathbf{x}) = \rho(\mathbf{x}), \quad \overline{\psi}(0, \mathbf{x})P_- = \bar{\rho}(\mathbf{x}),$$
$$P_-\psi(T, \mathbf{x}) = \rho'(\mathbf{x}), \quad \overline{\psi}(T, \mathbf{x})P_+ = \bar{\rho}'(\mathbf{x}).$$

These are, however, not to be considered as specifying boundary conditions, but describe couplings of sources for the undefined field components near the boundary. For example, defining

$$\xi(\mathbf{x}) = P_- \frac{\delta}{\delta\bar{\rho}(\mathbf{x})}, \quad \bar{\xi}(\mathbf{x}) = -\frac{\delta}{\delta\rho(\mathbf{x})} P_+, \tag{2.171}$$

we can consider correlation functions of the form

$$\langle \mathcal{O}^a A^a(x) \rangle \sim \int [\mathrm{d}U \, \mathrm{d}\psi \, \mathrm{d}\bar{\psi}] \, \mathcal{O}^a A^a(x) \mathrm{e}^{-S}\big|_{\rho=\bar{\rho}=\rho'=\bar{\rho}'=0}, \tag{2.172}$$

where all sources are set to zero after differentiating and

$$\mathcal{O}^a \equiv -\sum_{\mathbf{y},\mathbf{z}} \bar{\xi}(\mathbf{y}) \frac{1}{2}\tau^a \gamma_5 \xi(\mathbf{z}). \tag{2.173}$$

In this setting the extra boundary counterterms[28] appearing in Sint's original paper amount to a renormalization of the sources $\xi(\mathbf{x}) \to Z_\xi^{1/2}\xi(\mathbf{x})$.

An important point is that the SF fermion boundary conditions imply a gap in the spectrum of the Dirac operator at least for g_0 small enough. This has the consequence that simulations at zero quark mass $m_q = 0$ with the Schrödinger functional are not problematic.

Also, an extra option is to impose quasi-periodic boundary conditions in the spatial direction of the form

$$\psi(x + L\hat{k}) = \mathrm{e}^{i\theta_k}\psi(x), \quad \overline{\psi}(x + L\hat{k}) = \mathrm{e}^{-i\theta_k}\overline{\psi}(x), \tag{2.174}$$

which are equivalent to modifying the covariant derivative to

$$(\nabla_k \psi)(x) = \frac{1}{a}\left[\mathrm{e}^{ia\theta_k/L}U(x,k)\psi(x+a\hat{k}) - \psi(x)\right], \tag{2.175}$$

and similarly for ∇_k^*. Such boundary conditions with various choices of the θ_k serve as extra probes of the system.

2.4   Lattice artifacts

Probably in the future, computers will be so powerful that physically large enough lattices will be measurable with very small lattice spacing a, such that lattice artifacts become numerically irrelevant. Even so, the question of the nature of lattice artifacts is of theoretical interest. However, at present it is important in practice to gain insight in the form of the artifacts in order to make reliable extrapolations of numerical data to the continuum limit.

Usually, we make extrapolations of, e.g. ratios of masses of the form (2.2) assuming leading artifacts are predominantly polynomial in the lattice spacing. Lattice artifacts are non-universal, e.g. the exponent p and the coefficient C_{12} in Eq. (2.2) depend on the lattice action. This can be used in various ways, e.g. if we simulate different actions and the data for glueball masses looks as in Fig. 2.12, this would be a support of the (expected) *universality* of the continuum limit and one could make a constrained joint fit.

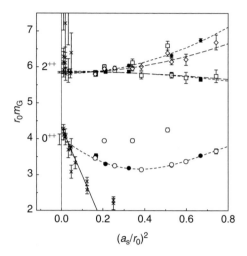

Fig. 2.12 An old plot of glueball masses in pure YM measured with different actions (Morningstar and Peardon, 1996); the results indicate universality. The points with large lattice artifacts for $(d_s/r_0)^2 < 02$ are the plaquette action and the others from improved actions.

More ambitious ways of using the non-universality involve designing actions with larger values of the exponent p, which are called *Symanzik improved actions* (Symanzik, 1979, (1981*b*)) or even constructing *perfect actions* having in principle $p = \infty$ (see Section 2.6).

Again, most of our knowledge concerning lattice artifacts comes from studies of perturbation theory. Some non-perturbative support for the validity of the structure found there comes from investigations in the $1/n$-expansion of QFT in 2 dimensions (see (Knechtli *et al.*, 2005), (Wolff *et al.*, 2006) and references therein), and also from many numerical simulations.

2.4.1 Free fields

Let us first consider a free scalar field theory on an infinite 4-dimensional hypercubic lattice with standard action:

$$S_0 = a^4 \sum_{x,\mu} \frac{1}{2} \left[\partial_\mu \phi(x) \partial_\mu \phi(x) + m^2 \phi(x)^2 \right], \tag{2.176}$$

where $\partial_\mu f(x) = \left[f(x + a\hat{\mu}) - f(x) \right]/a$. The 2-point function is

$$\widetilde{G}(k) = \frac{1}{\hat{k}^2 + m^2}, \tag{2.177}$$

where $\hat{k}_\mu = \frac{2}{a} \sin \frac{ak_\mu}{2}$. Noting that for small a,

$$\hat{k}^2 = k^2 - \frac{1}{12}a^2 \sum_\mu k_\mu^4 + O(a^4), \tag{2.178}$$

we can write (ϕ_0 the corresponding field in the continuum theory)

$$\tilde{G}(k) = \tilde{G}_{\text{cont}}(k) - a^2 \langle S_1^{\text{eff}} \tilde{\phi}_0(k)\phi_0(0)\rangle + \ldots \tag{2.179}$$

with

$$S_1^{\text{eff}} = -\int d^4x \sum_\mu \frac{1}{24}\partial_\mu^2\phi_0(x)\partial_\mu^2\phi_0(x). \tag{2.180}$$

On-shell information is obtained from the (lattice) two-point function $G(x)$ when x is separated from 0 by a physical distance. Performing the integral over k_0 we obtain the representation

$$G(x) = \int_{-\pi/a}^{\pi/a} \frac{d^3k}{(2\pi)^3} e^{ikx} \frac{e^{-\epsilon(\mathbf{k},a,m)|x_0|}}{R(\mathbf{k},m)}, \tag{2.181}$$

where the energy spectrum $\epsilon(\mathbf{k},a,m)$ is given by

$$\cosh(a\epsilon(\mathbf{k},a,m)) - 1 = \frac{1}{2}a^2\left(\hat{\mathbf{k}}^2 + m^2\right). \tag{2.182}$$

Defining the pole mass by $m_{\text{p}} = \epsilon(0,a,m)$ we obtain in the continuum limit $a \to 0$, m_{p} fixed:

$$\epsilon(\mathbf{k},a,m)^2 = m_{\text{p}}^2 + \mathbf{k}^2 - \frac{a^2}{12}T(\mathbf{k},m_{\text{p}}) + O(a^4); \quad T(\mathbf{k},m_{\text{p}}) = \sum_j k_j^4 + \mathbf{k}^2\left(\mathbf{k}^2 + 2m_{\text{p}}^2\right). \tag{2.183}$$

The cutoff effects are (for $m_{\text{p}} = 0$) of order $> 10\%$ for $k > 2\pi/(5a)$. One can improve this situation by adding $O(a^2)$ terms to the action. This can be done in many ways, the simplest possibility being

$$S = S_0 + cS_1, \tag{2.184}$$

$$S_1 = a^4 \sum_{x,\mu} \frac{a^2}{2}\partial_\mu^2\phi(x)\partial_\mu^2\phi(x), \tag{2.185}$$

the latter involving interaction of next-to-nearest neighbors. The energy spectrum is now given by the solution to

$$\cosh(a\epsilon(\mathbf{k}, a, m)) - 1 - 2c\left[\cosh(a\epsilon(\mathbf{k}, a, m)) - 1\right]^2 = \frac{1}{2}a^2\left(\hat{\mathbf{k}}^2 + m^2 + ca^4\sum_j \hat{k}_j^4\right).$$

(2.186)

Now, for small lattice spacing a the energy spectrum has the form

$$\epsilon(\mathbf{k}, a, m)^2 = m_p^2 + \mathbf{k}^2 + \left(c - \frac{1}{12}\right)a^2 T(\mathbf{k}, m_p) + O(a^4),$$

(2.187)

from which we see that the energy is $O(a^2)$ improved if we chose $c = \frac{1}{12}$.

Note that for the improved action another energy level is present but its real part[29] always remains close to the cutoff, and hence it is irrelevant for the continuum limit.

Exercise 2.5 What is S_1^{eff} for the action

$$S_0 + cS_1 + d\sum_{x,\mu,\nu}\frac{a^2}{2}\partial_\mu^*\partial_\mu\phi(x)\partial_\nu^*\partial_\nu\phi(x).$$

Show that the energy is improved for $c = \frac{1}{12}$ for arbitrary d.

2.4.2 Symanzik's effective action

Based on low-order perturbative computations in various field theories one arrives at the following *conjecture*: In a large class of interacting lattice theories (in particular asymptotically free theories) there exists an (Symanzik's) effective continuum action

$$S_1^{\text{eff}} = \int d^d y\, \mathcal{L}_1(y),$$

(2.188)

such that a Green function of products of a multiplicatively renormalizable lattice field φ at widely separated points x_i takes the form

$$Z_\varphi^{r/2}\langle\varphi(x_1)\ldots\varphi(x_r)\rangle_{\text{latt}} = \langle\varphi_0(x_1)\ldots\varphi_0(x_r)\rangle_{\text{cont}}$$

$$-a^p\int d^d y\,\langle\mathcal{L}_1(y)\varphi_0(x_1)\ldots\varphi_0(x_r)\rangle_{\text{cont}}$$

$$+a^p\sum_{k=1}^r\langle\varphi_0(x_1)\ldots\varphi_1(x_k)\ldots\varphi_0(x_r)\rangle_{\text{cont}} + \ldots,$$

(2.189)

where φ_0, φ_1 are renormalized continuum fields, in particular φ_1 is a sum of local operators of dimension $d_\varphi + p$ depending on the specific operator φ and having the same lattice quantum numbers as φ. The effective Lagrangian is a sum

$$\mathcal{L}_1 = \sum_i c_i(g(\mu), a\mu) \mathcal{O}_{\mathrm{R}i}(\mu), \qquad (2.190)$$

of local operators $\mathcal{O}_{\mathrm{R}i}$ of dimension $d + p$ having the symmetries of the lattice action. μ is the renormalization scale, e.g. of the dimensional regularization used in the continuum.

Note a) in the integral over y one in general encounters singularities at points $y = x_k$. A subtraction prescription must thus be applied, but the arbitrariness in this procedure amounts to a redefinition of φ_1. b) The coefficients c_i are, as indicated, functions of the lattice spacing a, but the dependence is thought to be weak (logarithmic).

If the conjecture is true then one generically expects $O(a^2)$ artifacts in pure Yang–Mills theory and $O(a)$ effects with Wilson fermions. All present numerical data seems consistent with these expectations but until now only a small range of a is available.

2.4.3   Logarithmic corrections to $O(a^2)$ lattice artifacts

In 2D lattice models, e.g. the non-linear $O(n)$ sigma model, which is perturbatively asymptotically free, one can simulate lattices with very large correlation lengths ($>$ $200a$). In these theories the expectation is also $O(a^2)$ artifacts. Hence, it came as a surprise, as mentioned by Hasenfratz in his LATT2001 plenary talk (Hasenfratz, 2002), that data on a step-scaling function in this model seemed to show $O(a)$ effects as illustrated in Fig. 2.13! This was rather unsettling and motivated Balog, Niedermayer and myself (2010, 2009) to investigate the logarithmic corrections to the $O(a^2)$ in the framework of renormalized perturbation theory. We found that generic artifacts in the $O(n)$ sigma model are of the form $a^2 \ln^s(a^2)$ with $s = n/(n-2)$. For $n = 3$ the exponent is $s = 3$, and such strong logarithmic corrections to the $O(a^2)$ effects can explain the peculiar behavior, and yield good fits of the data for various actions (Balog *et al.*, 2010). For $n = \infty$ the exponent is $s = 1$ that is consistent with what is found in leading orders of the $1/n$ expansion (Wolff *et al.*, 2006).

The steps involved in obtaining the result above are as follows.

1) Classify operators of dimension 4 (recall for this case $d = 2$ and $p = 2$) that appear in the Symanzik effective Lagrangian (2.190).

2) Compute the c_i at tree level (the coefficients normalized such that $c_i = c_i^{(0)} + O(g^2)$). Although finally interested only in on-shell observables, it is sometimes convenient to work off-shell and compute a sufficient number of correlation functions $G^{(r)}$ with a product of r basic fields, in the continuum and on the lattice. For the lattice Green function,

$$G_{\mathrm{latt}}^{(r)} = G_{\mathrm{cont}}^{(r)} + a^2 \sum_i c_i(g) G_i^{(r)} + \dots, \qquad (2.191)$$

where $G_i^{(r)}$ are continuum correlation functions involving additional insertion of a composite field $\mathcal{O}_{\mathrm{R}i}$.

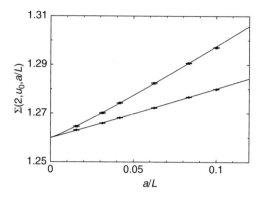

Fig. 2.13 Monte Carlo measurements of the O(3) σ-model step scaling function at $u_0 = 1.0595$ for two lattice actions. The fits shown contain a and $a \ln a$ terms. Fits of the form $k_1 a^2 + k_2 a^2 \ln a + k_3 a^4$ have unacceptably high χ^2/dof.

3) The ratios $\delta_i^{(r)} = G_i^{(r)}/G^{(r)}$ that characterize the lattice artifacts (but are themselves independent of the lattice regularization) obey an RG equation of the form

$$\left\{ \left(-a \frac{\partial}{\partial a} + \beta \frac{\partial}{\partial g} \right) \delta_{ij} + \nu_{ij} \right\} \delta_j^{(r)} = 0, \qquad (2.192)$$

where ν_{ij} is obtained from the mixing of the O_{Ri} to 1-loop (see Section 2.2.3). If we have a basis where the renormalization is diagonal to one loop $\nu_{ij} = -2b_0 \triangle_i \delta_{ij} g^2 + \ldots$ then the operator associated to the largest value of $\triangle_i$ generically dominates the artifacts if the corresponding tree level coefficient $c_i(0) \neq 0$.

The program should be carried out for lattice actions used for large-scale simulations of QCD, when technically possible, in order to check if potentially large logarithmic corrections to lattice artifacts predicted by perturbative analysis appear.

2.4.4 Symanzik improved lattice actions

If Symanzik's conjecture is true it practically follows that it is possible to find a Symanzik improved lattice action such that $S_1^{\text{eff}} = 0$, i.e. for this action there are no lattice artifacts $O(a^p)$. The conjecture is generally accepted for AF theories, but I should mention that a rigorous proof of the existence of a Symanzik improved lattice action for any theory (including ϕ^4) to all orders PT[30] is, to my knowledge, not complete. But there is an all-order proof by Keller (1993) for the existence of Symanzik improved actions for ϕ_4^4 and QED in the framework of a continuum regularization (using flow equations).[31] A subtle point is that the continuum limit of lattice ϕ_4^4 theory is probably trivial, i.e. a free theory. The renormalized coupling goes to zero as $c/\ln(a\mu)$ and hence the continuum limit is actually reached only logarithmically! Treating the renormalized coupling g effectively as a constant for a range of cutoffs one has for small g a perturbative Lagrangian description of the low-energy physics, and in this case the Symanzik effective Lagrangian describes the leading cutoff corrections to this.

An important ingredient of a lattice proof (to all orders PT) would presumably need a proof of the small a expansion of an arbitrary ℓ-loop Feynman diagram of the form

$$F(p,a) \sim a^{-\omega} \sum_{n=0}^{\infty} a^n \sum_{r=0}^{\ell} (\ln am)^r F_{nr}(p), \tag{2.193}$$

which we are quite confident holds and hence often stated in the literature, but which again has not, to my knowledge, been proven for $\ell \geq 2$.

Exercise 2.6 As an example of an expansion of the form (2.193), consider a lattice ϕ_4^4 theory with free propagator $1/[R(k,a) + m^2]$ with

$$R(q/a, a) = a^{-2}r(q), \quad r(q) \sim_{q=0} q^2 + c(q^2)^2 + d\sum_{\mu} q_{\mu}^4 + \dots .$$

Show that the tadpole integral $J_1 = \int_{-\pi/a}^{\pi/a} d^4k \, [R(k,a) + m^2]^{-1}$ has an expansion of the form

$$J_1 \sim a^{-2} \left\{ r_0 + a^2 m^2 [r_1 + s_1 \ln(am)] + a^4 m^4 [r_2 + s_2 \ln(am)] + O(a^6) \right\}$$

with $s_2 = 0$ if $c = d = 0$.

2.4.5 On-shell improved action for pure Yang–Mills theory

With the insight gained from our previous discussion we are now prepared to consider improved actions for Yang–Mills theory in 4 dimensions (Luscher, 1984). As is by now familiar, the first step is to classify the independent (up to total derivatives) gauge-invariant operators of dimension 6, which are scalars under lattice rotations; there are three such operators (Weisz, 1983):

$$O_1 = \sum_{\mu,\nu} \mathrm{tr} D_\mu F_{\mu\nu} D_\mu F_{\mu\nu}, \quad O_2 = \sum_{\mu,\nu,\rho} \mathrm{tr} D_\mu F_{\nu\rho} D_\mu F_{\nu\rho}, \quad O_3 = \sum_{\mu,\nu,\rho} \mathrm{tr} D_\mu F_{\mu\rho} D_\nu F_{\nu\rho}, \tag{2.194}$$

where $D_\mu F_{\nu\rho} = \partial_\mu F_{\nu\rho} + g_0 [A_\mu, F_{\nu\rho}]$. Candidates for Symanzik improved actions have the form

$$S_{\mathrm{imp}} = \frac{2}{g_0^2} \sum_i c_i(g_0) \sum_{\mathcal{C}_i \in \mathcal{S}_i} \mathcal{L}(\mathcal{C}_i), \tag{2.195}$$

$$\mathcal{L}(\mathcal{C}) = \Re \, \mathrm{tr} \, [1 - U(\mathcal{C})], \tag{2.196}$$

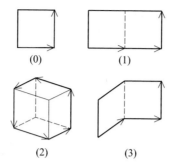

(0) (1)

(2) (3)

Fig. 2.14 4- and 6-link closed curves on the lattice

where $U(\mathcal{C})$ is the ordered product of link variables around the closed curves $\mathcal{C}$, and $\mathcal{S}_i$ are sets with a given topology e.g. $\mathcal{S}_0 =$ the set of plaquettes,

$\mathcal{S}_1 =$ the set of 2×1 rectangles, $\mathcal{S}_2 =$ the set of "twisted chairs", $\mathcal{S}_3 =$ the set of "chairs", as depicted in Fig. 2.14.

Identifying the $U(x, \mu)$ with phase factors in the continuum associated with the links as in Eq. (2.164) (with C_k replaced by $g_0 A_\mu$), the classical small a expansion of the local lattice operators is given by[32]

$$\mathcal{O}_i(x) \equiv \frac{1}{4} \sum_{\mathcal{C} \in \mathcal{S}_i, x \in \mathcal{C}} \mathcal{L}(\mathcal{C}) \tag{2.197}$$

$$= a^4 z_i \operatorname{tr} F_{\mu\nu} F_{\mu\nu} + a^6 \sum_{j=1}^{3} p_{ij} O_j(x). \tag{2.198}$$

We need only 4 lattice operators to represent the 4 continuum operators of dimension 4,6 appearing in the effective action. We chose the sets of curves with smallest perimeter ≤ 6 mentioned above, but many other choices are admissible (and have appeared in the literature). In order that the coefficient of F^2 in the classical expansion has the usual normalization the coefficients must satisfy

$$c_0(g_0) + 8c_1(g_0) + 8c_2(g_0) + 16c_3(g_0) = 1. \tag{2.199}$$

Further, one finds that improvement of the classical action requires

$$c_1(0) = -\frac{1}{12}, \quad c_2(0) = c_3(0) = 0. \tag{2.200}$$

On the other hand, improvement of on-shell quantities only requires two conditions among the coefficients, e.g. for the static 2-quark potential at tree level one finds

$$V(R) = -C \frac{g_0^2}{4\pi R} \left[1 + 3 \left(c_1(0) - c_2(0) - c_3(0) - \frac{1}{12} \right) \frac{a^2}{R^2} + \dots \right]. \tag{2.201}$$

Spectral quantities in finite volumes, e.g. with twisted boundary conditions, can also be computed, and reproduces the improvement condition above and in addition the condition $c_2(0) = 0$ (Luscher and Weisz 1985b,c). There are no other independent relations. This is as expected because on-shell the operator O_3 can be dropped from the effective action since it vanishes when using the equations of motion. For on-shell improvement we conclude that we can choose

$$c_3(g_0) = 0 \quad \text{for all } g_0. \tag{2.202}$$

The coefficients c_i can be computed to higher orders in PT. To 1-loop they have been computed from the same observables as mentioned above (Luscher and Weisz, 1985a).

   In principle, the $c_i(g_0)$ could be computed non-perturbatively by demanding cutoff effects to vanish from some spectral levels, e.g. requiring the $J^{\rm PC} = 2^{++}$ states to be degenerate. This has not been done yet, the reason being that the physical goal is QCD and to achieve full $O(a^2)$ improvement in that theory is much more difficult because there are more dimension-6 operators to be taken into account, in particular those involving products of 4 quark fields. Nevertheless improved gauge actions, based on various considerations, are used in practical large-scale numerical simulations.

2.5 $O(a)$ improved Wilson fermions

As discussed in Hernandez' lectures, there are many ways of putting fermions on the lattice, each having their own particular advantages and disadvantages. In this section I will only discuss lattice artifacts with Wilson fermions. From the tree-level coupling of quarks to one gluon

$$\sim g_0 T^a \left\{ \gamma_\mu - \frac{1}{2} a(p + p')_\mu + O(a^2) \right\}, \tag{2.203}$$

we immediately see that there are $O(a)$ off-shell effects, and some persist on-shell. The on-shell $O(a)$ improvement program (Luscher *et al.*, 1997) proceeds on the same lines as for the pure gauge fields in the last section. The first step is thus to classify the independent local gauge-invariant operators of dimension 5 that can occur in the effective Lagrangian $\mathcal{L}_1$:

$$O_1 = g_0 \overline{\psi} i \sigma_{\mu\nu} F_{\mu\nu} \psi, \quad \sigma_{\mu\nu} = \frac{i}{2} [\gamma_\mu, \gamma_\nu], \tag{2.204}$$

$$O_2 = \overline{\psi} D_\mu D_\mu \psi + \overline{\psi} \overleftarrow{D}_\mu \overleftarrow{D}_\mu \psi, \quad \overline{\psi} \overleftarrow{D}_\mu = \overline{\psi} \left(\overleftarrow{\partial}_\mu - g_0 A_\mu \right), \tag{2.205}$$

$$O_3 = m g_0^2 \mathrm{tr} F_{\mu\nu} F_{\mu\nu}, \quad O_4 = m \overline{\psi} \left(D_\mu - \overleftarrow{D}_\mu \right) \gamma_\mu \psi, \quad O_5 = m^2 \overline{\psi} \psi. \tag{2.206}$$

On-shell we can use the equations of motion $(\gamma D + m)\psi = 0$ to derive relations

$$O_1 - O_2 + 2O_5 \simeq 0, \quad O_4 + 2O_5 \simeq 0, \tag{2.207}$$

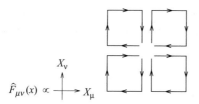

Fig. 2.15 Lattice representative of $g_0 F_{\mu\nu}$

which can be used to eliminate O_2, O_4. A Symanzik–improved action should then be constructible by adding a linear combination of lattice representations of O_1, O_3, O_5

$$\delta S = a^5 \sum_x \left\{ c_1(g_0)\widehat{O}_1(x) + c_3(g_0)\widehat{O}_3(x) + c_5(g_0)\widehat{O}_5(x) \right\}, \qquad (2.208)$$

to the Wilson fermion action. Now, $\widehat{O}_3, \widehat{O}_5$ are already present in the original lattice action, so adding these terms merely corresponds to a rescaling of the bare coupling and masses by terms $\sim 1 + O(am)$, i.e. they can be dropped until we discuss renormalization. We conclude that the on-shell improved Wilson action has only one extra term

$$S_{\text{imp}} = S_W + a^5 \sum_x c_{\text{SW}}(g_0)\overline{\psi}(x)\frac{i}{4}\sigma_{\mu\nu}\widehat{F}_{\mu\nu}(x)\psi(x), \qquad (2.209)$$

where the lattice representative $\widehat{F}_{\mu\nu}(x)$ of $g_0 F_{\mu\nu}(x)$ depicted in Fig. 2.15 has a "clover-leaf form". This action was first written down by Sheikholeslami and Wohlert (1985). The coefficient $c_{\text{SW}}(g_0)$ is known to 1-,2-loop order of PT for various gauge actions.

Two comments are in order here. First, as for any improvement, $O(a)$ improvement is only an asymptotic concept; it could happen that at some g_0 we have bad luck and $O(a^2)$ effects are bigger than $O(a)$ effects for some quantities. Secondly, a rather nice feature is that the extra overhead CPU cost in simulations of adding a SW term is not very substantial.

2.5.1 O(a) improvement of operators

In order to eliminate lattice artifacts in correlation functions involving composite operators, we must also improve the operators themselves, by adding local terms of higher dimension having the same quantum numbers. For example, for the axial isovector current A_μ^a the dimension-4- operator $(A_\mu^a)_1$ in the effective Lagrangian description could involve terms

$$O_{1\mu}^a = \overline{\psi}\left(D_\nu - \overleftarrow{D}_\nu\right)\sigma_{\mu\nu}\gamma_5\frac{1}{2}\tau^a\psi, \qquad (2.210)$$

$$O_{2\mu}^a = \partial_\mu \overline{\psi}\gamma_5\frac{1}{2}\tau^a\psi = \partial_\mu P^a, \qquad O_{3\mu}^a = mA_\mu^a. \qquad (2.211)$$

On-shell one can eliminate $O^a_{1\mu}$ in favor of $O^a_{2\mu}, O^a_{3\mu}$; moreover $O^a_{3\mu}$ is just a renormalization of the original operator. An ansatz for an improved lattice bare operator is then

$$(A^a_I)_\mu = A^a_\mu + a c_A \frac{1}{2} \left(\partial_\mu + \partial^*_\mu\right) P^a, \qquad (2.212)$$

and similarly for other operators, e.g.

$$P^a_I = P^a, \quad (V^a_I)_\mu = V^a_\mu + a c_V \frac{1}{2} \left(\partial_\nu + \partial^*_\nu\right) i\bar\psi \sigma_{\mu\nu} \frac{1}{2} \tau^a \psi. \qquad (2.213)$$

The improvement coefficients c_V, c_A appearing above are $O(g_0^2)$.

2.5.2   Mass-independent renormalization scheme

In lattice QCD it is often advantageous to use a mass-independent renormalization scheme as discussed in Section 2.2. Without $O(a)$ improvement this would involve renormalization constants independent of the quark masses and take the form

$$g_R^2 = g_0^2 Z_g(g_0^2, a\mu), \qquad (2.214)$$

$$m_R = m_q Z_m(g_0^2, a\mu), \quad m_q = m_0 - m_c. \qquad (2.215)$$

To obtain correlation functions at physical distances to approach their continuum limit at a rate $O(a^2)$ with improved Wilson fermions, we must modify the form of the renormalized parameters (and thereby account for the terms $\hat O_3, \hat O_5$ we dropped previously) in order to avoid uncanceled $O(am_q)$ effects (Luscher *et al.*, 1997). We do this by introducing modified bare parameters

$$\tilde g_0^2 = g_0^2 \left(1 + b_g a m_q\right), \qquad (2.216)$$

$$\tilde m_q^2 = m_q \left(1 + b_m a m_q\right), \qquad (2.217)$$

and define renormalized parameters through

$$g_R^2 = \tilde g_0^2 Z_g(\tilde g_0^2, a\mu), \qquad (2.218)$$

$$m_R = \tilde m_q Z_m(\tilde g_0^2, a\mu). \qquad (2.219)$$

We can compute the new improvement coefficients in perturbation theory, e.g. from the "pole mass" at tree level

$$m_P = \frac{1}{a} \ln(1 + am_0) = m_q - \frac{1}{2} am_q + \cdots, \qquad (2.220)$$

we deduce

$$b_m = -\frac{1}{2} + O(g_0^2). \qquad (2.221)$$

The lowest contribution to the coefficient b_g is $O(g_0^2)$ and first obtained by Sint and Sommer (1996) from the 1-loop computation of the SF coupling with fermions at fixed $z = m_q L$:

$$\bar{g}_{\text{SF}}^2(L) = g_0^2 + g_0^4 \left[2b_0 \ln(L/a) + C(z) + N_f k a m_q + \ldots\right], \qquad (2.222)$$

from which one deduces $b_g = N_f k g_0^2 + \ldots$.

Similar factors are required for renormalized composite operators

$$A_{\text{R}\mu}^a = Z_A \left(1 + b_A a m_q\right) (A_I^a)_\mu, \qquad (2.223)$$

$$P_R^a = Z_P \left(1 + b_P a m_q\right) P^a. \qquad (2.224)$$

The coefficients $b_A, b_P = 1 + O(g_0^2)$, which were first computed to 1-loop order by (Gabrielli *et al.*, 1991), don't depend on renormalization conditions – the latter are applied at $m_q = 0$.

2.5.3 Determination of RGI masses and running masses

We recall from Section 2.2 that renormalization-group-invariant masses are scheme independent. Running quark masses can be defined non-perturbatively in various ways; a most efficient way makes use of the PCAC relation:

$$\overline{m}(\mu) = \frac{Z_A \langle \partial A^a \mathcal{O} \rangle}{2 Z_P(\mu) \langle P^a \mathcal{O} \rangle}, \qquad (2.225)$$

where we have not yet specified the source $\mathcal{O}$ (the running mass should be practically independent of this) nor the precise definition of the expectation value. The running of a PCAC mass is determined by the running of the renormalization constant of the pseudoscalar density Z_P.

One scheme for computing scale-dependent renormalization constants, e.g. Z_P, is the regularization-independent (RI) MOM scheme described for couplings in Section 2.3.1. Originally introduced by (Martinelli *et al.*, 1994), it is now quite popular and used by many collaborations. Progress using this scheme was reported by Y. Aoki at LATT09 (Aoki, 2008), (Sturm *et al.*, 2009) and will be covered in these lectures by Vladikas.

Again, an alternative is to apply finite-size recursion techniques here too, and defining Z_P running with the volume. Then, in the continuum we can define an associated step scaling function σ_P through

$$\frac{\overline{m}(1/L)}{\overline{m}(1/2L)} = \frac{Z_P(2L)}{Z_P(L)} = \sigma_P(u), \quad u = \bar{g}^2(L). \qquad (2.226)$$

In perturbation theory we have

$$\sigma_P(u) = 1 - d_0 \ln(2) u + O(u^2). \qquad (2.227)$$

After having chosen a definition of Z_P on the lattice for a given $g_0, L/a$, we can proceed to compute $\sigma_P(u)$ via

$$\sigma_P(u) = \lim_{a \to 0} \Sigma_P(u, a/L), \qquad (2.228)$$

$$\Sigma_P(u, a/L) = \frac{Z_P(g_0, 2L/a)}{Z_P(g_0, L/a)}\Big|_{\bar{g}^2(L) = u}. \qquad (2.229)$$

Suppose we have already determined the step-scaling function for the running coupling (as described in Section 2.3) at say 8 points:

$$L_k = 2^{-k} L_{\max}, \quad u_k = \bar{g}^2(L_k), \quad k = 1, \dots, 8. \tag{2.230}$$

Then, at each step k we measure $\Sigma_P(u_k, a/L)$ for a sequence of values of L/a and extrapolate these to the continuum limit:

$$\Sigma_P(u_k, a/L) = \sigma_P(u_k) + O(a^p), \tag{2.231}$$

where $p = 2$ for the $O(a)$ improved theory (up to boundary terms). Thereby, we have

$$\frac{M}{\overline{m}(1/L_{\max})} = \frac{M}{\overline{m}(1/L_8)} \prod_{k=1}^{8} \sigma_P(u_k). \tag{2.232}$$

Since L_8 is a very small physical length we can safely use PT to determine the first factor $\frac{M}{\overline{m}(1/L_8)}$. To relate the RGI mass M to a low-energy scale it then remains to compute $L_{\max}$ and $\overline{m}(1/L_{\max})$ in physical units e.g. f_π.

2.5.4 The Alpha Collaboration project

It has been an Alpha Collaboration goal since ~ 1998 to measure running couplings, running quark masses and renormalization constants in QCD to a high precision. The program uses $O(a)$ improved Wilson fermions in the SF framework. As we have tried to emphasize this procedure is comparatively clean and care is taken to control systematic errors at each stage, but it involves a lot of preparatory work and attention to details.

First, there are extra improvement coefficients needed to cancel $O(a)$ boundary effects, even for the pure gauge theory $\sim \int_{\partial\Lambda} F_{0k}^2, \int_{\partial\Lambda} F_{ij}^2$. This involves adjusting weights for the plaquettes at the boundary

$$w(p) = \left\{ \begin{array}{c} \frac{1}{2}c_s(g_0^2), \text{ for } p \in \Lambda \\ c_t(g_0^2) \text{ for } p \text{ temporal touching } \Lambda \end{array} \right\}, \tag{2.233}$$

with $c_{s,t} = 1 + O(g_0^2)$. There are similar coefficients $\tilde{c}_s, \tilde{c}_t$ for fermionic boundary terms. Moreover, there is a b_ξ-coefficient for the renormalized boundary operator $\xi_R = Z_\xi(\bar{g}_0^2, a\mu)(1 + b_\xi a m_q)$ although these factors usually cancel in ratios defining observables of interest. It has been observed that these coefficients have small coefficients, and thus in practice it is considered safe to use the 1-loop perturbative approximation instead of determining them non-perturbatively.

Many of the steps must be carried out in a definite order. For example, the SF running coupling is defined as before and at zero quark masses $m_q = 0$. This means that before measuring the coupling the critical mass m_c must be determined but this in turn depends on $c_{\rm csw}$. In this project m_c is defined through the vanishing of the PCAC mass – other definitions, e.g. by vanishing pion mass will differ by $O(a^2)$ in the improved theory.

In an improved theory we expect (Luscher *et al.*, 1997)

$$\langle (A_R)^a_\mu(x)\mathcal{J}^a \rangle = 2m_R \langle P_R^a(x)\mathcal{J}^a \rangle + O(a^2) \tag{2.234}$$

for arbitrary sources $\mathcal{J}^a$. In the framework of the SF we can take $\mathcal{J}^a = \mathcal{O}^a$ as defined in Eq. (2.173). In terms of the bare correlation functions

$$f_A(x_0) = -\frac{1}{3}\langle A_0^a(x)\mathcal{O}^a \rangle, \tag{2.235}$$

$$f_P(x_0) = -\frac{1}{3}\langle P^a(x)\mathcal{O}^a \rangle, \tag{2.236}$$

we then define improved bare PCAC masses by

$$m(x_0, \theta_k, C, C') = \frac{\frac{1}{2}(\partial_0 + \partial_0^*)f_{A_I}(x_0)}{2f_p(x_0)}. \tag{2.237}$$

Finally, renormalized SF PCAC masses are given by

$$\overline{m}_{SF}(L) = \frac{(1 + b_A am_q)Z_A(g_0)}{(1 + b_P am_q)Z_P(g_0, L/a)}m, \tag{2.238}$$

where to complete the definitions we must specify the arguments of m.

It is clear that the coefficients c_A, c_{csw} can be determined by demanding that m is independent of the sources up to terms $O(a^2)$. This can be done by appropriately choosing independent configurations of x_0, θ_k, C, C'. One can proceed as follows: noting that $m(x_0)$ is linear in c_A, $m(x_0) = r(x_0) + c_A s(x_0)$ we can form from two choices of boundary conditions $(C_1, C_1'), (C_2, C_2')$ at say $\theta_k = 0$, a linear combination

$$M_1(x_0, y_0) = m_1(x_0) - s_1(x_0)\frac{[m_1(y_0) - m_2(y_0)]}{s_1(y_0) - s_2(y_0)}, \tag{2.239}$$

which is independent of c_A and equal to $m_1(x_0)$ up to $O(a^2)$ corrections (in the improved theory). We can now give a condition determining, c_{csw} e.g.

$$0 = M_1(T/4, 3T/4) - M_2(T/4, 3T/4) \quad \text{(for } M_1(T/2, T/4) = 0). \tag{2.240}$$

A condition such as the latter in brackets is necessary to complete the specification; to ensure that the results are relatively insensitive to the precise choice one should check that the dependence on m is weak for $m \simeq m_c$ (which is only known approximately at this stage). Once c_{csw} has been determined, we can specify a computation of c_A, e.g. by comparing values of m at $C = C' = 0$ and varying θ_k.

The data for c_{csw}, c_A thus obtained are usually fitted in the measured range to (rational) functions of g_0 that incorporate the known perturbative coefficients. These functions can now be considered as definitions of the improved theory, which can then be used in other simulations (not necessarily in the SF framework).

It is important to appreciate that c_{csw}, c_A have $O(a)$ ambiguities; there is no way to remove these outside of PT. Once these coefficients have been specified, we can use the vanishing of m_{PCAC} for a determination of m_c. Again, the precise value depends on all specific definitions involved.

Knowing c_{csw}, m_c we can now compute $\bar{g}_{\text{SF}}^2(L)$ and obtain the step-scaling function in the continuum limit.

Next, one can compute the pseudoscalar density scaling function σ_P as described in Section 2.5.3 in a given SF scheme, e.g.

$$Z_P(g_0, L/a) = \frac{c\sqrt{f_1}}{f_P(T/2)}\Big|_{T=L, m=m_c, C=C'=0, \theta_k=1/2}, \tag{2.241}$$

$$f_1 = -\frac{1}{3L^6}\langle \mathcal{O}'^a \mathcal{O}^a\rangle, \tag{2.242}$$

and c is chosen st $Z_P|_{g_0=0} = 1$.

Finally, one can compute the vector isovector current renormalization constant Z_V using the WI (the derivation is left as an exercise)

$$(1 + b_V a m_q) Z_V f_V(x_0) = f_1 + O(a^2), \tag{2.243}$$

where

$$f_V(x_0) = \frac{1}{6L^3}\langle (V_I)_0^a(x) \mathcal{O}_{\text{ext}}^a\rangle, \tag{2.244}$$

$$\mathcal{O}_{\text{ext}}^a = \epsilon^{abc}\mathcal{O}^b\mathcal{O}'^c, \tag{2.245}$$

and for Z_A:

$$(1 + b_A a m_q) Z_A^2 f_{A_I A_I}(x_0, y_0) = f_1 + O(a^2), \tag{2.246}$$

where

$$f_{AA}(x_0, y_0) = \frac{a^6}{L^6} \sum_{\mathbf{x}, \mathbf{y}} \epsilon^{abc}\langle A_0^a(x) A_0^b(y) \mathcal{O}_{\text{ext}}^c\rangle, \tag{2.247}$$

the conditions for determining Z_A, Z_V being at $m_q = 0$. No explicit results are shown here but the program outlined above is practically complete for $N_f = 2$ species of dynamical fermions (Della Morte *et al.* 2005*a*, 2006, 2005*b*), (Della Morte *et al.*, 2005*c*), (Della Morte *et al.*, 2005*d*), and for $N_f = 3$ (Aoki *et al.*, 2009). In particular, the SF coupling is measured over a wide range of energies (Della Morte *et al.*, 2005*a*) with a result similar in quality to that for $N_f = 0$. Finally, some results for the Alpha program are recently available for $N_f = 4$ (Tekin *et al.*, 2010).

2.6   Other improved actions

Many types of improved actions have been used in large-scale simulations. As mentioned before this is also useful for numerically verifying universality. Often due to the pressure of completing the simulation routines before a new supercomputer is delivered, pragmatic choices of the action to be used have to be made. For example, the CPPACS collaboration decided to employ the Iwasaki action (Iwasaki, 1983) that has the form (2.195) with $c_0 = 3.648, c_1 = -0.331, c_2 = c_3 = 0$ independent of g_0, but optimized for values of g_0 in the range to be simulated (e.g. by demanding good rotational symmetry properties of the static potential).

2.6.1   Perfect actions

An action used by Hasenfratz *et al.* (2005) is based on Wilson's renormalization group approach (see Niedermayer's review (1998)). The intriguing realization is that in the huge class of lattice actions that have the same universal continuum limit there are "perfect actions" for which the physical quantities that can be measured have no lattice artifacts at all even when the correlation lengths are $O(a)$!

To understand how this comes about consider an RG transformation in configuration space of a pure gauge theory of the form

$$\mathrm{e}^{-\beta' A'(V)} = \int [dU] \, \mathrm{e}^{-\beta[A(U) + T(U,V)]} \qquad (2.248)$$

with kernel

$$T(U,V) = -\frac{\kappa}{N} \sum_{n_B,\mu} \Re \, \mathrm{tr} \left[V(n_B,\mu) Q^\dagger(n_B,\mu) - N_\mu^\beta \right], \qquad (2.249)$$

where $Q(n_B,\mu)$ is a sum over products of link variables U on paths from $2n_B$ to $2(n_B + \hat{\mu})$ on the original lattice (e.g. a sum over staples). For physical quantities that can be measured with action $\beta' A'$ we will get the same results as those measured with βA. The main problem, however, is that we are now dealing with an infinite coupling parameter space.

The situation is simplified for the case of asymptotically free theories with one relevant direction labelled by β, with the critical surface where the correlation length diverges given by $\beta = \infty$. If we start the RG transformation with β very large then we expect that A' will have an expansion of the form

$$A'(V) = A_0(V) + \frac{1}{\beta} A_1(V) + \ldots, \qquad (2.250)$$

(see Fig. 2.16) where

$$A_0(V) = \min_U \left[A(U) + T(U,V) \right]. \qquad (2.251)$$

Performing this minimization with $A_0(U)$ on the critical surface $\beta = \infty$ we stay on that surface. Repeating it infinitely many times we finally reach the "fixed-point action" A^{FP} associated with the kernel T. It satisfies the equation

$$A^{\mathrm{FP}}(V) = \min_U \left[A^{\mathrm{FP}}(U) + T(U,V) \right]. \qquad (2.252)$$

This action is classically perfect; there exist solutions of the classical field equations (e.g. instanton configurations) with no lattice artifacts. If one starts the RG

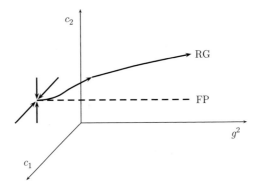

Fig. 2.16   RG flow

transformation close to the FP one obtains an RG trajectory along which one has very small lattice artifacts even for very small correlation lengths.

The proposal of Hasenfratz and Niedermayer (1994) is to simulate the action βA^{FP} for β large. In practice, it is too time consuming to determine A^{FP} configuration by configuration. Instead, one does it for a large number of configurations and parameterizes the results with a relatively small number of loops. Also, they tune the parameter κ in Eq. (2.249) to weaken the strength of the spread of loops in lattice units.

Fermions can be included along similar lines. A very nice feature is that fermion actions that are obtained by blocking from a continuum action that has chiral symmetry have Dirac operators that satisfy the Ginsparg–Wilson relation (Ginsparg and Wilson, 1982)

$$\{\mathcal{D}, \gamma_5\} = (1 + s)\mathcal{D}\gamma_5\mathcal{D}, \tag{2.253}$$

(or some mild local generalization thereof), where s is a parameter $|s| \leq 1$. Consider, for example, lattice fermions χ obtained from blocking a free massless continuum field ψ. The lattice Dirac operator can be obtained by a minimization procedure over classical fields:

$$\overline{\chi}_n \mathcal{D}_{nn'} \chi_{n'} = \min_{\overline{\psi}, \psi} \left\{ \overline{\psi} D \psi + \left(\overline{\chi} - \overline{\psi}\omega^\dagger \right) \left(\chi - \omega\psi \right) \right\}, \tag{2.254}$$

where

$$D_{xx'} = (\gamma_\mu \partial_\mu)_{xx'}, \tag{2.255}$$

and

$$\omega_{nx} = 1 \ \ \text{if} \ \ x \in \ \text{block} \ n$$
$$= 0 \ \ \ \text{otherwise.} \tag{2.256}$$

The minimizing field is given by

$$\psi_0(\chi) = A^{-1}\omega^\dagger\chi, \quad \overline{\psi}_0(\chi) = \overline{\chi}\omega A^{-1}, \tag{2.257}$$

with

$$A = D + \omega^\dagger\omega. \tag{2.258}$$

Exercise 2.7 Show that the lattice Dirac operator

$$\mathcal{D} = 1 - \omega A^{-1}\omega^\dagger \tag{2.259}$$

satisfies the GW relation (2.253) with $s = 1$.

Moreover, Hasenfratz *et al.* show in a very illuminating paper (2006) the general result that lattice actions induced by a RG procedure inherit all the symmetries of the continuum theory, and they give a general procedure that delivers the corresponding symmetry transformation on the lattice, e.g. for the U(1) axial continuum transformation

$$\delta\psi = i\epsilon\gamma_5\psi, \quad \delta\overline{\psi} = i\epsilon\overline{\psi}\gamma_5, \tag{2.260}$$

one obtains Lüscher's lattice transformation (Lüscher, 1998)

$$\delta\chi = i\epsilon\gamma_5(1 - \mathcal{D})\chi, \quad \delta\overline{\chi} = i\epsilon\overline{\chi}(1 - \mathcal{D})\gamma_5. \tag{2.261}$$

2.6.2 Neuberger's action

There are also classes of Dirac operators satisfying the GW relation that are not obtained by RG considerations, e.g. Neuberger's overlap massless Dirac operator (Neuberger, 1998*a*, 1998*b*)

$$D_N = \frac{(1+s)}{a}\left[1 - A\left(A^\dagger A\right)^{-1/2}\right], \tag{2.262}$$

$$A = 1 + s - aD_W, \tag{2.263}$$

which is discussed in Kaplan's lectures (see also Niedermayer's review (1999)). Simulations with associated actions are highly desirable because of their excellent chiral properties, however, at present they are very (CPU) expensive and hence only pursued by a minority of groups such as JLQCD (Ohki *et al.*, 2009). Here, I only want to summarize a few points concerning their renormalization and their lattice artifacts.

The Feynman rules are straightforward to derive but a bit more complicated than other actions because the action is not ultralocal. The free massless propagator is rather simple and takes the form ($\mathring{p}_\mu = a^{-1}\sin(ap_\mu)$)

$$S_0(p) = \frac{1}{2(1+s)} \left\{ -\frac{i\gamma\mathring{p}}{\mathring{p}^2} [\omega(p) + b(p)] + 1 \right\}, \tag{2.264}$$

$$b(p) = \frac{(1+s)}{a} - \frac{a}{2}\mathring{p}^2, \quad \omega(p) = \sqrt{\mathring{p}^2 + b(p)^2}. \tag{2.265}$$

The quark–antiquark 1-gluon vertex $V(p,q)$ has a relatively familiar structure, however, the quark–antiquark 2-gluon vertex also involves an integral over the product of two such vertices $\sim \int_r V(r,p)V(r,q)K(p,q,r)$ with some kernel K. With such a structure it is not immediately clear that the Reisz conditions for power counting are met, however, Reisz and Rothe (2000) have proven the renormalizability of Neuberger fermions with 2 species[33] of massless quarks.

As explained in Kaplan's talk the massless GW action can be decomposed in a sum of left- and right-handed parts

$$\overline{\psi}\mathcal{D}\psi = \overline{\psi}_L\mathcal{D}\psi_L + \overline{\psi}_R\mathcal{D}\psi_R, \tag{2.266}$$

with

$$\psi_{R/L} = \widehat{P}_\pm\psi, \quad \overline{\psi}_{R/L} = \overline{\psi}P_\mp, \tag{2.267}$$

where the projectors are given by

$$P_\pm = \frac{1}{2}(1 \pm \gamma_5), \quad \widehat{P}_\pm = \frac{1}{2}(1 \pm \widehat{\gamma}_5), \tag{2.268}$$

$$\widehat{\gamma}_5 = \gamma_5(1 - a\mathcal{D}). \tag{2.269}$$

It follows that scalar and pseudoscalar densities are naturally defined by

$$S = \overline{\psi}_L\psi_R + \overline{\psi}_R\psi_L = \overline{\psi}\left(1 - \frac{a}{2}\mathcal{D}\right)\psi, \tag{2.270}$$

$$P = \overline{\psi}_L\psi_R - \overline{\psi}_R\psi_L = \overline{\psi}\gamma_5\left(1 - \frac{a}{2}\mathcal{D}\right)\psi, \tag{2.271}$$

which transform into one another under the chiral transformation

$$\delta\psi_L = -\psi_L, \quad \delta\overline{\psi}_L = \overline{\psi}_L, \quad \delta\psi_R = 0 = \delta\overline{\psi}_R. \tag{2.272}$$

Mass terms are naturally introduced into GW actions via

$$\mathcal{D} \to \mathcal{D} + mS. \tag{2.273}$$

The mass is renormalized multiplicatively as in the continuum. The pattern of renormalization of quark bilinears and 4-quark operators is also the same as in the continuum, and for Neuberger fermions computed to 1-loop perturbation theory by Capitani and Giusti (2000).

Exactly conserved flavor vector currents that do not require renormalization have been constructed (Chandrasekharan, 1999), (Kikukawa and Yamada, 1999), (Hasenfratz *et al.*, 2002). Since these are relatively complicated other currents of the form $\overline{\psi}\gamma_\mu\frac{1}{2}\left(1 - \frac{a}{2}\mathcal{D}\right)\psi$ are usually used as observables. They also have nice chiral transformation properties but are not exactly conserved and hence require finite renormalization (Alexandrou *et al.*, 2000).

The discussions of lattice artifacts are again based on accepting the validity of arguments based on Symanzik's effective action. First, for $m = 0$ there are no O(a) artifacts for spectral quantities since in this case any such effects would be described by operators of dimension 5 having symmetries of the lattice action. No such operator can be constructed for GW fermions (a SW-like term breaks the chiral symmetry). When $m \neq 0$ the GW action is invariant under the transformation $\psi \to \hat{\gamma}_5 \psi$, $\overline{\psi} \to -\overline{\psi}\gamma_5$ together with $m \to -m$. This symmetry is, however, not quite sufficient to disqualify O(am) artifacts in correlation functions since these transformations change the measure for topological non-trivial configurations. One can argue, however, that in large volumes one can, for spectral quantities, restrict attention to the topologically trivial sector and hence also expect no O(am) artifacts.

Matrix elements of composite operators mentioned above do, however, have O(a) effects. The improvement of bilinears is obtained by the prescription ($m = 0$) (Capitani *et al.*, 1999*a*, 2000):

$$(\overline{\psi}\Gamma\psi)_I = \overline{\psi}\left(1 - \frac{a}{2}D\right)\Gamma\left(1 - \frac{a}{2}D\right)\psi. \tag{2.274}$$

As a last point I would like to mention that Schrödinger functionals for overlap fermions have been constructed by Lüscher (2006) and in a different way by Sint (for N_{f} even) (2007*b*). In order to obtain the natural SF boundary conditions (2.170) in the continuum limit, Lüscher simply modifies the Neuberger operator so that it satisfies the GW relation up to terms on the boundary

$$\{\gamma_5, \mathcal{D}\} = (1 + s)a\mathcal{D}\gamma_5\mathcal{D} + \triangle_B, \tag{2.275}$$

where $\triangle_B\psi(x) = 0$, unless $x_0 = 0, T$. A simple proposal is

$$\mathcal{D} = \frac{(1 + s)}{a}\left[1 - \frac{1}{2}\left(U + \gamma_5 U^\dagger \gamma_5\right)\right], \tag{2.276}$$

with

$$U = A\left(A^\dagger A + caP\right)^{-1/2}, \tag{2.277}$$

$$P\psi(x) = \frac{1}{a}\left\{\delta_{x_0,a}P_-\psi(x)|_{x_0=a} + \delta_{x_0,T-a}P_+\psi(x)|_{x_0=T-a}\right\}, \tag{2.278}$$

where c is a parameter that can be tuned for O(a) improvement, and has been computed to 1-loop order PT by Takeda (2008).

2.6.3 Twisted mass lattice QCD

As the last topic, I would like to mention Wilson twisted mass lattice QCD; and refer the reader to excellent reviews by Sint (2007*a*) and Shindler (2008). For the case of two flavors ($N_{\mathrm{f}} = 2$) the action takes the form

$$S_{\text{TM}} = \int \mathrm{d}\chi \mathrm{d}\overline{\chi}\,\overline{\chi} D_{\text{TM}}\chi, \tag{2.279}$$

$$D_{\text{TM}} = D_W + m_0 + i\mu_{\text{q}}\gamma_5\tau^3. \tag{2.280}$$

It appeared first in papers by Aoki and Gosch (1989, 1990) who were considering the extra twisted mass term as an external probe to study the phase diagram of Wilson fermions. In these works it was realized that the associated hermitian operator $Q \equiv \gamma_5 D_{TM}$ has no zero modes for $\mu_{\text{q}} > 0$:

$$Q = Q_W + i\mu_{\text{q}}\tau^3, \quad \text{and} \quad Q_W = Q_W^\dagger \;\Rightarrow\; Q^\dagger Q = Q_W Q_W^\dagger + \mu_{\text{q}}^2. \tag{2.281}$$

Thus, adding a twisted mass term provides a local field theoretic solution to the problem of treating exceptional configurations[34] in quenched simulations. The TM formulation also has some advantages in dynamical HMC simulations.

Soon, it was realized (Frezzotti *et al.* 2001*a,b*) that TMQCD is a viable alternative regularization of QCD, and it is, for an additional reason that I will soon explain, now used in some large-scale simulations, This may at first sight seem strange because parity is apparently strongly violated, but this is not the case. To see this, consider first TMQCD in the formal continuum limit

$$S_{\text{TMQCD}} = \int \overline{\chi} \left(\gamma D + m_{\text{q}} + i\mu_{\text{q}}\gamma_5\tau^3\right)\chi. \tag{2.282}$$

Now, making a change of variables in the functional integral

$$\psi = V(\omega)\chi, \quad \overline{\psi} = \overline{\chi}V(\omega), \tag{2.283}$$

$$V(\omega) = \exp\left(i\omega\gamma_5\frac{\tau^3}{2}\right), \quad \text{with} \quad \tan\omega = \frac{\mu_{\text{q}}}{m_{\text{q}}}, \tag{2.284}$$

the action takes the usual form

$$S_{\text{TMQCD}} = \int \overline{\psi}\left(\gamma D + M\right)\psi, \tag{2.285}$$

with

$$M = \sqrt{m_{\text{q}}^2 + \mu_{\text{q}}^2}. \tag{2.286}$$

Hence, continuum twisted mass QCD is formally equivalent to QCD. The symmetries in one basis are in one-to-one correspondence with symmetries in the other basis. The ψs are called the "physical basis" because in this basis the physical interpretation is clearer and χs are called the TM basis.

The formal equivalence can be made more rigorous (Frezzotti *et al.*, 2001*a*) by considering a lattice regularization of TMQCD that preserves a chiral symmetry such as GW fermions. In such a case the regularized correlation functions in the different bases are equal:

$$\langle \mathcal{O}[\psi,\overline{\psi}]\rangle_{(M,0)} = \langle \mathcal{O}[\chi,\overline{\chi}]\rangle_{(m_{\text{q}},\mu_{\text{q}})}. \tag{2.287}$$

Operators in one basis are equivalent to associated operators in the other; this requires some familiarization, e.g. for fermion bilinears in the TM basis define

$$V_\mu^a = \overline{\chi}\gamma_\mu \frac{\tau^a}{2}\chi, \quad \text{similarly} \quad A_\mu^a, S^a, P^a \quad \text{and} \quad S^0 = \overline{\chi}\chi, \tag{2.288}$$

and in the physical basis

$$\mathcal{V}_\mu^a = \overline{\psi}\gamma_\mu \frac{\tau^a}{2}\psi, \quad \text{similarly} \quad \mathcal{A}_\mu^a, \mathcal{S}^a, \mathcal{P}^a \quad \text{and} \quad \mathcal{S}^0 = \overline{\psi}\psi. \tag{2.289}$$

Then, these are related by

$$\mathcal{A}_\mu^a = \cos\omega\, A_\mu^a + \epsilon^{3ab}\sin\omega\, V_\mu^a, \quad a = 1, 2, \tag{2.290}$$

$$= A_\mu^a, \quad a = 3, \tag{2.291}$$

and similarly for $\mathcal{V}_\mu^a$, and

$$\mathcal{P}^a = P^a, \quad a = 1, 2 \tag{2.292}$$

$$= \cos\omega\, P^3 + 2i\sin\omega\, S^0, \quad a = 3, \tag{2.293}$$

$$\mathcal{S}^0 = \cos\omega S^0 + 2i\sin\omega P^3. \tag{2.294}$$

The familiar PCVC and PCAC relations

$$\partial_\mu \mathcal{V}_\mu^a = 0, \tag{2.295}$$

$$\partial_\mu \mathcal{A}_\mu^a = 2M\mathcal{P}^a, \tag{2.296}$$

translate in the TM basis to

$$\partial_\mu V_\mu^a = -2\mu_q \epsilon^{3ab} P^b, \tag{2.297}$$

$$\partial_\mu A_\mu^a = 2m_q P^a + i\mu_q \delta^{3a} S^0. \tag{2.298}$$

For the case of a regularization that breaks chiral symmetry, such as Wilson TM fermions, one expects that near the continuum limit

$$\langle \mathcal{O}_R[\chi, \overline{\chi}]\rangle_{(m_q, \mu_q)} = \langle \mathcal{O}_R[\psi, \overline{\psi}]\rangle^{\text{cont}}_{(M,0)} + O(a). \tag{2.299}$$

The lattice artifacts depend on m_q and on the twist angle ω. In this sense, ordinary Wilson fermions are just a special case of the WTM regularization of QCD.

To discuss the lattice artifacts the determination of the structure of the Symanzik effective action again plays a central role. This first requires listing the symmetries of the WTM action; among these are

1. gauge invariance, lattice rotations and translations, and charge conjugation;
2. $\mathcal{P}_F^a$-symmetry that is conventional parity $\mathcal{P}$ combined with a flavor rotation:
   $\mathcal{P}_F^a : \chi(x) \to i\tau^a \mathcal{P}\chi(x), \quad a = 1, 2 \quad \mathcal{P}\chi(x_0, \mathbf{x}) = \gamma_0\chi(x_0, -\mathbf{x});$
3. and $\widetilde{\mathcal{P}} \equiv \mathcal{P} \times \{\mu_q \to -\mu_q\};$

4. similarly for time reversal;
5. also the $U_V(1)$ subgroup of $SU_V(2)$ is maintained.

With this knowledge one can again show that the only terms in the on-shell effective action (apart from ones that can be incorporated in the renormalizations of the bare parameters) is the SW term

$$S_1^{\text{eff}} \propto \int d^4x\, \overline{\chi}(x)\sigma_{\mu\nu}F^{\mu\nu}(x)\chi(x). \tag{2.300}$$

The $O(a)$ improvement program of the Wilson fermions can be extended to the case of arbitrary twist angle $\omega \neq 0$. In general, more b-coefficients are required, e.g. for the mass-independent renormalization one has to introduce improved bare mass parameters

$$\widetilde{m}_q = m_q(1 + b_m a m_q) + \tilde{b}_m a\mu_q^2, \tag{2.301}$$

$$\widetilde{\mu}_q = \mu_q(1 + b_\mu a m_q), \tag{2.302}$$

and for the renormalized axial current

$$(A_R)_\mu^a = Z_A(1 + b_A a m_q)[A_\mu^a + ac_A \partial_\mu P^a + a\mu_q \tilde{b}_A \epsilon^{3ab} V_\mu^b. \tag{2.303}$$

An obvious simplification of the regularization at maximal twist angle $\omega = \pi/2$ was realized early on, but TMQCD only became more popular as a practical regularization after the realization by Frezzotti and Rossi (2004) of a related property called "automatic $O(a)$ improvement". This is the property that properly renormalized physical expectation values have only $O(a^2)$ artifacts provided that the untwisted mass is tuned to its critical value, e.g. by tuning the PCAC mass $\propto \langle \partial_\mu A_\mu^1 \mathcal{O} \rangle$ to 0. The original demonstration (Frezzotti and Rossi, 2004) of this was not complete, but the discoverers and many other authors developed independent proofs (Sharpe and Wu, 2005), (Shindler, 2006), (Sint, 2006), (Aoki and Bar, 2004, 2006), (Frezzotti *et al.*, 2006).

Working in the TM basis with

$$S^{\text{eff}} = S_0 + aS_1^{\text{eff}}, \tag{2.304}$$

$$S_0 = \int d^4x\, \overline{\chi}\left[\gamma D + i\mu_q \gamma_5 \tau^3\right]\chi \tag{2.305}$$

we have

$$\langle \mathcal{O}_R \rangle = \langle \mathcal{O}_R \rangle^{\text{cont}} - a\langle \mathcal{O}_R S_1^{\text{eff}} \rangle^{\text{cont}} + a\langle \mathcal{O}_1^{\text{eff}} \rangle^{\text{cont}} + O(a^2). \tag{2.306}$$

One proof is based on the following symmetry of S_0:

$$\mathcal{R}_5^1: \quad \chi \to i\tau^1 \gamma_5 \chi, \quad \overline{\chi} \to \overline{\chi} i\tau^1 \gamma_5. \tag{2.307}$$

$\mathcal{R}_5^1$ alone is not a symmetry of WTM, but $\mathcal{R}_5^1 \times \mathcal{D}$ is, where $\mathcal{D}$ essentially measures the parity of the dimensions of operators

$$\mathcal{D}: \quad U(x,\mu) \rightarrow U^\dagger(-x - a\hat{\mu}, \mu), \tag{2.308}$$

$$\chi(x) \rightarrow e^{3i\pi/2}\chi(-x), \quad \overline{\chi}(x) \rightarrow e^{3i\pi/2}\overline{\chi}(-x). \tag{2.309}$$

Consider now operators $\mathcal{O}_\pm$ that are even(odd) under $\mathcal{R}_5^1$. Now, S_1^{eff} is odd under $\mathcal{R}_5^1$; also (using $\mathcal{D}$) the effective operators $O_{\pm 1}^{\text{eff}}$ are odd(even) under $\mathcal{R}_5^1$, and so it follows

$$\langle \mathcal{O}_{+R} \rangle = \langle \mathcal{O}_{+R} \rangle^{\text{cont}} + O(a^2), \tag{2.310}$$

$$\langle \mathcal{O}_{-R} \rangle = O(a). \tag{2.311}$$

One can extend the TM regularization to include non-degenerate quarks and also further flavors in various ways.

Another advantage of using the TM formulation is in the simplification of some renormalization properties of composite operators. For example, in the computation of f_π with Wilson fermions from the correlator $\langle A_{\text{R}0}^1(x) P_{\text{R}}^1(y) \rangle$ one needs Z_{A}, but using the equivalence

$$\langle A_{\text{R}0}^1(x) P_{\text{R}}^1(y) \rangle_{(M_{\text{R}},0)} \simeq \langle V_0^2(x) P_{\text{R}}^1(y) \rangle_{(M_{\text{R}},\pi/2)} \tag{2.312}$$

no renormalization constant for the current on the rhs is needed.

Another example is the 4-fermion operator appearing in the description of $\triangle S = 2$ transitions. In the physical basis

$$\mathcal{O}_{(\mathcal{V}-\mathcal{A})(\mathcal{V}-\mathcal{A})}^{\triangle S = 2} = \sum_\mu \left(\bar{s}\gamma_\mu(1 - \gamma_5)d \right)^2 \tag{2.313}$$

is a sum of two parts, one parity even $\mathcal{O}_{\mathcal{VV}+\mathcal{AA}}$ and the other parity odd $-\mathcal{O}_{\mathcal{VA}+\mathcal{AV}}$. As mentioned in Section 2.2.7 for Wilson fermions the parity-even operator mixes with many other operators. However, with WTM regularization at maximal twist in the u, d sector and Wilson for the s quarks one obtains

$$\mathcal{O}_{\mathcal{VV}+\mathcal{AA}} \simeq \mathcal{O}_{VA+AV}. \tag{2.314}$$

Using $C, \widetilde{P}$ symmetry one can show that $\mathcal{O}_{VA+AV}$ renormalizes diagonally, which simplifies the problem considerably (Pena *et al.*, 2004).

In conclusion, TMQCD is a competitive practical regularization of QCD, especially at maximal twist. The price to be paid, however, is a breaking of parity and flavor symmetries. This is usually at the $O(a^2)$ level but these effects can in some situations be uncomfortably large.

Acknowledgments

I would like to thank Tassos Vladikas and the organizing committee for inviting me to present these lectures. I especially thank Rainer Sommer for his many constructive suggestions for improvement of this manuscript. Most of all I would like to thank my wife Teo for all her support during my career and for her patience while preparing these lectures.

Notes

1. which are equivalent to the Wightman axioms in Minkowski space (Wightman, 1956, Streater and Wightman, 1989).
2. see also (Parasiuk, 1960).
3. including my first lecturer (in 1966) on QFT, who began his lectures by saying "There are only about 3 people in the world who really understand renormalization theory, and I am not one of them!"
4. A (sub)diagram is connected if all pairs of its vertices are connected by a path following internal lines.
5. the usual perturbative assumption that a finite coupling exists in the continuum limit is not fulfilled.
6. Some involving exotic mathematics such as quantum groups, p-adic numbers, . . .
7. In the case of dimensional regularization the subtraction is simpler since the subtractions are in this case ϵ-pole parts of subintegrals.
8. It is amusing to hear Dyson's account of his realization of the problem at http:webstories.com
9. It has been noted that the BPH R-operation has the structure of a Hopf algebra; see, e.g. Kreimer's book (2000).
10. For slightly simplified proofs of BPH renormalization see (Caswell and Kennedy, 1982).
11. For gauge-invariant correlation functions it is in principle possible to do PT in finite volumes without gauge fixing by solving systems of coupled Schwinger–Dyson equations. But in practice this is not usually done because the system is difficult to solve analytically (without hints of the structure of the associated Feynman diagrams).
12. According to Muta (1987), these generalized WI were first discussed by 't Hooft (1971). In his discussion only a part of the system of identities was taken into account.
13. since the symmetry transformations are multiplicatively renormalized.
14. Clearly $\mathcal{E}_\mathrm{R}$ isn't determined uniquely by this condition.
15. Degrees of divergence are defined as before. The superficial degree of divergence of a diagram with N_G, N_{FP}, N_F external gluon, ghost, and fermion lines is: $d = 4 - N_G - \frac{3}{2}(N_{FP} + N_F)$ have 8 cases with $d \geq 0$: $(N_G, N_{FP}, N_F) = (0,0,0), (2,0,0), (0,2,0), (0,0,2), (3,0,0), (1,2,0), (1,0,2), (4,0,0)$. The vacuum diagram can be dropped because it is absorbed in the normalization of the generating functional. The rest have degrees $2, 1, 1, 1, 1, 0, 0, 0$, respectively. Although superficial degrees > 0 the full amplitudes are all only log divergent due to the gauge symmetry and Lorentz invariance.
16. where for simplicity of notation we consider the case of just one mass parameter m.
17. This remains to be done, but Giedt anticipates no additional principle difficulties (private communication).

18. To prove this start using the rules to obtain $D\mathrm{tr}\gamma_5\gamma_\mu\gamma_\nu = \mathrm{tr}\gamma_5\gamma_\rho^2\gamma_\mu\gamma_\nu = -\mathrm{tr}\gamma_\rho\gamma_5\gamma_\rho\gamma_\mu\gamma_\nu = -\mathrm{tr}\gamma_5\gamma_\rho\gamma_\mu\gamma_\nu\gamma_\rho = (-D+4)\mathrm{tr}\gamma_5\gamma_\mu\gamma_\nu$ so that $\mathrm{tr}\gamma_5\gamma_\mu\gamma_\nu = 0$ unless $D = 2$. Next, repeat the manipulations with the trace involving γ_5 and 4 gamma matrices.

19. problems with naive γ_5 algebra do not come in at this stage.

20. It has been shown non-perturbatively in some 2D models e.g. in the 2D Ising field theory (Wu *et al.*, 1976) and also in the massless Thirring model (Wilson, 1970).

21. Because of the difficulty with treating γ_5 in the framework of dimensional regularization, during the renormalization procedure one has to include mixing with "evanescent operators" (ones that vanish for $D = 4$).

22. see also (Martinelli *et al.*, 1982).

23. Note on a given $(L/a)^4$ lattice k can only take certain allowed integer values.

24. The RFST was first submitted by Lüscher as an (unaccepted) project proposal for the European Monte-Carlo Collaboration (EMC^2) formed after an early realization that large-scale simulations required large collaborations.

25. It was fortunate that M. Lüscher was already informed about the SF in scalar theories (Luscher, 1985), and that U. Wolff had already suggested a similar construction for the 2 D non-linear O(n) sigma model.

26. This particular SF coupling runs similarly to PT down to low energies, but this is not a universal property of non-perturbative running couplings.

27. Some additional evidence for the existence of non-perturbatively asymptotically free theories comes from studies of integrable models in 2D (see e.g. (Balog and Weisz, 2004)).

28. of the form $\propto \int_{x_0=0} \mathrm{d}^3x\, \overline{\psi}(x)P_-\psi(x) + \int_{x_0=T} \mathrm{d}^3x\, \overline{\psi}(x)P_+\psi(x)$.

29. A spectral representation exists but energy levels can be complex.

30. Symanzik's published papers dealt with lower-order PT probably he considered the generalization straightforward.

31. Here, the improvement refers to effects involving the UV cutoff Λ occurring in the definition of the bare propagators.

32. For this computation it is convenient to chose an axial gauge, and it is sufficient to consider abelian fields.

33. their proof uses the fact that for $N_{\mathrm{f}} \geq 2$ GW lattice fermions possess an exact chiral (flavor mixing) symmetry.

34. These are configurations that have exceptionally small eigenvalues of Q_W and hence give huge contributions to some correlation functions, e.g. $\langle P^a(x)P^a(y)\rangle$ in quenched simulations.

References

Adler, Stephen L. (1969). *Phys. Rev.*, **177**, 2426–2438.

Alexandrou, C., Follana, E., Panagopoulos, H., and Vicari, E. (2000). *Nucl. Phys.*, **B580**, 394–406.

Alles, B. et al. (1997). *Nucl. Phys.*, **B502**, 325–342.

Altarelli, Guido and Maiani, L. (1974). *Phys. Lett.*, **B52**, 351–354.

Aoki, S. et al. (2009). *JHEP*, **10**, 053.

Aoki, Sinya and Bar, Oliver (2004). *Phys. Rev.*, **D70**, 116011.

Aoki, Sinya and Bar, Oliver (2006). *Phys. Rev.*, **D74**, 034511.

Aoki, Sinya and Gocksch, Andreas (1989). *Phys. Lett.*, **B231**, 449.

Aoki, Sinya and Gocksch, Andreas (1990). *Phys. Lett.*, **B243**, 409–412.

Aoki, Yasumichi (2008). *PoS*, **LATTICE2008**, 222.

Baikov, P. A., Chetyrkin, K. G., and Kuhn, J. H. (2009). *Nucl. Phys. Proc. Suppl.*, **189**, 49–53.

Balog, Janos, Niedermayer, Ferenc, and Weisz, Peter (2009). *Phys. Lett.*, **B676**, 188–192.

Balog, Janos, Niedermayer, Ferenc, and Weisz, Peter (2010). *Nucl. Phys.*, **B824**, 563–615.

Balog, Janos and Weisz, Peter (2004). *Phys. Lett.*, **B594**, 141–152.

Becchi, C., Rouet, A., and Stora, R. (1976). *Annals Phys.*, **98**, 287–321.

Bjorken, J. D. and Drell, S. D. (1965). McGraw Hill Chapter 19.

Bogoliubov, N. N. and Parasiuk, O. S. (1957). *Acta Math.*, **97**, 227–266.

Bogoliubov, N. N. and Shirkov, D. V. (1959). New York, Interscience, 1959.

Breitenlohner, P. and Maison, D. (1977a). *Commun. Math. Phys.*, **52**, 39.

Breitenlohner, P. and Maison, D. (1977b). *Commun. Math. Phys.*, **52**, 55.

Callan, Jr., Curtis G. (1970). *Phys. Rev.*, **D2**, 1541–1547.

Capitani, Stefano and Giusti, Leonardo (2000). *Phys. Rev.*, **D62**, 114506.

Capitani, S., Gockeler, M., Horsley, R., Rakow, Paul E. L., and Schierholz, G. (1999a). *Phys. Lett.*, **B468**, 150–160.

Capitani, S., Gockeler, M., Horsley, R., Rakow, Paul E. L., and Schierholz, G. (2000). *Nucl. Phys. Proc. Suppl.*, **83**, 893–895.

Capitani, Stefano, Luscher, Martin, Sommer, Rainer, and Wittig, Hartmut (1999b). *Nucl. Phys.*, **B544**, 669–698.

Caracciolo, Sergio, Menotti, Pietro, and Pelissetto, Andrea (1992). *Nucl. Phys.*, **B375**, 195–242.

Caswell, William E. and Kennedy, A. D. (1982). *Phys. Rev.*, **D25**, 392.

Celmaster, William and Gonsalves, Richard J. (1979a). *Phys. Rev. Lett.*, **42**, 1435.

Celmaster, William and Gonsalves, Richard J. (1979b). *Phys. Rev.*, **D20**, 1420.

Chandrasekharan, Shailesh (1999). *Phys. Rev.*, **D60**, 074503.

Collins, John C. (1984). Cambridge, University Press (1984) 380p.

Constantinou, M. and Panagopoulos, H. (2008). *Phys. Rev.*, **D77**, 057503.

David, F. (1986). *Nucl. Phys.*, **B263**, 637–648.

Della Morte, Michele et al. (2005a). *Nucl. Phys.*, **B713**, 378–406.

Della Morte, Michele et al. (2005b). *Nucl. Phys.*, **B729**, 117–134.

Della Morte, Michele et al. (2006). *PoS*, **LAT2005**, 233.

Della Morte, Michele, Hoffmann, Roland, Knechtli, Francesco, Sommer, Rainer, and Wolff, Ulli (2005c). *JHEP*, **07**, 007.

Della Morte, Michele, Hoffmann, Roland, and Sommer, Rainer (2005d). *JHEP*, **03**, 029.

Donoghue, John F. (1995).

Dyson, F. J. (1949a). *Phys. Rev.*, **75**, 486–502.

Dyson, F. J. (1949b). *Phys. Rev.*, **75**, 1736–1755.

Feldman, J. S. (1975). In Erice 1975, Proceedings, Renormalization Theory, Dordrecht 1976, 435–460.

Frezzotti, Roberto, Grassi, Pietro Antonio, Sint, Stefan, and Weisz, Peter (2001a). *JHEP*, **08**, 058.

Frezzotti, R., Martinelli, G., Papinutto, M., and Rossi, G. C. (2006). *JHEP*, **04**, 038.

Frezzotti, R. and Rossi, G. C. (2004). *JHEP*, **08**, 007.

Frezzotti, Roberto, Sint, Stefan, and Weisz, Peter (2001b). *JHEP*, **07**, 048.

Gabrielli, E., Martinelli, G., Pittori, C., Heatlie, G., and Sachrajda, Christopher T. (1991). *Nucl. Phys.*, **B362**, 475–486.

Gaillard, M. K. and Lee, Benjamin W. (1974). *Phys. Rev. Lett.*, **33**, 108.

Gell-Mann, Murray and Low, Francis (1951). *Phys. Rev.*, **84**, 350–354.

Giedt, Joel (2007). *Nucl. Phys.*, **B782**, 134–158.

Ginsparg, Paul H. and Wilson, Kenneth G. (1982). *Phys. Rev.*, **D25**, 2649.

Glimm, J. and Jaffe, Arthur M. (1968). *Phys. Rev.*, **176**, 1945–1951.

Gross, D. J. and Wilczek, Frank (1973). *Phys. Rev. Lett.*, **30**, 1343–1346.

Hahn, Y. and Zimmermann, W. (1968). *Commun. Math. Phys.*, **10**, 330.

Hart, A., von Hippel, G. M., Horgan, R. R., and Muller, E. H. (2009). *Comput. Phys. Commun.*, **180**, 2698–2716.

Hasenfratz, Anna and Hasenfratz, Peter (1980). *Phys. Lett.*, **B93**, 165.

Hasenfratz, Anna, Hasenfratz, Peter, and Niedermayer, Ferenc (2005). *Phys. Rev.*, **D72**, 114508.

Hasenfratz, P. (2002). *Nucl. Phys. Proc. Suppl.*, **106**, 159–170.

Hasenfratz, P., Hauswirth, S., Jorg, T., Niedermayer, F., and Holland, K. (2002). *Nucl. Phys.*, **B643**, 280–320.

Hasenfratz, P. and Niedermayer, F. (1994). *Nucl. Phys.*, **B414**, 785–814.

Hasenfratz, Peter, Niedermayer, Ferenc, and von Allmen, Reto (2006). *JHEP*, **10**, 010.

Hepp, Klaus (1966). *Commun. Math. Phys.*, **2**, 301–326.

Iwasaki, Y. (1983). UTHEP-118.

Jauch, J. M. and Rohrlich, F. (1955). Addison Wesley (1955) chapters 9 and 10.

Johnson, R. W. (1970). *J. Math. Phys.*, **11**, 2161–2165.

Karsten, Luuk H. and Smit, Jan (1981). *Nucl. Phys.*, **B183**, 103.

Keller, Georg (1993). *Helv. Phys. Acta*, **66**, 453–470.

Keller, G., Kopper, Christoph, and Salmhofer, M. (1992). *Helv. Phys. Acta*, **65**, 32–52.

Kikukawa, Yoshio and Yamada, Atsushi (1999). *Nucl. Phys.*, **B547**, 413–423.

Knechtli, Francesco, Leder, Bjorn, and Wolff, Ulli (2005). *Nucl. Phys.*, **B726**, 421–440.

Kreimer, D. (2000). Cambridge, University Press (2000) 259 p.

Lee, B. W. and Zinn-Justin, Jean (1972). *Phys. Rev.*, **D5**, 3121–3137.

Lepage, G. Peter and Mackenzie, Paul B. (1993). *Phys. Rev.*, **D48**, 2250–2264.

Lowenstein, John H. (1972). University of Maryland Technical Report No. 73-068, 1972 (unpublished); Lectures given at Int. School of Mathematical Physics, Erice, Sicily, Aug 17–31, 1975.

Lowenstein, J. H. and Zimmermann, W. (1975). *Commun. Math. Phys.*, **44**, 73–86.

Luscher, M. (1984). In Les Houches 1984, Proceedings, Critical Phenomena, Random Systems, Gauge Theories, 359–374.

Luscher, M. (1985). *Nucl. Phys.*, **B254**, 52–57.

Luscher, M. (1988). Lectures given at Summer School "Fields, Strings and Critical Phenomena", Les Houches, France, Jun 28–Aug 5, 1988, 451–528.

Luscher, Martin (1991*a*). *Nucl. Phys.*, **B364**, 237–254.

Luscher, Martin (1991*b*). *Nucl. Phys.*, **B354**, 531–578.

Luscher, Martin (1998). *Phys. Lett.*, **B428**, 342–345.

Luscher, Martin (2006). *JHEP*, **05**, 042.

Luscher, Martin, Narayanan, Rajamani, Weisz, Peter, and Wolff, Ulli (1992). *Nucl. Phys.*, **B384**, 168–228.

Luscher, Martin, Sint, Stefan, Sommer, Rainer, and Weisz, Peter (1996). *Nucl. Phys.*, **B478**, 365–400.

Luscher, Martin, Sint, Stefan, Sommer, Rainer, Weisz, Peter, and Wolff, Ulli (1997). *Nucl. Phys.*, **B491**, 323–343.

Luscher, M. and Weisz, P. (1985*a*). *Phys. Lett.*, **B158**, 250.

Luscher, M. and Weisz, P. (1985*b*). *Commun. Math. Phys.*, **97**, 59.

Luscher, M. and Weisz, P. (1985*c*). *E: Commun. Math. Phys.*, **98**, 433.

Luscher, Martin and Weisz, Peter (1995). *Nucl. Phys.*, **B452**, 234–260.

Magnen, Jacques, Rivasseau, Vincent, and Seneor, Roland (1992). *Phys. Lett.*, **B283**, 90–96.

Magnen, Jacques, Rivasseau, Vincent, and Seneor, Roland (1993). *Commun. Math. Phys.*, **155**, 325–384.

Martinelli, G. (1984). *Phys. Lett.*, **B141**, 395.

Martinelli, G. et al. (1994). *Nucl. Phys. Proc. Suppl.*, **34**, 507–509.

Martinelli, G., Parisi, G., and Petronzio, R. (1982). *Phys. Lett.*, **B114**, 251.

Matthews, P. T. and Salam, Abdus (1951). *Rev. Mod. Phys.*, **23**, 311–314.

Mills, R. L. and Yang, Chen-Ning (1966). *Prog. Theor. Phys. Suppl.*, **37**, 507–511.

Morningstar, Colin and Peardon, Mike J. (1996).

Muta, T. (1987). *World Sci. Lect. Notes Phys.*, Singapore **5**, 1–409.

Nakanishi, N. (1971). New York, Gordon and Breach 1971.

Neuberger, Herbert (1998*a*). *Phys. Lett.*, **B417**, 141–144.

Neuberger, Herbert (1998*b*). *Phys. Lett.*, **B427**, 353–355.

Niedermayer, F. (1998). *Nucl. Phys. Proc. Suppl.*, **60A**, 257–266.

Niedermayer, Ferenc (1999). *Nucl. Phys. Proc. Suppl.*, **73**, 105–119.

Ohki, H. *et al.* (2009).

Osterwalder, Konrad and Schrader, Robert (1973). *Commun. Math. Phys.*, **31**, 83–112.

Osterwalder, Konrad and Schrader, Robert (1975). *Commun. Math. Phys.*, **42**, 281.

Parasiuk, O. S. (1960). *Ukrainskii Math. Jour.*, **12**, 287.

Parisi, G. (1985). *Nucl. Phys.*, **B254**, 58–70.

Patrascioiu, Adrian and Seiler, Erhard (2000). *hep-th/0002153*.

Pena, Carlos, Sint, Stefan, and Vladikas, Anastassios (2004). *JHEP*, **09**, 069.

Polchinski, Joseph (1984). *Nucl. Phys.*, **B231**, 269–295.

Politzer, H. David (1973). *Phys. Rev. Lett.*, **30**, 1346–1349.

Reisz, T. (1988*a*). *Commun. Math. Phys.*, **116**, 573.

Reisz, Thomas (1988*b*). *Commun. Math. Phys.*, **116**, 81.

Reisz, T. (1988*c*). *Commun. Math. Phys.*, **117**, 639.

Reisz, T. (1988*d*). MPI-PAE/PTh-79/88.

Reisz, T. (1989). *Nucl. Phys.*, **B318**, 417.

Reisz, T. and Rothe, H. J. (2000). *Nucl. Phys.*, **B575**, 255–266.

Rossi, G. C. and Testa, M. (1980*a*). *Nucl. Phys.*, **B163**, 109.

Rossi, G. C. and Testa, M. (1980*b*). *Nucl. Phys.*, **B176**, 477.

Salam, Abdus (1951*a*). *Phys. Rev.*, **84**, 426–431.

Salam, Abdus (1951*b*). *Phys. Rev.*, **82**, 217–227.

Salmhofer, M. (1999). Berlin, Germany, Springer (1999) p. 231.

Seiler, E. (1975). In Erice 1975, Proceedings, Renormalization Theory, Dordrecht 1976, 415–433.

Sharatchandra, H. S. (1978). *Phys. Rev.*, **D18**, 2042.

Sharpe, Stephen R. and Wu, Jackson M. S. (2005). *Phys. Rev.*, **D71**, 074501.

Sheikholeslami, B. and Wohlert, R. (1985). *Nucl. Phys.*, **B259**, 572.

Shindler, Andrea (2006). *PoS*, **LAT2005**, 014.

Shindler, A. (2008). *Phys. Rept.*, **461**, 37–110.

Sint, Stefan (1994). *Nucl. Phys.*, **B421**, 135–158.

Sint, Stefan (1995). *Nucl. Phys.*, **B451**, 416–444.

Sint, Stefan (2006). *PoS*, **LAT2005**, 235.

Sint, Stefan (2007*a*). Lectures given at Workshop on Perspectives in Lattice QCD, Nara, Japan, 31 Oct – 11 Nov 2005, hep-lat/0702008.

Sint, Stefan (2007*b*). *PoS*, **LAT2007**, 253.

Sint, Stefan and Sommer, Rainer (1996). *Nucl. Phys.*, **B465**, 71–98.

Slavnov, A. A. (1972). *Theor. Math. Phys.*, **10**, 99–107.

Sommer, R. (1994). *Nucl. Phys.*, **B411**, 839–854.

Streater, R. F. and Wightman, A. S. (1989). Redwood City, USA, Addison-Wesley (1989) 207 p. (Advanced book classics).

Stueckelberg, E. C. G. and Green, T. A. (1951). *Helv. Phys. Acta.*, **24**, 153.

Sturm, C. et al. (2009). *Phys. Rev.*, **D80**, 014501.

Symanzik, K. (1961). pp 485–517 in Lecture on High Energy Physics II, Hecegnovi, 1961, Ed. B. Jakśić.

Symanzik, K. (1970). *Commun. Math. Phys.*, **18**, 227–246.

Symanzik, K. (1971). *Commun. Math. Phys.*, **23**, 49.

Symanzik, K. (1979). DESY 79/76 (Cargèse lecture, 1979).

Symanzik, K. (1981*a*). *Nucl. Phys.*, **B190**, 1.

Symanzik, K. (1981*b*). in Mathematical problems in theoretical physics, eds. R. Schrader, R. Seiler, D. A. Uhlenbrock, Springer Lecture Notes in Physics, Berlin, Heidelberg, New York Vol. 153 (1982) 47.

't Hooft, Gerard (1971). *Nucl. Phys.*, **B33**, 173–199.

't Hooft, Gerard and Veltman, M. J. G. (1972). *Nucl. Phys.*, **B44**, 189–213.

Takahashi, Y. (1957). *Nuovo Cim.*, **6**, 371.

Takeda, Shinji (2008). *PoS*, **LATTICE2008**, 218.

Taylor, J. C. (1971). *Nucl. Phys.*, **B33**, 436–444.

Tekin, Fatih, Sommer, Rainer, and Wolff, Ulli (2010). *Phys. Lett.*, **B683**, 75–79.

Ward, John Clive (1950). *Phys. Rev.*, **78**, 182.

Ward, John Clive (1951). *Proc. Phys. Soc. Lond.*, **A64**, 54.

Weinberg, Steven (1960). *Phys. Rev.*, **118**, 838–849.

Weisz, P. (1983). *Nucl. Phys.*, **B212**, 1.

Wightman, A. S. (1956). *Phys. Rev.*, **101**, 860–866.

Wightman, A. S. (1975). In Erice 1975, Proceedings, Renormalization Theory, Dordrecht 1976, 1–24.

Wilson, Kenneth G. (1970). *Phys. Rev.*, **D2**, 1473.

Wilson, Kenneth G. (1971*a*). *Phys. Rev.*, **B4**, 3174–3183.

Wilson, Kenneth G. (1971*b*). *Phys. Rev.*, **B4**, 3184–3205.

Wilson, K. G. and Kogut, John B. (1974). *Phys. Rept.*, **12**, 75–200.

Wilson, K. G. and Zimmermann, W. (1972). *Commun. Math. Phys.*, **24**, 87–106.

Wolff, Ulli, Knechtli, Francesco, Leder, Bjorn, and Balog, Janos (2006). *PoS*, **LAT2005**, 253.

Wu, Tai Tsun (1962). *Phys. Rev.*, **125**, 1436–1450.

Wu, Tai Tsun, McCoy, Barry M., Tracy, Craig A., and Barouch, Eytan (1976). *Phys. Rev.*, **B13**, 316–374.

Zimmermann, W. (1968). *Commun. Math. Phys.*, **11**, 1–8.

Zimmermann, Wolfhart (1973*a*). *Ann. Phys.*, **77**, 536–569.

Zimmermann, Wolfhart (1973*b*). *Ann. Phys.*, **77**, 570–601.

Zinn-Justin, Jean. Lectures given at Int. Summer Inst. for Theoretical Physics, Jul 29–Aug 9, 1974, Bonn, West Germany.

Zinn-Justin, Jean (2002). *Int. Ser. Monogr. Phys.*, **113**, 1–1054.

3

Three topics in renormalization and improvement

Anastassios VLADIKAS

INFN - "Tor Vergata",
c/o Department of Physics,
University of Rome "Tor Vergata",
Via della Ricerca Scientifica 1,
I-00133 Rome, Italy

To Paulette, Michael and Sofia

3.1 Introduction

The important topic of lattice renormalization and improvement, covered by P. Weisz in this School, is quite complicated, even for students who have considerable expertise in calculations based on continuum regularization (e.g. Pauli-Villars, dimensional regularization). The reader is advised to familiarize himself with the basic concepts, before tackling the present chapter, which is meant to be a complement to Chapter 2 of this book (Weisz, 2010). Our aim is to present three very specific topics:

- The consequences of the loss of chiral symmetry in the Wilson lattice regularization of the fermionic action and its recovery in the continuum limit. The treatment of these arguments involves lattice Ward identities.

- The definition and properties of mass-independent renormalization schemes, which are suitable for a non-perturbative computation of various operator renormalization constants.

- The modification of the Wilson fermion action, by the introduction of a chirally twisted mass term (known as twisted mass QCD and abbreviated as tmQCD), which results to improved (re)normalization and scaling properties for physical quantities of interest.

Loss of chiral symmetry implies that the quark mass is not multiplicatively renormalizable, having also an additive counterterm, proportional to the ultraviolet cutoff (i.e. the inverse lattice spacing). Moreover, the no-renormalization theorem of the partially conserved vector and axial currents (PCVC and PCAC) turns out to be somewhat intricate. Wilson fermions preserve flavor (vector) symmetry, with an exactly conserved current, which is a point-split discretization of the corresponding continuum operator. The vector local current $V_\mu(x)$, familiar to us from continuum QCD, is to be normalized by a scale-independent normalization factor $Z_V(g_0) \neq 1$, which is a function of the bare coupling g_0, tending to unity in the continuum limit. On the other hand, axial symmetry is broken by the Wilson term. Thus, *any* discretization of the axial current on the lattice is accompanied by its scale-independent normalization, which tends to unity in the continuum limit. For instance, the familiar axial current $A_\mu(x)$ has a normalization $Z_A(g_0) \neq 1$. Another consequence of the loss of chiral symmetry is that operators belonging to the same chiral multiplet do not renormalize by the same renormalization factor. Although they all have identical anomalous dimensions (i.e. the same scale dependence), ratios of their renormalization constants are scale-independent quantities that become unity only in the continuum limit. These properties are established with the aid of Ward identities. The very same Ward identities may also be used in practical computations for the non-perturbative determination of these scale-invariant current normalizations and ratios of renormalization constants, at fixed lattice spacing.

It is a well-established fact that lattice perturbation theory is characterized by poor convergence. Renormalization constants calculated perturbatively are therefore sources of rather large systematic errors in lattice predictions and postdictions of many interesting physical quantities. The second topic of this chapter is the discussion of a family of renormalization schemes (RI/MOM, RI/SMOM) that are used for the non-pertubative renormalization of fermionic composite operators. The two main sources of systematic error (unwanted low-energy effects in the infrared and discretization effects in the ultraviolet) necessitate the existence of a "renormalization window", in which these errors are controlled. The infrared problems due to a Goldstone pole in some renormalization constants is discussed in considerable detail. Although our presentation is based on Wilson fermions, these renormalization methods are equally well applied in other lattice regularizations, such as domain-wall fermions.

The last topic concerns the benevolent effects of tmQCD in the renormalization and improvement properties of physical quantities. For this modified Wilson fermion action the Wilson term (responsible for a hard breaking of chiral symmetry) and the twisted mass term (which breaks the symmetry softly) are in some sense orthogonal in chiral space. This results in some fermion operators having simpler renormalization patterns in tmQCD than in the standard Wilson lattice theory. In these circumstances, the computation of their matrix elements is under better control in the tmQCD framework. Moreover, discrete symmetries in tmQCD imply that a particular tuning of the twisted mass action results in $\mathcal{O}(a)$ improvement of physical quantities, without the need for Symanzik counterterms. This is the so-called "automatic improvement" of Wilson fermions.

In these lecture notes we have striven to give explicit derivations of the most important results. Having opted for a detailed treatment of the theoretical issues, we have not addressed the many interesting numerical results, which have appeared in the literature over the years. Moreover, since these are meant to be paedagogical lectures addressing this advanced subject at an introductory level, we have only included the essential references that would facilitate the student's further study. This inevitably introduces some bias, for which we apologize.

3.2 Basics

Although we understand that the student is familiar with the basics of quark mass and composite operator renormalization, we recapitulate them here for completeness and in order to fix our notation. We also collect several useful definitions; although this is somewhat tedious, it is important to spell out the not-so-standard notation right from the beginning.

Concerning notation, we have preferred economy to mathematical rigor. Since the various bare quantities discussed below are defined in the lattice regularization, integrals like say, $\int d^4x_1 d^4x_2$ of Eq. (3.28), are really sums ($a^8 \sum_{x_1,x_2}$), which run over all lattice sites, labelled by x_1 and x_2, etc. The occasional use of integrals instead of sums, partial derivatives instead of finite differences, and Dirac functions instead of Kronecker symbols, simplifies notation, hopefully avoiding any confusion. Moreover, a space-time function and its Fourier transform will be indicated by the same symbol with a different argument (e.g. $f(x)$ and $f(p)$); again mathematical rigor is being sacrificed in favor of notational economy.

3.2.1   The Wilson lattice action and its symmetries

We opt for the lattice regularization scheme, proposed by Wilson, which consists of a gluonic action[1] (Wilson, 1974),

$$S_G = \frac{6}{g_0^2} \sum_P \left[1 - \frac{1}{6} \mathrm{Tr}\left[U_P + U_P^\dagger \right] \right] \tag{3.1}$$

and a fermionic one (Wilson, 1977),

$$S_F = a^4 \sum_x \bar\psi(x) \left[\frac{1}{2} \sum_\mu \left\{ \gamma_\mu (\vec{\mathcal{D}}_\mu^* + \vec{\mathcal{D}}_\mu) - a\, \vec{\mathcal{D}}_\mu^* \vec{\mathcal{D}}_\mu \right\} + M_0 \right] \psi(x)$$

$$= -a^4 \sum_{x,\mu} \frac{1}{2a} \left[\bar\psi(x)(1 - \gamma_\mu) U_\mu(x)\psi(x + a\hat\mu) + \bar\psi(x + a\hat\mu)(1 + \gamma_\mu) U_\mu^\dagger(x)\psi(x) \right]$$

$$+ a^4 \sum_x \bar\psi(x) \left(M_0 + \frac{4}{a} \right) \psi(x). \tag{3.2}$$

In standard notation, $U_\mu(x) \equiv \exp[i g_0 a G_\mu(x)]$ is the lattice gauge link, depending on the gauge (gluon) field $G_\mu(x)$, g_0 is the bare coupling constant, a the lattice spacing (i.e. our UV cutoff) and U_P the Wilson oriented plaquette. Two discretizations of the covariant derivative are used in the above definition:

$$a\, \vec{\mathcal{D}}_\mu\, \psi(x) = U_\mu(x)\psi(x + a\hat\mu) - \psi(x) \tag{3.3}$$

$$a\, \vec{\mathcal{D}}_\mu^*\, \psi(x) = \psi(x) - U_\mu^\dagger(x - a\hat\mu)\psi(x - a\hat\mu). \tag{3.4}$$

Later, we will also make use of the backward covariant derivative, defined as:

$$a\bar\psi(x) \overleftarrow{\mathcal{D}}_\mu = \bar\psi(x + a\hat\mu)U_\mu^\dagger(x) - \bar\psi(x). \tag{3.5}$$

The quark field $\psi(x)$ is a vector in flavor space. Its components are denoted by ψ_f ($f = 1, 2, 3, \ldots, N_F$). The diagonal bare mass matrix is denoted by M_0 and its elements by m_{0f} ($f = 1, \ldots, N_F$). Two continuous symmetries of particular interest are the flavor and chiral symmetries. The flavor group $SU(N_F)_V$ consists in the global vector transformations of the fermion field:

$$\psi(x) \to \psi'(x) = \exp\left[i\,\alpha_V^a \frac{\lambda^a}{2} \right] \psi(x),$$

$$\bar\psi(x) \to \bar\psi'(x) = \bar\psi(x) \exp\left[-i\,\alpha_V^a \frac{\lambda^a}{2} \right], \tag{3.6}$$

with $a = 1, \cdots, N_F^2 - 1$. The group generators in the fundamental representation are the flavor matrices $\lambda^a/2$ ($a = 1, \ldots, N_F^2 - 1$), satisfying:

$$\text{Tr}[\lambda^a \lambda^b] = 2\delta^{ab},$$

$$\left[\frac{\lambda^a}{2}, \frac{\lambda^b}{2}\right] = i f^{abc} \frac{\lambda^c}{2},$$

$$\left\{\frac{\lambda^a}{2}, \frac{\lambda^b}{2}\right\} = d^{abc} \frac{\lambda^c}{2} + \frac{\delta^{ab}}{N_F} I, \tag{3.7}$$

where I represents the $N_F \times N_F$ unit matrix in flavor space; f^{abc} are the $SU(N_F)$ structure constants and d^{abc} are totally symmetric. For $N_F = 3$ the generators λ^a are the eight Gell-Mann matrices, while for $N_F = 2$ they are the three Pauli matrices. Under these transformations the Wilson lattice action (3.2) is invariant, provided all quark masses are degenerate ($m_{01} = \cdots = m_{0N_F}$). Vector (or flavor) symmetry is preserved by the Wilson lattice regularization.

Next, we consider the axial transformation of the fermion field:

$$\psi(x) \to \psi'(x) = \exp\left[i \alpha_A^a \frac{\lambda^a}{2} \gamma_5\right] \psi(x),$$

$$\bar{\psi}(x) \to \bar{\psi}'(x) = \bar{\psi}(x) \exp\left[i \alpha_A^a \frac{\lambda^a}{2} \gamma_5\right], \tag{3.8}$$

with $a = 1, \cdots, N_F^2 - 1$. Even in the absence of a mass matrix ($M_0 = 0$), the presence of the Wilson term ($a \bar{\psi} \overset{*}{\vec{\mathcal{D}}}_\mu \vec{\mathcal{D}}_\mu \psi$) in the action (3.2) ensures that it is not invariant under these transformations. Consequently, the chiral group $SU(N_F)_L \otimes SU(N_F)_R$, related to axial and vector transformations, is not a symmetry of Wilson fermions. An interesting generalization of chiral symmetries is described in Appendix 3.7. For completeness, we also collect the discrete symmetries of the Wilson action in Appendix 3.8.

We close this subsection with a general observation. Many important quantities we will be using, such as the quark propagator and the various correlation functions G_O, Λ_O, Γ_O (to be defined below), are gauge dependent. This implies that some gauge-fixing procedure has been applied in their calculation. The gauge of choice is usually the Landau gauge, which on the lattice is understood as imposing a discrete version of the condition $\partial_\mu G_\mu = 0$ on all lattice gauge fields G_μ. The reader is advised to consult a review of lattice gauge fixing for more details (Giusti *et al.*, 2001).

3.2.2 Quark-mass renormalization

We recapitulate the general renormalization properties of the quark masses with Wilson fermions. It is convenient (Bhattacharya *et al.*, 2006) to separate non-singlet and singlet quark-mass contributions of the bare mass matrix:

$$M_0 = \sum_d \widetilde{m}^d \lambda^d + m^{\text{av}} I. \tag{3.9}$$

The sum runs over diagonal group generators only (e.g. $d = 3$ for $SU(2)_V$ and $d = 3, 8$ for $SU(3)_V$) and m^{av} is the average of the diagonal elements of M_0. Note that in the flavor symmetric theory, where all masses are degenerate, we have $\widetilde{m}^d = 0$ and $M_0 = m^{\mathrm{av}} I$; cf. Eq. (3.195). In terms of the mass decomposition (3.9), the bare mass of a specific quark of flavor f is

$$m_{0f} = m^{\mathrm{av}} + \sum_d \widetilde{m}^d \lambda^d_{ff}, \qquad (3.10)$$

with λ^d_{ff} the fth diagonal entry of the matrix λ^d.

As shown in Appendix 3.7, spurionic flavor symmetry implies that all components $\widetilde{m}^a$ transform as a multiplet in the adjoint representation of $SU(N_F)_V$, while m^{av} is a singlet. Thus, as far as this symmetry is concerned, members of the adjoint multiplet renormalize in the same way, while the singlet m^{av} could renormalize differently. In the same Appendix we have also seen that spurionic axial transformations mix flavor non-singlets $\widetilde{m}^a$ and singlets m^{av}. If that were a symmetry of the theory, then $\widetilde{m}^a$ and m^{av}, being members of the same chiral multiplet would share a common renormalization constant. This is not the case, however, as the Wilson term of the lattice action breaks chiral spurionic symmetry. Moreover, flavor symmetry also suggests that the $\widetilde{m}^d$ components renormalize multiplicatively, while the lack of chiral symmetry allows m^{av} to mix with the identity. Thus, the singlet mass m^{av} is also subject to additive renormalization. The bottom line is the following renormalization pattern:

$$\left[\widetilde{m}^d(g_{\mathrm{R}}) \right]_{\mathrm{R}} = \lim_{a \to 0} Z_m(g_0, a\mu) \, \widetilde{m}^d(g_0),$$

$$\left[m^{\mathrm{av}}(g_{\mathrm{R}}) \right]_{\mathrm{R}} = \lim_{a \to 0} Z_{m^0}(g_0, a\mu) \left[m^{\mathrm{av}}(g_0) - m_{\mathrm{cr}} \right]. \qquad (3.11)$$

We denote by Z_m, Z_{m^0} the multiplicative renormalization constants and μ the renormalization scale; $g_{\mathrm{R}}(\mu)$ is the renormalized gauge coupling. One must also keep in mind the dependence of the bare coupling on the lattice spacing; $g_0^2(a) \sim 1/\ln(a)$. The additive renormalization factor $m_{\mathrm{cr}} = [C(g_0)/a]$, with $C(g_0)$ a coefficient bearing the dependence on the bare coupling, is a flavor independent, linearly divergent counterterm. At tree level, $m_{\mathrm{cr}} = -4/a$. This counterterm is a well-known feature of Wilson fermions. As already pointed out, it is a consequence of the loss of chiral symmetry by this regularization (even when the bare quark masses are switched off). The fact that there are two mass-renormalization constants, $Z_m \neq Z_{m^0}$, albeit with the same scale-μ dependence, is also due to the loss of chiral symmetry. We will establish useful relations between them in section 3.3.

It is common practice to omit, in expressions like (3.11), the continuum limit on the r.h.s.. The resulting equations relate quantities at fixed lattice spacing and are true up to discretization effects.

Given that Eq. (3.10) relates the bare mass with flavor f to the singlet and non-singlet bare masses, we *define* the renormalized mass of this quark as the same combination of the renormalized singlet and non-singlet masses (Bhattacharya *et al.*, 2006):

$$[m_f]_{\mathrm{R}} = [m^{\mathrm{av}}]_{\mathrm{R}} + \sum_d \widetilde{[\tilde{m}^d]}_{\mathrm{R}} \lambda^d_{ff}$$

$$= Z_{m^0}[m^{\mathrm{av}}(g_0) - m_{\mathrm{cr}}] + Z_m \sum_d \widetilde{\tilde{m}^d} \lambda^d_{ff}$$

$$= Z_{m^0}[m^{\mathrm{av}}(g_0) - m_{\mathrm{cr}}] + Z_m[m_{0f} - m^{\mathrm{av}}]. \tag{3.12}$$

The second line is obtained by using Eqs. (3.11). The third line is obtained from the second, by substituting the sum in its last term from Eq. (3.10). This result can also be conveniently rewritten as

$$[m_f]_{\mathrm{R}} = Z_m \left[m_{0f} - m_{\mathrm{cr}} + \left(\frac{Z_{m^0}}{Z_m} - 1 \right)(m^{\mathrm{av}} - m_{\mathrm{cr}}) \right]. \tag{3.13}$$

Note that it is not enough to take $m_{0f} \to m_{\mathrm{cr}}$ in order to obtain the chiral limit $[m_f]_{\mathrm{R}} \to 0$; this is achieved only when *all bare quark masses* $m_{01}, \cdots m_{0N_F}$ tend to m_{cr}. Also, note that if the regularization scheme were chirally invariant (it isn't!) then m_{cr} would be absent and $Z_m = Z_{m^0}$; in this case the quark mass would be multiplicatively renormalizable ($[m_f]_{\mathrm{R}} = Z_m m_{0f}$), for each flavor.

It is instructive to consider the case of degenerate masses ($\tilde{m}^d = 0$ and $m^{\mathrm{av}} = m_{0f}$). Then, Eq. (3.13) reduces to the more familiar

$$[m_f(g_{\mathrm{R}})]_{\mathrm{R}} = \lim_{a \to 0} [Z_{m^0}(g_0, a\mu) \, m_f(g_0, m_{0f})], \tag{3.14}$$

where the subtracted quark mass is defined as

$$m_f(g_0, m_{0f}) = m_{0f}(g_0) - m_{\mathrm{cr}}. \tag{3.15}$$

The chiral limit is now simply $m_f \to 0$; i.e. $m_{0f} \to m_{\mathrm{cr}}$. In the mass-degenerate case, or whenever flavor dependence is unimportant, the flavor index will be dropped, leaving us with m_0, m, m_{R}, etc.

We also point out that combining Eq. (3.13) for two flavors, say $f = 1, 2$ we obtain

$$[m_1]_{\mathrm{R}} - [m_2]_{\mathrm{R}} = Z_m \left[m_{01} - m_{02} \right]$$

$$= Z_m \left[m_1 - m_2 \right]; \tag{3.16}$$

i.e. bare (and subtracted) mass differences renormalize with Z_m rather than Z_{m^0}. As already anticipated, the two renormalization constants Z_m and Z_{m^0} have the same scale dependence. We will show this in Section 3.3, and also prove that their ratio is a finite function of the bare coupling g_0, which tends to unity in the continuum limit.

3.2.3 Quark propagator

The bare quark propagator in coordinate space, $\mathcal{S}_f(x_1 - x_2; m) = \langle \psi_f(x_1) \bar{\psi}_f(x_2) \rangle$, has the Fourier transform

$$\mathcal{S}(p; m) = \int d^4 x \exp(-ipx) \mathcal{S}(x, m), \tag{3.17}$$

where the flavor index f has been dropped (it will reappear wherever necessary). The renormalized propagator is then given by

$$[\mathcal{S}(p; g_{\mathrm{R}}, m_{\mathrm{R}}, \mu)]_{\mathrm{R}} = \lim_{a \to 0} [Z_\psi(g_0, a\mu)\, \mathcal{S}(p; g_0, m)], \tag{3.18}$$

where $Z_\psi^{1/2}$ is the wave-function renormalization; $\psi_{\mathrm{R}} = Z_\psi^{1/2}\psi$. Note that in the bare quantities, the bare mass m_0 has been traded off for the simpler (from the point of view of the chiral limit) subtracted mass m. For later convenience, we define two "projections" (i.e. traces in spin and colour space) of $\mathcal{S}^{-1}(p, m)$ and its derivative with respect to momentum:

$$\Gamma_\Sigma(p; m) = \frac{-i}{48}\ \mathrm{Tr}\left[\gamma_\mu \frac{\partial \mathcal{S}^{-1}(p; m)}{\partial p_\mu}\right], \tag{3.19}$$

$$\Gamma_m(p; m) = \frac{1}{12}\ \mathrm{Tr}\left[\frac{\partial \mathcal{S}^{-1}(p; m)}{\partial m}\right]. \tag{3.20}$$

These quantities have been defined so that their tree-level values are equal to unity. From Eqs (3.14) and (3.16) we see that $\Gamma_\Sigma(p)$ and $\Gamma_m(p)$ renormalize like

$$[\Gamma_\Sigma(p)]_{\mathrm{R}} = \lim_{a \to 0}\left[Z_\psi^{-1}\Gamma_\Sigma(p, m)\right], \tag{3.21}$$

$$[\Gamma_m(p)]_{\mathrm{R}} = \lim_{a \to 0}\left[Z_\psi^{-1}Z_m^{-1}\Gamma_m(p, m)\right]. \tag{3.22}$$

It is natural to fix the quark-field and quark-mass renormalization Z_ψ and Z_m by imposing renormalization conditions on $[\Gamma_\Sigma(p)]_{\mathrm{R}}$ and $[\Gamma_m(p)]_{\mathrm{R}}$ at a given momentum $p = \mu$. Such a scheme will be the subject of Section 3.4. It is clearly not a unique choice.

3.2.4   Quark bilinear operators (composite fields)

In this chapter we will make extensive use of local bilinear quark operators of the form:

$$O_\Gamma^f(x) = \bar\psi(x)\Gamma\frac{\lambda^f}{2}\psi(x), \tag{3.23}$$

where Γ stands for a generic Dirac matrix and $f = 1, \cdots, N_F^2 - 1$ (flavor non-singlet case). Specific bilinear operators will be denoted according to their Lorentz group transformations: the scalar, pseudoscalar and tensor densities are

$$S^f(x) = \bar\psi(x)\frac{\lambda^f}{2}\psi(x), \qquad P^f(x) = \bar\psi(x)\frac{\lambda^f}{2}\gamma_5\psi(x),$$

$$T_{\mu\nu}^f(x) = \bar\psi(x)\frac{\lambda^f}{2}\gamma_\mu\gamma_\nu\psi(x), \tag{3.24}$$

respectively, whereas the vector and axial vector currents are

$$V_\mu^f(x) = \bar\psi(x)\frac{\lambda^f}{2}\gamma_\mu\psi(x), \qquad A_\mu^f(x) = \bar\psi(x)\frac{\lambda^f}{2}\gamma_\mu\gamma_5\psi(x). \qquad (3.25)$$

In the above, implicit color and spin indices are contracted. Besides the non-singlet bilinear quark operators defined in Eqs. (3.24) and (3.25), we will also be using the singlet ones

$$S^0(x) = \bar\psi(x)\frac{\lambda^0}{2}\psi(x), \qquad\qquad P^0(x) = \bar\psi(x)\frac{\lambda^0}{2}\gamma_5\psi(x), \qquad (3.26)$$

with $\lambda^0 \equiv \sqrt{2/N_F}\,I$.

We will sometimes simplify the notation by using the symbols S, V_μ, etc. (dropping the flavor superscript f). These are meant to be non-singlet operators of the form $O_\Gamma = \bar\psi_1\Gamma\psi_2$. For example if we substitute λ^f with $\lambda^+ = \lambda_1 + i\lambda_2$ in Eqs. (3.24) and (3.25), we end up with, say, $\bar u d$, $\bar u\gamma_\mu d$, etc. The renormalization and improvement of four-fermion operators, although very important, is beyond the scope of the present chapter.

The insertion of the operator $O_\Gamma = \bar\psi_1\Gamma\psi_2$ in the 2-point fermionic Green function gives the vertex function

$$G_O(x_1 - x, x_2 - x) = \langle \psi_1(x_1)\, O_\Gamma(x)\bar\psi_2(x_2)\rangle, \qquad (3.27)$$

with translational invariance explicitly taken into account in the notation. Placing, for simplicity, the operator at the origin $x = 0$, the above correlation function becomes, in momentum space, the vertex function

$$G_O(p_1, p_2) = \int d^4x_1 d^4x_2 \exp(-ip_1x_1)\exp(ip_2x_2)\, G_O(x_1, x_2). \qquad (3.28)$$

The corresponding amputated Green function is given by

$$\Lambda_O(p_1, p_2) = \mathcal{S}_1^{-1}(p_1)\, G_O(p_1, p_2)\, \mathcal{S}_2^{-1}(p_2), \qquad (3.29)$$

where the propagator subscripts denote flavor. Note that G_O and Λ_O are rank-2 tensors in color and spin space; the color and spin indices are suppressed in the notation. Since it is preferable to work with scalar, rather than tensor-like correlation functions, we define the projected amputated Green function $\Gamma_O(p)$ as follows:

$$\Gamma_O(p) = \frac{1}{12}\mathrm{Tr}\,[P_O\Lambda_O(p, p)]. \qquad (3.30)$$

The trace is over spin and color indices. In some cases Lorentz indices need to be added; e.g. the axial current A_μ has vertex functions G_A^μ, Γ_A^μ, etc. Thus, with the normalization factor $1/12$, P_O is the Dirac matrix that renders the tree-level value of $\Gamma_O(p)$ equal to unity; i.e. it projects out the nominal Dirac structure of the Green function $\Lambda_O(p)$. It is easy to work out these so-called "projectors" for each of the operators of Eqs (3.24) and (3.25):

$$P_S = I; \qquad P_P = \gamma_5; \qquad P_T^{\mu\nu} = \frac{1}{12}\gamma_\mu\gamma_\nu$$

$$P_V^\mu = \frac{1}{4}\gamma_\mu; \qquad\qquad P_A^\mu = \frac{1}{4}\gamma_5\gamma_\mu. \qquad\qquad (3.31)$$

Repeated Lorenz indices, appearing in the definition of these projectors and the corresponding Λ_{OS} are meant to be summed over. For simplicity of notation, specific Green functions will be denoted, for instance, as G_S, Γ_P, Λ_V, etc.

The dimension-3 bilinear operators S^f, P^f and $T_{\mu\nu}^f$ (with $a = f, \ldots, N_F^2 - 1$), are non-singlet in flavor space. As there are no other operators of equal or smaller dimension with the same quantum numbers, they renormalize multiplicatively; their renormalization constants are denoted as Z_S, Z_P and Z_T. Thus, the renormalized operator is formally given by

$$[O_\Gamma(g_R, m_R, \mu)]_R = \lim_{a \to 0} [Z_O(g_0, a\mu)O_\Gamma(g_0, m)]. \qquad (3.32)$$

For the currents V_μ^f and A_μ^f there is a lattice version of the no-renormalization theorem, which will be the subject of the following sections on Ward identities. Essentially, we will show in Section 3.3 that these currents, once normalized by appropriate finite, scale-independent factors Z_V and Z_A, go over to their continuum limit up to discretization effects that are proportional to positive powers of the lattice spacing. We then write the continuum currents as:

$$[O_{V,A}(g_R, m_R)]_R = \lim_{a \to 0} [Z_{V,A}(g_0)O_{V,A}(g_0, m)]. \qquad (3.33)$$

Combining the above wih Eqs. (3.18) and (3.27) – (3.29), we find for the renormalized Green functions:

$$[G_O(p_1, p_2)]_R = \lim_{a \to 0} [Z_\psi Z_O G_O(p_1, p_2)], \qquad (3.34)$$

$$[\Lambda_O(p_1, p_2)]_R = \lim_{a \to 0} [Z_\psi^{-1} Z_O \Lambda_O(p_1, p_2)]. \qquad (3.35)$$

The renormalization of $\Gamma_O(p)$ is identical to that of $\Lambda_O(p)$:

$$[\Gamma_O(p)]_R = \lim_{a \to 0} \left[Z_\psi^{-1} Z_O \Gamma_O(p) \right]. \qquad (3.36)$$

The singlet pseudoscalar density P^0 renormalizes multiplicatively in a similar fashion; its renormalization constant is denoted by Z_{P^0}. The singlet scalar density S^0, besides its multiplicative renormalization constant Z_{S^0}, also requires a power subtraction, as we will discuss in Section 3.3.6.

The above definitions of renormalization constants are purely formal. By this we mean that some arbitrary (still unspecified), choice of renormalization scheme and scale is implied. No reference to a specific scheme has been made so far. However, it is implicit in the notation that only mass-independent schemes are taken into consideration (otherwise Z_m, Z_ψ and Z_O would also depend on quark masses). In

Section 3.4.1 we will discuss a specific mass-independent renormalization scheme, known as RI/MOM (Martinelli *et al.*, 1995).

3.3 Lattice Ward identities

Ward identities are relations between Green functions; essentially they express how a classical continuum symmetry is realized at the quantum level. Since lattice actions break chiral symmetry[2] the resulting Ward identities are not trivial discretized transcriptions of the formal continuum ones. In fact, they dictate the proper normalization of (partially) conserved currents on the lattice and relate the renormalization parameters of quantities that in the continuum belong to the same chiral multiplet, such as the scalar and pseudoscalar densities. In this way, the question of recovering chiral symmetry in the continuum limit (up to discretization effects generically denoted here as $\mathcal{O}(a)$-effects) is linked to the appropriate (re)normalization of lattice operators in the lattice regularization. In particular, we will discuss the following topics in detail:

- the conservation of vector symmetry $SU(N_F)_V$ on the lattice with Wilson fermions;

- the loss of chiral $SU(N_F)_L \otimes SU(N_F)_R$ symmetry on the lattice, even in the chiral limit, and its recovery in the continuum limit;

- the derivation of vector and axial lattice Ward identities and the finite normalization of the vector and axial currents, resulting from them;

- the relations between quark-mass renormalization and the renormalization of the scalar and pseudoscalar densities, arising from lattice Ward identities.

Let us first recall that at the classical level, Ward identities connect the four-divergence of vector and axial currents to the scalar and pseudoscalar densities. The so-called "partially conserved vector current" (PCVC) relation

$$\partial_\mu V_\mu^f(x) + \bar{\psi}(x)\left[\frac{\lambda^f}{2}, M_0\right]\psi(x) = 0, \tag{3.37}$$

results, through the standard Noether construction, from the invariance of the QCD (Euclidean) fermionic action under $SU(N_F)_V$ transformations. Vector current conservation holds *in the degenerate mass case* $(M_0 \propto I)$. The "partially conserved axial current" (PCAC) relation

$$\partial_\mu A_\mu^f(x) - \bar{\psi}(x)\left\{\frac{\lambda^f}{2}, M_0\right\}\psi(x) = 0, \tag{3.38}$$

results from the invariance of the QCD action under axial transformations. Axial current conservation holds *in the chiral limit* $(M_0 = 0)$.[3] Similar, possibly more familiar, expressions are obtained in the specific case with λ^a replaced by $\lambda^+ \equiv \lambda^1 + i\lambda^2$. This involves transformations on two flavors only, ψ_1 and ψ_2 (with corresponding masses m_1 and m_2). With this choice, the quantities $S, P, T_{\mu\nu}, V_\mu, A_\mu$ of Eqs. (3.24) and (3.25) have the form $O_\Gamma = \bar{\psi}_1 \Gamma \psi_2$ and the above Ward identities become

$$\partial_\mu V_\mu(x) = (m_1 - m_2)S(x), \tag{3.39}$$

$$\partial_\mu A_\mu(x) = (m_1 + m_2)P(x). \tag{3.40}$$

Beyond the classical level, these Ward identities are to be understood as statements between operators. More precisely, they are insertions in expectation values of multilocal operators $O(x_1, \ldots, x_n)$, consisting of a product of quark and gluon fields at different space-time points $(x_1 \neq x_2 \neq \cdots \neq x_n)$. In the bare (lattice) theory these expectation values are defined in the path-integral formalism as

$$\langle O(x_1, \ldots, x_n) \rangle = \frac{1}{Z} \int [\mathcal{D}U][\mathcal{D}\psi][\mathcal{D}\bar{\psi}] O(x_1, \ldots, x_n) \exp[-S_G - S_F]. \tag{3.41}$$

Lattice Ward identities are obtained in a way analogous to the continuum case. We recall that the basic trick is to consider, instead of the usual *global* $SU(N_F)_L \otimes SU(N_F)_R$ chiral transformations, infinitesimal *local* vector and axial transformations of the fermionic fields; these are

$$\delta\psi(x) = i\left[\delta\alpha_V^a(x)\frac{\lambda^a}{2}\right]\psi(x) \quad ; \quad \delta\bar{\psi}(x) = -i\bar{\psi}(x)\left[\delta\alpha_V^a(x)\frac{\lambda^a}{2}\right] \tag{3.42}$$

and

$$\delta\psi(x) = i\left[\delta\alpha_A^a(x)\frac{\lambda^a}{2}\gamma_5\right]\psi(x) \quad ; \quad \delta\bar{\psi}(x) = i\bar{\psi}(x)\left[\delta\alpha_A^a(x)\frac{\lambda^a}{2}\gamma_5\right], \tag{3.43}$$

respectively. The operator expectation value of Eq. (3.41) is invariant under any change of the fermionic fields ψ and $\bar{\psi}$, since they are integration variables. This implies that

$$\frac{\delta}{\delta\alpha^a(x)}\langle O(x_1, \ldots, x_n) \rangle = 0, \tag{3.44}$$

where $\delta\alpha^a$ may be either $\delta\alpha_V^a$ or $\delta\alpha_A^a$, depending on the transformation under consideration (vector or axial). The above expression leads to the relation

$$\left\langle \frac{\delta O(x_1, \ldots, x_n)}{\delta\alpha^a(x)} \right\rangle = \left\langle O(x_1, \ldots, x_n)\frac{\delta S_F}{\delta\alpha^a(x)} \right\rangle. \tag{3.45}$$

This is the quantum field theoretic expression leading to a Ward identity: for vector transformations $\delta\alpha_V^a$, the variation of the fermionic action on the r.h.s. corresponds to the PCVC relation of Eq. (3.37), while for axial transformations $\delta\alpha_V^a$, it gives rise to the PCAC relation of Eq. (3.38). The exact meaning of this statement and the subtleties that accompany it in the case of Wilson fermions is the subject of the rest of this section.

3.3.1   Lattice vector Ward identities

It is easy to check that the *global* vector transformations corresponding to those of Eq. (3.42) are a symmetry of the action (3.2) when quarks are degenerate in mass (i.e. M_0 is proportional to the unit matrix). In common lore, "vector symmetry is a lattice symmetry". From the *local* transformations of Eq. (3.42) the following vector Ward identity can be derived (Karsten and Smit, 1981):

$$
i\left\langle \frac{\delta O(x_1, \ldots, x_n)}{\delta \alpha_V^a(x)} \right\rangle = a^4 \sum_\mu \nabla_x^\mu \left\langle \widetilde{V}_\mu^a(x) O(x_1, \ldots, x_n) \right\rangle
$$

$$
+ a^4 \left\langle \bar\psi(x) \left[\frac{\lambda^a}{2}, M_0 \right] \psi(x) O(x_1, \ldots, x_n) \right\rangle, \tag{3.46}
$$

where $a\nabla_x^\mu f(x) = (f(x) - f(x - \mu))$ is an asymmetric lattice derivative, and

$$
\widetilde{V}_\mu^a(x) = \frac{1}{2}\left[\bar\psi(x)(\gamma_\mu - 1) U_\mu(x) \frac{\lambda^a}{2} \psi(x + a\hat\mu) \right.
$$

$$
\left. + \bar\psi(x + a\hat\mu)(\gamma_\mu + 1) U_\mu^\dagger(x) \frac{\lambda^a}{2} \psi(x) \right] \tag{3.47}
$$

is a point-split vector current. With $x \neq x_1, \ldots, x_n$ and in the limit of degenerate bare quark masses ($M_0 \propto I$), Eq. (3.46) leads to the conservation of the point-split lattice vector current, $\nabla_x^\mu \widetilde{V}_\mu^a(x) = 0$. The conservation of the standard local vector current $V_\mu^a(x) = \bar\psi(x) \frac{\lambda^a}{2} \gamma_\mu \psi(x)$ on the lattice is somewhat more intricate, as we will see below.

Without loss of generality, we again simplify matters by combining two versions of Eq. (3.46), one with flavor index $a = 1$ and one with $a = 2$, into a single Ward identity, involving the raising Gell-Mann matrix λ^+ of $SU(N_F)_V$ and the two flavors ψ_1 and ψ_2 (with corresponding bare masses m_{01} and m_{02}). The vector current and scalar density are then bilinear operators of the form $O_\Gamma(x) = \bar\psi_1 \Gamma \psi_2$. We also chose the specific operator $O(x_1, x_2) = \psi_1(x_1)\bar\psi_2(x_2)$. In terms of these quantities the lattice vector Ward identity (3.46) becomes

$$
\nabla_x^\mu \left\langle \psi_1(x_1) \widetilde{V}_\mu(x) \bar\psi_2(x_2) \right\rangle = [m_{01} - m_{02}]\left\langle \psi_1(x_1) S(x) \bar\psi_2(x_2) \right\rangle
$$

$$
- \delta(x_2 - x)\left\langle \psi_1(x_1)\bar\psi_1(x_2) \right\rangle - \delta(x_1 - x)\left\langle \psi_2(x_1)\bar\psi_2(x_2) \right\rangle. \tag{3.48}
$$

Recalling the definition (3.27) of the vertex functions G_O, this is written as

$$
\sum_\mu \nabla_x^\mu G_{\widetilde{V}}^\mu (x_1 - x, x_2 - x; m_{01}, m_{02}) = (m_{01} - m_{02}) G_S (x_1 - x, x_2 - x; m_{01}, m_{02})
$$

$$
+ \delta(x_2 - x) S_1 (x_1 - x_2; m_{01})
$$

$$
- \delta(x_1 - x) S_2 (x_1 - x_2; m_{02}). \tag{3.49}
$$

This expression relates bare quantities that are divergent in the continuum limit. Let us first work in the degenerate mass limit ($m_{01} = m_{02}$), in which the first term on the r.h.s. vanishes. A glance at Eq. (3.18) shows that multiplying the resulting expression by Z_ψ renders the r.h.s. finite. Thus, the l.h.s., which is now $Z_\psi \sum_\mu \nabla_x^\mu G_{\widetilde{V}}^\mu$, is also finite and does not require any further renormalization. Consequently, the point-split conserved vector current (3.47) satisfies the vector Ward identity with its normalization given by the trivial factor

$$Z_{\widetilde{V}} = 1. \tag{3.50}$$

This is the no-renormalization theorem on the lattice with Wilson fermions (Karsten and Smit, 1981; Bochicchio *et al.*, 1985). It is an exact analog of the familiar one in continuum QCD. The only difference is that the lattice vector current satisfying it is not the familiar one $V_\mu = \bar{\psi}_1 \gamma_\mu \psi_2$, but the point-split one defined in Eq. (3.47). This result guarantees a proper definition of the vector charge and the validity of current algebra.

We will now show that the correlation functions of the local vector current $V_\mu(x) = \bar{\psi}_1(x)\gamma_\mu\psi_2(x)$ differ from those of the conserved current by finite contributions, which vanish in the continuum limit. The argument goes as follows: first we express the conserved current as

$$\widetilde{V}_\mu(x) = V_\mu(x) + \frac{a}{2}\left[\bar{\psi}_1(x)(\gamma_\mu - 1)\overrightarrow{\mathcal{D}}_\mu \psi_2(x) + \bar{\psi}_1(x)\overleftarrow{\mathcal{D}}_\mu (\gamma_\mu + 1)\psi_2(x)\right]$$

$$= V_\mu(x) + a\Delta_\mu(x), \tag{3.51}$$

where we have used the lattice asymmetric covariant derivatives of Eqs. (3.3) and (3.5). The second term on the r.h.s. of Eq. (3.51) is a dimension-4 operator Δ_μ. For definitiveness, we now consider the vertex function $G_{\widetilde{V}}(p)$; from Eq. (3.51) it follows that

$$G_{\widetilde{V}}(p) = G_V(p) + a\, G_\Delta(p), \tag{3.52}$$

which we have shown to be finite, up to an overall quark field renormalization Z_ψ; cf. Eqs. (3.34) and (3.50). The term $aG_\Delta(p)$ vanishes at tree level in the continuum limit. Beyond tree level, however, this term contributes, due to the power divergence induced by mixing with a lower-dimensional operator. To see this, write the renormalized dimension-4 operator as

$$[\Delta_\mu]_{\mathrm{R}} = Z_\Delta\left[\Delta_\mu + \frac{c(g_0, am)}{a}V_\mu\right]. \tag{3.53}$$

This renormalization pattern is dictated by dimensional arguments and the quantum numbers of Δ_μ. The dimensionless renormalization constant Z_Δ is at worst logarithmically divergent. The dimensionless mixing coefficient $c(g_0, am)$ depends on the gauge coupling and the bare quark masses (collectively denoted by am). It has been shown that it cannot depend on a renormalization scale $a\mu$ (Testa, 1998). From the point of view of dimensional analysis, $c(g_0, am)$ could contain logarithms of the form $\ln(am)$. However, the absence of such a logarithmic dependence has been explicitly shown for the axial current at all orders in perturbation theory; the situation is analogous for the vector current (Curci, 1986). This is not unexpected, as $\ln(am)$ terms would lead to an ill-defined chiral limit at fixed lattice spacing. Thus, we are left with a regular dependence of $c(g_0, am)$ on the quark mass (i.e. positive powers of am), which may be dropped in a mass-independent scheme leaving us with $c(g_0)$.

From these considerations we deduce that, up to regular $\mathcal{O}(a)$ terms (i.e. discretization effects), Eq. (3.52) may be written as

$$G_{\widetilde{V}}(p) = G_V(p) + \frac{a}{Z_\Delta}[G_\Delta(p)]_\mathrm{R} - c(g_0)G_V(p)$$

$$= [1 - c(g_0)]G_V(p) + \cdots . \tag{3.54}$$

The term proportional to $[G_\Delta]_\mathrm{R}$ has been dropped in the last expression, being proportional to the lattice spacing (the renormalized correlation is finite by construction, while Z_Δ is at most logarithmically divergent). Consequently, the local current V_μ has a finite normalization:

$$Z_V(g_0) \equiv 1 - c(g_0) \neq 1. \tag{3.55}$$

Note that Eq. (3.49) (and any other vector Ward identity) can be expressed in terms of the local current, by substituting $\widetilde{V}_\mu$ with $[Z_V V_\mu]$.

Next, we go back to Eq. (3.49) and, keeping the two masses distinct, integrate it w.r.t. x; the surface trerm on the l.h.s. vanishes and we obtain

$$(m_{02} - m_{01}) \int d^4x G_S (x_1 - x, x_2 - x; m_{01}, m_{02}) = \mathcal{S}_1 (x_1 - x_2; m_{01})$$

$$- \mathcal{S}_2 (x_1 - x_2; m_{02}). \tag{3.56}$$

Again, the r.h.s. of above expression becomes finite (renormalized) once we multiply through by Z_ψ. This means that, up to this quark field renormalization, the l.h.s. is also finite. In other words, the product of the bare quark mass difference and the integrated scalar operator $\int d^4x S(x)$ is a scale-independent (or renormalization-group-invariant) quantity. As there is no operator of dimension $d \leq 3$ that could mix with S and vanish identically under the integral $\int d^4x$, also the product of the bare quark mass difference and the (unintegrated) scalar density S is scale independent. With the difference of renormalized quark masses formally given in Eq. (3.16) and the operator renormalization given by Eq. (3.32), this means that

$$Z_S(g_0, a\mu) = Z_m^{-1}(g_0, a\mu). \tag{3.57}$$

Thus, vector Ward identities imply an exact relation between the quark-mass renormalization Z_m and that of the scalar density. In other words, once a renormalization condition is imposed on the quark mass in order to determine Z_m, the scalar density renormalization constant Z_S is also known (or vice versa). A corollary is that the mass anomalous dimension is the opposite of that of the scalar operator ($\gamma_m = -\gamma_S$).

3.3.2 Lattice axial Ward identities

Far less obvious are the consequences of axial transformations with Wilson fermions. This is because even at vanishing quark masses (i.e. $M_0 = 0$), the Wilson term of the lattice action (3.2) is not invariant under the *global* version of the axial transformations (3.43). However, by imposing suitable renormalization conditions, PCAC is recovered in the continuum limit (Karsten and Smit, 1981; Bochicchio *et al.*, 1985; Testa, 1998). The axial WI, namely Eq. (3.45) for the axial transformations (3.43), is

$$i\left\langle \frac{\delta O(x_1,\ldots,x_n)}{\delta\alpha_A^a(x)} \right\rangle = a^4 \sum_\mu \nabla_x^\mu \left\langle \widetilde{A}_\mu^a(x) O(x_1,\ldots,x_n) \right\rangle$$

$$- a^4 \left\langle \bar\psi(x)\left\{ \frac{\lambda^a}{2}, M_0 \right\}\gamma_5\psi(x)O(x_1,\ldots,x_n) \right\rangle - a^4 \left\langle X^a(x)O(x_1,\ldots,x_n) \right\rangle, \quad (3.58)$$

where $\widetilde{A}_\mu^a(x)$ is a bilinear point-split axial current given by

$$\widetilde{A}_\mu^a(x) = \frac{1}{2}\left[\bar\psi(x)\gamma_\mu\gamma_5 U_\mu(x)\frac{\lambda^a}{2}\psi(x+a\hat\mu) + \bar\psi(x+a\hat\mu)\gamma_\mu\gamma_5 U_\mu^\dagger(x)\frac{\lambda^a}{2}\psi(x) \right]. \quad (3.59)$$

The term X^a in the above Ward identity is the variation of the Wilson term under axial transformations:

$$X^a(x) = -\frac{1}{2}a\left[\bar\psi(x)\frac{\lambda^a}{2}\gamma_5 \overrightarrow{D}^2\, \psi(x) + \bar\psi(x) \overleftarrow{D}^2\, \frac{\lambda^a}{2}\gamma_5\psi(x) \right], \quad (3.60)$$

where

$$a^2 \overrightarrow{D}^2\, \psi(x) = \sum_\mu \left[U_\mu(x)\psi(x+a\hat\mu) + U_\mu^\dagger(x-a\hat\mu)\psi(x-a\hat\mu) - 2\psi(x) \right]$$

$$a^2 \bar\psi(x) \overleftarrow{D}^2 = \sum_\mu \left[\bar\psi(x+a\hat\mu)U_\mu^\dagger(x) + \bar\psi(x-a\hat\mu)U_\mu(x-a\hat\mu) - 2\bar\psi(x) \right]. \quad (3.61)$$

Since X^a cannot be cast in the form of a four-divergence, it cannot be absorbed in a redefinition of the axial current $\widetilde{A}_\mu^a$. This is of course a corollary of the fact that the Wilson term in the fermionic action breaks chiral symmetry. It is a dimension-4 operator that, in the naive continuum limit vanishes, being of the form $X^a = aO_5^a$, with O_5^a a dimension-5 operator. However, when inserted in correlation functions such as Eq.(3.58), O_5^a will generate power divergences that will cancel the prefactor a. It is therefore mandatory to understand its renormalization properties. Its mixing with other dimension-5 operators may be ignored, since the corresponding mixing coefficients are at most logarithmically divergent and their contribution to $X^a = aO_5^a$ vanishes in the continuum limit. Mixing with lower-dimensional operators is, on the other hand, relevant, as the corresponding coefficients are inverse powers of the lattice spacing (Testa, 1998). The only possibility for such operator mixing, respecting the symmetry properties of O_5^a is:

$$[O_5^a(x)]_{\rm R} = Z_5\left[O_5^a(x) + \bar\psi(x)\left\{ \frac{\lambda^a}{2}, \frac{\overline{M}}{a} \right\}\gamma_5\psi(x) + \frac{(Z_{\widetilde{A}}-1)}{a}\nabla_x^\mu \widetilde{A}_\mu^a(x) \right]. \quad (3.62)$$

The functional dependence of the mixing coefficients $Z_{\widetilde{A}}$ and $\overline{M}$ is established by arguments, very similar to the ones of Section 3.3.1, concerning Z_V. These coefficients involve subtractions of operators of lower dimension than that of O_5^a, which implies that they are independent of any renormalization scale μ (Testa, 1998). Any dependence on aM cannot be divergent (e.g. $\ln(aM)$) because this would compromise the chiral limit of $[O_5^a(x)]_{\mathrm{R}}$-insertions in renormalized correlation functions. The absence of $\ln(aM)$-dependence of $Z_{\widetilde{A}}$ has been explicitly proved in perturbation theory (Curci, 1986). Moreover, any regular dependence of $Z_{\widetilde{A}}$ on aM will vanish in a mass-independent renormalization scheme and may therefore be considered a lattice artefact. We conclude that the dimensionless axial current coefficient only depends on the gauge coupling, $Z_{\widetilde{A}}(g_0)$. Similarly, the dimension-1 mass coefficient has the functional form $\overline{M} = w(g_0, aM)/a$, with $w(g_0, aM)$ a regular function of aM.

Using Eq. (3.62), we substitute $X^a = a \times O_5^a$ in Eq. (3.58), obtaining

$$
i\Big\langle \frac{\delta O(x_1, \ldots, x_n)}{\delta \alpha_A^a(x)} \Big\rangle = a^4 \sum_\mu \nabla_x^\mu \Big\langle Z_{\widetilde{A}} \widetilde{A}_\mu^a(x) O(x_1, \ldots, x_n) \Big\rangle
$$

$$
- a^4 \Big\langle \bar{\psi}(x) \Big\{ \frac{\lambda^a}{2}, [M_0 - \overline{M}] \Big\} \gamma_5 \psi(x) O(x_1, \ldots, x_n) \Big\rangle
$$

$$
- a^4 \Big\langle \frac{X_{\mathrm{R}}^a(x)}{Z_5} O(x_1, \ldots, x_n) \Big\rangle. \tag{3.63}
$$

In order to simplify the presentation, we again specify the operator $O(x_1, x_2) = \psi_1(x_1)\bar{\psi}_2(x_2)$, and the Gell-Mann matrix λ^+ of $SU(N_F)_A$, in place of λ^a. The above expression becomes

$$
\nabla_x^\mu \Big\langle \psi_1(x_1) Z_{\widetilde{A}} \widetilde{A}_\mu(x) \bar{\psi}_2(x_2) \Big\rangle = [m_{01} + m_{02} - \overline{m}_1 - \overline{m}_2] \Big\langle \psi_1(x_1) P(x) \bar{\psi}_2(x_2) \Big\rangle
$$

$$
- \delta(x - x_2) \Big\langle \psi_1(x_1)\bar{\psi}_1(x_2)\gamma_5 \Big\rangle - \delta(x - x_1) \Big\langle \gamma_5 \psi_2(x_1)\bar{\psi}_2(x_2) \Big\rangle
$$

$$
+ \frac{a}{Z_5} \Big\langle \psi_1(x_1)[O_5(x)]_{\mathrm{R}} \bar{\psi}_2(x_2) \Big\rangle, \tag{3.64}
$$

with $\overline{m}_1, \overline{m}_2$ the elements of the diagonal[4] matrix $\overline{M}$. Note that these quantities depend on the bare coupling and all bare quark masses; i.e. $\overline{m}_f(g_0, m_{01}, \cdots, m_{0N_F})$. The last term on the r.h.s. is genuinely $\mathcal{O}(a)$. Since it vanishes in the continuum limit, it may be safely dropped as a discretization effect. This leaves us with the axial Ward identity

$$
Z_{\widetilde{A}} \sum_\mu \nabla_x^\mu G_{\widetilde{A}}^\mu (x_1 - x, x_2 - x; m_{01}, m_{02}) \tag{3.65}
$$

$$
= \big(m_1^{\mathrm{PCAC}} + m_2^{\mathrm{PCAC}}\big) G_P (x_1 - x, x_2 - x; m_{01}, m_{02})
$$

$$
- \delta(x_2 - x)\mathcal{S}_1 (x_1 - x_2; m_{01}) \gamma_5 - \delta(x_1 - x)\gamma_5 \mathcal{S}_2 (x_1 - x_2; m_{02}) + \mathcal{O}(a),
$$

where the PCAC (bare) quark mass is defined by

$$m_f^{\text{PCAC}} \equiv m_{0f} - \overline{m}_f(g_0, m_{01}, \cdots, m_{0N_F}) \qquad (f = 1, 2, 3, \cdots, N_F). \qquad (3.66)$$

We may now define the chiral limit on the basis of Ward identity (3.65): first we take for simplicity al masses m_0 to be degenerate, which allows us to simply the notation by writing $\overline{m}_f$ as $\overline{m}(g_0, m_0)$. Then, the chiral limit is defined as the value m_{cr} of m_0 for which $\overline{m}(g_0, m_{\text{cr}}) = m_{\text{cr}}$ (i.e. m^{PCAC} vanishes). A relation between the PCAC quark mass defined in Eq. (3.66) and the subtracted bare mass $m = m_0 - m_{\text{cr}}$ is readily obtained by expanding the former around the chiral point m_{cr}:

$$m^{\text{PCAC}} = m_0 - \overline{m}(m_{\text{cr}}) - \left. \frac{\partial \overline{m}(m)}{\partial m} \right|_{m_{\text{cr}}} (m_0 - m_{\text{cr}}) + \cdots$$

$$= (m_0 - m_{\text{cr}}) \left[1 - \left. \frac{\partial \overline{m}(m)}{\partial m} \right|_{m_{\text{cr}}} + \cdots \right]. \qquad (3.67)$$

In the chiral limit, the first term on the r.h.s. of Eq. (3.65) vanishes. Multiplying the resulting expression by Z_ψ renders the r.h.s. finite. Thus, also the l.h.s. is finite and does not require any further renormalization. Now, all terms of Eq. (3.65) differ from their continuum expressions by $\mathcal{O}(a)$ discretization effects. The point-split axial current (3.59) with the normalization

$$Z_{\widetilde{A}} \neq 1, \qquad (3.68)$$

satisfies the continuum axial Ward identity up to $\mathcal{O}(a)$. In practice, it turns out to be more convenient to work with the lattice local axial current $A_\mu^a(x)$. We can show, in a fashion analogous to the case of the vector current (c.f. the power-counting argument based on Eqs. (3.51)-(3.55)), that $A_\mu^a(x)$ has a finite normalization constant Z_A. Thus, we have $[A_\mu^a]_{\text{R}} = \lim_{a \to 0}[Z_{\widetilde{A}} \widetilde{A}_\mu^a] = \lim_{a \to 0}[Z_A A_\mu^a]$. From now on, the combination $Z_{\widetilde{A}} \widetilde{A}_\mu^a$ will always be substituted by $Z_A A_\mu^a$ wherever it appears in a WI. Also analogous to the vector current case is the lack of mass dependence of these constants. We therefore have:

$$Z_{\widetilde{A}}(g_0^2), Z_A(g_0^2) \neq 1. \qquad (3.69)$$

Note that the tree-level value of both $Z_{\widetilde{A}}$ and Z_A is unity. In perturbation theory $Z_A = 1 + z_1 g_0^2 + \cdots$ (and similarly for $Z_{\widetilde{A}}$). Thus, in the continuum limit ($a \to 0$, $g_0^2(a) \to 0$) we have $Z_A, Z_{\widetilde{A}} \to 1$.

From here on things proceed very much like the case of vector Ward identities. Integrating Eq. (3.64) over x eliminates the surface term with the axial current divergence, leaving us with

$$(m_1^{\text{PCAC}} + m_2^{\text{PCAC}}) \int d^4x G_P (x_1 - x, x_2 - x; m_{01}, m_{02})$$

$$= \mathcal{S}(x_1 - x_2; m_{01}) \gamma_5 + \gamma_5 \mathcal{S}(x_1 - x_2; m_{02}). \qquad (3.70)$$

Upon multiplying through with the quark field renormalization constant Z_ψ, the r.h.s. becomes finite. Thus, also the l.h.s. is finite, which means that the product of the PCAC quark mass times the pseudoscalar density is a renormalization-group-invariant quantity (i.e. it is scale independent). The anomalous dimension of the pseudoscalar density is the opposite to that of the quark mass ($\gamma_m = -\gamma_P$).[5] Given that the pseudoscalar operator P is multiplicatively renormalized by Z_P, this suggests the following renormalization pattern for the PCAC quark mass:

$$[m_f(g_R, \mu)]_R \equiv Z_P^{-1}(g_0, a\mu) m_f^{\text{PCAC}}(g_0, m_{01}, \cdots m_{0N_F}). \tag{3.71}$$

This expression amounts to a definition of the renormalized quark mass, in a scheme that fixes the renormalization of the pseudoscalar density Z_P. Identifying this definition of $[m_f]_R$ with the generic expression (3.13) gives a closed expression for $\overline{m}_f$.

In conclusion, lattice axial Ward identities lead to: (i) the definition of the chiral limit as the solution of the equation $m^{\text{PCAC}}(m_{\text{cr}}) = m_{\text{cr}}$; (ii) the normalization of the axial current A_μ by Z_A. The last requirement may raise the following question: the properly normalized matrix element, from which the pion decay constant f_π is computed, is given by

$$Z_A \langle \pi | A_\mu | 0 \rangle = \langle \pi | [A_\mu]_R | 0 \rangle + \mathcal{O}(a). \tag{3.72}$$

Since we know that in the continuum limit $Z_A \to 1$, why bother to normalize the axial current by Z_A? The answer provided by axial Ward identities like Eq. (3.65) (and Eq. (3.74) below) is that close to the continuum limit the normalized matrix element in the above expression differs from its continuum limit by discretization effects that are positive powers of the lattice spacing. Consequently, omitting Z_A (a regular function of the bare coupling g_0^2), implies that the bare matrix element converges to its continuum limit like $g_0^2 \sim 1/\ln(a)$. This is much slower than the convergence of $Z_A A_\mu$, shown in Eq. (3.72).

3.3.3 Hadronic Ward identities for Z_V and Z_A

So far we have seen how Ward identities fix the normalization of lattice partially conserved currents and the ratio of the scalar and pseudoscalar renormalization constants. We have mostly used vector and axial Ward identities based on the variation of the operator $O(x_1, x_2) = \psi(x_1)\bar{\psi}(x_2)$, commonly referred to as "Ward identities on quark states". This terminology refers to the fact that at large time separations, say $x_1^0 \ll x^0 \ll x_2^0$, expressions like (3.49) and (3.65) eventually produce PCVC and PCAC relations between matrix elements of quark states. These Ward identities can form a basis for the determination of the current normalizations Z_V, Z_A and the ratio Z_S/Z_P, but this is usually not very practical (for instance they require gauge fixing, which is affected by the Gribov ambiguity, etc.). It is therefore preferable to write down Ward identities involving correlation functions of gauge-invariant operators, based on suitable choices of $O(x_1, x_2)$. These are called hadronic Ward identities because they give rise, in terms of the LSZ procedure, to relations between matrix elements of hadronic states. Here, we will briefly describe some examples of Ward identities leading to nonperturbative computations of the scale-independent parameters Z_V, Z_A. A similar discussion concerning ratios of scalar and pseudoscalar renormalization constants is the subject of the next section.

The computation of Z_V is based on the observation that, up to discretization effects, $\widetilde{V}_\mu = Z_V V_\mu + \mathcal{O}(a)$. It is then straightforward to obtain, at fixed UV cutoff (i.e. fixed g_0) estimates of Z_V by comparing the insertion of the point-split vector current in a correlation function, to that of the local current. For example, equations

$$\sum_{k=1}^{3} \int d^3\mathbf{x} \langle \widetilde{V}_k^a(x) V_k^b(z) \rangle = Z_V \sum_{k=1}^{3} \int d^3\mathbf{x} \langle V_k^a(x) V_k^b(z) \rangle, \tag{3.73}$$

$$\int d^3\mathbf{x} \int d^3\mathbf{y} \langle P^a(x) \widetilde{V}_0^b(z) P^c(y) \rangle = Z_V \int d^3\mathbf{x} \int d^3\mathbf{y} \langle P^a(x) V_0^b(z) P^c(y) \rangle,$$

are valid up to $\mathcal{O}(a)$. Thus they may be solved for Z_V, providing two independent estimates for this quantity, which differ by discretization effects. Strictly speaking, integration over space dimensions is not an essential feature, but it is useful in practice, as it averages out statistical fluctuations. In order to avoid singularities from contact terms, timeslices are kept distinct ($x^0 \neq y^0 \neq z^0$). Note that the flavor superscripts a, b, c must be chosen so as to give, through Wick contractions of the fermion fields, connected diagrams of valence quarks with non-vanishing expectation values.

A non-perturbative determination of Z_A is based on two axial Ward identities. We write down Eq. (3.63) for the operator $O(y) = P^b(y)$; with $y \neq x$ the l.h.s. vanishes. As we have argued above, the term containing X_R^a may also be dropped, being a discretization effect. We also work with the local, rather than the point-split current. Thus, up to $\mathcal{O}(a)$, we have

$$\sum_\mu \nabla_x^\mu \langle A_\mu^a(x) P^b(y) \rangle = \left\langle \left[\bar{\psi}(x) \left\{ \frac{\lambda^a}{2}, \frac{[M_0 - \overline{M}]}{Z_A} \right\} \gamma_5 \psi(x) \right] P^b(y) \right\rangle. \tag{3.74}$$

Like before, flavor indices a, b must be appropriately chosen. The above Ward identity may be solved for the ratio $[M_0 - \overline{M}_0]/Z_A$; this is one of the most standard methods for the determination of the PCAC quark mass, up to the axial current normalization factor Z_A. At fixed gauge coupling and degenerate quark masses, the quantity $[m_0 - \overline{m}_0]/Z_A$ is computed at several values of the bare quark mass m_0 and subsequently extrapolated to zero, in order to obtain a non-perturbative estimate of $m_{\rm cr}$; cf. Eq. (3.66) and the comment that follows it. Note that in order to increase the signal stability, the above Ward identity is usually integrated over all space coordinates of y (with $x^0 \neq y^0$).

The next step is to work out the axial Ward identity (3.58), with the operator choice $O(x, y) = A_\nu^b(x) V_\rho^c(y)$ (for $N_F > 2$):

$$Z_A \nabla_z^\mu \langle A_\mu^a(z) A_\nu^b(x) V_\rho^c(y) \rangle \tag{3.75}$$

$$= \left\langle \left[\bar{\psi}(z) \left\{ \frac{1}{2} \lambda^a, (M_0 - \overline{M}) \right\} \gamma_5 \psi(z) \right] A_\nu^b(x) V_\rho^c(y) \right\rangle + \left\langle \overline{X}^a(z) A_\nu^b(x) V_\rho^c(y) \right\rangle$$

$$+ i f^{abd} \delta(z - x) \langle V_\nu^d(x) V_\rho^c(y) \rangle + i f^{acd} \delta(z - y) \langle A_\nu^b(x) A_\rho^d(y) \rangle,$$

where again use of Eq. (3.62) has been made and the shorthand notation $\overline{X}^a = a[O_5^a]_R/Z_5$ has been introduced. Once more, flavor superscripts a, b, c are chosen so as to give, upon Wick contraction, non-vanishing expectation values of connected diagrams of valence quarks. The correlation function with the $\overline{X}^a$ operator gives rise to contact terms. Symmetry arguments impose that these contact terms must have the form

$$\langle \overline{X}^a(z) A_\nu^b(x) V_\rho^c(y) \rangle = - ik_1(g_0^2) f^{abd} \delta(z-x) \langle V_\nu^d(x) V_\rho^c(y) \rangle \qquad (3.76)$$
$$+ ik_2(g_0^2) f^{acd} \delta(z-y) \langle A_\nu^b(x) A_\rho^d(y) \rangle + \dots,$$

where the ellipsis stands for localized Schwinger terms, which will vanish after integration over z, to be performed below (we always keep $x \neq y$). For $x, y \neq z$ the l.h.s. vanishes in the continuum limit as discussed in Section 3.3.2.

We now proceed as follows: (I) rewrite Eq. (3.75), using Eq. (3.76) and expressing all bare quantities in terms of the renormalized ones; (II) recall that the product of the PCAC quark mass and the pseudoscalar density in the first term of the r.h.s. of Eq. (3.75) is renormalization-group invariant (c.f. Eq. (3.71)); (III) require that the renormalized quantities obey the corresponding continuum (nominal) axial Ward identity. Thus, we obtain

$$k_1(g_0^2) = 1 - \frac{Z_V}{Z_A}$$
$$k_2(g_0^2) = \frac{Z_A}{Z_V} - 1, \qquad (3.77)$$

and the Ward identity in terms of properly normalized currents is:

$$Z_A^2 Z_V \left\langle \left[\nabla_z^\mu A_\mu^a(z) - \bar{\psi}(z) \left\{ \frac{\lambda^a}{2}, \frac{\overline{M} - M_0}{Z_A} \right\} \gamma_5 \psi(z) \right] A_\nu^b(x) V_\rho^c(y) \right\rangle$$
$$= +if^{abd} \delta(z-x) Z_V^2 \langle V_\nu^d(x) V_\rho^c(y) \rangle$$
$$+if^{acd} \delta(z-y) Z_A^2 \langle A_\nu^b(x) A_\rho^d(y) \rangle. \qquad (3.78)$$

Performing the integration over z kills off the first term on the l.h.s.[6] We also integrate over $\mathbf{x}$ in order to improve the signal-to-noise ratio in practical computations (recall that $x \neq y$ is necessary in order to eliminate Schwinger terms; thus $x^0 \neq y^0$):

$$\int d^4z \int d^3\mathbf{x} \left\langle \left[\bar{\psi}(z) \left\{ \frac{1}{2} \lambda^a, \frac{\overline{M} - M_0}{Z_A} \right\} \gamma_5 \psi(z) \right] A_\nu^b(x) V_\rho^c(y) \right\rangle \qquad (3.79)$$
$$= -i \frac{Z_V}{Z_A^2} f^{abd} \int d^3\mathbf{x} \langle V_\nu^d(x) V_\rho^c(y) \rangle - i \frac{1}{Z_V} f^{acd} \int d^3\mathbf{x} \langle A_\nu^b(x) A_\rho^d(y) \rangle.$$

Since Z_V is known from Eqs. (3.73) and $[\overline{M} - M_0]/Z_A$ is known from Eq. (3.74), the above Ward identity may be solved for Z_A. For $N_F = 2$, this determination of Z_A is clearly not viable. A method for obtaining Z_A for any $N_F \geq 2$ is based on an axial Ward identity with quark external states (Martinelli *et al.*, 1993).

3.3.4   Hadronic Ward identity for the ratio Z_S/Z_P

A further application of hadronic lattice Ward identities concerns the renormalization of the scalar and pseudoscalar densities. In a regularization that respects chiral symmetry, these operators, belonging to the same chiral multiplet, are renormalized by the same parameters $Z_S = Z_P$. This is not the case of Wilson fermions, which break chiral symmetry. Following a line of reasoning similar to the one of the previous section, applied to the operator $O(x, y) = S^g(x)P^h(y)$ (with $g \neq h$), we obtain the Ward identity (for $N_F > 2$)

$$
\int d^4z \int d^3\mathbf{x} \Big\langle \Big[\bar{\psi}(z) \Big\{ \frac{1}{2}\lambda^f, \frac{\overline{M} - M_0}{Z_A} \Big\} \gamma_5 \psi(z) \Big] S^g(x)P^h(y) \Big\rangle \tag{3.80}
$$

$$
= \frac{Z_P}{Z_A Z_S} d^{fgl} \int d^3\mathbf{x} \langle P^l(x)P^h(y) \rangle + \frac{Z_S}{Z_A Z_P} d^{fhl} \int d^3\mathbf{x} \langle S^g(x)S^l(y) \rangle
$$

where the ds are defined in Eq. (3.7). Once Z_A has been determined through, say, Eq. (3.79), the above Ward identity may be used to compute the ratio Z_S/Z_P.

Note that Ward identities such as (3.79) and (3.80) may be regarded as explicit demonstrations of the fact that Z_V, Z_A and Z_S/Z_P are scale-independent functions of the bare gauge coupling: upon solving them for these Z-factors, we obtain solutions that are combinations of *bare* correlation functions. These bear no dependence on renormalization schemes and scales and remain finite in the continuum limit. At fixed gauge coupling (i.e. fixed UV cutoff), they provide non-perturbative estimates of Z_V, Z_A and Z_S/Z_P.

Practical simulations of these Ward identities are often performed on lattices with periodic and/or antiperiodic boundary conditions for the gauge and fermion fields. For a given gauge coupling, simulations are carried out at several non-zero quark masses and the resulting Z-factors are extrapolated to the chiral limit. Alternatively, Schrödinger functional (Dirichlet) boundary conditions and suitably chosen correlation functions enable us to perform computations of Z_V, Z_A and Z_S/Z_P directly in the chiral limit.[7] Possible differences among numerical results for the same Z-factors, obtained from different formulations and/or Ward identities, are due to discretization effects and provide an estimate of this source of systematic error.

A determination of the ratio Z_S/Z_P, valid also for $N_F = 2$, is based on the expression

$$
[m_1]_\mathrm{R} - [m_2]_\mathrm{R} = Z_S^{-1} \big[m_{01} - m_{02} \big]
$$

$$
= Z_P^{-1} \big[m_1^{\mathrm{PCAC}} - m_2^{\mathrm{PCAC}} \big]. \tag{3.81}
$$

The first line is obtained from Eqs. (3.16) and (3.57), while the second one is from Eq. (3.71). The ratio of bare to PCAC mass differences gives an estimate of Z_S/Z_P. Such a computation must be carried out for pairs of non-degenerate quark masses.

Subsequently results are to be extrapolated to the chiral limit. To the best of our knowledge, this method has never been applied in practical computations.

3.3.5 Singlet scalar and pseudoscalar operators

It is sometimes convenient to work with Ward identities involving matrix elements of gauge-invariant operators between hadronic states, obtained from Ward identities between correlations functions such as Eq. (3.75), and the use of the standard LSZ procedure. The reader should have no difficulty performing these steps. We focus on the following Ward identity, based on the axial variation of the pseudoscalar density P^g amongst hadronic states $|h_1\rangle \neq |h_2\rangle$:

$$\nabla^\mu_x \langle h_1 | \mathcal{T}[Z_A A^f_\mu(x) P^g(0)] | h_2 \rangle - 2[m_0 - \overline{m}] \langle h_1 | \mathcal{T}[P^f(x) P^g(0)] | h_2 \rangle$$

$$= \langle h_1 | \mathcal{T}[\overline{X}^f(x) P^g(0)] | h_2 \rangle + i \left\langle h_1 \left| \frac{\delta P^g(0)}{\delta \alpha^f_A(x)} \right| h_2 \right\rangle. \tag{3.82}$$

The flavor index $f = 1, \cdots, N_F^2 - 1$ characterizes non-singlet currents, pseudoscalar densities, etc., while $g = 0, \cdots, N_F^2 - 1$ also covers the case of the flavor singlet pseudoscalar density P^0. For simplicity we assume exact flavor symmetry; i.e. $M_0 = m_0 I$ and thus $\{\lambda^f, M_0\} = 2m_0\lambda^f$. The axial variation of the pseudoscalar density is

$$\frac{\delta P^g(0)}{\delta \alpha^f_A(x)} = i\delta(x) \left[d^{fgh} S^h(0) + \delta^{fg} \sqrt{\frac{2}{N_F}} S^0(0) + \delta^{g0} \sqrt{\frac{2}{N_F}} S^f(0) \right]. \tag{3.83}$$

The axial current normalization Z_A and the mass subtraction $\overline{m}$ in Eq. (3.82) are generated in standard fashion by the counterterms of the operator $O^f_5 = X^f/a$; cf. Eq. (3.62). This is not, however, the end of the story as we now allow for the case in which the space-time point x comes close to the origin ($x \approx 0$). In this case, the insertion of $\overline{X}^f$ in off-shell Green functions does not vanish as we approach the continuum limit. Rather, it generates local contact terms that are constrained, by flavor symmetry, to have the form:

$$\langle h_1 | \mathcal{T}[\overline{X}^f(x) P^g(0)] | h_2 \rangle = -\delta(x) \left[c_1(g_0^2) d^{fgh} \langle h_1 | S^h(0) | h_2 \rangle + c_2(g_0^2) \delta^{fg} \sqrt{\frac{2}{N_F}} \langle h_1 | S^0(0) | h_2 \rangle \right.$$

$$\left. + c_3(g_0^2) \delta^{g0} \sqrt{\frac{2}{N_F}} \langle h_1 | S^f(0) | h_2 \rangle \right]. \tag{3.84}$$

Putting it all together, and going over to the chiral limit, yields the Ward identity

$$\nabla^\mu_x \langle h_1 | \mathcal{T}[Z_A A^f_\mu(x) P^g(0)] | h_2 \rangle = -\delta(x) \left\{ [1 + c_1(g_0^2)] d^{fgh} \langle h_1 | S^h(0) | h_2 \rangle \right.$$

$$+ \delta^{fg} \sqrt{\frac{2}{N_F}} [1 + c_2(g_0^2)] \langle h_1 | S^0(0) | h_2 \rangle \tag{3.85}$$

$$\left. + \delta^{g0} \sqrt{\frac{2}{N_F}} [1 + c_3(g_0^2)] \langle h_1 | S^f(0) | h_2 \rangle \right\}.$$

For a non-singlet pseudoscalar density P^g (i.e. for $g \neq 0$) and for $f \neq g$, the last terms in the r.h.s. of the above expression vanish. Since P^g is multiplicatively renormalizable, it is adequate to multiply the above expression by Z_P to render it finite. This results in the following relation between the renormalization factor of the non-singlet scalar density and Z_P:

$$Z_S = [1 + c_1(g_0^2)]Z_P. \tag{3.86}$$

If, on the the hand, $f = g \neq 0$, the second term on the r.h.s. of Eq. (3.85) survives and we obtain the relation:

$$Z_{S^0} = [1 + c_2(g_0^2)]Z_P. \tag{3.87}$$

Finally, we consider the case in which P^g is a singlet pseudoscalar density (i.e. $g = 0$), for which only the last term on the r.h.s. of Eq. (3.85) survives. Multiplying through by the factor Z_{P^0} we obtain the relation

$$Z_S = [1 + c_3(g_0^2)]Z_{P^0}. \tag{3.88}$$

Thus, it has been established (Bochicchio *et al.*, 1985; Rossi *et al.*, 2003) that the $[N_F, \overline{N}_F] \oplus [\overline{N}_F, N_F]$ representation of the chiral group $SU(N_F)_L \otimes SU(N_F)_R$ is formed by the rescaled operators

$$P^f \qquad\qquad [1 + c_1]S^f \qquad\qquad f \neq 0, \tag{3.89}$$

$$\frac{1 + c_1}{1 + c_3}P^0 \qquad\qquad [1 + c_2]S^0, \tag{3.90}$$

which renormalize by a common factor Z_P. Conversely, the renormalized operators $P_R^f = Z_P P^f$, $S_R^f = Z_S P^f$ (for $f \neq 0$) and $P_R^0 = Z_{P^0} P^f$ have a common anomalous dimension (arising from, say, Z_P) but renormalization constants Z_P, Z_S and Z_{P^0} that differ by finite normalization factors. These factors tend to unity in the continuum limit. Their presence is yet another consequence of chiral-symmetry breaking by the Wilson term. Analogous statements are true for the renormalization factor Z_{S^0} of the singlet scalar density S^0, which, however, is also subject to power subtractions, as we will see below.

Another important result (Maiani *et al.*, 1987; Rossi *et al.*, 2003) is the relation between the ratio Z_P/Z_{S^0} and the ratio of PCAC to subtracted quark masses. The first step towards its derivation consists in a Ward identity analogous to that of Eq. (3.82), but with S^0 in place of P^g:

$$\nabla_y^\mu \langle h_1 | T[Z_A A_\mu^f(y) S^0(x)] | h_2 \rangle - 2[m_0 - \overline{m}] \langle h_1 | T[P^f(y) S^0(x)] | h_2 \rangle$$

$$= \langle h_1 | T[\overline{X}^f(y) S^0(x)] | h_2 \rangle + i \left\langle h_1 \left| \frac{\delta S^0(x)}{\delta \alpha_A^f(y)} \right| h_2 \right\rangle$$

$$= -\delta(x - y)[1 + t(g_0^2)]\sqrt{\frac{2}{N_F}} \langle h_1 | P^f(y) | h_2 \rangle. \tag{3.91}$$

The last equation is the result of performing the functional derivative $\delta S^0(x)/\delta \alpha_A^f(y)$ and establishing that the contact term, arising when $\overline{X}^f(y)$ is in the vicinity of $S^0(x)$, is proportional to P^f, with $t(g_0^2)$ the proportionality factor. Multiplying through by Z_{S^0} gives the renormalized Ward identity

$$Z_{S^0}\nabla_y^\mu\langle h_1|\mathcal{T}[Z_A A_\mu^f(y)S^0(x)]|h_2\rangle - 2[m_0 - \overline{m}]Z_{S^0}\langle h_1|\mathcal{T}[P^f(y)S^0(x)]|h_2\rangle$$

$$= -\delta(x-y)\sqrt{\frac{2}{N_F}}Z_P\langle h_1|P^f(y)|h_2\rangle, \tag{3.92}$$

with the identification $Z_P = [1 + t(g_0^2)]Z_{S^0}$.

The next step consists in differentiating the axial Ward identity between hadronic states with respect to the bare quark mass:

$$\frac{\partial}{\partial m_0}\nabla_y^\mu\langle h_1|Z_A A_\mu^f(y)|h_2\rangle = \frac{\partial}{\partial m_0}2[m_0 - \overline{m}]\langle h_1|P^f(y)|h_2\rangle. \tag{3.93}$$

The mass derivative on the r.h.s. gives

$$\frac{\partial}{\partial m_0}2[m_0 - \overline{m}]\langle h_1|P^f(y)|h_2\rangle = 2\left[1 - \frac{\partial\overline{m}}{\partial m_0}\right]\langle h_1|P^f(y)|h_2\rangle$$

$$+ 2[m_0 - \overline{m}]\frac{\partial}{\partial m_0}\langle h_1|P^f(y)|h_2\rangle. \tag{3.94}$$

Now, differentiating a Green's function with respect to the quark mass m_0 amounts to inserting the operator $\int d^4x\,\overline{\psi}(x)\psi(x) = \sqrt{2N_F}\int d^4x\,S^0(x)$. Performing this insertion on the l.h.s. of Eq. (3.93) and the last term on the r.h.s. of Eq. (3.94) results in the following identity

$$\nabla_y^\mu\langle h_1|Z_A\mathcal{T}\left[A_\mu^f(y)\int d^4x\,S^0(x)\right]|h_2\rangle = \sqrt{\frac{2}{N_F}}\left[1 - \frac{\partial\overline{m}}{\partial m_0}\right]\langle h_1|P^f(y)|h_2\rangle$$

$$+ 2[m_0 - \overline{m}]\langle h_1|\mathcal{T}\left[P^f(y)\int d^4x\,S^0(x)\right]|h_2\rangle. \tag{3.95}$$

Multiplying both sides of the above by Z_{S^0} gives a finite expression. Upon comparing it with Eq. (3.92), and taking the chiral limit $m_0 \to m_{\rm cr}$, we obtain a relation between renormalization factors

$$\frac{Z_P}{Z_{S^0}} = 1 - \frac{\partial\overline{m}}{\partial m_0}\Big|_{m_{\rm cr}}, \tag{3.96}$$

Compared with Eqs. (3.66) and (3.67), this gives a relation between the PCAC and the subtracted quark masses:

$$\lim_{m_0\to m_{\rm cr}}\left[\frac{m^{\rm PCAC}}{m_0 - m_{\rm cr}}\right] = \frac{Z_P}{Z_{S^0}}. \tag{3.97}$$

Combining this expression with the renormalized quark mass definition (3.71), we obtain

$$m_{\mathrm{R}} = Z_{S^0}^{-1}[m_0 - m_{\mathrm{cr}}], \qquad (3.98)$$

which, upon comparison with Eqs. (3.14) and (3.15) leads to

$$Z_{S^0} = Z_{m^0}^{-1} \qquad (3.99)$$

We see that the above expression, as well as Eq. (3.57), relate the singlet and non-singlet mass renormalization constants to the corresponding scalar density renormalization constants.

In perturbation theory, the difference between singlet and non-singlet operators arises from extra diagrams in the singlet case, involving fermion loops. As these drop out in the quenched approximation, the quenched renormalization constants obey $Z_S = Z_{S^0}$ and $Z_P = Z_{P^0}$.

3.3.6   The chiral condensate

Our last topic related to lattice Ward identities is the derivation of the proper definition of the chiral condensate with Wilson fermions (Bochicchio *et al.*, 1985). The starting point is the Ward identity (3.82) for the vacuum expectation values of the non-singlet axial current A_μ^f and the pseudoscalar density P^g (i.e. $f, g \neq 0$ and $|h_1\rangle = |h_2\rangle = |0\rangle$):

$$\nabla_x^\mu \langle 0 | \mathcal{T}[Z_A A_\mu^f(x) P^g(0)]|0\rangle - 2[m_0 - \overline{m}]\langle 0|\mathcal{T}[P^f(x)P^g(0)]|0\rangle$$

$$= \langle 0|\mathcal{T}[\overline{X}^f(x)P^g(0)]|0\rangle + \delta^{fg}\delta(x)\frac{1}{N_F}\langle 0|\bar{\psi}\psi|0\rangle. \qquad (3.100)$$

Note that only the term proportional to S^0 survives in the vacuum expectation value of the axial variation of the pseudoscalar density (3.83). Moreover, the contact terms arising in the limit of vanishing lattice spacing from the insertion of $\overline{X}^f(x)$ with $P^g(0)$ at $x \approx 0$ are not exactly those of Eq. (3.84): at first sight it appears that the only contact term surviving in Eq. (3.84), sandwiched between vacuum states is the one proportional to $\langle 0|S^0|0\rangle$. But this is not the whole story, since we should also take into account terms that vanish in Eq. (3.84), when hadronic states $|h_1\rangle \neq |h_2\rangle$, but survive when these states are equal (e.g. the vacuum). The reader can easily convince himself that in the chiral limit the contact term structure is

$$\langle 0|\mathcal{T}[\overline{X}^f(x)P^g(0)]|0\rangle = \delta^{fg}\left[\frac{1}{N_F}\delta(x)\frac{b_0(g_0^2)}{a^3} + \Box\delta(x)\frac{b_1(g_0^2)}{a}\right], \qquad (3.101)$$

where the term proportional to $\langle 0|S^0|0\rangle$ is subdominant (it may be considered as incorporated in b_0). Away from the chiral limit we also have terms proportional to the mass (e.g. $\propto m^{\mathrm{PCAC}}/a^2$) but these are subdominant compared to the cubic divergence and irrelevant for the present discussion.

Combining the last two equations in the chiral limit we obtain

$$\nabla_x^\mu \langle 0|T[Z_A\, A_\mu^f(x)P^g(0)]|0\rangle - \delta^{fg}\Box\delta(x)\frac{b_1(g_0^2)}{a} = \delta^{fg}\delta(x)\frac{1}{N_F}\left[\langle 0|\bar\psi\psi|0\rangle + \frac{b_0(g_0^2)}{a^3}\right].$$

(3.102)

The subtraction on the l.h.s. compensates a divergent term, arising in the expectation value $\langle 0|T[A_\mu^f(x)P^g(0)]|0\rangle$ when $x \approx 0$. This term is proportional to $\partial_\mu\delta(x)$. Multiplying the Ward identity by Z_P renders both sides finite. This leads to the following definition of the renormalized chiral condensate with Wilson fermions

$$\langle\bar\psi\psi\rangle_{\rm R} \equiv Z_P \frac{1}{N_f}\left[\langle 0|\bar\psi\psi|0\rangle + \frac{b_0(g_0^2)}{a^3}\right].$$

(3.103)

Note that a chirally symmetric regularization would lead to the much simpler result $\langle\bar\psi\psi\rangle_{\rm R} = Z_{S^0}\langle 0|\bar\psi\psi|0\rangle/N_f$. The more complicated renormalization pattern of the last expression is due to the loss of chiral symmetry by Wilson fermions.

Useful information may also be obtained from Ward identity (3.100), by integrating it over all space-time. We distinguish two cases. In the first case we take the chiral limit *before* performing the integration. Then, the term proportional to $(m_0 - \overline{m})$ vanishes. The integrated $\nabla_x^\mu \hat A_\mu^f$-term would also vanish, being the integral of a four-divergence, *provided chiral symmetry were not broken in QCD* (absence of Goldstone bosons). Since upon integration, the Schwinger b_1-term of Eq. (3.101) also vanishes, the integrated Ward identity (3.102) would then imply a vanishing chiral condensate. However, in QCD the symmetry is broken, and the presence of Goldstone bosons guarantees a non-vanishing surface term upon integrating $\nabla_x^\mu \hat A_\mu^f$. Thus, the r.h.s. of the Ward identity (i.e. the chiral condensate) is also non-zero.

The second case of interest consists in integrating Ward identity (3.100) over all space-time, *before* going to the chiral limit. Now, this integration will kill off the first term on the l.h.s. (it is an integral of the total derivative of the axial current in the presence of massive states) while the second term survives. Upon integration, the Schwinger b_1-term of Eq. (3.101) also vanishes. Taking the chiral limit *after* the integration, we obtain

$$\frac{\delta^{fg}}{N_f}\left[\langle 0|\bar\psi\psi|0\rangle + \frac{b_0(g_0^2)}{a^3}\right] = -\lim_{m_0\to m_{\rm cr}} 2[m_0 - \overline{m}]\int d^4x\,\langle 0|T[P^f(x)P^g(0)]|0\rangle. \quad (3.104)$$

From the above equation, we can derive two other expressions for the chiral condensate, by inserting, in standard fashion, a complete set of states in the time-ordered product of pseudoscalar densities. The spatial integration $\int d^3x$ projects zero-momentum states. Contributions from higher-mass states vanish in the chiral limit, leaving us with a zero-momentum pion state $|\pi(\vec 0)\rangle$. Upon performing the time integration $\int dx^0$, we find

$$\frac{1}{N_f}\left[\langle 0|\bar\psi\psi|0\rangle + \frac{b_0(g_0^2)}{a^3}\right] = \lim_{m_0\to m_{\rm cr}} \frac{(m_0 - \overline{m})}{m_\pi^2}\left|\langle 0|P(0)|\pi(\vec 0)\rangle\right|^2,$$

(3.105)

where m_π is the mass of the pseudoscalar state $|\pi\rangle$ and the flavor indices f, g have been suppressed for simplicity.

Next, recall that the definition of the pion decay constant is given by

$$\langle 0|[A_\mu(x)]_{\mathrm{R}}|\pi(\vec{p})\rangle = if_\pi p_\mu \exp(ipx). \tag{3.106}$$

We also know that

$$Z_A \nabla^\mu_x \langle 0|A_\mu(x)|\pi(\vec{p})\rangle = 2[m_0 - \overline{m}]\langle 0|P(x)|\pi(\vec{p})\rangle = -f_\pi m^2_\pi \exp(ipx), \tag{3.107}$$

where the first equation is the axial Ward identity between states $|0\rangle$ and $|\pi\rangle$, while the second one is obtained by derivation of Eq. (3.106). Taking the square of the second equation at $\vec{p} = \vec{0}$ and combining it with Eq. (3.105) we find

$$\frac{1}{N_f}\left[\langle 0|\bar{\psi}\psi|0\rangle + \frac{b_0(g_0^2)}{a^3}\right] = \lim_{m_0 \to m_{\mathrm{cr}}} \frac{f_\pi^2 m_\pi^2}{4(m_0 - \overline{m})}. \tag{3.108}$$

Multiplying both sides by Z_P yields the familiar Gell-Mann–Oakes–Renner relation (Gell-Mann *et al.*, 1968):

$$\langle \bar{\psi}\psi \rangle_{\mathrm{R}} = \lim_{m_{\mathrm{R}} \to 0} \frac{f_\pi^2 m_\pi^2}{4m_{\mathrm{R}}}. \tag{3.109}$$

Note that the non-vanishing of the chiral condensate in the last two equations implies the well-known linear dependence of the pseudoscalar mass squared on the quark mass, close to the chiral limit. Provided that numerical simulations can be performed close to the chiral limit, Eqs. (3.105) and (3.108) may be used for the computation of the chiral condensate with Wilson fermions.[8]

This concludes our discussion of lattice Ward identities with Wilson fermions. A related subject is that of Ward identities of the singlet axial current, related to the question of $U(1)_A$ anomaly and the η' mass. Although extremely important, this topic is beyond the scope of the present chapter.

3.4   Momentum-subtraction (MOM) schemes

We have seen that Ward identities may be used in order to determine non-perturbatively the normalization of partially conserved currents and finite ratios of renormalization constants of operators of the same chiral multiplet. These quantities would be identically equal to unity if chiral symmetry were preserved by the lattice regularization. On the other hand, local operators like S^a, P^a and $T^a_{\mu\nu}$ are subject to multiplicative renormalization by scale-dependent renormalization parameters Z_S, Z_P and Z_T, which must be fixed by a more or less arbitrary renormalization condition. We now present a renormalization scheme, known as the RI/MOM scheme, which is suitable for the non-perturbative evaluation of these parameters.

Before presenting the RI/MOM scheme explicitly, we will discuss a few general features of momentum-subtraction schemes. For definitiveness, let us consider the dimension-3, multiplicatively renormalizable operator O_Γ^a defined in Eq. (3.23).

Mimicking what is usually done with operator renormalization in continuum momentum subtraction (MOM) schemes, we impose that a suitable renormalized vertex function between say, quark states, at a given momentum scale μ, be equal to its tree-level value. For instance,

$$\left[\langle p'|O_\Gamma^a|p\rangle\right]_R\Big|_{p'^2=p^2=\mu^2} = Z_O(a\mu)\langle p'|O_\Gamma^a|p\rangle_{\text{bare}} = \langle p'|O_\Gamma^a|p\rangle_{\text{tree}}, \qquad (3.110)$$

where $|p\rangle$, $|p'\rangle$ are single quark states of four-momenta p, p', respectively. This is a familiar example of a momentum-subtraction scheme (MOM scheme). Some important properties of such schemes are:

- The renormalization condition, imposed on quark states, is gauge dependent. Gauge fixing is required and one typically opts for the Landau gauge. Care should be taken that no problems arise from such a choice. For instance, dependence of the results on Gribov copies of a given configuration ensemble is obviously undesirable.

- MOM is a mass-independent, infinite-volume renormalization scheme. This means that in principle the condition (3.110) is written at infinite lattice volumes and vanishing quark masses. In practice, numerical simulations are performed at large but finite volumes and non-zero quark masses. Results are then extrapolated to the chiral limit; their independence on the volume must be carefully checked.

- The above renormalization condition is imposed in the chiral limit. By working at non-vanishing exceptional momenta, any infrared problems associated with vanishing quark masses are avoided (Itzykson and Zuber, 1980).

- Once the bare matrix element in the above MOM condition is regularized on the lattice, the renormalization constant $Z_O(a\mu)$. computed non-perturbatively, is subject to systematic errors due to discretization effects that are typically $\mathcal{O}(a\mu)$, $\mathcal{O}(a\Lambda_{\text{QCD}})$ and $\mathcal{O}(am)$, for non-improved Wilson fermions. Effects that are $\mathcal{O}(am)$ are extrapolated away by going to the chiral limit, while those that are $\mathcal{O}(a\mu)$ and $\mathcal{O}(a\Lambda_{\text{QCD}})$ are entangled with the non-perturbative definition of Z_O.

The reliability of non-perturbative Z_O estimates, computed in a MOM scheme, depend on the existence of a so-called "renormalization window" for the renormalization scale μ, defined through the inequalities

$$\Lambda_{\text{QCD}} \ll \mu \ll \mathcal{O}(a^{-1}). \qquad (3.111)$$

The upper bound guarantees that $\mathcal{O}(a\mu)$ discretization effects are under control (and so are $\mathcal{O}(a\Lambda_{\text{QCD}})$ ones). The lower bound is imposed for two reasons. The first reason is that in many cases, related to operator weak matrix elements, the physical amplitude of interest $\mathcal{A}$ is expressed in terms of an operator product expansion (OPE), schematically written as:

$$\mathcal{A} = \langle f|\mathcal{H}_{\text{eff}}|i\rangle = C_W\left(\frac{\mu}{M_W}\right)\langle f|O(\mu)|i\rangle_R, \qquad (3.112)$$

where $|i\rangle$ and $|f\rangle$ are physical initial and final states. The scale-dependent Wilson coefficient C_W comprises all short-distance effects of the physical process under consideration. It is known (typically to NLO) in perturbation theory. The renormalized weak matrix element

$$\langle f|O(\mu)|i\rangle_{\mathrm{R}} = \lim_{a\to 0} Z[g_0(a), a\mu]\langle f|O[g_0(a)]|i\rangle_{\mathrm{bare}}, \tag{3.113}$$

is a long-distance quantity, computed non-perturbatively. Its scale dependence cancels that of C_W (to say, NLO), so as to have a scale-independent physical amplitude $\mathcal{A}$. Since C_W is a perturbative quantity, it should be calculated at a scale μ, well above Λ_{QCD}. But μ is also the scale at which Z_O must be computed non-perturbatively, which leads to the lower bound of the renormalization window.

The second reason for requiring that $\Lambda_{\mathrm{QCD}} \ll \mu$ is related to the existence of unwanted non-perturbative effects, which affect the determination of some Z_Os at scales close to the infrared. This so-called "Goldstone pole contamination" will be treated extensively below.

Moreover, the UV cutoff a^{-1} must be well above the energy scales of the problem, such as Λ_{QCD}, μ (and any quark masses of active flavors). Finally, all these scales must also be higher than the lattice extension L, which acts as an infrared cutoff. Thus, we must tune the bare coupling $g_0(a)$ so that

$$L^{-1} \ll \Lambda_{\mathrm{QCD}} \ll \mu \ll a^{-1}. \tag{3.114}$$

With present-day computer resources, this hierarchy of scales is not always easy to satisfy in practice. One could then turn to finite-size scaling methods, similar to the ones based on the Schrödinger functional (Weisz, 2010). Although these methods have been described in general terms for the RI/MOM scheme (Donini *et al.*, 1999), they have only been systematically formulated and developed very recently (Arthur and Boyle, 2011), with extremely promising preliminary results.

3.4.1   The RI/MOM scheme

A specific renormalization scheme, which is particularly well suited for non-perturbative computations of renormalization constants is the RI/MOM scheme. As the name betrays, it is a momentum-subtraction scheme, similar to the one described above (the reason for the acronym "RI" is described in Appendix 3.9). The difference lies in the fact that the renormalization condition is not imposed on operator matrix elements of single-quark states, but on the amputated projected correlation function defined in Eq. (3.30). Since the renormalization pattern of this correlation function is given by Eq. (3.36), it immediately follows that the renormalization condition

$$\left[\Gamma_O(\mu, g_{\mathrm{R}}, m_{\mathrm{R}} = 0)\right]_{\mathrm{R}} = \lim_{a\to 0} \left[Z_\psi^{-1}(a\mu, g_0) Z_O(a\mu, g_0)\Gamma_O(p, g_0, m) \right]_{\substack{p^2 = \mu^2 \\ m \to 0}} = 1, \tag{3.115}$$

fixes the combination $Z_\psi^{-1} Z_O$. Note that unity on the r.h.s. stands for the tree-level value of the correlation function Γ_O, which has been constructed so as to satisfy this property.

There are several ways to disentangle the two renormalization constants from the product $Z_\psi^{-1} Z_O$. A conceptually straightforward method is based on the conservation of the vector current $\tilde{V}_\mu$. We *assume*[9] that the RI/MOM condition for the conserved current is compatible with the Ward identity result of Eq. (3.50), $Z_{\tilde{V}} = 1$. This means that

$$[\Gamma_{\tilde{V}}(\mu, g_R, m_R = 0)]_R = \lim_{a \to 0} \left[Z_\psi^{-1}(a\mu, g_0) \Gamma_{\tilde{V}}(p, g_0, m) \right]_{\substack{p^2 = \mu^2 \\ m \to 0}} = 1. \qquad (3.116)$$

The quark field renormalization Z_ψ cancels in the ratio $\Gamma_O(\mu)/\Gamma_{\tilde{V}}(\mu)$, which can be solved for Z_O. Alternatively, one often uses the product $Z_V V_\mu$ in place of $\tilde{V}_\mu$, with Z_V known from some Ward identity.

Another method is to first compute Z_ψ from the RI/MOM condition of the quark propagator, based on Eqs. (3.19) and (3.21):

$$[\Gamma_\Sigma(\mu)]_R = \lim_{a \to 0} \left[Z_\psi^{-1}(a\mu, g_0) \Gamma_\Sigma(p, g_0, m) \right]_{\substack{p^2 = \mu^2 \\ m \to 0}} = 1. \qquad (3.117)$$

Once Z_ψ is known, Eq. (3.115) gives us Z_O. This method is avoided in practice, because the definition of Γ_Σ involves derivatives, which look rather ugly with discretized momenta. A slightly different renormalization condition is preferred for the fermion field renormalization:

$$\frac{i}{12} \text{Tr} \left[\frac{\slashed{p}[\mathcal{S}^{-1}]_R}{p^2} \right] = \lim_{a \to 0} (Z_\psi')^{-1} \frac{i}{12} \text{Tr} \left[\frac{\slashed{p}\mathcal{S}^{-1}}{p^2} \right]_{\substack{p^2 = \mu^2 \\ m \to 0}} = 1. \qquad (3.118)$$

Operator renormalization constants Z_O, computed with Z_ψ' rather than Z_ψ, are said to be in the RI'/MOM scheme. In general, we expect that $Z_\psi' = Z_\psi + \mathcal{O}(g_0^2)$. However, in the Landau gauge it is known that $Z_\psi' = Z_\psi + \mathcal{O}(g_0^4)$. As this is the gauge of preference, differences between RI/MOM and RI'/MOM results are expected to be small.

For the quark mass, an RI/MOM renormalization condition, based on Eqs. (3.20) and (3.22) would be:

$$[\Gamma_m(\mu)]_R = \lim_{a \to 0} \left[Z_\psi^{-1}(a\mu, g_0) Z_m^{-1}(a\mu, g_0) \Gamma_m(p, g_0, m) \right]_{\substack{p^2 = \mu^2 \\ m \to 0}} = 1. \qquad (3.119)$$

In numerical simulations, this is also inconvenient, as Γ_m requires derivation of w.r.t the mass. The following condition is used in practice:

$$[\Gamma_m'(\mu)]_R = \lim_{a \to 0} \left[Z_\psi^{-1}(a\mu, g_0) Z_m^{-1}(a\mu, g_0) \Gamma_m'(p, g_0, m) \right]_{\substack{p^2 = \mu^2 \\ m \to 0}} = 1 \qquad (3.120)$$

with

$$\Gamma_m'(p; m) = \frac{1}{12m} \text{Tr} \left[\mathcal{S}^{-1}(p; m) \right]. \qquad (3.121)$$

Note that the renormalization pattern of Γ'_m is analogous to that of Eq. (3.22) for Γ_m. The two conditions are equivalent, as differentiation of Eq. (3.120) w.r.t. the quark mass m results in Eq. (3.119).)

In practical simulations, the RI/MOM' conditions are solved for Z'_ψ and Z_O at fixed bare coupling. The lattices are big with periodic and/or antiperiodic boundary conditions. At each g_0, one has to work at several small quark masses and then extrapolate the results to the chiral limit. The whole procedure is repeated for several lattice momenta ap in order to empirically establish a range of renormalization scales $ap = a\mu$, within the renormalization window of Eq. (3.114).

3.4.2   Goldstone pole contamination

It is clear from the previous section that RI/MOM is a mass-independent renormalization scheme, i.e. the renormalization conditions are imposed in the chiral limit. For field theories with vertices with degree four (such as QCD), renormalized at some fixed Euclidean point, it is well known (Itzykson and Zuber, 1980), that if all masses of a Feynman diagram go to zero, infrared singularities may appear, unless the external momenta are non-exceptional.[10] This is not our case: the renormalization condition (3.115) is imposed on the amputated-projected Green function $\Gamma_O(p)$, derived from the momentum-space correlation function $G_O(p_1, p_2)$ of Eq. (3.28) for $p_1 = p_2$. In other words, the operator O carries zero momentum ($q \equiv p_1 - p_2 = 0$) and quark-field external momenta are exceptional. In order to investigate which correlation functions are subject to dangerous infrared singularities, we apply the LSZ reduction formula (Peskin and Schroeder, 1995) to the correlation functions $G_O(p_1, p_2)$ (Papinutto, 2001). Returning momentarily to Minkowski space-time, we obtain for the pseudoscalar operator P:

$$G_P(p_1, p_2) \sim \int d^4x_2 \exp(ip_2 x_2) \Big[\langle 0|T[\psi(0)\bar\psi(x_2)]|\pi(\vec{q})\rangle \frac{i\theta(q^0)}{q^2 - m_\pi^2} \langle \pi(\vec{q})|[P(0)]_R|0\rangle$$

$$+ \langle 0|[P(0)]_R|\pi(-\vec{q})\rangle \frac{i\theta(-q^0)}{q^2 - m_\pi^2} \langle \pi(-\vec{q})|T[\psi(0)\bar\psi(x_2)]|0\rangle \Big] + \dots, \qquad (3.122)$$

where the ellipsis indicates terms without pion poles; isospin indices are implicit.[11]

Combining Eqs. (3.107) and (3.109) we obtain for the vacuum-to-pion pseudoscalar matrix element

$$\langle 0|[P(0)]_R|\pi(p)\rangle = -2\frac{\langle \bar\psi\psi\rangle_R}{f_\pi} \exp(ipx). \qquad (3.123)$$

Since $\langle \bar\psi\psi\rangle_R \neq 0$ (spontaneous symmetry breaking), the above matrix element survives the chiral limit. This last result, applied to Eq. (3.122) in the limit $q_\mu \to 0$, implies that the correlation function $G_P(p, p)$ (and thus also $\Gamma_P(p)$) have a $1/m_\pi^2$ pole. The presence of this so-called Goldstone pole means that the RI/MOM renormalization condition, which strictly speaking is valid in the chiral limit, can only be applied to Z_P with some precaution (see below for details).

The situation is less dramatic in the case of the axial current, for which the LSZ reduction formula is

$$G_A^\mu(p_1, p_2) \sim \int d^4x_2 \exp(ip_2x_2) \Big[\langle 0|T[\psi(0)\bar\psi(x_2)]|\pi(\vec q)\rangle \frac{i\theta(q^0)}{q^2 - m_\pi^2} \langle\pi(\vec q)|[A_\mu(0)]_R|0\rangle$$

$$+ \langle 0|[A_\mu(0)]_R|\pi(-\vec q)\rangle \frac{i\theta(-q^0)}{q^2 - m_\pi^2} \langle\pi(-\vec q)|T[\psi(0)\bar\psi(x_2)]|0\rangle \Big] + \ldots \qquad (3.124)$$

Since the relevant matrix element behaves like $\langle 0|[A_\mu(0)]_R|\pi(q)\rangle \sim q_\mu$ (cf. Eq (3.106)) in the limit $q_\mu \to 0$ the Goldstone pole contribution of $G_A^\mu(p,p)$ vanishes. However, as we will see below, the pole of $G_A^\mu(p,p)$ still plays an important role in axial Ward identities, used for the determination of Z_A. Other correlation functions such as G_V^μ, G_S and G_T (as well as Γ_V, Γ_S and Γ_T) have analogous LSZ reduction formulae. In these cases, however, the interpolating operators V_0, S and $T_{\mu\nu}$ have Lorentz properties which ensure that the corresponding intermediate single-particle state is not a pion. As the mass of these particles does not vanish in the chiral limit, these correlation functions are free of Goldstone pole singularities.[12]

It is important to realize that the Goldstone pole term of $G_P(p,p)$ vanishes in the limit of large external momentum $p^2 \to \infty$. This has been discussed in general terms in the original RI/MOM paper (Martinelli *et al.*, 1995), where the following correlation function was considered:

$$F_O(p) = \int d^4x \, d^4y \, \exp(-ipx) \langle [\bar\psi(0)\Gamma\psi(x)]O_\Gamma(y) \rangle, \qquad (3.125)$$

with $O_\Gamma(y)$ defined in Eq. (3.23). In the limit of large p^2, the dominant contribution to $F_O(p)$ comes from regions of integration where the integrand is singular; i.e. $x \sim 0$ and $x \sim y$. Based on the OPE and dimensional arguments, it is shown that (Martinelli *et al.*, 1995)

$$F_O(p) = c_\Gamma \frac{\ln^{\gamma_\Gamma}(p^2/\mu^2)}{p^2} + d_\Gamma \frac{\ln^{\delta_\Gamma}(p^2/\mu^2)}{p^4} \tilde\Delta_\Gamma(0), \qquad (3.126)$$

where c_Γ, γ_Γ, d_Γ and δ_Γ can be calculated in perturbation theory, while

$$\tilde\Delta_\Gamma(0) = \int d^4y \langle O_\Gamma(0)O_\Gamma(y) \rangle \qquad (3.127)$$

is a non-perturbative quantity. This means that, in the large p^2 limit, the perturbative contribution of $F_O(p)$ (i.e. the first term, proportional to c_Γ) dominates by one power of p^2 over the non-perturbative one (i.e. the term proportional to d_Γ). This is why $F_O(p)$ is infrared safe in perturbation theory, when p^2 is large. In other words, in the large external momentum limit, the contamination by the Goldstone pole residing in $\tilde\Delta_\Gamma(0)$, as well as other infrared chiral effects, vanishes. At finite momenta however, the Green function Γ_P, defined in Eq. (3.30), is not regular in the chiral limit. By requiring that $\Lambda_{QCD} \ll \mu$ (the lower bound of the renormalization window (3.114)) one is attempting to reduce, in practical simulations, the contamination due to infrared chiral effects, such as the Goldstone pole.

3.4.3   RI/MOM scheme and Ward identities

Once the wave-function renormalization Z_ψ is computed, the RI/MOM renormalization condition (3.115) fixes the renormalization parameter Z_O; examples are Z_S, Z_P, Z_T but also Z_V and Z_A. However, Z_V, Z_A and the ratio of Z_S/Z_P do not depend on any renormalization scheme (or renormalization scale), as they are fixed by Ward identities. It is then important that their determination through the RI/MOM renormalization condition be compatible with such identities.[13] In this section we will establish that the RI/MOM determination of Z_S/Z_P agrees with that from Ward identities only in the limit of large renormalization scales $\mu \gg \Lambda_{\mathrm{QCD}}$, where Goldstone pole contaminations die off.[14] On the other hand, the RI/MOM determination of Z_V is compatible with Ward identities at all momentum scales. Finally, Goldstone poles do not affect the RI/MOM determination of Z_A, provided the chiral limit is approached with care.

Let us start from the vector Ward identity (3.49), with $\tilde{V}_\mu$ replaced by $Z_V V_\mu$. We Fourier transform and amputate Eq. (3.49), obtaining (at small aq_μ)

$$
\sum_\mu iq_\mu\, Z_V\, \Lambda_V^\mu \left(p + \frac{q}{2}, p - \frac{q}{2}; m_{01}, m_{02}\right) - (m_{02} - m_{01})\Lambda_S \left(p + \frac{q}{2}, p - \frac{q}{2}; m_{01}, m_{02}\right)
$$

$$
= \mathcal{S}^{-1}\left(p + \frac{q}{2}; m_{01}\right) - \mathcal{S}^{-1}\left(p - \frac{q}{2}; m_{02}\right). \tag{3.128}
$$

We next go through the following steps: (i) derive the above expression w.r.t. q_μ; (ii) let $m_{01} = m_{02}$, so as to dispose of the term with the scalar vertex function Λ_S; (iii) trace the resulting equation with γ_μ, (iv) take the limit $q_\mu \to 0$. This gives

$$
Z_V\, \Gamma_V(p) + Z_V \lim_{q_\mu \to 0} \frac{q_\mu}{48} \mathrm{Tr}\left[\gamma_\rho \frac{\partial \Lambda_V(p + q/2, p - q/2)}{\partial q_\rho}\right] = \Gamma_\Sigma(p). \tag{3.129}
$$

The LSZ reduction formula for the vector current is analogous to Eq. (3.124), but the denominator does not involve a Goldstone pole m_π. This implies that, upon taking its derivative w.r.t. q_ρ, and in the limit $q_\mu \to 0$, the second term on the l.h.s. of the above Ward identity vanishes (there are no infrared singularities). Thus, we obtain

$$
Z_V \Gamma_V(p) = \Gamma_\Sigma(p). \tag{3.130}
$$

Solved for Z_V this is yet another Ward identity determination of the vector current normalization. On the other hand, if the above expression is multiplied by Z_ψ^{-1}, determined in the RI/MOM scheme through Eq. (3.117), we obtain at momenta $p = \mu$

$$
Z_\psi^{-1} Z_V \Gamma_V(\mu) = 1, \tag{3.131}
$$

which is the RI/MOM condition for Z_V. We have shown that for the vector current the RI/MOM scheme is consistent with Ward identities.

The situation is not quite the same for axial Ward identities. Starting from Eq. (3.65), written for degenerate quark masses, we perform the same steps that led from Eq. (3.128) to Eq. (3.129) and obtain

$$Z_A \Gamma_A(p) + Z_A \lim_{q_\mu \to 0} \frac{q_\mu}{48} \text{Tr}\left[\gamma_5 \gamma_\rho \frac{\partial \Lambda_A^\mu(p+q/2, p-q/2)}{\partial q_\rho}\right] \tag{3.132}$$

$$+ \frac{2m^{\text{PCAC}}}{48} \lim_{q_\mu \to 0} \text{Tr}\left[\gamma_5 \gamma_\rho \frac{\partial \Lambda_P(p+q/2, p-q/2)}{\partial q_\rho}\right] = \Gamma_\Sigma(p).$$

The LSZ reduction formulae (i.e. Eq. (3.124) for the axial current and Eq. (3.122) for the pseudoscalar density) may now be used in order to study the Goldstone pole contributions to the second and third term on the l.h.s. It can be easily shown that, away from the chiral limit, as $q_\mu \to 0$, these terms vanish. Approaching subsequently the chiral limit, just like in the vector current case, the axial Ward identity determination of Z_A is equivalent to the RI/MOM condition. If however, the chiral limit $m^{\text{PCAC}} \to 0$ is taken *before* $q_\mu \to 0$, the third term vanishes but the second one survives. This term is essential to the saturation of the Ward identity, in the presence of a massless Goldstone boson (i.e. any divergences due to the presence of massless pions in the first and second term will cancel). In this regime, the determination of Z_A from the RI/MOM condition is not feasible, unless one considers the $p \to \infty$ case. As we have seen in Section 3.4.2, in this limit these Goldstone pole terms become negligible.

The simplest way to see a discrepancy between Ward identities and RI/MOM is through Z_S/Z_P (Giusti and Vladikas, 2000). We start from the Ward identities (3.56) and (3.70), which we Fourier transform to momentum space; cf. Eqs. (3.17) and (3.28). We next amputate the resulting correlation functions as in Eq. (3.29) and project them as in Eqs. (3.30) and (3.31), obtaining the following vector and axial Ward identities:

$$(m_{02} - m_{01})\Gamma_S(p; m_{01}, m_{02}) = \frac{1}{12}\text{Tr}\left[\mathcal{S}^{-1}(p; m_{02}) - \mathcal{S}^{-1}(p; m_{01})\right], \tag{3.133}$$

$$(m_2^{\text{PCAC}} + m_1^{\text{PCAC}})\Gamma_P(p; m_{01}, m_{02}) = \frac{1}{12}\text{Tr}\left[\mathcal{S}^{-1}(p; m_{01}) + \mathcal{S}^{-1}(p; m_{02})\right]. \tag{3.134}$$

In the mass-degenerate limit these become

$$\Gamma_S(p) = \frac{1}{12}\text{Tr}\left[\frac{\partial \mathcal{S}^{-1}(p; m_0)}{\partial m_0}\right], \tag{3.135}$$

$$m^{\text{PCAC}}\Gamma_P(p) = \frac{1}{12}\text{Tr}\left[\mathcal{S}^{-1}(p; m_0)\right]. \tag{3.136}$$

The vector Ward identity (3.135) and the definition (3.20) imply that $\Gamma_S(p) = \Gamma_m(p)$. Once again we see from this and Eq. (3.22) that if Z_S is determined by the RI/MOM condition, then compatibility with this Ward identity requires that the mass renormalization satisfy $Z_m = Z_S^{-1}$.

We now turn to the axial Ward identity (3.136). The inverse quark propagator may be considered a function of either bare quark mass, m_0 or m^{PCAC}. We differentiate the above axial Ward identity w.r.t. m^{PCAC}, obtaining

$$\Gamma_P(p) + m^{\text{PCAC}} \frac{\partial \Gamma_P}{\partial m^{\text{PCAC}}} = \frac{1}{12} \text{Tr}\left[\frac{\partial \mathcal{S}^{-1}(p; m^{\text{PCAC}})}{\partial m^{\text{PCAC}}}\right]$$

$$= \frac{\partial m_0}{\partial m^{\text{PCAC}}} \frac{1}{12} \text{Tr}\left[\frac{\partial \mathcal{S}^{-1}(p; m_0)}{\partial m_0}\right]$$

$$= \frac{Z_S}{Z_P} \frac{1}{12} \text{Tr}\left[\frac{\partial \mathcal{S}^{-1}(p; m_0)}{\partial m_0}\right]. \tag{3.137}$$

In the last step, the substitution of the derivative $\partial m_0 / \partial m^{\text{PCAC}}$ by the ratio Z_S/Z_P is a consequence of Eq. (3.81). With the aid of the LSZ expression (3.122) at zero momentum transfer ($q_\mu = 0$) we easily confirm that the second term of the l.h.s. of Eq. (3.137) diverges in the chiral limit, due to the presence of a Goldstone pole. This term is essential for the cancellation of a similar contribution of the first term on the l.h.s. Again as $p \to \infty$ these contributions become negligible.

The ratio of the above vector and axial Ward identities (3.135) and (3.137) gives the following result for Z_P/Z_S:

$$\frac{Z_P}{Z_S} = \frac{\dfrac{\Gamma_S}{\Gamma_P}}{1 + \dfrac{m^{\text{PCAC}}}{\Gamma_P} \dfrac{\partial \Gamma_P}{\partial m^{\text{PCAC}}}}. \tag{3.138}$$

But this is not what the RI/MOM renormalization condition (3.115) gives for the same ratio:

$$\frac{Z_P}{Z_S} = \frac{\Gamma_S}{\Gamma_P}. \tag{3.139}$$

The two determinations differ by a factor that, as we have already discussed, is necessary in order to cancel Goldstone pole divergences in Γ_P and that becomes negligible in the limit $p^2 \to \infty$; in practice this limit is realized as $p^2 \gg \Lambda_{\text{QCD}}$. The absence of this factor from the RI/MOM determination is commonly referred to as Goldstone pole contamination.

It is also instructive to identify the origin of the Goldstone pole in the nonperturbative part of the quark propagator:

$$\mathcal{S}^{-1}(p; m) = i\not{p}\Sigma_1(p^2; m; \mu^2) + m\Sigma_2(p^2; m; \mu^2) + \Sigma_3(p; m; \mu^2), \tag{3.140}$$

where Σ_k (with $k = 1, \cdots, 3$) are form factors. The functional form of the first two terms is dictated by general symmetry arguments; the form factors Σ_1 and Σ_2 may be calculated in perturbation theory. The last term is a non-perturbative form factor, known to several orders in the OPE; to $\mathcal{O}(1/p^2)$ it is given by (Politzer, 1976; Pascual and de Rafael, 1982):

$$\Sigma_3(p; \mu; m) = g_0^2 K \langle \bar{\psi}\psi \rangle \frac{1}{p^2} + \mathcal{O}\left(p^{-4}\right), \tag{3.141}$$

where K is a mass-independent, gauge-dependent factor, in which logarithmic divergences have also been absorbed. Note the non-perturbative nature of this term: it is proportional to the chiral condensate $\langle \bar{\psi}\psi \rangle$ and vanishes in the large-scale limit $p^2 \to \infty$. Upon inserting the quark propagator expressions (3.140) and (3.141) into Ward identities (3.135) and (3.136), we find[15]

$$\Gamma_S(p; m) = \Sigma_2(p; m) + m \frac{\partial \Sigma_2(p; m)}{\partial m} + \mathcal{O}\left(p^{-4}\right) \tag{3.142}$$

$$\Gamma_P(p; m) = \frac{m}{m^{\mathrm{PCAC}}} \Sigma_2(ap, am) + g_0^2 K \frac{\langle \bar{\psi}\psi \rangle}{m^{\mathrm{PCAC}} p^2} + \mathcal{O}\left(p^{-4}\right)$$

$$= \frac{Z_{S^0}}{Z_P}\left[\Sigma_2(ap, am) + g_0^2 K \frac{\langle \bar{\psi}\psi \rangle}{mp^2}\right] + \mathcal{O}\left(p^{-4}\right). \tag{3.143}$$

Thus, with the aid of an axial Ward identity, we see that, to $\mathcal{O}(1/p^2)$ in the OPE, the correlation function Γ_P has a pole in the quark mass, proportional to the chiral symmetry-breaking order parameter $\langle \bar{\psi}\psi \rangle$. On the contrary, using a vector Ward identity we see that such a pole is absent, to $\mathcal{O}(1/p^2)$, from the scalar correlation function Γ_S.[16]

As stated repeatedly, this non-perturbative contribution to Γ_P vanishes like $1/p^2$ at large momenta. At finite momenta, however, Γ_P is not regular in the chiral limit and $p^2 m^{\mathrm{PCAC}} \gg \Lambda_{\mathrm{QCD}}^3$ must be enforced. In practical simulations it is usually difficult to satisfy this requirement. A remedy consists in fitting Γ_P, at fixed momenta p^2, by $A(p^2) + B(p^2)/m$. Once the fit parameters $A(p^2)$ and $B(p^2)$ are determined, the Goldstone pole contamination is removed from Z_P by imposing the RI/MOM condition on the subtracted correlation function $A(p^2)$ (Cudell *et al.*, 1999, 2001).

As we have shown to $\mathcal{O}(p^{-2})$, Γ_S does not suffer from Goldstone pole contaminations and therefore the RI/MOM determination of Z_S should be reliable even at relatively low renormalization scales. The same is true for the ratio Z_P/Z_S, computed from Ward identities, which are completely free of these problems. Thus, a reliable evaluation of Z_P consists in multiplying the ratio of Z_P/Z_S, obtained from a Ward identity, by Z_S in the RI/MOM scheme. The ratio Z_P/Z_S may be computed from hadronic Ward identities, cf. Eq. (3.80). Alternatively, Ward identities in momentum space, based on the correlation functions Γ_S and Γ_P have also been used with satisfactory results (Giusti and Vladikas, 2000).

Besides the single Goldstone pole, other, more complicated non-perturbative effects are of course also present. Some of these could be revealed by higher orders in the OPE, which involve higher-dimension condensates (Pascual and de Rafael, 1982; Lavelle and

Oleszczuk, 1992). In any case, all these contributions ought to disappear as $p^2 \gg \Lambda_{\text{QCD}}$.

3.4.4   The RI/SMOM scheme

In practice, it is not easy to satisfy the bounds imposed by the renormalization window of Eq. (3.111). At least one fine example where everything falls into place has been provided in the literature (Becirevic *et al.*, 2004). However, several cases exist in which the renormalization window turns out to be quite narrow: upon attempting to compute RI/MOM renormalization parameters at very high momenta, so as to reduce contamination due to the infrared chiral-symmetry-breaking effects, one is faced with the problem of discretization errors, which start flawing the data, even in an $\mathcal{O}(a)$-improved setup. The current precision of simulations is such that the infrared non-perturbatve effects are a significant source of error. Clearly, it would be advantageous to renormalize the operators in a scheme that is free of this problem, suppressing these effects by choosing kinematics without channels of exceptional momenta. One such choice, which for quark bilinears is characterized by very simple and convenient kinematics is the recently proposed variant of the RI/MOM scheme, called RI/SMOM (SMOM stands for *symmetric* momentum subtraction scheme). Although the new scheme has been introduced in the framework of domain-wall fermions (Aoki *et al.*, 2008; Sturm *et al.*, 2009), its definition and the advantages derived from it are independent of the regularization details.

Essentially, the scheme consists in the choice of new kinematics for the vertex function $G_O(p_1, p_2)$ of Eq. (3.28). In the standard RI/MOM scheme the renormalization conditions for quark bilinear operators are imposed on Green functions with the operator inserted between equal incoming and outgoing momenta, satisfying (in Euclidean space-time)

$$p_1^2 = p_2^2 = \mu^2; \qquad q = p_1 - p_2 = 0. \tag{3.144}$$

In the above momentum configuration, the momentum inserted at the operator is therefore $q = 0$ so that there is an *exceptional* channel, i.e. one in which the squared momentum transfer q^2 is much smaller than the typical large-scale μ^2. These kinematics define an asymmetric subtraction point. The renormalization procedure for RI/SMOM is very similar, but with the incoming and outgoing quarks having different momenta, p_1 and p_2, satisfying

$$p_1^2 = p_2^2 = q^2 = \mu^2; \qquad q = p_1 - p_2 \neq 0. \tag{3.145}$$

There are now no exceptional channels. An example of such a symmetric momentum configuration on a lattice of linear extension L is

$$p_1 = \frac{2\pi}{L}(0, 1, 1, 0); \qquad p_2 = \frac{2\pi}{L}(1, 1, 0, 0), \tag{3.146}$$

where $p = 2\pi/L(n_x, n_y, n_z, n_t)$. With this choice of a symmetric subtraction point, renormalized quantities such as $[G_O]_{\text{R}}$, $[\Lambda_O]_{\text{R}}$ and $[\Gamma_O]_{\text{R}}$ depend only on a single scale μ^2. The amputated Green function Λ_O is obtained once more as in Eq. (3.29), whereas for the projected amputated vertex function Γ_O we introduce the following set of projectors:

$$P_S = I; \qquad P_P = \gamma_5; \qquad P_T = \frac{1}{12}\gamma_\mu\gamma_\nu$$

$$P_V = \frac{1}{q^2}q_\mu \not{q}; \qquad P_A = \frac{1}{q^2}q_\mu\gamma_5 \not{q}. \tag{3.147}$$

The RI/SMOM renormalization condition is that of Eq. (3.115), imposed on the vertex functions $[\Gamma_O]_R$ (implicitly defined though the above projectors), in the symmetric momentum configuration of Eq. (3.145). We will use the shorthand notation $\{\mu^2\}$ for these kinematics.

Comparing with Eq. (3.31), we see that the RI/SMOM definitions of Γ_S, Γ_P and Γ_T are identical to those of the standard RI/MOM scheme, while Γ_V and Γ_A have been redefined through new projectors. The reason behind these changes is that we must ensure that the new scheme is consistent with Ward identities. If we wish to maintain condition (3.118) for the wave-function renormalization Z_ψ, we must modify the definitions of Γ_V and Γ_A, as implied by Eq. (3.147). These modifications must be accompanied by a redefinition of the quark mass renormalization: instead of condition (3.22), defined through Eq. (3.20), we must impose, in a symmetric momentum configuration, that

$$\lim_{a \to 0} \left\{ [\Gamma'_m(p)]_R - \frac{1}{2}\text{Tr}\left[q_\mu[\Lambda_A^\mu]_R\gamma_5 \right] \right\}\Big|_{\{\mu^2\}, m_R \to 0} = 1, \tag{3.148}$$

where $\Gamma'_m(p)$ is defined in Eq. (3.120) and Λ_A^μ in Eq. (3.29). Following a line of reasoning similar to that of Section 3.4.3, it has been shown that this RI/SMOM scheme is consistent with Ward identities (Sturm *et al.*, 2009). Clearly these choices are not unique. It is also possible to maintain the standard RI/MOM definitions for Γ_V and Γ_A but then the wave-function renormalization condition must be modified in order to ensure compatibility with Ward identities (Sturm *et al.*, 2009). This is a different RI/SMOM scheme.

Compared to the standard RI/MOM, the RI/SMOM scheme displays an improved infrared behavior. We have explicitly shown in Sections 3.4.2 and 3.4.3 how the Goldstone pole contaminates the chiral limit of certain correlation functions as their quark external momenta become exceptional ($p_1 = p_2$) and the vertex operator does not inject any momentum ($q = 0$). In the RI/SMOM scheme the kinematics are arranged so that this situation is avoided. In fact, using Weinberg's theorem it has been shown that for the asymmetric subtraction point, chiral-symmetry-breaking effects vanish like $1/p^2$ for large external momenta p^2. On the other hand, with bilinear operator renormalization constants defined at a symmetric subtraction point (with non-exceptional kinematics), unwanted infrared effects are better behaved and vanish with larger asymptotic powers, which are $\mathcal{O}(1/p^6)$ (Aoki *et al.*, 2008). Numerical evidence, in the framework of domain-wall fermion discretization, provides strong support to these arguments (Aoki *et al.*, 2008).

3.5 Twisted mass QCD (tmQCD), renormalization, and improvement

In recent years, there has been a newcomer to the family of lattice actions, known as lattice QCD with a chirally twisted mass term (tmQCD for short). It consists in a modification of the mass term of the lattice Wilson fermion action. The new regularization has several advantages, but these come at a price. On the positive side, the tmQCD fermion matrix is free of the spurious zero modes that plague quenched simulations with Wilson fermions at small quark mass, as well as dynamical fermion algorithms. Moreover, the renormalization pattern of certain physical quantities is much simpler with tmQCD. Finally, the bare mass parameters of the theory may be tuned in such a way that improvement is "automatic"; i.e. physical quantities such as masses and matrix elements are free of $\mathcal{O}(a)$-effects, without having to introduce Symanzik counterterms. The price to pay is loss of symmetry: parity and time reversal are only recovered in the continuum limit. Moreover, flavor symmetry is also affected and this leads to loss of degeneracy between some hadrons. An important example concerns pions: the neutral pion mass differs from that of the two charged pions. Since this is due to discretization effects, degeneracy is restored in the continuum. A complete presentation of tmQCD is beyond our scope. The interested reader may consult existing review articles (Sint, 2007; Shindler, 2008). Here, we will concentrate on the renormalization and improvement properties of the theory.

3.5.1   Classical tmQCD

It is instructive to begin by writing down the tmQCD classical Lagrangean density in the continuum. For simplicity, we will consider QCD with two degenerate flavors (called up and down quarks). The Dirac spinor in flavor space is then given by the doublet $\bar{\chi} = (\bar{u} \ \bar{d})$ and the fermionic Lagrangian density is defined as

$$\mathcal{L}_{\text{tm}} = \bar{\chi} \left[\slashed{D} + m_0 + i\mu_q \tau^3 \gamma_5 \right] \chi.$$

(3.149)

Compared to the familiar standard QCD Lagrangian density, we now have an additional mass term that, being a pseudoscalar, is "twisted" in chiral space. Apparently, this is not classical QCD, as the extra twisted mass term breaks parity and $SU(2)$ flavor symmetry. However, this is illusory. To see this, we redefine fermionic fields through chiral transformations (chiral rotations) in the third direction of flavor (isospin) space:

$$\chi \to \chi' = \exp\left[i\frac{\alpha}{2} \gamma_5 \tau^3 \right] \chi$$

$$\bar{\chi} \to \bar{\chi}' = \bar{\chi} \exp\left[i\frac{\alpha}{2} \gamma_5 \tau^3 \right].$$

(3.150)

We also redefine the two mass parameters through spurionic transformations

$$m_0 \to m_0' = \cos(\alpha)\, m_0 + \sin(\alpha)\mu_q$$

$$\mu_q \to \mu_q' = \cos(\alpha)\, \mu_q - \sin(\alpha)m_0.$$

(3.151)

Under these transformations, the Lagrangian density transforms as follows:

$$\mathcal{L}_{\text{tm}} \to \mathcal{L}'_{\text{tm}} = \bar{\chi}' \left[\slashed{D} + m'_0 + i\mu'_q \tau^3 \gamma_5 \right] \chi'. \tag{3.152}$$

This means that the theory is form invariant under the chiral rotations (3.150), provided they are accompanied by the mass spurionic transfrormations (3.151). In other words, these changes of field variables and mass definitions do not change the content of the theory. In fact tmQCD is a family of equivalent Lagrangian densities, connected by chiral and spurionic transformations. A member of this family of equivalent theories is standard QCD, obtained through a specific transformation angle α, chosen so that $\mu_q \to \mu'_q = 0$.

The above argument may be reformulated by defining an *invariant mass* M_{inv} and a *twist angle* ω:

$$M_{\text{inv}} = \sqrt{m_0^2 + \mu_q^2} \quad ; \quad \tan(\omega) = \frac{\mu_q}{m_0}. \tag{3.153}$$

The reason for this terminology will be clarified shortly. The Lagrangian density may be rewritten as

$$\mathcal{L}_{\text{tm}} = \bar{\chi} \left[\slashed{D} + M_{\text{inv}} \exp\left(i\omega \tau^3 \gamma_5 \right) \right] \chi; \tag{3.154}$$

i.e. it is a function of a single mass parameter M_{inv} and a dimensionless angle ω. The transformation (3.152) of the Lagrangian density may also be written as

$$\mathcal{L}_{\text{tm}} \to \mathcal{L}'_{\text{tm}} = \bar{\chi}' \left[\slashed{D} + M'_{\text{inv}} \exp\left(i\omega' \tau^3 \gamma_5 \right) \right] \chi', \tag{3.155}$$

with the invariant mass and twist angle defined as follows:

$$M'_{\text{inv}} = \sqrt{[m'_0]^2 + [\mu'_q]^2} \quad ; \quad \tan(\omega') = \frac{\mu'_q}{m'_0}. \tag{3.156}$$

Using Eqs. (3.151), it is easy to show that $M'_{\text{inv}} = M_{\text{inv}}$; i.e. the invariant mass is indeed invariant under spurionic transformations. In physical terms, this is the quark mass, which in tmQCD is seen to be a combination of both the standard mass m_0 and the twisted mass μ_q. Morevoer, the new twist angle ω' can easily be expressed in terms of the old twist angle ω and the rotation angle α:

$$\tan(\omega') = \tan(\omega - \alpha). \tag{3.157}$$

Again, we see that tmQCD may be regarded as a family of equivalent theories, parametrized by an invariant mass M_{inv} and a twist angle ω. Starting with a specific tmQCD theory (i.e. a given value of ω, defined through $\tan(\omega) = \mu_q/m_0$), we obtain standard QCD by performing chiral rotations with a transformation angle $\alpha = \omega$, which brings us to $\omega' = 0$ (i.e. $\mu'_q = 0$).

It is worth noting two values of the twist angle that are of special interest to us. First, as already explained, for chiral rotations with $\alpha = \omega$, we obtain standard QCD (i.e. $\omega' = 0$, $\mu'_q = 0$ and $m'_0 = M_{\text{inv}}$). In this case, the quark mass is carried entirely by the standard mass m'_0. The second case of interest is when the chiral rotations are $\alpha = \omega - \pi/2$ (i.e. $\omega' = \pi/2$, $\mu'_q = M_{\text{inv}}$ and $m'_0 = 0$). In this case, the quark mass is carried entirely by the twisted mass μ'_q. This is known as the *maximally twisted* theory.

Since QCD and tmQCD at the classical level are equivalent theories, they should share the same symmetries. This means, for example, that loss of parity due to the presence of the twisted mass term in the tmQCD Lagrangian density is only apparent. It is indeed straightforward to confirm that Lagrangian density (3.149) is invariant under the parity transformations:

$$
\begin{aligned}
x = (x^0, \mathbf{x}) &\to x^{\mathcal{P}} = (x^0, -\mathbf{x}), \\
U_0(x) &\to U_0(x^{\mathcal{P}}), \\
U_k(x) &\to U_k(x^{\mathcal{P}} - \hat{k})^{\dagger}, \\
\chi(x) &\to \gamma_0 \exp[i\omega\gamma_5\tau^3]\chi(x^{\mathcal{P}}), \\
\bar{\chi}(x) &\to \bar{\chi}(x^{\mathcal{P}}) \exp[i\omega\gamma_5\tau^3]\gamma_0.
\end{aligned}
\tag{3.158}
$$

This means that in the classical tmQCD framework, parity is still a symmetry, albeit with modified transformations of the fermion fields (cf. Eq (3.199)). It should come as no surprise that the new parity transformations of χ and $\bar{\chi}$ involve the very same chiral rotations that connect tmQCD to standard QCD.

Analogous results may be obtained for the vector (isospin) symmetry. It is not really lost, as the twisted mass term of Eq. (3.149) may lead us to believe at first sight. Rather, it is simply transcribed as follows:

$$
\begin{aligned}
\chi(x) &\to \exp\left[-i\frac{\omega}{2}\gamma_5\tau^3\right] \exp\left[i\frac{\theta^a}{2}\tau^a\right] \exp\left[i\frac{\omega}{2}\gamma_5\tau^3\right]\chi(x) \\
\bar{\chi}(x) &\to \bar{\chi}(x) \exp\left[i\frac{\omega}{2}\gamma_5\tau^3\right] \exp\left[-i\frac{\theta^a}{2}\tau^a\right] \exp\left[-i\frac{\omega}{2}\gamma_5\tau^3\right],
\end{aligned}
\tag{3.159}
$$

with θ^a $(a = 1, 2, 3)$ the three rotation angles. We denote this symmetry group as $SU(2)_V^\omega$. Recall that from the beginning tmQCD has been formulated for a mass-degenerate isospin doublet.

Axial transformations are transcribed in the following form:

$$
\begin{aligned}
\chi(x) &\to \exp\left[-i\frac{\omega}{2}\gamma_5\tau^3\right] \exp\left[i\frac{\theta^a}{2}\tau^a\gamma_5\right] \exp\left[i\frac{\omega}{2}\gamma_5\tau^3\right]\chi(x) \\
\bar{\chi}(x) &\to \bar{\chi}(x) \exp\left[i\frac{\omega}{2}\gamma_5\tau^3\right] \exp\left[i\frac{\theta^a}{2}\tau^a\gamma_5\right] \exp\left[-i\frac{\omega}{2}\gamma_5\tau^3\right].
\end{aligned}
\tag{3.160}
$$

It is easy to verify that these transformations are a symmetry of the Lagrangian density (3.154), when $M_{\text{inv}} = 0$.

The field rotations, relating standard QCD and tmQCD also relate composite field operators, defined in the two theories. Since tmQCD (with twist angle ω) and standard QCD (with twist angle $\omega' = 0$) are related by chiral rotations (3.150) with $\alpha = \omega$, the following relations hold between the vector and axial currents of the two theories:

$$[V_\mu^a]_{\text{tmQCD}} = \cos(\omega)[V_\mu^a]_{\text{QCD}} - \epsilon^{3ab}\sin(\omega)[A_\mu^b]_{\text{QCD}} \qquad a,b = 1,2,$$

$$[A_\mu^a]_{\text{tmQCD}} = \cos(\omega)[A_\mu^a]_{\text{QCD}} - \epsilon^{3ab}\sin(\omega)[V_\mu^b]_{\text{QCD}} \qquad a,b = 1,2,$$

$$[V_\mu^3]_{\text{tmQCD}} = [V_\mu^3]_{\text{QCD}},$$

$$[A_\mu^3]_{\text{tmQCD}} = [A_\mu^3]_{\text{QCD}}. \tag{3.161}$$

Recall that a,b are flavour indices. Similarly, for the pseudoscalar density and the isospin singlet scalar density $S^0 \equiv \bar\chi\chi$ we obtain

$$[P^a]_{\text{tmQCD}} = [P^a]_{\text{QCD}} \qquad a,b = 1,2,$$

$$[P^3]_{\text{tmQCD}} = \cos(\omega)[P^3]_{\text{QCD}} - \frac{i}{2}\sin(\omega)[S^0]_{\text{QCD}},$$

$$[S^0]_{\text{tmQCD}} = \cos(\omega)[S^0]_{\text{QCD}} - 2i\sin(\omega)[P^3]_{\text{QCD}}. \tag{3.162}$$

Finally, the familiar PCVC and PCAC Ward identities, have the following form in the tmQCD formulation:

$$\partial_\mu [V_\mu^a]_{\text{tmQCD}} = -2\mu_q \epsilon^{3ab}[P^b]_{\text{tmQCD}} \tag{3.163}$$

$$\partial_\mu [A_\mu^a]_{\text{tmQCD}} = 2m_0[P^a]_{\text{tmQCD}} + i\mu_q \delta^{3a}[S^0]_{\text{tmQCD}}. \tag{3.164}$$

3.5.2 Lattice tmQCD

So far we have achieved precious little! The relation between classical QCD and tmQCD is simply a change of fermionic field variables, accompanied by a redefinition of the masses. A change of variables cannot bring about new physics. Thus, classical tmQCD is simply an intricate way of writing down QCD. It is upon passing over to the field-theoretic formulation that the equivalence between the two theories is less trivial and this has important consequences, both in renormalization and improvement properties of many physical quantities.

The proof that standard Wilson fermion QCD and tmQCD are equivalent field theories will only be shown schematically here. For a detailed demonstration, the reader is advised to consult the original reference (Frezzotti *et al.*, 2001). The starting point is that the equivalence in question proceeds through linear relations between renormalized Green functions of the two theories. As a first step, we consider QCD and tmQCD, regularized on the lattice with Ginsparg–Wilson fermions. The exact form of the lattice actions is not important in this discussion; for example in both lattice actions the discretization of $\slashed{D}$ may be the Neuberger operator (Neuberger, 1998*a,b,c*). What is important is that chiral symmetry is not broken, owing to the Ginsparg–Wilson relation obeyed by the regularized Dirac operator. This implies that the chiral rotations (3.150) map the lattice tmQCD bare action to the lattice QCD one, just like in the classical case. On the other hand, the variation of composite operators transforms members of a given multiplet among each other, plus some extra terms. It has been shown that the presence of these extra terms does not affect the main line of reasoning (Frezzotti *et al.*, 2001). Thus, they will be ignored in this discussion. The fact that chiral rotations transform both the tmQCD action and the composite fields into those of standard QCD implies that bare Green functions of the two theories are related. For example, the insertion of the scalar operator S^0 in a Green function gives rise to the following relation

$$\langle \cdots S^0 \cdots \rangle^{\text{GW}}_{\text{QCD}} = \cos(\omega)\langle \cdots S^0 \cdots \rangle^{\text{GW}}_{\text{tmQCD}} + i\sin(\omega)\langle \cdots P^3 \cdots \rangle^{\text{GW}}_{\text{tmQCD}}, \qquad (3.165)$$

where the ellipses stand for other operators, which have the same form in QCD and tmQCD (e.g. V^3_μ, P^1, etc.) and the superscript GW indicates bare Green functions, regularized in a Ginsparg-Wilson lattice framework. The subscripts QCD and tmQCD denote the mass term regularization (standard or twisted) in which these Green functions are defined.

The next step is to pass from bare to renormalized Green functions. Since chiral symmetry is preserved by the Ginsparg–Wilson regularization, members of the same chiral multiplet, such as S^0 and P^3, renormalize with the same renormalization factor $Z_{S^0} = Z_P$. In a mass-independent renormalization scheme, this factor is the same both for standard QCD and tmQCD. This is easy to understand, since the two theories only differ in the way mass terms are introduced in the action. The bottom line is that when both sides of the last equation are multiplied by the same factor $Z_{S^0} = Z_P$, we obtain a linear mapping between renormalized quantities, calculated in two frameworks (QCD and tmQCD):

$$\langle \cdots [S^0]_{\text{R}} \cdots \rangle_{\text{QCD}} = \cos(\omega)\langle \cdots [S^0]_{\text{R}} \cdots \rangle_{\text{tmQCD}} + i\sin(\omega)\langle \cdots [P^3]_{\text{R}} \cdots \rangle_{\text{tmQCD}}.$$
$$(3.166)$$

Thus, based on a specific lattice regularization that respects chiral symmetry, we have shown the equivalence of QCD and tmQCD beyond the classical level. So far the line of reasoning has been similar to the one of the classical theories: a change of variables and mass definitions in the actions and the path integrals maps QCD lattice (bare) expectation values into those of tmQCD. Multiplicative renormalization subsequently ensures that the same relations hold for the continuum (renormalized) Green functions. Just like in the classical case, we do not expect anything interesting out of these changes of variables.

Things become non-trivial, however, once we realize that the last expression, being true in the continuum, does not depend on the regularization details. This is a consequence of the principle of universality, which is a generally accepted assumption. In particular, universality implies that Eq. (3.166) is also true, up to discretization effects, for renormalized (continuum) Green functions, obtained from any other regularization; e.g. lattice Wilson fermions. Since chiral symmetry is broken by the Wilson term, the renormalization patterns of bare operators (which in the continuum belong to the same chiral multiplet) are now very different. This has important consequences. For example, as will be discussed below, the renormalization of S^0 with Wilson fermions is fairly complicated, and this renders the direct computation of the l.h.s. of Eq. (3.166) rather cumbersome. On the other hand, the renormalization of P^3 with Wilson fermions is much simpler (be it in standard QCD or in tmQCD). One may tune the mass parameters of tmQCD so that the twist angle is $\omega = \pi/2$ and the first term in the r.h.s. of Eq. (3.166) vanishes. Rather than computing the l.h.s. using standard lattice QCD with Wilson fermions, one may then compute the r.h.s. with the tmQCD lattice regularization of Wilson fermions, which has simpler renormalization properties. There are several interesting cases, besides the one sketched above, in which it is preferable, from the renormalization point of view, to work in a tmQCD framework.

Having explained the main idea, we now present Wilson fermion tmQCD in some detail. The standard Wilson action (3.2) is modified by the addition of the twisted mass term:

$$S_F^{\text{tm}} = a^4 \sum_x \bar{\chi}(x) \left[\frac{1}{2} \sum_\mu \{ \gamma_\mu (\overset{\rightarrow}{\mathcal{D}}_\mu^* + \overset{\rightarrow}{\mathcal{D}}_\mu) - a\, \overset{\rightarrow}{\mathcal{D}}_\mu^* \overset{\rightarrow}{\mathcal{D}}_\mu \} + m_0 + i\mu_q \tau^3 \gamma_5 \right] \chi(x). \quad (3.167)$$

The difference between the classical case and the present formulation is that the Wilson term breaks several symmetries. We saw, for example, that in the classical case parity is preserved in the modified form of Eqs. (3.158). This symmetry is now broken by the Wilson term of the action. It does, however, survive, when combined with flavor exchange

$$\chi(x) \to i\gamma_0 \tau^1 \chi(x^{\mathcal{P}}),$$
$$\bar{\chi}(x) \to -i\bar{\chi}(x^{\mathcal{P}}) \tau^1 \gamma_0, \quad (3.168)$$

where the gauge field transformations are the usual ones.[17] Equivalently, instead of flavor exchange we can combine parity with a flip of the twisted mass sign,

$$\chi(x) \to i\gamma_0 \chi(x^{\mathcal{P}}),$$
$$\bar{\chi}(x) \to -i\bar{\chi}(x^{\mathcal{P}}) \gamma_0,$$
$$\mu_q \to -\mu_q, \quad (3.169)$$

leaving the action S_F^{tm} invariant. Analogous considerations can be made for time-reversal. The consequence of this loss of symmetry is that matrix elements such as $\langle 0 | V_0^3(0) | \pi^0 \rangle$, which, due to parity, vanish in the standard Wilson theory, are non-zero in tmQCD. Of course, since we have argued in the previous section that tmQCD is a legitimate regularization of continuum QCD, such matrix elements will vanish in the continuum limit. They are lattice artefacts proportional to (some power of) the lattice spacing. Also note that often we study the asymptotic behavior of correlation functions such as $\langle P(x) P^\dagger(0) \rangle$ at large time separations, by introducing a complete set of states between the operators. In the standard QCD case these would be pseudoscalar states; in tmQCD loss of parity implies that more states (e.g. scalars) are also allowed.

We now proceed to examine continuous symmetries. The twisted vector symmetry of Eq. (3.159) is hard-broken by the Wilson term. Axial symmetry (cf. Eq. (3.160)) is softly broken by the mass term M_{inv} as expected, but also by the Wilson term; the latter breaking is hard. Once the continuum and chiral limits are taken, these symmetries are expected to be restored. It is important to note, however, that the subgroup of transformations

$$\chi(x) \rightarrow \exp\left[i\frac{\theta^3}{2}\tau^3\right]\chi(x)$$

$$\bar{\chi}(x) \rightarrow \bar{\chi}(x)\exp\left[-i\frac{\theta^3}{2}\tau^3\right], \tag{3.170}$$

obtained from Eq. (3.159) by setting $\theta^1 = \theta^2 = 0$, remains a vector symmetry of the Wilson tmQCD action. We denote this group as $U(1)_V^3$. Thus, due to the Wilson term we have the symmetry-breaking pattern $SU(2)_V^\omega \rightarrow U(1)_V^3$. The reduced symmetry causes a lack of degeneracy between the neutral pion π^0 and the two degenerate charged pions $\pi^\pm$. The mass difference between charged and neutral pions is a discretization effect, proportional to (some power of) the lattice spacing, which vanishes in the continuum limit, where the full symmetry is restored.

Another interesting symmetry is derived from the axial transformations (3.160) which, upon setting $\omega = \pi/2$ and $\theta^2 = \theta^3 = 0$, reduce to

$$\chi(x) \rightarrow \exp\left[i\frac{\theta^1}{2}\tau^2\right]\chi(x)$$

$$\bar{\chi}(x) \rightarrow \bar{\chi}(x)\exp\left[-i\frac{\theta^1}{2}\tau^2\right]. \tag{3.171}$$

This group of transformations is denoted by $U(1)_A^1$, though it has the appearance of the usual vector symmetry. Analogously, for $\theta^1 = \theta^3 = 0$ we obtain a similar group of transformations, called $U(1)_A^2$. Indeed, these are axial symmetries, being symmetries of the massless Wilson action, softly broken by the twisted mass term. Thus, in the chiral limit the twisted theory is symmetric under $U(1)_V^3 \otimes U(1)_A^1 \otimes U(1)_A^2$, which amounts to an $SU(2)$ group, with one "vector" and two "axial" generators. It is not surprising that in the chiral limit, where the action (3.167) reduces to the standard Wilson one, the full $SU(2)$ symmetry is recovered. The interpretation of this symmetries as vector (in the standard case) or vector/axial (in tmQCD), depends on how the soft mass term is introduced in the action (Sint, 2007).

3.5.3   Renormalization with tmQCD

So far, we have seen that the introduction of a twisted mass term in the Wilson fermion regularization breaks some discrete and continuous symmetries. We have also argued that this lattice theory goes over to QCD as the lattice cutoff is removed, so any effects due to loss of symmetry by the twisted lattice action vanish in the continuum limit. Having seen the shortcomings of tmQCD, it is high time we discuss some of the advantages. First, it is straightforward to see that the fermion determinant corresponding to the tmQCD action (3.167) is positive definite, as long as $\mu_q^2 \neq 0$. This is an important advantage for lattice simulations close to the chiral regime, but will not be further discussed here. A second advantage concerns renormalization. We have already argued, in rather general terms, that some operator renormalization is simpler in tmQCD than in standard Wilson fermion lattice theory. Here, we will discuss some important examples in detail.

Before we do so, we should understand the renormalization properties of the two bare mass parameters of tmQCD, m_0 and μ_q (or equvalently, M_{inv} and ω). The symmetries of the tmQCD action suggest that, for degenerate flavors, the standard quark mass m_0 renormalizes as in Eqs. (3.14) and (3.15), while the twisted mass μ_q renormalizes multiplicatively. Moreover, the PCVC relation (3.163) is exact in lattice tmQCD, with the local vector current replaced by the point-split one (3.47). This in turn implies that the product of twisted mass μ_q and pseudoscalar density P^b (i.e. the r.h.s. of Eq. (3.163)) is renormalization-group invariant. Thus, Z_P^{-1} renormalizes μ_q:

$$[\mu_q]_{\text{R}} = Z_P^{-1}\mu_q. \tag{3.172}$$

The above statements are generalizations of the arguments of Section 3.3, concerning standard lattice QCD with Wilson quarks. The reader should have no problem in convincing himself of their validity for tmQCD.

Given these mass renormalizations, we define the twist angle through the following ratio of renormalized quantities:

$$\tan(\omega) = \frac{[\mu_q]_{\text{R}}}{m_{\text{R}}} = \frac{Z_P^{-1}\mu_q}{Z_{S^0}^{-1}[m_0 - m_{\text{cr}}]}. \tag{3.173}$$

Note that with Ginsparg–Wilson fermions, chiral symmetry ensures that the two bare masses m_0 and μ_q renormalize multiplicatively with the same renormalization factor, and thus it makes no difference whether the twist angle is defined through a ratio of bare or renormalized masses. It is the loss of chiral symmetry with Wilson fermions that leads to the above redefinition of the twist angle w.r.t. Eq. (3.153).

We stress that for mass-independent renormalization, carried out in the chiral limit, there is no distinction between standard QCD and tmQCD, since the mass terms are absent from the action. Therefore, all renormalization constants computed in standard QCD simulations (e.g. Z_m, Z_P) may also be used for tmQCD quantities. For instance, the renormalization factors of m and μ_q in the last equation are the very same Z_{S^0} and Z_P that renormalize the densities S^0 and P in the standard Wilson theory.[18]

With the above observations in mind, we can immediately generalize the tree-level expressions (3.161) for the full theory. In particular, the continuum axial current A^a (for $a = 1, 2$) is expressed as linear combinations of renormalized tmQCD ones:

$$[A_\mu^a]_{\text{cont}} = Z_A[A_\mu^a]_{\text{QCD}} = \cos(\omega)Z_A[A_\mu^a]_{\text{tmQCD}} + \epsilon^{3ab}\sin(\omega)[\tilde{V}_\mu^b]_{\text{tmQCD}}. \tag{3.174}$$

The subscripts QCD and tmQCD indicate bare quantities in the respective lattice regularizations. The first equation in the above expression links the continuum axial current to the lattice one in the standard formulation, as detailed in Section 3.3.2. The second equation does the same job in the tmQCD framework. As previously explained, the normalization factor Z_A is the same in both regularizations. The last term of the second equation contains the exactly conserved point-split vector current; thus the absence of a normalization factor. We may of course use, instead of $[\tilde{V}_\mu^a]_{\text{tmQCD}}$, the local current $[V_\mu^a]_{\text{tmQCD}}$, multiplied by Z_V. Similarly for the pseudoscalar density P^a (for $a = 1, 2$) we have

$$[P^a]_{\text{cont}} = Z_P[P^a]_{\text{QCD}} = Z_P[P^a]_{\text{tmQCD}}. \qquad (3.175)$$

Equalities (3.174) and (3.175) are valid up to discretization errors.

We are finally ready to see a first advantage of maximally twisted tmQCD. To ensure maximal twist (i.e. $\omega = \pi/2$), the standard bare mass m_0 of the tmQCD regularization is tuned to its critical value m_{cr}; cf. Eq. (3.173). Note that knowledge of the ratio Z_P/Z_{S^0} is not required for this tuning. Then, according to Eq. (3.174) the axial current in the continuum limit is the vector current in tmQCD, which requires no normalization. Using Eqs. (3.174) and (3.175), we see that the standard PCAC Ward identity in the continuum corresponds to the PCVC relation

$$\sum_{\vec{x}} \nabla_x^\mu \langle \tilde{V}_\mu^1(x) P^2(0) \rangle_{\text{tmQCD}} = -2\mu_q \sum_{\vec{x}} \langle P^1(x) P^2(0) \rangle_{\text{tmQCD}}. \qquad (3.176)$$

This is an exact Ward identity in lattice tmQCD. Inserting, in standard fashion, a complete set of states between operators in the above expectations values, we obtain, in the limit of large time separations

$$m_\pi \langle 0|[\tilde{V}_0(0)]_{\text{tmQCD}}|\pi \rangle = 2\mu_q \langle 0| \, [P(0)]_{\text{tmQCD}} \, |\pi \rangle. \qquad (3.177)$$

Flavor indices have been dropped for simplicity. Combining these expressions and Eq. (3.174) we get, for the pion decay constant f_π

$$
\begin{aligned}
f_\pi &\equiv \frac{1}{m_\pi} \langle 0| \, [A_0(0)]_{\text{cont}} \, |\pi \rangle \\[6pt]
&= \lim_{a \to 0} \frac{1}{m_\pi} \langle 0| \, [\tilde{V}_0(0)]_{\text{tmQCD}} \, |\pi \rangle \\[6pt]
&= \lim_{a \to 0} \frac{2\mu_q}{m_\pi^2} \langle 0| \, [P(0)]_{\text{tmQCD}} \, |\pi \rangle.
\end{aligned}
\qquad (3.178)
$$

The last expression provides, within the tmQCD framework, a definition of f_π that is free of any normalization or renormalization factors. Recall that in the standard Wilson fermion case, the computation of f_π from the matrix element $\langle 0|A_0|\pi \rangle$ requires knowledge of the axial current normalization Z_A. Thus, the systematic error due to Z_A has been eliminated in tmQCD.

Another example of the advantages of tmQCD is provided by the renormalization of the chiral condensate operator S^0. In the standard Wilson quark case, the symmetries of the theory imply the renormalization pattern (cf. Eq. (3.103)):

$$[S^0]_{\text{R}} = Z_{S^0}\left[[S^0]_{\text{QCD}} + \frac{c_S(g_0^2)}{a^3}\right] + \cdots, \qquad (3.179)$$

where the ellipsis stands for less vigorous power divergences (quadratic and linear) which, for dimensional reasons, are proportional to powers of the quark mass m (i.e. they are absent in the chiral limit). In tmQCD with maximal twist ($\omega = \pi/2$), this operator is mapped on P^3 (cf. Eq. (3.162)). The symmetries of lattice tmQCD with Wilson quarks imply the renormalization pattern

$$[S^0]_{\mathrm{R}} = Z_P \left[[P^3]_{\mathrm{tmQCD}} + \frac{\mu_q c_P(g_0^2)}{a^2} \right] + \cdots , \qquad (3.180)$$

where the ellipsis again stands for less vigorous linear power divergences, which depend on the quark mass μ_q. Thus, in the chiral limit the chiral condensate, regularized in tmQCD with maximal twist, is multiplicatively renormalizable. It is well known that power divergences are hard to control, whether calculated perturbatively or computed non-perturbatively (i.e. numerically). Suppose that the dimensionless coefficients c_S and c_P are calculated perturbatively. Such a calculation misses out dimensionless non-perturbative contributions, which are $\mathcal{O}(a\Lambda_{\mathrm{QCD}})$. Since these contributions are to be multiplied by the power divergences $1/a^3$ and $1/a^2$ of the power subtractions, the perturbative calculation of c_S and c_P is inaccurate by terms that diverge in the continuum limit! On the other hand, if the coefficients c_S and c_P are calculated non-perturbatively, they are bound to be subject to discretization effects. These are expected to be $\mathcal{O}(a)$.[19] So again the error of the power subtractions diverges in the continuum limit. Nevertheless, it is remarkable that in tmQCD the chiral condensate suffers from less-severe power divergences. If somehow one could work in the chiral limit of maximally tmQCD, determined non-perturbatively at very high accuracy (i.e. $m_{\mathrm{cr}} \sim \mathcal{O}(a^3)$), then the power subtraction of Eq. (3.180) could be ignored altogether and the chiral condensate computation would be greatly simplified.

These considerations have been extended to the calculation of matrix elements of dimension-6 four fermion operators, which determine the non-perturbative contributions in say, neutral Kaon oscillations and Kaon non-leptonic decays into two pions. In lattice QCD with standard Wilson quarks, the renormalization pattern of these operators is very complicated, due to the introduction of counterterms that would have been absent, if chiral symmetry were preserved. In tmQCD it is possible to map these operators into their partners of opposite parity, which have a much simpler renormalization pattern, in spite of the loss of chiral symmetry. Explaining this in detail is beyond the scope of the present chapter. The interested reader is advised to consult the relevant literature, where these results are presented in considerable detail (Frezzotti *et al.*, 2001; Dimopoulos *et al.*, 2006; Pena *et al.*, 2004; Frezzotti and Rossi, 2004).

3.5.4 "Automatic" improvement in maximally twisted QCD

Besides simplified renormalization patterns, tmQCD in its maximally twisted version has another nice property, commonly known as "automatic improvement". By this we mean that physical quantities (e.g. masses, matrix elements, etc.) have discretization effects that are subleading; i.e. $\mathcal{O}(a^2)$. Improvement of these quantities comes about by carefully tuning the twist angle $\omega = \pi/2$, without the need of introducing Symanzik counterterms.[20] Following the original proof of this statement (Frezzotti and Rossi, 2004*a*), several simplified versions have appeared in the literature. In order to understand the line of reasoning of this section, the reader is strongly advised to acquaint himself with the relevant sections of Chapter 2 of the present volume (Weisz, 2010).

The study of discretization effects of lattice observables is based on the Symanzik expansion, in the light of which the lattice action, close to the continuum, is described in terms of an effective theory

$$S_F^{\mathrm{tm}} = \int d^4 y \mathcal{L}_0 + a \int d^4 y \mathcal{L}_1 + \cdots . \tag{3.181}$$

As the notation in the above expression implies, our starting point is the tmQCD Wilson fermion action (3.167) at maximal twist (i.e. with $m_0 = m_{\mathrm{cr}}$). Thus, the Symanzik counterterms reflect the symmetries of this action:

$$\mathcal{L}_0 = \bar{\chi} \left[\slashed{D} + i \, [\mu_q]_{\mathrm{R}} \, \gamma_5 \tau^3 \right] \chi \tag{3.182}$$

$$\mathcal{L}_1 = i \, b_{\mathrm{sw}} \left[\bar{\chi} \sigma \cdot F \chi \right] + b_\mu \, [\mu_q]_{\mathrm{R}}^2 \, [\bar{\chi} \chi]. \tag{3.183}$$

Besides the above expansion for the action, Symanzik improvement implies that d-dimensional composite lattice fileds Φ_{latt} are also subject to an effective theory description of the form

$$\Phi_{\mathrm{latt}} = \Phi_0 + a \, \Phi_1, \tag{3.184}$$

where Φ_0, Φ_1 are continuum composite fields of dimensions d and $d+1$ respectively. Actually, Φ_{latt} is a lattice transcription of Φ_0. Combining Eqs. (3.181) and (3.184) gives, the lowest-order Symanzik expansion for the vacuum expectation value of Φ_{latt}

$$\langle \Phi_{\mathrm{latt}} \rangle_{\mathrm{tm}} = \langle \Phi_0 \rangle_0 + a \, \langle \Phi_1 \rangle_0 - a \int d^4 y \langle \Phi_0 \mathcal{L}_1 \rangle_0 + \mathcal{O}(a^2). \tag{3.185}$$

The l.h.s. is the expectation value of the lattice operator in maximally twisted lattice QCD. The expectation values of the r.h.s. are continuum quantities; their subscript $\langle \cdots \rangle_0$ indicates that they are defined in terms of the continuum tree-level action $S_0 = \int d^4 y \mathcal{L}_0$. Automatic improvement means that the last two terms on the r.h.s. vanish identically, due to some symmetries of the maximally twisted continuum theory.

To prove this, we first explain what the relevant lattice symmetries are. We define the *discrete chiral transformations in the first isospin direction* $\mathcal{R}_5^1$ as:

$$\chi \to i \, \gamma_5 \, \tau^1 \, \chi,$$
$$\bar{\chi} \to i \, \bar{\chi} \, \gamma_5 \, \tau^1. \tag{3.186}$$

Next, we define the *operator dimensionality transformations* $\mathcal{D}$:

$$\chi(x) \to \exp[\frac{3i\pi}{2}] \, \chi(-x),$$
$$\bar{\chi}(x) \to \bar{\chi}(-x) \exp[\frac{3i\pi}{2}]$$
$$U_\mu(x) \to U_\mu^\dagger(-x - a\hat{\mu}). \tag{3.187}$$

The gauge lattice action is invariant under the above transformations of the link fields $U_\mu(x)$. Composite operators of even (odd) dimension d are even (odd) under these

Table 3.1 "Parity" of the various terms of the fermion tmQCD action under discrete symmetries.

$\mathcal{S}_F^{\text{tm}}$	$\mathcal{R}_5^1$	$\mathcal{D}$	$[\mu_q \to -\mu_q]$
$\sum_x \sum_\mu \bar{\chi}\gamma_\mu(\overset{\to}{\mathcal{D}}_\mu^* + \overset{\to}{\mathcal{D}}_\mu)\chi$	$+$	$+$	$+$
$\sum_x \sum_\mu \bar{\chi}\,\overset{\to}{\mathcal{D}}_\mu^*\overset{\to}{\mathcal{D}}_\mu\,\chi$	$-$	$-$	$+$
$m_{\text{cr}}\sum_x \bar{\chi}\chi$	$-$	$-$	$+$
$i\sum_x \bar{\chi}\mu_q\tau^3\gamma_5\chi$	$+$	$-$	$-$

transformations, up to the flip of sign of the space-time argument x. Finally, we define the *twisted mass sign-flip transformation*

$$\mu_q \to -\mu_q. \tag{3.188}$$

Each term of the lattice tmQCD action is even or odd under these transformations; i.e. it has definitive "parity" with respect to these symmetries. In Table 3.1 we list these properties explicitly. It is then clear that the tmQCD fermion action is invariant under the combined transformations $\mathcal{R}_5^1 \otimes \mathcal{D} \otimes [\mu_q \to -\mu_q]$, which is therefore a symmetry of the lattice theory. On the other hand, the continuum tmQCD action $\mathcal{S}_0 = \int d^4y\mathcal{L}_0$ has positive $\mathcal{R}_5^1$-parity (cf. Eq. (3.182) and the first and last rows of Table 3.1). Thus, $\mathcal{R}_5^1$ is a symmetry of the continuum theory.

Let us now examine each term of Eq. (3.185) in the light of these symmetries. It is important to keep in mind that the three vacuum expectation values on the r.h.s. are continuum quantities, determined by the continuum tmQCD action $\mathcal{S}_0 = \int d^4y\mathcal{L}_0$, which is symmetric under $\mathcal{R}_5^1$.

1. Φ_{latt}: Without loss of generality, we assume this operator to be even under $\mathcal{R}_5^1$ and to have even dimension d. Then, its vacuum expectation value $\langle\Phi_{\text{latt}}\rangle_{\text{tm}}$ is invariant under the combined transformations $\mathcal{R}_5^1 \otimes \mathcal{D} \otimes [\mu_q \to -\mu_q]$, which is a symmetry of the lattice theory. This implies that the terms on the r.h.s. of Eq. (3.185) are also invariant under the same combined transformations.
2. Φ_0: This operator has positive $\mathcal{R}_5^1$-parity and even dimension d, being the continuum counterpart of the lattice operator Φ_{latt}.
3. Φ_1: This operator has dimension $d+1$, so it is odd under $\mathcal{D}$. Since its expectation value $\langle\Phi_1\rangle_0$ is even under $\mathcal{R}_5^1 \otimes \mathcal{D} \otimes [\mu_q \to -\mu_q]$, the operator Φ_1 must be odd under $\mathcal{R}_5^1$. This implies that $\langle\Phi_1\rangle_0$ vanishes, being the expectation value of an $\mathcal{R}_5^1$-odd observable Φ_1, weighted by an $\mathcal{R}_5^1$-even action $\mathcal{S}_0$.
4. $\mathcal{L}_1$: This is the $\mathcal{O}(a)$-counterterm of the continuum Lagrangian $\mathcal{L}_0$. It is a continuum dimension-5 operator. From Eqs. (3.183) and (3.186) we deduce that $\mathcal{L}_1$ is odd under $\mathcal{R}_5^1$ transformations.
5. $\Phi_0\mathcal{L}_1$: This is a product of two composite operators. The operator Φ_0 is $\mathcal{R}_5^1$-even while the operator $\mathcal{L}_1$ is $\mathcal{R}_5^1$-odd. It follows that the $\langle\Phi_0\,\mathcal{L}_1\rangle_0$ vanishes, being the expectation value of an $\mathcal{R}_5^1$-odd observable Φ_1, weighted by an $\mathcal{R}_5^1$-even action $\mathcal{S}_0$.

We have shown that the lattice expectation value $\langle\Phi_{\text{latt}}\rangle_{\text{tm}}$ of Eq. (3.185) is equal to the leading-order term of the Symanzik expansion $\langle\Phi_0\rangle_0$ up to $\mathcal{O}(a^2)$ counterterms. The $\mathcal{O}(a)$ counterterms of the expansion are identically zero. Therefore, without introducing Symanzik counterterms, we find that expectation values of lattice operators in the maximally twisted theory are $\mathcal{O}(a)$-improved. Note that it is the vacuum expectation values, rather than the maximally twisted action itself, which are automatically improved.

The previous arguments have overlooked the following subtlety. The proof rests on the vanishing of the continuum vacuum expectation values $\langle\Phi_1\rangle_0$ and $\langle\Phi_0\,\mathcal{L}_1\rangle_0$, due to their breaking of the discrete "chiral" symmetry $\mathcal{R}_5^1$, which is a symmetry of the continuum tmQCD action $\mathcal{S}_0$. But do these "chiral condensates" vanish in a theory that, being a regularization of QCD, is characterized by spontaneous symmetry breaking? The answer is affirmative, because the term generating spontaneous symmetry breaking is the twisted mass term, while possible $\mathcal{O}(a)$ counterterms are generated by the "chirally orthogonal" Wilson term, which in tmQCD breaks vector symmetry. In other words, the $\mathcal{R}_5^1$-symmetry on which the automatic improvement is based is not really a chiral symmetry. It is a discrete subgroup of the flavor symmetry $SU(2)_V$ of Eq. (3.159), for $\omega = \pi/2$ (i.e. maximal twist) and transformation angles $(\alpha_1, \alpha_2, \alpha_3) = (0, \pi, 0)$. Flavor symmetry is not spontaneously broken in the continuum, and thus the quantities $\langle\Phi_1\rangle_0$ and $\langle\Phi_0\,\mathcal{L}_1\rangle_0$ do indeed vanish by symmetry arguments. The situation is not altered by simulations, provided we take extra care that the continuum limit is approached before the chiral limit. In this way the chiral phase of the vacuum is driven by the mass term and not by the Wilson term. This is ensured by imposing the condition $\mu_q \gg a\Lambda_{\text{QCD}}^2$.

We have shown that tmQCD is a variant of Wilson fermion regularization that, at the price of sacrificing some symmetries, enjoys simpler renormalization properties of composite operators and automatic improvement. The advantages on renormalization patterns are typically obtained by tuning the twist angle to a value that maps the original operator to its counterpart of opposite parity (e.g. axial current to vector current, scalar density to pseudoscalar density, etc.). In some cases, this twist angle turns out to be different from $\pi/2$, the maximal twist value that ensures automatic improvement. Thus, the renromalization and improvement advantages are not always satisfied simultaneously. Such difficulties arise with the four-fermion operators related to neutral Kaon oscillations and $K \to \pi\pi$ decays (Pena *et al.*, 2004). A way out of these problems has been proposed (Frezzotti and Rossi, 2004), based on mixed actions with maximally twisted valence quarks of the so-called Osterwalder–Seiler variety. Once again, the gains are accompanied by certain disadvantages, but these are issues beyond the scope of this chapter.

3.6 Conclusions

The present chapter focuses on three specific topics: (i) lattice Ward identities; (ii) the non-perturbative RI/MOM renormalization scheme; (iii) renormalization and improvement of twisted mass QCD. From the study of these topics, significant theoretical and practical advantages may be obtained for the renormalization and improvement of lattice composite operators. Throughout we have used the Wilson regularization of the fermionic action. However, at least as far as the first two topics are concerned, the issues discussed go beyond a specific regularization.

Once chiral symmetry is lost by the regularization, Ward identities ensure its recovery in the renormalized theory, as the UV cutoff is removed. In the Wilson fermion case chiral symmetry is broken by an irrelevant operator. The consequences are immediately seen in the power subtraction of the quark mass, the finite normalization of the axial current, the different renormalization constants of operators belonging to the same chiral multiplet, etc. In other regularizations, such as domain-wall fermions, the loss of chirality only appears at the "practical" level (i.e. when the fifth dimension of the domain wall is not infinite in simulations). The resulting loss of chiral symmetry is less manifest in such cases, but its effects are analogous to those of Wilson fermions (albeit quantitatively less important).

The RI/MOM scheme has been presented in considerable detail. Emphasis has been laid on the theoretical aspects of the scheme, the conceptual problems and their resolution, without presenting any results, which may be easily dug out in the literature. Again, the language used was that of Wilson fermions, but it is quite clear that non-perturbative operator renormalization is nowadays a requisite for all lattice regularization schemes.

Finally, the variant of Wilson fermions known as twisted mass QCD has been discussed. Only selected aspects of this formulation have been exposed. The fact that the tmQCD fermion determinant is positive definite at non-zero twisted mass μ_q has only been touched upon, although it is of great importance to simulations close to the chiral regime. Moreover, we have limited our discussion to tmQCD with two light degenerate quarks, although it is possible to generalize the theory so as to include heavy flavors (Frezzotti and Rossi, 2004c). This subject has many faces, but the most important properties of tmQCD are arguably those connected to renormalization and improvement. Significant simplifications of the standard Wilson formulation are obtained, which are, however, accompanied by loss of symmetry, recoverable only in the continuum limit. This is hardly surprising, being yet another example of the well-known fact that most significant theoretical gains in quantum field theory come at a price.

3.7 Appendix: spurionic chiral symmetry

Fermion fields are decomposed into left- and right-components $\psi = \psi_L + \psi_R$, with

$$\psi_L = \frac{1 - \gamma_5}{2} \psi, \qquad \bar{\psi}_L = \bar{\psi} \frac{1 + \gamma_5}{2},$$

$$\psi_R = \frac{1 + \gamma_5}{2} \psi, \qquad \bar{\psi}_R = \bar{\psi} \frac{1 - \gamma_5}{2}. \tag{3.189}$$

They transform under the chiral group $SU(N_F)_L \otimes SU(N_F)_R$ as follows:

$$\psi_L \to \psi_L' = U_L \psi_L, \qquad \bar{\psi}_L \to \bar{\psi}_L' = \bar{\psi}_L U_L^\dagger,$$
$$\psi_R \to \psi_R' = U_R \psi_R, \qquad \bar{\psi}_R \to \bar{\psi}_R' = \bar{\psi}_R U_R^\dagger, \qquad (3.190)$$

with

$$U_{L,R} = \exp\left[i\alpha_{L,R}^a \frac{\lambda^a}{2} \right] \qquad a = 1, \cdots, N_F^2 - 1. \qquad (3.191)$$

The vector transformations (3.6) are recovered for $U_L = U_R$ (i.e. $\alpha_L^a = \alpha_R^a$), while the axial transformations (3.8) are recovered for $U_L = U_R^\dagger$ (i.e. $\alpha_L^a = -\alpha_R^a$).

The kinetic term of the lattice action (3.2) is invariant under these transformations, while the mass and Wilson terms break chiral symmetry. However, the symmetry may be enforced on the mass term, once the mass matrix M_0 is generalized, so that it is neither real nor diagonal (the physically relevant case corresponds to $M_0 = M_0^\dagger$ and diagonal). In order for the action to remain Hermitian, the mass term must be modified as follows:

$$\bar{\psi}_L M_0 \psi_R + \bar{\psi}_R M_0 \psi_L \to \bar{\psi}_L M_0 \psi_R + \bar{\psi}_R M_0^\dagger \psi_L. \qquad (3.192)$$

We also impose the following chiral transformation of the mass matrix:

$$M_0 \to M_0' = U_L M_0 U_R^\dagger. \qquad (3.193)$$

Transformations of the mass parameters are called spurionic. Under the transformations (3.190) and (3.193) the modified mass term of the action is chirally invariant. Note that the Wilson term in the action remains a (spurionic) chiral-symmetry-breaking term.

The generalized, non-diagonal bare mass matrix M_0 may be decomposed into non-singlet and singlet quark mass contributions (this terminology will become clear shortly):

$$M_0 = \sum_{a=1}^{N_F^2 - 1} \tilde{m}^a \lambda^a + m^{\mathrm{av}} I. \qquad (3.194)$$

Using the trace property of Eq. (3.7) for the group generators λ^a we see that

$$\tilde{m}^a = \frac{1}{2} \mathrm{Tr}[M_0 \lambda^a], \qquad m^{\mathrm{av}} = \frac{1}{N_F} \mathrm{Tr}[M_0]. \qquad (3.195)$$

Clearly m^{av} is the average of the diagonal elements of M_0. It is straightforward to show that under infinitesimal chiral transformations these mass components transform as follows:

$$\tilde{m}^a \to [\tilde{m}^a]' = \tilde{m}^a - f^{abc} \frac{\alpha_L^b + \alpha_R^b}{2} \tilde{m}^c + id^{abc} \frac{\alpha_L^b - \alpha_R^b}{2} \tilde{m}^c + i \frac{\alpha_L^a - \alpha_R^a}{2} m^{\mathrm{av}},$$

$$m^{\mathrm{av}} \to [m^{\mathrm{av}}]' = m^{\mathrm{av}} + \frac{i}{N_F} (\alpha_L^a - \alpha_R^a) \tilde{m}^a. \qquad (3.196)$$

In the special case of vector transformations $(\alpha_L^a = \alpha_R^a = \alpha^a)$, the above simplify to:

$$\tilde{m}^a \to \left[\tilde{m}^a\right]' = \tilde{m}^a - f^{abc}\alpha^b\tilde{m}^c;$$

$$m^{\mathrm{av}} \to \left[m^{\mathrm{av}}\right]' = m^{\mathrm{av}}. \qquad (3.197)$$

Thus, the non-singlet mass components $\tilde{m}^a$ are indeed $SU(N_F)_V$ multiplets in the adjoint representation (triplets for $N_F = 2$, octets for $N_F = 3$, etc.). On the other hand, m^{av} is a flavor singlet. Under infinitesimal axial transformations $(\alpha_L^a = -\alpha_R^a = \alpha^a)$, Eqs. (3.196) reduce to

$$\tilde{m}^a \to \left[\tilde{m}^a\right]' = \tilde{m}^a + d^{abc}\alpha^b\tilde{m}^c + i\alpha^a m^{\mathrm{av}},$$

$$m^{\mathrm{av}} \to \left[m^{\mathrm{av}}\right]' = m^{\mathrm{av}} + \frac{2i}{N_F}\alpha^a\tilde{m}^a. \qquad (3.198)$$

Therefore, flavor non-singlets $\tilde{m}^a$ and the singlet m^{av} transform into each other under axial transformations; i.e. they belong to the same chiral multiplet.

3.8 Appendix: lattice discrete symmetries

In this Appendix we gather the discrete symmetry transformations of the Wilson action (3.1), (3.2) (Bernard, 1989). Although of rather limited value to the present chapter, they are generally useful for the classification of physical states, the mixing of operators under renormalization, etc. We start with parity, which is defined as follows:

$$x = (x^0, \mathbf{x}) \to x^{\mathcal{P}} = (x^0, -\mathbf{x}),$$

$$U_0(x) \to U_0(x^{\mathcal{P}}),$$

$$U_k(x) \to U_k(x^{\mathcal{P}} - \hat{k})^\dagger,$$

$$\psi(x) \to \gamma_0\psi(x^{\mathcal{P}}),$$

$$\bar{\psi}(x) \to \bar{\psi}(x^{\mathcal{P}})\gamma_0. \qquad (3.199)$$

Time reversal is given by

$$x = (x^0, \mathbf{x}) \to x^{\mathcal{T}} = (-x^0, \mathbf{x}),$$

$$U_0(x) \to U_0(x^{\mathcal{T}} - \hat{0})^\dagger,$$

$$U_k(x) \to U_k(x^{\mathcal{T}}),$$

$$\psi(x) \to \gamma_0\gamma_5\psi(x^{\mathcal{T}}),$$

$$\bar{\psi}(x) \to \bar{\psi}(x^{\mathcal{T}})\gamma_5\gamma_0. \qquad (3.200)$$

Charge conjugation is defined as

$$U_\mu(x) \to U_\mu(x)^*,$$
$$\psi(x) \to \mathcal{C}\bar{\psi}^T(x),$$
$$\bar{\psi}(x) \to -\psi^T(x)\mathcal{C}^{-1}. \tag{3.201}$$

The T superscript denotes transpose vectors and matrices. The charge conjugation matrix satisfies

$$\mathcal{C}\gamma_\mu\mathcal{C} = -\gamma_\mu^* = -\gamma_\mu^T. \tag{3.202}$$

A realization of $\mathcal{C}$ is $\mathcal{C} = \gamma_0\gamma_2$.

3.9   Appendix: regularization-dependent scheme

The acronym "RI" stands for "regularization independent" and the reason for this is mostly historical: before the advent of lattice non-perturbative renormalization, Z_O-factors were calculated in lattice perturbation theory, with $\overline{\text{MS}}$ as the scheme of preference. This means that two perturbative calculations had to be carried out; one in the continuum and one on the lattice. Schematically, in dimensional regularization (DR) the continuum perturbative expression for the bare correlation function Γ_O, at one loop is

$$\Gamma_O^{\text{DR}}(p, g_0, \epsilon) = \left[1 + \frac{g_0^2(\mu)}{(4\pi)^2}\left(\gamma_\Gamma^{(0)}\frac{1}{\hat{\epsilon}} - \gamma_\Gamma^{(0)}\ln(\frac{p^2}{\mu^2}) + C_\Gamma^{\text{DR}}\right)\right], \tag{3.203}$$

where $\gamma_\Gamma^{(0)}$ is the LO anomalous dimension of the correlation function Γ_O. To this order it is a universal, scheme-independent quantity. As implied by the notation, the finite constant C_Γ^{DR} depends on the regularization scheme and the chosen gauge. The factor $1/\hat{\epsilon}$ is an abbreviation for

$$\frac{1}{\hat{\epsilon}} = \frac{1}{\epsilon} + \ln(4\pi) - \gamma_E, \tag{3.204}$$

where $\epsilon = (4 - D)/2$ and γ_E stands for Euler's constant. In DR we work in D dimensions, where the original bare coupling g_0 has dimension ϵ. Here, the scale μ is introduced to render the bare coupling $g_0(\mu)$ dimensionless. The μ dependence of the r.h.s. of Eq. (3.203) is only apparent.

Imposing the $\overline{\text{MS}}$ renormalization condition amounts to removing the $1/\hat{\epsilon}$ divergence. Since the renormalization constant of the projected Green function Γ_O is given by $Z_\Gamma = Z_\psi^{-1}Z_O$ (see Eq. (3.36)), this implies for Z_Γ the value

$$Z_\Gamma^{\overline{\text{MS}},\text{DR}}(g_0(\mu), \epsilon) = 1 - \frac{g_0^2(\mu)}{(4\pi)^2}\gamma_\Gamma^{(0)}\frac{1}{\hat{\epsilon}}. \tag{3.205}$$

Consequently, the renormalized Green function is given by

$$\left[\Gamma_O^{\overline{\text{MS}}}(p, g_{\overline{\text{MS}}}(\mu), \mu)\right]_R = \lim_{\epsilon \to 0} \left[Z_\Gamma^{\overline{\text{MS}},\text{DR}}(g_0(\mu), \epsilon)\Gamma_O^{\text{DR}}(p, g_0, \epsilon)\right]$$

$$= 1 + \frac{g_{\overline{\text{MS}}}^2(\mu)}{(4\pi)^2}\left[-\gamma_\Gamma^{(0)}\ln(p/\mu)^2 + C_\Gamma^{\text{DR}}\right], \qquad (3.206)$$

where, to this order, we are free to replace g_0 by $g_{\overline{\text{MS}}}(\mu)$, the $\overline{\text{MS}}$ renormalized coupling constant.

The same calculation can be repeated on the lattice. Now, the UV cutoff is provided by the inverse finite lattice spacing a^{-1} and thus the 1-loop calculation yields:

$$\Gamma_O^{\text{LAT}}(p, g_0(a), a) = 1 + \frac{g_0(a)^2}{(4\pi)^2}\left(-\gamma_\Gamma^{(0)}\ln(pa)^2 + C_\Gamma^{\text{LAT}}\right) + \mathcal{O}(a), \qquad (3.207)$$

where $g_0(a)$ is the bare coupling of the lattice action. The renormalization scheme can again be chosen at will; as previously stated, the $\overline{\text{MS}}$ is often chosen also on the lattice.[21] Thus, we must satisfy the renormalization condition

$$\left[\Gamma_O^{\overline{\text{MS}}}(p, g_{\overline{\text{MS}}}(\mu), \mu)\right]_R = \lim_{a \to 0}\left[Z_\Gamma^{\overline{\text{MS}},\text{LAT}}(\mu a, g_0(a))\Gamma_O^{\text{LAT}}(p, g_0(a), a)\right], \qquad (3.208)$$

where again the lattice coupling $g_0(a)$ should be traded for the $\overline{\text{MS}}$ renormalized coupling constant $g_{\overline{\text{MS}}}(\mu)$. This point of principle is of limited relevance for a 1-loop calculation. From Eqs. (3.206), (3.207) and (3.208), the following renormalization constant is obtained:

$$Z_\Gamma^{\overline{\text{MS}},\text{LAT}}(\mu a, g_0(a)) = 1 + \frac{g_0(a)^2}{(4\pi)^2}\left[\gamma_\Gamma^{(0)}\ln(\mu a)^2 + C_\Gamma^{\text{DR}} - C_\Gamma^{\text{LAT}}\right]. \qquad (3.209)$$

We now recall that the renormalization constant of the amputated vertex Γ_O is $Z_\Gamma = Z_\psi^{-1}Z_O$. The quark field renormalization Z_ψ can be calculated, from the quark propagator $\mathcal{S}(p)$, with an analogous procedure; c.f. Eq. (3.22). The result is

$$Z_\psi^{\overline{\text{MS}},\text{LAT}}(\mu a, g_0(a)) = 1 + \frac{g_0(a)^2}{(4\pi)^2}\left[\gamma_\Sigma^{(0)}\ln(\mu a)^2 + C_\Sigma^{DR} - C_\Sigma^{\text{LAT}}\right]. \qquad (3.210)$$

Combining Eqs. (3.209) and (3.210) we obtain

$$Z_O^{\overline{\text{MS}},\text{LAT}}(\mu a, g_0(a)) = 1 + \frac{g_0(a)^2}{(4\pi)^2}\left[\gamma_O^{(0)}\ln(\mu a)^2 + \Delta_\Gamma + \Delta_\Sigma\right], \qquad (3.211)$$

where

$$\gamma_O = \gamma_\Gamma + \gamma_\Sigma,$$
$$\Delta_\Gamma = C_\Gamma^{\text{DR}} - C_\Gamma^{\text{LAT}}, \qquad (3.212)$$
$$\Delta_\Sigma = C_\Sigma^{\text{DR}} - C_\Sigma^{\text{LAT}}.$$

It is this renormalization constant (with this choice of renormalization condition) that is usually denoted by Z_O in lattice perturbation theory calculations. The dependence of $Z_O^{\overline{\text{MS}},\text{LAT}}$ on the coefficients C_Γ^{DR} and C_Σ^{DR} comes from the choice of the $\overline{\text{MS}}$ renormalization condition (see Eqs. (3.206) and (3.208)), whereas its dependence on C_Γ^{LAT} and C_Σ^{LAT} from the lattice regularization (see Eq. (3.207)). Two perturbative calculations are thus necessary, one in the continuum for the C^{DR}s and one on the lattice for C^{LAT}s. The presence of the C^{DR}s on the r.h.s. of Eq. (3.209) is sometimes referred to as the "regularization dependence" of the renormalization scheme.

On the other hand, the RI/MOM scheme is obtained by imposing condition (3.115) on the lattice correlation function (3.207). It follows that the renormalization constant of interest, in one-loop perturbation theory, is given by

$$Z_\Gamma^{\text{RI/MOM}}(\mu a, g_0(a)) = 1 + \frac{g_0(a)^2}{(4\pi)^2} \left[\gamma_\Gamma^{(0)} \ln(\mu a)^2 - C_\Gamma^{\text{LAT}} \right]. \tag{3.213}$$

Similarly, condition (3.22), related to the quark field renormalization, gives

$$Z_\psi^{\text{RI/MOM}}(\mu a, g_0(a)) = 1 + \frac{g_0(a)^2}{(4\pi)^2} \left[\gamma_\Sigma^{(0)} \ln(\mu a)^2 - C_\Sigma^{\text{LAT}} \right]. \tag{3.214}$$

The last two expressions combine to give

$$Z_O^{\text{RI/MOM}}(\mu a, g_0(a)) = 1 + \frac{g_0(a)^2}{(4\pi)^2} \left[\gamma_O^{(0)} \ln(\mu a)^2 - C_\Gamma^{\text{LAT}} - C_\Sigma^{\text{LAT}} \right]. \tag{3.215}$$

Clearly, the above RI/MOM result does not depend on a continuum regularization (such as DR), but only on the lattice. This is referred to as "regularization independence".

Acknowledgments

The present chapter is based on a short course, delivered at the XCIII Les Houches Summer School (August 2009). The exciting atmosphere of the School and the students' enthusiasm have encouraged me to write up a significantly expanded version of the original course. I thank my colleagues at Les Houches, students and lecturers alike, for their stimulating and supportive attitude. I also thank Ben Svetitsky for his constant (pure malt) spiritual support during the preparation of my lectures at Les Houches.

I am grateful to Giancarlo Rossi, Rainer Sommer and Chris Sachrajda for carefully reading the manuscript, for their numerous useful suggestions, and for their constructive criticism. I am especially indebted to Massimo Testa, for his patience and constant advise during several long encounters, throughout the various phases of preparation of this manuscript. If these lectures prove useful to young lattice researchers, it is largely thanks to the help provided by these colleagues. Naturally, responsibility for any remaining shortcomings rests entirely with the author.

Notes

1. Clearly, renormalization schemes, constants, etc. depend on the lattice actions. However, the choice of a specific gluonic action, such as that of Eq. (3.1), is of no particular consequence to the topics discussed in this chapter.
2. This statement is usually reserved for the Wilson fermion action only; staggered fermions are known to display a reduced chiral symmetry, while Ginsparg–Wilson ones preserve a lattice chiral symmetry. Yet, in practice even with the latter actions there is some loss of chirality: for domain-wall fermions the extension of the fifth dimension is never quite infinite.
3. Note that unless otherwise stated, flavor indices run over $f = 1, \cdots, N_F^2 - 1$. The singlet case ($f = 0$), related to anomalous Ward identities, will not be discussed in these chapter.
4. If the matrix $\overline{M}$ were not diagonal, quantum numbers like strangeness would be violated by the theory.
5. This is not unexpected: anomalous dimensions are continuum quantities and since in the continuum scalar and pseudoscalar operators belong to the same chiral multiplet, they have the same anomalous dimension $\gamma_S = \gamma_P$. We have already derived from vector Ward identities that $\gamma_S = -\gamma_m$.
6. Note that Eq. (3.79) is only valid away from the chiral limit, where the integral over z of the total divergence $\nabla_z^\mu A_\mu^a(z)$ vanishes. At zero quark mass, the term containing the total divergence of the axial current contributes, because of the presence of massless Goldstone bosons.
7. For scale-dependent renormalization parameters, such as Z_P, renormalization group running is naturally performed non-perturbatively in the Schrödinger functional scheme, for a large range of renormalization scales. As this topic is beyond the scope of the present chapter the reader is advised to consult the literature on non-pertrubative renormalization and the Schrödinger functional (Lüscher, 1998; Sommer, 1997; Weisz, 2010).
8. Computing the chiral condensate directly form the trace of the quark propagator is not viable with Wilson fermions, due to the presence of the cubic divergence proportional to b_0.
9. This assumption will be proved in Section 3.4.3.
10. A correlation function has non-exceptional momenta if no partial sum of the incoming momenta p_i vanishes.
11. Contrary to what we have done so far, we find it convenient to identify x_1 with the origin (i.e. $x_1 = 0$) in the correlation function $G_P(x_1 - x, x_2 - x)$ of Eq. (3.27). Then, the Fourier transform, analogous to that of Eq. (3.27), is performed w.r.t to q (conjugate variable of x) and p_2 (conjugate variable of x_2), in order to obtain $G(p_1, p_2)$, with $p_2 \equiv p_1 - q$.

12. Intermediate two-pion states are of course allowed. These, however, give rise not to poles, but to branch cuts with less severe, logarithmic singularities in the chiral limit.

13. Strictly speaking, it is of course conceivable that one renormalizes, say S and P, without taking Ward identities into account. This is acceptable from the point of view of renormalization (i.e. removal of divergences) but distorts, by terms finite in the bare coupling g_0, the recovery of symmetry as the regularization is removed.

14. For simplicity, we limit our discussion to the renormalization properties of quark bilinear operators. Analogous conclusions may be drawn for other more complicated cases. For example, the scale-independent mixing coefficients of the four-fermion operators, such as the ones involved in $K \to \pi\pi$ decays, are also fixed by Ward identities (Bochicchio *et al.*, 1985; Donini *et al.*, 1999). When computed in the RI/MOM scheme (Donini *et al.*, 1999), they are also subject to Goldstone pole contamination (Becirevic *et al.*, 2004).

15. We have swapped, in Eq. (3.135) the dependence from the bare quark mass m_0 with that of the subtracted bare mass m.

16. An instructive exercise consists in plugging-in the above OPE expressions for Γ_S and Γ_P on the r.h.s. of Eq. (3.138) in order to explicitly confirm that they combine to give the ratio Z_P/Z_S. Contributions proportional to the chiral condensate cancel out.

17. The transformations of Eq. (3.168), with τ^1 replaced by τ^2, are also a symmetry.

18. The same is true of the power subtraction m_{cr}, being the counterterm of the standard bare mass m_0. However, non-perturbatove determinations of m_{cr}, computed in a tmQCD framework, require extra care (Frezzotti *et al.*, 2006).

19. If Symanzik improvement is used, these discretization effects are $O(a^2)$.

20. In the realistic case of QCD with non-degenerate quark masses, the non-pertubative determination of Symanzik counterterms is a non-trivial and intricate task (Bhattacharya *et al.*, 2006).

21. This seemingly unnatural choice (the $\overline{\mathrm{MS}}$ is closely linked to continuum DR) has a few advantages. For example, matrix elements of effective Hamiltonians, once calculated non-perturbatively on the lattice, must be renormalized and combined with perturbatively calculated Wilson coefficients, in order to obtain physical amplitudes; cf. Eqs. (3.112) and (3.113). The renormalization-group invariance of these amplitudes is guaranteed only if the Wilson coefficients and the renormalization constants are calculated in the same renormalization scheme. Since the former are often known in the $\overline{\mathrm{MS}}$ scheme, this scheme is also preferred for the calculation of the latter.

References

Aoki, Y. *et al.* (2008). *Phys. Rev.*, **D78**, 054510.

Arthur, R. and Boyle, P. A. (2011). arXiv:1006.0422 [hep-lat], to appear in Phys. Rev. D.

Becirevic, D. *et al.* (2004). *JHEP*, **08**, 022.

Bernard, C.W. (1989). *Weak matrix elements on and off the lattice*, 233. Lectures given at TASI '89, Boulder, CO, Jun 4–30, 1989.

Bhattacharya, T. *et al.* (2006). *Phys. Rev.*, **D73**, 034504.

Bochicchio, M. *et al.* (1985). *Nucl. Phys.*, **B262**, 331.

Cudell, J.-R., Le Yaouanc, A., and Pittori, C. (1999). *Phys. Lett.*, **B454**, 105.

Cudell, J.-R., Le Yaouanc, A., and Pittori, C. (2001). *Phys. Lett.*, **B516**, 92.

Curci, G. (1986). *Phys. Lett.*, **B167**, 425.

Dimopoulos, P. *et al.* (2006). *Nucl. Phys.*, **B749**, 69.

Donini, A. *et al.* (1999). *Eur. Phys. J.*, **C10**, 121.

Frezzotti, R. *et al.* (2001). *JHEP*, **08**, 058.

Frezzotti, R. *et al.* (2006). *JHEP*, **04**, 038.

Frezzotti, R. and Rossi, G. C. (2004a). *JHEP*, **08**, 007.

Frezzotti, R. and Rossi, G. C. (2004b). *JHEP*, **10**, 070.

Frezzotti, R. and Rossi, G. C. (2004c). *Nucl. Phys. Proc. Suppl.*, **128**, 193.

Gell-Mann, M., Oakes, R.J., and Renner, B. (1968). *Phys. Rev.*, **175**, 2195.

Giusti, L. *et al.* (2001). *Int. J. Mod. Phys.*, **A16**, 3487.

Giusti, L. and Vladikas, A. (2000). *Phys. Lett.*, **B488**, 303.

Itzykson, C. and Zuber, J.-B. (1980). *Quantum field theory*. McGraw-Hill International Book Company, New york section.8.3.

Karsten, L.H. and Smit, J. (1981). *Nucl. Phys.*, **B183**, 103.

Lavelle, M. J. and Oleszczuk, M. (1992). *Phys. Lett.*, **B275**, 133.

Lüscher, M. (1998). Lectures given at Les Houches Summer School in Theoretical Physics, Session 68: Probing the Standard Model of Particle Interactions, Les Houches, France, 28 Jul–5 Sep 1997, edited by R.Gupta *et al.* (Elsevier Science, Amsterdam,1999).

Maiani, L. *et al.* (1987). *Nucl. Phys.*, **B293**, 420.

Martinelli, G. *et al.* (1993). *Phys. Lett.*, **B311**, 241.

Martinelli, G. *et al.* (1995). *Nucl. Phys.*, **B445**, 81.

Neuberger, H. (1998a). *Phys. Lett.*, **B417**, 141.

Neuberger, H. (1998b). *Phys. Lett.*, **B427**, 353.

Neuberger, H. (1998c). *Phys. Rev.*, **D57**, 5417.

Papinutto, M. (2001). *Ph.D. thesis*. unpublished.

Pascual, P. and de Rafael, E. (1982). *Zeit. Phys.*, **C12**, 127.

Pena, C., Sint, S., and Vladikas, A. (2004). *JHEP*, **09**, 069.

Peskin, M.E. and Schroeder, V.D. (1995). *An introduction to quantum field theory*. Addison-Wesley Publishing Company, Reading, MA section.7.2.

Politzer, H.D. (1976). *Nucl. Phys.*, **B117**, 397.

Rossi, G.C., Martinelli, G., and Testa, M. (2003). Unpublished notes.

Shindler, A. (2008). *Phys. Rep.*, **461**, 37.

Sint, S. (2007). Lectures given at the Nara Workshop, Japan, 31 Oct–11 Nov 2005; in Perspectives in Lattice QCD, edited by Y. Kuramashi, World Scientific, Singapore, 2008; hep-lat/0702008.

Sommer, R. (1997). Lectures given at the 36th Internationale Universitätswochen Für Kernphysik und Teilchenphysik, Schladming, Austria, 1–8 Mar 1997. In Computing particle properties; edited by H. gausterer and Ch. Lang (Springer Publishing Company, Berlin, 1998).

Sturm, C. *et al.* (2009). *Phys. Rev.*, **D80**, 014501.

Testa, M. (1998). *JHEP*, **04**, 002.

Weisz, P. (2010). Chapter 2 of this volume.

Wilson, K.G. (1974). *Phys. Rev.*, **D10**, 2445.

Wilson, K.G. (1977). New Phenomena In Subnuclear Physics. Part A. Proceedings of the First Half of the 1975 International School of Subnuclear Physics, Erice, Sicily, July 11–August 1, 1975, ed. A. Zichichi, Plenum Press, New York, 1977, p. 69.

4

Chiral symmetry and lattice fermions

David B. KAPLAN

Institute for Nuclear Theory, University of Washington, Seattle, WA 98195-1550
Lectures delivered at the Les Houches École d'Été de Physique Thńeorique

4.1   Chiral symmetry

4.1.1   Introduction

Chiral symmetries play an important role in the spectrum and phenomenology of both the Standard Model and various theories for physics beyond the Standard Model. In many cases chiral symmetry is associated with non-perturbative physics that can only be quantitatively explored in full on a lattice. It is therefore important to implement chiral symmetry on the lattice, which turns out to be less than straightforward. In this chapter I discuss what chiral symmetry is, why it is important, how it is broken, and ways to implement it on the lattice. There have been many hundreds of papers on the subject and this is not an exhaustive review; the limited choice of topics I cover reflects on the scope of my own understanding and not the value of the omitted work.

4.1.2   Spinor representations of the Lorentz group

To understand chiral symmetry one must understand Lorentz symmetry first. Since we will be discussing fermions in various dimensions of space-time, consider the generalization of the usual Lorentz group to d dimensions. The Lorentz group is defined by the real matrices Λ that preserve the form of the d-dimensional metric

$$\Lambda^T \eta \, \Lambda = \eta, \qquad \eta = \mathrm{diag}\,(1, -1, \ldots -1). \tag{4.1}$$

With this definition, the inner product between two 4-vectors, $v^\mu \eta_{\mu\nu} w^\mu = v^T \eta w$, is preserved under the Lorentz transformations $v \to \Lambda v$ and $w \to \Lambda w$. This defines the group $SO(d-1, 1)$, which—like $SO(d)$—has $d(d-1)/2$ linearly independent generators, which may be written as $M^{\mu\nu} = -M^{\nu\mu}$, where the indices $\mu, \nu = 0, \ldots, (d-1)$ and

$$\Lambda = e^{i\theta_{\mu\nu} M^{\mu\nu}}, \tag{4.2}$$

with $\theta_{\mu\nu} = -\theta_{\nu\mu}$ being $d(d-1)/2$ real parameters. Note that μ, ν label the $d(d-1)/2$ generators, while in a representation R each M is a $d_R \times d_R$ matrix, where d_R is the dimension of R. By expanding Eq. (4.1) to order θ one sees that the generators M must satisfy

$$(M^{\mu\nu})^T \eta + \eta M^{\mu\nu} = 0. \tag{4.3}$$

From this equation it is straightforward to write down a basis for the $M^{\mu\nu}$ in the d-dimensional defining representation and determine the commutation relations for the algebra,

$$\left[M^{\alpha\beta}, M^{\gamma\delta} \right] = i \left(\eta^{\beta\gamma} M^{\alpha\delta} - \eta^{\alpha\gamma} M^{\beta\delta} - \eta^{\beta\delta} M^{\alpha\gamma} + \eta^{\alpha\delta} M^{\beta\gamma} \right). \tag{4.4}$$

A Dirac spinor representation can be constructed as

$$M^{\alpha\beta} \equiv \Sigma^{\alpha\beta} = \frac{i}{4} \left[\gamma^\alpha, \gamma^\beta \right], \tag{4.5}$$

where the gamma matrices satisfy the Clifford algebra:

$$\{\gamma^\alpha, \gamma^\beta\} = 2\eta^{\alpha\beta}. \tag{4.6}$$

Solutions to the Clifford algebra are easy to find by making use of direct products of Pauli matrices. In a direct product space we can write a matrix as $M = a \otimes A$, where a and A are matrices of dimension d_a and d_A, respectively, acting in different spaces; the matrix M then has dimension $(d_a \times d_A)$. Matrix multiplication is defined as $(a \otimes A)(b \otimes B) = (ab) \otimes (AB)$. It is usually much easier to construct a representation when you need one rather than to look one up and try to keep the conventions straight! One finds that solutions for the γ matrices in d dimensions obey the following properties:

1. For both $d = 2k$ and $d = 2k + 1$, the γ-matrices are 2^k dimensional.
2. For even space-time dimension $d = 2k$ (such as our own with $k = 2$) one can define a generalization of γ_5 to be

$$\Gamma = i^{k-1} \prod_{\mu=0}^{2k-1} \gamma^\mu \tag{4.7}$$

with the properties

$$\{\Gamma, \gamma^\mu\} = 0, \quad \Gamma = \Gamma^\dagger = \Gamma^{-1}, \quad \text{Tr}(\Gamma\gamma^{\alpha_1} \cdots \gamma^{\alpha_{2k}}) = 2^k i^{-1-k} \epsilon^{\alpha_1 \ldots \alpha_{2k}}, \tag{4.8}$$

where $\epsilon_{012\ldots 2k-1} = +1 = -\epsilon^{012\ldots 2k-1}$.
3. In $d = 2k + 1$ dimensions one needs one more γ-matrix than in $d = 2k$, and one can take it to be $\gamma^{2k} = i\Gamma$.

Sometimes, it is useful to work in a specific basis for the γ-matrices; a particulary useful choice is a "chiral basis", defined to be one where Γ is diagonal. For example, for $d = 2$ and $d = 4$ (Minkowski space-time) one can choose

$$d = 2: \quad \gamma^0 = \sigma_1, \quad \gamma^1 = -i\sigma_2, \quad \Gamma = \sigma_3 \tag{4.9}$$

$$d = 4: \quad \gamma^0 = -\sigma_1 \otimes 1, \quad \gamma^i = i\sigma_2 \otimes \sigma_i, \quad \Gamma = \sigma_3 \otimes 1. \tag{4.10}$$

4.1.2.1 γ-matrices in Euclidian space-time

In going to Euclidian space-time with metric $\eta^{\mu\nu} = \delta_{\mu\nu}$, one takes

$$\partial_0^M \to i\partial_0^E, \quad \partial_i^M \to \partial_i^E \tag{4.11}$$

and defines

$$\gamma_M^0 = \gamma_E^0, \quad \gamma_M^i = i\gamma_E^i, \tag{4.12}$$

so that

$$(\gamma_E^\mu)^\dagger = \gamma_E^\mu, \quad \{\gamma_E^\mu, \gamma_E^\nu\} = 2\delta_{\mu\nu} \tag{4.13}$$

and $\not{D}_M \to i\not{D}_E$, with

$$\not{D}_E = -\not{D}_E^\dagger \tag{4.14}$$

and the Euclidian Dirac operator is $(\not{D}_E + m)$. The matrix $\Gamma^{(2k)}$ in $2k$ dimensions is taken to equal γ_E^{2k} in $(2k+1)$ dimensions:

$$\Gamma_E^{(2k)} = \gamma_E^{2k} = \Gamma_M^{(2k)}, \qquad \text{Tr}(\Gamma_E\, \gamma_E^{\alpha_1} \cdots \gamma_E^{\alpha_{2k}}) = -2^k i^k \epsilon^{\alpha_1 \ldots \alpha_{2k}}, \tag{4.15}$$

where $\epsilon_{012\ldots 2k-1} = +1 = +\epsilon^{012\ldots 2k-1}$.

4.1.3   Chirality in even dimensions

This section on chiral fermions follow from the properties of the matrix Γ in even dimensions. The existence of Γ means that Dirac spinors are reducible representations of the Lorentz group, which in turn means we can have symmetries ("chiral symmetries") that transform different parts of Dirac spinors in different ways. To see this, define the projection operators

$$P_\pm = \frac{(1 \pm \Gamma)}{2}, \tag{4.16}$$

which have the properties

$$P_+ + P_- = \mathbf{1}, \qquad P_\pm^2 = P_\pm, \qquad P_+P_- = 0. \tag{4.17}$$

Since in odd spatial dimensions $\{\Gamma, \gamma^\mu\} = 0$ for all μ, it immediately follows that Γ commutes with the Lorentz generators $\Sigma^{\mu\nu}$ in Eq. (4.5): $[\Gamma, \Sigma^{\mu\nu}] = 0$. Therefore, we can write $\Sigma^{\mu\nu} = \Sigma_+^{\mu\nu} + \Sigma_-^{\mu\nu}$, where

$$\Sigma_\pm^{\mu\nu} = P_\pm \Sigma^{\mu\nu} P_\pm, \qquad \Sigma_+^{\alpha\beta}\Sigma_-^{\mu\nu} = \Sigma_-^{\mu\nu}\Sigma_+^{\alpha\beta} = 0. \tag{4.18}$$

Thus, $\Sigma^{\mu\nu}$ is reducible: spinors $\psi_\pm$, which are eigenstates of Γ with eigenvalue ± 1, respectively, transform independently under Lorentz transformations.

The word "chiral" comes from the Greek word for hand, $\chi\epsilon\iota\rho$. The projection operators P_+ and P_- are often called P_R and P_L, respectively; what does handedness have to do with the matrix Γ? Consider the Lagrangian for a free massless Dirac fermion in $1+1$ dimensions and use the definition of $\Gamma = \gamma^0\gamma^1$:

$$\begin{aligned}
\mathcal{L} &= \bar{\Psi} i\not{\partial}\Psi \\
&= \Psi^\dagger i \left(\partial_t + \Gamma\partial_x\right)\Psi \\
&= \Psi_+^\dagger i \left(\partial_t + \partial_x\right)\Psi_+ + \Psi_-^\dagger i \left(\partial_t - \partial_x\right)\Psi_-,
\end{aligned} \tag{4.19}$$

where

$$\Gamma\Psi_\pm = \pm\Psi_\pm. \tag{4.20}$$

We see then that the solutions to the Dirac equation for $m = 0$ are

$$\Psi_\pm(x, t) = \Psi_\pm(x \mp t)$$

so that Ψ_+ corresponds to right-moving solutions, and Ψ_- corresponds to left-moving solutions. This is possible since the massless particles move at the speed of light and the direction of motion is invariant under proper Lorentz transformations in $(1+1)$ dimensions.

In the chiral basis (4.9), the positive energy plane-wave solutions to the Dirac equation are

$$\Psi_+ = e^{-iE(t-x)} \begin{pmatrix} 1 \\ 0 \end{pmatrix}, \qquad \Psi_- = e^{-iE(t+x)} \begin{pmatrix} 0 \\ 1 \end{pmatrix}. \tag{4.21}$$

It is natural to write P_+ and P_- as P_R and P_L, respectively.

Exercise 4.1 You should perform the same exercise in $3+1$ dimensions and find that solutions $\Psi_\pm$ to the massless Dirac equation satisfying $\Gamma\Psi_\pm = \pm\Psi_\pm$ must also satisfy $|\vec{p}| = E$ and $(2\vec{p}\cdot\vec{S}/E)\Psi_\pm = \pm\Psi_\pm$, where $S_i = \frac{1}{2}\epsilon_{0ijk}\Sigma^{jk}$ are the generators of rotations. Thus, $\Psi_\pm$ correspond to states with positive or negative helicity, respectively, and are called right- and left-handed particles.

4.1.4   Chiral symmetry and fermion mass in four dimensions

Consider the Lagrangian for a single flavor of Dirac fermion in $3+1$ dimensions coupled to a background gauge field,

$$\mathcal{L} = \bar{\Psi}(i\slashed{D} - m)\Psi = \left(\bar{\Psi}_L i\slashed{D}\Psi_L + \bar{\Psi}_R i\slashed{D}\Psi_R\right) - m\left(\bar{\Psi}_L\Psi_R + \bar{\Psi}_R\Psi_L\right), \tag{4.22}$$

where I have defined

$$\Psi_L = P_-\Psi, \quad \bar{\Psi}_L = \Psi_L^\dagger\gamma^0 = \bar{\Psi}P_+, \quad \Psi_R = P_+\Psi, \quad \bar{\Psi}_R = \bar{\Psi}P_-. \tag{4.23}$$

For now, I am assuming that $\Psi_{L,R}$ are in the same complex representation of the gauge group, where D_μ is the gauge covariant derivative appropriate for that representation. It is important to note that the property $\{\gamma_5, \gamma^\mu\} = 0$ ensured that the kinetic terms in Eq. (4.23) do not couple left-handed and right-handed fermions; on the other hand, the mass terms do.[1] The above Lagrangian has an exact $U(1)$ symmetry, associated with fermion number, $\Psi \to e^{i\alpha}\Psi$. Under this symmetry, left-handed and right-handed components of Ψ rotate with the same phase; this is often called a "vector symmetry". In the case where $m = 0$, it apparently has an additional symmetry where the left- and right-handed components rotate with the opposite phase, $\Psi \to e^{i\alpha\gamma_5}\Psi$; this is called an "axial symmetry", $U(1)_A$.

Symmetries are associated with Noether currents, and symmetry violation appears as a non-zero divergence for the current. Recall the Noether formula for a field ϕ and infinitesimal transformation $\phi \to \phi + \epsilon\delta\phi$:

$$J^\mu = -\frac{\partial\mathcal{L}}{\partial(\partial_\mu\phi)}\delta\phi, \qquad \partial_\mu J^\mu = -\delta\mathcal{L}. \tag{4.24}$$

In the Dirac theory, the vector symmetry corresponds to $\delta\Psi = i\Psi$, and the axial symmetry transformation is $\delta\Psi = i\gamma_5\Psi$, so that the Noether formula yields the vector and axial currents:

$$U(1): \qquad J^\mu = \overline{\Psi}\gamma^\mu\Psi, \qquad\qquad \partial_\mu J^\mu = 0 \tag{4.25}$$

$$U(1)_A: \qquad J_A^\mu = \overline{\Psi}\gamma^\mu\gamma_5\Psi, \qquad\qquad \partial_\mu J_A^\mu = 2im\overline{\Psi}\gamma_5\Psi. \tag{4.26}$$

Some comments are in order:

- Equation (4.26) is not the whole story! We will soon talk about additional contributions to the divergence of the axial current from the regulator, called anomalies, which do not decouple as the regulator is removed.

- The fact that the fermion mass breaks chiral symmetry means that fermion masses get multiplicatively renormalized, which means that fermions can naturally be light (unlike scalars in most theories); more on this later.

- The variation of a general fermion bilinear $\overline{\Psi}X\Psi$ under chiral symmetry is

$$\delta\overline{\Psi}X\Psi = i\overline{\Psi}\{\gamma_5, X\}\Psi. \tag{4.27}$$

This will vanish if X can be written as the product of an odd number of gamma matrices. In any even dimension the chirally invariant bilinears include currents, with $X = \gamma^\mu$ or $X = \gamma^\mu\Gamma$, while the bilinears that transform non-trivially under the chiral symmetry include mass terms, $X = \mathbf{1}, \Gamma$. Thus, gauge interactions can be invariant under chiral symmetry, while fermion masses are always chiral-symmetry violating. In $(3 + 1)$ dimensions, anomalous electromagnetic moment operators corresponding to $X = \sigma_{\mu\nu}, \sigma_{\mu\nu}\gamma_5$ are also chiral-symmetry violating.

- A more general expression for the classical divergence of the axial current for a bilinear action $\int \overline{\Psi}D\Psi$ in any even dimension is

$$\partial_\mu J_A^\mu = i\overline{\Psi}\{\Gamma, D\}\Psi. \tag{4.28}$$

On the lattice one encounters versions of the fermion operator D that violate chiral symmetry even for a massless fermion.

The Lagrangian for N_f flavors of massive Dirac fermions in odd d, coupled to some background gauge field may be written as

$$\mathcal{L} = \left(\overline{\Psi}_L^a i\slashed{D}\Psi_L^a + \overline{\Psi}_R^a i\slashed{D}\Psi_R^a\right) - \left(\overline{\Psi}_L^a M_{ab}\Psi_R^b + \overline{\Psi}_R^a M_{ab}^\dagger\Psi_L^b\right). \tag{4.29}$$

The index on Ψ denotes flavor, with $a, b = 1, \dots N_f$, and M_{ab} is a general complex mass matrix (no distinction between upper and lower flavor indices). Again, assuming the fermions to be in a complex representation of the gauge group, this theory is invariant under independent chiral transformations if the mass matrix vanishes:

$$\Psi_R^a \to U_{ab}\Psi_R^b, \quad \Psi_L^a \to V_{ab}\Psi_L^b, \quad U^\dagger U = V^\dagger V = \mathbf{1}, \tag{4.30}$$

where U and V are independent $U(N_f)$ matrices. Since $U(N_f) = SU(N_f) \times U(1)$, it is convenient to write

$$U = e^{i(\alpha+\beta)} R, \quad V = e^{i(\alpha-\beta)} L, \quad R^\dagger R = L^\dagger L = \mathbf{1}, \quad |R| = |L| = 1, \quad (4.31)$$

so that the symmetry group is $SU(N_f)_L \times SU(N_f)_R \times U(1) \times U(1)_A$ with $L \in SU(N_f)_L$, $R \in SU(N_f)_R$.

If we turn on the mass matrix, the chiral symmetry is explicitly broken, since the mass matrix couples left- and right-handed fermions to each other. If $M_{ab} = m\delta_{ab}$ then the "diagonal" or "vector" symmetry $SU(N_f) \times U(1)$ remains unbroken, where $SU(N_f) \subset SU(N_f)_L \times SU(N_f)_R$ corresponding to the transformation (4.30), (4.31) with $L = R$. If M_{ab} is diagonal but with unequal eigenvalues, the symmetry may be broken down as far as $U(1)^{N_f}$, corresponding to independent phase transformations of the individual flavors. With additional flavor-dependent interactions, these symmetries may be broken as well.

4.1.5 Weyl fermions

4.1.5.1 Lorentz group as $SU(2) \times SU(2)$ and Weyl fermions

We have seen that Dirac fermions in even dimensions form a reducible representation of the Lorentz group. Dirac notation is convenient when both LH and RH parts of the Dirac spinor transform as the same complex representation under a gauge group, and when there is a conserved fermion number. This sounds restrictive, but applies to QED and QCD. For other applications—such as chiral gauge theories (where LH and RH fermions carry different gauge charges, as under $SU(2) \times U(1)$), or when fermion number is violated (as is the case for neutrinos with a Majorana mass), or when fermions transform as a real representation of gauge group—it is much more convenient to use irreducible fermion representations, called Weyl fermions.

The six generators of the Lorentz group may be chosen to be the three Hermitian generators of rotations J_i, and the three anti-Hermitian generators of boosts K_i, so that an arbitrary Lorentz transformation takes the form

$$\Lambda = e^{i(\theta_i J_i + \omega_i K_i)}. \quad (4.32)$$

In terms of the $M_{\mu\nu}$ generators in Section 4.1.2,

$$J_i = \tfrac{1}{2}\epsilon_{0i\mu\nu} M^{\mu\nu}, \quad K_i = M^{0i}. \quad (4.33)$$

These generators have the commutation relations

$$[J_i, J_j] = i\epsilon_{ijk} J_k, \quad [J_i, K_j] = i\epsilon_{ijk} K_k, \quad [K_i, K_j] = -i\epsilon_{ijk} J_k. \quad (4.34)$$

It is convenient to define different linear combinations of generators

$$A_i = \frac{J_i + iK_i}{2}, \quad B_i = \frac{J_i - iK_i}{2}, \quad (4.35)$$

satisfying an algebra that looks like $SU(2) \times SU(2)$, except for the fact that the generators Eq. (4.35) are not Hermitian and therefore the group is non-compact:

$$[A_i, A_j] = i\epsilon_{ijk}A_k, \quad [B_i, B_j] = i\epsilon_{ijk}B_k, \quad [A_i, B_j] = 0. \tag{4.36}$$

Thus, Lorentz representations may be labelled with two $SU(2)$ spins $j_{A,B}$, corresponding to the two $SU(2)$s: (j_A, j_B), transforming as

$$\Lambda(\vec{\theta}, \vec{\omega}) = D^{j_A}(\vec{\theta} - i\vec{\omega}) \times D^{j_B}(\vec{\theta} + i\vec{\omega}), \tag{4.37}$$

where the D^j is the usual $SU(2)$ rotation in the spin j representation; boosts appear as imaginary parts to the rotation angle; the D^{j_A} and D^{j_B} matrices act in different spaces and therefore commute. For example, under a general Lorentz transformation, a LH Weyl fermion ψ and a RH Weyl fermion χ transform as $\psi \to L\psi$, $\chi \to R\chi$, where

$$L = e^{i(\vec{\theta} - i\vec{\omega})\cdot\vec{\sigma}/2}, \qquad R = e^{i(\vec{\theta} + i\vec{\omega})\cdot\vec{\sigma}/2}. \tag{4.38}$$

Evidently the two types of fermions transform the same way under rotations, but differently under boosts.

The dimension of the (j_A, j_B) representation is $(2j_A + 1)(2j_B + 1)$. In this notation, the smaller irreducible Lorentz representations are labelled as:

$$(0,0): \quad \text{scalar}$$

$$(\tfrac{1}{2},0), (0,\tfrac{1}{2}): \quad \text{LH and RH Weyl fermions}$$

$$(\tfrac{1}{2},\tfrac{1}{2}): \quad \text{four-vector}$$

$$(1,0), (0,1): \quad \text{self-dual and anti-self-dual antisymmetric tensors}$$

A Dirac fermion is the reducible representation $(\tfrac{1}{2},0) \oplus (0,\tfrac{1}{2})$ consisting of a LH and a RH Weyl fermion.

Parity interchanges the two $SU(2)$s, transforming a (j_1, j_2) representation into (j_2, j_1). Similarly, charge conjugation effectively flips the sign of K_i in Eq. (4.37) due to the factor of i implying that if a field ϕ transforms as (j_1, j_2), then $\phi^\dagger$ transforms as (j_2, j_1).[2] Therefore, a theory of N_L flavors of LH Weyl fermions ψ_i and N_R flavors of RH Weyl fermions χ_a may be recast as a theory of $(N_L + N_R)$ LH fermions by defining $\chi_a \equiv \omega_a^\dagger$. The fermion content of the theory can be described entirely in terms of LH Weyl fermions then, $\{\psi_i, \omega_a\}$; this often simplifies the discussion of parity-violating theories, such as the Standard Model or Grand Unified Theories. Note that if the RH χ_a transformed under a gauge group as representation R, the conjugate fermions ω_a transform under the conjugate representation $\bar{R}$.

For example, QCD written in terms of Dirac fermions has the Lagrangian:

$$\mathcal{L} = \sum_{i=u,d,s\ldots} \overline{\Psi}_n(i\slashed{D} - m_n)\Psi_n, \tag{4.39}$$

where D_μ is the $SU(3)_c$ covariant derivative, and the Ψ_n fields (both LH and RH components) transform as a **3** of $SU(3)_c$. However, we could just as well write the theory in terms of the LH quark fields ψ_n and the LH antiquark fields χ_n. Using the γ-matrix basis in Eq. (4.10), we write the Dirac spinor Ψ in terms of two-component LH spinors ψ and χ as

$$\Psi = \begin{pmatrix} -\sigma_2\chi^\dagger \\ \psi \end{pmatrix}. \tag{4.40}$$

Note that ψ transforms as a **3** of $SU(3)_c$, while χ transforms as a $\bar{\mathbf{3}}$. Then the kinetic operator becomes (up to a total derivative)

$$\overline{\Psi}i\slashed{D}\Psi = \psi^\dagger iD_\mu\sigma^\mu\psi + \chi^\dagger iD_\mu\sigma^\mu\chi, \qquad \sigma^\mu \equiv \{\mathbf{1}, -\vec{\sigma}\}, \tag{4.41}$$

and the mass terms become

$$\begin{aligned} \overline{\Psi}_R\Psi_L &= \chi\sigma_2\psi = \psi\sigma_2\chi \\ \overline{\Psi}_L\Psi_R &= \psi^\dagger\sigma_2\chi^\dagger = \chi^\dagger\sigma_2\psi^\dagger, \end{aligned} \tag{4.42}$$

where I used the fact that fermion fields anticommute. Thus, a Dirac mass in terms of Weyl fermions is just

$$m\overline{\Psi}\Psi = m(\psi\sigma_2\chi + h.c), \tag{4.43}$$

and preserves a fermion number symmetry, where ψ has charge $+1$ and χ has charge -1. On the other hand, one can also write down a Lorentz-invariant mass term of the form

$$m(\psi\sigma_2\psi + h.c.), \tag{4.44}$$

which violates fermion number by two units; this is a Majorana mass, which is clumsy to write in Dirac notation. Experimentalists are trying to find out which form neutrino masses have—Dirac or Majorana? If the latter, lepton number is violated by two units and could show up in neutrinoless double beta decay, where a nucleus decays by emitting two electrons and no antineutrinos.

The Standard Model is a relevant example of a chiral gauge theory. Written in terms of LH Weyl fermions, the quantum numbers of a single family under $SU(3) \times SU(2) \times U(1)$ are:

$$\begin{aligned} Q &= (3,2)_{+\frac{1}{6}} & L &= (1,2)_{-\frac{1}{2}} \\ U^c &= (\bar{3},1)_{-\frac{2}{3}} & E^c &= (1,1)_{+1} \\ D^c &= (\bar{3},1)_{+\frac{1}{3}}. \end{aligned} \tag{4.45}$$

Evidently this is a complex representation and chiral. If neutrino masses are found to be Dirac in nature (i.e. lepton-number preserving) then a partner for the neutrino must be added to the theory, the "right-handed neutrino", that can be described by a LH Weyl fermion which is neutral under all Standard Model gauge interactions, $N = (1,1)_0$.

If unfamiliar with two-component notation, you can find all the details in Appendix A of Wess and Bagger's classic book on supersymmetry (Wess and Bagger, (1992); the notation used here differs slightly as I use the metric and γ-matrix conventions of Itzykson and Zuber (Itzykson and Zuber, (1980), and write out the σ_2 matrices explicitly.

Exercise 4.2 Consider a theory of N_f flavors of Dirac fermions in a real or pseudoreal representation of some gauge group. [Real representations combine symmetrically to form an invariant, such as a triplet of $SU(2)$; pseudoreal representations combine antisymmetrically, such as a doublet of $SU(2)$.] Show that if the fermions are massless the action exhibits a $U(2N_f) = U(1) \times SU(2N_f)$ flavor symmetry at the classical level [the $U(1)$ subgroup being anomalous in the quantum theory]. If the fermions condense as in QCD, what is the symmetry-breaking pattern? How do the resultant Goldstone bosons transform under the unbroken subgroup of $SU(2N_f)$?

Exercise 4.3 To see how the $(\frac{1}{2}, \frac{1}{2})$ representation behaves like a four-vector, consider the 2×2 matrix $P = P_\mu \sigma^\mu$, where σ^μ is given in Eq. (4.41). Show that the transformation $P \rightarrow LPL^\dagger$ (with $\det L = 1$) preserves the Lorentz-invariant inner product $P_\mu P^\mu = (P_0^2 - P_i P_i)$. Show that with L given by Eq. (4.38), P_μ transforms properly like a four-vector.

Exercise 4.4 Is it possible to write down an anomalous electric or magnetic moment operator in a theory of a single charge-neutral Weyl fermion?

4.1.6 Chiral symmetry and mass renormalization

Some operators in a Lagrangian suffer from additive renormalizations, such as the unit operator (cosmological constant), and scalar mass terms, such as the Higgs mass in the Standard Model, $|H|^2$. Therefore, the mass scales associated with such operators will naturally be somewhere near the UV cutoff of the theory, unless the bare couplings of the theory are fine tuned to cancel radiative corrections. Such fine-tuning problems have obsessed particle theorists since the work of Wilson and 't Hooft on renormalization and naturalness in the 1970s. However, such intemperate behavior will not occur for operators that violate a symmetry respected by the rest of the theory: if the bare couplings for such operators were set to zero, the symmetry would ensure they could not be generated radiatively in perturbation theory. Fermion mass operators generally fall into this benign category.

Consider the following toy model: QED with a charge-neutral complex scalar field coupled to the electron,

$$\mathcal{L}\overline{\Psi}(i\slashed{D} - m)\Psi + |\partial\phi|^2 - \mu^2|\phi|^2 - g|\phi|^4 + y\left(\overline{\Psi}_R \phi \Psi_L + \overline{\Psi}_L \phi^* \Psi_R\right). \quad (4.46)$$

Note that in the limit $m \rightarrow 0$ this Lagrangian respects a chiral symmetry $\Psi \rightarrow e^{i\alpha\gamma_5}\Psi$, $\phi \rightarrow e^{-2i\alpha}\phi$. The symmetry ensures that if $m = 0$, a mass term for the fermion would not be generated radiatively in perturbation theory. With $m \neq 0$, this means that any renormalization of m must be proportional to m itself (i.e. m is "multiplicatively renormalized"). This is evident if one traces chirality through the Feynman diagrams;

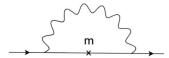

Fig. 4.1 One-loop renormalization of the electron mass in QED due to photon exchange. A mass operator flips chirality, while gauge interactions do not. A contribution to the electron mass requires an odd number of chirality flips, and so there has to be at least one insertion of the electron mass in the diagram: the electron mass is multiplicatively renormalized. A scalar interaction flips chirality when the scalar is emitted, and flips it back when the scalar is absorbed, so replacing the photon with a scalar in the above graph again requires a fermion mass insertion to contribute to mass renormalization.

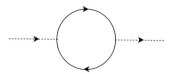

Fig. 4.2 One-loop additive renormalization of the scalar mass due to a quadratically divergent fermion loop.

see Fig. 4.1. Multiplicative renormalization implies that the fermion mass can at most depend logarithmically on the cutoff (by dimensional analysis): $\delta m \sim (\alpha/4\pi)m \ln m/\Lambda$.

In contrast, the scalar mass operator $|\phi|^2$ does not violate any symmetry and therefore suffers from additive renormalizations, such as through the graph in Fig. 4.2. By dimensional analysis, the scalar mass operator can have a coefficient that scales quadratically with the cutoff: $\delta\mu^2 \sim (y^2/16\pi^2)\Lambda^2$. This is called an additive renormalization, since $\delta\mu^2$ is not proportional to μ^2. It is only possible in general to have a scalar in the spectrum of this theory with mass much lighter than $y\Lambda/4\pi$ if the bare couplings are finely tuned to cause large radiative corrections to cancel. When referring to the Higgs mass in the Standard Model, this is called the hierarchy problem.

If chiral symmetry is broken by operators other than the mass term, then the fermion mass will no longer in general be multiplicatively renormalized, and fine tuning may be necessary. This is particularly true if chiral symmetry is broken by "irrelevant" operators. Consider adding to QED a dimension-five operator of the form $W = \frac{r}{\Lambda}\overline{\Psi}D_\mu D^\mu \Psi$, where r is a dimensionless coupling. This operator breaks chirality and therefore one can substitute it for the mass operator in Fig. 4.1; an estimate of the diagram then gives an additive renormalization of the fermion mass, $\delta m \sim (\alpha/4\pi)r\Lambda$. Thus, unless r is extremely small (*e.g.* $r \lesssim m/\Lambda$) the chiral-symmetry-breaking effects of this operator will be important and fine tuning will be necessary to ensure a light fermion in the spectrum. This example is relevant to Wilson's method for putting fermions on the lattice, which does not respect chiral symmetry and entails adding to the action a lattice version of W, where $a \sim 1/\Lambda$ is the lattice spacing and $r \sim 1$. Therefore, Wilson fermions acquire an $O(1/a)$ correction to their mass that needs to be cancelled by a bare contribution in order to describe a world with light fermions.

4.1.7 Chiral symmetry in QCD

4.1.7.1 Chiral-symmetry breaking and Goldstone bosons

So far the discussion of chiral symmetry in terms of the effect on fermion masses has been appropriate for a weakly coupled theory. As was presented in M. Golterman's lectures, the low-energy spectrum of QCD is described by a chiral Lagrangian, encoding the interactions of the meson octet that are the approximate Nambu–Goldstone bosons of spontaneously broken $SU(3)_L \times SU(3)_R$ chiral symmetry.[3] The Goldstone bosons would be massless in the limit of exact chiral symmetry, and so tuning away the leading finite-lattice-spacing correction for Wilson fermions can be accomplished by tuning the bare quark mass to eliminate the $1/a$ dependence of the square of the pion mass.

In contrast to QCD, $N = 1$ super-Yang–Mills theory has a single Weyl fermion (the gaugino) transforming as an adjoint under the gauge group; the theory has a $U(1)_A$ symmetry at the classical level—phase rotations of the gaugino—but it is broken by anomalies to a discrete symmetry. This discrete symmetry is then spontaneously broken by a gluino condensate, but, without any continuous symmetries, no Goldstone bosons are produced. What should the spectrum of this theory look like? Presumably a bunch of massive boson and fermion glueball-like states. They will form degenerate supersymmetric multiplets when the gluino mass is tuned to zero, but there is no particle that becomes massless in the chiral limit in this case, and therefore tuning the bare mass is difficult.

After tuning away the $O(1/a)$ mass correction, there remain for non-chiral lattice fermions the dimension-5 chiral-symmetry-violating operators in the Symanzik action that require $O(a)$ tuning, as discussed by Golterman. In contrast, chiral fermions receive finite lattice corrections only at $O(a^2)$, simply because one cannot write down a dimension-5 chiral-symmetry-preserving operator in QCD.

4.1.7.2 Operator mixing

One encounters additional factors of $1/a$ when computing weak processes. One of the most curious feature of the strong interactions is the $\Delta I = 1/2$ rule, which is the observation that $\Delta S = 1$ transitions in nature are greatly enhanced when they change isospin by $\Delta I = 1/2$, in comparison to $\Delta I = 3/2$. For example, one requires for the amplitudes for kaon decay $K \to \pi\pi$:

$$\frac{\mathcal{A}(\Delta I = 1/2)}{\mathcal{A}(\Delta I = 3/2)} \simeq 20. \tag{4.47}$$

To compute this in the Standard Model, one starts with four-quark operators generated by W-exchange, which can be written as the linear combination of two operators

$$\mathcal{L}_{\Delta S=1} = -V_{ud}V_{us}^* \frac{G_F}{\sqrt{2}} \left[C_+(\mu, M_w)\mathcal{O}^+ + C_-(\mu, M_w)\mathcal{O}^- \right],$$
$$\mathcal{O}^\pm = [(\bar{s}d)_L(\bar{u}u)_L \pm (\bar{s}u)_L(\bar{u}d)_L] - [u \leftrightarrow c], \tag{4.48}$$

where $(\bar{q}q')_L \equiv (\bar{q}\gamma^\mu P_L q')$. If one ignores the charm-quark contribution, the $\mathcal{O}^-$ transforms as an **8** under $SU(3)_f$, while $\mathcal{O}^+$ transforms as a **27**; therefore $\mathcal{O}^-$ is pure $I = 1/2$, while $\mathcal{O}^+$ is a mix of $I = 3/2$ and $I = 1/2$. The full $\mathcal{O}^\pm$ operators are in $SU(4)_f$ multiplets; while $SU(4)_f$ is not a good symmetry of the spectrum, it is only broken by quark masses that do not affect the log divergences of the theory. Thus, the running of the operators respect $SU(4)_f$ down to $\mu = m_c$, and there is no mixing between $\mathcal{O}^\pm$. At the weak scale $\mu = M_W$, one finds $|C^+/C^-| = 1 + O(\alpha_s(M_W))$, showing that there is no $\Delta I = 1/2$ enhancement intrinsic to the weak interactions. One then scales these operators down to $\mu \sim 2$ GeV in order to match onto the lattice theory; using the renormalization group to sum up leading $\alpha_s \ln \mu/M_W$ corrections gives an enhancement $|C^+/C^-| \simeq 2$—which is in the right direction, but not enough to explain Eq. (4.47), which should then come either from QCD at long distances, or else from new physics! This is a great problem for the lattice to resolve.

A wonderful feature about using dimensional regularization and $\overline{MS}$ in the continuum is that an operator will never mix with another operator of lower dimension. This is because there is no UV mass scale in the scheme that can make up for the mismatch in operator dimension. This is not true on the lattice, where powers of the inverse lattice spacing $1/a$ can appear. In particular, the dimension-6 four-fermion operators $\mathcal{O}^\pm$ could in principle mix with dimension-3 two-fermion operators. The only $\Delta S = 1$ dimension-3 operator that could arise is $\bar{s}\gamma_5 d$, which is also $\Delta I = \frac{1}{2}$.[4] If the quarks were massless, the lattice theory would possess an exact discrete "$\hat{S}CP$" symmetry under which one interchanges $s \leftrightarrow d$ and performs a CP transformation to change LH quarks into LH antiquarks; the operators $\mathcal{O}_\pm$ are even under SCP, while $\bar{s}\gamma_5 d$ is odd, so the operator that could mix on the lattice is

$$\mathcal{O}_p = (m_s - m_d)\,\bar{s}\gamma_5 d. \tag{4.49}$$

In a theory where the quark masses are the only source of chiral-symmetry breaking, we have $\mathcal{O}_p = \partial_\mu A_\mu^{\bar{s}d}$, the divergence of the $\Delta S = 1$ axial current. Therefore, on-shell matrix elements of this operator vanish, since the derivative gives $(p_K - p_{2\pi}) = 0$, i.e. no momentum is being injected by the weak interaction. We can ignore $\mathcal{O}_p$ then when the $K \to \pi\pi$ amplitude is measured with chiral lattice fermions with on-shell momenta.

For a lattice theory without chiral symmetry, $\mathcal{O}_p = \partial_\mu A_\mu^{\bar{s}d} + O(a)$ and so it has a non-vanishing $O(a)$ matrix element. In this case operators $\mathcal{O}_\pm$ from Eq. (4.48) in the continuum match onto the lattice operators,

$$\mathcal{O}^\pm(\mu) = Z^\pm(\mu a, g_0^2) \left[\mathcal{O}^\pm(a) + \frac{C_p^\pm}{a^2} \mathcal{O}_p \right] + \mathcal{O}(a). \tag{4.50}$$

In general, then one would need to determine the coefficient $C_p^\pm$ to $O(a)$ in order to determine the $\Delta I = \frac{1}{2}$ amplitude for $K \to \pi\pi$ to leading order in an a-expansion, which is not really feasible. Other weak matrix elements such as B_K and ϵ'/ϵ similarly benefit from the use of lattice fermions with good chiral symmetry.

4.1.8   Fermion determinants in Euclidian space

Lattice computations employ Monte Carlo integration, which requires a positive integrand that can be interpreted as a probability distribution. In order that the lattice action yield a positive measure, it is certainly necessary (though not sufficient) that the continuum theory one is approximating have this property. Luckily, the fermion determinant for vector-like gauge theories (such as QCD), $\det(\slashed{D} + m)$, has this property in Euclidian space. Since $\slashed{D}^\dagger = -\slashed{D}$ and $\{\Gamma, \slashed{D}\} = 0$, it follows that there exist eigenstates ψ_n of $\slashed{D}$ such that

$$\slashed{D}\psi_n = i\lambda_n\psi_n, \qquad \slashed{D}\Gamma\psi_n = -i\lambda_n\Gamma\psi_n, \qquad \lambda_n \text{ real.} \tag{4.51}$$

For non-zero λ, ψ_m and $\Gamma\psi_n$ are all mutually orthogonal and we see that the eigenvalue spectrum contains $\pm i\lambda_n$ pairs. On the other hand, if $\lambda_n = 0$ then ψ_n can be an eigenstate of Γ as well, and $\Gamma\psi_n$ is not an independent mode. Therefore,

$$\det(\slashed{D} + m) = \prod_{\lambda_n > 0} (\lambda_n^2 + m^2) \times \prod_{\lambda_n = 0} m, \tag{4.52}$$

which is real and for positive m is positive for all gauge fields.

What about a chiral gauge theory? The fermion Lagrangian for a LH Weyl fermion in Euclidian space looks like $\bar{\psi}D_L\psi$ with $D_L = D_\mu\sigma_\mu$ and (in the chiral basis Eq. (4.10), continued to Euclidian space) $\sigma_\mu = \{1, i\vec{\sigma}\}$. Note that D_L has no nice hermiticity properties, which means its determinant will be complex, its right eigenvectors and left eigenvectors will be different, and its eigenvectors will not be mutually orthogonal. Furthermore, D_L is an operator that maps vectors from the space $\mathcal{L}$ of LH Weyl fermions to the space $\mathcal{R}$ of RH Weyl fermions. In Euclidian space, these spaces are unrelated and transform independently under the $SU(2) \times SU(2)$ Lorentz transformations. Suppose we have an orthonormal basis $|n, \mathcal{R}\rangle$ for the RH Hilbert space and $|n, \mathcal{L}\rangle$ for the LH Hilbert space; we can expand our fermion integration variables as

$$\psi = \sum_n c_n |n, \mathcal{L}\rangle, \qquad \bar{\psi} = \sum_n \bar{c}_n \langle n, \mathcal{R}|, \tag{4.53}$$

so that

$$\int [d\psi][d\bar{\psi}] \, e^{-\int \bar{\psi}D_L\psi} = \det_{mn} \langle m, \mathcal{R}|D_L|n, \mathcal{L}\rangle. \tag{4.54}$$

However, the answer we get will depend on the basis we choose. For example, we could have chosen a different orthonormal basis for the $\mathcal{L}$ space $|n', \mathcal{L}\rangle = \mathcal{U}_{n'n}|n, \mathcal{L}\rangle$ that differed from the first by a unitary transformation $\mathcal{U}$; the resultant determinant would differ by a factor $\det\mathcal{U}$, which is a phase. If this phase were a number, it would not be an issue—but it can in general be a functional of the background gauge field, so that different choices of phase for $\det D_L$ lead to completely different theories.

We do know that if D_R is the fermion operator for RH Weyl fermions in the same gauge representation as D_L, then $\det D_R = \det D_L^*$ and so $\det D_R \det D_L = \det\slashed{D}$. Therefore, the norm of $|\det D_L|$ can be defined as

$$\det D_L = \sqrt{|\det \slashed{D}|}\, e^{iW[A]},\qquad\qquad (4.55)$$

where the phase $W[A]$ is a functional of the gauge fields. What do we know about $W[A]$?

1. Since $\det\slashed{D}$ is gauge invariant, $W[A]$ should be gauge invariant unless the fermion representation has a gauge anomaly, in which case it should correctly reproduce that anomaly.
2. It should be analytic in the gauge fields, so that the computations of gauge field correlators (or the gauge current) are well defined.
3. It should be a local functional of the gauge fields.

In Fig. 4.3 I show two possible ways to define $\det D_L$, neither of which satisfy the above criteria. The naive choice of just setting $W[A] = 0$ not only fails to reproduce the anomaly (if the fermion representation is anomalous) but is also non-analytic and non-local. It corresponds to taking the product of all the positive eigenvalues λ_n of $\slashed{D}$ (up to an uninteresting overall constant phase). This definition is seen to be non-analytic where eigenvalues cross zero. Another definition might be to take the product of positive eigenvalues at some reference gauge field A_0, following those eigenvalues as they cross zero; this definition is analytic, but presumably not local, and is always gauge invariant.

There have been quite a few papers on how to proceed in defining this phase $W[A]$ in the context of domain-wall fermions, including a rather complicated explicit construction for $U(1)$ chiral gauge theories on the lattice (Lüscher, 1999; 2000a); however, even if a satisfactory definition of $W[A]$ is devised, it could be impossible to simulate using Monte Carlo algorithms due to the complexity of the fermion determinant.

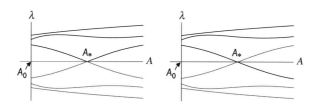

Fig. 4.3 The eigenvalue flow of the Dirac operator as a function of gauge fields, and two unsatisfactory ways to define the Weyl fermion determinant $\det D_L$ as a square root of $\det\slashed{D}$. The expression $\sqrt{|\det\slashed{D}|}$ corresponds to the picture on the left, where $\det D_L$ is defined as the product of positive eigenvalues of $\slashed{D}$; this definition is non-analytic at A_*. The picture on the right corresponds to the product of half the eigenvalues, following those that were positive at some reference gauge field A_0. This definition is analytic, but not necessarily local. Both definitions are gauge invariant, which is incorrect for an anomalous fermion representation.

4.1.9   Parity and fermion mass in odd dimensions

In this section I will be discussing fermions in $(2k + 1)$ dimensions with a spatially varying mass term that vanishes in some $2k$-dimensional region; in such cases we find chiral modes of a $2k$-dimensional effective theory bound to this mass defect. Such an example could arise dynamically when fermions have a Yukawa coupling to a real scalar ϕ that spontaneously breaks a discrete symmetry, where the surface with $\phi = 0$ forms a domain wall between two different phases; for this reason such fermions are called domain-wall fermions, even though we will be putting the spatially dependent mass in by hand and not through spontaneous symmetry breaking.

To study domain-wall fermions it is useful to say a few words about fermions in odd dimensions where there is no analoy of Γ and therefore there is no such thing as chiral symmetry. Nevertheless, fermion masses still break a symmetry: parity. In a theory with parity symmetry one has extended the Lorentz group to include improper rotations: spatial rotations R for which the determinant of R is negative. Parity can be defined as a transformation where an odd number of the spatial coordinates flip sign. In even dimensions parity can be the transformation $\mathbf{x} \to -\mathbf{x}$ and

$$\Psi(\mathbf{x}, t) \to \gamma^0 \Psi(-\mathbf{x}, t) \qquad \text{(parity, } d \text{ even).} \tag{4.56}$$

Note that under this transformation Ψ_L and Ψ_R are exchanged and that a Dirac mass term is parity invariant.

However, in odd dimensions the transformation $\mathbf{x} \to -\mathbf{x}$ is just a rotation; instead we can define parity as the transformation that just flips the sign of one coordinate x^1, and

$$\Psi(\mathbf{x}, t) \to \gamma^1 \Psi(\tilde{\mathbf{x}}, t), \qquad \tilde{\mathbf{x}} = (-x^1, x^2, \dots, x^{2k}). \tag{4.57}$$

Remarkably, a Dirac mass term flips sign under parity in this case; and since there is no chiral symmetry in odd d to rotate the phase of the mass matrix, the sign of the quark mass is physical, and a parity-invariant theory of massive quarks must have them come in pairs with masses $\pm M$, with parity interchanging the two.

4.1.10   Fermion masses and regulators

We have seen that theories of fermions in any dimension can possess symmetries that forbid masses—chiral symmetry in even dimensions and parity in odd dimensions. This property obviously can have a dramatic impact on the spectrum of a theory. Supersymmetry ingeniously puts fermions and bosons in the same supermultiplet, which allows scalars to also enjoy the benefits of chiral symmetry, which is one reason theorists have been so interested in having supersymmetry explain why the Higgs boson of the Standard Model manages to be so much lighter than the Planck scale. However, precisely because mass terms violate these symmetries, it is difficult to maintain them in a regulated theory. After all, one regulates a theory by introducing a high mass scale in order to eliminate UV degrees of freedom in the theory, and this mass scale will typically violate chiral symmetry in even dimensions, or parity in odd. This gives rise to "anomalous" violation of the classical fermion symmetries, my next topic.

4.2 Anomalies

4.2.1 The $U(1)_A$ anomaly in $1 + 1$ dimensions

One of the fascinating features of chiral symmetry is that sometimes it is not a symmetry of the quantum field theory even when it is a symmetry of the Lagrangian. In particular, Noether's theorem can be modified in a theory with an infinite number of degrees of freedom: the modification is called an "anomaly." Anomalies turn out to be very relevant both for phenomenology, and for the implementation of lattice field theory. The reason anomalies affect chiral symmetries is that regularization requires a cutoff on the infinite number of modes above some mass scale, while chiral symmetry is incompatible with fermion masses.[5]

Anomalies can be seen in many different ways. I think the most physical is to look at what happens to the ground state of a theory with a single flavor of massless Dirac fermion in $(1 + 1)$ dimensions in the presence of an electric field. Suppose one adiabatically turns on a constant positive electric field $E(t)$, then later turns it off; the equation of motion for the fermion is[6] $\frac{dp}{dt} = eE(t)$ and the total change in momentum is

$$\Delta p = e \int E(t)\, dt. \qquad (4.58)$$

Thus, the momenta of both left- and right-moving modes increase; if one starts in the ground state of the theory with a filled Dirac sea, after the electric field has turned off, both the right-moving and left-moving sea has shifted to the right, as in Fig. 4.4. The final state differs from the original by the creation of particle–antiparticle pairs: right-moving particles and left-moving antiparticles. Thus, while there is a fermion current in the final state, fermion number has not changed. This is what one would expect from conservation of the $U(1)$ current,

$$\partial_\mu J^\mu = 0. \qquad (4.59)$$

However, recall that right-moving and left-moving particles have positive and negative chirality, respectively; therefore the final state in Fig. 4.4 has net axial charge, even though the initial state did not. This is peculiar, since the coupling of the

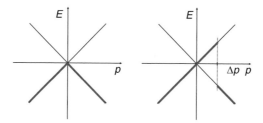

Fig. 4.4 On the left: the ground state for a theory of a single massless Dirac fermion in $(1 + 1)$ dimensions; on the right: the theory after application of an adiabatic electric field with all states shifted to the right by Δp, given in Eq. (4.58). Filled states are indicated by the heavier lines.

electromagnetic field in the Lagrangian does not violate chirality. We can quantify the effect. If we place the system in a box of size L with periodic boundary conditions, momenta are quantized as $p_n = 2\pi n/L$. The change in axial charge is then

$$\Delta Q_A = 2\frac{\Delta p}{2\pi/L} = \frac{e}{\pi}\int d^2x\, E(t) = \frac{e}{2\pi}\int d^2x\, \epsilon_{\mu\nu}F^{\mu\nu}, \qquad (4.60)$$

where I expressed the electric field in terms of the field strength F, where $F^{01} = -F^{10} = E$. This can be converted into the local equation using $\Delta Q_A = \int d^2x\, \partial_\mu J_A^\mu$, a modification of Eq. (4.26):

$$\partial_\mu J_A^\mu = 2im\overline{\Psi}\Gamma\Psi + \frac{e}{2\pi}\epsilon_{\mu\nu}F^{\mu\nu}, \qquad (4.61)$$

where in the above equation I have included the classical violation due to a mass term as well. The second term is the axial anomaly in $1+1$ dimensions; it would vanish for a non-abelian gauge field, due to the trace over the gauge generator.

So how did an electric field end up violating chiral charge? Note that this analysis relied on the Dirac sea being infinitely deep. If there had been a finite number of negative energy states, then they would have shifted to higher momentum, but there would have been no change in the axial charge. With an infinite number of degrees of freedom, though, one can have a "Hilbert Hotel": the infinite hotel that can always accommodate another visitor, even when full, by moving each guest to the next room and thereby opening up a room for the newcomer. This should tell you that it will not be straightforward to represent chiral symmetry on the lattice. A lattice field theory approximates quantum field theory with a finite number of degrees of freedom—the lattice is a big hotel, but quite conventional. In such a hotel there can be no anomaly.

We can derive the anomaly in other ways, such as by computing the anomaly diagram Fig. 4.5, or by following Fujikawa (Fujikawa, 1979; 1980) and carefully accounting for the Jacobian from the measure of the path integral when performing a chiral transformation. It is particularly instructive for our later discussion of lattice fermions to compute the anomaly in perturbation theory using Pauli–Villars regulators of mass M. We replace our axial current by a regulated current

$$J_{A,\mathrm{reg}}^\mu = \overline{\Psi}\gamma^\mu\Gamma\Psi + \overline{\Phi}\gamma^\mu\Gamma\Phi, \qquad (4.62)$$

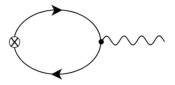

Fig. 4.5   The anomaly diagram in $1+1$ dimensions, with one Pauli–Villars loop and an insertion of $2iM\overline{\Phi}\Gamma\Phi$ at the X.

where Φ is our Pauli–Villars field; it follows then that

$$\partial_\mu J^\mu_{A,\text{reg}} = 2im\overline{\Psi}\Gamma\Psi + 2iM\overline{\Phi}\Gamma\Phi. \tag{4.63}$$

We are interested in matrix elements of $J^\mu_{A,\text{reg}}$ in a background gauge field between states without any Pauli–Villars particles, and so we need to evaluate $\langle 2iM\overline{\Phi}\Gamma\Phi\rangle$ in a background gauge field and take the limit $M \to \infty$ to see if $\partial_\mu J^\mu_{A,\text{reg}}$ picks up any anomalous contributions that do not decouple as we remove the cutoff.

To compute $\langle 2iM\overline{\Phi}\Gamma\Phi\rangle$ we need to consider all Feynman diagrams with a Pauli–Villars loop, and insertion of the $\overline{\Phi}\Gamma\Phi$ operator, and any number of external $U(1)$ gauge fields. By gauge invariance, a graph with n external photon lines will contribute n powers of the field strength tensor $F^{\mu\nu}$. For power counting, it is convenient that we normalize the gauge field so that the covariant derivative is $D_\mu = (\partial_\mu + iA_\mu)$; then the gauge field has mass dimension 1, and $F^{\mu\nu}$ has dimension 2. In $(1+1)$ dimensions $\langle 2iM\overline{\Phi}\Gamma\Phi\rangle$ has dimension 2, and so simple dimensional analysis implies that the graph with n photon lines must make a contribution proportional to $(F^{\mu\nu})^n/M^{2(n-1)}$. Therefore, only the graph in Fig. 4.5 with one photon insertion can make a contribution that survives the $M \to \infty$ limit (the graph with zero photons vanishes). Calculation of this diagram yields the same result for the divergence of the regulated axial current as we found in Eq. (4.61).

Exercise 4.5 Compute the diagram in Fig. 4.5 using the conventional normalization of the gauge field $D_\mu = (\partial_\mu + ieA_\mu)$ and verify that $2iM\langle\overline{\Phi}\Gamma\Phi\rangle = \frac{e}{2\pi}\epsilon_{\mu\nu}F^{\mu\nu}$ when $M \to \infty$.

Note that in this description of the anomaly we (i) effectively rendered the number of degrees of freedom finite by introducing the regulator; (ii) the regulator explicitly broke the chiral symmetry; (iii) as the regulator was removed, the symmetry-breaking effects of the regulator never decoupled, indicating that the anomaly arises when the two vertices in Fig. 4.5 sit at the same space-time point. While we used a Pauli–Villars regulator here, the use of a lattice regulator will have qualitatively similar features, with the inverse lattice spacing playing the role of the Pauli–Villars mass, and we can turn these observations around: A lattice theory will not correctly reproduce anomalous symmetry currents in the continuum limit, unless that symmetry is broken explicitly by the lattice regulator. This means we would be foolish to expect to construct a lattice theory with exact chiral symmetry. But can the lattice break chiral symmetry just enough to explain the anomaly, without losing the important consequences of chiral symmetry at long distances (such as protecting fermion masses from renormalization)?

4.2.2 Anomalies in $3 + 1$ dimensions

4.2.2.1 The $U(1)_A$ anomaly

An analogous violation of the $U(1)_A$ current occurs in $3 + 1$ dimensions as well.[7] One might guess that the analog of $\epsilon_{\mu\nu}F^{\mu\nu} = 2E$ in the anomalous divergence Eq. (4.4) would be the quantity $\epsilon_{\mu\nu\rho\sigma}F^{\mu\nu}F^{\rho\sigma} = 8\vec{E}\cdot\vec{B}$, which has the right dimensions and properties under parity and time reversal. So we should consider the behavior a massless Dirac fermion in $(3+1)$ dimensions in parallel constant E and B fields. First, turn on a B field pointing in the $\hat{z}$ direction: this gives rise to Landau levels, with energy levels E_n characterized by non-negative integers n as well as spin in the $\hat{z}$ direction S_z and momentum p_z, where

$$E_n^2 = p_z^2 + (2n+1)eB - 2eBS_z. \tag{4.64}$$

The number density of modes per unit transverse area is defined to be g_n, which can be derived by computing the zero-point energy in Landau modes and requiring that it yield the free-fermion result as $B \to 0$. We have $g_n \to p_\perp dp_\perp/(2\pi)$ with $[(2n+1)eB - 2eBS_z] \to p_\perp^2$, implying that

$$g_n = eB/2\pi. \tag{4.65}$$

Th dispersion relation Eq. (4.64) looks like that of an infinite number of one-dimensional fermions of mass $m_{n,\pm}$, where

$$m_{n\pm}^2 = (2n+1)eB - 2eBS_z, \quad S_z = \pm\tfrac{1}{2}. \tag{4.66}$$

The state with $n=0$ and $S_z = +\tfrac{1}{2}$ is distinguished by having $m_{n,+} = 0$; it behaves like a *massless* one-dimensional Dirac fermion (with transverse density of states g_0) moving along the $\hat{z}$-axis with dispersion relation $E = |p_z|$. If we now turn on an electric field also pointing along the $\hat{z}$-direction we know what to expect from our analysis in $1+1$ dimensions: we find an anomalous divergence of the axial current equal to

$$g_0 eE/\pi = e^2 EB/2\pi^2 = \left(\frac{e^2}{16\pi^2}\right)\epsilon_{\mu\nu\rho\sigma}F^{\mu\nu}F^{\rho\sigma}. \tag{4.67}$$

If we include an ordinary mass term in the $(3+1)$-dimensional theory, then we get

$$\partial_\mu J_A^\mu = 2im\overline{\Psi}\Gamma\Psi + \left(\frac{e^2}{16\pi^2}\right)\epsilon_{\mu\nu\rho\sigma}F^{\mu\nu}F^{\rho\sigma}. \tag{4.68}$$

One can derive this result by computing $\langle M\overline{\Phi}i\Gamma\Phi\rangle$ for a Pauli–Villars regulator as in the $(1+1)$-dimensional example; now the relevant graph is the triangle diagram of Fig. 4.6.

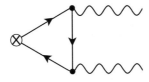

Fig. 4.6 The $U(1)_A$ anomaly diagram in $3+1$ dimensions, with one Pauli–Villars loop and an insertion of $2iM\overline{\Phi}\Gamma\Phi$.

If the external fields are non-abelian, the analog of Eq. (4.68) is

$$\partial_\mu J_A^\mu = 2im\overline{\Psi}\Gamma\Psi + \left(\frac{g^2}{16\pi^2}\right)\epsilon_{\mu\nu\rho\sigma}F_a^{\mu\nu}F_b^{\rho\sigma}\mathrm{Tr}T_aT_b. \tag{4.69}$$

If the fermions transform in the defining representation of $SU(N)$, it is conventional to normalize the coupling g so that $\mathrm{Tr}T_aT_b = \frac{1}{2}\delta_{ab}$. This is still called an "abelian anomaly", since J_A^μ generates a $U(1)$ symmetry.

4.2.2.2 Anomalies in Euclidian space-time

Continuing to Euclidian space-time by means of Eqs. (4.11)–(4.15) changes the anomaly equations simply by eliminating the factor of i from in front of the fermion mass:

$$2d: \quad \partial_\mu J_A^\mu = 2m\overline{\Psi}\Gamma\Psi + \frac{e}{2\pi}\epsilon_{\mu\nu}F^{\mu\nu} \tag{4.70}$$

$$4d: \quad \partial_\mu J_A^\mu = 2m\overline{\Psi}\Gamma\Psi + \left(\frac{g^2}{16\pi^2}\right)\epsilon_{\mu\nu\rho\sigma}F_a^{\mu\nu}F_b^{\rho\sigma}\mathrm{Tr}T_aT_b. \tag{4.71}$$

4.2.2.3 The index theorem in four dimensions

For non-abelian gauge theories the quantity on the far right of Eq. (4.71) is a topological charge density, with

$$\nu = \frac{g^2}{64\pi^2}\int d^4x_E\,\epsilon_{\mu\nu\rho\sigma}F_a^{\mu\nu}F_a^{\rho\sigma} \tag{4.72}$$

the winding number associated with $\pi_3(G)$, the homotopy group of maps of S_3 (spacetime infinity) into the gauge group G.

Consider then continuing the anomaly equation Eq. (4.69) to Euclidean space and integrating over space-time its vacuum expectation value in a background gauge field (assuming the fermions to be in the N-dimensional representation of $SU(N)$ so that $\mathrm{Tr}T_aT_b = \frac{1}{2}\delta_{ab}$). The integral of $\partial_\mu\langle J_A^\mu\rangle$ vanishes because it is a pure divergence, so we get

$$\int d^4x_E\, m\langle\overline{\Psi}\Gamma\Psi\rangle = -\nu. \tag{4.73}$$

The matrix element is

$$\int [d\Psi][d\overline{\Psi}]\,e^{-S_E}\,(m\,\overline{\Psi}\Gamma\Psi)\Big/\int[d\Psi][d\overline{\Psi}]\,e^{-S_E}, \tag{4.74}$$

where $S_E = \int\overline{\Psi}(\slashed{D}_E + m)\Psi$. We can expand Ψ and $\overline{\Psi}$ in terms of eigenstates of the anti-Hermitian operator $\slashed{D}_E$, where

$$\slashed{D}_E\psi_n = i\lambda_n\psi_n, \qquad \int d^4x_E\,\psi_m^\dagger\psi_n = \delta_{mn}, \tag{4.75}$$

with

$$\Psi = \sum c_n \psi_n, \qquad \overline{\Psi} = \sum \bar{c}_n \psi_n^\dagger. \tag{4.76}$$

Then,

$$\int d^4 x_E \, m \, \langle \overline{\Psi} \Gamma \Psi \rangle = \left(\sum_n \int d^4 x_E \, m \, \psi_n^\dagger \Gamma \psi_n \prod_{k \neq n} (i\lambda_k + m) \right) \bigg/ \prod_k (i\lambda_k + m)$$

$$= m \sum_n \int d^4 x_E \psi_n^\dagger \Gamma \psi_n / (i\lambda_n + m). \tag{4.77}$$

Recall that $\{\Gamma, \slashed{D}\} = 0$; thus

$$\slashed{D}\psi_n = i\lambda_n \psi_n \quad \text{implies} \quad \slashed{D}(\Gamma\psi_n) = -i\lambda_n(\Gamma\psi_n). \tag{4.78}$$

Thus, for $\lambda_n \neq 0$, the eigenstates ψ_n and $(\Gamma\psi_n)$ must be orthogonal to each other (they are both eigenstates of $\slashed{D}$ with different eigenvalues), and so $\int \psi_n^\dagger \Gamma \psi_n$ vanishes for $\lambda_n \neq 0$ and does not contribute to the sum in Eq. (4.77). In contrast, modes with $\lambda_n = 0$ can simultaneously be eigenstates of $\slashed{D}$ and of Γ. Let n_+, n_- be the number of RH and LH zero-modes respectively. The sum in Eq. (4.77) then just equals $(n_+ - n_-) = (n_R - n_L)$, and combining with Eq. (4.73) we arrive at the index equation

$$n_- - n_+ = \nu, \tag{4.79}$$

which states that the difference in the numbers of LH and RH zero-mode solutions to the Euclidian Dirac equation in a background gauge field equals the winding number of the gauge field. With N_f flavors, the index equation is trivially modified to read

$$n_- - n_+ = N_f \nu. \tag{4.80}$$

This link between eigenvalues of the Dirac operator and the topological winding number of the gauge field provides a precise definition for the topological winding number of a gauge field on the lattice, where there is no topology—provided we have a definition of a lattice Dirac operator that exhibits exact zero modes. We will see that the overlap operator is such an operator.

4.2.2.4 More general anomalies

Even more generally, one can consider the 3-point correlation function of three arbitrary currents as in Fig. 4.7,

$$\langle J_a^\alpha(k) J_b^\beta(p) J_c^\gamma(q) \rangle, \tag{4.81}$$

and show that the divergence with respect to any of the indices is proportional to a particular group theory factor,

$$k_\mu \langle J_a^\mu(k) J_a^\alpha(p) J_c^\beta(q) \rangle \propto \text{Tr} Q_a \{Q_b, Q_c\} \Big|_{R-L} \epsilon^{\alpha\beta\rho\sigma} k_\rho k_\sigma, \tag{4.82}$$

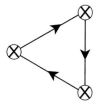

Fig. 4.7 Anomalous three-point function of three currents.

where the Qs are the generators associated with the three currents in the fermion representation, with the symmetrized trace computed as the difference between the contributions from RH and LH fermions in the theory. The anomaly $\mathcal{A}$ for the fermion representation is defined by the group-theory factor,

$$\mathrm{Tr}\left(Q_a\{Q_b, Q_c\}\right)\Big|_{R-L} \equiv \mathcal{A}\, d_{abc}, \qquad (4.83)$$

with d_{abc} the totally symmetric-invariant tensor of the symmetry group. For a *simple* group G (meaning G is not $U(1)$ and has no factor subgroups), d_{abc} is only non-zero for $G = SU(N)$ with $N \geq 3$; even in the case of $SU(N)$, d_{abc} will vanish for real irreducible representations (for which $Q_a = -Q_a^*$), or for judiciously chosen reducible complex representations, such as $\mathbf{\bar 5} \oplus \mathbf{10}$ in $SU(5)$. For a semisimple group $G_1 \times G_2$ (where G_1 and G_2 are themselves simple) there are no mixed anomalies since the generators are all traceless, implying that if $Q \in G_1$ and $\mathcal{Q} \in G_2$ then $\mathrm{Tr}\left(Q_a\{\mathcal{Q}_b, \mathcal{Q}_c\}\right) \propto \mathrm{Tr} Q_a = 0$. When considering groups with $U(1)$ factors there can be non-zero mixed anomalies of the form $U(1)G^2$ and $U(1)^3$ where G is simple; the $U(1)^3$ anomalies can involve different $U(1)$ groups. With a little group theory it is not difficult to compute the contribution to the anomaly of any particular group representation.

If a current with an anomalous divergence is gauged, then the theory does not make sense. That is because a divergenceless current is required for the unphysical modes in the gauge field A_μ to decouple; if they do not decouple, their propagator has a piece that goes as $k_\mu k_\nu / k^2$ that does not fall off at large momentum, and the theory is not renormalizable.

When global $U(1)$ currents have anomalous divergences, that is interesting. We have seen that the $U(1)_A$ current is anomalous, which explains the η' mass; the divergence of the axial isospin current explains the decay $\pi^0 \to \gamma\gamma$; the anomalous divergence of the baryon number current in abackground $SU(2)$ potential in the Standard Model predicts baryon violation in the early Universe and the possibility of weak-scale baryogenesis.

Exercise 4.6 Verify that all the gauge currents are anomaly free in the Standard Model with the representation in Eq. (4.45). The only possible G^3 anomalies are for $G = SU(3)$ or $G = U(1)$; for the $SU(3)^3$ anomaly use the fact that a LH Weyl fermion contributes $+1$ to $\mathcal{A}$ if it transforms as a $\mathbf{3}$ of $SU(3)$, and contributes -1 to $\mathcal{A}$ if it is a $\mathbf{\bar 3}$. There are two mixed anomalies to check as well: $U(1)SU(2)^2$ and $U(1)SU(3)^2$.

This apparently miraculous cancellation suggests that each family of fermions may be unified into a spinor of $SO(10)$, since the vanishing of anomalies that happens automatically in $SO(10)$ is of course maintained when the symmetry is broken to a smaller subgroup, such as the Standard Model.

Exercise 4.7   Show that the global B (baryon number) and L (lepton number) currents are anomalous in the Standard Model (4.45), but that $B - L$ is not.

4.2.3   Strongly coupled chiral gauge theories

Strongly coupled chiral gauge theories are particularly intriguing, since they can contain light composite fermions, which could possibly describe the quarks and leptons we see. A nice toy example of a strongly coupled chiral gauge theory is $SU(5)$ with LH fermions

$$\psi = \bar{5}, \qquad \chi = 10. \tag{4.84}$$

It so happens that the ψ and the χ contribute with opposite signs to the $SU(5)^3$ anomaly $\mathcal{A}$ in Eq. (4.83), so this seems to be a well-defined gauge theory. Furthermore, the $SU(5)$ gauge interactions are asymptotically free, meaning that interactions becomes strong at long distances. One might therefore expect the theory to confine as QCD does. However, unlike QCD, there are no gauge-invariant fermion bilinear condensates that could form, which in QCD are responsible for baryon masses. That being the case, might there be any massless composite fermions in the spectrum of this theory? 't Hooft came up with a nice general argument involving global anomalies that suggests there will be.

In principle, there are two global $U(1)$ chiral symmetries in this theory, corresponding to independent phase rotations for ψ and χ; however, both of these rotations have global $\times SU(5)^2$ anomalies, similar to the global $\times SU(3)^2$ of the $U(1)_A$ current in QCD. This anomaly can only break one linear combination of the two $U(1)$ symmetries, and one can choose the orthogonal linear combination that is anomaly free. With a little group theory you can show that the anomaly-free global $U(1)$ symmetry corresponds to assigning charges as

$$\psi = \bar{5}_3, \qquad \chi = 10_{-1}, \tag{4.85}$$

where the subscript gives the global $U(1)$ charge. This theory has a non-trivial global $U(1)^3$ anomaly, $\mathcal{A} = 5 \times (3)^3 + 10 \times (-1)^3 = 125$. 't Hooft's argument is that this (global)^3 anomaly restricts—and helps predict—the low-energy spectrum of the theory. Applied to the present model, his argument goes as follows. Imagine weakly gauging this $U(1)$ symmetry. This would be bad news as the theory stands, since a (gauge)^3 anomaly leads to a sick theory, but one can add a LH "spectator fermion" $\omega = 1_{-5}$ that is a singlet under $SU(5)$ but has charge -5 under this $U(1)$ symmetry, cancelling the $U(1)^3$ anomaly. This weak $U(1)$ gauge interaction plus the $SU(5)$-singlet ω fermion should not interfere with the strong $SU(5)$ dynamics. If that dynamics leads to confinement and no $U(1)$ symmetry breaking, then the weak $U(1)$ gauge theory must remain anomaly free at low energy, implying that there has to be one or more massless composite fermion to cancel the $U(1)^3$ anomaly of the ω. A good candidate massless composite LH fermion is $(\psi\psi\chi)$, that is an $SU(5)$-singlet (as required by confinement), and that has $U(1)$ charge of $(3+3-1)=5$, exactly cancelling the $U(1)^3$ anomaly of the ω. Now forget the thought experiment: do not gauge the $U(1)$ and do not include the ω spectator fermion. It should still be true that this $SU(5)$ gauge theory produces a single massless composite fermion $(\psi\psi\chi)$.[8]

While it is hard to pin down the spectrum of general strongly coupled chiral gauge theories using 't Hooft's anomaly-matching condition alone, a lot is known about strongly coupled supersymmetric chiral gauge theories, and they typically have a very interesting spectrum of massless composite fermions, which can be given small masses and approximate the quarks and leptons we see by tweaking the theory. See for example (Kaplan *et al.*, 1997), which constructs a theory with three families of massless composite fermions, each with a different number of constituents.

Chiral gauge theories would be very interesting to study on the lattice, but pose theoretical problems that have not been solved yet—and that, if they were, might then be followed by challenging practical problems related to complex path-integral measures and massless fermions. Perhaps one of you will crack this interesting problem.

4.2.4 The non-decoupling of parity violation in odd dimensions

We have seen that chiral symmetry does not exist in odd space dimensions, but that a discrete parity symmetry can forbid a fermion mass. One would then expect a regulator—such as Pauli–Villars fields—to break parity. Indeed they do: on integrating the Pauli–Villars field out of the theory, one is left with a Chern–Simons term in the Lagrangian with coefficient $M/|M|$, which does not decouple as $M \to \infty$. In $2k+1$ dimensions the Chern–Simons form for an abelian gauge field is proportional to

$$\epsilon^{\alpha_1 \cdots \alpha_{2k+1}} A_{\alpha_1} F_{\alpha_2 \alpha_3} \cdots F_{\alpha_{2k}\alpha_{2k+1}}, \tag{4.86}$$

which violates parity; the Chern–Simons form for non-abelian gauge fields is more complicated.

For domain-wall fermions we will be interested in a closely related but slightly different problem: the generation of a Chern–Simons operator on integrating out a heavy fermion of mass m. In $1+1$ dimensions with an abelian gauge field one computes the graph in Fig. 4.8, which gives rise to the Lagrangian

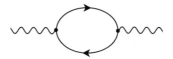

Fig. 4.8 Integrating out a heavy fermion in three dimensions gives rise to the Chern–Simons term in the effective action of Eq. (4.87).

$$\mathcal{L}_{CS} = \frac{e^2}{8\pi} \frac{m}{|m|} \epsilon^{\alpha\beta\gamma} A_\alpha \partial_\beta A_\gamma.$$ (4.87)

What is interesting is that it implies a particle number current

$$J_\mu = \frac{1}{e} \frac{\partial \mathcal{L}_{CS}}{\partial A_\mu} = \frac{e}{8\pi} \frac{m}{|m|} \epsilon^{\mu\alpha\beta} F_{\alpha\beta},$$ (4.88)

which we will see is related to the anomaly Eq. (4.61) in $1 + 1$ dimensions.

Exercise 4.8   Verify the coefficient in Eq. (4.87) by computing the diagram Fig. 4.8. By isolating the part that is proportional to $\epsilon_{\mu\nu\alpha} p^\alpha$ before performing the integral, one can make the diagram very easy to compute.

4.3   Domain-wall fermions

4.3.1   Chirality, anomalies, and fermion doubling

You have heard of the Nielsen–Ninomiya theorem. It states that a fermion action in $2k$ Euclidian space-time dimensions

$$S = \int_{\pi/a}^{\pi/a} \frac{d^{2k}p}{(2\pi)^4} \overline{\Psi}_{-\mathbf{p}} \tilde{D}(\mathbf{p}) \Psi(\mathbf{p})$$ (4.89)

cannot have the operator $\tilde{D}$ satisfy all four of the following conditions simultaneously:

1. $\tilde{D}(\mathbf{p})$ is a periodic, analytic function of p_μ;
2. $D(\mathbf{p}) \propto \gamma_\mu p_\mu$ for $a|p_\mu| \ll 1$;
3. $\tilde{D}(\mathbf{p})$ invertible everywhere except $p_\mu = 0$;
4. $\{\Gamma, \tilde{D}(\mathbf{p})\} = 0$.

   The first condition is required for locality of the Fourier transform of $\tilde{D}(\mathbf{p})$ in coordinate space. The next two state that we want a single flavor of conventional Dirac fermion in the continuum limit. The last item is the statement of chiral symmetry. One can try keeping that and eliminating one or more of the other conditions; for example, the SLAC derivative took $\tilde{D}(\mathbf{p}) = \gamma_\mu p_\mu$ within the Brillouin zone (BZ), which violates the first condition—if taken to be periodic, it is discontinuous at the edge of the BZ. This causes problems—for example, the QED Ward identity states that the photon vertex Γ_μ is proportional to $\partial \tilde{D}(\mathbf{p})/\partial p_\mu$, which is infinite at the BZ boundary. Naive fermions satisfy all the conditions except (3): there $\tilde{D}(\mathbf{p})$ vanishes at the 2^4 corners of the BZ, and so we have 2^4 flavors of Dirac fermions in the continuum. Staggered fermions are somewhat less redundant, producing four flavors in the continuum for each lattice field; Creutz fermions are the least redundant, giving rise to two copies for each lattice field. The discussion in any even space-time dimension is analogous.

This roadblock in developing a lattice theory with chirality is obviously impossible to get around when you consider anomalies. Remember that anomalies do occur in the continuum but that with a UV cutoff on the number of degrees of freedom there are no anomalies, and the exact symmetries of the regulated action are the exact symmetries of the quantum theory. The only way a symmetry current can have a non-zero divergence is if either the original action or the UV regulator explicitly violates that symmetry. The implication for lattice fermions is that any symmetry that is exact on the lattice will be exact in the continuum limit, while any symmetry anomalous in the continuum limit must be broken explicitly on the lattice.

A simple example to analyze is the case of a "naive" lattice action for a single RH fermion,

$$S = \frac{1}{2a} \sum_{\mathbf{n},\mu} \overline{\Psi}_+(\mathbf{n}) \gamma_\mu \left[\Psi_+(\mathbf{n} + \hat{\mu}) - \Psi_+(\mathbf{n} - \hat{\mu}) \right]$$
$$= \frac{1}{2a} \sum_{\mathbf{p},\mu} 2i \sin a p_\mu \overline{\Psi}_+(-\mathbf{p}) \gamma_\mu \Psi_+(\mathbf{p}),$$
$$\Gamma \Psi_+ = \Psi_+, \tag{4.90}$$

so that $\tilde{D}(\mathbf{p}) = i\gamma_\mu \sin ak_\mu/a$. This vanishes at every corner of the BZ; expanding about these points we write $p_\mu = q_\mu + n_\mu \pi/a$ with $n_\mu \in \{0,1\}$ and $aq_\mu \ll 1$ and find

$$\tilde{D}(\mathbf{p}) \simeq i \sum_\mu (-1)^{n_\mu} \gamma_\mu q_\mu. \tag{4.91}$$

These zeroes of $\tilde{D}(\mathbf{p})$ (at $q_\mu = 0$) correspond to the 2^d doublers in d-dimensional Euclidian space-time, violating condition (3) in the Nielsen–Ninomiya theorem. However, $\tilde{D}(\mathbf{p})$ does satisfy condition (4) and the action S is invariant under the symmetry

$$\Psi_+(\mathbf{p}) \to e^{i\alpha} \Psi_+(\mathbf{p}) = e^{i\alpha \Gamma} \Psi_+(\mathbf{p}), \tag{4.92}$$

which looks like a chiral symmetry—yet for a continuum theory with 2^d RH fermions, a phase symmetry would be anomalous, which we know cannot result from a symmetric lattice theory!

The resolution is that the continuum theory does *not* have 2^d RH fermions, but rather 2^{d-1} Dirac fermions, and the exact lattice $U(1)$ symmetry corresponds to an exact fermion number symmetry in the continuum, which is not chiral and not anomalous. To show this, note that $\tilde{D}(\mathbf{p})$ in Eq. (4.91) has funny signs near the corners of the BZ. We can convert back to our standard gamma matrix basis using the similarity transformation $P(n)\gamma_\mu P(n)^{-1} = (-1)^{n_\mu}\gamma_\mu$; but then $P(n)\Gamma P(n)^{-1} = (-1)^{\sigma(n)}\Gamma$, where $\sigma(n) = \sum_\mu n_\mu$ (since Γ is the product of all the γ^μ). Therefore,

$$\Gamma[P(n)\Psi_+(\mathbf{q})] = (-1)^{\sigma(n)}[P(n)\Psi_+(\mathbf{q})] \tag{4.93}$$

and in the continuum we have 2^{d-1} RH fermions and 2^{d-1} LH fermions, and the exact and apparently chiral symmetry of the lattice corresponds to an exact and anomaly-free fermion number symmetry in the continuum. The redundancy of staggered and

Creutz fermions serves the same purpose, ensuring that all exact lattice symmetries become anomaly-free vector symmetries in the continuum limit.

4.3.2 Domain-wall fermions in the continuum

4.3.2.1 Motivation

What we would like is a realization of chiral symmetry on the lattice that (i) is not exact, so we can correctly recover anomalies, but that (ii) retains all the good features of chiral symmetry in the continuum, such as protection from additive renormalization of fermion masses. There is a curious example in the continuum of such a system, which gave a clue on how to achieve this. The example has to do with fermions in odd dimension that interact with a domain wall. To be concrete, consider a system in three dimensions [coordinates (x_0, x_1, x_2)], where the fermion has a mass that depends on x_2 and switches sign at $x_2 = 0$. For simplicity, I will take $m(x_2) = m\epsilon(x_2) = mx_2/|x_2|$. Curiously enough, we will show that a massless fermion mode exists bound to this 2-dimensional surface, and that it is chiral: there exists a RH mode, and not a LH one. Thus, the low-energy limit of this theory looks like a 2-dimensional theory of a Weyl fermion with a chiral symmetry, even though we started with a 3-dimensional theory in which there can be no chirality.

Yet we know that the low-energy effective theory is anomalous. Recall that for a massless Dirac fermion coupled to photons in two Euclidian dimensions, the vector current is conserved ($\partial_\mu J^\mu = 0$), while the axial current is not in general [$\partial_\mu J^\mu_A = (e^2/\pi E)$]. Thus, the fermion current for a RH Weyl current satisfies

$$\partial_\mu J^\mu_R = \tfrac{1}{2}\partial_\mu(J^\mu + J^\mu_A) = \frac{e}{2\pi}E. \tag{4.94}$$

If we turn on an electric field pointing in the x_1 direction, then the charge on the mass defect must increase with time. Yet, in the full 3-dimensional theory, there is only one fermion current $J^\mu = \overline{\Psi}\gamma^\mu\Psi$, and it is conserved. So even though only massive states live off the mass defect, we see that they must somehow know about the anomaly and allow chiral symmetry to be violated as the anomaly requires.

You might be suspicious that there is some hidden fine tuning here to keep the chiral mode massless, but that cannot be: the low-energy effective theory would need a LH mode as well in order for there to be a mass. Distortions of the domain-wall mass function cannot change this result, unless the mass $m(x_2)$ becomes small enough somewhere to change the spectrum of the low-energy effective theory. But even in that case there is an index theorem that requires there to be a massless Weyl fermion as the sign of $m(x_2)$ changes at an odd number of locations.

This looks like a useful trick to apply to the lattice: an anomalous chiral symmetry emerging at low energy from a full theory with no fundamental chiral symmetry, and without fine tuning. In this section I show how the continuum theory works, and then how it can be transcribed to the lattice. In the next section I will discuss how the effective theory can be described directly using the overlap formulation, without any reference to the higher-dimensional parent theory.

4.3.2.2 The model

Even though fermions in even and odd dimensions look quite different, one finds an interesting connection between them when considering the Dirac equation with a space-dependent mass term. One can think of a space-dependent mass as arising from a Higgs mechanism, for example, where there is a topological defect trapped in the classical Higgs field, such as a domain wall or a vortex. A domain wall can naturally arise when the Higgs field breaks a discrete symmetry; a vortex arises when the Higgs field breaks a $U(1)$ symmetry [see John Preskill's lectures "Vortices and Monopoles" at the 1985 Les Houches Summer School (Preskill (1985))]. Domain-wall defects are pertinent to putting chiral fermions on the lattice, so I will consider that example.

Consider a fermion in Euclidian space-time with dimension $d = 2k + 1$, where the coordinates are written as $\{x_0, x_1, \ldots x_{2k-1}, s\} \equiv \{x_\mu, s\}$, where $\mu = 0, \ldots, 2k - 1$ and s is what I call the coordinate x_{2k}. The $(2k + 1)$ γ matrices are written as $\{\gamma_0, \ldots, \gamma_{2k}, \Gamma\}$. This fermion is assumed to have an s-dependent mass with the simple form

$$m(s) = m\epsilon(s) = \begin{cases} +m & s > 0 \\ -m & s < 0 \end{cases}, \qquad m > 0. \tag{4.95}$$

This mass function explicitly breaks the Poincaré symmetry of $(2k + 1)$-dimensional space-time, but preserves the Euclidian. Poincaré symmetry of $2k$-dimensional space-time. The fermion is also assumed to interact with $2k$-dimensional background gauge fields $A_\mu(x_\mu)$ that are independent of s. The Dirac equation may be written as

$$\left[\slashed{D} + \Gamma \partial_s + m(s) \right] \Psi(x_\mu, s) = 0, \tag{4.96}$$

where $\slashed{D}$ is the lower-dimension ($d = 2k$) covariant Dirac operator. The spinor Ψ can be factorized as the product of functions of s times spinors $\psi(x_\mu)$,

$$\Psi(x_\mu, s) = \sum_n [b_n(s)P_+ + f_n(s)P_-] \psi_n(x_\mu), \qquad P_\pm = \frac{1 \pm \Gamma}{2}, \tag{4.97}$$

satisfying the equations

$$[\partial_s + m(s)]b_n(s) = \mu_n f_n(s), \\ [-\partial_s + m(s)]f_n(s) = \mu_n b_n(s), \tag{4.98}$$

and

$$(\slashed{D} + \mu_n)\Psi_n(x) = 0. \tag{4.99}$$

One might expect all the eigenvalues in Eq. (4.98) to satisfy $|\mu_n| \gtrsim O(m)$, since that is the only scale in the problem. However, there is also a solution to Eq. (4.98) with eigenvalue $\mu = 0$ given by

$$b_0 = N e^{-\int_0^s m(s')ds'} = N e^{-m|s|}. \tag{4.100}$$

This solution is localized near the defect at $s = 0$, falling off exponentially fast away from it. There is no analogous solution to Eq. (4.98) of the form

$$f_0 \sim e^{+\int_0^s m(s')ds'},$$

since that would be exponentially growing in $|s|$ and not normalizable. Therefore, as seen from Eq. (4.99) the spectrum consists of an infinite tower of fermions satisfying the $d = 2k$ Dirac equation: massive Dirac fermions with mass $O(m)$ and higher, plus a single massless right-handed chiral fermion. The massless fermion is localized at the defect at $s = 0$, whose profile in the transverse extra dimension is given by Eq. (4.100); the massive fermions are not localized. Because of the gap in the spectrum, at low energy the accessible part of the spectrum consists only of the massless RH chiral fermion.

Exercise 4.9   Construct a $d = 2k + 1$ theory whose low-energy spectrum possesses a single light $d = 2k$ Dirac fermion with mass arbitrarily lighter than the domain-wall scale m. There is more than one way to do this.

Some comments are in order:

- It is not a problem that the low-energy theory of a single right-handed chiral fermion violates parity in $d = 2k$ since the mass for Ψ breaks parity in $d = 2k + 1$;

- Furthermore, nothing is special about right-handed fermions, and a left-handed mode would have resulted if we had chosen the opposite sign for the mass in Eq. (4.95). This makes sense because choosing the opposite sign for the mass can be attained by flipping the sign of all the space coordinates: a rotation in the $(2k + 1)$-dimensional theory, but a parity transformation from the point of view of the $2k$-dimensional fermion zero mode.

- The fact that a chiral mode appeared at all is a consequence of the normalizability of $\exp(-\int_0^s m(s')ds')$, which in turn follows from the two limits $m(\pm\infty)$ being non-zero with opposite signs. Any function $m(s)$ with that boundary condition will support a single chiral mode, although in general there may also be a number of very light fermions localized in regions wherever $|m(s)|$ is small—possibly extremely light if $m(s)$ crosses zero a number of times, so that there are widely separated defects and antidefects.

- Gauge boson loops will generate contributions to the fermion mass function that are even in s. If the coupling is sufficiently weak, it cannot affect the masslessness of the chiral mode. However, if the gauge coupling is strong, or if the mass m is much below the cutoff of the theory, the radiative corrections could cause the fermion mass function to never change sign, and the chiral mode would not exist. Or it could still change sign, but become small in magnitude in places, causing the chiral mode to significantly delocalize. An effect like this can cause trouble with lattice simulations at finite volume and lattice spacing; more later.

4.3.3 Domain-wall fermions and the Callan–Harvey mechanism

Now turn on the gauge fields and see how the anomaly works, following Callan and Harvey, (1985). To do this, I integrate out the heavy modes in the presence of a background gauge field. Although I will be interested in having purely $2k$-dimensional gauge fields in the theory, I will for now let them be arbitrary $(2k+1)$-dimensional fields. And since it is hard to integrate out the heavy modes exactly, I will perform the calculation as if their mass were constant, and then substitute $m(s)$; this is not valid where $m(s)$ is changing rapidly (near the domain wall) but should be adequate farther away. Also, in departure from the work of (Callan and Harvey, 1985), I will include a Pauli–Villars field with constant mass $M < 0$, independent of s; this is necessary to regulate fermion loops in the wave-function renormalization for the gauge fields, for example.

When one integrates out the heavy fields, one generates a Chern–Simons operator in the effective Lagrangian, as discussed in Section 4.2.4:

$$\mathcal{L}_{\text{CS}} = \left(\frac{m(s)}{|m(s)|} + \frac{M}{|M|} \right) \mathcal{O}_{CS} = \left(\epsilon(s) - 1 \right) \mathcal{O}_{CS}. \tag{4.101}$$

Note that with $M < 0$, the coefficient of the operator equals -2 on the side where $m(s)$ is negative, and equals zero on the side where it is positive. For a background $U(1)$ gauge field one finds in Euclidian space-time:

$$d = 3: \quad \mathcal{O}_{CS} = -\frac{e^2}{8\pi} \epsilon_{abc} (A_a \partial_a A_c), \tag{4.102}$$

$$d = 5: \quad \mathcal{O}_{CS} = -\frac{e^3}{48\pi^2} \epsilon_{abcde} (A_a \partial_b A_c \partial_d A_e). \tag{4.103}$$

Differentiating $\mathcal{L}_{\text{CS}}$ by A_μ and dividing by e gives the particle number current:

$$J_a^{(CS)} = (\epsilon(s) - 1) \begin{cases} -\frac{e}{8\pi} \epsilon_{abc} (F_{bc}) & d = 3 \\[2mm] -\frac{e^2}{64\pi^2} \epsilon_{abcde} (F_{bc} F_{de}) & d = 5, \end{cases} \tag{4.104}$$

where I use Latin letters to denote the coordinates in $2k+1$ dimensions, while Greek letters will refer to indices on the $2k$-dimensional defect. So when we turn on background $2k$-dimensional gauge fields, particle current flows either onto or off of the domain wall along the transverse s direction on the left side (where $m(s) = -m$). If we had regulated with a positive mass Pauli–Villars field, the current would flow on the right side. But in either case, this bizarre current exactly accounts for the anomaly. Consider the case of a 2-dimensional domain wall embedded in 3 dimensions. If we turn on an E field we know that from the point of view of a 2D creature, RH Weyl particles are created, where from Eq. (4.102),

$$\partial_\mu J_{\mu,R} = \tfrac{1}{2} \partial_\mu J_{\mu,A} = \frac{e}{4\pi} \epsilon_{\mu\nu} F_{\mu\nu}. \tag{4.105}$$

We see from Eq. (4.104) this current is exactly compensated for by the Chern–Simons current $J_2^{(CS)} = \frac{e}{4\pi}\epsilon_{2\mu\nu}F_{\mu\nu}$ that flows onto the domain wall from the $-s = -x_2$ side. The total particle current is divergenceless.

This is encouraging: (i) we managed to obtain a fermion whose mass is zero due to topology and not fine tuning; (ii) the low-energy theory therefore has a chiral symmetry even though the full 3D theory does not; (iii) the only remnant of the explicit chiral-symmetry breaking of the full theory is the anomalous divergence of the chiral symmetry in the presence of gauge fields. One drawback though is the infinite dimension in the s direction, since we will eventually want to simulate this on a finite lattice; besides, it is always disturbing to see currents streaming in from $s = -\infty$! One solution is to work in finite $(2k + 1)$ dimensions, in which case we end up with a massless RH mode stuck to the boundary on one side and a LH mode on the other (which is great for a vector-like theory of massless Dirac fermions, but not for chiral gauge theories). This is what one does when simulating domain-wall fermions. The other solution is more devious, leads to the "overlap operator", and is the subject of Section 4.4.

4.3.3.1   *Domain-wall fermions on a slab*

To get a better understanding of how the theory works, it is useful to consider a compact extra dimension. In particular, consider the case of periodic boundary conditions $\Psi(x_\mu, s + 2s_0) = \Psi(x_\mu, s)$; we define the theory on the interval $-s_0 \le s \le s_0$ with $\Psi(x_\mu, -s_0) = \Psi(x_\mu, s_0)$ and mass $m(s) = m\frac{s}{|s|}$. Note that the mass function $m(s)$ now has a domain-wall kink at $s = 0$ and an antikink at $s = \pm s_0$. There are now two exact zero-mode solutions to the Dirac equation,

$$b_0(s) = Ne^{-\int_{-s_0}^{s} m(s')ds'}, \qquad f_0(s) = Ne^{+\int_{-s_0}^{s} m(s')ds'}. \qquad (4.106)$$

Both solutions are normalizable since the transverse direction is finite; b_0 corresponds to a right-handed chiral fermion located at $s = 0$, and f_0 corresponds to a left-handed chiral fermion located at $s = \pm s_0$. However, in this case the existence of exactly massless modes is a result of the fact that $\int_{-s_0}^{+s_0} m(s)\,ds = 0$ that is not a topological condition and not robust. For example, turning on weakly coupled gauge interactions will cause a shift in the mass by $\delta m(s) \propto \alpha m$ (assuming m is the cutoff) that ruins this property. However: remember that to get a mass in the $2k$-dimensional defect theory, the RH and LH chiral modes have to couple to each other. The induced mass will be

$$\delta\mu_0 \sim \delta m \int ds\, b_0(s)\, f_0(s) = \delta m\, N^2 \sim \alpha m \times \frac{2ms_0}{\cosh[ms_0]} \equiv m_{\text{res}}, \qquad (4.107)$$

which vanishes exponential fast as $(Ms_0) \to \infty$. Nevertheless, at finite s_0 there will always be some chiral-symmetry breaking, in the form of a residual mass, called m_{res}. If, however, one wants to work on a finite line segment in the extra dimension instead of a circle, we can take an asymmetric mass function,

$$m(s) = \begin{cases} -m & -s_0 \le s \le 0 \\ +\infty & 0 < s < s_0. \end{cases} \tag{4.108}$$

This has the effect of excluding half the space, so that the extra dimension has boundaries at $s = -s_0$ and $s = 0$. Now, even without extra interactions, one finds

$$m_{\text{res}} \sim 2m e^{-2ms_0}. \tag{4.109}$$

Any matrix element of a chiral-symmetry-violating operator will be proportional to the overlap of the LH and RH zero-mode wave functions, which is proportional to m_{res}. On the lattice the story of m_{res} is more complicated—as discussed in Section 4.3.4—both because of the discretization of the fermion action, and because of the presence of rough gauge fields. Lattice computations with domain-wall fermions need to balance the cost of simulating a large extra dimension versus the need to make m_{res} small enough to attain chiral symmetry.

4.3.3.2   The (almost) chiral propagator

Before moving to the lattice, I want to mention an illuminating calculation by Lüscher (Lüscher, 2000a) who considered non-interacting domain-wall fermions with a semi-infinite fifth dimension, negative fermion mass, and LH Weyl fermion zero-mode bound to the boundary at $s = 0$. He computed the Green function for propagation of the zero mode from $(x, s = 0)$ to $(y, s = 0)$ and examined the chiral properties of this propagator. The differential operator to invert should be familiar now:

$$D_5 = \slashed{\partial}_4 + \gamma_5 \partial_s - m, \qquad s \ge 0. \tag{4.110}$$

We wish to look at the Green function G that satisfies

$$D_5 G(x, s; y, t) = \delta^{(4)}(x - y)\delta(s - t), \qquad P_+ G(x, 0; y, t) = 0. \tag{4.111}$$

The solution Lüscher found for propagation along the boundary was

$$G(x, s; y, t)\Big|_{s=t=0} = 2P_- D^{-1} P_+, \tag{4.112}$$

where D is the peculiar-looking operator

$$D = [1 + \gamma_5 \epsilon(H)], \qquad H \equiv \gamma_5(\slashed{\partial}_4 - m) = H^\dagger, \qquad \epsilon(\mathcal{O}) \equiv \frac{\mathcal{O}}{\sqrt{\mathcal{O}^\dagger \mathcal{O}}}. \tag{4.113}$$

This looks pretty bizarre! Since H is Hermitian, in a basis where H is diagonal, $\epsilon(H) = \pm 1$! But don't conclude that in this basis the operator is simply $D = (1 \pm \gamma_5)$—you must remember, that in the basis where H is diagonal, $\epsilon(H)\gamma_5$ is not (by which I mean $\langle m|\epsilon(H)\gamma_5|n\rangle$ is in general non-zero for $m \ne n$ in the H eigenstate basis). In fact, Eq. (4.113) looks very much like the overlap operator discovered some years earlier and that we will be discussing soon.

A normal Weyl fermion in four dimensions would have a propagator $P_-(\slashed{\partial}_4)^{-1}P_+$; here we see that the domain-wall fermion propagator looks like the analogous object arising from the fermion action $\bar{\Psi} D \Psi$, with D playing the role of the four-dimensional Dirac operator $\slashed{\partial}_4$. So what are the properties of D?

- For long-wavelength modes (e.g. $k \ll m$) we can expand D in powers of $\slashed{\partial}_4$ and find

$$D = \frac{1}{m} \left(\slashed{\partial}_4 - \frac{\partial_4^2}{2m} + \cdots \right), \tag{4.114}$$

which is reassuring: we knew that at long wavelengths we had a garden variety Weyl fermion living on the boundary of the extra dimension (the factor of $1/m$ is an unimportant normalization).

- A massless Dirac action is chirally invariant because $\{\gamma_5, \slashed{\partial}_4\} = 0$. However, the operator D does not satisfy this relationship, but rather:

$$\{\gamma_5, D\} = D \gamma_5 D, \tag{4.115}$$

or equivalently,

$$\{\gamma_5, D^{-1}\} = \gamma_5. \tag{4.116}$$

This is the famous Ginsparg–Wilson equation, first introduced in the context of the lattice (but not solved) many years ago (Ginsparg and Wilson, 1982). Note the right-hand side of the above equations encodes the violation of chiral symmetry that our Weyl fermion experiences; the fact that the right side of Eq. (4.116) is local in space-time implies that violations of chiral symmetry will be seen in Green functions *only* when operators are sitting at the same space-time point. We know from our previous discussion that the only chiral-symmetry violation that survives to low energy in the domain-wall model is the anomaly, and so it must be that the chiral-symmetry violation in Eqs. (4.115)–(4.116) encode the anomaly and nothing else at low energy.[9]

4.3.4   Domain-wall fermions on the lattice

The next step is to transcribe this theory onto the lattice. If you replace continuum derivatives with the usual lattice operator $D \to \frac{1}{2}(\nabla^* + \nabla)$ (where ∇ and ∇^* are the forward and backward lattice difference operators, respectively) then one discovers...doublers! Not only are the chiral modes doubled in the $2k$ dimensions along the domain wall, but there are two solutions for the transverse wave function of the zero mode, $b_0(s)$, one of which alternates sign with every step in the s direction and that is a LH mode. So this ends up giving us a theory of naive fermions on the lattice, only in a much more complicated and expensive way!

However, when we add Wilson terms $\frac{r}{2}\nabla^*\nabla$ for each of the dimensions, things get interesting. You can think of these as mass terms that are independent of s but that are dependent on the wave number k of the mode, vanishing for long wavelength. What happens if we add a k-dependent spatially constant mass $\Delta m(k)$ to the step function mass $m(s) = m\epsilon(s)$? The solution for $b_0(s)$ in Eq. (4.100) for an infinite extra dimension becomes

$$b_0 = Ne^{-\int_0^s [m(s') + \Delta m(k)] ds'}, \tag{4.117}$$

which is a normalizable zero-mode solution, albeit distorted in shape, so long as $|\Delta m(k)| < m$. However, for $|\Delta m(k)| > m$, the chiral mode vanishes. What happens to it? It becomes more and more extended in the extra dimension until it ceases to be normalizable. What is going on is easier to grasp for a finite extra dimension: as $|\Delta m(k)|$ increases with increasing k, eventually the b_0 zero-mode solution extends to the opposing boundary of the extra dimension, when $|\Delta m(k)| \sim (m - 1/s_0)$. At that point it can pair up with the LH mode and become heavy.

So the idea is: add a Wilson term, with strength such that the doublers at the corners of the Brillouin zone have $|\Delta m(k)|$ too large to support a zero-mode solution. Under separation of variables, one looks for zero-mode solutions with $\Psi(x, s) = e^{ipx} \phi_\pm(s) \psi_\pm$ with $\Gamma \psi_\pm = \pm \psi$. One then finds (for $r = 1$)

$$\slashed{\partial}_4 \psi_\pm = 0, \qquad -\phi_\pm(s \mp 1) + (m_{\text{eff}}(s) + 1)\phi_\pm(s) = 0, \tag{4.118}$$

where

$$m_{\text{eff}}(s) = m\epsilon(s) + \sum_\mu (1 - \cos p_\mu) \equiv m\epsilon(s) + F(p). \tag{4.119}$$

Solutions of the form $\phi_\pm(s) = z_\pm^s$ are found with

$$z_\pm = (1 + m_{\text{eff}}(s))^{\mp 1} = (1 + m\epsilon(s) + F(p))^{\mp 1}; \tag{4.120}$$

they are normalizable if $|z|^{\epsilon(s)} < 1$. Solutions are found for ψ_+ only, provided that m is in the range $F(p) < m < F(p) + 2$. (For $r \neq 1$, this region is found by replacing $m \to m/r$.) However, even though the solution is only found for ψ_+, the chirality of the solutions will alternate with corners of the Brillouin zone, just as we found for naive fermions, Eq. (4.93). The picture for the spectrum in 2D is shown in Fig. 4.9. It was first shown in (Kaplan, 1992) that doublers could be eliminated for domain-wall fermions on the lattice; the rich spectrum in Fig. 4.9 was worked out in (Jansen and Schmaltz, 1992), where for 4D they found the number of zero-mode solutions to be the Pascal numbers $(1, 4, 6, 4, 1)$ with alternating chirality, the critical values for $|m/r|$ being $0, 2, \ldots, 10$. One implication of their work is that the Chern–Simons currents must also change discontinuously on the lattice at these critical values of $|m|/r$; indeed that is the case, and the lattice version of the Callan–Harvey mechanism was verified analytically in (Golterman *et al.*, 1993).

Figure 4.9 suggests that chiral fermions will exist in two space-time dimensions so long as $0 < |m/r| < 6$, with critical points at $|m/r| = 0, 2, \ldots, 6$ where the numbers of massless flavors and their chiralities change discontinuously. In four space-time dimensions a similar calculation leads to chiral fermions for $0 < |m/r| < 10$ with critical points at $|m/r| = 0, 2, \ldots, 10$. However, this reasoning ignores the gauge fields. In perturbation theory one would expect the bulk fermions to obtain a radiative mass correction of size $\delta m \sim O(\alpha)$ in lattice units, independent of the extra dimension s. Extrapolating shamelessly to strong coupling, one then expects the domain-wall form of the mass to be ruined when $\alpha \sim 1$ for $|m/r| \sim (2n + 1)$, $n = 0, \ldots, 4$ causing a loss

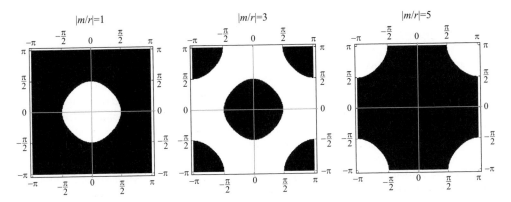

Fig. 4.9 Domain-wall fermions in $d = 2$ on the lattice: dispersion relation plotted in the Brillouin zone. Chiral modes exist in white regions only. For $0 < |m/r| < 2$ there exists a single RH mode centered at $(k_1, k_2) = (0, 0)$. for $2 < |m/r| < 4$ there exist two LH modes centered at $(k_1, k_2) = (\pi, 0)$ and $(k_1, k_2) = (0, \pi)$; for $4 < |m/r| < 6$ there exists a single RH mode centered at $(k_1, k_2) = (\pi, \pi)$. For $|m/r| > 6$ there are no chiral mode solutions.

of chiral symmetry; near the critical points in $|m/r|$ the critical gauge coupling that destroys chiral symmetry will be smaller.

While qualitatively correct, this argument ignores the discrete nature of the lattice. On the lattice, the exponential suppression $m_{\text{res}} \sim \exp(-2ms_0)$ found in Eq. (4.107) is replaced by $\hat{T}^{L_s} = \exp(-L_s \hat{h})$, where $\hat{T}$ is a transfer matrix in the fifth dimension that is represented by L_s lattice sites. Good chiral symmetry is attained when $\hat{h}$ exhibits a "mass gap," i.e. when all its eigenvalues are positive and bounded away from zero. However, one finds that, at strong coupling, rough gauge fields can appear that give rise to near-zero modes of $\hat{h}$, destroying chiral symmetry, with $m_{\text{res}} \propto 1/L_s$ (Christ, 2006; Antonio *et al.*, 2008). To avoid this problem, one needs to work at weaker coupling and with an improved gauge action that suppresses the appearance of rough gauge fields.

At finite lattice spacing the phase diagram is expected to look something like in Fig. 4.10 where I have plotted m versus g^2, the strong-coupling constant. On this diagram, $g^2 \to 0$ is the continuum limit. Domain-wall fermions do not require fine tuning so long as the mass is in one of the regions marked by an "X", which yield $\{1, 4, 6, 4, 1\}$ chiral flavors from left to right. The shaded region is a phase called the Aoki phase (Aoki, 1984); it is presently unclear whether the phase extends to the continuum limit (left side of Fig. 4.10) or not (right side) (Golterman *et al.*, 2005). In either case, the black arrow indicates how for Wilson fermions one tunes the mass from the right to the boundary of the Aoki phase to obtain massless pions and chiral symmetry; if the Aoki phase extends down to $g^2 = 0$ then the Wilson program will work in the continuum limit, but not if the RH side of Fig. 4.10 pertains. See (Golterman and Shamir, 2000; 2003) for a sophisticated discussion of the physics behind this diagram.

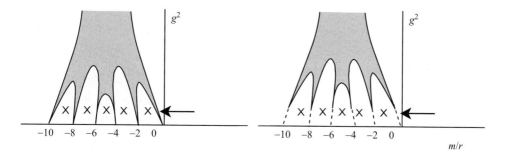

Fig. 4.10 A sketch of the possible phase structure of QCD with Wilson fermions where the shaded region is the Aoki phase—pictured extending to the continuum limit (left) or not (right). When using Wilson fermions one attempts to tune the fermion mass to the phase boundary (arrow) to obtain massless pions; this is only possible in the continuum limit if the picture on the left is correct. For domain-wall fermions chiral symmetry results at infinite L_s when one simulates in any of the regions marked with an "X". There are six "fingers" in this picture instead of five due to the discretization of the fifth dimension.

Of course, in the real world we do not see exact chiral symmetry, since quarks and leptons do have mass. A mass for the domain-wall fermion can be included as a coupling between the LH mode at $s = 1$, and the RH mode at $s = N_s$:

$$m_q \left[\overline{\psi}(\mathbf{x}, 1) P_+ \psi(\mathbf{x}, N_s) + \overline{\psi}(\mathbf{x}, N_s) P_- \psi(\mathbf{x}, 1) \right] \tag{4.121}$$

and correlation functions are measured by sewing together propagators from one boundary to itself for chiral-symmetry-preserving operators, or from one boundary to the other for operators involving a chiral flip. The latter will require insertions of the mass operator above to be non-zero (assuming a neglible m_{res})—just as it should be in the continuum.

4.3.4.1 Shamir's formulation

Domain-wall fermions are used by a number of lattice collaborations these days, using the formulation of Shamir (Shamir, 1993; Furman and Shamir, 1995), which is equivalent to the continuum version of domain-wall fermions on a slab described above. The lattice action is given by:

$$\sum_{b=1}^{5} \sum_{\mathbf{x}} \sum_{s=1}^{N_s} \left[\tfrac{1}{2} \overline{\psi} \gamma_b (\partial_b^* + \partial_b) \psi - m \overline{\psi} \psi - \frac{r}{2} \overline{\psi} \partial_b^* \partial_b \psi \right], \tag{4.122}$$

where the lattice coordinate on the 5D lattice is $\mathbf{n} = \{\mathbf{x}, s\}$, $\mathbf{x}$ and s being the 4D and fifth-dimension lattice coordinates, respectively. The difference operators are

$$\partial_b \psi(\mathbf{n}) = \psi(\mathbf{n} + \hat{\mu}_b) - \psi(\mathbf{n}), \qquad \partial_b^* \psi(\mathbf{n}) = \psi(\mathbf{n}) - \psi(\mathbf{n} - \hat{\mu}_b), \tag{4.123}$$

where $\hat{\mu}_b$ is a unit vector in the x_b direction. In practice, of course, these derivatives are gauged in the usual way by inserting gauge-link variables. The boundary conditions are defined by setting fields to zero on sites with $s = 0$ and $s = N_s + 1$. I have reversed the sign of m and r from Shamir's original paper, since the above sign for r appears to be relatively standard now. For domain-wall fermions, m has the opposite sign from standard Wilson fermions, which is physics, not convention. The above action gives rise to a RH chiral mode bound to the $s = 1$ boundary of the lattice, and a LH chiral mode bound at the $s = N_s$ boundary.

4.3.5   The utility of domain-wall fermions

Theoretically, chiral symmetry can be as good a symmetry as one desires if one is close enough to the continuum limit (to avoid delocalization of the zero mode due to large gauge-field fluctuations) and large extra dimension. In practical simulations, the question is whether the residual-mass term can be small enough to warrant the simulation cost. This was reviewed dispassionately and at length in (Sharpe, 2007), and I refer you to that article if you are interested in finding out the details. Currently, domain-wall fermions are being extensively applied to QCD; for some diverse examples from the past year see (Yamazaki, 2009; Chiu *et al.*, 2009*b*; Gavai and Sharma, 2009; Torok *et al.*, 2009; Ohta, RBC and UKQCD Collaborations, 2009; Cheng *et al.*, 2010; Chiu *et al.*, 2009*a*), and a recent overview (Jansen, 2008). Another recent application has been to $N = 1$ supersymmetric Yang–Mills theory (Giedt *et al.*, 2008; 2009; Endres, 2008; 2009*a*;*b*) based on a domain-wall formulation for Majorana fermions (Kaplan and Schmaltz, 2000) and earlier numerical work (Fleming *et al.*, 2001).

4.4   Overlap fermions and the Ginsparg–Wilson equation

4.4.1   Overlap fermions

We have seen that the low-energy limit of a domain-wall fermion in the limit of large extra dimension is a single massless Dirac fermion, enjoying the full extent of the chiral symmetry belonging to massless fermions in the continuum. In this low-energy limit, the effective theory is four-dimensional if the original domain-wall fermion lived in five dimensions. One might wonder whether one could dispense with the whole machinery of the extra dimension and simply write down the low-energy four-dimensional theory to start with. Furthermore, one would like a four-dimensional formulation with exact chiral symmetry, which could only occur for domain-wall fermions with infinite extent in the time direction, which is not very practical numerically!

Neuberger and Narayanan found an extremely clever way to do this, leading to the four-dimensional "overlap operator" that describes lattice fermions with perfect chiral symmetry. The starting point is to consider a five-dimensional fermion in the continuum with a single domain wall, and to consider the fifth dimension to be time (after all, it makes no difference in Euclidian space). Then, $\gamma_5(\slashed{D}_4 + m(s))$ looks like the Hamiltonian, where s is the new time coordinate, and $m(-\infty) = -m_1$, $m(\infty) = +m_2$, where $m_{1,2} > 0$. The path integral projects onto ground states, and so the partition function for this system is $Z = \langle \Omega, -m_1 | \Omega, +m_2 \rangle$, where the state $|\Omega, m\rangle$ is the ground state of $\mathcal{H}_4(m) = \gamma_5(\slashed{D}_4 + m)$ We know that this should describe a massless Weyl fermion. Note that the partition function is in general complex with an ill-defined phase (we can redefine the phase of $|\Omega, -m_1\rangle$ and $|\Omega, m_2\rangle$ separately and arbitrarily). If we now instead imagine that the fermion mass function $m(s)$ exhibits a wall–antiwall pair, with the two defects separated infinitely far apart, we recognize a system that will have a massless Dirac fermion in the spectrum, and $Z = \left|\langle \Omega, -m_1|\Omega, +m_2\rangle\right|^2$, which is real, positive, and independent of how we chose the phase for the ground states.

We can immediately transcribe this to the lattice, where we replace $\not{D}_4$ with the four-dimensional Wilson operator,

$$\mathcal{H}(m) = \gamma_5(D_w + m) = \gamma_5\left(D_\mu\gamma_\mu - \frac{r}{2}D_\mu^2 + m\right), \qquad (4.124)$$

with D_μ being the symmetric covariant derivative on the lattice, and D_μ^2 being the covariant lattice Laplacian. Note that $\mathcal{H}(m)$ is Hermitian, and so its eigenvalues are real. Furthermore, one can show that it has equal numbers of positive and negative eigenvalues.

We can account for the γ-matrix structure of $\mathcal{H}(m)$ explicitly on a chiral basis where $\gamma_5 = \sigma_3 \otimes 1$:

$$\mathcal{H}(m) = \begin{pmatrix} B + m & C \\ C^\dagger & -B - m \end{pmatrix}, \qquad (4.125)$$

where $B = -\frac{r}{2}\nabla^2$ is the Wilson operator and $C = D_\mu\sigma_\mu$, where $\sigma_\mu = \{i, \vec{\sigma}\}$. For simplicity for $\langle\Omega, m_2|$ one can take $m_2 \to \infty$, in which case $\mathcal{H} \sim +m_2\gamma_5$.

We know that $Z = \left|\langle \Omega, -m_1|\Omega, +m_2\rangle\right|^2$ will represent a massless Dirac fermion on the lattice, so long as $0 < m_1 < 2r$, with m_2 arbitrary. The ground states of interest may be written as Slater determinants of all the one-particle wave functions with negative energy. Let us designate the one-particle energy eigenstates of $\mathcal{H}(-m_1)$ and $\mathcal{H}(m_2)$ to be $|n, -m_1\rangle$ and $|n, m_2\rangle$, respectively, with

$$\langle n, m_2|n', -m_1\rangle \equiv U_{nn'} = \begin{pmatrix} \alpha & \beta \\ \gamma & \delta \end{pmatrix}_{nn'}, \qquad U^\dagger U = 1, \qquad (4.126)$$

where the block structure of U is in the same γ-matrix space that we introduced in writing $\mathcal{H}$ in block form, Eq. (4.125). Now, we want to only fill negative energy eigenstates, so it is convenient to introduce the sign function

$$\varepsilon(\lambda) \equiv \frac{\lambda}{\sqrt{\lambda^\dagger\lambda}}. \qquad (4.127)$$

With $m_2 \to \infty$ we have

$$\varepsilon(\mathcal{H}(m_2)) \xrightarrow[m_2 \to \infty]{} \gamma_5 = \begin{pmatrix} 1 & 0 \\ 0 & -1 \end{pmatrix}. \qquad (4.128)$$

Assuming $\mathcal{H}(-m_1)$ has no exact zero modes then, it follows that all eigenvalues come in $\pm$ pairs (just like the operator γ_5) and we can choose our basis $|n, -m_1\rangle$ so that

$$\varepsilon(\mathcal{H}(m_2)) = U\gamma_5 U^\dagger = U\begin{pmatrix} 1 & 0 \\ 0 & -1 \end{pmatrix}U^\dagger. \qquad (4.129)$$

Therefore, the Slater determinant we want is

$$
\begin{aligned}
Z &= \left| \langle \Omega, m_2 | \Omega, -m_1 \rangle \right|^2 \\
&= \left| \det U_{22} \right|^2 \\
&= \det \delta^\dagger \det \delta \\[2mm]
&= \det \left(\frac{1 + \gamma_5 \varepsilon(\mathcal{H}(-m_1))}{2} \right).
\end{aligned}
\tag{4.130}
$$

Some steps have been omitted from ths derivation (Narayanan, 2001); see Exercise 4.4.1.

Exercise 4.10 Prove the assertion that if $\mathcal{H}(-m_1)$ has no zero modes, it has equal numbers of positive and negative eigenvalues.

Exercise 4.11 You should prove the last step in Eq. (4.130), breaking it down to the following steps:

(a) Show that $\det \delta^\dagger = \det \alpha \det U^\dagger$;
(b) ... so that $\det \delta^\dagger \det \delta = \det \delta \det \alpha \det U^\dagger = \det \left[\frac{1}{2}(U + \gamma_5 U \gamma_5) U^\dagger \right]$;
(c) ... which combines with Eq. (4.129) to yield Eq. (4.130).

On the other hand, $Z \propto \det D$, where D is the fermion operator. So, we arrive at the overlap operator (dropping the subscript from m_1):

$$
\begin{aligned}
D &= 1 + \gamma_5 \varepsilon(\mathcal{H}(-m)) \\[2mm]
&= 1 + \gamma_5 \frac{\mathcal{H}(-m)}{\sqrt{\mathcal{H}(-m)^2}} \\[2mm]
&= 1 + \frac{D_w - m}{\sqrt{(D_w - m)^\dagger (D_w - m)}},
\end{aligned}
\tag{4.131}
$$

a remarkable result. It was subsequently shown explicitly that this fermion operator can be derived directly from lattice domain-wall fermions at infinite wall separation (Neuberger, 1998c; Kikukawa and Noguchi, 1999). Recall from our discussion of domain-wall fermions that at least for weak gauge fields, we need $0 < m < 2r$ in order to obtain one flavor of massless Dirac fermion (where I have set the lattice spacing $a = 1$).

4.4.1.1 *Eigenvalues of the overlap operator*

Recall that the eigenvalues of the Dirac operator in the continuum are $\pm i\lambda_n$ for real non-zero λ_n, plus n_+ RH and n_- LH zero modes, where the difference is constrained by the index theorem to equal the topological winding number of the gauge field. Thus, the spectrum looks like a line on the imaginary axis. What does the spectrum of the overlap operator look like? Consider

$$
(D - 1)^\dagger (D - 1) = \epsilon(\mathcal{H})^2 = 1.
\tag{4.132}
$$

Thus, $(D - 1)$ is a unitary matrix and the eigenvalues of D are constrained to lie on a circle of unit radius in the complex plane, with the center of the circle at $z = 1$. If you put the lattice spacing back into the problem, $D \to aD$ in the above expression to get the dimensions right, and so the eigenvalues sit on a circle of radius $1/a$ centered at $1/a$. Thus, as $a \to 0$ the circle gets bigger, and the eigenvalues with small magnitude almost lie on the imaginary axis, like the continuum eigenvalues. See the problem below, where you are to show that the eigenfunctions of D with real eigenvalues are chiral.

4.4.1.2 Locality of the overlap operator

If just presented with the overlap operator Eq. (4.130) without knowing how it was derived, one might worry that its unusual structure could entail momentum-space singularities corresponding to unacceptable non-local behavior in coordinate space. (From its derivation from domain-wall fermions this would be very surprising for sufficiently weakly coupled gauge fields, since the domain-wall theory looks well defined and local with a mass gap.) The locality of the overlap operator (i.e. that it falls off exponentially in coordinate space) was proven analytically in (Hernández *et al.*, 1999), under the assumption of sufficiently smooth gauge-link variables, namely that $|1 - U| < 1/30$. They also claimed numerical evidence for locality that was less restrictive.

4.4.1.3 The value of m and the number of fermions

For domain-wall fermions we found the interesting phase structure as a function of m/r, where in the intervals between the critical values $m/r = \{0, -2, -4 \ldots, -2(2k + 1)\}$ there were $\{1, 2k, \ldots, 1\}$ copies of chiral fermions with alternating chirality, as shown in Fig. 4.9. One would expect something analogous then for the overlap operator, since it is equivalent to a domain-wall fermion on a $2k$-dimensional lattice with infinite continuous extra dimension. The equation of motion for the domain-wall modes is slightly different from that found in Eq. (4.118) due to the continuous dimension:

$$\slashed{p}_{2k}\psi_{\pm} = 0, \qquad \pm\phi'_{\pm}(s) + (m_{\text{eff}}(s) + 1)\phi_{\pm}(s) = 0 \qquad (4.133)$$

where

$$m_{\text{eff}}(s) = m\epsilon(s) + r\sum_{\mu}(1 - \cos p_{\mu}) \equiv m\epsilon(s) + rF(p). \qquad (4.134)$$

Solutions are of the form $\phi_{\pm}(s) = e^{\mp \int^s m_{\text{eff}}(t)\,dt}$. With $rF(p) > 0$, ϕ_{-} is never normalizable, while for ϕ_{+}, normalizability requires $m/r > F(p)$. Thus, the overlap operator Eq. (4.131) represents $\{1, 1 + 2k, \ldots, 2^{2k}\}$ massless Dirac fermions for m/r in the intervals $(0, 2), (2, 4), \ldots, (4k, \infty)$. In $2k = 4$ dimensions, these flavor numbers are $\{1, 5, 11, 15, 16\}$.

4.4.1.4 Simulating the overlap operator

The overlap operator has exact chiral symmetry, in the sense that it is an exact solution of the Ginsparg–Wilson relation, which cannot be said for domain-wall fermions at finite N_s; furthermore, it is a four-dimensional operator, which would seem to be easier to simulate than a 5D theory. However, the inverse square root of an operator is expensive to compute, and requires some approximations. The algorithms for computing it are described in detail in an excellent review by Kennedy (Kennedy, 2006). Amusingly, he explains that the method for computing the overlap operator can be viewed as simulating a five-dimensional theory, albeit one with more general structure than the domain-wall theory. For a recent review comparing the computational costs of different lattice fermions, see (Jansen, 2008).

Exercise 4.12 Show that the overlap operator in Eq. (4.131) has the following properties:

(a) At zero gauge field and acting on long-wavelength fermion modes, $D \simeq \partial_4$, the ordinary Dirac operator for a massless fermion.
(b) It satisfies the Ginsparg–Wilson equation, Eq. (4.115):

$$\{\gamma_5, D\} = D\gamma_5 D. \tag{4.135}$$

Exercise 4.13

(a) Show that one can write $D = 1 + V$, where $V^\dagger V = 1$, and that therefore D can be diagonalized by a unitary transformation, with its eigenvalues lying on the circle $z = 1 + e^{i\phi}$.
(b) Show that, despite D being non-Hermitian, normalized eigenstates satisfying $D|z\rangle = z|z\rangle$ with different eigenvalues are orthogonal, satisfying $\langle z'|z\rangle = \delta_{z'z}$.
(c) Show that if $D|z\rangle = z|z\rangle$ then $D^\dagger|z\rangle = z^*|z\rangle$.
(d) Assuming that $\gamma_5 D\gamma_5 = D^\dagger$, show that $\langle z|\gamma_5|z\rangle = 0$ unless $z = 0$ or $z = 2$, in which   case $\langle z|\gamma_5|z\rangle = \pm 1$.

4.4.2 The Ginsparg–Wilson equation and its consequences

In 1982 Paul Ginsparg and Kenneth Wilson wrote a paper about chiral lattice fermions that was immediately almost completely forgotten, accruing 10 citations in the first ten years and none in the subsequent five; today it is marching toward 700 citations. The reason for this peculiar history is that they wrote down an equation they speculated should be obeyed by a fermion operator in the fixed-point action of a theory tuned to the chiral point—but they did not solve it. After domain-wall and overlap fermions were discovered in the early 1990s, it was realized that they provided a solution to this equation (the domain-wall solution only being exact in the limit of infinite extra dimension). Shortly afterward, M. Lüscher elaborated on how the salient features of chirality flowed from the Ginsparg-Wilson equation—in particular, how anomalies and multiplicative mass renormalization were consequences of the equation, which provided a completely explicit four-dimensional explanation for the success of the overlap and domain-wall fermions.

4.4.2.1 Motivation

A free Wilson fermion with its mass tuned to the critical value describes a chiral fermion in the continuum. As we have seen, chiral symmetry does not exist on the lattice, but its violation is not evident at low energy, except through correctly reproducing the anomaly. However, imagine studying this low-energy effective theory by repeatedly performing block spin averages. One would eventually have a lattice theory with all the properties one would desire: chiral fermions and chiral anomalies. What is the fermion operator in this low-energy theory, and how does it realize chiral symmetry? Motivated by this question, Ginsparg and Wilson performed a somewhat simpler calculation: they took a continuum theory with chiral symmetry and anomalies, and performed an average over space-time cells to create a lattice theory, and asked how the chiral symmetry in the original theory was expressed in the resulting lattice theory.

The starting point is the continuum theory

$$Z = \int [d\psi]\,[d\bar{\psi}]\,e^{-S(\psi,\bar{\psi})}. \tag{4.136}$$

I assume there are N_f identical flavors of fermions, and that S is invariant under the full $U(N_f) \times U(N_f)$ chiral symmetry. We define $\psi_{\mathbf{n}}$ to be localized averages of ψ,

$$\psi_{\mathbf{n}} = \int d^4x\,\psi(x) f(\mathbf{x} - a\mathbf{n}), \tag{4.137}$$

where $f(\mathbf{x})$ is some function with support in the region of $|\mathbf{x}| \lesssim a$. Then, up to an irrelevant normalization, we can rewrite

$$Z = \int [d\psi]\,[d\bar{\psi}] \int \prod_{\mathbf{n}} d\chi_{\mathbf{n}}\,d\bar{\chi}_{\mathbf{n}}\,e^{-\left[\sum_{\mathbf{n}} \alpha(\bar{\chi}_{\mathbf{n}} - \bar{\psi}_{\mathbf{n}})(\chi_{\mathbf{n}} - \psi_{\mathbf{n}}) - S(\psi,\bar{\psi})\right]}$$

$$\equiv \int \prod_{\mathbf{n}} d\chi_{\mathbf{n}}\,d\bar{\chi}_{\mathbf{n}}\,e^{-S_{\mathrm{lat}}(\bar{\chi}_{\mathbf{n}},\chi_{\mathbf{n}})} \equiv e^{-\bar{\chi} D \chi}, \tag{4.138}$$

where α is a dimensionful parameter and D is the resulting lattice fermion operator. Since there are N_f copies of all the fields, the operator D is invariant under the vector $U(N_f)$ symmetry, so that if T is a $U(N_f)$ generator, $[T, D] = 0$. The lattice action is therefore defined as

$$e^{-\bar{\chi} D \chi} = \int [d\psi]\,[d\bar{\psi}]\,e^{-\left[\sum_{\mathbf{n}} \alpha\,(\bar{\chi}_{\mathbf{n}} - \bar{\psi}_{\mathbf{n}})(\chi_{\mathbf{n}} - \psi_{\mathbf{n}}) - S(\psi,\bar{\psi})\right]}. \tag{4.139}$$

Note that explicit chiral-symmetry breaking has crept into our definition of S_{lat} through the fermion bilinear we have introduced in the Gaussian in order to change variables.

Now consider a chiral transformation on the lattice variables, $\chi_{\mathbf{n}} \to e^{i\epsilon\gamma_5 T}\chi_{\mathbf{n}}$, $\bar{\chi}_{\mathbf{n}} \to \bar{\chi}_{\mathbf{n}} e^{i\epsilon\gamma_5 T}$, where T is a generator for a $U(N_f)$ flavor transformation. This is accompanied by a corresponding change of integration variables ψ, $\bar{\psi}$:

$$e^{-\bar{\chi}e^{i\epsilon\gamma_5 T}De^{i\epsilon\gamma_5 T}\chi} = \int [d\psi]\,[d\bar{\psi}]\, e^{i\int \epsilon\mathcal{A}\,\mathrm{Tr}\,T}\, e^{-[\sum_n \alpha\,(\bar{\chi}_n-\bar{\psi}_n)e^{2i\epsilon\gamma_5 T}(\chi_n-\psi_n)-S(\psi,\bar{\psi})]}.$$

$$(4.140)$$

where $\mathcal{A}$ is the anomaly due to the non-invariance of the measure $[d\psi]\,[d\bar{\psi}]$ as computed by Fujikawa (Fujikawa (1979):

$$\mathcal{A} = \frac{1}{16\pi^2}\epsilon_{\epsilon\beta\gamma\delta}\mathrm{Tr}F_{\alpha\beta}F_{\gamma\delta} \qquad (4.141)$$

with

$$\int \mathcal{A} = 2\nu, \qquad (4.142)$$

ν being the topological charge of the gauge field.

Expanding to linear order in ϵ gives

$$-\bar{\chi}\{\gamma_5, D\}T\chi\, e^{-\bar{\chi}D\chi} = \int [d\psi]\,[d\bar{\psi}]\, \left(2\nu\,\mathrm{Tr}\,T + \sum_n \left[(\bar{\chi}_n - \bar{\psi}_n)2\alpha\gamma_5 T(\chi_n - \psi_n)\right]\right)$$

$$\times \exp\left[-\sum_m (\bar{\chi}_m - \bar{\psi}_m)(\chi_m - \psi_m) - S(\psi,\bar{\psi})\right]$$

$$= \sum_n \left(2\nu\,\mathrm{Tr}\,T - \frac{2}{\alpha}\frac{\delta}{\delta\chi_n}\gamma_5 T\frac{\delta}{\delta\bar{\chi}_n}\right)e^{-\bar{\chi}D\chi}$$

$$= \left(\mathrm{Tr}\gamma_5 DT + 2\nu\,\mathrm{Tr}\,T - \frac{2}{\alpha}\bar{\chi}_n D\gamma_5 DT\chi_n\right)e^{-\bar{\chi}D\chi}. \qquad (4.143)$$

Defining $\alpha \equiv 2/a$ this yields the operator identity

$$(\{\gamma_5, D\} - a\,D\gamma_5 D)\,T = (\mathrm{Tr}\,\gamma_5 DT + 2\nu\,\mathrm{Tr}\,T). \qquad (4.144)$$

If T is taken to be a traceless generator of $U(N_f)$, multiplying both sides by T and taking the trace yields the Ginsparg–Wilson equation:

$$\{\gamma_5, D\} = a\,D\gamma_5 D. \qquad (4.145)$$

If, on the other hand, we take T to be the unit matrix and use Eq. (4.145) we find

$$\mathrm{Tr}\gamma_5 D = 2N_f\nu. \qquad (4.146)$$

This latter equation was not derived in the original Ginsparg–Wilson paper; from our discussion of the index theorem Eq. (4.80), it follows from Eq. (4.146) that $\mathrm{Tr}\gamma_5 D = (n_- - n_+)$, where $n_\pm$ are the number of $\pm$ chirality zero modes. We will see that this is indeed the case.

Note that the GW relation Eq. (4.145) is the same equation satisfied by the overlap operator (Neuberger, 1998b)—and therefore by the domain-wall propagator at infinite wall separation on the lattice, being equivalent as shown in (Neuberger, 1998a;c)—as well as by the infinitely separated domain-wall propagator in the continuum (Lüscher,

$2000a$). In fact, the general overlap operator derived by Neuberger

$$D = 1 + \gamma_5 \epsilon(\mathcal{H}) \tag{4.147}$$

is the only explicit solution to the GW equations that is known.

4.4.2.2 Exact lattice chiral symmetry

Missing from the discussion so far is how the overlap operator is able to ensure multiplicative renormalization of fermion masses (and similarly, multiplicative renormalization of pion masses). In the continuum, both phenomena follow from the fact that fermion masses are the only operators breaking an otherwise good symmetry. The GW relation states exactly how chiral symmetry is broken on the lattice, but does not specify a symmetry that *is* exact on the lattice and capable of protecting fermion masses from additive renormalization.

Lüscher was able to solve this problem by discovering that the GW relation implied the existence of an exact symmetry of the lattice action: $\int \bar\psi D\psi$ is invariant under the transformation

$$\delta\psi = \gamma_5\left(1 + \frac{a}{2}D\right)\psi, \qquad \delta\bar\psi = \bar\psi\left(1 - \frac{a}{2}D\right)\gamma_5. \tag{4.148}$$

Note that this becomes ordinary chiral symmetry in the $a \to 0$ limit, and that it is broken explicitly by a mass term for the fermions.

4.4.2.3 Anomaly

If this symmetry were an exact symmetry of the path integral, we would run foul of all the arguments we have made so far: it becomes the anomalous $U(1)_A$ symmetry in the continuum, so it cannot be an exact symmetry on the lattice! The answer is that this lattice chiral transformation is not a symmetry of the measure of the lattice path integral:

$$\delta[d\psi][d\bar\psi] = [d\psi][d\bar\psi]\left(\text{Tr}\left[\gamma_5\left(1 + \frac{a}{2}D\right)\right] + \text{Tr}\left[\left(1 - \frac{a}{2}D\right)\gamma_5\right]\right)$$
$$= [d\psi][d\bar\psi] \times a\text{Tr}\gamma_5 D, \tag{4.149}$$

where I used the relation $d\det M/dx = \det[M]\text{Tr}M^{-1}dM/dx$. Unlike the tricky non-invariance of the fermion measure in the continuum under a $U(1)_A$ transformation—which only appears when the measure is properly regulated—here we have a perfectly ordinary integration measure and a transformation that gives rise to a Jacobean with a non-trivial phase (unless, of course, $\text{Tr}\gamma_5 D = 0$). To make sense, $\text{Tr}\gamma_5 D$ must map into the continuum anomaly ... and we have already seen that it does, from Eq. (4.146).

What remains is to prove the index theorem (Hasenfratz *et al.*, 1998; Lüscher, 1998), the lattice equivalent of Eq. (4.79). From Exercise 4.4.1.4 it follows that for states $|z\rangle$ satisfying $D|z\rangle = z|z\rangle$

$$\mathrm{Tr}\gamma_5 D = \sum_z \langle z|\gamma_5 D|z\rangle = 2N_f(n_+^{(2)} - n_-^{(2)}), \tag{4.150}$$

where $n_\pm^{(2)}$ are the number of positive and negative chirality states with eigenvalue $z = 2$. We also know that

$$0 = \mathrm{Tr}\gamma_5 = \sum_z \langle z|\gamma_5|z\rangle = (n_+ - n_-) + (n_+^{(2)} - n_-^{(2)}), \tag{4.151}$$

where $n_\pm$ are the number of $\pm$ chirality zero modes at $z = 0$. Therefore, we can write

$$\mathrm{Tr}\gamma_5 D = 2(n_- - n_+). \tag{4.152}$$

Substituting into Eq. (4.146) we arrive at the lattice index theorem,

$$(n_- - n_+) = \nu N_f, \tag{4.153}$$

which is equivalent to the continuum result Eq. (4.80), and provides an interesting definition for the topological charge of a lattice gauge field. A desirable feature of the overlap operator is the existence of exact zero-mode solutions in the presence of topology; it is also a curse for realistic simulations, since the zero modes make it difficult to sample different global gauge topologies. And while it cannot matter what the global topology of the Universe is, fixing the topology in a lattice QCD simulation gives rise to spurious effects that only vanish with a power of the volume (Edwards, 2002).

4.4.3   Chiral gauge theories: the challenge

Chiral fermions on the lattice make an interesting story whose final chapter, on chiral gauge theories, has barely been begun. It is a story that is both theoretically amusing and of practical importance, given the big role chiral symmetry plays in the Standard Model. I have tried to stress that the understanding of anomalies has been the key to both understanding the puzzling doubling problem and its resolution. In terms of practical application chiral formulations are more expensive than other fermion formulations, but they have advantages when studying physics where chirality plays an important role. For a recent review comparing different fermion formulations, see (Jansen, 2008).

While domain-wall and overlap fermions provide a way to represent any global chiral symmetry without fine tuning, it may be possible to attain these symmetries by fine tuning in theories with either staggered or Wilson fermions. In contrast, there is currently no practical way to regulate general non-abelian chiral gauge theories on the lattice. [There have been a lot of papers in this area, however, in the context of domain-wall/overlap/Ginsparg–Wilson fermions; for a necessarily incomplete list of references that gives you a flavor of the work in this direction, see (Kaplan, 1992; 1993; Narayanan and Neuberger, 1993; 1995; 1996; Kaplan and Schmaltz, 1996; Lüscher, 1999; Aoyama and Kikukawa, 1999; Lüscher, 2000b; Kikukawa and Nakayama, 2001; Kikukawa, 2002; Kadoh and Kikukawa, 2008; Hasenfratz and von Allmen, 2008)]. Thus, we lack a non-perturbative regulator for the Standard Model—but then again, we think perturbation theory suffices for understanding the Standard Model in the real world. If a solution to putting chiral gauge theories on the lattice proves to be a complicated and not especially enlightening enterprise, then it probably is not worth the effort (unless the LHC finds evidence for a strongly coupled chiral gauge theory!). However, if there is a compelling and physical route to such theories, that would undoubtedly be very interesting.

Even if eventually a lattice formulation of the Standard Model is achieved, we must be ready to address the sign problem associated with the phase of the fermion determinant in such theories. A sign problem has for years plagued attempts to compute properties of QCD at finite baryon chemical potential; the same physics is responsible for poor signal-to-noise ratio experienced when measuring correlators in multibaryon states. To date there have not been any solutions that solve this problem. We can at least take solace in the fact that the sign problems encountered in chiral gauge theories and in QCD at finite baryon density are not independent: After all, the Standard Model at fixed non-trivial $SU(2)$ topology with a large winding number can describe a transition from the QCD vacuum to a world full of iron atoms and neutrinos!

Acknowledgments

I have benefitted enormously in my understanding of lattice fermions from conversations with Maarten Golterman, Pilar Hernández, Yoshio Kikukawa, Martin Lüscher, Rajamani Narayanan, Herbert Neuberger, and Stephen Sharpe. Many thanks to them (and no blame for the content of this chapter). This work is supported in part by U.S. DOE grant No. DE-FG02-00ER41132.

Notes

1. I will use the familiar γ_5 in $3+1$ dimensions instead of Γ when there is no risk of ambiguity.
2. For this reason, the combined symmetry CP does not alter the particle content of a chiral theory, so that CP violation must arise from complex coupling constants.
3. With six flavors of quarks in QCD, one might ask why an $SU(3)$ chiral symmetry instead of $SU(2)$ or $SU(6)$. The point is that the chiral symmetry is broken by quark masses, and whether the breaking is large or small depends on the ratio m_q/Λ_{QCD}, where here Λ_{QCD} is some strong interaction scale in the 100s of MeV. The u and d quarks are much lighter than Λ_{QCD}, the strange quark is borderline, and the c, b, t quarks are much heavier. Therefore, $SU(2) \times SU(2)$ is a very good symmetry of QCD; $SU(3) \times SU(3)$ is a pretty good symmetry of QCD; but assuming chiral symmetry for the heavier quarks is not justified. Radiative corrections in the baryon sector go as $\sqrt{m_q/\Lambda_{QCD}}$ and so there even $SU(3) \times SU(3)$ does not appear to be very reliable.
4. The operator $\bar{s}d$ is removed by rediagonalizing the quark mass matrix and does not give rise to $K \to \pi\pi$.

5. Dimensional regularization is not a loophole, since chiral symmetry cannot be analytically continued away from odd space dimensions.
6. While in much of this chapter I will normalize gauge fields so that $D_\mu = \partial_\mu + iA_\mu$, in this section I need to put the gauge coupling back in. If you want to return to the nicer normalization, set the gauge coupling to unity, and put a $1/g^2$ factor in front of the gauge action.
7. Part of the content of this section comes directly from John Preskill's class notes on the strong interactions, available at his web page: http://www.theory. caltech.edu/~preskill/notes.html.
8. You may wonder about whether fermion condensates form to break the global $U(1)$ symmetry. Perhaps, but it seems unlikely. The lowest-dimension gauge invariant fermion condensates involve four fermion fields—such as $\langle \chi\chi\chi\psi \rangle$ or $\langle (\chi\psi)(\chi\psi)^\dagger \rangle$— which are all neutral under the $U(1)$ symmetry. Furthermore, there are arguments that a Higgs phase would not be distinguishable from a confining phase for this theory.
9. A lattice solution to Eq. (4.115) (the only solution in existence) is the overlap operator discovered by Neuberger (Neuberger, 1998*a*;*b*); it was a key reformulation of earlier work (Narayanan and Neuberger, 1993; 1995) on how to represent domain-wall fermions with an infinite extra dimension (and therefore exact chiral symmetry) in terms of entirely lower-dimensional variables. We will discuss overlap fermions and the Ginsparg–Wilson equation further in the next section.

References

Antonio, David J. *et al.* (2008). *Phys. Rev.*, **D77**, 014509.

Aoki, Sinya (1984). *Phys. Rev.*, **D30**, 2653.

Aoyama, Tatsumi, and Kikukawa, Yoshio (1999). hep-lat/9905003.

Callan, Curtis G., Jr., and Harvey, Jeffrey A. (1985). *Nucl. Phys.*, **B250**, 427.

Cheng, M. *et al.* (2010). *Phys. Rev.*, **D81**, 054510.

Chiu, Ting-Wai *et al.* (2009*a*). *PoS*, **LAT2009**, 034.

Chiu, Ting-Wai, Hsieh, Tung-Han, and Tseng, Po-Kai (2009*b*). *Phys. Lett.*, **B671**, 135–138.

Christ, N. (2006). *PoS*, **LAT2005**, 345.

Edwards, Robert G. (2002). *Nucl. Phys. Proc. Suppl.*, **106**, 38–46.

Endres, Michael G. (2008). *PoS*, **LATTICE2008**, 025.

Endres, Michael G. (2009*a*). *Phys. Rev.*, **D79**, 094503.

Endres, Michael G. (2009*b*). arXiv:0912.0207 [hep-lat].

Fleming, George Tamminga, Kogut, John B., and Vranas, Pavlos M. (2001). *Phys. Rev.*, **D64**, 034510.

Fujikawa, Kazuo (1979). *Phys. Rev. Lett.*, **42**, 1195.

Fujikawa, Kazuo (1980). *Phys. Rev.*, **D21**, 2848.

Furman, Vadim and Shamir, Yigal (1995). *Nucl. Phys.*, **B439**, 54–78.

Gavai, R. V. and Sharma, Sayantan (2009). *Phys. Rev.*, **D79**, 074502.

Giedt, Joel, Brower, Richard, Catterall, Simon, Fleming, George T., and Vranas, Pavlos (2008). arXiv:0807.2032 [hep-lat].

Giedt, Joel, Brower, Richard, Catterall, Simon, Fleming, George T., and Vranas, Pavlos (2009). *Phys. Rev.*, **D79**, 025015.

Ginsparg, Paul H. and Wilson, Kenneth G. (1982). *Phys. Rev.*, **D25**, 2649.

Golterman, Maarten and Shamir, Yigal (2000). *JHEP*, **09**, 006.

Golterman, Maarten and Shamir, Yigal (2003). *Phys. Rev.*, **D68**, 074501.

Golterman, Maarten, Sharpe, Stephen R., and Singleton, Robert L., Jr. (2005). *Phys. Rev.*, **D71**, 094503.

Golterman, Maarten F. L., Jansen, Karl, and Kaplan, David B. (1993). *Phys. Lett.*, **B301**, 219–223.

Hasenfratz, Peter, Laliena, Victor, and Niedermayer, Ferenc (1998). *Phys. Lett.*, **B427**, 125–131.

Hasenfratz, Peter and von Allmen, Reto (2008). *JHEP*, **02**, 079.

Hernández, Pilar, Jansen, Karl, and Lüscher, Martin (1999). *Nucl. Phys.*, **B552**, 363–378.

Itzykson, C. and Zuber, J. B. (1980). *Quantum field theory.* : McGrawHill, New York 705 pp. (International Series in Pure and Applied Physics).

Jansen, Karl (2008). *PoS*, **LATTICE2008**, 010.

Jansen, Karl, and Schmaltz, Martin (1992). *Phys. Lett.*, **B296**, 374–378.

Kadoh, Daisuke, and Kikukawa, Yoshio (2008). *JHEP*, **05**, 095.

Kaplan, David B. (1992). *Phys. Lett.*, **B288**, 342–347.

Kaplan, D. B. (1993). *Nucl. Phys. Proc. Suppl.*, **30**, 597–600.

Kaplan, David B., Lepeintre, Francois, and Schmaltz, Martin (1997). *Phys. Rev.*, **D56**, 7193–7206.

Kaplan, David B., and Schmaltz, Martin (1996). *Phys. Lett.*, **B368**, 44–52.

Kaplan, David B., and Schmaltz, Martin (2000). *Chin. J. Phys.*, **38**, 543–550.

Kennedy, A. D. (2006). hep-lat/0607038.

Kikukawa, Yoshio (2002). *Phys. Rev.*, **D65**, 074504.

Kikukawa, Yoshio, and Nakayama, Yoichi (2001). *Nucl. Phys.*, **B597**, 519–536.

Kikukawa, Yoshio, and Noguchi, Tatsuya (1999). hep-lat/9902022.

Lüscher, Martin (1998). *Phys. Lett.*, **B428**, 342–345.

Lüscher, Martin (1999). *Nucl. Phys.*, **B549**, 295–334.

Lüscher, Martin (2000*a*). hep-th/0102028.

Lüscher, Martin (2000*b*). *Nucl. Phys.*, **B568**, 162–179.

Narayanan, Rajamani (2001). *Int. J. Mod. Phys.*, **A16S1C**, 1203–1206.

Narayanan, Rajamani, and Neuberger, Herbert (1993). *Phys. Rev. Lett.*, **71**, 3251–3254.

Narayanan, Rajamani, and Neuberger, Herbert (1995). *Nucl. Phys.*, **B443**, 305–385.

Narayanan, R., and Neuberger, H. (1996). *Nucl. Phys. Proc. Suppl.*, **47**, 591–595.

Neuberger, Herbert (1998*a*). *Phys. Lett.*, **B417**, 141–144.

Neuberger, Herbert (1998*b*). *Phys. Lett.*, **B427**, 353–355.

Neuberger, Herbert (1998*c*). *Phys. Rev.*, **D57**, 5417–5433.

Ohta, Shigemi, for the RBC and UKQCD Collaborations, (2009). arXiv:0910.5686 [hep-lat].

Preskill, John (1985). Lectures presented at the 1985 Les Houches Summer School, Les Houches, France, Jul 1–Aug 8, 1985.

Shamir, Yigal (1993). *Nucl. Phys.*, **B406**, 90–106.

Sharpe, Stephen R. (2007). arXiv:0706.0218 [hep-lat].

Torok, Aaron *et al.* (2009). arXiv:0907.1913 [hep-lat].

Wess, J., and Bagger, J. (1992). *Supersymmetry and supergravity.* princeton Unverscity press Princeton, USA (1992) 259 pp.

Yamazaki, Takeshi (2009). *Phys. Rev.*, **D79**, 094506.

5
Lattice QCD at non-zero temperature and baryon density

Owe PHILIPSEN

Institut für Theoretische Physik, Westfälische Wilhelms-Universität Münster,
Wilhelm-Klemm-Str.9, 48149 Münster, Germany

Introduction

One of the key features of QCD is asymptotic freedom. While the theory is amenable to perturbation theory at large momenta, it is non-perturbative for energy scales $\lesssim 1$ GeV and lattice QCD is the only known method for first-principles calculations in this regime. The running of the coupling with momentum scale immediately implies the existence of different states of nuclear matter at asymptotically high densities or temperatures: when the coupling on the scales of temperature T or baryon chemical potential μ_B is sufficiently weak, confinement gets lost. At a critical temperature $T_c \sim 200$ MeV, QCD predicts a transition between the familiar confined hadron physics and a deconfined phase of quark gluon plasma (QGP). At the same temperature, chiral symmetry gets restored. A thermal environment with sufficiently high temperatures for a QCD plasma certainly existed during the early stages of the Universe, which passed through the quark–hadron transition on its way to its present state. On the other hand, for high densities and low temperatures, there is an attractive interaction between quarks to form Cooper pairs, and exotic non-hadronic phases such as a color superconductor have been predicted. Such physics might be realized in the cores of compact stars.

Current and future heavy-ion collision experiments are attempting to create hot and dense quark-gluon plasma at RHIC (BNL), LHC (CERN) and FAIR (GSI). These studies will have a bearing far beyond QCD in the context of early Universe and astro-particle physics. Many other prominent features of the observable Universe, such as the baryon asymmetry or the seeding for structure formation, have been determined primordially in hot plasmas described by non-abelian gauge theories. The QCD plasma serves as a prototype also for those, since it is the only one we can hope to produce in laboratory experiments. There is thus ample motivation to provide controlled theoretical predictions for the physics of hot and dense QCD. As we shall see, this turns out to be even more challenging than QCD in the vacuum.

In this chapter the focus is on basic concepts and methods, with a few exemplary results for illustration only and a very incomplete list of references. For summaries of the latest results and literature lists, see the annual proceedings of the LATTTICE conferences.

5.1 Aspects of finite-temperature field theory in the continuum

5.1.1 Statistical mechanics reminder

We wish to describe a system of particles in some volume V that is in thermal contact with a heat bath at temperature T. Associated with the particles may be a set of conserved charges $N_i, i = 1, 2, \ldots$ (such as particle number, electric charge, baryon number, etc.). In statistical mechanics there is a choice of ensembles to describe this situation. In the canonical ensemble, V and the N_i are kept fixed while the system exchanges energy with the heat bath. In the grand canonical description, exchange of particles with the heat bath is also allowed. Of course, both ensembles provide the same description for physical observables in the thermodynamic limit, $V \to \infty$, but for specific situations one or the other may be more appropriate. In quantum field theory, the most direct description is in terms of the grand canonical ensemble. Its density operator and partition function are given as

$$\rho = e^{-\frac{1}{T}(H - \mu_i N_i)}, \qquad Z = \hat{\mathrm{Tr}}\rho, \qquad \hat{\mathrm{Tr}}(\ldots) = \sum_n \langle n|(\ldots)|n\rangle, \tag{5.1}$$

where μ_i are chemical potentials for the conserved charges, and the quantum-mechanical trace is a sum over all energy eigenstates of the Hamiltonian. Thermodynamic averages for an observable O are then obtained as $\langle O \rangle = Z^{-1}\hat{\mathrm{Tr}}(\rho O)$.

From the partition function, all other thermodynamic equilibrium quantities follow by taking appropriate derivatives. In particular, the (Helmholtz) free energy, pressure, entropy, mean values of charges and energy are obtained as

$$F = -T \ln Z,$$
$$p = \frac{\partial (T \ln Z)}{\partial V},$$
$$S = \frac{\partial (T \ln Z)}{\partial T},$$
$$\bar{N}_i = \frac{\partial (T \ln Z)}{\partial \mu_i},$$
$$E = -pV + TS + \mu_i \bar{N}_i. \tag{5.2}$$

Since the free energy is known to be an extensive quantity, $F = fV$, and we are interested in the thermodynamic limit, it is often more convenient to consider the corresponding densities,

$$f = \frac{F}{V}, \quad p = -f, \quad s = \frac{S}{V}, \quad n_i = \frac{\bar{N}_i}{V}, \quad \epsilon = \frac{E}{V}. \tag{5.3}$$

5.1.2 QCD at finite temperature and quark density

Let us now consider the grand canonical partition function of QCD. The derivation of its path-integral representation is discussed in detail in the textbooks (Kapusta and Gale, 2006) and requires discretized Euclidean time followed by a continuum limit. We shall thus derive it more conveniently in its lattice version in Section 5.2.1 and just quote the result here,

$$Z(V, \mu_f, T; g, m_f) = \hat{\mathrm{Tr}}\left(e^{-(H - \mu_f Q_f)/T}\right) = \int DA\, D\bar{\psi}\, D\psi\, e^{-S_g[A_\mu]}\, e^{-S_f[\bar{\psi}, \psi, A_\mu]}, \tag{5.4}$$

with the Euclidean gauge and fermion actions

$$S_g[A_\mu] = \int_0^{1/T} dx_0 \int_V d^3x\, \frac{1}{2}\mathrm{Tr}\, F_{\mu\nu}(x)F_{\mu\nu}(x),$$

$$S_f[\bar{\psi}, \psi, A_\mu] = \int_0^{1/T} dx_0 \int_V d^3x \sum_{f=1}^{N_f} \bar{\psi}_f(x)\left(\gamma_\mu D_\mu + m_f - \mu_f \gamma_0\right)\psi_f(x). \tag{5.5}$$

The index f labels the different quark flavors, and the covariant derivative contains the gauge coupling g,

$$D_\mu = (\partial_\mu - igA_\mu), \quad A_\mu = T^a A^a_\mu(x), \quad a = 1, \ldots N^2 - 1,$$

$$F_{\mu\nu}(x) = \frac{i}{g}[D_\mu, D_\nu]. \tag{5.6}$$

The thermodynamic limit is obtained by sending the spatial three-volume $V \to \infty$. The difference to the Euclidean path integral at $T = 0$ is that the temporal direction is compactified and kept finite, i.e. the theory lives on a torus whose compactification radius defines the inverse temperature, $1/T$. The path integral is to be evaluated with periodic and antiperiodic boundary conditions in the temporal direction for bosons and fermions, respectively,

$$A_\mu(\tau, \mathbf{x}) = A_\mu(\tau + \frac{1}{T}, \mathbf{x}), \qquad \psi(\tau, \mathbf{x}) = -\psi(\tau + \frac{1}{T}, \mathbf{x}), \tag{5.7}$$

which ensures Bose–Einstein statistics for bosons and the Pauli principle for fermions. Clearly, the path integral for vacuum QCD on infinite four-volume is smoothly recovered from this expression for $T \to 0$.

The partition function depends on the external macroscopic parameters T, V, μ_f, as well as on the microscopic parameters like quark masses and the coupling constant. The conserved quark numbers corresponding to the chemical potentials μ_f are

$$Q_f = \bar{\psi}_f \gamma_0 \psi_f. \tag{5.8}$$

Since the QCD phase transition happens on a scale ~ 200 MeV, we neglect the c, b, t quarks or treat them non-relativistically when needed. We will thus consider mostly two and three flavors of quarks, and always take $m_u = m_d$. The case $m_s = m_{u,d}$ is then denoted by $N_f = 3$, while $N_f = 2 + 1$ implies $m_s \neq m_{u,d}$. Furthermore, we will couple all flavors to the same chemical potential μ unless otherwise stated.

The only, but fundamental, difference compared to the $T = 0$ path integral is due to the compactness of the time direction. For example, the Fourier expansion of the fields on a finite volume $V = L^3$ is

$$A_\mu(\tau, \mathbf{x}) = \frac{1}{\sqrt{VT}} \sum_{n=-\infty}^{\infty} \sum_{\mathbf{p}} e^{i(\omega_n \tau + \mathbf{p} \cdot \mathbf{x})} A_{\mu,n}(p), \quad \omega_n = 2n\pi T,$$

$$\psi(\tau, \mathbf{x}) = \frac{1}{\sqrt{V}} \sum_{n=-\infty}^{\infty} \sum_{\mathbf{p}} e^{i(\omega_n \tau + \mathbf{p} \cdot \mathbf{x})} \psi_n(p), \quad \omega_n = (2n+1)\pi T. \tag{5.9}$$

The allowed momenta are $p_i = (2\pi n_i)/L$, where $n_i \in \mathbb{Z}$. The normalization factors in front are chosen such that the Fourier modes, called the Matsubara modes, are dimensionless. In the thermodynamic limit the momenta become continuous

$$\frac{1}{V} \sum_{n_1, n_2, n_3} \xrightarrow{V \to \infty} \int \frac{d^3 p}{(2\pi)^3}, \tag{5.10}$$

but the Matsubara frequencies ω_n stay discrete due to the compactness of the time direction. Note that bosons/fermions have even/odd Matsubara frequencies, respectively, to ensure the bondary conditions Eq. (5.7). We thus obtain modified Feynman rules (Kapusta and Gale, 2006) compared to the vacuum. The zero components of four vectors contain the Matsubara frequencies ω_n, and the four-momentum integration associated with internal lines gets replaced by a 3D integral and a Matsubara sum,

$$\sum_{n=-\infty}^{\infty} \int \frac{d^3p}{(2\pi)^3}. \tag{5.11}$$

5.1.3 Perturbative expansion

Similar to $T = 0$, we can take the path integral as the starting point for a perturbative expansion. For some generic field ϕ, we proceed just as in the vacuum and split the action in a free and an interacting part, $S = (S_0 + S_i)$, in order to expand in powers of the interaction, and hence the coupling constant,

$$Z = N \int D\phi \, e^{-(S_0 + S_i)} = N \int D\phi \, e^{-S_0} \sum_{l=0}^{\infty} \frac{(-1)^l}{l!} S_i^l. \tag{5.12}$$

Thus, we get for the log of the partition function

$$\ln Z = \ln Z_0 + \ln Z_i = \ln \left(N \int D\phi \, e^{-S_0} \right) + \ln \left(1 + \sum_{l=1}^{\infty} \frac{(-1)^l}{l!} \frac{\int D\phi \, e^{-S_0} S_i^l}{\int D\phi \, e^{-S_0}} \right). \tag{5.13}$$

We shall evaluate the ideal gas part, $\ln Z_0$, in the next section. The corrections to the non-interacting case are the sum of all loop diagrams without external legs. When evaluating loop diagrams, UV divergences are encountered and the renormalization program has to be performed. Whatever regularization and renormalization is necessary and sufficient at zero temperature is also necessary and sufficient at finite temperature. This is not surprising, since the ultraviolet structure of the theory, i.e. the microscopic short-distance regime, is unchanged by the introduction of macroscopic parameters like temperature and baryon chemical potential.

On the other hand, the infrared structure of the theory does get changed. This is easily seen by considering the inverse propagator for a bosonic degree of freedom,

$$p^2 + m^2 = \omega_n^2 + \mathbf{p}^2 + m^2 = (2n\pi T)^2 + \mathbf{p}^2 + m^2. \tag{5.14}$$

The Matsubara frequencies act like effective thermal masses $\sim T$ for all modes with $n \neq 0$. In the case of fermions, there are only non-zero modes. Hence, all fermionic and all non-zero bosonic modes are infrared-safe even in the limit of vanishing bare mass, $m = 0$. By contrast, for the bosonic zero mode the inverse propagator is $\mathbf{p}^2 + m^2$, which is identical to that of a 3D field theory. Thus, 4D Yang–Mills theory at finite temperature contains in the zero-mode sector the 3D Yang–Mills theory, which is a confining theory. Its propagator is infrared divergent for $m = 0$, and the divergence is worse than in 4D. This points at non-perturbative behavior and is at the heart of the Linde problem of finite-temperature perturbation theory, Section 5.1.6.

Let us briefly discuss qualitatively the self-energy corrections to the gluon propagator. When computing the self-energy diagrams, one finds that for the color electric field A_0 a gluon mass is generated, the electric or Debye mass. To leading order it is

$$m_E^{LO} = \left(\frac{N}{3} + \frac{N_f}{6}\right)^{1/2} gT. \tag{5.15}$$

Thus, color electric fields get screened by the medium at finite temperature, whereas for color magnetic fields A_i one finds the corresponding magnetic mass $m_M = 0$ at this order. However, there are contributions to $m_M \sim g^2 T$ starting at the two-loop level. As we shall see in Section 5.1.6, all loops contribute equally to the coefficient, such that the calculation of the magnetic screening mass is an entirely non-perturbative problem.

5.1.4   Ideal gases

Ideal, i.e. non-interacting, gases of particles are important model systems to guide our intuition. It is therefore instructive to see how their thermal properties are derived from the path integral. Let us consider a bosonic system of real scalar fields, for simplicity. After Fourier transformation, the action reads

$$S_0 = \frac{1}{2T^2} \sum_{n=-\infty}^{\infty} \sum_{\mathbf{p}} (\omega_n^2 + \omega^2)\phi_n(p)\phi_n^*(p), \tag{5.16}$$

where we abbreviate $\omega = \sqrt{\mathbf{p}^2 + m^2}$ and $\phi_n^*(p) = \phi_{-n}(p)$ for a real scalar field. The partition function then factorizes into a product over all Matsubara modes,

$$Z_0 = N \prod_{n=-\infty}^{\infty} \prod_{\mathbf{p}} \int d\phi_n \, \exp[-\frac{1}{2T^2}(\omega_n^2 + \omega^2)\phi_n(p)\phi_n^*(p)]$$

$$= N\prod_{\mathbf{p}} \int d\phi_0 \exp[-\frac{1}{2T^2}(\omega_0^2 + \omega^2)\phi_0^2(p)] \prod_{n>0} \int d\phi_n \, d\phi_n^* \exp[-\frac{1}{2T^2}(\omega_n^2 + \omega^2)\phi_n(p)\phi_n^*(p)]$$

$$= N' \prod_{\mathbf{p}} (2\pi)^{1/2} \left(\frac{\omega_0^2 + \omega^2}{T^2}\right)^{-\frac{1}{2}} \prod_{n>0} \int d|\phi_n| \, |\phi_n| \exp[-\frac{1}{2T^2}(\omega_n^2 + \omega^2)|\phi_n|^2]$$

$$= N' \prod_{\mathbf{p}} (2\pi)^{1/2} \left(\frac{\omega_0^2 + \omega^2}{T^2}\right)^{-\frac{1}{2}} \prod_{n>0} \frac{1}{4} \frac{\omega_n^2 + \omega^2}{T^2} = N'' \prod_{n=-\infty}^{\infty} \prod_{\mathbf{p}} \left(\frac{\omega_n^2 + \omega^2}{T^2}\right)^{-\frac{1}{2}}, \tag{5.17}$$

where we have changed to polar coordinates in the third line and absorbed constant numerical factors into the normalization in front. Thus, just as for zero temperature, the partition function of the free theroy can be formally written as

$$Z_0 = N \int D\phi \, e^{-S(\phi)} = N''(\det\Delta^{-1})^{-1/2}, \tag{5.18}$$

where $\Delta^{-1} = (\omega_n^2 + \omega^2)/T^2$ is the inverse propagator in momentum space.

Since N'' is (V, T) independent, it will not contribute to thermodynamics, Eqs. (5.2), and may be dropped. (It will contribute, e.g., to the entropy as an additive constant, which we are allowed to set to zero by the third law of thermodynamics). Thus, we obtain

$$\ln Z_0 = -\frac{1}{2} \sum_{n=-\infty}^{\infty} \sum_{\mathbf{p}} \ln \frac{\omega_n^2 + \omega^2}{T^2}. \tag{5.19}$$

The Matsubara sum is performed using the formulae

$$\ln\left[(2\pi n)^2 + \frac{\omega^2}{T^2}\right] = \int_1^{\omega^2/T^2} \frac{d\theta^2}{\theta^2 + (2\pi n)^2} + \ln(1 + (2\pi n)^2),$$

$$\sum_{n=-\infty}^{\infty} \frac{1}{n^2 + (\frac{\theta}{2\pi})^2} = \frac{2\pi^2}{\theta}\left(1 + \frac{2}{e^\theta - 1}\right), \tag{5.20}$$

leading to the expression

$$\ln Z_0 = -\sum_{\mathbf{p}} \int_1^{\omega/T} d\theta \left(\frac{1}{2} + \frac{1}{e^\theta - 1}\right) + \text{T-indep.}$$

$$\stackrel{V\to\infty}{\longrightarrow} V \int \frac{d^3 p}{(2\pi)^3} \left[\frac{-\omega}{2T} - \ln\left(1 - e^{-\frac{\omega}{T}}\right)\right]. \tag{5.21}$$

One observes that the integral over the first term diverges in the UV since $\omega \sim |\mathbf{p}|$. This, however, is familiar from the quantum-mechanical harmonic oscillator. The term corresponds to the zero-point energy and gives a divergent vacuum contribution to the energy and pressure for $T \to 0$. The renormalization condition is that the vacuum has zero pressure,

$$p_{\text{phys}}(T) = p(T) - p(T = 0). \tag{5.22}$$

The final result then is the familar form of the partition function for a free gas of spinless bosons,

$$\ln Z_0 = -V \int \frac{d^3 p}{(2\pi)^3} \ln\left(1 - e^{-\frac{\omega}{T}}\right). \tag{5.23}$$

For the massless case, $m = 0$, the momentum integral can be done exactly and one finds the famous result for the pressure of one bosonic degree of freedom at zero chemical potential,

$$p = \frac{\pi^2}{90} T^4. \tag{5.24}$$

Doing the same calculation for non-abelian gauge fields A_μ^a, each field component corresponds to one bosonic mode and we have to sum over $a = 1, \ldots N^2 - 1$ and $\mu = 1 \ldots 4$ in Eq. (5.16). Hence, we get a factor of $4(N^2 - 1)$ in front of Eq. (5.23). Going through a similar sequence of steps for a free Dirac field, one finds instead

$$\ln Z_0 = 2V \int \frac{d^3 p}{(2\pi)^3} \left[\ln \left(1 + e^{-\frac{\omega - \mu}{T}} \right) + \ln \left(1 + e^{-\frac{\omega + \mu}{T}} \right) \right]. \tag{5.25}$$

The factor of two in front accounts for the two spin states of a fermion, and there are two terms from a fermion and an antifermion in the brackets. Recall that for gauge theories gauge fixing is necessary in order to invert the two-point function. Thus, for the free photon or Yang–Mills gas, two of the four bosonic Lorentz degrees of freedom get cancelled by the corresponding ghost contributions, such that we obtain two polarization states per massless vector particle, as it should be.

All of this can be summarized by the one-particle partition function

$$\ln Z_i^1(V, T) = \eta V \nu_i \int \frac{d^3 p}{(2\pi)^3} \ln(1 + \eta e^{-(\omega_i - \mu_i)/T}), \tag{5.26}$$

where $\eta = -1$ for bosons and $\eta = 1$ for fermions and ν_i gives the number of degrees of freedom for the particle i. The momentum integration for one massless fermionic degree of freedom gives the pressure

$$p = \frac{7}{8} \frac{\pi^2}{90} T^4. \tag{5.27}$$

To compute the pressure of an ideal gas of gluons and massless quarks, we now simply have to count the degrees of freedom to obtain (two polarization states for each of $(N^2 - 1)$ gluons, two polarization states per quark, three colors per quark and the same for antiquarks)

$$\frac{p}{T^4} = \left(2(N^2 - 1) + 4NN_f \frac{7}{8} \right) \frac{\pi^2}{90}. \tag{5.28}$$

This is the Stefan–Boltzmann limit of the QCD pressure that is valid for vanishing coupling, i.e. in the limit $T \to \infty$.

5.1.5   The hadron resonance gas model

There is another ideal gas system that is useful to model the low-temperature behavior of QCD. The modelling assumption is that, at low temperatures when we still have confinement, the QCD partition function is close to that of a free gas of hadrons. While strong coupling effects are responsible for confinement of the quarks and gluons, the interactions between the hadrons are considerably weaker and may even be neglected if the gas is sufficiently dilute. In this case, the QCD partition function factorizes into one-particle partition functions $Z_i^1(V, T)$,

$$\ln Z(V, T) \approx \sum_i \ln Z_i^1(V, T). \tag{5.29}$$

The particles to be inserted are the known hadrons and hadron resonances (for QCD purposes electroweak decays are to be neglected). Taking the ν_i and the energies from the particle data booklet, one can supply this formula with hundreds of hadron resonances, do the momentum integral and obtain a thermodynamic pressure that compares remarkably well with Monte Carlo simulations of QCD (Karsch *et al.*, 2003). We shall see in Section 5.2.9 that this is no accident, but that the hadron resonance gas model can actually be derived from lattice QCD as an effective theory for the strong-coupling regime.

5.1.6   The Linde problem

As an example to illustrate the Linde problem, consider the $l+1$-loop contribution to the QCD pressure, Fig. 5.1, with $2l$ three gluon vertices and $3l$ propagators. The Matsubara sums over the internal lines, Eq. (5.11), contain a term coming exclusively from the zero modes. Dispensing with the index structures, its contribution is given by the 3D loop integral

$$I \sim g^{2l} \left(T \int d^3 p \right)^{l+1} p^{2l}(p^2 + m^2)^{-3l}, \tag{5.30}$$

where we have introduced a mass m by hand as an infrared regulator. The momentum integrals have to be performed up to a scale T, which appears as an effective UV cut-off after doing the Matsubara sum over the other modes. Parametrically, the integral then evaluates to

$$I \sim \begin{cases} g^6 T^4 \ln(T/m) & \text{for} \quad l = 3 \\ g^6 T^4 (g^2 T/m)^{l-3} & \text{for} \quad l > 3. \end{cases} \tag{5.31}$$

It is therefore infrared divergent for $m \to 0, l > 2$ and the usual bare perturbation theory breaks down. However, as we discussed before, mass scales are generated dynamically by loop corrections to the propagators. Evaluating the integral with those mass scales corresponds to an effective resummation of the perturbative series (the diagram now contains loop insertions that would make it formally higher order in bare perturbation theory). If our internal lines contain A_0 fields, we need to do so with $m_E \sim gT$ and observe that as a consequence our series contains odd powers of g in addition to the logarithms. Summing up mass corrections for the A_i amounts

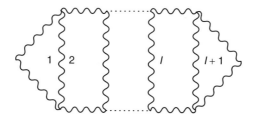

Fig. 5.1 $l+1$-loop Feynman diagram contributing to the pressure.

to an insertion of $m_M \sim g^2 T$, and we see that the gauge coupling drops out of the effective expansion parameter entirely for $l > 3$. Hence,   *all* loop orders contribute to the pressure at order g^6, which thus is a fully non-perturbative problem.

The same problem occurs when calculating other quantities, with the order of the breakdown depending on the observable. For example, for the electric mass m_E it appears already at NLO and for the magnetic mass m_M even at the leading non-zero order. Thus, at finite temperature, perturbation theory only works to some finite order that depends on the observable. Note that this is true no matter how weak the coupling g. Even electroweak physics at finite temperatures is inherently non-perturbative, and perturbative answers are only useful to the extent that the calculable orders are sufficient for a good approximation of physical results.

5.2   The lattice formulation for zero baryon density

5.2.1   Action and partition function

Let us now consider the lattice formulation of $SU(N)$ pure gauge theory on a hypercubic lattice, $N_s^3 \times N_\tau$, with lattice spacing a and the standard Wilson gauge action

$$S_g[U] = \sum_x \sum_{1 \le \mu < \nu \le 4} \beta \left(1 - \frac{1}{3} \mathrm{ReTr} U_p \right), \tag{5.32}$$

where $U_p = U_\mu(x) U_\nu(x + a\hat{\mu}) U_\mu^\dagger(x + a\hat{\nu}) U_\nu^\dagger(x)$ is the elementary plaquette, and the bare lattice and continuum gauge couplings are related by $\beta = 2N/g^2$. As usual, we impose periodic boundary conditions in all directions, $U_\mu(\tau, \mathbf{x}) = U_\mu(\tau + N_\tau, \mathbf{x}), U_\mu(\tau, \mathbf{x}) = U_\mu(\tau, \mathbf{x} + N_s)$. The connection between zero-and finite-temperature physics is most easily exhibited by the transfer matrix, which relates the path-integral representation of a Euclidean lattice field theory to the Hamiltonian formulation. A transfer matrix element between two timeslices τ and $\tau + 1$ is given by (Montvay and Münster, 1994),

$$T[U_i(\tau + 1), U_i(\tau)] = e^{-aH} = \int DU_0(\tau) \; \exp -L[U_i(\tau + 1), U_0(\tau), U_i(\tau)], \tag{5.33}$$

where the action is written as a sum over timeslices,

$$S_g = \sum_\tau L[U_i(\tau + 1), U_0(\tau), U_i(\tau)],$$

$$L[U_i(\tau + 1), U_0(\tau), U_i(\tau)] = \frac{1}{2} L_1[U_i(\tau + 1)] + \frac{1}{2} L_1[U_i(\tau)]$$

$$+ L_2[U_i(\tau + 1), U_0(\tau), U_i(\tau)]$$

$$L_1[U_i(\tau)] = -\frac{\beta}{N} \sum_{p(\tau)} \mathrm{ReTr} U_p,$$

$$L_2[U_i(\tau+1), U_0(\tau), U_i(\tau)] = -\frac{\beta}{N} \sum_{p(\tau,\tau+1)} \mathrm{Re}\,\mathrm{Tr}\, U_p, \tag{5.34}$$

and $p(\tau), U_i(\tau)$ denote all spatial plaquettes and links contained in the timeslice τ with arguments $\mathbf{x}$ suppressed, while $p(\tau+1,\tau), U_0(\tau)$ are time-like plaquettes and links connecting the timeslices τ and $\tau+1$. The partition function is now conveniently expressed as

$$Z = \int \prod_\tau (DU_i(\tau, \mathbf{x})\, T[U_i(\tau+1), U_i(\tau)]) = \hat{\mathrm{Tr}}(T^{N_\tau}) = \hat{\mathrm{Tr}}(e^{-N_\tau a H}). \tag{5.35}$$

Note that the periodic boundary condition in the temporal direction is necessary for the trace operation in order to have identical states $|n\rangle$ on the timeslices 1 and N_τ. In this discretized form we can immediately see that Z is equivalent to the partition function of a thermal system if we identify

$$\frac{1}{T} \equiv a N_\tau. \tag{5.36}$$

The thermal expectation value of an observable is then

$$\langle O \rangle = Z^{-1} \hat{\mathrm{Tr}}(e^{-\frac{H}{T}} O) = Z^{-1} \sum_n \langle n|T^{N_\tau} O|n\rangle = \frac{\sum_n \langle n|O|n\rangle\, e^{-a N_\tau E_n}}{\sum_n e^{-a N_\tau E_n}}. \tag{5.37}$$

As in the continuum, we are interested in the thermodynamic limit and hence $N_s \to \infty$, while keeping $a N_\tau = 1/T$ finite.

In this form we easily see the connection to $T = 0$ physics: projection on the vacuum expectation value is achieved by taking N_τ to infinity,

$$\langle 0|O|0\rangle = \lim_{N_\tau \to \infty} \frac{\sum_n \langle n|O|n\rangle\, e^{-a N_\tau (E_n - E_0)}}{\sum_n e^{-a N_\tau (E_n - E_0)}}. \tag{5.38}$$

In order to describe our gauge theory at finite temperatures, we simply need to dispense with this step. In that case the expectation value receives contributions from all eigenstates $|n\rangle$ with Boltzmann weights $e^{-\frac{E_n}{T}}$, Eq. (5.37). Hence, all lattices with finite N_τ (in particular those used for numerical simulations!) correspond to a finite temperature $T = 1/(a N_\tau)$, and for a description of vacuum physics sufficiently large N_τ is required.

On a Euclidean lattice, space and time directions are in principle indistinguishable as long as we provide them with equal boundary conditions. For some applications it is useful to define a Hamiltonian that translates states spatially, say in the z-direction, defined through a transfer matrix between adjacent z-slices, and write the partition function in terms of that Hamiltonian (with $U(z)$ denoting $\{U_\mu(z)|\mu \neq 3\}$),

$$T[U(z+1), U(z)] \equiv e^{-a H_z}, \quad Z = \mathrm{Tr}(e^{-a N_z H_z}). \tag{5.39}$$

For thermal physics, we want to take $N_{x,y,z} \to \infty$ and keep N_τ, which is now hidden in the definition of the Hamiltonian, finite. It is thus equivalent to the "zero temperature" ($N_z \to \infty$) physics of the Hamiltonian H_z, which acts on states defined on a space with two infinite and one finite, compactified directions. Clearly, H_z has reduced symmetry compared to H. In either description, from a calculational point of view finite temperature physics on a Euclidean space-time lattice is simply a finite-size effect: the Boltzmann-weighted sums, i.e. thermal effects, become noticeable once the temporal lattice size is small enough for the system to be sensitive to the boundary.

Adding fermions is now the same as for zero temperature and does not change this picture. Once a suitable action

$$S_f = \sum_{x,y} \bar{\psi}(x) M_{xy}(m_f)\, \psi(y) \tag{5.40}$$

has been selected, the Gauss integral can be done and, minding the appropriate boundary conditions in the temporal direction, we end up with the partition function

$$Z(N_s, N_\tau; \beta, m_f) = \int DU \prod_f \det M(m_f)\, \mathrm{e}^{-S_g[U]},$$

$$U_\mu(\tau, \mathbf{x}) = U_\mu(\tau + N_\tau, \mathbf{x}),$$

$$\psi(\tau, \mathbf{x}) = -\psi(\tau + N_\tau, \mathbf{x}). \tag{5.41}$$

For definiteness, the lattice action for N_f degenerate Wilson fermions is given by

$$S_f^W = \frac{1}{2a} \sum_{x,\mu,f} a^4\, \bar{\psi}_f(x)[(\gamma_\mu - r)U_\mu(x)\psi_f(x+\hat{\mu}) - (\gamma_\mu + r)U_\mu^\dagger(x-\hat{\mu})\psi_f(x-\hat{\mu})]$$

$$+ \left(m + 4\frac{r}{a}\right)\sum_{x,f} a^4\, \bar{\psi}_f(x)\psi_f(x). \tag{5.42}$$

5.2.2   Tuning temperature and the continuum limit

According to Eq. (5.36), one way of tuning temperature on the lattice is by choosing N_τ. But this is not satisfactory as this is only possible in discrete steps, and for realistic lattice spacings these are much too coarse. Hence, common practice in numerical simulations is to keep N_τ fixed and instead vary the lattice spacing a via the lattice coupling, $\beta = 2N/g^2(a)$, thus affecting temperature. This is a marked difference from simulations at zero temperature. In particular, simulation points at different temperature correspond to different lattice spacings and thus have different cutoff effects.

The relation between a and β is given by the renormalization group. For example, for the lambda parameter in lattice regularization we have to leading order in perturbation theory for $SU(3)$

$$a\Lambda_L = \left(\frac{6b_0}{\beta}\right)^{-b_1/2b_0^2} e^{-\frac{\beta}{12b_0}},$$

$$b_0 = \frac{1}{16\pi^2}\left(11 - \frac{2}{3}N_f\right), \quad b_1 = \left(\frac{1}{16\pi^2}\right)^2\left[102 - \left(10 + \frac{2}{3}\right)N_f\right]. \quad (5.43)$$

However, perturbation theory is not convergent for accessible lattice spacings. The way out is to express the calculated observables in terms of known physical quantities of the same mass dimension. For example, if we want to compute the critical temperature of the QCD phase transition, T_c, by keeping N_τ fixed and tuning β, the location of a phase transition will be given as a critical coupling β_c. The critical temperature in units of a hadron mass is then

$$\frac{T_c}{m_H} = \frac{1}{a_c m_H N_\tau} = \frac{1}{a(\beta_c)m_H N_\tau}. \quad (5.44)$$

In order to set the physical scale for T_c, we then have to calculate the zero-temperature hadron mass in lattice units at the value of the critical coupling, $(am_H)(\beta_c)$.

This procedure is good as long as we are able to simulate physical quark masses. For most practitioners, this is not yet the case, and furthermore there are many interesting theoretical questions concerning regimes with unphysical quark masses, such as the quenched and the chiral limits. In order to set a scale in those cases, one uses quantities that display only little sensitivity to the quark mass values. Examples are the string tension or the Sommer scale,

$$\frac{T}{\sqrt{\sigma}} = \frac{1}{a\sqrt{\sigma}N_\tau}, \quad \sigma \approx 425 \text{ MeV}; \quad Tr_0 = \frac{r_0}{aN_\tau}, \quad r^2\frac{dV(r)}{dr} = 1.65. \quad (5.45)$$

In order to take the continuum limit, we now have to compute expectation values of our observables of interest, $\langle O \rangle(\beta, m_f)$, in the thermodynamic limit for various lattice spacings a. Then, we can extrapolate to $a \to 0$. Since the continuum limit has to be taken along lines of constant physics, i.e. keeping temperature and mass ratios fixed, this is equivalent to taking $N_\tau \to \infty$. A continuum extrapolation therefore requires simulations on a sequence of lattices with different N_τ. For example, at a temperature $T = 200$ MeV ~ 1 fm^{-1}, Eq. (5.36) with $N_\tau = 4, 8, 12$ implies $a \approx 0.25, 0.125, 0.083$ fm.

5.2.3 Constraints on lattice simulations and systematic errors

It is important to realize from the outset that current lattice simulations at finite temperature and density are still hampered by sizeable systematic errors and uncertainties. Let us discuss the origins of those. The Compton wavelength of a hadron is proportional to its inverse mass m_H^{-1}, and the largest of those constitutes the correlation length of the statistical system. To keep finite-size as well as discretization errors small, we need to require

$$a \ll m_H^{-1} \ll aN_s. \quad (5.46)$$

For the low-T phase we thus need a lattice size of several inverse pion masses. The push to do physical quark masses is only just beginning to be a possibility on the most powerful machines and with the cheapest actions (i.e. staggered, with Wilson rapidly catching up). On the other hand, at high T screening masses scale as $m_H \sim T$, thus

$$N_\tau^{-1} \ll 1 \ll N_s N_\tau^{-1}. \tag{5.47}$$

Hence, while we desire large N_τ, the spatial lattice size should be significantly larger than the temporal one. This limits the directly accessible temperatures to several T_c.

Another important subject are cutoff effects. Similar to zero temperature, one can formally expand lattice observables in powers of the lattice spacing about their continuum limit. According to the discussion above, this is tantamount to expanding in the dimensionless N_τ^{-1} about zero. In order to reduce cutoff effects, one can then apply improved actions and operators, again in complete analogy to QCD in the vacuum.

In order to study thermodynamic behavior, one is often interested in the temperature dependence of certain quantities. This already implies sequences of simulations at many β-values for just one N_τ. If one is interested in a phase transition, one is moreover confronted with fluctuations between different phases, increasing correlation lengths and critical slowing down, i.e. increased autocorrelation times. For these reasons most thermal simulations require orders of magnitude more Monte Carlo trajectories than typical zero-temperature problems. Simulations are thus necessarily run on comparatively coarse lattices, implying larger systematic errors that are so far less controlled than for vacuum physics.

5.2.4 The ideal gas on the lattice

Similar to Section 5.1.4, we can evaluate the ideal gas for a bosonic field on the lattice. The starting point is the equation employing the free propagator,

$$\ln Z_0 = -\frac{1}{2} \ln \det\Delta = \frac{1}{2} \mathrm{Tr} \ln \Delta^{-1}$$

$$= V \sum_{n=-N_\tau/2}^{N_\tau/2-1} \int_{\frac{\pi}{a}}^{\frac{\pi}{a}} \frac{d^3 p}{(2\pi)^3} \ln(\hat{p}^2 + (am)^2)$$

$$= V \sum_{n=-N_\tau/2}^{N_\tau/2-1} \int_{\frac{\pi}{a}}^{\frac{\pi}{a}} \frac{d^3 p}{(2\pi)^3} \ln\left(4\sin^2(\frac{a\omega_n}{2}) + 4\hat{\omega}^2\right), \tag{5.48}$$

where now we have to deal with the lattice momenta,

$$\hat{p}^2 = 4\sin^2(\frac{a\omega_n}{2}) + 4\sum_{j=1}^{3} \sin^2(\frac{ap_j}{2}),$$

$$4\hat{\omega}^2 = 4\sum_{j=1}^{3} \sin^2(\frac{ap_j}{2}) + (am)^2. \tag{5.49}$$

Due to the finite lattice spacing the Matsubara sum is only from $-N_\tau/2, \ldots, N_\tau/2 - 1$. In order to perform the sum, we employ the formula (Kaste and Rothe, 1997)

$$\frac{1}{N_\tau} \sum_{n=-N_\tau/2}^{N_\tau/2-1} g(e^{i\omega_n}) = -\sum_{z_i} \frac{\text{Res}(\frac{g(z_i)}{z_i})}{z_i^{N_\tau} - 1} \tag{5.50}$$

to the derivative of the sum of logs,

$$L(\omega^2) \equiv \frac{1}{N_\tau} \sum_n \ln\left(4\sin^2(\frac{a\omega_n}{2}) + 4\hat{\omega}^2\right),$$

$$\frac{dL}{d\hat{\omega}^2} = \frac{1}{N_\tau} \sum_n \frac{4}{4\hat{\omega}^2 + 2 - e^{i\omega_n} - e^{-i\omega_n}},$$

$$\frac{g(z)}{z} \equiv \frac{4}{4\hat{\omega}^2 + 2 - z - z^{-1}}. \tag{5.51}$$

This function has simple poles at $z = 2\hat{\omega}^2 + 1 \pm 2\hat{\omega}\sqrt{\hat{\omega}^2 + 1}$. It is now convenient to change variables by $\hat{\omega} = \sinh(aE/2)$ to obtain

$$\frac{dL}{d(aE)} = \frac{dL}{d\hat{\omega}^2}\frac{d\hat{\omega}^2}{d(aE)} = 2\left(\frac{1}{e^{N_\tau aE} - 1} + 1\right),$$

$$L = \frac{2}{N_\tau} \ln\left(1 - e^{-N_\tau aE}\right) + 2aE. \tag{5.52}$$

We recognize again the vacuum-energy contribution, which needs to be subtracted to arrive at the final result for the pressure of a bosonic gas on the lattice,

$$\ln Z_0 = -V \int_{-\frac{\pi}{a}}^{\frac{\pi}{a}} \frac{d^3p}{(2\pi)^3} \ln(1 - e^{-N_\tau aE}). \tag{5.53}$$

One can now study the approach to the continuum by extending the integration range to infinity and expanding the log in small lattice spacing. The leading mistake we make by the first step in the positive integration range of the integral is

$$I(a) = \int_{\frac{\pi}{a}}^{\infty} \frac{d^3p}{(2\pi)^3} \ln(1 - e^{-N_\tau aE}) = I(0) + \frac{dI}{da}a + \ldots, \tag{5.54}$$

and similarly for the negative range. With $I(0) = 0$ and $I'(a) \propto \exp{-N_\tau \pi}$, with $N_\tau = 1/(aT)$, this is exponentially small and can be neglected compared to the power corrections of the integrand. For the expansion of the integrand, we need the corrections to the dispersion relation. Writing

$$E(\mathbf{p}) = E^{(0)}(\mathbf{p}) + a^2 E^{(2)}(\mathbf{p}) + \ldots, \qquad E^{(0)}(\mathbf{p}) = \sqrt{\mathbf{p}^2 + m^2}, \tag{5.55}$$

and expanding both sides of the equation

$$\sinh^2\left(\frac{aE}{2}\right) = \sum_{j=1}^{3} \sin^2\left(\frac{ap_j}{2}\right) + \frac{(am)^2}{4},$$

$$\frac{(aE)^2}{4} + \frac{(aE)^4}{48} = \sum_{j=1}^{3}\left(\frac{(ap_j)^2}{4} - \frac{(ap_j)^4}{48}\right) + \frac{(am)^2}{4} + O(a^6), \tag{5.56}$$

one finds for the a^2 correction to the dispersion relation

$$E^{(2)}(\mathbf{p}) = -\frac{1}{24E^{(0)}(\mathbf{p})}\left(\sum_{j=1}^{3} p_j^4 + E^{(0)}(\mathbf{p})\right). \tag{5.57}$$

Changing to dimensionless variables, $x = p/T, \varepsilon = E/T$, we then have for the expansion of the pressure

$$\frac{p}{T^4} = \left(\frac{p}{T^4}\right)_{\text{cont}} - a^2 \int \frac{d^3x}{(2\pi)^3} \frac{\varepsilon^{(2)}(x)}{e^{-\varepsilon^{(0)}(x)} - 1} + \cdots. \tag{5.58}$$

This means that the pressure of a free gas of bosons has leading lattice corrections of $O(a^2)$. For the massless case the integral can again be done and one arrives at

$$\frac{p}{p_{\text{cont}}} = 1 + \frac{8\pi^2}{21}\frac{1}{N_\tau^2} + O\left(\frac{1}{N_\tau^4}\right), \tag{5.59}$$

where p_{cont} is the continuum result Eq. (5.24).

One can repeat a similar calculation for various fermion actions and discuss their cutoff effects. For the free gas in the chiral limit this can be found in (Hegde *et al.*, 2008), leading-order interactions and mass effects have been evaluated in (Philipsen and Zeidlewicz, 2010).

5.2.5   The quenched limit of QCD and $Z(N)$-symmetry

In the quenched limit, i.e. when quarks are infinitely heavy, $m_f \to \infty$, the QCD partition function reduces to that of a pure gauge theory plus static quark fields. Let us examine the consequences of the compact temporal direction for gauge symmetry. The action is of course invariant under standard gauge transformations,

$$S_g[U^g] = S_g[U] \quad \text{with} \quad U_\mu^g(x) = g(x)U_\mu(x)g^{-1}(x+\hat{\mu}), \quad g(x) \in SU(N), \tag{5.60}$$

with our earlier periodic boundary conditions

$$U_\mu(\tau, \mathbf{x}) = U_\mu(\tau + N_\tau, \mathbf{x}), \quad g(\tau, \mathbf{x}) = g(\tau + N_\tau, \mathbf{x}). \tag{5.61}$$

However, with the temporal boundary in place, we can also consider gauge transformations with topologically non-trivial matrices $g'(x)$, i.e. matrices that cannot be taken to unity by a smooth change of the parameters characterizing the group elements.

Consider a boundary condition on the transformation matrices, which is only periodic up to a constant matrix h,

$$g'(\tau + N_\tau, \mathbf{x}) = hg'(\tau, \mathbf{x}), \quad h \in SU(N). \tag{5.62}$$

Such a $g'(x)$ does not go into itself when winding around the torus once, but picks up a "twist" factor h. After a gauge transformation with g', the gauge links behave across the boundary as

$$U_\mu^{g'}(\tau + N_\tau, \mathbf{x}) = h\, U_\mu^{g'}(N_\tau, \mathbf{x})\, h^{-1}. \tag{5.63}$$

They are only consistent with the periodicity requirement if $[h, U_i^{g'}] = 0$. This is satisfied if h is proportional to the unit matrix, i.e. it is in the center of the group,

$$h = z\mathbf{1} \in Z(N), \quad z = \exp i\frac{2\pi n}{N}, \quad n \in \{0, 1, 2, \ldots N - 1\}. \tag{5.64}$$

Thus, pure gauge theory at finite temperature is invariant under gauge transformations with non-trivial winding through the temporal boundary, for any global twist factor in $Z(N)$. Note that this is *not* a symmetry of the QCD Hamiltonian. The latter is acting on a particular timeslice and does not know about temporal boundary conditions. The Hamiltonian generates translations in Euclidean time and its symmetries are defined on the space orthogonal to that. Rather, the center symmetry discussed here is a symmetry of the space-wise Hamiltonian H_z, Eq. (5.39), which acts on a z-slice and thus *is* sensitive to the boundary in the temporal direction.

Since the center symmetry is related to non-trivial gauge transformations winding through the temporal boundary, gauge-invariant observables can only be sensitive to it if they wind too. Such an observable is a Wilson line in the temporal direction closing onto itself, a Polyakov loop,

$$L(\mathbf{x}) = \prod_{x_0}^{N_\tau} U_0(x). \tag{5.65}$$

Physically, it corresponds to the propagator of a static quark. Under gauge transformations,

$$L^g(\mathbf{x}) = g(\mathbf{x})L(\mathbf{x})g^{-1}(\mathbf{x}),$$

$$L^{g'}(\mathbf{x}) = g'(1, \mathbf{x})L(\mathbf{x})g'^{-1}(1 + N_\tau, \mathbf{x}) = g'(1, \mathbf{x})L(\mathbf{x})g'^{-1}(1, \mathbf{x})h^{-1}. \tag{5.66}$$

Thus, we see that the traced Polyakov loop is gauge invariant under topologically trivial gauge transformations, while it picks up a center element when transformed with a winding transformation,

$$\mathrm{Tr}L^g = \mathrm{Tr}L, \quad \mathrm{Tr}L^{g'} = z^*\mathrm{Tr}L. \tag{5.67}$$

The Polyakov loop emerges naturally in the QCD path integral to leading order in the hopping expansion. For example, the partition function for a pure gauge theory with a static quark sitting at $\mathbf{x}$ is

$$Z_Q = \int DU \ \mathrm{Tr}L(\mathbf{x}) \ e^{-S_g[U]}. \tag{5.68}$$

Hence,

$$\langle \mathrm{Tr}L \rangle = \frac{1}{Z} \int DU \ \mathrm{Tr}L \ e^{-S_g} = \frac{Z_Q}{Z} = e^{-(F_Q - F_0)/T}, \tag{5.69}$$

and the expectation value of the Polyakov loop gives the free-energy difference between a Yang–Mills plasma with and without the static quark (McLerran and Svetitsky, 1981). From this we can immediately infer two limiting cases: For $T \to 0$ Yang–Mills theory is confining and it would cost infinite energy to remove the quark to infinity, i.e. $F_Q = \infty$ and therefore $\langle \mathrm{Tr}L \rangle = 0$. On the other hand, $T \to \infty$ corresponds to $\beta \to \infty$, for which $U_0 \to 1$ and $\langle \mathrm{Tr}L \rangle \to \mathrm{Tr}\mathbf{1} = N$. Clearly, a non-zero expectation value is no longer invariant under center transformations and signals the spontaneous breaking of center symmetry. Therefore, QCD in the quenched limit has a true (non-analytic) deconfinement phase transition corresponding to the breaking of the global center symmetry, and the average of the Polyakov loop is the corresponding order parameter.

If we add dynamical quark fields to the theory, they will behave as

$$\psi^g(x) = g(x)\psi(x), \quad \psi(\tau + N_\tau, \mathbf{x}) = -\psi(\tau, \mathbf{x}), \quad \psi^{g'}(\tau + N_\tau, \mathbf{x}) = -h\psi(\tau, \mathbf{x}). \tag{5.70}$$

Since statstical mechanics requires antiperiodic boundary conditions for fermions, the trivial $h = 1$ is the only permissible choice, i.e. there is *no* center symmetry in the presence of dynamical quarks. Physically, if there are dynamical quarks, their pair production screens the confining force (it leads to string breaking) and F_Q is finite. Correspondingly, $\langle \mathrm{Tr}L \rangle \neq 0$ for all temperatures and the Polyakov loop is no longer a true order parameter. In this case, a non-analytic phase transition as a function of temperature is not necessary, confined and deconfined regions may also be analytically connected by a smooth crossover.

5.2.6   The chiral limit

Chiral symmetry and its breaking on the lattice is the subject of another chapter. Let us merely summarize what we need for the finite-temperature discussion. In the limit of zero quark masses the classical QCD Lagrangian in the continuum is invariant under global chiral-symmetry transformations, the total symmetry being $U_A(1) \times SU_L(N_f) \times SU_R(N_f)$. The axial $U_A(1)$ is anomalous, quantum corrections break it down to $Z(N_f)$. The remainder gets spontaneously broken to the diagonal subgroup, $SU_L(N_f) \times SU_R(N_f) \to SU_V(N_f)$, giving rise to $N_f^2 - 1$ massless Goldstone bosons, the pions. The order parameter signalling chiral symmetry is the chiral condensate,

$$\langle \bar{\psi}\psi \rangle = \frac{1}{N_s^3 N_\tau} \frac{\partial}{\partial m_f} \ln Z. \tag{5.71}$$

It is non-zero for $T < T_c$, when chiral symmetry is spontaneously broken, and zero for $T > T_c$. Hence, there is a non-analytic finite-temperature phase transition corresponding to chiral-symmetry restoration. For non-zero quark masses, chiral symmetry is broken explicitly and the chiral condensate $\langle \bar{\psi}\psi \rangle \neq 0$ for all temperatures. Again, in this case there is no need for a non-analytic phase transition. Of course, for most lattice fermions chiral symmetry is either reduced (staggered) or broken completely (Wilson), rendering the behavior in the chiral limit a most difficult subject of study.

5.2.7 Physical QCD

QCD with physical quark masses obviously does not correspond to either the chiral or quenched limit. The $Z(3)$ symmetry as well as the chiral symmetry are explicitly broken. Nevertheless, physical QCD displays confinement as well as three very light pions as "remnants" of those symmetries. In the presence of mass terms there is no true order parameter, i.e. the expectation values of the Polyakov loop as well as the chiral condensate are non-zero everywhere. Hence, the deconfined or chirally symmetric phase is analytically connected with the confined or chirally broken phase, and there is no need for a non-analytic phase transition. The following questions then arise, which should be answered by numerical simulations: for which parameter values of QCD is there a true phase transition, and what is its order? Are confinement and chirality changing across the same single transition or are there different transitions? If there is just one transition, which is the driving mechanism? If there is only a smooth crossover, how do the properties of matter change in the different regions?

5.2.8 Strong-coupling expansions at finite T

Strong-coupling expansions are well known from spin models and QCD at zero temperature. In contrast to the asymptotic series obtained by weak coupling expansions, they yield convergent series in the lattice gauge coupling $\beta = 2N/g^2$ within a finite radius of convergence. The series approximates the true answer the better the lower β, i.e. for a fixed N_τ the lower the temperature. In pure gauge theory, the convergence radius is bounded from above by the critical coupling of the deconfinement transition, β_c. Hence, strong-coupling series are analytical low-temperature results, complementary to weak-coupling perturbation theory that is valid at high temperatures.

A detailed introduction to strong-coupling methods in the vacuum can be found in (Montvay and Münster, 1994). Here, we merely summarize the main formulae for Yang–Mills theory and discuss the modifications for finite-temperature applications (Langelage *et al.*, 2008).

The Wilson gauge action can be written as a sum over all single plaquette actions,

$$S_g[U] = \sum_x \sum_{1 \leq \mu < \nu \leq 4} \beta \left(1 - \frac{1}{3} \mathrm{Re} \mathrm{Tr} U_p \right) \equiv \sum_p S_p. \qquad (5.72)$$

Note that analytically we are able to consider an infinite spatial volume, $N_s \to \infty$. The plaquette action has an expansion in terms of characters $\chi_r(U) = \mathrm{Tr} D_r(U)$ of the representation matrices $D_r(U)$ of the group elements U,

$$\exp -S_p = c_0(\beta)[1 + \sum_{r \neq 0} d_r c_r(\beta) \chi_r(U_p)], \qquad (5.73)$$

and d_r is the dimension of the representation r. With these ingredients the free-energy density can be written as

$$\tilde{f} \equiv -\frac{1}{\Omega} \ln Z = -6 \ln c_0(\beta) - \frac{1}{\Omega} \sum_{C=(X_i^{n_i})} a(C) \prod_i \Phi(X_i)^{n_i}, \qquad (5.74)$$

where $\Omega = V \cdot N_\tau$ is the lattice volume and c_0 is the expansion coefficient of the trivial representation, which has been factored out. The combinatorial factor $a(C)$ is introduced via a moment-cumulant formalism, and equals 1 for clusters C that consist of only one graph or so-called polymer X_i. The contribution of a graph X_i is

$$\Phi(X_i) = \int DU \prod_{p \in X_i} d_r c_{r_p} \chi_{r_p}(U_p). \qquad (5.75)$$

The quantity in Eq. (5.74) is customarily called a free energy, even at zero physical temperature, because the path integral corresponds to a partition function if one formally identifies the lattice coupling β with $1/T$. One may use the fundamental representations of the gauge groups, $c_f \equiv u$, as the effective expansion parameter, which together with some higher ones can be expressed as series in the lattice coupling

$$\text{SU(2):} \quad u = \frac{\beta}{4} + O(\beta^2) \qquad \text{SU(3):} \quad u = \frac{\beta}{18} + O(\beta^2). \qquad (5.76)$$

Here, we are interested in a physical temperature $T = 1/(aN_\tau)$, realized by compactifying the temporal extension of the lattice. The physical free energy is then obtained by subtracting the formal ($N_\tau = \infty$) free energy, which removes the divergent vacuum energy as in the continuum. Thus, the physical free-energy density reads

$$f(N_\tau, u) = \tilde{f}(N_\tau, u) - \tilde{f}(\infty, u). \qquad (5.77)$$

The contributing polymers X_i have to be objects with a closed surface, since

$$\int dU \, \chi_r(U) = \delta_{r,0}. \qquad (5.78)$$

This means the group integration projects out the trivial representation at each link. To calculate the group integrals one uses the integration formula

$$\int dU \, \chi_r(UV) \chi_r(WU^{-1}) = \frac{1}{d_r} \chi_r(VW). \qquad (5.79)$$

Because of the difference in Eq. (5.77), those graphs contributing in the same way to $\tilde{f}(N_\tau)$ and $\tilde{f}(\infty)$ drop out of the physical free energy. This is true for all polymers with time extent less than N_τ. The calculation thus reduces to graphs with a temporal size of N_τ on the finite N_τ lattice, and graphs spanning or extending N_τ on the infinite

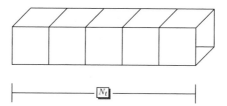

Fig. 5.2 Graph X_1 contributing to the lowest order $f(N_\tau, u)$ of the expansion of the physical free-energy density at finite temperature.

lattice. Such graphs contribute either to $\tilde{f}(N_\tau)$ or to $\tilde{f}(\infty)$ (and in some cases to both). It is therefore clear that the strong-coupling series for the physical free energy starts at a higher order than the formal zero-temperature free energy. Moreover, the order of the leading contribution depends on N_τ.

The lowest-order graph existing due to the boundary condition on the finite N_τ lattice, but not on the infinite lattice, is a tube of length N_τ with a cross-section of one single plaquette, as shown in Fig. 5.2. It forms a closed torus through the periodic boundary and thus gives a non-vanishing contribution, which is easily calculated to be $\Phi(X_1) = u^{4N_\tau}$. We need to sum up all such graphs on the lattice. There are three spatial directions for the cross-section of the tube, giving a factor of 3. Translations in time take the graph into itself and do not give a new contribution, while we get $V\Phi(X_1)$ from all spatial translations. Together with the $1/\Omega$ in Eq. (5.74) this gives a factor of $1/N_\tau$. The contribution of all tubes with all plaquettes in the fundamental representation for $SU(2)$ is thus

$$\Phi(X_1) = \frac{3}{N_\tau} u^{4N_\tau}, \tag{5.80}$$

which is – up to a sign – also the leading-order result for the physical free energy. For $SU(N)$ with $N \geq 3$ we have an additional factor of 2 because there are also complex conjugate fundamental representations. Thus, in the strong-coupling limit $u \to 0$ the free-energy density and pressure are zero.

The leading correction comes from tubes with inner plaquettes, higher orders have local decorations of additional plaquettes either in the fundamental or in higher representations. For the interesting case of $SU(3)$, these contributions up to the calculated orders are (Langelage *et al.*, 2007)

$$f(N_\tau, u) = -\frac{3}{N_\tau} u^{4N_\tau} c^{N_\tau} \left[1 + 12N_\tau u^4 + 42N_\tau u^5 - \frac{115343}{2048} N_\tau u^6 - \frac{597663}{2048} N_\tau u^7 \right]$$

$$- \frac{3}{N_\tau} u^{4N_\tau} b^{N_\tau} \left[1 + 12N_\tau u^4 + 30N_\tau u^5 - \frac{17191}{256} N_\tau u^6 - 180N_\tau u^7 \right], \tag{5.81}$$

with $b = 1 - 3u - 6v + 8w$, $c = 1 + 3u + 6v + 8w - 18u^2$ and $v = \beta^2/432 + O(\beta^4)$, $w = \beta^2/288 + O(\beta^4)$. This series is valid only for $N_\tau \geq 5$, for smaller N_τ there are modifications coming from polymers with cross-sections larger than one plaquette.

5.2.9 The strong-coupling regime as an ideal hadron gas

It is now interesting to ask how the QCD pressure can be interpreted in the strong-coupling regime. From the Wilson action it is clear that the strong-coupling limit is also non-interacting. However, as we have noted already, in this limit the pressure is zero. Considering strong but finite couplings, let us recall the first orders of the $T = 0$ glueball mass calculations (Münster, 1981; Seo, 1982),

$$
m(A_1^{++}) = -4\ln u - 3u + 9u^2 - \frac{27}{2}u^3 - 7u^4 - \frac{297}{2}u^5 + \frac{858827}{10240}u^6 + \frac{47641149}{71680}u^7,
$$

$$
m(E^{++}) = -4\ln u - 3u + 9u^2 - \frac{27}{2}u^3 + 17u^4 - \frac{153}{2}u^5 + \frac{1104587}{10240}u^6 + \frac{29577789}{71680}u^7,
$$

$$
m(T_1^{+-}) = -4\ln u + 3u + \frac{9}{2}u^3 - \frac{98}{4}u^4 + \frac{33}{4}u^5 - \frac{36771}{1280}u^6 + \frac{117897}{448}u^7, \tag{5.82}
$$

where the arguments on the left side denote the representations of the point group, which map into the different spin states (Montvay and Münster, 1994). We observe that the expansion of the free energy can be written

$$
f(N_\tau, u) = -\frac{1}{N_\tau}\left[e^{-m(A_1^{++})N_\tau} + 2e^{-m(E^{++})N_\tau} + 3e^{-m(T_1^{+-})N_\tau}\right]\left(1 + O(u^4)\right). \tag{5.83}
$$

The prefactors before the exponentials correspond to the number of polarizations of the respective glueball states. Note that higher spin states start with $\sim 6\ln u$ (Schor, 1983), thus contributing to the order $\sim u^{6N_t}$ or higher in the free energy. Hence, through two non-trivial orders our result is that of a free glueball gas, modified by higher-order corrections. By employing a leading-order hopping expansion, the same conclusion can be reached when heavy quarks are added (Langelage and Philipsen, 2010b). This is a rather remarkable result. It allows us to see from a first-principles calculation that the pressure is exponentially small in the confined phase, and that it is well approximated by an ideal gas of quasi-particles that correspond to the $T = 0$ hadron excitations.

5.3 Some applications

5.3.1 The equation of state

Energy density $\epsilon(T)$ and pressure $p(T)$ as a function of temperature are certainly among the most fundamental thermodynamic quantities of QCD governing the expansion of the plasma in the early Universe as well as in heavy-ion collisions. Let us use the ideal-gas results to develop some intuition about what will happen in QCD. In the high-temperature limit $T \to \infty$, we have a gas of non-interacting gluons and quarks with

$$
\frac{p}{T^4} = \left(16 + \frac{7}{8}12N_f\right)\frac{\pi^2}{90}. \tag{5.84}
$$

On the other hand, as $T \to 0$ we consider a hadron resonance gas model. For temperatures significantly below ~ 200 MeV, only the pions are relativistic, which come in three charge states with spin zero, $\nu = 3$. A gas of non-interacting pions has pressure

$$\frac{p}{T^4} = 3\frac{\pi^2}{90}. \tag{5.85}$$

Thus, for QCD we expect the pressure to change as a function of temperature from a small to a large value as a signal of deconfinement.

For the fully interacting case, we need to compute the free-energy density and the pressure from the QCD partition function. A technical obstacle here is that, in a Monte Carlo simulation, one cannot compute the partition function directly, since all expectation values are normalized to Z,

$$\langle O \rangle = Z^{-1}\mathrm{Tr}(\rho O). \tag{5.86}$$

So, we need to specify some observable O to access the partition function. The most frequently used detour is called the integral method (Engels *et al.*, 1990), in which a derivative of the free energy is calculated and then integrated,

$$\left.\frac{f}{T^4}\right|_{T_o}^{T} = -\frac{1}{V}\int_{T_o}^{T} \mathrm{d}x\, \frac{\partial\; x^{-3}\ln Z(V,x)}{\partial x}. \tag{5.87}$$

On the lattice, it is convenient to take the derivative with respect to the gauge coupling instead of temperature,

$$\left.\frac{f}{T^4}\right|_{\beta_o}^{\beta} = -\frac{N_\tau^3}{N_s^3}\int_{\beta_o}^{\beta} \mathrm{d}\beta' \left(\left\langle \frac{\partial \ln Z}{\partial \beta'} \right\rangle - \left\langle \frac{\partial \ln Z}{\partial \beta'} \right\rangle_{T=0}\right),$$

$$= -N_\tau^4 \int_{\beta_0}^{\beta} \mathrm{d}\beta'\; \left(3\langle \mathrm{Tr}U_p^t + \mathrm{Tr}U_p^s\rangle - 6\langle \mathrm{Tr}U_p\rangle_{T=0}\right). \tag{5.88}$$

Now, we simply need to measure expectation values of temporal (U_P^t) and spatial (U_p^s) plaquettes with sufficient accuracy, so they can be integrated numerically. Note that this introduces a lower integration constant, which needs to be fixed for the result to be meaningful. While we do not know $f(\beta_0)$ from first principles, we can choose β_0 corresponding to a temperature below the phase transition, where the free energy should be well modelled by a weakly interacting hadron gas. For $T \lesssim 130$ MeV, all hadrons become non-relativistic and we can approximate the pressure by zero. Note, however, that this procedure is not good for the chiral limit, when pions become massless and stay relativistic down to very low temperatures.

Another difficulty is that strong discretization effects are to be expected. At high temperature the relevant partonic degrees of freedom have momenta of order $\pi T \sim \pi/(aN_t)$ on the scale of the lattice spacing, by which they are strongly affected. For the equation of state it is therefore particularly important to gain control over these effects and carry out the continuum limit $a \to 0$. This motivates the use of improved actions, designed to minimize cutoff effects in the approach to the continuum.

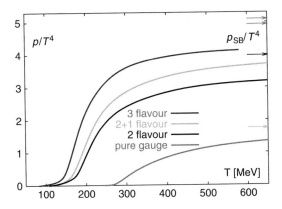

Fig. 5.3 Flavor dependence of the pressure for $N_\tau = 4$ lattices compared to a continuum extrapolated pure gauge result. From (Karsch *et al.*, 2000).

The results of a computation of the pressure with an improved action (Karsch *et al.*, 2000) are shown in Fig. 5.3. The data have been obtained for $N_f = 2, 3$ with (bare) mass $m_q/T = 0.4$ as well as for $N_f = 2 + 1$ with a heavier mass $m_q^s/T = 1$ on a very coarse lattice, $N_\tau = 4$. Nevertheless, interesting qualitative features can be observed. For comparison, continuum extrapolated pure gauge results are also included. The figure shows a rapid rise of the pressure in a narrow transition region. The critical temperature as well as the magnitude of p/T^4 reflect the number of degrees of freedom liberated at the transition. This last conclusion is firm, since the pressure also rises for fixed temperature when light quarks are added to the theory, consistent with the behavior in the Stefan–Boltzmann limit. Another interesting feature is that the curves fall short of the ideal-gas values, i.e. interactions are still strong just above T_c. An important question then is whether these features survive in the continuum limit. In pure gauge theory this can be firmly established by numerical extrapolation, while for light dynamical quarks this is the subject of ongoing simulation programs.

5.3.2 Screening masses

Essentially, all static equilibrium properties of a thermal quantum field theory are encoded in its equal time correlation functions. These are quantities that are well defined and calculable to good precision by lattice methods. Unfortunately, these quantities are not directly accessible in heavy-ion-collision experiments. Nevertheless, their theoretical knowledge provides us with the relevant length scales in the plasma, from which conclusions about the active degrees of freedom and their dynamics may be drawn.

The concept of screening is most easily introduced in a QED plasma. Suppose we insert a static external charge in a plasma of freely moving charges. These will be attracted or repelled, arranging themselves such that the polarization cancels out the field of the static charge, which therefore is screened. Beyond a certain distance called a screening length, a particle in the plasma will not feel the presence of the external charge. The screening length is defined by the spatial exponential decay of the equal time correlator of the electric field,

$$\lim_{\mathbf{x}\to\infty} \langle E_i(\mathbf{x},t)E_j(\mathbf{0},t)\rangle = const.\ e^{-m_D|\mathbf{x}|}, \tag{5.89}$$

and the inverse screening length m_D is the Debye mass. In perturbation theory, m_D corresponds to the pole of the A_0 propagator at Matsubara frequency $\omega_n = 0$, i.e. it corresponds to the electric gluon mass, $m_D = m_E$, Eq. (5.15). Its Fourier transformation leads to the Debye-screened potential of a static charge,

$$V(r) = Q \int \frac{d^3p}{(2\pi)^3} \frac{e^{i\mathbf{p}\cdot\mathbf{x}}}{\mathbf{p}^2 + \Pi_{00}(0,\mathbf{p})} = \frac{e^{-m_D r}}{4\pi r}. \tag{5.90}$$

The same concept has been carried over to QCD where it is speculated to be responsible for the loss of confinement for heavy quarkonium states by screening of color charges in the plasma (Matsui and Satz, 1986). However, there are some conceptual difficulties with the definition of the Debye mass in QCD. First, color-electric fields are no physical observables since they are gauge dependent, and so is their correlator. One might argue that the pole mass of this correlator is still gauge-invariant order by order in perturbation theory and consider the perturbative series (Rebhan, 1993),

$$m_D = m_D^{LO} + \frac{3}{4\pi}g^2 T \ln\frac{m_D}{g^2 T} + cg^2 T + O(g^3). \tag{5.91}$$

In this case, one encounters the Linde problem already in next-to-leading order: the coefficient c receives contributions from all loop orders and cannot be evaluated in perturbation theory.

For a non-perturbative evaluation of screening, one therefore generalizes the concept to gauge-invariant sources, averaged over Euclidean time,

$$\bar{O}(\mathbf{x}) = \frac{1}{N_\tau}\sum_\tau O(\mathbf{x},\tau), \quad \langle \bar{O}(\mathbf{x})\bar{O}(\mathbf{0})\rangle_c = \sum_n c_n\, e^{-M_n|\mathbf{x}|}. \tag{5.92}$$

These are correlations in space, and the connected parts therefore fall off with the eigenvalues of the space-wise Hamiltonian H_z, Eq. (5.39). In analogy to QED, these masses correspond to the inverse length scale over which the equilibrated medium is sensitive to the insertion of a static source carrying the quantum numbers of O. Because of the compact Euclidean time direction at $T > 0$, the continuum rotation symmetry of the hypertorus orthogonal to the correlation direction is broken down from $O(3)$ to $O(2) \times Z(2)$, corresponding to rotations in the (x,y)-plane and reflections of τ. Correspondingly, screening masses are classified by quantum numbers J_R^{PC}, where J, P, C are standard spin, parity and charge conjugation, while R is associated with the Euclidean time reflection. The appropriate subgroup for the lattice theory is $D^4 \times Z(2) = D_h^4$. The irreducible representations and the classification of operators have been worked out for pure gauge theory (Grossman *et al.*, 1994; Datta and Gupta, 1998) as well as for staggered quarks (Gupta, 1999).

Just as in $T = 0$ spectrum calculations, we can consider either glueball or meson and baryon-like operators. Once again it is useful to develop some intuition by considering the limits of low and asymptotically high temperature. In the latter case, standard perturbation theory should apply and we can evaluate the correlators using quark and gluon propagators. For a mesonic operator the exponential decay at large distances is then dominated by two quark lines, with an effective mass given by their lowest Matsubara frequencies, $\sim \pi T$. Neglecting bare quark masses, a meson correlator in the non-interacting limit then decays with $M_{\bar{q}q} \sim 2\pi T$. Perturbative corrections lift the degeneracy and appear to generally shift this towards larger values (Laine and Vepsäläinen, 2004). For purely gluonic operators, the lowest Matsubara frequency is zero and the exponential decay is determined by dynamically generated mass scales. For operators like $\mathrm{Tr} A_0^2$ the propagators feature the perturbative Debye mass, and hence $M_{A_0^2} \sim 2m_E$. However, for operators involving A_i, we are faced with the Linde problem and a perturbative evaluation is not possible.

On the other hand, at low temperatures we have small β and a calculation by strong-coupling methods is feasible. As an example, let us consider the color-electric field correlator $\langle \mathrm{Tr} F_{0i}^a(\mathbf{x}) \, \mathrm{Tr} F_{0i}^a(\mathbf{y}) \rangle$, which is in the $J_T^{PC} = 0_+^{++}$ channel (T denotes reflection in Euclidean time) containing the ground state and the mass gap. On the lattice, this corresponds to a correlation of temporal plaquettes, and the quantum numbers under the point group D_h^4 are A_1^{++}. Temporarily assigning separate gauge couplings to all plaquettes, the correlator can be defined as

$$C(z) = \langle \mathrm{Tr}\, U_{p_1}(0) \, \mathrm{Tr}\, U_{p_2}(z) \rangle = N^2 \frac{\partial^2}{\partial\beta_1 \partial\beta_2} \ln Z(\beta, \beta_1 \beta_2) \Big|_{\beta_{1,2}=\beta}. \tag{5.93}$$

At zero temperature the exponential decay is the same as for correlations in the time direction, and thus determined by the glueball masses, the lowest of which may be extracted as

$$m = -\lim_{z\to\infty} \frac{1}{z} \ln C(z). \tag{5.94}$$

The leading-order graphs for the strong-coupling series at zero temperature are shown in Fig. 5.4 (left). This leads to the lowest-order contribution:

$$C(z) = A\, u^{4z} = A e^{-m_s z}. \tag{5.95}$$

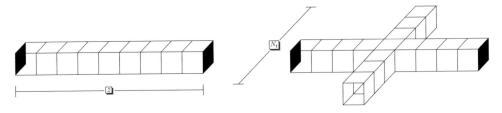

Fig. 5.4 Left: Leading graph contributing to the plaquette correlation function. Right: Graph contributing to the finite temperature effect of the correlator.

Thus, to leading order the glueball mass is $m_s = -4 \ln u(\beta)$.

Now we switch on a physical temperature, i.e. keep the lattice volume compact in the time direction. As in the case of the free energy, we are here only interested in the temperature effects, i.e. in the mass difference

$$\Delta m(T) = m(T) - m(0) = -\lim_{z \to \infty} \frac{1}{z} [\ln C(T; z) - \ln C(0; z)] \qquad (5.96)$$

$$= -\lim_{z \to \infty} \left[\ln \left(1 + \frac{\Delta C(T; z)}{C(0; z)} \right) \right], \qquad (5.97)$$

with $\Delta C(T; z) = C(T; z) - C(0; z)$. A typical graph contributing in lowest order to this difference is shown in Fig. 5.4 (right). Summing up all leading-order graphs gives (Langelage *et al.*, 2008)

$$\Delta m(T) = -\frac{2}{3} N_\tau \, u^{4N_t - 6}, \qquad (5.98)$$

i.e. the screening masses decrease compared to their $T = 0$ values. Again, the leading-order result is generic for all $SU(N)$ and quantum number channels. As in the case of the free energy, the difference only receives N_τ-dependent higher orders of u, leading to weak temperature effects. We conclude that in the confinement phase the lowest screening masses in each quantum number channel should be close to the corresponding zero-temperature particle masses, with a significant temperature dependence showing up only near T_c. This explains the findings of numerical investigations of the lowest screening mass in $SU(3)$ gauge theory, which for temperatures as high as $T = 0.97T_c$ see very little temperature dependence, $\Delta m(T)/m(0) \gtrsim 0.83$ (Datta and Gupta, 1998).

Away from the high-and the low-temperature limits, numerical simulations have to be performed. One particularly interesting aspect of screening masses is that they permit study of chiral-symmetry restoration across the quark–hadron transition by looking at degeneracy patterns. For example, consider the lowest lying scalar, pseudoscalar, vector and axial-vector mesons. Chiral $SU(2) \times SU(2)$ implies degeneracy between the scalar and vector, while an intact $U_A(1)$ implies degeneracy between the parity-flipped states. A calculation of these masses for temperatures around T_c will thus allow insight into the details of chiral-symmetry restoration.

meson	σ	$\vec{\pi}$	$\vec{\delta}$	η'
operator	$\bar{\psi}\psi$	$\bar{\psi}\gamma^5\vec{\tau}\psi$	$\bar{\psi}\vec{\tau}\psi$	$\bar{\psi}\gamma^5\psi$

$$\begin{array}{ccc} & \leftarrow \quad SU(2) \times SU(2) \quad \rightarrow & \\ \uparrow & \sigma & \vec{\pi} \\ U_A(1) & & \\ \downarrow & \eta' & \vec{\delta} \end{array}$$

5.3.3 The free energy of a static quark–antiquark pair

Similar to a single static quark, Eq. (5.68) and the discussion following it, we can consider a static quark–antiquark pair in a Yang–Mills plasma,

$$\langle \mathrm{Tr} L^\dagger(\mathbf{x}) \mathrm{Tr} L(\mathbf{0}) \rangle = \frac{Z_{\bar{Q}Q}}{Z} = \exp -\frac{F_{\bar{Q}Q}(\mathbf{x}, T) - F_0(T)}{T}. \qquad (5.99)$$

(In the following we drop the subtraction of F_0 from the notation, which is always implied.) Such a system is particularly interesting because it represents the non-relativistic limit of heavy quarkonia in the plasma, which are routinely produced in heavy-ion collisions. In order to clarify the meaning of this quantity, let us consider its spectral decomposition. We use again the transfer-matrix formalism, but with a slight modification accounting for the static sources. Let us fix to temporal gauge, $U_0(\tau, \mathbf{x}) = 1, \tau = 1, \ldots N_\tau - 1$, which we are allowed to do on all temporal links but those in one timeslice. On the gauge-fixed timeslices, we have the Kogut–Susskind Hamiltonan H_0, which acts on the Hilbert space of states with static sources, from Eq. (5.33),

$$(T_0)_{\tau+1,\tau} \equiv e^{-aH_0} = \exp -L[U_i(\tau + 1), 1, U_i(\tau)]. \tag{5.100}$$

Now, we can be rewrite the Polyakov loop correlator exactly as

$$\langle \mathrm{Tr}L^\dagger(\mathbf{x})\mathrm{Tr}L(\mathbf{0}) \rangle = \tag{5.101}$$

$$\frac{1}{Z}\hat{\mathrm{Tr}}\left(T_0^{N_\tau-1} \int DU_0(N_\tau)\ U_{0\alpha\alpha}^\dagger(N_\tau, \mathbf{x})U_{0\beta\beta}(N_\tau, \mathbf{0})\ e^{-L[U_i(1),U_0(N_\tau),U_i(N_\tau)]} \right).$$

Next, we employ gauge invariance of the kernel of the transfer matrix, Eq. (5.34), under gauge transformation in the upper timeslice,

$$R(g)L = L[U_i^g(\tau + 1), U_0(\tau)g^{-1}, U_i(\tau)] = L[U_i(\tau + 1), U_0(\tau), U_i(\tau)], \tag{5.102}$$

where $R(g)|\psi\rangle = |\psi^g\rangle$ imposes a gauge transformation on the states it acts on. Choosing $g(\mathbf{x}) = U_0(N_\tau, \mathbf{x})$, we then obtain

$$\langle \mathrm{Tr}L^\dagger(\mathbf{x})\mathrm{Tr}L(\mathbf{0}) \rangle = \frac{1}{Z}\hat{\mathrm{Tr}}(T_0^{N_\tau} P_{\alpha\alpha\beta\beta}), \tag{5.103}$$

where we have introduced the projection operator

$$P_{\alpha\beta\mu\nu} = \int Dg\ g_{\alpha\beta}^\dagger(\mathbf{x})g_{\mu\nu}(\mathbf{0})R(g). \tag{5.104}$$

This operator annihilates all wave functions not transforming as

$$\psi_{\beta\mu}[U^g] = g(\mathbf{x})_{\beta\gamma}\ \psi_{\gamma\delta}[U]\ g_{\delta\mu}^\dagger(\mathbf{0}). \tag{5.105}$$

Specifically,

$$P_{\alpha\beta\mu\nu}|\psi_{\gamma\delta}\rangle = \frac{1}{N^2}\delta_{\beta\gamma}\delta_{\mu\delta}|\psi_{\alpha\nu}\rangle. \tag{5.106}$$

Thus, P projects onto the subspace of states with a color triplet sitting at $\mathbf{x}$ and an anti-triplet at $\mathbf{0}$. Inserting a complete set of eigenstates of the Hamiltonian H_0, we then find

$$\langle \mathrm{Tr}L^\dagger(\mathbf{x})\mathrm{Tr}L(\mathbf{0}) \rangle = \frac{1}{N^2 Z}\sum_{n,\alpha,\beta}\langle n_{\alpha\beta}|n_{\beta\alpha}\rangle\ e^{-\frac{E_n^{Q\bar{Q}}(|\mathbf{x}|)}{T}} = \frac{1}{Z}\sum_n e^{-\frac{E_n^{\bar{Q}Q}(|\mathbf{x}|)}{T}}. \tag{5.107}$$

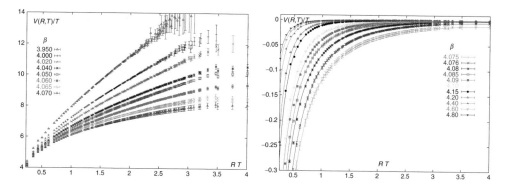

Fig. 5.5 Static quark–antiquark free energy/potential, Eq. (5.99), for $T < T_c$ (left) and $T > T_c$ (right), on 32×4 with Symanzik improved action (Kaczmarek *et al.*, 2000).

We easily recognize this as a ratio of partition functions, the numerator being just the Boltzmann sum over all energy levels $E_n^{\bar{Q}Q}$, which are eigenvalues of H_0 in the presence of the $\bar{Q}Q$ pair. Let us now consider the zero-temperature limit of this expression,

$$\lim_{T \to 0} \frac{\sum_n e^{-\frac{E_n^{\bar{Q}Q}}{T}}}{\sum_n e^{-\frac{E_n}{T}}} = \lim_{T \to 0} e^{-(E_0^{\bar{Q}Q} - E_0)/T} \frac{1 + e^{-(E_1^{\bar{Q}Q}(|\mathbf{x}|) - E_0^{\bar{Q}Q}(|\mathbf{x}|))/T} + \dots}{1 + e^{-(E_1 - E_0)/T} + \dots} \to e^{-\frac{V(|\mathbf{x}|)}{T}}.$$

(5.108)

Here, we have identified the static potential with the ground-state energy of a static quark–antiquark pair above the vacuum, $V(|\mathbf{x}|) = E_0^{\bar{Q}Q}(|\mathbf{x}|) - E_0$. This is precisely the same quantity that at zero temperature determines the ground state of the exponential fall-off of the Wilson loop. Thus, we can extract the same quantity from a Polyakov loop correlator, Eq. (5.107), in the limit $N_\tau \to \infty$. At finite temperatures instead, the Polyakov loop correlator corresponds to the Boltzmann-weighted sum over all excited states of the static potential, and hence to a free energy. This free energy is often called a T-dependent potential, $V(r, T) \equiv F_{\bar{Q}Q}(r, T)$, even though this is not quite justified, as we will discuss below.

The Polyakov loop correlator is readily simulated, with results as in Fig. 5.5. It gives a linearly rising free energy in the confined phase, whose effective string tension reduces with temperature, while in the deconfined phase the potential is screened, thus exhibiting clearly the phase transition from a confined to deconfined phase. However, the screened free energy does not correspond to the Debye-screened potential from Eq. (5.90), as becomes apparent when considering its spatial decay at high T. Fitting the screened free energy to

$$\frac{F_{\bar{Q}Q}}{T} = -\frac{c(T)}{(rT)^d} e^{-m(T)r}, \qquad (5.109)$$

gives $d \approx 1.5$ and $m = M_{0^{++}}$, i.e. the screening mass corresponds to the lightest glueball channel (Hart *et al.*, 2000). This can already be seen in perturbation theory, where the leading term is by two-gluon exchange and thus $m = 2m_D$. Non-perturbatively it follows from the fact that the traced Polyakov loop is a gauge-invariant operator coupling to all powers of gauge fields, hence its correlator at large distance is dominated by the lightest gauge-invariant gluonic screening mass.

For the following, let us consider explicitly $N = 3$ colors. In order to find a thermal potential in analogy to QED Debye screening, attempts have been made to decompose the traced and gauge-invariant Polyakov loop into color singlet and octet configurations of the static sources, $\mathbf{3} \times \bar{\mathbf{3}} = \mathbf{1} + \mathbf{8}$, (McLerran and Svetitsky, 1981; Nadkarni, 1986)

$$e^{-F_{\bar{Q}Q}(r,T)/T} = \frac{1}{9}\, e^{-F_1(r,T)/T} + \frac{8}{9}\, e^{-F_8(r,T)/T}, \tag{5.110}$$

$$e^{-F_1(r,T)/T} = \frac{1}{3}\langle \mathrm{Tr}L^\dagger(\mathbf{x})L(\mathbf{y})\rangle,$$

$$e^{-F_8(r,T)/T} = \frac{1}{8}\langle \mathrm{Tr}L^\dagger(\mathbf{x})\mathrm{Tr}L(\mathbf{y})\rangle - \frac{1}{24}\langle \mathrm{Tr}L^\dagger(\mathbf{x})L(\mathbf{y})\rangle. \tag{5.111}$$

Note that the correlators in the singlet and octet channels are gauge dependent, and the color decomposition only holds perturbatively in a fixed gauge, which has motivated many gauge-fixed lattice simulations. However, both options are unphysical at a non-perturbative level. To understand this, let us start from something physical and consider a meson operator in an octet state, $O^a = \bar{\psi}(\mathbf{x})U(\mathbf{x},\mathbf{x}_0)T^aU(\mathbf{x}_0,\mathbf{y})\psi(\mathbf{y})$, with $\mathbf{x}_0$ the meson's center of mass. In the plasma the color charge can always be neutralized by a gluon. In the correlators for the singlet and octet operators, we integrate out the heavy quarks, replacing them by Wilson lines,

$$\langle O(\mathbf{x},\mathbf{y};0)O^\dagger(\mathbf{x},\mathbf{y};N_\tau)\rangle \propto \langle \mathrm{Tr}L^\dagger(\mathbf{x})U(\mathbf{x},\mathbf{y};0)L(\mathbf{y})U^\dagger(\mathbf{x},\mathbf{y};N_\tau)\rangle,$$

$$\langle O^a(\mathbf{x},\mathbf{y};0)O^{a\dagger}(\mathbf{x},\mathbf{y};N_\tau)\rangle \propto \left[\frac{1}{8}\langle \mathrm{Tr}L^\dagger(\mathbf{x})\mathrm{Tr}L(\mathbf{y})\rangle \right.$$
$$\left. - \frac{1}{24}\langle \mathrm{Tr}L^\dagger(\mathbf{x})U(\mathbf{x},\mathbf{y};0)L(\mathbf{y})U^\dagger(\mathbf{x},\mathbf{y};N_\tau)\rangle\right]. \tag{5.112}$$

We have now arrived at gauge-invariant expressions, because we used a gauge string between the sources. The singlet correlator corresponds to a periodic Wilson loop that wraps around the boundary. The connection to the gauge-fixed correlators is readily established, replacing the gauge string by gauge-fixing functions, $U(\mathbf{x},\mathbf{y}) = g^{-1}(\mathbf{x})g(\mathbf{y})$. Thus, in axial gauge, $U(\mathbf{x},\mathbf{y}) = 1$ (and only there), the gauge-fixed correlators are identical to the gauge-invariant ones.

Repeating the spectral analysis the full correlators take the form (Jahn and Philipsen, 2004)

$$e^{-F_1(r,T)/T} = \frac{1}{Z}\frac{1}{9}\sum_n \langle n_{\delta\gamma}|U_{\gamma\delta}(\mathbf{x},\mathbf{y})U^\dagger_{\alpha\beta}(\mathbf{x},\mathbf{y})|n_{\beta\alpha}\rangle\, e^{-E_n^{\bar{Q}Q}(r)/T},$$

$$e^{-F_8(r,T)/T} = \frac{1}{Z}\frac{1}{9}\sum_n \langle n_{\delta\gamma}|U^a_{\gamma\delta}(\mathbf{x},\mathbf{y})U^{\dagger a}_{\alpha\beta}(\mathbf{x},\mathbf{y})|n_{\beta\alpha}\rangle\, e^{-E_n^{\bar{Q}Q}(r)/T}. \tag{5.113}$$

The energy levels in the exponents are identically the same in Eqs. (5.107) and (5.113) and correspond to the familiar gauge-invariant static potential at zero temperature and its excitations. However, while Eq. (5.107) is purely a sum of exponentials and thus a true free energy, the singlet and octet correlators contain matrix elements that do depend on the operators used, thus giving a path/gauge-dependent weight to the exponentials contributing to F_1, F_8. Since the spectral information contained in the average and gauge-fixed singlet and octet channels is the same, we must conclude that any difference between those correlators is entirely gauge dependent and thus unphysical.

The definition of a temperature-dependent static potential for bound states with the appropriate perturbative Debye-screened limit has recently been achieved. However, it requires a Wilson-loop correlator in real time evaluated in a thermal ensemble (Laine *et al.*, 2007; Brambilla *et al.*, 2008). Since this cannot be done by straightforward lattice simulations, we shall not further discuss it here.

5.3.4 Phase transitions and phase diagrams

A question of utmost importance for experimental programs ranging from heavy-ion collisions to astroparticle physics are the different forms of nuclear matter for various conditions specified by temperature and baryon chemical potential, and whether these are separated by phase transitions. In statistical mechanics, phase transitions are defined as singularities, or non-analyticities, in the free energy as a function of its thermodynamic parameters. However, on finite volumes, free energies are always analytic functions and the theorem of Lee and Yang (Yang and Lee, 1952; Lee and Yang, 1952) states that singularities only develop in the thermodynamic limit of infinitely many particles, or $V \to \infty$. This is particularly obvious in the case of lattice QCD, whose partition function is a functional integral over a compact group with a bounded exponential as an integrand and without zero-mass excitations. It is thus a perfectly analytic function of T, μ, V for any finite V. Hence, a theoretical establishment of a true phase transition requires finite-size scaling (FSS) studies on a series of increasing and sufficiently large volumes to extrapolate to the thermodynamic limit.

Three different situations can emerge: a first-order phase transition is characterized by coexistence of two phases, and hence a discontinuous jump of the order parameter (and other quantities), while a second-order transition shows a continuous transition of the order parameter accompanied by a divergence of the correlation length and some other quantities, like the heat capacity. Finally, a marked change in the physical properties of a system may also occur without any non-analyticity of the free energy, in which case it is called an analytic crossover. A familiar system featuring all these possibilities is water, with a weakening first-order liquid–gas phase transition terminating in a critical endpoint with Z(2) universality, as well as a triple point where the first-order liquid–gas and solid–liquid transitions meet. Similar structures are also conjectured to be present in the QCD phase diagram (Halasz *et al.*, 1998; Rajagopal and Wilczek, 2000), Fig. 5.6, where asymptotic freedom suggests the existence of at

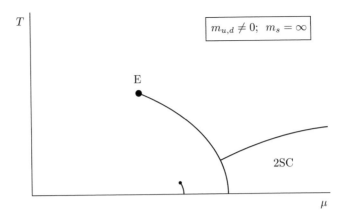

Fig. 5.6   Conjectured phase diagram for $N_f = 2$ QCD with finite light-quark masses. The physical case $N_f = 2 + 1$ is believed to qualitatively look the same, based on universality and continuity arguments as well as input from QCD-like models (Rajagopal and Wilczek, 2000).

least three different regimes: the usual hadronic matter, a quark gluon plasma at high T and low μ, as well as a color superconducting state at low T and high μ.

5.3.5   The (pseudo-) critical coupling and temperature

The first task when investigating a potential phase transition is to locate the phase boundary. Thus, for QCD with a fixed quark content, we are interested in the critical (or pseudo-critical) temperature where the transition from the confining regime to the plasma regime takes place. The method to locate a transition in statistical mechanics usually is to look for rapid changes of suitable observables $O(\mathbf{x})$ and peaks in their susceptibilities. Typical examples are the Polyakov loop, the chiral condensate or the plaquette, $O \in \{\mathrm{Tr}L, \bar{\psi}\psi, \mathrm{Tr}U_p, \ldots\}$. Generalized susceptibilities for $O(\mathbf{x})$ in statistical mechanics are defined as the volume integral over the connected correlation function,

$$\chi_O = \int d^3x \left(\langle O(\mathbf{x})O(0)\rangle - \langle O(\mathbf{x})\rangle\langle O(0)\rangle \right). \tag{5.114}$$

(Note that on the lattice this can be generalized to 4D. However, the thermal equilibrium system likewise lives in three dimensions, which are the ones to be used for finite-size scaling analyses in the following section). If we instead consider the corresponding volume average,

$$\bar{O} = \frac{1}{V} \int d^3x \, O(\mathbf{x}), \tag{5.115}$$

then because of its translation invariance the integration gives trivial volume factors and we obtain on the lattice

$$\chi_{\bar{O}} = N_s^3 (\langle \bar{O}^2 \rangle - \langle \bar{O} \rangle^2) = N_s^3 \langle (\delta \bar{O})^2 \rangle, \tag{5.116}$$

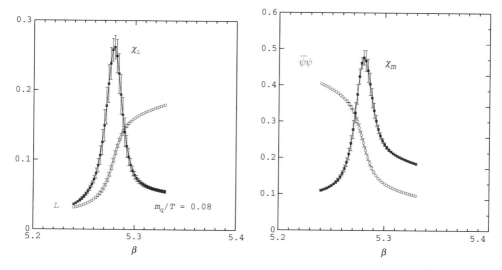

Fig. 5.7 Polyakov loop $\langle \mathrm{Tr} L \rangle$ and chiral condensate $\langle \bar{\psi}\psi \rangle$ together with their susceptibilities signal a transition in two-flavor QCD. From (Karsch, 2002).

with the fluctuations $\delta \bar{O} = \bar{O} - \langle \bar{O} \rangle$. At a phase transition fluctuations are maximal, hence the locations of the peaks of susceptibilities define (pseudo-) critical couplings, $\chi(\beta_c, m_f) = \chi_{\mathrm{max}} \Rightarrow \beta_c(m_f)$, which can be turned into temperatures as discussed in Section 5.2.2. In practice, often the two-loop beta function is used as a short cut, although this becomes valid only when the lattice spacing is fine enough to be in the perturbative regime.

Note that for an analytic crossover the pseudo-critical couplings defined from different observables do not need to coincide. The partition function is analytic everywhere and there is no uniquely specified "transition". This holds in particular for pseudo-critical couplings extracted from finite lattices. As the thermodynamic limit is approached, the couplings defined in different ways will merge where there is a non-analytic phase transition, and stay separate in the case of a crossover, as illustrated in Fig. 5.8 for a putative QCD phase diagram.

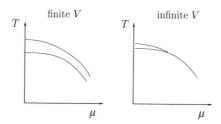

Fig. 5.8 Location of pseudo-critical parameters defined by different observables. In the infinite-volume limit, the lines merge for a true phase transition, but stay separate for a crossover.

5.3.6 Universality, finite-size scaling and signals for criticality

A fascinating phenomenon in physics is the "universality" exhibited by physical systems near critical points of second-order phase transitions. It is due to the divergence of the correlation length, which implies that the entire system acts as a coherent collective. Hence, microscopic physics becomes unimportant, the collective behavior is determined by the global symmetries and the number of dimensions of the system. With the divergence of the correlation length any characteristic length scale disappears from the problem, and thermodynamic observables in the critical region obey scale-invariant power laws. For example, at a second-order ferromagnetic phase transition, the magnetization (the order parameter) vanishes as $M \sim |t|^\beta$ with $t = (T - T_c)/T_c$, while the specific heat, the magnetic susceptibility and the correlation length diverge, $C \sim |t|^{-\alpha}$, $\chi \sim |t|^{-\gamma}$ and $\xi \sim |t|^{-\nu}$, respectively. The critical exponents $\alpha, \beta, \gamma, \nu$ and similar ones for other quantities are the same for all systems within a universality class. The latter are usually labelled by spin models, since those can be readily solved numerically. Unfortunately, for most systems, and particularly QCD, it is not obvious which global symmetries the system will exhibit at a critical point, since those are a dynamically determined subset of the total symmetry of the theory. Moreover, since there is no true order parameter in the case of finite quark masses, fields and parameters of QCD map into those of the effective model in a non-trivial way.

Nevertheless, general scaling properties can be derived, we follow (Karsch *et al.*, 2001). In the vicinity of a critical point, the dynamics of QCD will be governed by an effective Hamiltonian in analogy to a spin model, with energy-like and magnetization-like (extensive) operators E, M that couple to two relevant couplings,

$$\frac{H_{eff}}{T} = \tau E + hM. \tag{5.117}$$

In the case of the Ising model τ and h are proportional to temperature and magnetic field, respectively. The power-law behavior for thermodynamic functions near a critical point follows from the scaling form of the singular part of the free energy,

$$f_s(\tau, h) = b^{-d} f_s(b^{D_\tau} \tau, b^{D_h} h), \tag{5.118}$$

with a dimensionless scale factor $b = LT = N_s/N_\tau$, the spatial dimension d and the dimensions of the scaling fields D_τ, D_h. Using general scaling relations between those and the critical exponents,

$$D_\tau = \frac{1}{\nu}, \quad \gamma = \frac{2D_h - d}{D_\tau}, \quad \alpha = 2 - \frac{d}{D_\tau}, \tag{5.119}$$

one derives the scaling of the susceptibilities as

$$\chi_E = V^{-1}\langle(\delta E)^2\rangle = -\frac{1}{T}\frac{\partial^2 f}{\partial \tau^2} \sim b^{\alpha/\nu},$$

$$\chi_M = V^{-1}\langle(\delta M)^2\rangle = -\frac{1}{T}\frac{\partial^2 f}{\partial h^2} \sim b^{\gamma/\nu}. \tag{5.120}$$

(Note the different volume factors between Eqs. (5.116) and (5.120), which is due to the fact that Eq. (5.120) uses extensive variables, while in Eq. (5.116) we use averages.) In QCD, the operators E, M will be mixtures of the plaquette action and the chiral condensate (and possibly higher-dimension operators), while τ and h are functions of β, m_f, μ. Thus, when measuring susceptibilities constructed from operators as in Eq. (5.116), these will in turn be mixtures of E and M. Therefore, in the thermodynamic limit all of them will show identical FSS behavior, which is dominated by the larger of α/ν and γ/ν. For the universality classes relevant for QCD, $Z(2)$ and $O(4), O(2)$, this is γ/ν.

By contrast, at a first-order phase transition the fluctuations diverge proportional to the spatial volume,

$$\chi_{\bar{O}} \sim V, \tag{5.121}$$

while for an analytic crossover the peaks of susceptibilities saturate at some finite maximal value in the thermodynamic limit.

Another possibility to detect phase transitions is to evaluate the partition function for complex values of the coupling β and look for Lee–Yang zeroes of the partition function, $Z(\beta) = 0$. On a finite volume, these will be at complex values of the coupling, since the partition function is analytic for real parameter values. On larger volumes a Lee–Yang zero approaches the real axis if there is a true phase transition. However, such evaluations require multihistogram and reweighting techniques (cf. Section 5.4.3), where the values of the observable at different parameter values are calculated from one simulation point, a technique that has to be handled with care (Ejiri, 2006). Figure 5.9 (left) shows an example of Lee–Yang zeroes for $SU(3)$ pure gauge theory at complex values of the coupling.

A more straightforward but numerically very expensive observable to determine the order of a phase transition is the Binder cumulant constructed from moments of fluctuations of the order parameter,

$$B_4(m, \mu) = \frac{\langle (\delta \bar{O})^4 \rangle}{\langle (\delta \bar{O})^2 \rangle^2}. \tag{5.122}$$

It is to be evaluated on the phase boundary $\beta = \beta_c(m_f, \mu)$, where the third moment of the fluctuation vanishes, $\langle (\delta \bar{O})^3 \rangle (\beta_c) = 0$. This observable is particularly well suited to locate the change from a first-order transition to a crossover regime as a function of some parameter, like quark mass or chemical potential. In the infinite-volume limit, $B_4 \to 1$ or 3 for a first-order transition or crossover, respectively, whereas it approaches a value characteristic of the universality class at a critical point. For $Z(2)$ (or 3D Ising) universality one has $B_4 \to 1.604$. Hence, when examining the change of a phase transition from a weakening first-order transition to a critical end-point and a crossover as a function of $x = am_f, a\mu$, B_4 is a non-analytic step function, which gets smoothed out to an analytic curve on finite volumes, with a slope increasing with volume to gradually approach the step function, Fig. 5.9 (right). Near a critical point the correlation length diverges as $\xi \sim r^{-\nu}$, where r is the distance to the critical point in the plane of temperature and magnetic-field-like variables. Since the gauge coupling

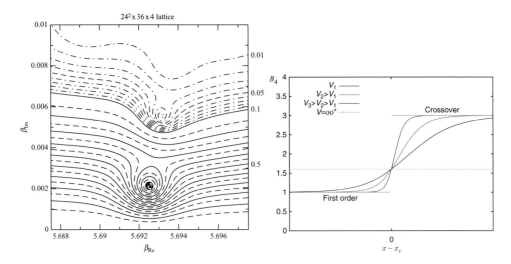

Fig. 5.9 Left: Lee–Yang zeroes in the complex β-plane for SU(3) pure gauge theory (Ejiri, 2006). Right: The Binder cumulant as a function of some parameter x that takes the phase transition from a first-order to a crossover regime, passing through a critical point of 3D Ising universality. (For example, $x = m_f$ in the case of $N_f = 3, \mu = 0$, cf. Fig. 5.10).

is tuned to $\beta = \beta_c(x)$, $r \sim |x - x_c|$ and B_4 is a function of the dimensionless ratio L/ξ, or equivalently, $(L/\xi)^{1/\nu}$. It can therefore be expanded as

$$B_4(x, N_s) = b_0 + b\, N_s^{1/\nu}(x - x^c) + \dots \qquad (5.123)$$

Close to the thermodynamic limit all curves intersect at a critical B_4 value, and moreover ν takes its universal value. Thus calculation of B_4 around a critical point gives access to two independent pieces of information governed by universality.

5.3.7 The nature of the QCD phase transition for $N_f = 2 + 1$ at $\mu = 0$

The qualitative picture for the order of the QCD phase transition at zero baryon density as a function of the quark masses is outlined schematically in Fig. 5.10. As discussed in Sections 5.2.5 and 5.2.6, for $N_f = 3$ in the limits of zero and infinite quark masses (lower left and upper right corners), order parameters corresponding to the breaking of a symmetry can be defined, implying true phase transitions. One finds numerically at small and large quark masses that a first-order transition takes place at a finite temperature T_c. On the other hand, one observes an analytic crossover at intermediate quark masses. Hence, each corner must be surrounded by a region of first-order transition, bounded by a second-order line. The line bounding the chiral transitions is commonly called the chiral critical line, the one bounding the deconfinement transition is accordingly the deconfinement critical line. We know from simulations on $N_\tau = 4$ with staggered fermions (de Forcrand and Philipsen, 2007) that this line is to the lower left of the physical point, and simulations on $N_\tau = 6$ (de Forcrand *et al.*, 2007; Endrődi *et al.*, 2007) show that it shrinks towards the lower

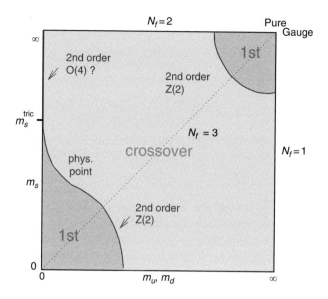

Fig. 5.10 Order of the QCD phase transition as a function of quark masses $(m_{u,d}, m_s)$ at $\mu = 0$.

left corner on finer lattices. A continuum extrapolation for the chiral critical line is not yet available. However, there are continuum extrapolated simulations with staggered quarks with physical masses that show that the $\mu = 0$ transition in QCD is a crossover (Aoki *et al.*, 2006).

The situation is not yet settled for the chiral limit of the two-flavor theory in the upper left corner. If the transition is second order, then chiral symmetry $SU(2)_L \times SU(2)_R \sim O(4)$ puts it in the universality class of 3d $O(4)$ spin models. In this case there must be a tricritical strange-quark mass m_s^{tric}, where the second-order chiral transition ends and the first-order region begins. The exponents at such a tricritical point would correspond to 3D mean field (Lawrie and Sarbach, 1984). On the other hand, a first-order scenario for the chiral limit of $N_f = 2$ so far has not been conclusively ruled out. In fact, it has been shown in a model with the same chiral symmetries as $N_f = 2$ QCD and a tunable $U_A(1)$ anomaly that both scenarios are possible and the order of the transition depends on the strength of the anomaly at the critical temperature (Chandrasekharan and Mehta, 2007).

5.4 Lattice QCD at non-zero baryon density

5.4.1 Implementing chemical potential

So far, we have considered lattice QCD at finite temperature with net baryon number zero, i.e. with a complete balance between baryons and antibaryons. This situation is realized during the evolution of the early Universe because the matter antimatter asymmetry is so tiny there. However, there are interesting questions about the nature of dense nuclear matter in the center of compact stars, and heavy-ion-collision experiments obviously also operate at finite baryon number. Thus, we now wish to use the grand canonical ensemble with a chemical potential μ for quark number Q (recall that each quark carries $1/3$ of baryon number, $Q = B/3, \mu = \mu_B/3$),

$$Z = \hat{\mathrm{Tr}}\, e^{-(H-\mu Q)}, \quad Q = \int d^3x\, \bar{\psi}(x)\gamma_0\psi(x) = \int d^3x\, \psi^\dagger(x)\psi(x). \tag{5.124}$$

For the following sections we recall that in Euclidean space-time $\gamma_\mu = \gamma_\mu^\dagger$, $\{\gamma_5, \gamma_\mu\} = 0$. Under charge conjugation, $C = \gamma_0\gamma_2$, the fermion number changes sign,

$$A_\mu^C = -A_\mu^*, \quad \psi^C = \gamma_0\gamma_2\bar{\psi}^T, \quad \bar{\psi}^C\gamma_0\psi^C = -\bar{\psi}\gamma_0\psi. \tag{5.125}$$

Thus, $\mu > 0$ represents a net baryon number and $\mu < 0$ a net antibaryon number. From this, one derives an important symmetry of the partition function. Since the measure and the fermion action for $\mu = 0$ are invariant under charge conjugation, we find that the partition function is an even function of chemical potential,

$$Z(\mu) = \int DA^C\, D\bar{\psi}^C D\psi^C \exp - \left[S_g^C + S_f^C(\mu = 0) - \mu \int_0^{1/T} dx_0\, Q^C \right]$$

$$= \int DA\, D\bar{\psi}D\psi \exp - \left[S_g + S_f(\mu = 0) + \mu \int_0^{1/T} dx_0\, Q \right]$$

$$= Z(-\mu). \tag{5.126}$$

In a straightforward implementation of chemical potential on the lattice we would simply replace the integral by a lattice sum,

$$S_f[M(\mu)] = S_f[M(0)] + a\mu \sum_x \psi(x)\gamma_0\psi(x). \tag{5.127}$$

However, when computing the energy density in lattice perturbation theory, one finds it diverges in the continuum limit,

$$\epsilon = \frac{1}{V}\frac{\partial}{\partial(\frac{1}{T})} \ln Z \xrightarrow{a \to 0} \infty. \tag{5.128}$$

This happens after duly subtracting the divergent vacuum contribution, and thus the chemical-potential term in Eq. (5.127) appears to spoil renormalizability. This would violate our earlier observation that renormalization of the vacuum theory is also sufficient for the theory at finite temperature and density. The reason is that the discretization violates a generalized symmetry of the continuum theory (Hasenfratz and Karsch, 1983). Note that the term μQ in the continuum looks like the zero component of the fermion current, $j^0 = \bar{\psi}\gamma^0\psi$, coupling to an external $U(1)$ gauge field,

$$\mu Q = -igA_0 j_0 \quad \text{with} \quad A_0 = i\frac{\mu}{g}. \tag{5.129}$$

Chemical potential for quark number is equivalent to a classical electromagnetic field A_0 with a constant imaginary value. The quark-number term can therefore be absorbed into the covariant derivative of the Dirac action, which then is invariant under $U(1)$ gauge transformations of the quark fields and A_0. This symmetry protects against new divergences and gets broken by a lattice implementation as in Eq. (5.127). The problem is solved by implementing chemical potential in the same way as an external $U(1)$ gauge field, namely as an additional temporal link variable

$$U_{0,\text{ext}} = e^{iagA_0} = e^{-a\mu}, \tag{5.130}$$

which multiplies all non-abelian temporal links. For example, the Wilson action with finite chemical potential then takes the form

$$
\begin{aligned}
S_f^W = a^3 \sum_x \Big(& \bar\psi(x)\psi(x) \\
& - \kappa \left[e^{a\mu} \bar\psi(x)(r - \gamma_0)U_0(x)\psi(x - \hat{0}) + e^{-a\mu}\bar\psi(x + \hat{0})(r + \gamma_0)U_0^\dagger(x)\psi(x) \right] \\
& - \kappa \sum_{j=1}^{3} \left[\bar\psi(x)(r - \gamma_j)U_j(x)\psi(x + \hat{j}) + \bar\psi(x + \hat{j})(r + \gamma_j)U_j^\dagger(x)\psi(x) \right] \Big).
\end{aligned} \tag{5.131}
$$

As usual on the lattice, this discretization is not unique, only the continuum limit is. For alternate ways of implementing chemical potential, see (Gavai, 1985). It is now easy to verify that

$$\det(\slashed{D}(U^\dagger) + m + \gamma_0\mu) = \det(\slashed{D}(U) + m - \gamma_0\mu), \tag{5.132}$$

and since $S_g[U^\dagger] = S_g[U]$ and $DU^\dagger = DU$, we have $Z(\mu) = Z(-\mu)$ on the lattice as well.

5.4.2 The sign problem

Let us consider what happens to the Dirac operator in the presence of chemical potential. Reality of the fermion determinant follows from the γ_5-hermiticity of the Dirac operator,

$$(\slashed{D} + m)^\dagger = \gamma_5(\slashed{D} + m)\gamma_5. \tag{5.133}$$

This relation is satisfied in the continuum as well as for all standard lattice Dirac operators. Now consider a complex chemical potential,

$$\gamma_5(\slashed{D} + m - \gamma_0\mu)\gamma_5 = (-\slashed{D} + m + \gamma_0\mu) = (\slashed{D} + m + \gamma_0\mu^*)^\dagger. \tag{5.134}$$

For the fermion determinant this implies,

$$\det(\slashed{D} + m - \gamma_0\mu) = \det^*(\slashed{D} + m + \gamma_0\mu^*), \tag{5.135}$$

which is complex unless $\text{Re}\,\mu = 0$. However, a complex fermion determinant cannot be interpreted as a probability measure for importance sampling and the evaluation of the path integral by Monte Carlo methods is spoiled. This is known as the "sign problem" of QCD.

It should be stressed that there is nothing wrong with the theory at finite μ. In particular, the partition function as well as the free energy and the other thermodynamic functions are all real after performing the path integral over the gauge fields, i.e. the imaginary parts of the fermion determinant in the background of gauge configurations average to zero. Being a property of the Dirac operator, the sign problem is generic for fermionic systems with particles and antiparticles or particles and holes, and is necessary for a correct description of the physics. For example, for the expectation value of the Polyakov loop we have

$$\langle \mathrm{Tr}L \rangle = e^{-\frac{F_Q}{T}} = \langle \mathrm{Re}\mathrm{Tr}L \, \mathrm{Re}\det M - \mathrm{Im}\mathrm{Tr}L \, \mathrm{Im}\det M \rangle_g,$$

$$\langle (\mathrm{Tr}L)^* \rangle = e^{-\frac{F_{\bar{Q}}}{T}} = \langle \mathrm{Re}\mathrm{Tr}L \, \mathrm{Re}\det M + \mathrm{Im}\mathrm{Tr}L \, \mathrm{Im}\det M \rangle_g, \tag{5.136}$$

where the angular brackets denote path integration with respect to the pure gauge action, $\langle \ldots \rangle_g = \int DU \ldots \exp -S_g[U]$. Thus, the free energy of a static quark or an antiquark in a plasma differ if and only if $\mathrm{Im}\det \neq 0$, i.e. for finite chemical potential μ. If there is no net quark or baryon number, it costs equal amounts of energy to insert a quark or an antiquark into the plasma, whereas this is different if the plasma already has a net baryon number. The origin of the sign problem may thus be traced to the behavior under charge conjugation,

$$\det(\slashed{D} + m - \gamma_0\mu) \xrightarrow{C} \det(\slashed{D} + m + \gamma_0\mu). \tag{5.137}$$

The "sign problem" thus is only a problem for a Monte Carlo evaluation of the partition function with importance-sampling methods. In some spin models, it can be solved by cluster algorithms, which are able to identify conjugate configurations so as to cancel imaginary parts during the simulation (Wiese, 2002). Unfortunately, these work only for special classes of Hamiltonians and QCD does not appear to be one of them. Another idea is to simply avoid importance sampling and evaluate expectation values by means of stochastical quantization, or Langevin algorithms (Karsch and Wyld, 1985). However, also in this case there are difficulties, for complex actions there is no proof that the Langevin method converges to the correct expectation value. While the method seems to work for various model systems, there appears to be no successful application to QCD yet.

Due to the symmetry properties of the determinant, there are a number of interesting special cases permitting a Monte Carlo treatment. As long as the fermion determinant is real, we are able to simulate even numbers of degenerate flavors with it. Eq. (5.135) shows that this is the case for purely imaginary $\mu = i\mu_i, \mu_i \in \mathbb{R}$, and we shall discuss later how this can be used to learn something about real μ.

Another interesting situation for $N_f = 2$ is when the chemical potential for u and d quarks are opposite, i.e. $\mu_u = -\mu_d \equiv \mu_I$, which corresponds to a chemical potential for isospin (Son and Stephanov, 2001). In this case we have for the determinants

$$\det(\slashed{D} + m - \gamma_0\mu_I)\det(\slashed{D} + m + \gamma_0\mu_I) = |\det(\slashed{D} + m - \gamma_0\mu_I)|^2 \geq 0. \tag{5.138}$$

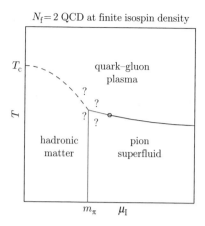

$N_\mathrm{f} = 2$ QCD at finite isospin density

Fig. 5.11 Schematic phase diagram for QCD at finite isosipin density.

Note that isospin is not conserved in the real world due to the electroweak interactions. Nevertheless, one can investigate interesting finite-density effects with this theory by actual simulations (Kogut and Sinclair, 2008). For example, one obtains a condensate of charged pions $\langle \pi^- \rangle \neq 0$ when $|\mu_I| > m_\pi$, Fig. 5.11.

Finally, the determinant is real for a two-color version of QCD with gauge group $SU(2)$. This is because this group has real representations, i.e. there is a matrix S such that $ST^a S^{-1} = -T^{a^*}$ with $S = \sigma^2$, the Pauli matrix. Now consider $S = C\gamma_5 \sigma^2$, and using $C\gamma_\mu C^{-1} = -\gamma_\mu^T$ we have

$$
\begin{aligned}
S[\slashed{D} + m - \gamma_0 \mu]S^{-1} &= C\gamma_5 \sigma^2 [\gamma_\mu(\partial_\mu - igA_\mu) + m - \gamma_0 \mu]\sigma^2 \gamma_5 C^{-1} \\
&= C\gamma_5 [\gamma_\mu(\partial_\mu + igA_\mu^*) + m - \gamma_0 \mu]\gamma_5 C^{-1} \\
&= [\gamma_\mu^T(\partial_\mu + igA_\mu^*) + m - \gamma_0^T \mu] \\
&= [\slashed{D} + m - \gamma_0 \mu^*]^*.
\end{aligned}
\tag{5.139}
$$

For real μ we have $\det M = \det^* M$, likewise permitting simulations to study qualitative density effects (Hands *et al.*, 2008).

In the following sections, the main concern will be with methods for the theory of immediate experimental interest, i.e. three-color QCD with real quark number chemical potential. As will become apparent, all those methods side-step the sign problem by introducing approximations that work only for sufficiently small chemical potential, empirically one finds these methods to be good as long as $\mu \lesssim T$.

5.4.3 Reweighting

Reweighting techniques are widely used in Monte Carlo simulations with importance sampling, in particular in order to interpolate data between simulation points when studying phase transitions (Ferrenberg and Swendsen, 1988). In the context of finite density physics, this technique is used to extrapolate to finite-chemical potential. The basic idea is to rewrite the path integral exactly,

$$Z(\mu) = \int DU \, \det M(\mu) \, e^{-S_g[U]} = \int DU \, \det M(0) \, \frac{\det M(\mu)}{\det M(0)} \, e^{-S_g[U]}$$

$$= Z(0) \left\langle \frac{\det M(\mu)}{\det M(0)} \right\rangle_{\mu=0}. \tag{5.140}$$

The angular brackets now denote averaging over an ensemble with the same measure $DU \, \det M(0)$ as for zero density, i.e. the path integral can be interpreted as an expectation value of a reweighting factor given by the ratio of determinants. While this is a mathematical identity, its practical evaluation by Monte Carlo methods turns out to be impressively difficult. First, a numerical evaluation of this path integral requires the calculation of the reweighting factor configuration by configuration and is expensive. This is aggravated by a signal-to-noise ratio that worsens exponentially with volume because the free energy is an extensive quantity $F = Vf$,

$$\frac{Z(\mu)}{Z(0)} = \exp -\frac{F(\mu) - F(0)}{T} = \exp -\frac{V}{T} \left(f(\mu) - f(0) \right). \tag{5.141}$$

Hence, the reweighting factor gets exponentially small as we increase V or μ/T, requiring exponentially increased statistics for its determination. The problem can be illustrated with a one-dimensional Gaussian integral with a complex "action",

$$Z(\mu) = \int dx \, e^{-x^2 + i\mu x}. \tag{5.142}$$

Figure 5.12 (left) shows the real part of the integrand for $\mu = 0$ and $\mu = 30$. It is obvious that the oscillatory bevahior of the integrand will rapidly become prohibitive for Monte Carlo integration with growing μ.

The second problem is known as the overlap problem. When we evaluate the integral with importance sampling, the integrand will be calculated the more frequently the larger its contribution to the integral. However, when we employ reweighting, the configurations will be generated with the measure $DU \, \det M(0)$, whose probability distribution is shifted compared to that of $DU \, \det M(\mu)$, Fig. 5.12 (right). If we had infinite statistics, this would not matter as long as our algorithm is ergodic and the entire configuration space is covered. The point of Monte Carlo methods is, however, to get away with a few "important" configurations on which we evaluate our determinant and observables. But if the difference between the integrands at $\mu = 0$ and $\mu \neq 0$ gets too large, the most frequent configurations are the unimportant ones. It is apparent that we only get a useful estimate for observables as long as there is sufficient overlap between the two integrands. The difficulty with this method is to realize when it fails. The statistical errors are determined by the fluctuations within the generated ensemble and will appear small, unless there are configurations that have tunneled far into the tail where the actual integrand is important.

When searching for a phase transition at some fixed $\beta_c(\mu)$, one-parameter reweighting uses an ensemble at $\beta_c(\mu), \mu = 0$, which is non-critical since $T_c(\mu) < T_c(0)$, thus missing important dynamics. Significant progress enabling finite-density simulations was made a few years ago, by a generalization to reweighting in two parameters (Fodor and Katz, 2002). The partition function is now rewritten as

$$Z(\mu, \beta) = Z(0, \beta_0) \left\langle \frac{e^{-S_g(\beta)} \det(M(\mu))}{e^{-S_g(\beta_0)} \det(M(\mu = 0))} \right\rangle_{\mu=0, \beta_0}, \qquad (5.143)$$

where the ensemble average is generated at $\mu = 0$ and a lattice gauge coupling β_0, while a reweighting factor takes us to the values μ, β of interest. In this way, one can use an ensemble on the phase boundary at $\mu = 0$, which actually samples both phases, and then reweight along the (pseudo-)critical line to the desired value of μ. On physical grounds, this is drastically improving the overlap when attempting to describe a phase transition.

Clearly, the choice of reweighting factors is not unique and one may ask for an optimal choice. A useful criterion is to minimize fluctuations in the reweighting factor, Eq. (5.140) (de Forcrand *et al.*, 2003). The optimal setup with a feasible Monte Carlo implementation is to split the determinant into modulus and phase, $\det M = |\det M| e^{i\theta}$, and employ the modulus for the generation of the ensemble. Note that this is equivalent to an ensemble with finite isospin chemical potential. Thus,

$$Z(\mu) = Z(\mu_I) \langle e^{i\theta} \rangle_{\mu_I}. \qquad (5.144)$$

Numerical experiments show that for small to moderate chemical potentials, this choice is advantageous (de Forcrand *et al.*, 2003) compared to the standard procedure. Moreover, it serves for interesting theoretical insights. There are indications in QCD (Splittorff, 2005) and firm results in random matrix models (Han and

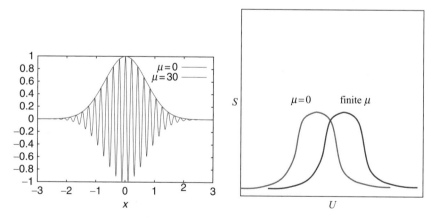

Fig. 5.12 Left: The integrand of Eq. (5.142) for different μ. Right: In order for reweighting to work, the probability distribution of the Monte Carlo ensemble must have sufficient "overlap" with the target integrand.

Stephanov, 2008) that the phase transition to the pion condensate in the theory with finite isospin chemical potential is directly related to the strength of the sign problem, which becomes maximal with an average sign near zero in the neighborhood of that transition, thus signalling a breakdown of reweighting methods there.

5.4.4   Finite density by Taylor expansion

Another straightforward way to employ Monte Carlo ensembles generated at $\mu = 0$ to learn about finite μ is by Taylor expanding the partition function in μ/T. The grand canonical partition function is an analytic function of its parameters away from phase transitions. The Lee–Yang theorem tells us that on finite volumes there are no non-analytic phase transitions, and so the partition function must be analytic everywhere. It is then natural to consider power series in μ/T for thermodynamic observables like the pressure (Allton *et al.*, 2002),

$$\frac{p}{T^4} = \sum_{n=0}^{\infty} c_{2n}(T) \left(\frac{\mu}{T}\right)^{2n} \equiv \Omega(T,\mu). \tag{5.145}$$

Note that there are only even powers of μ/T because of the reflection symmetry, Eq. (5.126). The leading Taylor coefficient is just the pressure at zero density, and all higher coefficients are derivatives evaluated at $\mu = 0$,

$$c_0(T) = \frac{p}{T^4}(T,\mu=0), \quad c_{2n}(T) = \frac{1}{(2n)!} \left.\frac{\partial^{2n}\Omega}{\partial(\frac{\mu}{T})^{2n}}\right|_{\mu=0}. \tag{5.146}$$

These can of course be calculated by standard Monte Carlo techniques. Once the coefficients are available, the series for other thermodynamic quantities follow, like the quark number density or the quark number susceptibility, respectively,

$$\frac{n}{T} = \frac{\partial\Omega}{\partial(\frac{\mu}{T})} = 2c_2\frac{\mu}{T} + 4c_4\left(\frac{\mu}{T}\right)^3 + \dots,$$

$$\frac{\chi_q}{T^2} = \frac{\partial^2\Omega}{\partial(\frac{\mu}{T})^2} = 2c_2 + 12c_4\left(\frac{\mu}{T}\right)^2 + 30c_6\left(\frac{\mu}{T}\right)^4 + \dots \tag{5.147}$$

Clearly, this strategy can be applied to any observable of interest and is well defined computationally. As one needs to calculate coefficient by coefficient, the full functions are only well approximated as long as the series converges sufficiently rapidly, i.e. for sufficiently small μ/T.

In practice, what needs to be evaluated during a simulation are the derivatives of a particular observable with respect to chemical potential,

$$\frac{\partial\langle O\rangle}{\partial\mu} = \left\langle\frac{\partial O}{\partial\mu}\right\rangle + N_f\left(\left\langle O\frac{\partial\ln\det M}{\partial\mu}\right\rangle - \langle O\rangle\left\langle\frac{\partial\ln\det M}{\partial\mu}\right\rangle\right). \tag{5.148}$$

Using $\det M = \exp \operatorname{Tr} \ln M$, this is converted into expressions like

$$\frac{\partial \ln \det M}{\partial \mu} = \operatorname{Tr}\left(M^{-1}\frac{\partial M}{\partial \mu}\right)$$

$$\frac{\partial^2 \ln \det M}{\partial \mu^2} = \operatorname{Tr}\left(M^{-1}\frac{\partial^2 M}{\partial \mu^2}\right) - \operatorname{Tr}\left(M^{-1}\frac{\partial M}{\partial \mu}M^{-1}\frac{\partial M}{\partial \mu}\right),$$

etc. $\qquad\qquad$ (5.149)

The advantage is that the derivatives of the fermion determinant are represented by local operators, which can be evaluated using random noise vectors. But it is clear that higher-order derivatives quickly turn into very complex expressions involving many cancellations, and hence are very cumbersome to evaluate with controlled accuracy. For a discussion of this point, see (Gavai and Gupta, 2005).

5.4.5 QCD at imaginary chemical potential

The hermiticity relation Eq. (5.135) tells us that the QCD fermion determinant is real for imaginary chemical potential $\mu = i\mu_i$. For such a parameter choice we can simulate without sign problem with no more complications than at $\mu = 0$, and moreover we can do so with an ensemble that is actually sensitive to chemical potential. In order to get back to real chemical potential, we can use once more the Taylor expansion,

$$\langle O \rangle(\mu_i) = \sum_{k=1}^{N} c_k \left(\frac{\mu_i}{T}\right)^{2k}, \qquad\qquad (5.150)$$

with only even powers for observables without explicit μ-dependence. Note that in this case the Monte Carlo results represent the left side of the equation and contain no approximation beyond the usual finite-size and cutoff effects. Thus, if there are sufficiently many and accurate data points, one can test whether the observable is well represented by a truncated Taylor expansion, and if this is the case one may analytically continue, $\mu_i \rightarrow -i\mu_i$.

In order to apply this technique, it is necessary to discuss a few general properties of the partition function. From the form of the grand canonical density operator, $Z = \hat{\operatorname{Tr}}\exp-(H - \mu Q)$, it is clear that the partition function is going to be periodic for imaginary μ. Is this a sensible theory? Modifying the discussion around Eq. (5.129), imaginary chemical potential is formally equivalent to a real external $U(1)$ field, so this parameter choice is well defined. Next, let us examine the periodicity (Roberge and Weiss, 1986). Since we always have a compactified temporal lattice direction, we can eliminate the quark-number term from the action and absorb it by the modified boundary condition,

$$Z^{(1)}(i\mu_i) = \int DU \det M(0)\mathrm{e}^{-S_g}, \quad \text{b.c.:}\quad \psi(\tau + N_\tau, \mathbf{x}) = -\mathrm{e}^{i\frac{\mu_i}{T}}\psi(\tau, \mathbf{x}). \qquad (5.151)$$

Now, let us consider gauge transformed fields $\psi^{g'}(x), U_\mu^{g'}(x)$, using the large gauge transformations discussed in Section 5.2.5, i.e.

$$g'(\tau + N_\tau, x) = e^{-i\frac{2\pi n}{N}} g'(\tau, \mathbf{x}).$$
(5.152)

The measure and the QCD action for $\mu = 0$ are invariant under such a transformation, so the new form of the partition function is

$$Z^{(2)}(i\mu_i) = \int DU \det M(0)e^{-S_g}, \quad \text{b.c.:} \quad \psi(\tau + N_\tau, \mathbf{x}) = -e^{-i\frac{2\pi n}{N}} e^{i\frac{\mu_i}{T}} \psi(\tau, \mathbf{x}).$$
(5.153)

We now observe that

$$Z^{(2)}\left(i\frac{\mu_i}{T} + i\frac{2\pi n}{N}\right) = Z^{(1)}\left(i\frac{\mu_i}{T}\right).$$
(5.154)

Since we have obtained one partition function from the other by a gauge transformation, the two give completely equivalent descriptions of physics, so we can conclude for the QCD partition function

$$Z\left(i\frac{\mu_i}{T} + i\frac{2\pi n}{N}\right) = Z\left(i\frac{\mu_i}{T}\right).$$
(5.155)

Comparing Eqs. (5.151) and (5.152), we see that for discrete values $\mu_i/T = 2\pi n/N$ an imaginary chemical potential is equivalent to a center transformation, and Eq. (5.155) tells us that the partition function is symmetric under such transformations, and therefore periodic. We have thus established two remarkable properties: the period of the QCD partition function in the presence of an imaginary chemcial potential is $2\pi/N$, and there is a good $Z(N)$ center symmetry even in the presence of finite mass quarks!

Let us now specialize to the physical case, $N = 3$. We recall from Section 5.2.5 that the Polyakov loop closes through the temporal boundary and hence picks up phases when changing the center sector, Eq. (5.67). Therefore the phase of the Polyakov loop is an observable to identify the different $Z(3)$-sectors as the imaginary chemical potential is increased. There are $Z(3)$ transitions between neighboring center sectors for all $(\mu_i/T)_c = \frac{2\pi}{3}\left(n + \frac{1}{2}\right), n = 0, \pm 1, \pm 2, \ldots$. It has been numerically verified that these transitions are first order for high temperatures and a smooth crossover for low temperatures (de Forcrand and Philipsen, 2002; D'Elia and Lombardo, 2003).

Figure 5.13 (left) shows a schematic phase diagram. The vertical lines represent the high-temperature, first-order $Z(3)$-transitions. On the other hand, at $\mu = 0$ we also have the chiral/deconfinement transition, which analytically continues to imaginary chemical potential, as indicated by the dotted lines. As discussed before, its order depends on the quark-mass combinations. It joins the $Z(3)$ transitions in their endpoint, whose nature has recently been investigated by explicit simulations for $N_f = 2, 3$ (D'Elia and Sanfilippo, 2009; de Forcrand and Philipsen, 2010). For small and large quark masses, where there is a first-order chiral and deconfinement transition, respectively, three first-order lines join up at this point, rendering it a triple point. For intermediate quark masses the quark–hadron transition is only a crossover, and the $Z(3)$ transition features an endpoint with 3D Ising universality.

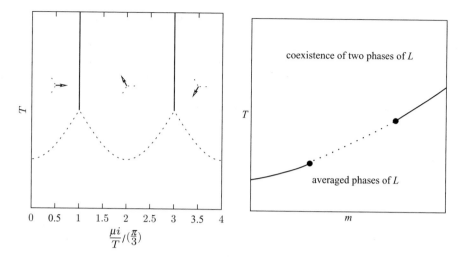

Fig. 5.13 Left: Periodic phase diagram for imaginary chemical potential. Vertical lines correspond to first-order transitions between the different $Z(3)$-sectors, the arrows indicate the corresponding phase of the Polyakov loop. The $\mu = 0$ chiral/deconfinement transition continues to imaginary chemical potential, its order depends on N_f and the quark masses. Right: Nature of the junction at fixed $\mu_i/T = \pi/3$ mod. $2\pi/3$. Solid lines are lines of triple points ending in tricritical points, connected by a $Z(2)$-line.

Tuning the quark masses interpolates between these situations, with tricritical points marking the changes, Fig. 5.13 (right). If we want to learn about certain observables for real μ by means of analytic continuation, the radius of convergence is given by the closest non-analyticity due to phase transitions. At high temperatures and for the chiral/deconfinement transition line itself, we are thus limited to the first sector, or $\mu/T < \pi/3$.

5.4.6 The canonical partition function

The grand canonical partition function for imaginary chemical potential appears again in the formulation of finite-density QCD by means of the canonical partition function (Roberge and Weiss, 1986). Consider the Fourier transform of $Z(i\mu_i/T)$,

$$
\begin{aligned}
\tilde{Z}(\bar{Q}) &= \frac{1}{2\pi} \int_{-\pi}^{\pi} d\left(\frac{\mu_i}{T}\right) \, \mathrm{e}^{-i\frac{\mu_i}{T}\bar{Q}} \, Z\left(i\frac{\mu_i}{T}\right) \\
&= \frac{1}{2\pi} \int_{-\pi}^{\pi} d\left(\frac{\mu_i}{T}\right) \, \mathrm{e}^{-i\frac{\mu_i}{T}\bar{Q}} \, Z\left(i\frac{\mu_i}{T} + i\frac{2\pi}{N}\right) \\
&= \frac{1}{2\pi} \int_{-\pi}^{\pi} d\left(\frac{\mu_i}{T}\right) \, \mathrm{e}^{-i\left(\frac{\mu_i}{T} - \frac{2\pi}{N}\right)\bar{Q}} \, Z\left(i\frac{\mu_i}{T}\right).
\end{aligned}
\tag{5.156}
$$

In the second line we have made use of the periodicity of the grand canonical partition function, Eq. (5.155), and in the third we have shifted $\mu_i/T \to \mu_i/T + 2\pi/N$. The first and the third line can only be equal with $\tilde{Z} \neq 0$ if

$$\frac{\bar{Q}}{N} = n, \quad n = 0, 1, \ldots \tag{5.157}$$

Inserting the grand canonical partition function with the quark number operator Q, we can write

$$\tilde{Z}(\bar{Q}) = \frac{1}{2\pi} \int_{-\pi}^{\pi} d\left(\frac{\mu_i}{T}\right) e^{-i\frac{\mu_i}{T}\bar{Q}} \mathrm{Tr}\left(e^{-\frac{(H - i\mu_i Q)}{T}}\right)$$

$$= \mathrm{Tr}\left(e^{-\frac{H}{T}} \delta(\bar{Q} - Q)\right), \tag{5.158}$$

where we have used the integral representation of the delta function,

$$\delta(\bar{Q} - Q) = \frac{1}{2\pi} \int d\left(\frac{\mu_i}{T}\right) \exp\left(i\frac{\mu_i(\bar{Q} - Q)}{T}\right). \tag{5.159}$$

We thus arrive at the conclusion that $\tilde{Z}(\bar{Q})$ corresponds to the QCD partition function evaluated with a delta-constraint fixing quark number to be $\bar{Q}$. This is just the canonical partition function for QCD,

$$\tilde{Z}(\bar{Q}) = Z_C(\bar{Q}). \tag{5.160}$$

Since quark number is constrained to be $\bar{Q} = nN$, baryon number $B = N\bar{Q}$ must be an integer.

The descriptions in terms of the canonical or the grand canonical partition function are identical in the thermodynamic limit. The grand canonical ensemble can be obtained from the canonical one by the fugacity expansion,

$$Z(V, T, \mu) = \sum_{B=1}^{\infty} e^{\frac{\mu_B B}{T}} Z_C(V, T, B)$$

$$Z(T, \mu) \overset{V \to \infty}{=} \int_{-\infty}^{\infty} d\rho\, e^{VN\rho\frac{\mu}{T}} Z_C(T, \rho)$$

$$= \lim_{V \to \infty} \int_{-\infty}^{\infty} d\rho\, e^{-\frac{V}{T}(f(\rho) - \mu\rho)}, \tag{5.161}$$

where $B = \rho V$ and f has now been defined by the canonical ensemble. It is then possible to convert back from baryon number to chemical potential by

$$\mu(\rho) = \frac{1}{N}\frac{\partial f(\rho)}{\partial \rho}. \tag{5.162}$$

Since the grand canonical partition function at imaginary chemical potential can be evaluated by Monte Carlo methods, this offers another approach to deal with finite density QCD (Alford *et al.*, 1999). Numerical data for the grand canonical partition function can be Fourier transformed numerically to give the canonical partition function at fixed baryon number. However, in this approach the sign problem re-enters at the stage of the Fourier transform. The latter clearly has an oscillatory integrand and thus the same problem we encountered before. The integration can only be done for sufficiently small quark numbers Q, whereas the thermodynamic limit requires $Q \to \infty$ with a fixed quark number density $Q/V = const$. Nevertheless, this is an interesting alternative approach that does not require reweighting or truncation of Taylor series, and thus is explored numerically (Kratochvila and de Forcrand, 2006; Alexandru *et al.*, 2005).

5.4.7   Plasma properties at finite density

Having developed computational tools for finite density, one can apply them to the studies discussed in the previous sections and see how finite baryon densities affect the screening masses, the equation of state or the static potential. In all those cases the influence of the chemical potential is found to be rather weak. These calculations appear to be well under control, and we will not further discuss them here. Instead, we outline recent attempts to determine the QCD phase diagram at finite density, where the order of the phase transition is expected to change as μ is increased, Fig. 5.13.

5.5   Towards the QCD phase diagram

5.5.1   The critical temperature at finite density

As in the case of zero density, let us first discuss the phase boundary, $T_c(\mu)$, before dealing with the order of the phase transition. The (pseudo-)critical line has been calculated for a variety of flavors and quark masses using different methods. Its computation is most straightforward by reweighting, where one directly evaluates susceptibilities of observables, Eq. (5.116), and locates their maxima to identify the critical couplings. With the Taylor expansion method, one computes instead the coefficients of such susceptibilites, cf. Eq. (5.147), which then is known to some order in μ/T. Similarly, one can evaluate susceptibilities at imaginary chemical potentials and locate their maxima as a function of μ_i.

On the other hand, the (pseudo-)critical gauge coupling can itself be expressed as a Taylor series. It was defined as an implicit function to be the coupling for which a generalized susceptibility peaks. The implicit function theorem then guarantees that, if $\chi(\beta, \mu)$ is an analytic function, so is $\beta_c(\mu)$, and hence

$$\beta_c(m_f, \frac{\mu}{T}) = \sum_n b_{2n}(m_f) \left(\frac{\mu}{T}\right)^{2n}. \tag{5.163}$$

Using some form of the renormalization group beta function, this can be converted into an expansion for the (pseudo-)critical temperature,

$$\frac{T_c(m_f, \mu)}{T_c(m_f, 0)} = 1 + t_2(m_f) \left(\frac{\mu}{T}\right)^2 + t_4(m_f) \left(\frac{\mu}{T}\right)^4 + \dots \tag{5.164}$$

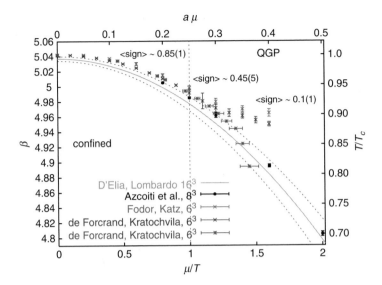

Fig. 5.14 Comparison of different methods to compute the critical couplings (Kratochvila and de Forcrand, 2006).

Thus, the Taylor expansion method as well as simulations at imaginary chemical potential followed by fits to polynomials allow for independent determinations of the coefficients, providing valuable cross-checks to control the systematics. For a quantitative comparison one needs data at one fixed parameter set and also eliminate the uncertainties of setting the scale. Such a comparison is shown for the critical coupling in Fig. 5.14, for $N_f = 4$ staggered quarks with the same action and quark mass $m/T \approx 0.2$. (For that quark mass the transition is first order along the entire curve.) One observes quantitative agreement up to $\mu/T \approx 1.3$, after which the different results start to scatter. Thus, we conclude that all methods discussed here appear to be reliable for $\mu/T \lesssim 1$.

For quark-mass values close to the physical ones, one observes that the critical temperature is decreasing only very slowly with μ on coarse lattices. These calculations can be repeated on finer lattices. While the cost for reweighting becomes formidable, the other methods should allow for a reliable determination of the leading coefficients in the continuum limit, and thus the physical phase boundary between the hadron and quark gluon plasma phase for $\mu \lesssim T$, in the near future.

5.5.2   The QCD phase diagram for $\mu \neq 0$ and the critical point

As in the case of $\mu = 0$, a determination of the order of the transition, and hence the search for the critical endpoint, is much harder, and we begin by discussing the qualitative picture. If a chemical potential is switched on for the light quarks, there is an additional parameter requiring an additional axis for our phase diagram characterizing the order of the transition, Fig. 5.10. This is shown in Fig. 5.15, where the horizontal plane is spanned by the $\mu = 0$ phase diagram in $m_s, m_{u,d}$ and the

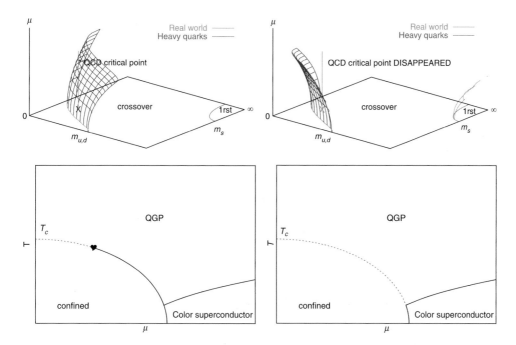

Fig. 5.15 Upper panel: The chiral critical surface in the case of positive (left) and negative (right) curvature. If the physical point is in the crossover region for $\mu = 0$, a finite μ chiral critical point will only arise in the scenario (left) with positive curvature, where the first-order region expands with μ. Note that for heavy quarks, the first-order region shrinks with μ, cf. Section 5.5.5. Lower panel: phase diagrams for fixed quark mass (here $N_f = 3$) corresponding to the two scenarios depicted above.

vertical axis represents μ. The critical line separating the first-order section from the crossover will now extend to finite μ and span a surface. *A priori* the shape of this surface is not known. However, the expected QCD phase diagram corresponds to the left scenario of Fig. 5.15. The first-order region expands as μ is turned on, so that the physical point, initially in the crossover region, eventually belongs to the chiral critical surface. At that chemical potential μ_E, the transition is second order: that is the QCD chiral critical point. Increasing μ further makes the transition first order. In this scenario the transition is generally strenghtened with real chemical potential. A completely different scenario arises if instead the transition weakens and the first-order region shrinks as μ is turned on. In that case, the physical point remains in the crossover region and there is no chiral critical point for moderate μ, Fig. 5.15 (right).

There are then two different strategies to learn about the QCD phase diagram. One can fix a particular set of quark masses and for that theory switch on and increase the chemical potential to see whether a critical surface is crossed or not, Section 5.5.3. Alternatively, Section 5.5.4 discusses how to start from the known critical line at $\mu = 0$ and study its evolution with a finite μ.

5.5.3 Critical point for fixed masses: reweighting and Taylor expansion

Reweighting methods at physical quark masses get a signal for a critical point at $\mu_B^E \sim 360$ MeV (Fodor and Katz, 2004). In this work $L^3 \times 4$ lattices with $L = 6 - 12$ were used, working with the standard staggered fermion action. Quark masses were tuned to $m_{u,d}/T_c \approx 0.037, m_s/T_c \approx 1$, corresponding to the mass ratios $m_\pi/m_\rho \approx 0.19, m_\pi/m_K \approx 0.27$, which are close to their physical values. A Lee–Yang zero analysis was employed in order to find the change from crossover behavior at $\mu = 0$ to a first-order transition for $\mu > \mu_E$. This is shown in Fig. 5.16. For a crossover the partition function has zeroes only off the real axis, whereas for a phase transition the zero moves to the real axis when extrapolated to infinite volume. For a critical discussion of the use of Lee–Yang zeros in combination with reweighting, see (Ejiri, 2006). A caveat of this calculation is the observation that the critical point is found in the immediate neighborhood of the onset of pion condensation in the phase-quenched theory, which is where the sign problem becomes maximally severe (Splittorff, 2005; Han and Stephanov, 2008), cf. the discussion in Section 5.4.3. Therefore, one would like to confirm this result with independent methods.

In principle, the determination of a critical point is also possible via the Taylor expansion. In this case true phase transitions will be signalled by an emerging non-analyticity, or a finite radius of convergence for the pressure series about $\mu = 0$, Eq. (5.145), as the volume is increased. The radius of convergence of a power series gives the distance between the expansion point and the nearest singularity, and may be extracted from the high-order behavior of the series. Possible definitions are

$$\rho, r = \lim_{n\to\infty} \rho_n, r_n \quad \text{with} \quad \rho_n = \left| \frac{c_0}{c_{2n}} \right|^{1/2n}, \quad r_n = \left| \frac{c_{2n}}{c_{2n+2}} \right|^{1/2}. \tag{5.165}$$

General theorems ensure that if the limit exists and asymptotically all coefficients of the series are positive, then there is a singularity on the real axis. More details as well as previous applications to strong-coupling expansions in various spin models can be found in (Guttman, 1989). In the series for the pressure such a singularity would correspond to the critical point in the (μ, T)-plane. The current best attempt is based

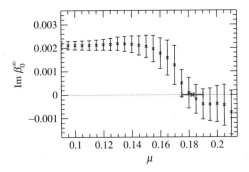

Fig. 5.16 Imaginary part of the Lee–Yang zero closest to the real axis as a function of chemical potential (Fodor and Katz, 2004).

on four consecutive coefficients, i.e. knowledge of the pressure to eighth order, and a critical endpoint for the $N_f = 2$ theory was reported in (Gavai and Gupta, 2005). There are also difficulties in this approach. First, there are different definitions for the radius of convergence, which are only unique in the asymptotic limit, but differ by numerical factors at finite n. Furthermore, the estimated ρ_n, r_n at a given order is neither an upper nor a lower bound on an actual radius of convergence. Finally, at finite orders the existence of a finite radius of convergence is a necessary, but not a sufficient condition for the existence of a critical point. For example, one also obtains finite estimates for a radius of convergence from the Taylor coefficients of the hadron resonance gas model, even though that model does not feature a non-analytic phase transition.

5.5.4   The change of the chiral critical line with μ

Rather than fixing one set of masses and considering the effects of μ, one may map out the critical surface in Fig. 5.15 by measuring how the $\mu = 0$ critical boundary line changes under the influence of μ. For example, for a given value of the strange-quark mass the corresponding critical $m_{u,d}$ has again a Taylor expansion,

$$\frac{m_{u,d}^c(\mu)}{m_{u,d}^c(0)} = 1 + c_1 \left(\frac{\mu}{T}\right)^2 + c_2 \left(\frac{\mu}{T}\right)^4 + \dots . \tag{5.166}$$

The curvature of the critical surface in lattice units is directly related to the behaviour of the Binder cumulant via the chain rule,

$$\frac{dam_c}{d(a\mu)^2} = -\frac{\partial B_4}{\partial(a\mu)^2} \left(\frac{\partial B_4}{\partial am}\right)^{-1}, \tag{5.167}$$

and similar expressions for the higher derivatives. While the second factor is sizeable and easy to evaluate, the μ dependence of the cumulant is excessively weak and requires enormous statistics to extract. In order to guard against systematic errors, this derivative can be evaluated in two independent ways. One is to fit the corresponding Taylor series of B_4 in powers of μ/T to data generated at imaginary chemical potential, the other is to compute the derivative directly and without fitting via the finite-difference quotient,

$$\frac{\partial B_4}{\partial(a\mu)^2} = \lim_{(a\mu)^2 \to 0} \frac{B_4(a\mu) - B_4(0)}{(a\mu)^2}. \tag{5.168}$$

Because the required shift in the couplings is very small, it is adequate and safe to use the original Monte Carlo ensemble for $am^c(0), \mu = 0$ and reweight the results by the standard Ferrenberg–Swendsen method. Moreover, by reweighting to imaginary μ the reweighting factors remain real positive and close to 1.

On coarse $N_\tau = 4$ lattices, the first two coefficients in Eq. (5.166) are found to be negative, (de Forcrand and Philipsen, 2008), hence the region featuring a first-order chiral transition is shrinking when a real chemical potential is turned on. This implies that, for moderate $\mu \lesssim T$, there is no critical point belonging to the chiral critical surface. An open questions here is what the higher-order corrections are. For example, a sudden change of behavior as in Fig. 5.16 would be difficult to capture with only few Taylor coefficients. Also, there might be a chiral critical point at larger chemical potentials and finally we do not yet know how the curvature of the critical surface behaves in the continuum limit. In addition, these findings do not exclude a critical point that is not associated with the chiral critical surface.

5.5.5 The change of the deconfinement critical line with μ

In light of these results, it is interesting to compare with the situation in the heavy-quark corner of the schematic phase diagram, Fig. 5.14. As the quark mass goes to infinity, quarks can be integrated out and QCD reduces to a gauge theory of Polyakov lines. At a second-order phase transition, universality allows us to neglect the details of gauge degrees of freedom, so the theory reduces to the 3D three-state Potts model, which is in the appropriate 3D Ising universality class. Hence, studying the three-state Potts model should teach us about the behavior of QCD in the neighborhood of the deconfinement critical line separating the quenched first-order region from the crossover region.

Here, we are interested in simulations at small chemical potential. In this case, the sign problem is mild enough for brute force-simulations at real μ to be feasible. In (Kim *et al.*, 2006), the change of the critical heavy-quark mass is determined as a function of real as well as imaginary μ, as shown in Fig. 5.17. Note that $M^c(\mu)$ rises with real chemical potential, such that the first-order region in Fig. 5.10 shrinks

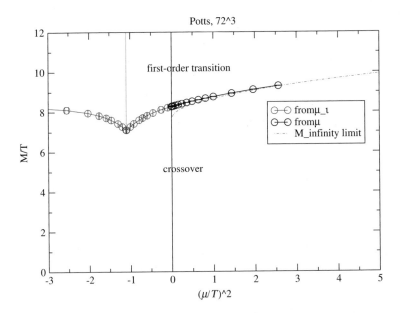

Fig. 5.17 The critical heavy quark mass separating first order from crossover as a function of μ^2 from the Potts model (Kim *et al.*, 2006).

as finite baryon density is switched on. The calculation also gives some insight onto the problem of analytic continuation: Fig. 5.17 clearly endorses the approach with $M_c(\mu)$ passing analytically through $\mu = 0$. However, high-order fits might be required in practice in order to reproduce the data on both sides of $\mu^2 = 0$.

The same qualitative behavior is observed in a combined strong-coupling and hopping-parameter expansion (Langelage and Philipsen, 2010a). While such a calculation is only accurate for coarse lattices, the universal features of the critical surface should be cutoff independent. Thus, QCD with heavy quarks is an example of the non-standard scenario discussed in the previous section.

5.5.6   The critical surfaces at imaginary μ

The chiral and deconfinement critical surfaces, upper Fig. 5.15 (right), continue to imaginary μ, which has been used to determine their curvature. Rather than studying the neighborhood of $\mu = 0$ with an aim to analytic continuation, it is also illuminating to follow those surfaces into the imaginary regime, until the phase boundary to the next $Z(3)$-sector, Fig. 5.13 (left), is hit. This has already been done in the case of the deconfinement transition within the Potts model, cf. the curve at negative μ^2, Fig. 5.17. The meeting point of the deconfinement critical line with the critical endpoint of the $Z(3)$ transition is then tricritical, and corresponds precisely to the tricritical point in the large-mass region of Fig. 5.13 (right).

For three flavors of quarks and varying quark masses, one can now draw a schematic diagram analogous to Fig. 5.6, showing the nature of the junction at $\mu = i\pi T/3$. At the time of writing, only the two tricritical points on the $N_f = 3$ diagonal and the light-mass one for $N_f = 2$ have been determined on coarse lattices. A natural extension would then be that there are two tricritical lines that delimit the quark-mass sections for which the junction corresponds to triple points. The area in between corresponds to a second-order endpoint of the $Z(3)$-transition, that no connection to the chiral/deconfinement transition which is merely a crossover there. The tricritical lines represent the boundaries in which the chiral and deconfinement critical surfaces end at imaginary $\mu = i\pi T/3$, before they periodically repeat. On coarse lattices they are found "outside" the corresponding $\mu = 0$ critical lines, which demonstrates an increase of the first-order area in the imaginary direction. In fact, for heavy quarks this change is monotonic. The functional form of the curve in Fig. 5.17, and hence the surface in Fig. 5.18 (right), was found to be determined by tricritical scaling for the whole range up to real $\mu \sim T$. This would explain the curvature of the critical surfaces, i.e. the weakening of the chiral and deconfinement transitions with real chemical potential.

5.5.7   Discussion

It thus appears that first-order transitions are weakened when a chemical potential for baryon number is switched on. The same observation is made for finite-isospin chemical potential (Kogut and Sinclair, 2008). Note, however, that most of these computations are done on $N_\tau = 4, 6$ lattices, where cutoff effects appear to be larger than finite-density effects. Hence, definite conclusions for continuum physics cannot yet

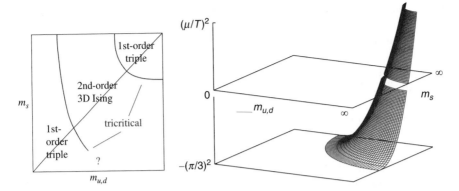

Fig. 5.18 Left: Nature of the $Z(3)$-transition endpoint at $\mu/T = i\pi/3$. (de Forcrand and Philipsen, 2010). Right: Deconfinement critical surface for real and imaginary μ.

be drawn. A general finding is the steepness of the critical surface, making the location of a possible critical endpoint extremely quark-mass sensitive, and hence difficult to determine accurately. Furthermore, the chiral first-order region is also observed to shrink with decreasing lattice spacing, such that in Fig. 5.10 $m_s^{tric} < m_s^{phys}$. In this case the chiral critical surface we have been discussing here might not be responsible for a possible critical point, regardless of its curvature, but another surface emanating from the putative $O(4)$-chiral limit, or one connected to finite-density physics, or.....
Possibilities and uncertainties abound! To weed them out we need to conclusively understand the situation in the $N_f = 2$ chiral limit. Finally, even if those issues get settled, all the methods discussed here are only valid for $\mu \lesssim T$. Thus, learning about high-density QCD, relevant for compact stars, requires entirely new theoretical methods. In summary, it remains a challenging but also most exciting task to settle even the qualitative features of the QCD phase diagram.

Acknowledgments

I would like to thank the organizers of the school for the invitation to lecture here, and all the students for the excellent working atmosphere, their interest and stimulating questions. I also thank Philippe de Forcrand for continued collaboration and availability for discussion over many years, as well as Jens Langelage and Lars Zeidlewicz for calculational input and checking parts of the manuscript.

References

Alexandru, A., Faber, M., Horvath, I., and Liu, K.-F. (2005). *Phys. Rev.*, **D72**, 114513.
Alford, Mark G., Kapustin, A., and Wilczek, F. (1999). *Phys. Rev.*, **D59**, 054502.
Allton, C. R. *et al.* (2002). *Phys. Rev.*, **D66**, 074507.
Aoki, Y., Endrodi, G., Fodor, Z., Katz, S. D., and Szabo, K. K. (2006). *Nature*, **443**, 675–678.

Brambilla, N., Ghiglieri, J., Vairo, A., and Petreczky, Peter (2008). *Phys. Rev.*, **D78**, 014017.

Chandrasekharan, S. and Mehta, A. C. (2007). *Phys. Rev. Lett.*, **99**, 142004.

Datta, S. and Gupta, S. (1998). *Nucl. Phys.*, **B534**, 392–416.

de Forcrand, Ph., Kim, S., and Philipsen, O. (2007). *PoS*, **LAT2007**, 178.

de Forcrand, Ph., Kim, S., and Takaishi, T. (2003). *Nucl. Phys. Proc. Suppl.*, **119**, 541–543.

de Forcrand, Ph. and Philipsen, O. (2002). *Nucl. Phys.*, **B642**, 290–306.

de Forcrand, Ph. and Philipsen, O. (2007). *JHEP*, **01**, 077.

de Forcrand, Ph. and Philipsen, O. (2008). *JHEP*, **11**, 012.

de Forcrand, Ph. and Philipsen, O. (2010). Phys. Rev. Lett., 105, 152001.

D'Elia, M. and Lombardo, M.-P. (2003). *Phys. Rev.*, **D67**, 014505.

D'Elia, M. and Sanfilippo, F. (2009). *Phys. Rev.*, **D80**, 111501.

Ejiri, S. (2006). *Phys. Rev.*, **D73**, 054502.

Endrödi, G., Fodor, Z., Katz, S. D., and Szabo, K. K. (2007). *PoS*, **LAT2007**, 182.

Engels, J., Fingberg, J., Karsch, F., Miller, D., and Weber, M. (1990). *Phys. Lett.*, **B252**, 625–630.

Ferrenberg, A. M. and Swendsen, R. H. (1988). *Phys. Rev. Lett.*, **61**, 2635–2638.

Fodor, Z. and Katz, S. D. (2002). *Phys. Lett.*, **B534**, 87–92.

Fodor, Z. and Katz, S. D. (2004). *JHEP*, **04**, 050.

Gavai, R. V. (1985). *Phys. Rev.*, **D32**, 519.

Gavai, R. V. and Gupta, S. (2005). *Phys. Rev.*, **D71**, 114014.

Grossman, B., Gupta, S., Heller, U. M., and Karsch, F. (1994). *Nucl. Phys.*, **B417**, 289–306.

Gupta, S. (1999). *Phys. Rev.*, **D60**, 094505.

Guttman, A. J. (1989). In *Phase transitions and critical phenomena* (ed. C. Domb and J. L. Lebowitz), Number 13, p. 1. Academic Press, London.

Halasz, A. M., Jackson, A. D., Shrock, R. E., Stephanov, M. A., and Verbaarschot, J. J. M. (1998). *Phys. Rev.*, **D58**, 096007.

Han, J. and Stephanov, M. A. (2008). *Phys. Rev.*, **D78**, 054507.

Hands, S., Sitch, P., and Skullerud, J.-I. (2008). *Phys. Lett.*, **B662**, 405.

Hart, A., Laine, M., and Philipsen, O. (2000). *Nucl. Phys.*, **B586**, 443–474.

Hasenfratz, P. and Karsch, F. (1983). *Phys. Lett.*, **B125**, 308.

Hegde, P., Karsch, F., Laermann, E., and Shcheredin, S. (2008). *Eur. Phys. J.*, **C55**, 423–437.

Jahn, O. and Philipsen, O. (2004). *Phys. Rev.*, **D70**, 074504.

Kaczmarek, O., Karsch, F., Laermann, E., and Lütgemeier, Martin (2000). *Phys. Rev.*, **D62**, 034021.

Kapusta, J. I. and Gale, C. (2006). *Finite-temperature field theory: Principles and applications.* University Press Cambridge, UK:(2006) 428 p.

Karsch, F. (2002). *Lect. Notes Phys.*, **583**, 209–249.

Karsch, F., Laermann, E., and Peikert, A. (2000). *Phys. Lett.*, **B478**, 447–455.

Karsch, F., Laermann, E., and Schmidt, C. (2001). *Phys. Lett.*, **B520**, 41–49.

Karsch, F., Redlich, K., and Tawfik, A. (2003). *Eur. Phys. J.*, **C29**, 549–556.

Karsch, F. and Wyld, H. W. (1985). *Phys. Rev. Lett.*, **55**, 2242.

Kaste, P. and Rothe, H. J. (1997). *Phys. Rev.*, **D56**, 6804–6815.

Kim, S., de Forcrand, Ph., Kratochvila, S., and Takaishi, T. (2006). *PoS*, **LAT2005**, 166.

Kogut, J. B. and Sinclair, D. K. (2008). *Phys. Rev.*, **D77**, 114503.

Kratochvila, S. and de Forcrand, Ph. (2006). *PoS*, **LAT2005**, 167.

Laine, M., Philipsen, O., Romatschke, P., and Tassler, M. (2007). *JHEP*, **03**, 054.

Laine, M. and Vepsäläinen, M. (2004). *JHEP*, **02**, 004.

Langelage, J., Münster, G., and Philipsen, O. (2007). *PoS*, **LAT2007**, 201.

Langelage, J., Münster, G., and Philipsen, O. (2008). *JHEP*, **07**, 036.

Langelage, J. and Philipsen, O. (2010*a*). *JHEP*, **01**, 089.

Langelage, J. and Philipsen, O. (2010*b*). *JHEP*, **04**, 055.

Lawrie, I. D. and Sarbach, S. (1984). In *Phase transitions and critical phenomena* (ed. C. Domb and J. L. Lebowitz), Number 9, p. 1. Academic Press, London.

Lee, T. D. and Yang, C. N. (1952). *Phys. Rev.*, **87**, 410–419.

Matsui, T. and Satz, H. (1986). *Phys. Lett.*, **B178**, 416.

McLerran, L. D. and Svetitsky, B. (1981). *Phys. Rev.*, **D24**, 450.

Montvay, I. and Münster, G. (1994). *Quantum fields on a lattice*. University Press Cambridge, UK. (1994) 491 p. (Cambridge monographs on mathematical physics).

Münster, G. (1981). *Nucl. Phys.*, **B190**, 439.

Nadkarni, S. (1986). *Phys. Rev.*, **D34**, 3904.

Philipsen, O. and Zeidlewicz, L. (2010). *Phys. Rev.*, **D81**, 077501.

Rajagopal, K. and Wilczek, F. (2000). In *At the frontier of particle physics* (ed. M. Shifman), World Scientific Number 3, p. 2061.

Rebhan, A. K. (1993). *Phys. Rev.*, **D48**, 3967–3970.

Roberge, A. and Weiss, N. (1986). *Nucl. Phys.*, **B275**, 734.

Schor, R. S. (1983). *Phys. Lett.*, **B132**, 161.

Seo, K. (1982). *Nucl. Phys.*, **B209**, 200–216.

Son, D. T. and Stephanov, M. A. (2001). *Phys. Rev. Lett.*, **86**, 592–595.

Splittorff, K. (2005). *hep-lat/0505001*.

Wiese, U. J. (2002). *Nucl. Phys.*, **A702**, 211–216.

Yang, Chen-Ning and Lee, T. D. (1952). *Phys. Rev.*, **87**, 404–409.

6

Computational strategies
in lattice QCD

Martin LÜSCHER

Physics Department, CERN, 1211 Geneva 23, Switzerland

Introduction

Numerical lattice QCD has seen many important innovations over the years. In this chapter an introduction to some of the basic techniques is provided, emphasizing their theoretical foundation rather than their implementation and latest refinements.

The development of computational strategies in lattice QCD requires physical insight to be combined with an understanding of modern numerical mathematics and of the capabilities of massively parallel computers. When a new method is proposed, it should ideally be accompanied by a theoretical analysis that explains why it is expected to work out. However, in view of the complexity of the matter, some experimenting is often required. The field thus retains a certain empirical character.

At present, numerical lattice QCD is still in a developing phase to some extent. The baryon spectrum, for example, remains difficult to compute reliably, because the signal-to-noise ratio of the associated two-point functions decreases exponentially at large distances. There is certainly ample room for improvements and it may also be necessary to radically depart from the known techniques in some cases. Hopefully, this chapter will encourage some of the problems to be studied and to be solved eventually.

6.1 Computation of quark propagators

Quark propagators in the presence of a specified SU(3) gauge field are fundamental building blocks in lattice QCD. Their computation amounts to solving the Dirac equation

$$D\psi(x) = \eta(x) \tag{6.1}$$

a number of times, where D denotes the massive lattice Dirac operator, $\eta(x)$ a given quark field (the source field) and $\psi(x)$ the desired solution.

Although the subject is nearly as old as lattice QCD itself, there have been important advances in the last few years that allow the quark propagators to be calculated much more rapidly than was possible before. As a consequence, the "measurement" of hadronic quantities and the simulation of the theory with light sea quarks are both accelerated significantly.

6.1.1 Preliminaries

6.1.1.1 *Accuracy and condition number*

The Dirac equation (6.1) is a large linear system that can only be solved iteratively, i.e. through some recursive procedure that generates a sequence $\psi_1, \psi_2, \psi_3, \ldots$ of increasingly accurate approximate solutions. A practical measure for the accuracy of an approximate solution ϕ is the norm of the associated residue

$$\rho = \eta - D\phi. \tag{6.2}$$

If, say, $\|\rho\| < \epsilon\|\eta\|$ for some small value ϵ, an important question is then by how much ϕ deviates from the exact solution ψ of the equation. Using standard norm estimates, the deviation is found to be bounded by

$$\|\psi - \phi\| < \epsilon \kappa(D)\|\psi\|, \tag{6.3}$$

where

$$\kappa(D) = \|D\|\|D^{-1}\| \tag{6.4}$$

is referred to as the condition number of D. The relative error of ϕ thus tends to be larger than ϵ by the factor $\kappa(D)$ (in this chapter, the standard scalar product of quark fields is used as well as the field and operator norms that derive from it).

The extremal eigenvalues of $D^\dagger D$, α_{min} and α_{max}, are proportional to the square of the quark mass m and the square of the inverse of the lattice spacing a, respectively. In particular, the condition number

$$\kappa(D) = (\alpha_{max}/\alpha_{min})^{1/2} \propto (am)^{-1} \tag{6.5}$$

can be very large at small quark masses and lattice spacings. One says that the Dirac operator is "ill-conditioned" in this case. The important point to keep in mind is that the accuracy of the solution of the Dirac equation that can be attained on a given computer is limited by the condition number.

6.1.1.2 *Iterative improvement*

If the residual error ϵ of the approximate solution ϕ is still well above the limit set by the machine precision and the condition number of the Dirac operator, an improved solution

$$\tilde\phi = \phi + \chi \tag{6.6}$$

may be obtained by approximately solving the residual equation

$$D\chi = \rho. \tag{6.7}$$

It is straightforward to show that the residue of the solution is reduced by the factor δ in this way if χ satisfies $\|\rho - D\chi\| < \delta\|\rho\|$. Note that χ is approximately equal to the deviation of ϕ from the exact solution of the Dirac equation and is therefore usually a small correction to ϕ.

Iterative improvement is used by all solvers that need to be restarted after a while. The GCR algorithm discussed below is an example of such a solver. Another application of iterative improvement, known as "single-precision acceleration", exploits the fact that modern processors perform 32-bit arithmetic operations significantly faster than 64-bit operations. The idea is to solve the Dirac equation to 64-bit precision by going through a few cycles of iterative improvement, where, in each cycle, the residual equation is solved to a limited precision using 32-bit arithmetic, while the residue ρ and the sum (6.6) are evaluated using 64-bit arithmetic (Giusti *et al.*, 2003).

6.1.2 Krylov-space solvers

The Krylov space $\mathcal{K}_n$ of dimension n is the complex linear space spanned by the fields

$$\eta,\, D\eta,\, D^2\eta,\, \ldots,\, D^{n-1}\eta. \tag{6.8}$$

Many popular solvers, including the CG, BiCGstab and GCR algorithms, explicitly or implicitly build up a Krylov space and search for the solution of the Dirac equation within this space. The very readable book of (Saad, 2003) describes these solvers in full detail. A somewhat simpler discussion of the CG (conjugate gradient) algorithm is given in the book of (Golub and van Loan, 1989) and useful additional references for the BiCGstab algorithm are (van der Vorst, 1992) and (Frommer *et al.*, 1994). Here, the GCR (generalized conjugate residual) algorithm is discussed as a representative case.

6.1.2.1 *The GCR algorithm*

The approximate solutions of the Dirac equation (6.1) generated by the GCR algorithm are the fields $\psi_k \in \mathcal{K}_k$, $k = 1, 2, 3, \ldots$, that minimize the norm of the residues

$$\rho_k = \eta - D\psi_k. \tag{6.9}$$

An equivalent requirement is that $D\psi_k$ coincides with the orthogonal projection of the source field η to the k-dimensional linear space $D\mathcal{K}_k$ (see Fig. 6.1). The algorithm proceeds somewhat indirectly by first constructing an orthonormal basis $\chi_0, \chi_1, \chi_2, \cdots$ of these spaces through a recursive process. Independently of the details of the construction, the orthogonality property mentioned above then implies that the fields

$$\rho_k = \eta - \sum_{l=0}^{k-1} c_l \chi_l, \qquad c_l = (\chi_l, \eta), \tag{6.10}$$

are the residues of the approximate solutions ψ_k. The residues are thus obtained before the latter are known.

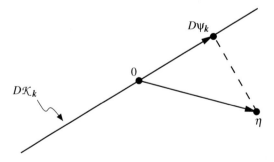

Fig. 6.1 The approximate solution $\psi_k \in \mathcal{K}_k$ of the Dirac equation constructed by the GCR algorithm is such that the distance $\|\eta - D\psi_k\|$ is minimized. The field $D\psi_k$ therefore coincides with the orthogonal projection of the source η to the space $D\mathcal{K}_k$.

In each iteration of the recursion, the next basis field χ_k is constructed from the previous fields $\chi_0, \ldots, \chi_{k-1}$ and the source η. First, the residue ρ_k is computed through Eq. (6.10) and χ_k is then taken to be a (properly orthonormalized) linear combination of $D\rho_k$ and the previous fields. Since χ_{k-1} is a linear combination of $D\rho_{k-1}$ and the fields $\chi_0, \ldots, \chi_{k-2}$, and so on, it is clear that the recursion also yields the coefficients a_{lj} in the equations

$$\chi_l = \sum_{j=0}^{l} a_{lj} D\rho_j, \qquad l = 0, 1, 2, \ldots, \tag{6.11}$$

in which $\rho_0 = \eta$.

Once the fields $\chi_0, \ldots, \chi_{n-1}$ are known for some n, the last solution ψ_n is obtained starting from the orthogonality condition

$$D\psi_n = \sum_{l=0}^{n-1} c_l \chi_l. \tag{6.12}$$

After substituting Eq. (6.11), the equation may then be divided by D and one finds that the solution is given by

$$\psi_n = \sum_{l=0}^{n-1} \sum_{j=0}^{l} c_l a_{lj} \rho_j. \tag{6.13}$$

Note that the right-hand side of this equation can be evaluated straightforwardly since all entries are known at this point.

In total, the computation of ψ_n requires n applications of the Dirac operator and the evaluation of some $\frac{1}{2} n^2$ linear combinations and scalar products of quark fields. Moreover, memory space for about $2n$ fields is needed. Choosing values of n from, say, 16 to 32 proves to be a reasonable compromise in practice, where the computational effort must be balanced against the reduction in the residue that is achieved. If the last solution is not sufficiently accurate, the algorithm can then always be restarted following the rules of iterative improvement.

6.1.2.2   Convergence properties

The convergence of the GCR algorithm and related Krylov-space solvers can be proved rigorously if the (complex) spectrum of the Dirac operator is contained in the half-plane on the right of the imaginary axis. In the case of the Neuberger–Dirac operator, for example, all eigenvalues of D lie on the circle shown in Fig. 6.2. The spectrum of the Wilson–Dirac operator is rather more complicated, but is usually contained in an ellipsoidal region in the right half-plane.

For simplicity, the Dirac operator is, in the following paragraphs, assumed to be diagonalizable and to have all its eigenvalues in the shaded disk $\mathbb{D}$ shown in Fig. 6.2. The convergence analysis of the GCR algorithm then starts from the observation that the residue ρ_k is given by

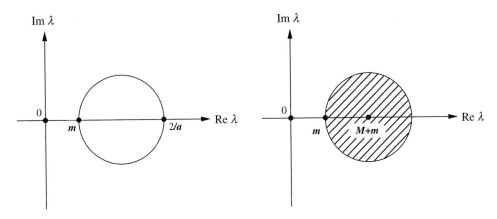

Fig. 6.2 The eigenvalues of the Neuberger–Dirac operator with bare quark mass m lie on a circle in the complex plane (drawing on the left). For the convergence analysis of the GCR algorithm, the spectrum of the Dirac operator is assumed to be contained in a disk $\mathbb{D}$ in the right-half plane with radius M and distance $m > 0$ from the origin (drawing on the right).

$$\rho_k = \eta - D\psi_k = p_k(D)\eta, \tag{6.14}$$

where $p_k(\lambda)$ is a polynomial of degree k that satisfies $p_k(0) = 1$. Since the algorithm minimizes the residue, it follows that

$$\|\rho_k\| = \min_{p_k} \|p_k(D)\eta\| \leq \min_{p_k} \|p_k(D)\|\|\eta\|, \tag{6.15}$$

the minimum being taken over all such polynomials.

Lattice quark fields are large arrays of complex numbers. In this language, the Dirac operator D is just a complex square matrix. The assumption that D is diagonalizable then implies the existence of a diagonal matrix Λ and of an invertible matrix V such that $D = V\Lambda V^{-1}$. As a consequence

$$\|p_k(D)\| = \|Vp_k(\Lambda)V^{-1}\| \leq \kappa(V)\|p_k(\Lambda)\| \tag{6.16}$$

and therefore

$$\|\rho_k\| \leq \kappa(V) \max_{\lambda \in \mathbb{D}} |p_k(\lambda)|\|\eta\| \tag{6.17}$$

for any polynomial $p_k(\lambda)$ of degree k satisfying $p_k(0) = 1$. One may, for example, insert

$$p_k(\lambda) = \left(1 - \frac{\lambda}{M+m}\right)^k, \tag{6.18}$$

in which case the inequality (6.17) leads to the bound[1]

$$\|\rho_k\| \leq \kappa(V)\left(1 + \frac{m}{M}\right)^{-k}\|\eta\|. \tag{6.19}$$

The GCR algorithm thus converges roughly like $\exp(-km/M)$ if $m/M \ll 1$. Note that the convergence rate $m/M \sim 2/\kappa(D)$ can be quite small in practice. For $m = 10$ MeV and $M = 2$ GeV, for example, the estimate (6.19) suggests that values of k as large as 4000 are required for a reduction of the residue by the factor 10^{-10}.

The GCR algorithm can also be applied to the so-called normal equation

$$D^\dagger D\psi = \eta \tag{6.20}$$

and to the Dirac equation

$$(i\gamma_5 D + \mu)\psi = \eta \tag{6.21}$$

in "twisted-mass" QCD. A notable difference with respect to the ordinary Dirac equation is that the operators on the left of these equations can be diagonalized through unitary transformations. Moreover, their spectra are contained in straight-line segments in the complex plane (see Fig. 6.3). Using Chebyshev polynomials in place of the power (6.18), the estimate

$$\|\rho_k\| \lesssim 2\mathrm{e}^{-rk}\|\eta\| \tag{6.22}$$

may be derived in these cases, where $r = m/M$ for the normal equation and $r = \mu/2M$ for the twisted-mass Dirac equation.

The convergence of Krylov-space solvers is thus mainly determined by the properties of the spectrum of the operator considered. In QCD the fact that the masses of the light quarks are much smaller than the inverse lattice spacing consequently tends to slow down the computations enormously. One can do better, however, by exploiting specific properties of the Dirac operator.

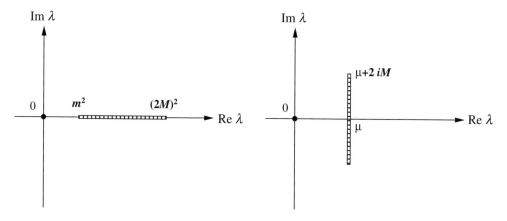

Fig. 6.3 The eigenvalues of the Hermitian operator $D^\dagger D$ and the twisted-mass Dirac operator $i\gamma_5 D + \mu$ occupy the shaded line segments shown in the left and right drawings, respectively.

6.1.2.3 *Preconditioning*

Preconditioning is a general strategy that allows such properties to be taken into account. Let L and R be some invertible operators acting on quark fields. Instead of the Dirac equation (6.1), one may then consider the so-called preconditioned equation

$$LDR\phi = L\eta. \tag{6.23}$$

Once this equation is solved, using a Krylov-space solver for example, the solution of the Dirac equation is obtained by setting $\psi = R\phi$. If $D \approx L^{-1}R^{-1}$, and if the application of L and R to a given quark field is not too time consuming, the total computer time required for the solution of the equation may be significantly reduced in this way.

In lattice QCD, a widely used preconditioning method for the Wilson–Dirac operator is "even–odd preconditioning". A lattice point $x \in \mathbb{Z}^4$ is referred to as even or odd depending on whether the sum of its coordinates x_μ is even or odd (see Fig. 6.4). If the points are ordered such that the even ones come first, the Dirac operator assumes the block form

$$D = \begin{pmatrix} D_{\text{ee}} & D_{\text{eo}} \\ D_{\text{oe}} & D_{\text{oo}}, \end{pmatrix} \tag{6.24}$$

where D_{eo}, for example, stands for the hopping terms that go from the odd to the even sites. The blocks on the diagonal, D_{ee} and D_{oo}, include the mass term and a Pauli term if the theory is $O(a)$-improved. Since they do not couple different lattice points, they can be easily inverted and it makes sense to consider the preconditioners

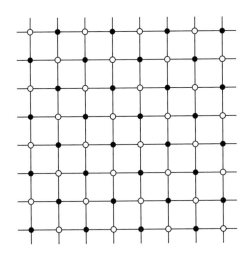

Fig. 6.4 Hypercubic lattices may be divided into the sublattices of the even and the odd sites (black and white points, respectively). Even–odd preconditioning effectively amounts to "integrating out" the quark field on the odd sublattice.

$$L = \begin{pmatrix} 1 & -D_{\text{eo}}D_{\text{oo}}^{-1} \\ 0 & 1, \end{pmatrix} \qquad R = \begin{pmatrix} 1 & 0 \\ -D_{\text{oo}}^{-1}D_{\text{oe}} & 1. \end{pmatrix} \qquad (6.25)$$

With this choice, the Dirac operator is block-diagonalized,

$$LDR = \begin{pmatrix} \hat{D} & 0 \\ 0 & D_{\text{oo}}, \end{pmatrix} \qquad \hat{D} = D_{\text{ee}} - D_{\text{eo}}D_{\text{oo}}^{-1}D_{\text{oe}}, \qquad (6.26)$$

and the solution of Eq. (6.23) thus amounts to solving a system in the space of quark fields on the even sublattice. The condition number of $\hat{D}$ is usually less than half the one of D and even–odd preconditioning consequently leads to an acceleration of the solver by a factor 2 to 3 or so.

Other preconditioners used in lattice QCD are the successive symmetric overrelaxation (SSOR) preconditioner (Frommer *et al.*, 1994) and a domain-decomposition preconditioner based on the Schwarz alternating procedure (Lüscher, 2004). The latter is an example of an "expensive" preconditioner, whose implementation involves an iterative procedure and is therefore inexact to some extent. Inaccuracies at this level of the algorithm are, however, not propagated to the final results if the GCR solver is used for the preconditioned equation (6.23). This algorithm actually always finds the best approximation to the solution in the space generated by applying the preconditioner to the residues of the previous solutions. For the same reason, the GCR algorithm is also safe of rounding errors.

6.1.3 Low-mode deflation

The low modes of the Dirac operator are intimately related to the spontaneous breaking of chiral symmetry and therefore play a special rôle in QCD. Treating them separately from the other modes seems appropriate from the physical point of view and is recommended for technical reasons at small quark masses.

6.1.3.1 *Textbook deflation*

In the case of the Hermitian system,

$$A\psi = \eta, \qquad A = D^{\dagger}D, \qquad (6.27)$$

there exists an orthonormal basis of eigenvectors v_k, $k = 1, 2, 3, \ldots$, such that

$$Av_k = \alpha_k v_k, \qquad 0 \le \alpha_1 \le \alpha_2 \le \ldots . \qquad (6.28)$$

The action on any quark field ψ of the orthonormal projector P to the N lowest modes is then given by

$$P\psi = \sum_{k=1}^{N} v_k (v_k, \psi). \qquad (6.29)$$

Since P commutes with A, the linear system (6.27) splits into the decoupled equations

$$A_\| \psi_\| = \eta_\|, \qquad \psi_\| = P\psi, \tag{6.30}$$

$$A_\perp \psi_\perp = \eta_\perp, \qquad \psi_\perp = (1 - P)\psi, \tag{6.31}$$

where $A_\| = PAP$ and $A_\perp = (1 - P)A(1 - P)$ are, respectively, referred to as the "little operator" and the "deflated operator".

If the eigenvectors $v_1, \dots, v_N$ are known, and if there are no zero modes, the solution of the little system (6.30) can be obtained exactly through

$$\psi_\| = \sum_{k=1}^{N} \frac{1}{\alpha_k} v_k (v_k, \eta). \tag{6.32}$$

The deflated system (6.31), on the other hand, can only be solved iteratively using the CG algorithm, for example. With respect to the full system, the associated condition number

$$\kappa(A_\perp) = \frac{\alpha_1}{\alpha_{N+1}} \kappa(A) \tag{6.33}$$

is, however, reduced and one therefore expects the solver to be accelerated by the factor $(\alpha_{N+1}/\alpha_1)^{1/2}$ or so.

The deflation of the Hermitian system (6.27) along these lines is straightforward to implement, but the method tends to be limited to small lattices, because the computer time required for the calculation of the low eigenvectors grows rapidly with the lattice volume. In the past few years, it was nevertheless further developed and improved in various directions (for a review, see (Wilcox, 2007), for example).

6.1.3.2   The Banks–Casher relation

In the continuum theory (which is considered here for simplicity), the eigenvalues of the Dirac operator D in presence of a given gauge field are of the form $m + i\lambda_k$, where m denotes the quark mass and $\lambda_k \in \mathbb{R}$, $k = 1, 2, 3, \dots$, the eigenvalues of the massless operator. The associated average spectral density,

$$\rho(\lambda, m) = \frac{1}{V} \sum_{k=1}^{\infty} \langle \delta(\lambda - \lambda_k) \rangle, \tag{6.34}$$

is conventionally normalized by the space-time volume V so that it has a meaningful infinite-volume limit.

In a now famous paper, (Banks and Casher, 1980) showed many years ago that the density at the origin,

$$\lim_{\lambda \to 0} \lim_{m \to 0} \lim_{V \to \infty} \rho(\lambda, m) = \frac{1}{\pi} \Sigma, \tag{6.35}$$

is proportional to the quark condensate

$$\Sigma = -\lim_{m \to 0} \lim_{V \to \infty} \langle \bar{u}u \rangle, \qquad u\text{: up-quark field,} \tag{6.36}$$

in the chiral limit. The spontaneous breaking of chiral symmetry in QCD is thus linked to the presence of a non-zero density of eigenvalues at the low end of the spectrum of the Dirac operator.

Since the eigenvalues $\alpha_k = m^2 + \lambda_k^2$ of $D^\dagger D$ are simply related to the eigenvalues of D, the Banks–Casher relation (6.35) immediately leads to the estimate

$$\nu(M, m) \simeq \frac{2}{\pi} \Lambda \Sigma V, \qquad \Lambda^2 = M^2 - m^2, \tag{6.37}$$

for the number of low modes of $D^\dagger D$ with eigenvalues $\alpha_k \leq M^2$. If one sets $m = 0$, $M = 100$ MeV and $\Sigma = (250\,\text{MeV})^3$, for example, and considers a space-time volume of size $V = 2L^4$, the mode numbers are estimated to be 21, 106 and 336 for $L = 2, 3$ and 4 fm, respectively. As illustrated by Fig. 6.5, the low-mode condensation is readily observed in numerical simulations of lattice QCD.

Since $\nu(M, m)$ increases proportionally to the space-time volume V, an effective deflation of the Dirac equation requires $O(V)$ modes to be deflated. The associated computational effort increases like V^2 and straightforward deflation consequently tends to become inefficient or impractical at large volumes. However, as explained in the following, the V^2-problem can be overcome using inexact deflation (Giusti *et al.*, 2003) and domain-decomposed deflation subspaces (Lüscher, 2007*a*).

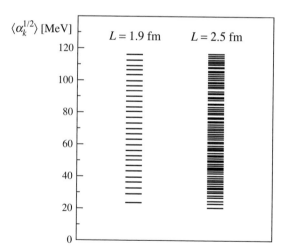

Fig. 6.5 Expectation value of the eigenvalues of $(D^\dagger D)^{1/2}$ below 116 MeV in $O(a)$-improved two-flavor QCD on a $2L \times L^3$ lattice at two values of L. Both spectra were obtained at lattice spacing $a = 0.08$ fm and renormalized sea-quark mass $m = 26$ MeV. From the smaller to the larger lattice, the number of modes per MeV increases approximately as predicted by the Banks–Casher relation.

6.1.3.3   Deflation w/o eigenvectors

Returning to the lattice Dirac equation (6.1), a more general form of deflation will now be described, which starts from an unspecified set $\phi_1, \ldots, \phi_N$ of N orthonormal quark fields. The linear space $\mathcal{S}$ spanned by these fields will play the rôle of the deflation subspace, but is not assumed to be an eigenspace of the Dirac operator.

The action on any quark field ψ of the orthogonal projector P to $\mathcal{S}$ is given by

$$P\psi = \sum_{k=1}^{N} \phi_k \left(\phi_k, \psi \right). \tag{6.38}$$

As before, the restriction PDP of the Dirac operator to $\mathcal{S}$ will be referred to as the "little Dirac operator". Its action is encoded in the complex $N \times N$ matrix

$$A_{kl} = (\phi_k, D\phi_l), \quad k, l = 1, \ldots, N, \tag{6.39}$$

through

$$PDP\psi = \sum_{k,l=1}^{N} \phi_k A_{kl} \left(\phi_l, \psi \right). \tag{6.40}$$

A technical assumption made in the following is that A (and thus PDP as an operator acting in $\mathcal{S}$) is invertible.

The form of inexact deflation discussed below is based on the projectors

$$P_L = 1 - DP(PDP)^{-1}P, \tag{6.41}$$

$$P_R = 1 - P(PDP)^{-1}PD. \tag{6.42}$$

It is not difficult to prove that

$$P_L^2 = P_L, \qquad P_R^2 = P_R, \tag{6.43}$$

$$PP_L = P_R P = (1 - P_L)(1 - P) = (1 - P)(1 - P_R) = 0, \tag{6.44}$$

$$P_L D = D P_R. \tag{6.45}$$

In particular, the operator P_L projects any quark field to the orthogonal complement $\mathcal{S}^{\perp}$ of the deflation subspace. Note, however, that P_L and P_R are not Hermitian and therefore not ordinary orthogonal projectors.

The Dirac equation (6.1) may now be split into two decoupled equations,

$$D\psi_{\parallel} = \eta_{\parallel}, \qquad D\psi_{\perp} = \eta_{\perp}, \tag{6.46}$$

where

$$\psi_{\parallel} = (1 - P_R)\psi, \qquad \psi_{\perp} = P_R \psi, \tag{6.47}$$

$$\eta_{\parallel} = (1 - P_L)\eta, \qquad \eta_{\perp} = P_L \eta. \tag{6.48}$$

The use of a different projector for the splitting of the solution ψ and the source η is entirely consistent in view of the commutator relation (6.45) and merely reflects the fact that the Dirac operator is not Hermitian.

The solution of the little system,

$$\psi_{\|} = P(PDP)^{-1}P\eta = \sum_{k,l=1}^{N} \phi_k (A^{-1})_{kl} (\phi_l, \eta), \qquad (6.49)$$

is easily found, but the other equation can only be solved using an iterative procedure such as the GCR algorithm. However, the operator on the left of the equation is the deflated operator

$$\hat{D} = DP_R = P_L D(1 - P), \qquad (6.50)$$

which acts on quark fields in $\mathcal{S}^{\perp}$ and that may have a much smaller condition number than D, particularly so at small quark masses. An acceleration of the computation is then achieved since the solution is obtained in fewer iterations than in the case of the unmodified Dirac equation.

The condition number $\kappa(\hat{D}) = \|\hat{D}\|\|\hat{D}^{-1}\|$ of $\hat{D}$ depends on the quark mass mainly through the factor

$$\|\hat{D}^{-1}\| = \|(1 - P)D^{-1}(1 - P)\| \leq \|(1 - P)(D^{\dagger}D)^{-1}(1 - P)\|^{1/2}. \qquad (6.51)$$

It is then quite obvious that $\kappa(\hat{D})$ will be much smaller than $\kappa(D)$ if $1 - P$ effectively "projects away" the low modes of $D^{\dagger}D$, i.e. if they are well approximated by the deflation subspace. However, contrary to what may be assumed, this requirement is fairly weak and does not imply that the deflation subspace must be spanned by approximate eigenvectors of $D^{\dagger}D$ (see Fig. 6.6).

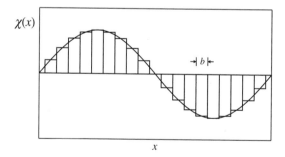

Fig. 6.6 The low modes $\chi(x)$ of the lattice Dirac operator in the free-quark theory are plane waves with small momenta p. These can be well approximated in the norm by functions that are constant on blocks of lattice points if the block size b satisfies $bp \ll 1$. Efficient deflation subspaces can thus be constructed using fields that are discontinuous and therefore far from being approximate eigenmodes of the Dirac operator.

6.1.3.4   *Domain-decomposed deflation subspaces*

In the following, a decomposition of the lattice into rectangular blocks Λ such as the one shown in Fig. 6.7 will be considered. A special kind of deflation subspace $\mathcal{S}$ is then obtained by choosing a set of orthonormal fields $\phi_1^\Lambda, \ldots, \phi_{N_s}^\Lambda$ on each block Λ and by taking $\mathcal{S}$ to be the linear span of all these fields. The associated projector is

$$P = \sum_\Lambda P_\Lambda, \qquad P_\Lambda \psi = \sum_{k=1}^{N_s} \phi_k^\Lambda \left(\phi_k^\Lambda, \psi \right). \qquad (6.52)$$

Evidently, this construction fits the general scheme discussed in Section 6.1.3.3 except perhaps for the labelling of the fields that span the deflation subspace.

The size of the blocks Λ is a tunable parameter of domain-decomposed subspaces. Usually the blocks are taken to be fairly small (4^4, 6^4 or 8×4^3, for example), but the exact choice should eventually be based on the measured performance of the deflated solver. A key feature of domain-decomposed subspaces is the fact that their dimension is proportional to the volume V of the lattice, while the application of the projector P to a given quark field requires only $O(N_s V)$ floating-point operations (and not $O(N_s V^2)$ operations, as would normally be the case). Such subspaces may thus allow the V^2-problem to be overcome, provided high deflation efficiencies can be achieved for some volume-independent number N_s of block modes.

In the case of the free-quark theory, Fig. 6.6 suggests that this strategy will work out if the block modes are chosen to be constant. Since quark fields have 12 complex components, one needs $N_s = 12$ such modes per block. In the presence of an arbitrary gauge field, the choice of the block modes is less obvious, however, because the notion of smoothness ceases to have a well-defined meaning.

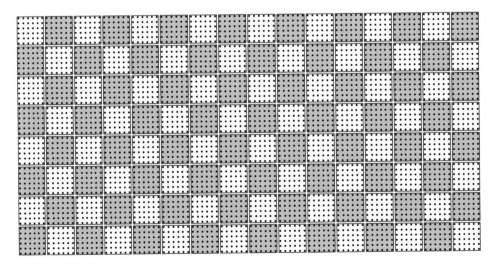

Fig. 6.7   Two-dimensional view of a 96×48^3 lattice, divided into 8192 non-overlapping blocks of size 6^4.

6.1.3.5 *Local coherence and subspace generation*

It is helpful to note at this point that the deflation deficits

$$\|(1-P)\chi\|^2 = \sum_\Lambda \|(1-P_\Lambda)\chi\|^2 \tag{6.53}$$

of the low modes χ of $D^\dagger D$ can only be small if they are small on all blocks Λ. Since the local deflation subspace has fixed dimension N_s, and since the number of low modes is proportional to V and thus tends to be much larger than N_s, this condition cannot in general be met unless the low modes happen to collapse to a lower-dimensional space on each block, i.e. unless they are "locally coherent" (see Fig. 6.8).

The local coherence of the low modes is numerically well established. On a 64×32^3 lattice with spacing $a = 0.08$ fm, for example, deflation deficits as small as a few per-cent are achieved when using 4^4 blocks and $N_s = 12$ block fields. Moreover, N_s does not need to be adjusted when the lattice volume increases. So far, however, no theoretical explanation of why the modes are locally coherent has been given. In particular, it is unclear whether the property has anything to do with chiral symmetry.

Efficient domain-decomposed deflation subspaces can now be constructed fairly easily (Lüscher, 2007a). One first notes that any quark field ψ satisfying $\|D\psi\| \leq M\|\psi\|$ for some sufficiently small value of M (say, $M = 100$ MeV) is well approximated by a linear combination of low modes and is therefore locally coherent with these. By generating a set $\psi_1, \ldots, \psi_{N_s}$ of independent fields of this kind, using inverse iteration, for example, and by applying the Gram–Schmidt orthonormalization process to the projected fields

$$\psi_k^\Lambda(x) = \begin{cases} \psi_k(x) & \text{if } x \in \Lambda, \\ 0 & \text{otherwise,} \end{cases} \tag{6.54}$$

one thus obtains a basis $\phi_1^\Lambda, \ldots, \phi_{N_s}^\Lambda$ of block fields with large projections to the low modes. [2] The generation of the deflation subspace along these lines requires a modest amount of computer time and certainly far less than would be needed for an

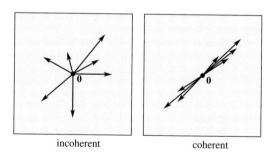

incoherent   coherent

Fig. 6.8 When restricted to a small block of lattice points, the $O(V)$ low modes of the Dirac operator tend to align to a relatively low-dimensional linear space, a property referred to as local coherence.

approximate computation of the low modes of the Dirac operator. Moreover, since N_s can be held fixed, the computational effort scales like V rather than V^2.

6.1.3.6   *Solving the deflated system*

Once the deflation subspace is constructed, the deflated equation $D\psi_\perp = \eta_\perp$ can in principle be solved using the GCR algorithm with D replaced by $\hat{D} = P_L D$. However, the application of the projector P_L requires the little system to be solved for a given source field, which is not a small task in general. Note that the little Dirac operator acts on fields on the block lattice with N_s complex components. An exact solution of the little system is therefore not practical on large lattices.

In the case of the Wilson–Dirac operator and its relatives, the little Dirac operator has only nearest-neighbor couplings among the blocks. The solution of the little system may then be obtained iteratively using the even–odd preconditioned GCR algorithm, for example. The effort required for the solution of the little system is nevertheless not completely negligible and it is advisable to consider solving the deflated right-preconditioned system

$$P_L D R \phi = \eta_\perp, \qquad \psi_\perp = P_R R \phi, \tag{6.55}$$

instead of the deflated equation directly, the operator R being a suitable preconditioner for D. A preconditioner that has been used in this context is the Schwarz alternating procedure (Lüscher, 2004). The important point to note is that the preconditioner tends to reduce the high-mode components of the residue of the current approximate solution, while the low-mode component of the residue is projected away by the projector P_L. Deflation and right-preconditioning thus tend to complement one another.

The performance figures plotted in Fig. 6.9 show that local deflation works very well in lattice QCD. In this study, the block size was taken to be 4^4 and N_s was set to 20. With respect to the even–odd preconditioned BiCGstab algorithm (points labelled EO+BiCGstab in the figure), the deflated Schwarz-preconditioned GCR algorithm (DFL+SAP+GCR) achieves an acceleration by more than an order of magnitude at the smallest quark masses considered.

On other lattices, the deflated Schwarz-preconditioned solver for the O(a)-improved Wilson–Dirac equation performs as well as in the case reported in Fig. 6.9. In particular, the accumulated experience unambiguously shows that the algorithm overcomes the V^2-problem and that the critical slowing down towards the chiral limit, which previously hampered quark-propagator computations, is nearly eliminated.

6.2   Simulation algorithms

Lattice QCD simulations are based on Markov chains and the concept of importance sampling. More specifically, most large-scale simulations performed today rely on some variant of the so-called Hybrid Monte Carlo algorithm (Duane *et al.*, 1987). An exception to this rule are simulations of the pure SU(3) gauge theory, where link-update algorithms are usually preferred for reasons of efficiency.

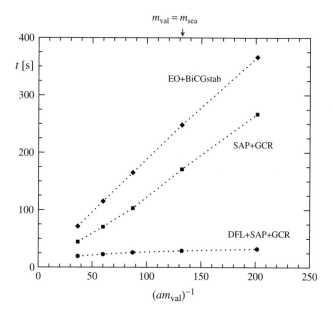

Fig. 6.9 Computer time needed for the solution of the O(a)-improved Wilson–Dirac equation in two-flavor QCD on a 64×32^3 lattice with spacing $a = 0.08$ fm. In these tests, the sea-quark mass m_{sea} was 26 MeV, the valence-quark mass m_{val} ranged from about 15 to 90 MeV and the relative residue of the solution was required to be 10^{-10}. All timings were taken on a PC cluster with 64 (single-core) processors.

QCD simulation algorithms have a long history and incorporate many ideas and improvements. Some important deficits remain, however, and further progress in algorithms will no doubt be required to be able to perform accurate simulations of a significantly wider range of lattices than is possible at present.

6.2.1 Importance sampling

6.2.1.1 *Statistical interpretation of the functional integral*

The quark fields in the QCD functional integral take values in a Grassmann algebra. So far, no practical method has been devised that would allow such fields to be simulated directly. The theory simulated is therefore always the one obtained after integrating out the quark fields.

In the case of two-flavor lattice QCD with mass-degenerate Wilson quarks, the partition function then assumes the form

$$\mathcal{Z} = \int \mathrm{D}[U] \left\{ \det D(U) \right\}^2 \mathrm{e}^{-S_g(U)}, \qquad \mathrm{D}[U] = \prod_{x,\mu} \mathrm{d}U(x,\mu), \qquad (6.56)$$

where $D(U)$ denotes the massive Wilson–Dirac operator in the presence of the gauge field U, $S_{\mathrm{g}}(U)$ the gauge action and $\mathrm{d}U(x,\mu)$ the SU(3)-invariant integration measure for the link variable $U(x, \mu)$. Since $D^\dagger = \gamma_5 D \gamma_5$, the quark determinant $\det D$ is real and the product

$$p(U) = \frac{1}{\mathcal{Z}} \{\det D(U)\}^2 \, e^{-S_g(U)} \tag{6.57}$$

is therefore a normalized probability density on the space of all gauge fields. The physics described by the theory is eventually extracted from expectation values

$$\langle \mathcal{O} \rangle = \int D[U] \, p(U) \mathcal{O}(U) \tag{6.58}$$

of observables $\mathcal{O}(U)$ such as a Wilson loop or a quark-line diagram. From this point of view, lattice QCD thus looks like a classical statistical system, where the states (the gauge-field configurations) occur with a certain probability and where one is interested in the expectation values of some properties of the states.

The simulation algorithms discussed in the following depend on the existence of a probabilistic representation of the theory. In particular, they do not apply to two-flavor QCD with non-degenerate quark masses and three-flavor QCD unless the product of the quark determinants is guaranteed to be non-negative (as is the case if the lattice Dirac operator preserves chiral symmetry).

6.2.1.2   Representative ensembles

Representative ensembles $\{U_1, \ldots, U_N\}$ of gauge fields are obtained by choosing the fields randomly with probability $D[U] \, p(U)$, i.e. such that

$$\text{no. of fields in } \mathfrak{R} = \int_{\mathfrak{R}} D[U] \, p(U) + O(N^{-1/2}) \tag{6.59}$$

for any open region $\mathfrak{R}$ in field space, the term of order $N^{-1/2}$ being a statistical error that depends on $\mathfrak{R}$ and the generated ensemble of fields. The high-probability regions thus contain many fields U_i and are therefore sampled well, while in other areas of field space there may be only a few fields or none at all (if N is not astronomically large).

Given a representative ensemble of fields, the expectation values of the observables of interest can be estimated through

$$\langle \mathcal{O} \rangle = \frac{1}{N} \sum_{i=1}^{N} \mathcal{O}(U_i) + O(N^{-1/2}). \tag{6.60}$$

A bit surprising may be the fact that results with small statistical errors can often be obtained in this way even if the ensemble contains only 100 or perhaps 1000 field configurations. Naive estimates, taking the dimension of field space into account, actually suggest that an accurate numerical evaluation of the functional integral requires some k^{32n} configurations, where n is the number of lattice points and k at least 10 or so.

The apparent paradox is resolved by noting that small ensembles of field configurations can only capture some aspects of the theory. That is, one should not expect to obtain the expectation values of all possible observables with small statistical errors. Typically, any quantity sensitive to the correlations of the field variables at large distances tends to have large statistical errors, sometimes to the extent that the results of the computation are completely useless.

6.2.1.3 *Translation symmetry and the infinite-volume limit*

In order to minimize finite-volume effects, periodic boundary conditions are usually imposed on all fields, an exception being the quark fields, which are often taken to be antiperiodic in time. Since the translation symmetry of the theory is preserved by this choice of boundary conditions, representative ensembles of gauge fields are expected "to look the same" in distant regions of a large lattice.

The meaning of this statement is best explained by dividing the lattice into blocks Λ, as in Section 6.1.3.4, and by considering an extensive quantity

$$\mathcal{E} = \sum_x \mathcal{O}(x), \tag{6.61}$$

where $\mathcal{O}(x)$ is some local gauge-invariant field. Translation symmetry then implies that the contributions $\langle \mathcal{O}_\Lambda \rangle$ to the sum

$$\langle \mathcal{E} \rangle = \sum_\Lambda \langle \mathcal{O}_\Lambda \rangle, \qquad \mathcal{O}_\Lambda = \sum_{x \in \Lambda} \mathcal{O}(x), \tag{6.62}$$

are all equal. Moreover, the statistical fluctuations of $\mathcal{O}_\Lambda$ and $\mathcal{O}_{\Lambda'}$ are practically uncorrelated and tend to cancel one another in the sum (6.62) if the blocks Λ and Λ' are separated by a distance larger than the range of the connected correlation function of $\mathcal{O}(x)$. For a fixed ensemble size, the statistical error of the density $\langle \mathcal{E} \rangle / V$ therefore decreases like $V^{-1/2}$ for $V \to \infty$, i.e. the statistics is effectively multiplied by a factor proportional to V.

The discussion also illustrates the fact that the efficiency of importance sampling depends on the observable considered and on how its expectation value is calculated. Rather than from Eq. (6.61), one might actually start from the identity $\langle \mathcal{E} \rangle = V \langle \mathcal{O}(0) \rangle$, in which case the information contained in the field ensemble away from the origin $x = 0$ remains unused. Both calculations yield the correct expectation value, but only the first profits from the available data on the full lattice and consequently obtains the result with much better statistical precision.

6.2.1.4 *Simulating Gaussian distributions*

Representative ensembles of fields can be easily constructed in the case of a complex field $\phi(x)$ distributed according to the Gaussian probability density

$$p_A(\phi) \propto \exp\left\{ -\sum_x \phi(x)^\dagger (A\phi)(x) \right\}, \tag{6.63}$$

where A is a Hermitian, strictly positive linear operator. If one sets

$$A = -\Delta + m_0^2, \qquad \Delta: \text{lattice Laplacian}, \tag{6.64}$$

for example, the theory describes a free scalar field with bare mass m_0. Another possible choice of A is

$$A = (DD^\dagger)^{-1}, \qquad D: \text{lattice Dirac operator}. \tag{6.65}$$

The field $\phi(x)$ must be a pseudo-fermion field in this case, i.e. a complex-valued and therefore bosonic quark field (see Section 6.2.5.1).

A representative ensemble of fields $\phi_1, \ldots, \phi_N$ may be generated for any Gaussian distribution by randomly choosing a set $\chi_1, \ldots, \chi_N$ of fields with normal distribution $p_1(\chi)$ and by setting

$$\phi_i(x) = (B\chi_i)(x), \qquad i = 1, \ldots, N, \tag{6.66}$$

where B is an operator satisfying

$$A = (BB^\dagger)^{-1}. \tag{6.67}$$

In the case of the pseudo-fermion fields, for example, one can simply take $B = D$, while a possible choice in the case of the free scalar field is

$$B = (-\Delta + m_0^2)^{-1/2}. \tag{6.68}$$

It is then straightforward to check that the fields ϕ_i generated in this way are correctly distributed.

Since the fields $\chi_i(x)$ are normally distributed, their components at different lattice points are decoupled and can be drawn randomly one after another. The generation of random numbers on a computer is a complicated subject, however, with many open ends and a vast literature (for an introduction, see (Knuth, 1997), for example). For the time being, one of the simulation-quality random number generators included in the GNU Scientific Library (http://www.gnu.org/software/gsl) may be used, among them the ranlux generator (Lüscher, 1994; James, 1994), which is based on a strongly chaotic dynamical system and thus comes with some theoretical understanding of why the generated numbers are random. An efficient ISO C code for this generator can be downloaded from http://cern.ch/luscher/ranlux.

6.2.2   Markov chains

Gaussian distributions are an exceptionally simple case where representative ensembles of fields can be generated instantaneously. In general, however, representative ensembles are generated through some recursive procedure (a Markov process) that obtains the field configurations one after another according to some stochastic algorithm.

This section is devoted to a theoretical discussion of such Markov processes. Rather than QCD or the SU(3) gauge theory, an abstract discrete system will be considered in order to avoid some technical complications, which might obscure the mechanism on which Markov-chain simulations are based. The discrete system is left unspecified, but is assumed to have the following properties:

(a) *There is a finite number n of states s.*
(b) *The equilibrium distribution $P(s)$ satisfies $P(s) > 0$ for all s and $\sum_s P(s) = 1$.*
(c) *The observables are real-valued functions $\mathcal{O}(s)$ of the states s.*

One is then interested in calculating the expectation values

$$\langle \mathcal{O} \rangle = \sum_s \mathcal{O}(s) P(s) \tag{6.69}$$

of the observables $\mathcal{O}(s)$.

6.2.2.1 *Transition probabilities*

A Markov chain is a random sequence $s_1, s_2, s_3, \ldots, s_N$ of states, where s_k is obtained from s_{k-1} through some stochastic algorithm. The chain thus depends on the initial state s_1 and the transition probability $T(s \to s')$ to go from the current state s to the next state s'. In the following, the basic idea is to choose the latter such that the Markov chain provides a representative ensemble of states for large N. The expectation value of any observable $\mathcal{O}(s)$ is then given by

$$\langle \mathcal{O} \rangle = \frac{1}{N} \sum_{k=1}^{N} \mathcal{O}(s_k) + \mathrm{O}(N^{-1/2}), \tag{6.70}$$

where the error term is dominated by the random fluctuations of the chain.

When trying to construct such transition probabilities, one may be guided by the following plausible requirements:

1. $T(s \to s') \geq 0$ for all s, s' and $\sum_{s'} T(s \to s') = 1$ for all s.
2. $\sum_s P(s) T(s \to s') = P(s')$ for all s'.
3. $T(s \to s) > 0$ for all s.
4. If S is a non-empty proper subset of states, there exist two states $s \in S$ and $s' \notin S$ such that $T(s \to s') > 0$.

Property 1 merely guarantees that $T(s \to s')$ is a probability distribution in s' for any fixed s, while property 2 says that the equilibrium distribution should be preserved by the update process. The other properties ensure that the Markov process does not get trapped in cycles (property 3, referred to as "aperiodicity") or in subsets of states (property 4, "ergodicity").

Later it will be shown that any transition probability $T(s \to s')$ satisfying $1 - 4$ generates Markov chains that simulate the system in the way explained above. These properties are thus sufficient to guarantee the correctness of the procedure.

6.2.2.2 *The acceptance–rejection method*

For illustration, an explicit example of a valid transition probability will now be constructed. In order to simplify the notation a little bit the states are labelled from

0 to $n-1$ and are thought to be arranged on a circle so that the neighbors of the state i are the states $i \pm 1 \mod n$. The construction then starts from the transition probability

$$T_0(i \to j) = \begin{cases} \frac{1}{3} & \text{if } j = i \text{ or } j = i \pm 1 \mod n, \\ 0 & \text{otherwise,} \end{cases} \tag{6.71}$$

which generates a random walk on the circle. This transition probability satisfies $1 - 4$, the equilibrium distribution being the flat distribution $P_0(i) = 1/n$.

Now, if the equilibrium distribution is not flat, as the one shown in Fig. 6.10, a valid transition probability is given by (Metropolis *et al.*, 1953)

$$T(i \to j) = T_0(i \to j)P_{\text{acc}}(i, j) + \delta_{ij} \sum_k T_0(i \to k)\left(1 - P_{\text{acc}}(i, k)\right), \tag{6.72}$$

where

$$P_{\text{acc}}(i, j) = \min\{1, P(j)/P(i)\} \tag{6.73}$$

is the so-called acceptance probability. In other words, starting from the current state i, the next state j is proposed with probability $T_0(i \to j)$. A random number $r \in [0, 1]$ is then chosen, with uniform distribution, and j is accepted if $P(j) \geq rP(i)$. If the proposed state is not accepted, the next state is taken to be the state i.

The acceptance–rejection method is widely used in various incarnations. In general, the challenge is to find an *a priori* transition probability $T_0(i \to j)$ where the proposed states are accepted with high probability, as otherwise the simulation will be very slow and therefore inefficient. The transition probability (6.72) incidentally satisfies

$$P(i)T(i \to j) = P(j)T(j \to i) \quad \text{for all} \quad i, j, \tag{6.74}$$

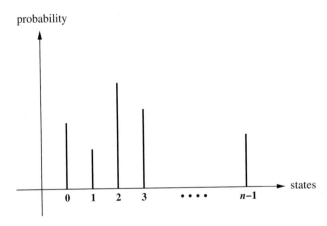

Fig. 6.10 Example of an equilibrium probability distribution of the abstract discrete system considered in this section. The update algorithm implementing the transition probability (6.72) generates a random walk in the space of states, where, in each step, one moves from the current state i to one of its neighbors j with probability $\frac{1}{3}P_{\text{acc}}(i, j)$ or else stays at i.

a property referred to as "detailed balance". In the literature, detailed balance is sometimes required for acceptable algorithms, but the condition is quite strong and not needed to ensure the correctness of the simulation.

6.2.2.3 *Implications of properties 1–4*

In the following, it is assumed that $T(s \to s')$ is a given transition probability satisfying the conditions 1–4 listed in Section 6.2.2.1. The associated Markov process is then shown to have certain mathematical properties, which will allow, in Section 6.2.2.4, to determine its asymptotic behavior at large times.

Let $\mathcal{H}$ be the linear space of real-valued functions $f(s)$ defined on the set of all states s. A useful norm of such functions is given by

$$\|f\|_1 = \sum_s |f(s)|. \tag{6.75}$$

The transition probability $T(s \to s')$ defines a linear operator T in $\mathcal{H}$ through

$$(Tf)(s') = \sum_s f(s)T(s \to s'). \tag{6.76}$$

Note that, by properties 1 and 2, the equilibrium distribution $P(s)$ has unit norm and is an eigenfunction of T with eigenvalue 1.

Lemma 6.1 *For all $f \in \mathcal{H}$, the bound $\|Tf\|_1 \leq \|f\|_1$ holds. Moreover, if $Tf = f$, there exists $c \in \mathbb{R}$ such that $f(s) = cP(s)$ for all states s.*

Proof: Any given function $f \in \mathcal{H}$ may be decomposed into positive and negative parts according to

$$f(s) = f_+(s) - f_-(s), \qquad f_{\pm}(s) = \frac{1}{2}\{|f(s)| \pm f(s)\} \geq 0. \tag{6.77}$$

Property 1 then implies

$$\|Tf_{\pm}\|_1 = \sum_{s'} |\sum_s f_{\pm}(s)T(s \to s')| = \sum_s f_{\pm}(s)\sum_{s'} T(s \to s') = \|f_{\pm}\|_1. \tag{6.78}$$

Using the triangle inequality, this leads to

$$\|Tf\|_1 = \|Tf_+ - Tf_-\|_1$$

$$\leq \|Tf_+\|_1 + \|Tf_-\|_1 = \|f_+\|_1 + \|f_-\|_1 = \|f\|_1, \tag{6.79}$$

which proves the first statement made in the lemma. Moreover, Eq. (6.79) shows that the equality $\|Tf\|_1 = \|f\|_1$ holds if and only if Tf_+ and Tf_- have disjoint support.

Now, let $f \in \mathcal{H}$ be such that $Tf = f$. The functions Tf_+ and Tf_- must have disjoint support in this case and are both non-negative. Since the decomposition $Tf = f_+ - f_-$ in positive and negative parts is unique, it follows that $Tf_+ = f_+$ and $Tf_- = f_-$. Properties 3 and 4, however, imply that the support of a non-negative function grows when T is applied, unless the support is empty or the whole set of states. Either f_+ or f_- must therefore be equal to zero. The function f thus has a definite sign.

Another function $g \in \mathcal{H}$ may now be defined through

$$g(s) = f(s) - cP(s), \qquad c = \sum_s f(s). \tag{6.80}$$

This function must also have a definite sign since $Tg = g$. Moreover, $\sum_s g(s) = 0$ by construction, which can only be true if $g = 0$, i.e. if $f(s) = cP(s)$. $\qquad\square$

Lemma 6.1 shows that the equilibrium distribution is the only distribution that is stationary under the action of the operator T. Some control over the complementary space

$$\mathcal{H}_0 = \{f \in \mathcal{H} \mid \sum_s f(s) = 0\} \tag{6.81}$$

of non-stationary functions will later be required as well. To this end, it is helpful to introduce a funny scalar product in $\mathcal{H}$,

$$(f, g) = \sum_s f(s) P(s)^{-1} g(s), \tag{6.82}$$

and the associated norm $\|f\| = (f, f)^{1/2}$.

Lemma 6.2 *There exists $0 \le \rho < 1$ such that $\|Tf\| \le \rho \|f\|$ for all $f \in \mathcal{H}_0$.*

Proof: With respect to the scalar product (6.82), the adjoint $T^\dagger$ of the operator T may be defined and thus also the symmetric operator $\hat{T} = T^\dagger T$. The action of $\hat{T}$ on any function $f \in \mathcal{H}$ is given by

$$(\hat{T}f)(s') = \sum_s f(s) \hat{T}(s \to s'), \tag{6.83}$$

$$\hat{T}(s \to s') = \sum_r T(s \to r) P(r)^{-1} T(s' \to r) P(s'). \tag{6.84}$$

A moment of thought then shows that $\hat{T}(s \to s')$ is a transition probability satisfying conditions 1–4. In particular, Lemma 6.1 holds for $\hat{T}$ as well.

Since $\hat{T}$ is symmetric, there exists a complete set of eigenfunctions $v_i \in \mathcal{H}$ of $\hat{T}$ with real eigenvalues λ_i $(i = 0, 1, \ldots, n-1)$. Without loss one may assume that

$$\lambda_0 \ge \lambda_1 \ge \ldots \ge \lambda_{n-1}, \qquad (v_i, v_j) = \delta_{ij}. \tag{6.85}$$

Moreover, noting $\lambda_i = (v_i, \hat{T}v_i) = \|Tv_i\|^2 \ge 0$, Lemma 6.1 (for $\hat{T}$) implies

$$\lambda_0 = 1, \quad v_0 = P \quad \text{and} \quad \lambda_1 < 1. \tag{6.86}$$

Now, since the states v_i, $i \ge 1$, satisfy

$$0 = (v_0, v_i) = \sum_s v_i(s), \tag{6.87}$$

they form a basis of the subspace $\mathcal{H}_0$, i.e. $\hat{T}$ maps $\mathcal{H}_0$ into itself and its largest eigenvalue in this subspace is λ_1. As a consequence,

$$\|Tf\|^2 = (f, \hat{T}f) \le \lambda_1 \|f\|^2 \quad \text{for all} \quad f \in \mathcal{H}_0. \tag{6.88}$$

This proves the lemma and also shows that the smallest possible value of the constant ρ is $\lambda_1^{1/2}$.

$\square$

6.2.2.4 *Statistical properties of Markov chains*

When analyzing Markov chains, an important conceptual point to note is that one cannot reasonably speak of the statistical properties of a single chain. However, one can ask what the average properties of the generated sequences of states are if many independent chains are considered. In practice, this corresponds to running several simulations in parallel, with different streams of random numbers.

In the following theoretical discussion, it will be assumed that an infinite number of independent simulations of length N have been performed, with the same initial state s_1 and a transition probability $T(s \to s')$ that has properties 1–4. As before, the expectation value of an observable $\mathcal{O}$ will be denoted by $\langle \mathcal{O} \rangle$, while

$$\overline{\mathcal{O}} = \frac{1}{N} \sum_{k=1}^{N} \mathcal{O}(s_k) \tag{6.89}$$

stands for its average over the states $s_1, s_2, \ldots, s_N$ generated in the course of a given simulation. One may then also consider functions $\phi(s_1, \ldots, s_N)$ of these sequences of states and their average $\langle\langle \phi \rangle\rangle$ over the infinitely many parallel simulations.

(a) Probability distribution of the states. The state s_k generated after $k-1$ steps changes randomly from one simulation to another. It is not difficult to show that the probability $P_k(s) = \langle\langle \delta_{ss_k} \rangle\rangle$ for s_k to coincide with s is given by

$$P_k(s) = \sum_{s_2, s_3, \ldots, s_{k-1}} T(s_1 \to s_2) T(s_2 \to s_3) \ldots T(s_{k-1} \to s)$$

$$= (T^{k-1} P_0)(s), \qquad P_0(s) = \delta_{ss_1}. \tag{6.90}$$

Noting $P_0 = P + f$, $f \in \mathcal{H}_0$, property 2 and Lemma 6.2 then imply

$$P_k(s) \underset{k \to \infty}{=} P(s) + \mathrm{O}\left(e^{-k/\tau}\right), \tag{6.91}$$

where $\tau = -1/\ln \rho > 0$ is the so-called exponential autocorrelation time of the Markov process. The state s_k is thus distributed according to the equilibrium distribution if k is much larger than τ. In particular, after so many steps, there is no longer any memory of the initial state s_1 and one says that the simulation has "thermalized".

(b) Calculation of expectation values. Together with the generated states, the average $\overline{\mathcal{O}}$ of the "measured" values of an observable $\mathcal{O}$ fluctuates randomly about the mean value

$$\langle\langle \overline{\mathcal{O}} \rangle\rangle = \frac{1}{N} \sum_{k=1}^{N} \langle\langle \mathcal{O}(s_k) \rangle\rangle = \sum_{s} \mathcal{O}(s) \frac{1}{N} \sum_{k=1}^{N} P_k(s). \tag{6.92}$$

Recalling Eq. (6.91), this formula shows that

$$\langle\!\langle \overline{\mathcal{O}} \rangle\!\rangle = \langle \mathcal{O} \rangle \tag{6.93}$$

if the first $k \gg \tau$ measurements of $\mathcal{O}$ are dropped. Averages of the measured values calculated after thermalization thus coincide with the expectation value $\langle \mathcal{O} \rangle$ up to statistical fluctuations.

(c) Autocorrelation functions. The states s_k in a Markov chain are statistically dependent to some extent, because they are generated one after another according to the transition probability $T(s \to s')$. As a consequence, the measured values of an observable $\mathcal{O}$ are statistically correlated, i.e. the autocorrelation function

$$\Gamma(t) = \langle\!\langle \mathcal{O}(s_k)\mathcal{O}(s_{k+t}) \rangle\!\rangle - \langle\!\langle \mathcal{O}(s_k) \rangle\!\rangle \langle\!\langle \mathcal{O}(s_{k+t}) \rangle\!\rangle \tag{6.94}$$

does not vanish. For $k \gg \tau$ and $t \geq 0$, the autocorrelation function is independent of k and given by

$$\Gamma(t) = \sum_{s_k, s_{k+1}, \dots, s_{k+t}} P(s_k)\mathcal{O}(s_k)T(s_k \to s_{k+1})\dots T(s_{k+t-1} \to s_{k+t})\mathcal{O}(s_{k+t}) - \langle \mathcal{O} \rangle^2. \tag{6.95}$$

Moreover, noting $P(s)\mathcal{O}(s) = P(s)\langle \mathcal{O} \rangle + f(s)$, $f \in \mathcal{H}_0$, it follows from this expression and Lemma 6.2 that $\Gamma(t)$ falls off exponentially, like $e^{-t/\tau}$, at large separations t. The measured values $\mathcal{O}(s_i)$ and $\mathcal{O}(s_j)$ are thus independently distributed if $|i - j| \gg \tau$.

(d) Statistical fluctuations. The statistical variance of the averages $\overline{\mathcal{O}}$ of an observable $\mathcal{O}$ is, after thermalization, given by

$$\langle\!\langle (\overline{\mathcal{O}} - \langle \mathcal{O} \rangle)^2 \rangle\!\rangle = \frac{1}{N^2} \sum_{l,j=1}^{N} \Gamma(|l - j|) = \Gamma(0)\frac{2\tau_{\mathcal{O}}}{N} + \mathrm{O}(N^{-2}), \tag{6.96}$$

where

$$\tau_{\mathcal{O}} = \frac{1}{2} + \sum_{t=1}^{\infty} \frac{\Gamma(t)}{\Gamma(0)} \tag{6.97}$$

denotes the so-called integrated autocorrelation time of $\mathcal{O}$. Up to terms of order $N^{-3/2}$, the standard deviation of $\overline{\mathcal{O}}$ from its expectation value $\langle \mathcal{O} \rangle$ is thus

$$\sigma = \sigma_0 \left(\frac{2\tau_{\mathcal{O}}}{N} \right)^{1/2}, \qquad \sigma_0 = \langle (\mathcal{O} - \langle \mathcal{O} \rangle)^2 \rangle^{1/2}, \tag{6.98}$$

i.e. the statistical error of $\overline{\mathcal{O}}$ decreases proportionally to $N^{-1/2}$. Note that σ_0 is just the standard deviation of $\mathcal{O}$ in equilibrium. In particular, σ_0 is a property of the system rather than of the Markov process. The integrated autocorrelation time $\tau_{\mathcal{O}}$, on the other hand, often strongly depends on the simulation algorithm.

The fact that the measured values of $\mathcal{O}$ are correlated leads to an increase of the variance of $\overline{\mathcal{O}}$ by the factor $2\tau_{\mathcal{O}}$ and thus lowers the efficiency of the simulation. In practice, one frequently chooses to measure the observables only on a subsequence of states separated by some fixed distance Δt in simulation time. As long as Δt is not much larger than $2\tau_{\mathcal{O}}$, the depletion of the measurements has no or little influence on the statistical error of $\overline{\mathcal{O}}$ and therefore helps reducing the computational load.

6.2.3 Simulating the SU(3) gauge theory

The theory of Markov chains developed in Section 6.2.2 can be extended to non-discrete systems like the pure SU(3) gauge theory on a finite lattice. Markov chains are sequences $U_1, U_2, \ldots, U_N$ of gauge-field configurations in this case, which are generated according to some transition probability. The latter must satisfy certain conditions analogous to those listed in Section 6.2.2.1 for the discrete system. The mathematics required at this point, however, tends to be quite heavy and no attempt will be made to prove of the correctness of the procedure (see (Tierney, 1994), for example).

6.2.3.1 Transition probability densities

The equilibrium probability density to be simulated is

$$p(U) = \frac{1}{\mathcal{Z}}e^{-S_{\mathrm{g}}(U)}, \qquad \mathcal{Z} = \int \mathrm{D}[U]\,e^{-S_{\mathrm{g}}(U)}, \tag{6.99}$$

where the gauge action $S_{\mathrm{g}}(U)$ is assumed to be a bounded function of the gauge field U. Markov processes in this theory are characterized by a transition probability density $T(U \to U')$ that specifies the probability $\mathrm{D}[U']\,T(U \to U')$ for the next configuration to be in the volume element $\mathrm{D}[U']$ at U' when the current configuration is U.

Note that transition probability densities may involve δ-function and other singularities, but the product $\mathrm{D}[U']\,T(U \to U')$ must be a well-defined measure on the field manifold for any given U. The obvious requirements are then

1. $T(U \to U') \geq 0$ *for all* U, U' *and* $\int \mathrm{D}[U']\,T(U \to U') = 1$ *for all* U.
2. $\int \mathrm{D}[U]\,p(U)T(U \to U') = p(U')$ *for all* U'.

Further conditions need to be added, however, in order to guarantee the aperiodicity, the ergodicity and thus the convergence of the Markov process. A sufficient but fairly strong condition is

3. *Every gauge field V has an open neighborhood $\mathcal{N}$ in field space such that $T(U \to U') \geq \epsilon$ for some $\epsilon > 0$ and all $U, U' \in \mathcal{N}$.*

This property ensures that, in every step, the Markov process spreads out in an open neighborhood of the current field. Moreover, using the compactness of the field manifold, it is possible to show that the process will reach any region in field space in a finite number of steps.

While properties 1–3 guarantee the asymptotic correctness of the simulations, the rigorous upper bounds on the exponential autocorrelation time obtained in the course of the convergence proofs tend to be astronomically large. In practice, simulations of lattice QCD therefore remain an empirical science to some extent, where one cannot claim, with absolute certainty, that the simulation results are statistically correct.

6.2.3.2 Link-update algorithms

If T_1 and T_2 are two transition probability densities satisfying 1 and 2, so does their composition

$$T(U \to U') = \int D[V]\, T_1(U \to V) T_2(V \to U'). \tag{6.100}$$

An update step according to the composed transition probability density first obtains the intermediate field V with probability $D[V]\, T_1(U \to V)$ and then generates U' with probability $D[U']\, T_2(V \to U')$. Composition allows simulation algorithms for fields to be constructed from elementary transitions, where a single field variable is changed at the time.

Link-update algorithms generate the next gauge field by updating the link variables one after another in some order. The Metropolis algorithm, for example, proceeds as follows (the SU(3) notation is summarized in Section 6.2.3.6):

(a) *Select a link (x, μ) and choose $X \in \mathfrak{su}(3)$ randomly in the ball $\|X\| \leq \epsilon$, with uniform distribution, where ϵ is some fixed positive number.*

(b) *Accept $U'(x, \mu) = e^X U(x, \mu)$ as the new value of the link variable on the selected link with probability $P_{\text{acc}} = \min\{1, e^{S_g(U) - S_g(U')}\}$.*

(c) *Leave the link variable unchanged if the new value proposed in step (b) is not accepted.*

It is not difficult to write down the transition probability density corresponding to the steps (a)–(c) and to check that it satisfies conditions 1 and 2. Moreover, the complete update cycle, where each link is visited once, satisfies condition 3 as well.

If the gauge action is local, the calculation of the action difference $S_g(U') - S_g(U)$ in step (b) involves only the field variables residing in the vicinity of the selected link. The computer time required per link update is then quite small. When all factors are taken into account, including the autocorrelation times, a more efficient link-update algorithm is, however, provided by the combination of the heatbath algorithm discussed below with a number of microcanonical moves (Section 6.2.3.5).

6.2.3.3 Heatbath algorithm

As a function of the field variable $U(x, \mu)$ residing on a given link (x, μ), the Wilson plaquette action is of the form

$$S_g(U) = -\operatorname{Re} \operatorname{tr}\{U(x, \mu) M(x, \mu)\} + \dots, \tag{6.101}$$

where $M(x, \mu)$ and the terms represented by the ellipsis do not depend on $U(x, \mu)$. Up to a constant factor involving the bare gauge coupling g_0, the complex 3×3 matrix

$$M(x, \mu) = \frac{2}{g_0^2} \sum_{\nu \neq \mu} \left\{ \begin{array}{c} \includegraphics \end{array} + \begin{array}{c} \includegraphics \end{array} \right\} \tag{6.102}$$

coincides with the "staple sum" of Wilson lines from $x + \hat{\mu}$ to x. Other popular gauge actions, including the $O(a^2)$-improved Symanzik action, are of the same form except that $M(x, \mu)$ gets replaced by a sum of more complicated Wilson lines.

The heatbath algorithm (Creutz, 1980) is a link-update algorithm, where the new value $U'(x, \mu)$ of the field variable on the selected link is chosen randomly with probability density proportional to $\exp\left(\operatorname{Re} \operatorname{tr}\{U'(x, \mu)M(x, \mu)\}\right)$. In other words, the link variable is updated according to its exact distribution in the presence of the other field variables.

While this algorithm fulfills conditions 1–3, it is difficult to implement in practice exactly as described here. However, for gauge group SU(2) (the case considered by (Creutz, 1980)), the situation is more favorable and there are highly efficient ways to generate random link variables with the required probability distribution (Fabricius and Haan, 1984; Kennedy and Pendleton, 1985).

6.2.3.4   The Cabibbo–Marinari method

The practical difficulties encountered when implementing the heatbath algorithm in the SU(3) theory can be bypassed as follows (Cabibbo and Marinari, 1982). Let $v \in$ SU(2) be embedded in SU(3) through

$$
v \to V = \begin{pmatrix} v_{11} & v_{12} & 0 \\ v_{21} & v_{22} & 0 \\ 0 & 0 & 1 \end{pmatrix}. \tag{6.103}
$$

A correct one-link update move, $U(x, \mu) \to U'(x, \mu)$, is then obtained by choosing v randomly with probability density proportional to $\exp\left(\operatorname{Re} \operatorname{tr}\{VU(x, \mu)M(x, \mu)\}\right)$ and by setting $U'(x, \mu) = VU(x, \mu)$. A moment of thought reveals that the distribution of v is of the same analytic form as the link distribution in the SU(2) theory. The highly efficient methods developed for the latter can thus be used here too.

In order to treat all color components of the link variables democratically, different embeddings of SU(2) in SU(3) should be used. A popular choice is to perform SU(2) rotations in the $(1, 2)$, $(2, 3)$ and $(3, 1)$ planes in color space before proceeding to the next link. The ergodicity of the algorithm is then again guaranteed.

6.2.3.5   Microcanonical moves

The link-update algorithms discussed so far tend to become inefficient when the lattice spacing a is reduced, a phenomenon known as "critical slowing down". Typically, the autocorrelation times of physical observables increase approximately like a^{-2}.

Microcanonical algorithms are based on field transformations that preserve the gauge action. Such transformations are valid transitions, satisfying conditions 1 and 2, provided the field integration measure is preserved too. The trajectories in field space generated by a microcanonical algorithm do not involve random changes of direction and are therefore quite different from the random walks performed by the other algorithms. An acceleration of the simulation is then often achieved when microcanonical moves are included in the update scheme.

A microcanonical link-update algorithm for the SU(3) theory is easily constructed following the steps taken in the case of the heatbath algorithm (Creutz, 1987; Brown and Woch, 1987). The link variable $U(x, \mu)$ on the selected link is again updated by applying Cabibbo–Marinari rotations, but the SU(2) matrix v is now set to

$$v = \frac{2w^2}{\operatorname{tr}\{w^\dagger w\}},\tag{6.104}$$

where w is a 2×2 matrix implicitly defined by

$$\operatorname{Re}\operatorname{tr}\{VU(x, \mu)M(x, \mu)\} = \operatorname{tr}\{vw^\dagger\} + \cdots\tag{6.105}$$

and the requirement that w is in SU(2) up to a real scale factor. The existence and uniqueness of w is implied by the reality and linearity properties of the expression on the left of Eq. (6.105) (no update is performed if w is accidentally equal to zero). The transformation $U(x, \mu) \to VU(x, \mu)$ defined in this way preserves the gauge action since $\operatorname{tr}\{vw^\dagger\} = \operatorname{tr}\{w^\dagger\}$, but it is less obvious that it also preserves the link integration measure, because w and therefore v depend on $U(x, \mu)$, i.e. the transformation is non-linear. It is straightforward to show, however, that

$$w|_{U(x,\mu) \to ZU(x,\mu)} = wz^\dagger\tag{6.106}$$

for all $z \in$ SU(2). Now, if $f(U)$ is any integrable function of $U = U(x, \mu)$, the substitution $U \to ZU$ leads to the identity

$$\int \mathrm{d}U f(VU) = \int \mathrm{d}U f(\tilde{V}U), \qquad \tilde{v} = \frac{2wz^\dagger w}{\operatorname{tr}\{w^\dagger w\}}.\tag{6.107}$$

Since this equation holds for any z, it remains valid when integrated over SU(2). Using the invariance of the group-integration measures under left- and right-multiplications, one then deduces that

$$\int \mathrm{d}U f(VU) = \int \mathrm{d}z \int \mathrm{d}U f(\tilde{V}U) = \int \mathrm{d}z \int \mathrm{d}U f(Z^\dagger U) = \int \mathrm{d}U f(U),\tag{6.108}$$

which proves that the transformation $U \to VU$ preserves the link-integration measure.

Microcanonical simulation algorithms are not ergodic and must therefore be combined with an ergodic one. A recommended scheme consists in updating all link variables once using the heatbath algorithm and subsequently n times using microcanonical moves. This combination is more efficient than the pure Metropolis or heatbath algorithm and it reportedly has an improved scaling behavior as a function of the lattice spacing if n is scaled roughly like a^{-1}.

6.2.3.6   *Appendix: SU(3) notation*

The Lie algebra $\mathfrak{su}(3)$ of SU(3) consists of all complex 3×3 matrices X that satisfy

$$X^\dagger = -X \quad \text{and} \quad \operatorname{tr}\{X\} = 0.\tag{6.109}$$

With this convention, the Lie bracket $[X,Y]$ maps any pair X, Y of $\mathfrak{su}(3)$ matrices to another element of $\mathfrak{su}(3)$. Moreover, the exponential series

$$\mathrm{e}^X = 1 + \sum_{k=1}^{\infty} \frac{X^k}{k!} \tag{6.110}$$

converges to an element of SU(3) (note the absence of factors of i in these and the following formulae).

One can always choose a basis T^a, $a = 1, \ldots, 8$, of $\mathfrak{su}(3)$ such that

$$\mathrm{tr}\{T^a T^b\} = -\frac{1}{2}\delta^{ab}. \tag{6.111}$$

With respect to such a basis, the elements $X \in \mathfrak{su}(3)$ are represented as

$$X = \sum_{a=1}^{8} X^a T^a, \qquad X^a = -2\,\mathrm{tr}\{X T^a\} \in \mathbb{R}. \tag{6.112}$$

Moreover, the natural scalar product on $\mathfrak{su}(3)$ is given by

$$(X, Y) = \sum_{a=1}^{8} X^a Y^a = -2\,\mathrm{tr}\{XY\}, \tag{6.113}$$

the associated matrix norm being $\|X\| = (X, X)^{1/2}$. If not specified otherwise, the Einstein summation convention is used for group indices.

6.2.4 The hybrid Monte Carlo (HMC) algorithm

The inclusion of the sea quarks in the simulations is difficult, because the quark determinants in the QCD functional integral depend non-locally on the gauge field. In particular, one-link update algorithms would require a computational effort proportional to the square of the lattice volume and are therefore not practical.

The HMC algorithm (Duane *et al.*, 1987) updates all link variables at once and has a much better scaling behavior with respect to the lattice volume. It will here be explained in general terms for an unspecified (possibly non-local) action $S(U)$, which is assumed to be real and differentiable.

6.2.4.1 *Molecular dynamics*

As an intermediate device, the HMC algorithm requires an $\mathfrak{su}(3)$-valued field

$$\pi(x, \mu) = \pi^a(x, \mu)T^a, \qquad \pi^a(x, \mu) \in \mathbb{R}, \tag{6.114}$$

to be added to the theory (the SU(3) notation is as in Section 6.2.3.6). The new field is interpreted as the canonical momentum of the gauge field, the associated Hamilton function being

$$H(\pi, U) = \frac{1}{2}(\pi, \pi) + S(U), \qquad (\pi, \pi) = \sum_{x,\mu} \pi^a(x, \mu)\pi^a(x, \mu). \qquad (6.115)$$

Evidently, since

$$\int D[U] \, \mathcal{O}(U) e^{-S(U)} = \text{constant} \times \int D[\pi]D[U] \, \mathcal{O}(U) e^{-H(\pi, U)}, \qquad (6.116)$$

the addition of the momentum field does not affect the physics content of the theory.

In the form (6.116), the theory is reminiscent of the classical statistical systems that describe a gas of molecules. Hamilton's equations,[3]

$$\dot{\pi}(x, \mu) = -F(x, \mu), \qquad F^a(x, \mu) = \left.\frac{\partial S(e^\omega U)}{\partial \omega^a(x, \mu)}\right|_{\omega=0}, \qquad (6.117)$$

$$\dot{U}(x, \mu) = \pi(x, \mu)U(x, \mu), \qquad (6.118)$$

are therefore often referred to as the "molecular dynamics equations". As usual, the dot on the left of these equations implies a differentiation with respect to time t, which is here a fictitious time unrelated to the time coordinate of space-time. The solutions of the molecular dynamics equations, $\pi_t(x, \mu)$ and $U_t(x, \mu)$, are uniquely determined by the initial values of the fields at $t = 0$. They may be visualized as trajectories in field space (or, more precisely, in phase space) parameterized by the time t.

6.2.4.2 The HMC strategy

The basic idea underlying the HMC algorithm is to pass from the original theory to the classical system (6.116) and to evolve the fields by integrating the molecular dynamics equations. Explicitly, the steps leading from the current gauge field $U(x, \mu)$ to the next field $U'(x, \mu)$ are the following:

(a) *A momentum field π is generated randomly with probability density proportional to $\exp\{-\frac{1}{2}(\pi, \pi)\}$.*

(b) *The molecular dynamics equations are integrated from time $t = 0$ to some later time $t = \tau$, taking π and U as the initial values of the fields.*

(c) *The new gauge field U' is set to the field U_τ obtained at time $t = \tau$ through the molecular dynamics evolution.*

If τ is set to a fixed value, as is usually done, the transition probability density corresponding to the steps (a)–(c) is given by

$$T(U \to U') = \frac{1}{\mathcal{Z}_\pi} \int D[\pi] \, e^{-\frac{1}{2}(\pi, \pi)} \prod_{x,\mu} \delta(U'(x, \mu), U_\tau(x, \mu)), \qquad (6.119)$$

where the partition function $\mathcal{Z}_\pi$ of the momentum field ensures the correct normalization and the Dirac δ-function is the one appropriate to the gauge-field integration measure.

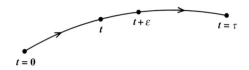

Fig. 6.11 The numerical integration of the molecular dynamics equations proceeds in time steps of size ϵ. In each step, the fields at time $t + \epsilon$ are computed from the fields at time t and possibly those at earlier times (if a higher-order scheme is used). The integration rule and the step size ϵ should evidently be such that the calculated fields at time $t = n\epsilon$, $n = 0, 1, \ldots, N_0$, closely follow the exact trajectory in phase space.

It is trivial to check that the transition probability density (6.119) satisfies the first of the three conditions listed in Section 6.2.3.1. The second condition is also fulfilled, but some work is required to show this. An important point to note is that the molecular dynamics equations are invariant under time reversal $\pi_t, U_t \to -\pi_{\tau-t}, U_{\tau-t}$. The molecular dynamics evolution $\pi_0, U_0 \to \pi_\tau, U_\tau$ therefore defines an invertible transformation of phase space. Moreover, it preserves the Hamilton function and, by Liouville's theorem, also the phase-space integration measure. It follows from these remarks that

$$\int D[U] \, e^{-S(U)} T(U \to U') = \frac{1}{Z_\pi} \int D[\pi] D[U] \, e^{-H(\pi,U)} \prod_{x,\mu} \delta(U'(x,\mu), U_\tau(x,\mu))$$

$$= \frac{1}{Z_\pi} \int D[\pi_\tau] D[U_\tau] \, e^{-H(\pi_\tau, U_\tau)} \prod_{x,\mu} \delta(U'(x,\mu), U_\tau(x,\mu)) = e^{-S(U')}, \qquad (6.120)$$

where the second equation is obtained by performing a change of variables from $\pi, U = \pi_0, U_0$ to π_τ, U_τ. Condition 2 is thus satisfied too.

For sufficiently small τ, the ergodicity of the algorithm (condition 3) can be proved as well. The proof is based on an expansion of π_τ, U_τ in powers of τ, from which one infers that the range of U_τ includes an open neighborhood of $U = U_0$ when the initial momentum $\pi = \pi_0$ varies over a neighborhood of the origin. In view of the non-linear nature of the theory, HMC simulations of lattice QCD are, however, expected to be ergodic at any (non-zero) value of τ, even if one is unable to show this. Ergodicity can, in any case, always be rigorously ensured by choosing $\tau \in [0, \tau_{\max}]$ randomly from one update step to the next.

6.2.4.3   Numerical integration of the molecular dynamics equations

In practice, the molecular dynamics equations cannot be integrated exactly and one must resort to some numerical integration method. As explained in Section 6.2.4.4, the integration error can be compensated by including an acceptance–rejection step in the HMC algorithm so that the correctness of the simulation is not compromised.

The numerical integration proceeds by dividing the time interval $[0, \tau]$ in N_0 steps of size ϵ and by applying a discrete integration rule that gives the correct result in the limit $\epsilon \to 0$ (see Fig. 6.11). Considering the Taylor expansions

$$\pi_{t+\epsilon} = \pi_t - \epsilon F|_{U=U_t} + O(\epsilon^2), \qquad (6.121)$$

$$U_{t+\epsilon} = U_t + \epsilon \pi_t U_t + O(\epsilon^2), \qquad (6.122)$$

it is clear that acceptable integration schemes can be built from the elementary operations

$$\mathcal{I}_0(\epsilon): \quad \pi, U \to \pi - \epsilon F, U, \qquad (6.123)$$

$$\mathcal{I}_U(\epsilon): \quad \pi, U \to \pi, e^{\epsilon \pi} U. \qquad (6.124)$$

The combination $\mathcal{I}_0(\frac{1}{2}\epsilon)\mathcal{I}_U(\epsilon)\mathcal{I}_0(\frac{1}{2}\epsilon)$, for example, takes π_t, U_t to $\pi_{t+\epsilon}, U_{t+\epsilon}$ up an error of order ϵ^3. The complete integration from time $t = 0$ to time $t = \tau$ then amounts to applying the product

$$\mathcal{J}_0(\epsilon, N_0) = \left\{ \mathcal{I}_0(\tfrac{1}{2}\epsilon)\mathcal{I}_U(\epsilon)\mathcal{I}_0(\tfrac{1}{2}\epsilon) \right\}^{N_0}, \qquad \epsilon = \frac{\tau}{N_0}, \qquad (6.125)$$

to the initial fields.

The "leap-frog integrator" (6.125) is remarkably simple and has a number of good properties. In particular, the integration is reversible,

$$\mathcal{J}_0(-\epsilon, N_0)\mathcal{J}_0(\epsilon, N_0) = 1, \qquad (6.126)$$

and $\mathcal{J}_0(\epsilon, N_0)$ is therefore an invertible mapping of phase space. Starting from the elementary integration steps (6.123) and (6.124), it is also trivial to show that the integrator preserves the field-integration measure $D[\pi]D[U]$. Through the numerical integration, these important properties of the molecular dynamics evolution are thus not lost.

The leap-frog integrator is widely used and appreciated for its simplicity, but there are many other integration schemes that can be employed (Leimkuhler and Reich, 2004; Hairer *et al.*, 2006). In particular, the so-called symplectic integrators all have the good properties mentioned above.

6.2.4.4   *Acceptance–rejection step*

For a fixed step size ϵ, the numerical integration of the molecular dynamics equations normally does not preserve the Hamilton function. In particular, the difference

$$\Delta H(\pi, U) = \{ H(\pi_\tau, U_\tau) - H(\pi_0, U_0) \}_{\pi_0 = \pi, U_0 = U} \qquad (6.127)$$

does not vanish in general. The HMC algorithm (as defined through steps (a)–(c) in Section 6.2.4.2) consequently violates condition 2 if the integration is performed numerically. It is possible to correct for this deficit by replacing step (c) through

(c′) *The new gauge field U' is set to the field U_τ obtained through the integration of the molecular dynamics equations with probability*

$$P_{\mathrm{acc}}(\pi, U) = \min\{1, e^{-\Delta H(\pi, U)}\}. \qquad (6.128)$$

Otherwise, i.e. if the proposed field is rejected, U' is set to U.

The transition probability density of the modified algorithm,

$$T(U \to U') = \frac{1}{Z_\pi} \int \mathrm{D}[\pi] \, \mathrm{e}^{-\frac{1}{2}(\pi,\pi)} \Big\{ P_{\mathrm{acc}}(\pi, U) \prod_{x,\mu} \delta(U'(x,\mu), U_\tau(x,\mu))$$

$$+ (1 - P_{\mathrm{acc}}(\pi, U)) \prod_{x,\mu} \delta(U'(x,\mu), U(x,\mu)) \Big\}, \tag{6.129}$$

can then again be shown to have the required properties, for any value of the integration step size ϵ, provided the integrator is reversible and measure-preserving.

The adjustable parameters of the HMC algorithm are then the trajectory length τ, the integration step size ϵ and further parameters of the integration scheme (if any). Evidently, the simulation will be inefficient if the average acceptance rate is low, i.e. if the numerical integration is not very accurate. The acceptance rate must, however, be balanced against the computer time required for the simulation, which grows roughly linearly with the step number $N_0 = \tau/\epsilon$. In the case of the leap-frog integrator, for example, $\langle P_{\mathrm{acc}} \rangle = 1 - \mathrm{O}(\epsilon^2)$ and the integration step size is then usually tuned so that acceptance rates of $70 - 80$ per cent are achieved.

It is more difficult to give a recommendation on the value of τ. Traditionally, the trajectory length is set to 1, but the choice of τ can have an influence on the autocorrelation times and the stability of the numerical integration. Some empirical studies are therefore required in order to determine the optimal value of τ in a given case.

6.2.5 Application to two-flavor QCD

In QCD with a doublet of mass-degenerate sea quarks, the probability density to be simulated is

$$p(U) = \frac{1}{Z} \mathrm{e}^{-S(U)}, \qquad S(U) = S_{\mathrm{g}}(U) - \ln |\det D(U)|^2, \tag{6.130}$$

where $D(U)$ denotes the lattice Dirac operator in presence of the gauge field U (cf. Section 6.2.1.1). The HMC algorithm can in principle be used to simulate this distribution, but a straightforward application of the algorithm is not possible in practice, because the calculation of the quark determinant and of the force deriving from it would require an unreasonable amount of computer time.

6.2.5.1 Pseudo-fermion fields

This difficulty can fortunately be overcome using the pseudo-fermion representation

$$|\det D(U)|^2 = \mathrm{constant} \times \int \mathrm{D}[\phi] \, \mathrm{e}^{-S_{\mathrm{pf}}(U,\phi)}, \tag{6.131}$$

$$S_{\mathrm{pf}}(U, \phi) = (D(U)^{-1}\phi, D(U)^{-1}\phi), \qquad \mathrm{D}[\phi] = \prod_{x,A,\alpha} \mathrm{d}\phi_{A\alpha}(x)\mathrm{d}\phi_{A\alpha}(x)^*, \tag{6.132}$$

of the quark determinant. The auxiliary field $\phi(x)$ introduced here carries a Dirac index A and a color index α, like a quark field, but its components are complex numbers rather than being elements of a Grassmann algebra. For the scalar product in Eq. (6.132) one can take the obvious one for such fields, with any convenient normalization. Step (a) of the HMC algorithm is then replaced by

(a′) *A momentum field π and a pseudo-fermion field ϕ are generated randomly with probability density proportional to* $\exp\{-\frac{1}{2}(\pi,\pi) - S_{\rm pf}(U,\phi)\}$.

Note that the pseudo-fermion action is quadratic in ϕ. The field can therefore be easily generated following the lines of Section 6.2.1.4. Once this is done, the algorithm proceeds as before, where the Hamilton function to be used in steps (b) and (c′) is

$$H(\pi, U) = \frac{1}{2}(\pi, \pi) + S_{\rm g}(U) + S_{\rm pf}(U, \phi). \tag{6.133}$$

Only the momentum π and the gauge field U are evolved by the molecular dynamics equations. The pseudo-fermion field ϕ remains unchanged and thus plays a spectator rôle at this point.

The steps (a′), (b) and (c′) implement the transition probability density

$$T(U \to U') = \frac{1}{\mathcal{Z}_\pi \mathcal{Z}_{\rm pf}(U)} \int D[\pi] D[\phi]\, e^{-\frac{1}{2}(\pi,\pi) - S_{\rm pf}(U,\phi)}$$

$$\times \left\{ P_{\rm acc}(\pi, U) \prod_{x,\mu} \delta(U'(x,\mu), U_\tau(x,\mu)) \right.$$

$$\left. + (1 - P_{\rm acc}(\pi, U)) \prod_{x,\mu} \delta(U'(x,\mu), U(x,\mu)) \right\}, \tag{6.134}$$

where $\mathcal{Z}_{\rm pf}(U)$ is the partition function of the pseudo-fermion field ϕ in the presence of the gauge field U. For simplicity, the dependence of U_τ and $P_{\rm acc}$ on ϕ has been suppressed. Starting from this formula, it is then not difficult to show that the algorithm correctly simulates the distribution (6.130).

6.2.5.2 *Performance of the HMC algorithm*

The force F that drives the molecular dynamics evolution in step (b) of the algorithm has two parts, F_0 and F_1, the first deriving from the gauge action and the other from the pseudo-fermion action in the Hamilton function (6.133). In the case of the Wilson theory, for example, the forces are

$$F_0^a(x, \mu) = -\mathrm{Re}\,\mathrm{tr}\{T^a U(x,\mu) M(x,\mu)\}, \tag{6.135}$$

$$F_1^a(x, \mu) = -2\,\mathrm{Re}\left(\gamma_5 D^{-1}\gamma_5 \psi, \delta_{x,\mu}^a D\psi\right), \qquad \psi = D^{-1}\phi, \tag{6.136}$$

where $M(x, \mu)$ is the staple sum (6.129) previously encountered in the pure gauge theory and

$$(\delta^a_{x,\mu} D\psi)(y) = \delta_{x+\hat\mu, y} \frac{1}{2}(1 + \gamma_\mu)U(x,\mu)^{-1}T^a\psi(x)$$

$$- \delta_{x,y}\frac{1}{2}(1 - \gamma_\mu)T^a U(x,\mu)\psi(x + \hat\mu). \tag{6.137}$$

Note that the computation of the pseudo-fermion force F_1 requires the Dirac equation to be solved twice. The by far largest fraction of the computer time is then usually spent in this part of the simulation program, particularly so at small sea-quark masses where the Dirac operator becomes increasingly ill-conditioned.

Traditionally, the performance of QCD simulation algorithms is measured by counting the number of floating-point operations required for the generation of a sample of statistically independent gauge-field configurations. Such performance estimations are quite primitive and tend to be subjective to some extent, because the term "statistically independent" is only loosely defined in this context. Moreover,

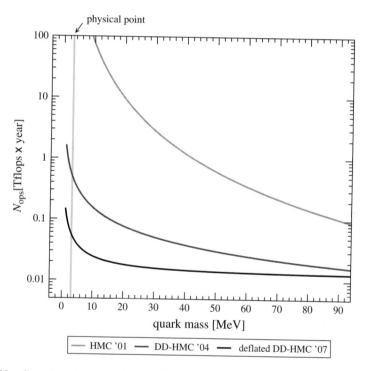

Fig. 6.12 Number of floating-point operations required for the generation of 100 statistically independent gauge-field configurations in $O(a)$-improved two-flavor QCD on a 64×32^3 lattice with spacing $a = 0.08$ fm. The top curve (Ukawa, 2002) represents the status reported at the memorable Berlin LATTICE conference in 2001, the middle one was obtained a few years later, using the so-called domain-decomposed HMC algorithm (Lüscher, 2005; Del Debbio *et al.*, 2007), and the lowest curve shows the performance of a recently developed deflated version of the latter (Lüscher, 2007*b*).

Fig. 6.13 Multiple time-step integration schemes divide the integration range $[0, \tau]$ in a hierarchy of intervals of increasing sizes $\epsilon_0, \epsilon_1, \dots$ such that ϵ_{k+1} is an integer multiple of ϵ_k.

the specific capabilities of the computers used for the simulation are not taken into account. The performance estimates plotted in Fig. 6.12 nevertheless clearly show that the simulations have become significantly faster since the beginning of the decade. In the following sections, some of the now widely used acceleration techniques are briefly described.

The curves shown in Fig. 6.12 refer to a particular choice of the lattice action and the lattice parameters. If the lattice volume V is increased at fixed lattice spacing and quark mass, and if the leap-frog integrator is used, the numerical effort required for the simulations is known to scale like $V^{5/4}$. As a function of the lattice spacing a, the scaling behavior of the HMC algorithm is more difficult to determine and is currently a debated issue, but there is little doubt that the required computer time increases at least like a^{-7}.

6.2.5.3 *Multiple time-step integration*

An acceleration of the HMC algorithm can often be achieved using adapted integration step sizes for different parts of the force F in the molecular dynamics equations (Sexton and Weingarten, 1992). The quark force F_1, for example, tends to be significantly smaller than the gauge force F_0 and can therefore be integrated with a larger step size than the latter.

If the integration step sizes for the forces F_0 and F_1 are taken to be

$$\epsilon_0 = \frac{\tau}{N_0 N_1}, \qquad \epsilon_1 = \frac{\tau}{N_1}, \tag{6.138}$$

where N_0, N_1 are some positive integers, the leap-frog integrators

$$\mathcal{J}_0(\epsilon_0, N_0) = \left\{ \mathcal{I}_0(\tfrac{1}{2}\epsilon_0)\mathcal{I}_U(\epsilon_0)\mathcal{I}_0(\tfrac{1}{2}\epsilon_0) \right\}^{N_0}, \tag{6.139}$$

$$\mathcal{J}_1(\epsilon_1, N_1) = \left\{ \mathcal{I}_1(\tfrac{1}{2}\epsilon_1)\mathcal{J}_0(\epsilon_0, N_0)\mathcal{I}_1(\tfrac{1}{2}\epsilon_1) \right\}^{N_1}, \tag{6.140}$$

integrate the molecular dynamics equations from time t to $t + \epsilon_1$ and $t + \tau$, respectively (see Fig. 6.13). In these equations,

$$\mathcal{I}_k(\epsilon) : \pi, U \to \pi - \epsilon F_k, U \tag{6.141}$$

are the elementary integration steps involving the force F_k. In particular, the application of $\mathcal{J}_0(\epsilon_0, N_0)$ consumes relatively little computer time, because a computation of the quark force is not required.

The hierarchical integration is profitable if the step size ϵ_1 can be set to a value larger than ϵ_0 without compromising the accuracy of the numerical integration too much. An acceleration of the simulation by a factor approximately equal to N_0 is then achieved.

6.2.5.4 *Frequency splitting of the quark determinant*

The factorization

$$|\det D|^2 = \det\{DD^\dagger + \mu^2\} \times \det\left\{\frac{DD^\dagger}{DD^\dagger + \mu^2}\right\} \tag{6.142}$$

of the quark determinant separates the contribution of the eigenvalues of $DD^\dagger$ larger than μ^2 from the contribution of the lower ones. In the HMC algorithm, the two factors may be represented by two pseudo-fermion fields, ϕ_1 and ϕ_2, with action

$$S_{\text{pf}}(U, \phi) = \left(\phi_1, (DD^\dagger + \mu^2)^{-1}\phi_1\right) + \left(\phi_2, \phi_2 + \mu^2(DD^\dagger)^{-1}\phi_2\right). \tag{6.143}$$

The quark force accordingly splits into two forces, F_1 and F_2, where the first is nearly insensitive to the quark mass, while the second involves the inverse of $DD^\dagger$ and is, in this respect, similar to the force derived from the full quark determinant.

When such determinant factorizations were first considered (Hasenbusch, 2001; Hasenbusch and Jansen, 2003), the main effect appeared to be that the fluctuations of the quark force along the molecular dynamics trajectories were reduced. The integration step size required for a given acceptance rate could consequently be increased by a factor 2 or so. (Urbach *et al.*, 2006) later noted that F_1 is the far dominant contribution to the quark force at small μ. Using a multiple time-step integrator, and after some tuning of μ, an important acceleration of the simulation was then achieved.

Another factorization of the quark determinant, leading to the DD-HMC algorithm (Lüscher, 2005), is obtained starting from a domain decomposition of the lattice like the ones previously considered in Section 6.1.3.4. The rôle of the scale μ that separates the high modes of the Dirac operator from low modes is here played by the block sizes, which are usually chosen to be in the range from 0.5 to 1 fm or so.

For the associated factorization of the quark determinant to work out, one needs to assume that the lattice Dirac operator has only nearest-neighbor hopping terms, as in the case of the (improved) Wilson–Dirac operator, and that the block division of the lattice is chessboard-colorable (see Fig. 6.14). With respect to the subset Ω of points contained in the black blocks and its complement Ω^*, the Dirac operator then naturally decomposes into four parts,

$$D = D_\Omega + D_{\Omega^*} + D_{\partial\Omega} + D_{\partial\Omega^*}, \tag{6.144}$$

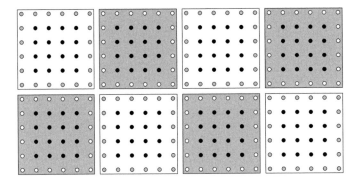

Fig. 6.14 Divisions of the lattice into non-overlapping blocks of lattice points can be chessboard-colored if there is an even number of blocks along each coordinate axis. The union of the sets of points contained in the black blocks is denoted by Ω and its complement (the points in the white blocks) by Ω^*. Their exterior boundaries, $\partial\Omega$ and $\partial\Omega^*$, consist of all points in the other set with minimal distance from the surfaces separating the blocks (open points).

where D_Ω includes all (diagonal and hopping) terms acting inside Ω and D_{Ω^*} those acting in Ω^*. The hopping terms from the white to the black blocks and the ones going in the opposite direction are included in $D_{\partial\Omega}$ and $D_{\partial\Omega^*}$, respectively.

Using the fact that Ω and Ω^* are disjoint sets of lattice points, the factorization

$$\det D = \det D_\Omega \det D_{\Omega^*} \det\{1 - D_\Omega^{-1} D_{\partial\Omega} D_{\Omega^*}^{-1} D_{\partial\Omega^*}\}, \tag{6.145}$$

may now be derived, where the first two factors further factorize into the determinants of the block Dirac operators. The factorization corresponds to a splitting of the quark force into an easy high-mode part (the block determinants) and an "expensive" but much smaller low-mode part. No fine tuning is required in this case and an acceleration of the simulation is again achieved using a multiple time-step integrator. Moreover, the domain decomposition is, as usual, favorable for the parallel processing of large lattices.[4]

6.2.5.5 *Chronological inversion and deflation acceleration*

In the course of the integration of the molecular dynamics equations, the Dirac equation must be solved many times. The exact details depend on which integrator and acceleration techniques are used, but the equations to be solved are usually of the form

$$D\psi = \phi, \qquad D\chi = \gamma_5\psi, \tag{6.146}$$

where the source field ϕ does not depend on the integration time t.

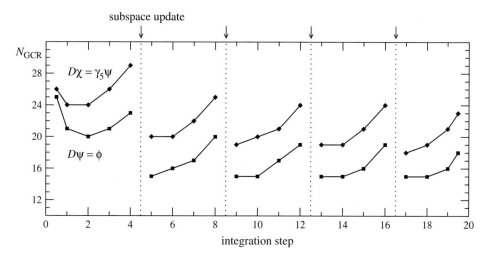

Fig. 6.15 History of the GCR solver iteration numbers required for the solution of the Dirac equation along a molecular dynamics trajectory, showing the effect of the refreshing of the deflation subspace. The data points were obtained at a sea-quark mass of 26 MeV on a 64×32^3 lattice with spacing $a = 0.08$ fm. On other lattices, the situation is practically the same.

For small integration step sizes, the gauge field changes only little from one step to the next and the solutions ψ_t and χ_t obtained at time t therefore tend to evolve smoothly. One may thus attempt to predict them from the previous solutions through a polynomial extrapolation in t, for example (Brower *et al.*, 1997). In general, the predicted solutions of the Dirac equation are not sufficiently accurate, but this deficit can easily be removed through iterative improvement (Section 6.1.1.2). The total number of iterations performed by the solver program is then often significantly reduced.

An important further reduction of the solver iteration numbers can be achieved using the low-mode deflation method described in Section 6.1.3 (Lüscher, 2007*b*). Since the gauge field changes along a molecular dynamics trajectory, the (domain-decomposed) deflation subspace must be refreshed from time to time in order to preserve its efficiency. An example illustrating this process is shown in Fig. 6.15. The subspace generation at the beginning of the trajectory and the periodic refreshing of the subspace requires some computer time, but this overhead is largely compensated by the fact that the solutions of the Dirac equations (6.146) are obtained much more rapidly than without deflation, particularly so at small quark masses (see Fig. 6.12).

6.2.5.6 *Improved integrators*

The use of integration schemes other than the leap-frog integrator can be profitable if fewer evaluations of the quark force are required for a given integration accuracy.

Higher-order schemes, for example, where the one-step error is reduced to $O(\epsilon^5)$, are not difficult to construct (Leimkuhler and Reich, 2004; Hairer *et al.*, 2006), but so far did not prove to be significantly faster than the leap-frog integrator.

Another possibility is to look for $O(\epsilon^3)$ integrators that minimize the integration error according to some criterion (Omelyan *et al.*, 2002; 2003; Takaishi and de Forcrand, 2006; Clark *et al.*, 2008*b*). An integrator of this kind is, in the case of a single time-step integration, given by

$$\widetilde{\mathcal{J}}_0(\epsilon, N_0) = \left\{ \mathcal{I}_0(\tfrac{1}{2}\tilde{\epsilon})\mathcal{I}_U(\tfrac{1}{2}\epsilon)\mathcal{I}_0(\epsilon - \tilde{\epsilon})\mathcal{I}_U(\tfrac{1}{2}\epsilon)\mathcal{I}_0(\tfrac{1}{2}\tilde{\epsilon}) \right\}^{N_0}, \qquad (6.147)$$

where $\tilde{\epsilon} \propto \epsilon$ is a tunable parameter. Experience suggests that the optimal values of $\tilde{\epsilon}/\epsilon$ are in the range $0.3 - 0.5$, depending a bit on the chosen accuracy criterion. Although the quark force needs to be computed twice per integration step, this integrator achieves a net acceleration by a factor 1.5 or so with respect to the leap-frog integrator, because fewer integration steps are required.

6.2.6 Inclusion of the strange quark

Unlike the light quarks, the strange and the heavy quarks do not pair up in approximately mass-degenerate doublets. For various technical reasons, single quarks are more difficult to include in the simulations than pairs and a special treatment is required.

6.2.6.1 *Strange-quark determinant*

One of the issues that needs to be addressed is the fact that the determinant

$$\det D_{\rm s} = \pm |\det D_{\rm s}| \qquad (6.148)$$

of the strange-quark Dirac operator $D_{\rm s}$ may not have the same sign for all gauge-field configurations. If chiral symmetry is exactly preserved on the lattice, the determinant is guaranteed to be positive, but sign changes from one configuration to another are not excluded in the case of the (improved) Wilson–Dirac operator, for example. The presence of positive and negative contributions to the QCD partition function potentially ruins the foundations on which numerical simulations are based. In particular, importance sampling ceases to have a clear meaning.

In all current simulations of QCD that include the strange quark, the regions in field space, where the strange-quark determinant is negative, are assumed (or can be shown) to have a totally negligible weight in the functional integral. The operator in the determinant may then be replaced by the non-negative Hermitian operator

$$|Q_{\rm s}| = \left(Q_{\rm s}^2\right)^{1/2}, \qquad Q_{\rm s} = \gamma_5 D_{\rm s}, \qquad (6.149)$$

without affecting the simulation results. As explained in the following sections, the so-modified theory can again be simulated using the HMC algorithm.

In the Wilson theory, the justification of this procedure rests on the observation that the physical strange quark is relatively heavy and that the Hermitian Dirac operator $Q_{\rm s}$ consequently tends to have a solid spectral gap in presence of the gauge

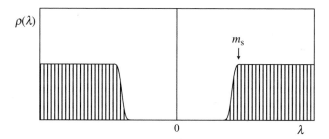

Fig. 6.16 Typical shape of the density $\rho(\lambda)$ of the eigenvalues λ of the Hermitian Wilson–Dirac operator Q_s near the origin. At lattice spacings $a \leq 0.1$ fm, and if the strange-quark mass m_s is greater than or equal to its physical value, the distribution of the eigenvalues with the smallest magnitude extends only slightly inside the spectral gap of the density in the continuum limit (Del Debbio *et al.*, 2006).

fields that dominate the QCD functional integral (see Fig. 6.16). Since the gap tends to widen when the strange-quark mass is increased, the sign of the determinant of Q_s cannot change and is therefore the same as the one at large masses, where it can be proved to be positive.

6.2.6.2   Pseudo-fermion representation

Although well defined, the operator $|Q_\mathrm{s}|$ is not directly accessible and one is forced to use approximations whenever its action on a quark field needs to be computed. In the present context, an approximation of its inverse is required, i.e. a tractable function R of Q_s^2 such that $|Q_\mathrm{s}|R \simeq$ constant. An exact representation of the strange-quark determinant can then be given using two pseudo-fermion fields,

$$\det|Q_\mathrm{s}| = \text{constant} \times \int \mathrm{D}[\phi_1]\mathrm{D}[\phi_2]\, \mathrm{e}^{-S_{\mathrm{pf,s}}(U,\phi_1,\phi_2)}, \qquad (6.150)$$

where only the second term in the pseudo-fermion action

$$S_{\mathrm{pf,s}} = (\phi_1,(|Q_\mathrm{s}|R)^{-1}\phi_1) + (\phi_2, R\phi_2) \qquad (6.151)$$

is included in the molecular dynamics Hamilton function. The first term is nearly independent of the gauge field and can be taken into account in the acceptance–rejection step at the end of the molecular dynamics evolution.

So far not many different approximations R were considered. The PHMC algorithm, for example, employs a polynomial approximation (Frezzotti and Jansen, 1997; 1999), while the RHMC algorithm is based on a rational approximation (Horvath *et al.*, 1999; Clark and Kennedy, 2007). There is also a complex version of the PHMC algorithm, where one starts from an approximation of D_s^{-1} by a polynomial in D_s (Takaishi and de Forcrand, 2002; Aoki *et al.*, 2002). None of these algorithms is completely trivial to implement or obviously preferable in view of its simplicity, accessibility to acceleration techniques or speed.

In the following, the RHMC algorithm is worked out as a representative case. The steps to be taken in the PHMC algorithm are similar, the main differences being issues of approximation accuracy and numerical stability.

6.2.6.3   *Optimal rational approximation*

The operator norm of the difference of two functions of Q_s^2 is bounded from above by the maximal absolute deviation of the functions in the range covered by the eigenvalues of Q_s^2. For the construction of optimal approximations of $|Q_s|^{-1}$, some information on the spectral range of Q_s^2 is therefore required as input.

Since Q_s^2 has a solid spectral gap, one can choose some fixed numbers M and $\epsilon > 0$ such that the spectrum of Q_s^2 is fully contained in the interval $[\epsilon M^2, M^2]$ with high probability (i.e. for most configurations in a representative ensemble of gauge fields). Now let

$$R(y) = A\frac{(y + a_1)(y + a_3)\ldots(y + a_{2n-1})}{(y + a_2)(y + a_4)\ldots(y + a_{2n})}, \tag{6.152}$$

be a rational function of degree $[n, n]$ in y, which approximates the function $1/\sqrt{y}$ in the range $[\epsilon, 1]$. The operator $R(Q_s^2/M^2)$ then approximates $M/|Q_s|$ up to a relative error (in the operator norm) less than or equal to

$$\delta = \max_{\epsilon \leq y \leq 1} |1 - \sqrt{y}R(y)| \tag{6.153}$$

if the spectrum of Q_s^2 is fully contained in $[\epsilon M^2, M^2]$ and a somewhat larger error if the spectrum extends slightly beyond the limits of this interval.

For a specified degree n, one would evidently like the coefficients a_r in Eq. (6.152) to be such that the error δ is minimized. It seems unlikely that the optimal coefficients can be worked out analytically, but the mathematician Zolotarev was able to do that a long time ago by relating the optimization problem to the theory of elliptic functions (see (Achiezer, 1992), for example; the coefficients are given explicitly in Section 6.2.6.5).[5]

The optimal rational approximation has a number of remarkable properties. One of them is that the error δ is a very rapidly decreasing function of the degree n. For $\epsilon = 10^{-5}$, for example, the approximation error is 5×10^{-4} if $n = 6$ and decreases to 1×10^{-7} and 8×10^{-15} if $n = 12$ and $n = 24$, respectively. The fact that the minimizing coefficients satisfy

$$a_1 > a_2 > \ldots > a_{2n} > 0, \qquad A > 0, \tag{6.154}$$

is also very important. In particular, the residues in the expansion in partial fractions,

$$R(y) = A\left\{1 + \frac{r_2}{y + a_2} + \ldots + \frac{r_{2n}}{y + a_{2n}}\right\}, \tag{6.155}$$

are all positive. The expansion is therefore well suited for the numerical evaluation of the action of the operator $R(Q_s^2/M^2)$ on the pseudo-fermion field ϕ_2. Note that this calculation essentially amounts to solving the equations

$$\left(Q_{\text{s}}^2 + \mu_{2k}^2\right)\psi_k = M^2\phi_2, \qquad \mu_r = M\sqrt{a_r}, \tag{6.156}$$

for $k = 1, \ldots, n$. Since the right-hand sides of these equations are the same, it is possible to solve all equations simultaneously using a so-called multimass solver (Jegerlehner, 1996). The total computational effort is then not very much larger than what would be required for the sequential solution of the equations that are the most difficult to solve (the ones at the largest values of k).

6.2.6.4   The RHMC algorithm

Having specified the operator R, the inclusion of the strange quark in the HMC algorithm is now straightforward. The algorithm proceeds according to the steps (a′), (b) and (c′), as before, and one merely has to add the contribution of the strange-quark pseudo-fermion fields. Some specific remarks on what exactly needs to be done may nevertheless be useful.

(a) Pseudo-fermion generation.   The random generation of the strange-quark pseudo-fermion fields in the first step requires two operators B and C to be found such that

$$BB^\dagger = |Q_{\text{s}}|R, \quad \text{and} \quad CC^\dagger = R^{-1} \tag{6.157}$$

(cf. Section 6.2.1.4). For C one can take the rational function

$$C = A^{-1/2}\frac{(Q_{\text{s}} + i\mu_2)\ldots(Q_{\text{s}} + i\mu_{2n})}{(Q_{\text{s}} + i\mu_1)\ldots(Q_{\text{s}} + i\mu_{2n-1})}, \tag{6.158}$$

but there is no similarly simple choice for the other operator. However, since

$$Z = \frac{Q_{\text{s}}^2 R^2}{M^2} - 1 \tag{6.159}$$

is of order δ, the series

$$B = \sqrt{M}(1 + Z)^{1/4} = \sqrt{M}\left\{1 + \frac{1}{4}Z - \frac{3}{32}Z^2 + \ldots\right\} \tag{6.160}$$

converges rapidly and may be truncated after the first few terms. The strange-quark pseudo-fermion fields can thus be generated with a computational effort equivalent to the one required for a few applications of the operator R to a given quark field.

(b) Strange-quark force.   As already mentioned, only the second term of the pseudo-fermion action (6.151) is included in the molecular dynamics Hamilton function. The computation of the associated force,

$$F_{\text{s}}^a(x, \mu) = -\frac{2A}{M^2}\sum_{k=1}^{n} r_{2k}\,\text{Re}\left(\psi_k, Q_{\text{s}}\gamma_5\delta_{x,\mu}^a D_{\text{s}}\psi_k\right), \tag{6.161}$$

requires the n linear systems (6.156) to be solved. The computer time needed for this calculation is therefore essentially the same as for one application of the operator R.

(c) Acceptance step. The acceptance probability is calculated as usual except for the fact that the change of the first term of the pseudo-fermion action (6.151),

$$\left(\phi_1, (|Q_s|R)^{-1}\phi_1\right)\big|_{U=U_\tau} - \left(\phi_1, (|Q_s|R)^{-1}\phi_1\right)\big|_{U=U_0}, \tag{6.162}$$

must be added to the difference (6.127) of the molecular-dynamics Hamilton function. Note that the action difference (6.162) can be computed by expanding $(|Q_s|R)^{-1}\phi_1$ in powers of Z.

6.2.6.5   *Appendix: Coefficients of the optimal rational approximation*

The analytic expressions for the coefficients of the rational function (6.152) that minimizes the approximation error (6.153) involve the Jacobi elliptic functions $\text{sn}(u, k)$, $\text{cn}(u, k)$ and the complete elliptic integral $K(k)$ (see (Abramowitz and Stegun, 1972), for example, for the definition of these functions). Explicitly, they are given by

$$a_r = \frac{\text{cn}^2(rv, k)}{\text{sn}^2(rv, k)}, \quad r = 1, 2, \ldots, 2n, \tag{6.163}$$

where

$$k = \sqrt{1 - \epsilon}, \quad v = \frac{K(k)}{2n + 1}. \tag{6.164}$$

The formulae for the amplitude A and the error δ,

$$A = \frac{2}{1 + \sqrt{1 - d^2}} \frac{c_1 c_3 \ldots c_{2n-1}}{c_2 c_4 \ldots c_{2n}}, \tag{6.165}$$

$$\delta = \frac{d^2}{\left(1 + \sqrt{1 - d^2}\right)^2}, \tag{6.166}$$

involve the coefficients

$$c_r = \text{sn}^2(rv, k), \quad r = 1, 2, \ldots, 2n, \tag{6.167}$$

$$d = k^{2n+1} \left(c_1 c_3 \ldots c_{2n-1}\right)^2. \tag{6.168}$$

All these expressions are free of singularities and can be programmed straightforwardly, using the well-known methods for the numerical evaluation of the Jacobi elliptic functions. An ISO C program that calculates the coefficients A, $a_1, \ldots, a_{2n}$ and the error δ to machine precision can be downloaded from `http://cern.ch/luscher/`.

6.3   Variance-reduction methods

Lattice QCD simulations produce ensembles $\{U_1, \ldots, U_N\}$ of gauge fields, which are representative of the functional integral at the specified gauge coupling and sea-quark masses. In this section it is taken for granted that autocorrelation effects can be safely neglected, i.e. that the separation in simulation time of subsequent field configurations is sufficiently large for this to be the case. As discussed in Section 6.2.2.4, the expectation value of any (real or complex) observable $\mathcal{O}(U)$ may then be calculated through

$$\langle \mathcal{O} \rangle = \frac{1}{N} \sum_{k=1}^{N} \mathcal{O}(U_k) + \mathrm{O}(N^{-1/2}), \tag{6.169}$$

where the statistical error is, to leading order in $1/N$, given by

$$\sigma(\mathcal{O}) = \frac{\sigma_0(\mathcal{O})}{N^{1/2}}, \qquad \sigma_0(\mathcal{O}) = \langle |\mathcal{O} - \langle \mathcal{O} \rangle|^2 \rangle^{1/2}. \tag{6.170}$$

In practice, the error is estimated from the variance of the "measured" values $\mathcal{O}(U_k)$, which is a correct procedure up to subleading terms.

For a given observable $\mathcal{O}$, another observable $\mathcal{O}'$ satisfying

$$\langle \mathcal{O}' \rangle = \langle \mathcal{O} \rangle, \qquad \sigma_0(\mathcal{O}') \ll \sigma_0(\mathcal{O}), \tag{6.171}$$

can sometimes be found. The desired expectation value is then obtained with a much smaller statistical error if $\mathcal{O}$ is replaced by $\mathcal{O}'$. Most variance-reduction methods are based on this simple observation. The construction of effective alternative observables is non-trivial, however, and may involve auxiliary stochastic variables and transformations of the functional integral.

The discussion in this section is often of a general nature and applies to most forms of lattice QCD, but if not specified otherwise, the Wilson formulation will be assumed, with or without O(a)-improvement and with two or more flavors of sea quarks.

6.3.1 Hadron propagators

The calculation of the properties of the light mesons and baryons is a central goal in lattice QCD. Finding good variance-reduction methods proves to be difficult in this field, but some important progress has nevertheless been made. In this section, the aim is to shed some light on the problem by discussing the statistical variance of hadron propagators.

6.3.1.1 The pion propagator

Once the sea-quark fields are integrated out, the correlation functions of local fields like the isospin pseudoscalar density

$$P^a = \bar{\psi} \frac{1}{2} \tau^a \gamma_5 \psi, \qquad \psi = \begin{pmatrix} u \\ d \end{pmatrix}, \qquad \tau^a: \text{Pauli matrices}, \tag{6.172}$$

become expectation values of quark-line diagrams. The diagrams are products of quark propagators, Dirac matrices and invariant color tensors, with all Dirac and color indices properly contracted (see Fig. 6.17).

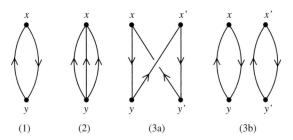

Fig. 6.17 Quark-line diagrams contributing to the pion propagator (1), the nucleon propagator (2) and the pion four-point function (3a and 3b). Directed lines from y to x stand for the light-quark propagator $S(x, y)$. At the vertices, the spinor indices are contracted with the appropriate color tensor and Dirac matrix (γ_5 in the case of the pion-field vertices).

For any given gauge field, the light-quark propagator is determined through the field equation

$$DS(x, y) = \delta_{xy} \tag{6.173}$$

and the chosen boundary conditions. The spinor indices have been suppressed in this equation, but one should keep in mind that $S(x, y)$ is, for fixed x and y, a complex 12×12 matrix in spinor space. In view of the γ_5-hermiticity of the Dirac operator, forward and backward propagators are related by

$$\gamma_5 S(y, x) \gamma_5 = S(x, y)^\dagger, \tag{6.174}$$

where the Hermitian conjugation refers to spinor space only.

In lattice QCD, the pion mass m_π is usually determined from the exponential decay of the "pion propagator"

$$\langle P^a(x) P^b(y) \rangle = -\frac{1}{2} \delta^{ab} \langle \text{tr}\{S(x, y) S(x, y)^\dagger\} \rangle \tag{6.175}$$

at large distances $|x - y|$. It is advantageous for this calculation to pass to the zero-momentum component of the propagator,

$$g_\pi(x_0 - y_0) = \langle \mathcal{O}_\pi(x_0, y) \rangle, \tag{6.176}$$

$$\mathcal{O}_\pi(x_0, y) = \sum_{\vec{x}} \text{tr}\{S(x, y) S(x, y)^\dagger\}, \tag{6.177}$$

whose asymptotic form at large times,

$$g_\pi(t) \underset{t \to \infty}{\propto} e^{-m_\pi t} + O(e^{-3m_\pi t}), \tag{6.178}$$

is dominated by a single exponential with exponent equal to the pion mass.

6.3.1.2 *Statistical error estimation*

For a given gauge field and source point y, the calculation of the propagator $S(x, y)$ allows the observable $\mathcal{O}_\pi(x_0, y)$ to be evaluated at all times x_0. The ensemble average of $\mathcal{O}_\pi(x_0, y)$ then provides a stochastic estimate of the zero-momentum pion propagator, from which one may be able to extract the pion mass.

The statistical error of the average of $\mathcal{O}_\pi(x_0, y)$ is proportional to the square root of the *a priori* variance

$$\sigma_0(\mathcal{O}_\pi)^2 = \langle \mathcal{O}_\pi(x_0, y)^2 \rangle - \langle \mathcal{O}_\pi(x_0, y) \rangle^2, \tag{6.179}$$

which may symbolically be written in the form

$$\sigma_0(\mathcal{O}_\pi)^2 = \sum_{\vec{x}, \vec{x}'} \left\{ \left\langle \vcenter \right\rangle - \left\langle \right\rangle \left\langle \right\rangle \right\}_{x_0' = x_0}. \tag{6.180}$$

An important point to note here is that the diagram in the first term also contributes to the $\pi\pi$ propagator $\langle P^a(x) P^b(x') P^a(y) P^b(y) \rangle$ (see Fig. 6.17). Moreover, the second term is just the square of the pion propagator. The variance is therefore expected to decay like $\mathrm{e}^{-2m_\pi |x_0 - y_0|}$ at large time separations, which implies that the pion propagator is obtained with a nearly time-independent relative statistical error.

Lattice QCD simulations tend to confirm this and they also show that the quark propagator typically falls off like

$$\mathrm{tr}\{S(x, y) S(x, y)^\dagger\}^{1/2} \propto \mathrm{e}^{-\frac{1}{2} m_\pi |x - y|} \tag{6.181}$$

at large distances, for every gauge field in a representative ensemble of fields. There is actually little room for a different behavior of the propagator, since both the mean and width of the distribution of $\mathrm{tr}\{S(x, y) S(x, y)^\dagger\}$ decay exponentially with the same exponent.

6.3.1.3 *Baryons and the exponential SNR problem*

The diagrams contributing to the nucleon two-point function involve 3 quark propagators. At zero spatial momentum and for every gauge-field configuration, the associated observable $\mathcal{O}_N(x_0, y)$ thus falls off roughly like $\mathrm{e}^{-\frac{3}{2} m_\pi |x_0 - y_0|}$ at large time separations. The nucleon propagator, however, decays much more rapidly, since the exponent (the nucleon mass m_N) is significantly larger than $\frac{3}{2} m_\pi$.

In the sum over all gauge fields, there must therefore be important cancellations among the measured values of $\mathcal{O}_N(x_0, y)$. More precisely, after averaging over N configurations, the nucleon propagator is obtained with a signal-to-noise ratio (SNR) proportional to $\sqrt{N} \mathrm{e}^{-(m_N - \frac{3}{2} m_\pi)|x_0 - y_0|}$. The number of measurements required for a specified statistical accuracy at a given time separation thus scales like

$$N \propto \mathrm{e}^{(2m_N - 3m_\pi)|x_0 - y_0|}. \tag{6.182}$$

Note that the exponent $2m_N - 3m_\pi$ is, in all cases of interest, not small and as large as $7\,\mathrm{fm}^{-1}$ at the physical point. The calculated values of the nucleon propagator therefore tend to rapidly disappear in the statistical noise at large time separations $|x_0 - y_0|$.

Computations of other hadron propagators are similarly affected by an exponential loss of significance, the only exception being the propagators of the stable pseudoscalar mesons. The masses and decay constants of the latter can thus be determined far more easily than those of the vector mesons and the baryons.

6.3.2 Using random sources

As already noted in Section 6.2.1.3, the link variables in distant regions of a large lattice are practically decoupled. Hadron propagators calculated at widely separated source points therefore tend to be sampled independently and their average consequently has smaller statistical fluctuations than the propagator at a single source point. Averaging over as many source points as possible is thus desirable, but tends to be expensive in terms of computer time, because the quark propagators must be recomputed at each point.

The random source method performs the sum over all or a selected set of source points stochastically (Michael and Peisa, 1998). In many cases, a significant acceleration of the computation can be achieved in this way. The idea of introducing random sources is also quite interesting from a purely theoretical point of view, because it extends the notion of an observable to stochastic observables (i.e. functions of the gauge field that depend on random auxiliary variables).

6.3.2.1 *Gaussian random fields*

Random sources may be thought of as a set of additional fields that are decoupled from the dynamical fields and therefore do not change the physics content of the theory. In the case considered here, the added fields are a multiplet

$$\eta_i(\vec{x}), \quad i = 1, \ldots, N_{\mathrm{src}}, \tag{6.183}$$

of pseudo-fermion fields on the fixed-time spatial lattice. Their action is taken to be

$$S_{\mathrm{src}}(\eta) = \sum_{i=1}^{N_{\mathrm{src}}} (\eta_i, \eta_i), \tag{6.184}$$

where the scalar product is the obvious one for such fields. For each gauge-field configuration in a representative ensemble of fields, the source fields are chosen randomly with probability density proportional to $e^{-S_{\mathrm{src}}(\eta)}$. Evidently, the set of fields obtained in this way is a representative ensemble for the joint probability density of the gauge field and the source fields.

One may now consider observables $\mathcal{O}(U, \eta)$ that depend on both the gauge field and the sources. When the latter are integrated out, such observables reduce to ordinary (non-stochastic) observables. Noting

$$\langle \eta_i(\vec{x}) \eta_j(\vec{y})^\dagger \rangle_{\mathrm{src}} = \delta_{ij} \delta_{\vec{x}\vec{y}}, \tag{6.185}$$

the integral over the source fields is easily worked out, using Wick's theorem, if $\mathcal{O}(U, \eta)$ is a polynomial in the source fields. In the case of the observable

$$\mathcal{O} = \frac{1}{N_{\text{src}}} \sum_{i=1}^{N_{\text{src}}} \sum_{\vec{x}, \vec{y}} \eta_i(\vec{x})^\dagger S(x, y)|_{x_0 = y_0} \eta_i(\vec{y}), \tag{6.186}$$

for example, the calculation yields

$$\langle \mathcal{O} \rangle_{\text{src}} = \sum_{\vec{x}} \text{tr}\{S(x, x)\}. \tag{6.187}$$

The random sources thus allow the trace (6.187) to be estimated stochastically.

6.3.2.2 *The pion propagator revisited*

For any fixed time y_0, the spinor fields

$$\phi_i(x, y_0) = \sum_{\vec{y}} S(x, y) \eta_i(\vec{y}), \qquad i = 1, \ldots, N_{\text{src}}, \tag{6.188}$$

can be computed by solving the Dirac equations $D\phi_i(x, y_0) = \delta_{x_0 y_0} \eta_i(\vec{x})$. It is then straightforward to show that the observable

$$\hat{\mathcal{O}}_\pi(x_0, y_0) = \frac{1}{N_{\text{src}} V_3} \sum_{i, \vec{x}} |\phi_i(x, y_0)|^2, \qquad V_3: \text{spatial lattice volume}, \tag{6.189}$$

has the same expectation value as $\mathcal{O}_\pi(x_0, y)$ and may therefore be used in place of the latter in a calculation of the pion propagator. Note that through the average over the source fields,

$$\langle \hat{\mathcal{O}}_\pi(x_0, y_0) \rangle_{\text{src}} = \frac{1}{V_3} \sum_{\vec{y}} \mathcal{O}_\pi(x_0, y), \tag{6.190}$$

one effectively sums over all source points at time y_0.

Whether the new observable is any better than the old one depends on whether it has smaller statistical fluctuations or not. A short calculation, using Wick's theorem, shows that the associated *a priori* variance is given by

$$\sigma_0(\hat{\mathcal{O}}_\pi)^2 = \sigma_0(\overline{\mathcal{O}}_\pi)^2 + \frac{1}{N_{\text{src}} V_3^2} \sum_{\vec{x}, \vec{x}'} \sum_{\vec{y}, \vec{y}'} \left\{ \left\langle \vcenter{\hbox{}} \right\rangle \right\}_{x_0' = x_0, y_0' = y_0}, \tag{6.191}$$

where $\overline{\mathcal{O}}_\pi(x_0, y_0)$ denotes the volume-averaged observable (6.190). Recalling the exponential decay (6.181) of the quark propagator, the second term is readily estimated to be of order $(N_{\text{src}} m_\pi^3 V_3)^{-1}$ at large volumes. Since the first term scales like $(m_\pi^3 V_3)^{-1}$, it tends to be the dominant contribution to the variance already for moderately large numbers of source fields (setting $N_{\text{src}} = 12$, for example, is often sufficient).

With respect to a calculation of the pion propagator at a single source point, the random source method described here thus achieves a reduction of the statistical error by a factor proportional to $(m_\pi^3 V_3)^{-1/2}$ for approximately the same computational cost. The use of random sources is therefore recommended in this case, particularly so on large lattices.

6.3.2.3 Further applications

Random sources are an interesting and useful tool. Here, only one particular application and variant of the method was discussed. Source fields on different subsets of points can be considered as well as random fields taking values in a group or, more generally, fields with any non-Gaussian distribution. The method combines well with low-mode averaging techniques (Neff *et al.*, 2001; Giusti *et al.*, 2004; DeGrand and Schaefer, 2004; Bali *et al.*, 2005; Foley *et al.*, 2005) and, to mention just one further example, it also played a central rôle in a recent calculation of the spectral density of the Hermitian Wilson–Dirac operator (Giusti and Lüscher, 2009).

In the case of the vector-meson and baryon propagators, the application of random source methods is complicated by the fact that the index contractions at the vertices of the quark-line diagrams couple different components of the quark propagators. Different kinds of random sources (one for each component, for example) then need to be introduced to be able to write down a correct random-source representation of the hadron propagator. The variance of such observables typically involves many diagrams, among them often also disconnected ones. As a consequence, the use of random sources for these propagators is not obviously profitable, at least as long as the spatial extent of the lattices considered is not significantly larger than 2 or 3 fm.

Random-source methods should preferably be applied only after a careful analysis of the variance of the proposed stochastic observables. Such an analysis can be very helpful in deciding which kind of random fields to choose and how exactly the stochastic observable must be constructed in order to achieve a good scaling of the statistical error with the lattice volume.

6.3.3 Multilevel simulations

Random-source methods can lead to an important reduction of statistical errors, but are unable to overcome the exponential SNR problem encountered in the case of the nucleon propagator, for example. In a multilevel simulation, the improved observables are constructed through a stochastic process, i.e. through a secondary or nested simulation. Exploiting the locality of the theory, an exponential reduction of the statistical error can then be achieved in certain cases.

So far, multilevel simulations have been limited to bosonic field theories, essentially because manifest locality is lost when the fermion fields are integrated out. Whether this limitation is a transient one is unclear at present, but it is certainly worth explaining the idea here and to show its impressive potential.

6.3.3.1 Statistical fluctuations of Wilson loops

In the following, a multilevel algorithm for the computation of the expectation values of Wilson loops in the pure SU(3) lattice gauge theory will be described. The lattice action is assumed to be the Wilson plaquette action.

Similarly to the hadron propagators, Wilson-loop expectation values suffer from an exponential SNR problem. Let $\mathcal{C}$ be a $T \times R$ rectangular loop in the (x_0, x_1)-plane, $U(\mathcal{C})$ the ordered product of the link variables around the loop and $\mathcal{W} = \mathrm{tr}\{U(\mathcal{C})\}$ its trace. For large loops, the expectation value of $\mathcal{W}$ satisfies the area law

$$\langle \mathcal{W} \rangle \sim \mathrm{e}^{-\sigma A}, \qquad A = TR, \tag{6.192}$$

where $\sigma \simeq 1\,\mathrm{GeV/fm}$. Since $\mathcal{W}$ is a number of order 1 for every gauge-field configuration, the *a priori* variance $\sigma_0(\mathcal{W})$ is practically equal to $\langle |\mathcal{W}|^2 \rangle$ and thus of order 1 too. One therefore needs to generate ensembles of at least

$$N = \langle \mathcal{W} \rangle^{-2} \sim \mathrm{e}^{2\sigma A} \tag{6.193}$$

configurations to be able to calculate the expectation value $\langle \mathcal{W} \rangle$ to a useful precision.

For loop areas A equal to 1, 2 and 4 fm^2, for example, the minimal ensemble sizes are thus estimated to be 2×10^4, 6×10^8 and 4×10^{17}, respectively. These figures may not be exactly right, because the Wilson loop is averaged over all possible translations in practice and since there are important subleading corrections to the area law (6.192). However, for large areas A, the minimal ensemble size is essentially determined by the rapidly growing exponential factor (6.193).

6.3.3.2 Factorization and sublattice expectation values

A multilevel algorithm invented many years ago is the "multihit method" (Parisi *et al.*, 1983). In this case, a reduction of the statistical error by an exponential factor with exponent proportional to T is achieved by replacing the time-like link variables along the Wilson loop through stochastic estimates of their average values in the presence of the other link variables. The method thus exploits the fact that the Wilson loop factorizes into a product of link variables.

If the loop is instead factorized into a product of two-link operators, an exponential error reduction is obtained with an exponent proportional to the area A (Lüscher and Weisz, 2001). A two-link operator is an object with four SU(3) indices,

$$\mathbb{T}(x_0)_{\alpha\beta\gamma\delta} = U(x,0)^*_{\alpha\beta} U(x + R\hat{1}, 0)_{\gamma\delta}, \tag{6.194}$$

residing at time x_0 and $\vec{x} = 0$ (see Fig. 6.18). Two-link operators at adjacent times can be multiplied and their product from time 0 to $T - a$ yields the tensor product of the time-like lines of the Wilson loop. The latter is then given by

$$\mathcal{W} = \mathrm{tr}\{\mathbb{L}(0)\mathbb{T}(0)\mathbb{T}(a)\ldots\mathbb{T}(T-a)\mathbb{L}(T)\}, \tag{6.195}$$

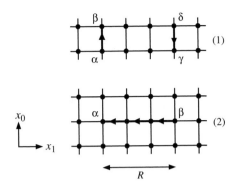

Fig. 6.18 A $T \times R$ Wilson loop in the (x_0, x_1)-plane with lower-left corner at $x = 0$ can be factorized into the product (6.195) of two-link and line operators. The SU(3) indices of the two-link operator (6.194) are assigned as shown in figure (1). The line operators (6.196) coincide with the spatial lines of the Wilson loop and have only two indices, as indicated in figure (2).

where the product

$$\mathbb{L}(x_0)_{\alpha\beta} = \left\{ U(x, 1)U(x + a\hat{1}, 1) \ldots U(x + (R - a)\hat{1}, 1) \right\}_{\alpha\beta} \tag{6.196}$$

denotes the spatial Wilson line at time x_0.

Consider now the sublattice bounded by the equal-time hyperplanes at some time y_0 and some later time z_0 (see Fig. 6.19). The Wilson action couples the link variables inside the sublattice (the "interior" link variables) to themselves and the spatial fields on the boundaries, but there is no interaction with the field variables elsewhere on the lattice. If $\mathcal{O}$ is any function of the interior link variables, its sublattice expectation value is defined by

$$[\mathcal{O}] = \frac{1}{Z_{\text{int}}} \int \mathrm{D}[U]_{\text{int}} \, \mathcal{O}(U) \, \mathrm{e}^{-S(U)}, \tag{6.197}$$

where one integrates over the interior variables. Note that the part of the action that does not depend on the latter drops out in the expectation value, which is thus a function of the spatial link variables at time y_0 and z_0 only.

If the lattice is decomposed into non-overlapping sublattices of this kind, the functional integral divides into an integral over the interior variables and an integral over the boundary fields. For even T/a, for example, the Wilson-loop expectation value may be rewritten in the form

$$\mathcal{W} = \langle \mathrm{tr}\{\mathbb{L}(0)[\mathbb{T}(0)\mathbb{T}(a)][\mathbb{T}(2a)\mathbb{T}(3a)] \ldots \mathbb{L}(T)\} \rangle. \tag{6.198}$$

The outer expectation value in this expression is the usual one involving an integration over all field variables. However, since the observable depends only on the spatial fields at time $x_0 = 0, 2a, 4a, \ldots, T$, the integral over the interior field variables yields the

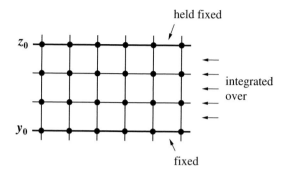

Fig. 6.19 Factorized representations of the Wilson-loop expectation value such as (6.198) are based on a division of the lattice into sublattices separated by equal-time hyperplanes. The sublattice expectation values [...] in these expressions involve an integration over the field variables residing on the links between the hyperplanes, all other link variables being held fixed.

product of the sublattice partition functions. This factor exactly cancels the product of the normalization factors of the sublattice expectation values.

The sublattice expectation value of a product of two-link operators is a correlation function of two segments of Wilson lines in presence of the boundary fields. A single segment transforms non-trivially under the center symmetry of the theory, where all time-like link variables at a given time are multiplied by an element of the center of SU(3). Barring spontaneous symmetry breaking, its sublattice expectation value therefore vanishes and the correlation function of two segments is consequently expected to go to zero at large separations R. Experience actually suggests that

$$[\mathbb{T}(y_0)\mathbb{T}(y_0 + a)\ldots\mathbb{T}(z_0 - a)] \sim e^{-\sigma(z_0 - y_0)R} \qquad (6.199)$$

if the time difference $z_0 - y_0$ is larger than, say, 0.5 fm or so. Recalling Eq. (6.198), the rapid decay of the Wilson-loop expectation value at large times T (and the area law if Eq. (6.199) holds) is thus seen to arise from a product of small factors.

6.3.3.3 *Multilevel update scheme*

The foregoing suggests that the expectation value of the Wilson loop may be accurately calculated, at any value of T, starting from a factorized representation like Eq.(6.198). An algorithm that implements the idea for this particular factorization proceeds in cycles consisting of the following steps:

(a) *Update the gauge field N_0 times using a combination of the heatbath and micro-canonical link-update algorithms.*

(b) *Estimate the expectation values $[\mathbb{T}(x_0)\mathbb{T}(x_0 + a)]$ by updating the field inside the associated sublattices N_1 times and by averaging the product of the two-link operators over the generated configurations.*

(c) *Compute the trace $\mathrm{tr}\{\mathbb{L}(0)[\mathbb{T}(0)\mathbb{T}(a)][\mathbb{T}(2a)\mathbb{T}(3a)]\ldots\mathbb{L}(T)\}$ using the estimates of the sublattice expectation values obtained in step (b).*

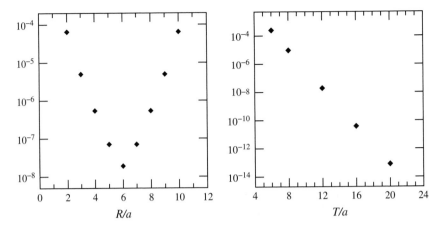

Fig. 6.20 Values of the correlation function of the Polyakov loop on a $(T/a) \times 12^3$ lattice with spacing $a \simeq 0.17$ fm and periodic boundary conditions, plotted as a function of the distance R at $T = 12a$ (left figure) and as a function of T at $R = 6a$ (right figure).

Each cycle thus yields an estimate of the trace $\text{tr}\{\mathbb{L}(0)[\mathbb{T}(0)\mathbb{T}(a)]\dots\mathbb{L}(T)\}$. A moment of thought shows that these estimates are averages of the Wilson loop over a particular set of field configurations generated through a valid simulation algorithm. Their average over many cycles therefore coincides with the expectation value of the Wilson loop up to statistical errors.

As explained in Section 6.3.3.2, the trace $\text{tr}\{\mathbb{L}(0)[\mathbb{T}(0)\mathbb{T}(a)]\dots\mathbb{L}(T)\}$ tends to decay exponentially, for every field configuration in a representative ensemble of fields. The statistical error of the stochastic estimates of the trace produced by the multilevel algorithm (a)–(c) is therefore guaranteed to fall off exponentially as well. With respect to a straightforward computation of the Wilson-loop expectation value, the sublattice averaging thus achieves an exponential error reduction.

For illustration, some results obtained in the course of an early application of the multilevel algorithm are shown in Fig. 6.20. In order to simplify the situation a little bit, the correlation function of two Polyakov loops (i.e. Wilson lines that wrap around the lattice in the time direction) was considered in this study. The line operators $\mathbb{L}(x_0)$ are then not needed and the observable coincides with the trace of a product of T/a two-link operators, T being the time-like extent of the lattice. As is evident from the data plotted in the figure, the multilevel algorithm allows the exponential decay of the correlation function to be followed over many orders of magnitude, which would not be possible (or only with an astronomical computer budget) using the standard simulation techniques.

6.3.3.4   *Remarks and further developments*

Multilevel algorithms of the kind described here have been employed in studies of the string behavior (Lüscher and Weisz, 2002; Kratochvila and de Forcrand, 2003; Pepe and Wiese, 2009), the glueball spectrum (Meyer, 2003; 2004) and the bulk viscosity

(Meyer, 2008) in pure gauge theories. Another, closely related multilevel algorithm has recently been proposed for the calculation of the energies of the lightest states with a specified (non-trivial) transformation behavior under the exact symmetries of the theory (Della Morte and Giusti, 2009). In all these cases, an important and sometimes impressive acceleration of the simulation was achieved.

The exponential SNR problem is, however, often not completely eliminated. Computations of Wilson-loop expectation values, for example, require sublattice expectation values of products of two-link operators to be determined to some accuracy. In view of Eq. (6.199), the number of sublattice updates that need to be performed in this part of the calculation is therefore expected to scale roughly like $e^{2\sigma(z_0 - y_0)R}$. With respect to a one-level simulation, the exponential growth of the required computational effort is nevertheless dramatically reduced, because the exponent now increases only proportionally to the distance R rather than the area A.

6.4 Statistical error analysis

The estimation of the statistical errors in numerical lattice QCD appears to be an easy topic. However, the physical quantities often need to be extracted from the simulation data through some non-linear procedure, which may involve complicated fits and extrapolations. The correct propagation of the errors becomes a non-trivial task under these conditions. Resampling techniques such as the jackknife and bootstrap methods (Efron and Tibshirani, 1993) allow the errors of the calculated physical quantities to be estimated with minimal effort, but are a bit of magic and are actually known to be incorrect in certain special cases (the jackknife method, for example, in general gives wrong results for the statistical error of the median of an observable).

The aim in this section is to build up a conceptually solid framework in which the error propagation is made transparent. In particular, the correctness of the jackknife method can then be established for a class of quantities, which includes most cases of interest.

6.4.1 Primary observables

Wilson loops, quark-line diagrams and any other function of the gauge field, including the stochastic ones considered in Section 6.3, are the primary observables in lattice QCD. Their expectation values are the first quantities calculated in a simulation project and all physical quantities are eventually obtained from these.

6.4.1.1 Correlation functions and data series

Let A_r be some real-valued primary observables labelled by an index r. As before, the lattice QCD expectation value of an observable $\mathcal{O}$ is denoted by $\langle \mathcal{O} \rangle$. The quantities of interest are then the n-point correlation functions $\langle A_{r_1} \ldots A_{r_n} \rangle$. Independently of any locality properties, their connected parts $\langle A_{r_1} \ldots A_{r_n} \rangle_c$ may be defined in the familiar way. In particular,

$$\langle A_r \rangle_c = \langle A_r \rangle, \tag{6.200}$$

$$\langle A_r A_s \rangle_c = \langle A_r A_s \rangle - \langle A_r \rangle \langle A_s \rangle, \tag{6.201}$$

$$\langle A_r A_s A_t \rangle_c = \langle A_r A_s A_t \rangle - \langle A_r \rangle \langle A_s A_t \rangle - \langle A_s \rangle \langle A_t A_r \rangle - \langle A_t \rangle \langle A_r A_s \rangle$$

$$+ 2 \langle A_r \rangle \langle A_s \rangle \langle A_t \rangle. \tag{6.202}$$

For an arbitrary number n of observables, the relation between the connected and the full correlation functions is given by the moment-cumulant transformation (see Section 6.4.1.4).

Suppose now that a representative ensemble of N statistically independent gauge-field configuration has been generated in the course of a simulation. The evaluation of the primary observable A_r on these configurations yields a series

$$a_{r,1}, a_{r,2}, \ldots, a_{r,N} \tag{6.203}$$

of "measured" values of this observable, whose average

$$\bar{a}_r = \frac{1}{N} \sum_{i=1}^{N} a_{r,i} \tag{6.204}$$

provides a stochastic estimate of its expectation value $\langle A_r \rangle$.

If $a_{r,i}$ and $a_{s,i}$, $i = 1, \ldots, N$, are the data series obtained for the observables A_r and A_s, the data series for the product $A_r A_s$ is $a_{r,i} a_{s,i}$. The product is actually just another primary observable, even in the case of the stochastic observables considered in Section 6.3.2 if the random sources are included as additional fields in the ensemble of fields generated by the simulation. A stochastic estimate of the n-point correlation function $\langle A_{r_1} \ldots A_{r_n} \rangle$ is thus obtained by calculating the average of $a_{r_1,i} \ldots a_{r_n,i}$.

6.4.1.2   *Simulation statistics*

As previously noted in Section 6.2.2.4, the question by how much $\bar{a}_r$ deviates from $\langle A_r \rangle$ can only be answered statistically when the simulation is repeated many times. In the following, the average over infinitely many simulations of any function ϕ of the measured values of the primary observables is denoted by $\langle\langle \phi \rangle\rangle$. The standard deviation of $\bar{a}_r$ from $\langle A_r \rangle$, for example, is given by $\langle\langle (\bar{a}_r - \langle A_r \rangle)^2 \rangle\rangle^{1/2}$.

Simulation statistics becomes a useful tool when the following two assumptions are made. First, it must be guaranteed that the measurements of the primary observables are unbiased, i.e. that the equality

$$\langle\langle a_{r,i} \rangle\rangle = \langle A_r \rangle \tag{6.205}$$

holds for all i and all primary observables A_r. The second assumption is that the fields generated by the simulations are statistically independent. In particular,

$$\langle\langle a_{r,i} a_{s,j} \rangle\rangle = \langle\langle a_{r,i} \rangle\rangle \langle\langle a_{s,j} \rangle\rangle \quad \text{if} \quad i \neq j. \tag{6.206}$$

More generally, the average $\langle\!\langle a_{r_1,i_1}\ldots a_{r_n,i_n}\rangle\!\rangle$ factorizes into a product of averages, one for each subset of the factors $a_{r,i}$ with a given value of the configuration index i. As discussed in Section 6.2.2.4, both conditions are fulfilled if the residual autocorrelations among the fields in the representative ensembles generated by the simulation algorithm are negligible.

In practice, the statistical independence of the measured values should be carefully checked by computing the associated autocorrelation functions. However, since the latter is estimated from the data and is therefore subject to statistical fluctuations, a reliable determination of the autocorrelation times can sometimes be difficult, particularly so if the available data series is not very long. The issue of "calculating the error of the error" was addressed by (Madras and Sokal, 1988) and in greater detail again by (Wolff, 2004) (see also (Lüscher, 2005), Appendix E, for some additional material).

6.4.1.3 *Distribution of the mean values*

The mean values $\bar{a}_r, \bar{a}_s, \ldots$ of the measured values of the primary observables are correlated to some extent, because the underlying ensemble of gauge-field configurations is the same. It is possible to work out the correlations

$$\langle\!\langle \bar{a}_{r_1}\ldots\bar{a}_{r_k}\rangle\!\rangle = \frac{1}{N^k}\sum_{i_1=1}^{N}\cdots\sum_{i_k=1}^{N}\langle\!\langle a_{r_1,i_1}\ldots a_{r_k,i_k}\rangle\!\rangle \tag{6.207}$$

in terms of the correlation functions $\langle A_{r_1}\ldots A_{r_n}\rangle$. In the case of the two-point correlation functions, for example, one obtains

$$\langle\!\langle \bar{a}_r\bar{a}_s\rangle\!\rangle = \frac{1}{N^2}\sum_{i,j=1}^{N}\langle\!\langle a_{r,i}a_{s,j}\rangle\!\rangle = \langle A_r\rangle\langle A_s\rangle + \frac{1}{N}\langle A_r A_s\rangle_c, \tag{6.208}$$

where use was made of Eqs. (6.205) and (6.206).

For any $k \geq 1$, the analogous expression for the correlation function (6.207) reads

$$\langle\!\langle \bar{a}_{r_1}\ldots\bar{a}_{r_k}\rangle\!\rangle = \sum_{l=1}^{k}\frac{1}{N^{k-l}l!}\sum_{P\in\mathcal{P}_{k,l}}\langle A_{P_1}\rangle_c\ldots\langle A_{P_l}\rangle_c. \tag{6.209}$$

The second sum in this formula runs over the set $\mathcal{P}_{k,l}$ of all partitions $P = (P_1,\ldots,P_l)$ of the set $\{1,\ldots,k\}$ into l non-empty subsets. Furthermore,

$$A_{P_i} = \prod_{j\in P_i} A_{r_j}. \tag{6.210}$$

Note that the subsets $P_1,\ldots,P_l$ are ordered and therefore distinguished. In particular, at large N the dominant term is

$$\langle\!\langle \bar{a}_{r_1}\ldots\bar{a}_{r_k}\rangle\!\rangle = \langle A_{r_1}\rangle\ldots\langle A_{r_k}\rangle + \mathrm{O}(N^{-1}). \tag{6.211}$$

The proof of Eq. (6.209) is a bit technical and is deferred to Section 6.4.1.4.

The statistical properties of the deviations

$$\delta\bar{a}_r = \bar{a}_r - \langle A_r \rangle \tag{6.212}$$

can now be determined as follows. First, note that Eq. (6.208) may be rewritten in the form

$$\langle\!\langle \delta\bar{a}_r \delta\bar{a}_s \rangle\!\rangle = \frac{1}{N} \langle A_r A_s \rangle_{\mathrm{c}}. \tag{6.213}$$

On average, the magnitude of the deviations $\delta\bar{a}_r$ is thus proportional to $N^{-1/2}$.

A more detailed characterization of the distribution of the deviations is obtained by working out their higher-order correlations. Starting from Eq. (6.209), it is possible to show (Section 6.4.1.4) that

$$\langle\!\langle \delta\bar{a}_{r_1} \ldots \delta\bar{a}_{r_k} \rangle\!\rangle = \sum_{l=1}^{k} \frac{1}{N^{k-l} l!} \sum_{P \in \tilde{\mathcal{P}}_{k,l}} \langle A_{P_1} \rangle_{\mathrm{c}} \ldots \langle A_{P_l} \rangle_{\mathrm{c}}, \tag{6.214}$$

where $\tilde{\mathcal{P}}_{k,l} \subset \mathcal{P}_{k,l}$ denotes the set of partitions of $\{1, \ldots, k\}$ into l subsets with two or more elements. In particular, for all even k

$$\langle\!\langle \delta\bar{a}_{r_1} \ldots \delta\bar{a}_{r_k} \rangle\!\rangle = \frac{1}{N^{k/2}} \left\{ \langle A_{r_1} A_{r_2} \rangle_{\mathrm{c}} \ldots \langle A_{r_{k-1}} A_{r_k} \rangle_{\mathrm{c}} + \text{permutations} \right\} + \ldots, \tag{6.215}$$

while for all odd k the leading terms are of order $N^{-(k+1)/2}$, because the admissible partitions contain at least one subset with 3 or more elements. Taken together, these results show that the joint probability distribution of the scaled deviations $\sqrt{N}\delta\bar{a}_r$ is Gaussian to leading order in $1/N$, with mean zero and variance $\langle A_r A_s \rangle_{\mathrm{c}}$.

6.4.1.4   Appendix: Proof of Eqs. (6.209) and (6.214)

Let J_r be real-valued sources for the selected primary observables A_r. The generating function of the correlation functions $\langle A_{r_1} \ldots A_{r_n} \rangle$ is a formal power series

$$\mathcal{Z}(J) = 1 + \sum_{n=1}^{\infty} \frac{1}{n!} \sum_{r_1} \ldots \sum_{r_n} \langle A_{r_1} \ldots A_{r_n} \rangle J_{r_1} \ldots J_{r_n} \tag{6.216}$$

in these sources. Similarly, the generating function of the connected parts of the correlation functions is given by

$$\mathcal{W}(J) = \sum_{n=1}^{\infty} \frac{1}{n!} \sum_{r_1} \ldots \sum_{r_n} \langle A_{r_1} \ldots A_{r_n} \rangle_{\mathrm{c}} J_{r_1} \ldots J_{r_n}. \tag{6.217}$$

The moment-cumulant transformation is then summarized by the identity

$$\mathcal{Z}(J) = e^{\mathcal{W}(J)} \tag{6.218}$$

among formal power series.

The left-hand side of Eq. (6.209) is related to the generating function $\mathcal{Z}(J)$ through

$$\langle\!\langle \bar{a}_{r_1} \ldots \bar{a}_{r_k} \rangle\!\rangle = \frac{1}{N^k} \frac{\partial^k \langle\!\langle \exp\{\sum_{r,i} a_{r,i} J_r\} \rangle\!\rangle}{\partial J_{r_1} \ldots \partial J_{r_k}} \bigg|_{J=0}$$

$$= \frac{1}{N^k} \frac{\partial^k \mathcal{Z}(J)^N}{\partial J_{r_1} \ldots \partial J_{r_k}} \bigg|_{J=0}. \qquad (6.219)$$

Use has here been made of the statistical independence of the data $a_{r,i}$ at different values of the index i and of the fact that their products at fixed i are unbiased estimators of the correlation functions of the primary observables. The insertion of Eq. (6.218) and the subsequent expansion of the exponential function $e^{\mathcal{W}(J)}$ now leads to the formula

$$\langle\!\langle \bar{a}_{r_1} \ldots \bar{a}_{r_k} \rangle\!\rangle = \sum_{l=1}^{k} \frac{1}{N^{k-l} l!} \frac{\partial^k \mathcal{W}(J)^l}{\partial J_{r_1} \ldots \partial J_{r_k}} \bigg|_{J=0}. \qquad (6.220)$$

Each derivative in this expression acts on the l factors $\mathcal{W}(J) \ldots \mathcal{W}(J)$ one by one. It is then not difficult to convince oneself that the possible distributions of the derivatives to the factors match the possible partitions $P \in \mathcal{P}_{k,l}$, thus proving Eq. (6.209).

The proof of Eq. (6.214) proceeds in the same way. In order to obtain the correlation functions of the deviations $\delta\bar{a}_r$, it suffices to substitute $a_{r,i} \to a_{r,i} - \langle A_r \rangle$ on the right of Eq. (6.219). The generating function in Eq. (6.220) then gets replaced by

$$\tilde{\mathcal{W}}(J) = \mathcal{W}(J) - \sum_r \langle A_r \rangle J_r. \qquad (6.221)$$

Since at least two derivatives must act on each factor $\tilde{\mathcal{W}}(J)$, the possible distributions of the derivatives to the factors now match the partitions $P \in \tilde{\mathcal{P}}_{k,l}$.

6.4.2 Physical quantities

The correlation functions of the primary observables only rarely have an immediate physical meaning. In the present context, any well-defined function of the expectation values $\langle A_r \rangle, \langle A_r A_s \rangle, \ldots$ is referred to as a physical quantity. Ratios of the expectation values of Wilson loops, for example, are considered to be physical quantities as well as the heavy-quark potential, which is a limit of such ratios.

6.4.2.1 *From the primary observables to the physical quantities*

In practice, the correlation functions required for the calculation of a physical quantity are approximated by the averages of the measured values of the appropriate primary observables. A prototype computation of the pion mass m_π, for example, starts from the data series for the observables $\mathcal{O}_\pi(x_0, y)$ at all times x_0 and, say, a single source point y (cf. Section 6.3.1.1). The so-called effective mass

$$m_{\text{eff}}(x_0) = -a^{-1} \ln \frac{\overline{\mathcal{O}}_\pi(x_0 + a, y)}{\overline{\mathcal{O}}_\pi(x_0, y)} \tag{6.222}$$

is then computed and fitted by a constant in a sensible range of x_0, where the contributions of the higher-energy states to the pion propagator are negligible with respect to the statistical errors. As long as this condition is satisfied, the fitted constant provides a stochastic estimate of the pion mass on the given lattice.

Once a definite fit procedure is adopted, the so-calculated values of the pion mass become a complicated but unambiguously defined function of the data series for the observables $\mathcal{O}_\pi(x_0, y)$. It should be noted, however, that this function is not simply a function of the average values of these observables. The fit also requires an estimate of the covariance matrix of the latter as input, which one obtains from the available data through the jackknife method or in some other way. Evidently, when fits of fitted quantities are considered, as may be the case in studies of the quark-mass dependence of the pion mass, the functional dependence of the calculated quantities on the primary data series is further obscured.

6.4.2.2   *Stochastic estimators*

A stochastic estimator of a physical quantity Q is any function ϕ of N and the measured values of the primary observables A_r such that

$$Q = \lim_{N \to \infty} \phi \quad \text{with probability 1.} \tag{6.223}$$

In the following, a class of stochastic estimators with some additional properties will be considered. More precisely, it will be assumed that the asymptotic expansion

$$\phi \underset{N \to \infty}{\sim} \sum_{k=0}^{\infty} N^{-k} \phi^{(k)}(\bar{a}_{r_1}, \bar{a}_{r_2}, \ldots) \tag{6.224}$$

holds, where the coefficients $\phi^{(k)}$ are smooth functions of their arguments. The number of arguments may grow with k but is required to be finite for all k.

It may not be obvious at this point why one needs to consider stochastic estimators with an explicit dependence on N. The leading coefficient in the expansion (6.224) is in fact a valid stochastic estimator for Q that depends only on the averages $\bar{a}_r$ of the primary data series. However, when a stochastic estimator is implicitly defined through a fit procedure, for example, its leading coefficient at large N may be inaccessible in practice. In these cases one does not really have the choice and is forced to deal with the operationally well-defined but N-dependent estimators.

A simple example of an admissible stochastic estimator is provided by the effective mass (6.222). The rôle of the physical quantity Q is here played by the effective mass calculated from the exact pion propagator $g_\pi(x_0 - y_0)$. More complicated, N-dependent stochastic estimators will be discussed in Section 6.4.3.

6.4.2.3   *Bias and covariance matrix*

Let Q_α be some physical quantities, labelled by an index α, and ϕ_α stochastic estimators of these. On average, the values of the estimators obtained in a simulation approximate the physical quantities up to a deviation given by

$$B_\alpha = \langle\langle \delta\phi_\alpha \rangle\rangle, \qquad \delta\phi_\alpha = \phi_\alpha - Q_\alpha. \tag{6.225}$$

B_α is referred to as the bias of the chosen estimators. The statistical fluctuations of the measured values are described by the covariance matrix

$$C_{\alpha\beta} = \langle\langle \delta\phi_\alpha \delta\phi_\beta \rangle\rangle \tag{6.226}$$

and are usually significantly larger than the bias.

For $N \to \infty$, the bias and the covariance matrix can be expanded in a series in inverse powers of N with coefficients depending on the correlation functions $\langle A_{r_1} \dots A_{r_n} \rangle$ of the primary observables. To show this, first recall that

$$\bar{a}_r = \langle A_r \rangle + \delta\bar{a}_r, \qquad \delta\bar{a}_r = \mathrm{O}(N^{-1/2}). \tag{6.227}$$

The Taylor expansion of the coefficient functions in eqn (6.224) in powers of the deviations $\delta\bar{a}_r$ then leads to the expression

$$\phi_\alpha = \hat{\phi}_\alpha^{(0)} + \sum_r \partial_r \hat{\phi}_\alpha^{(0)} \delta\bar{a}_r$$

$$+ \frac{1}{N}\hat{\phi}_\alpha^{(1)} + \frac{1}{2}\sum_{r,s} \partial_r \partial_s \hat{\phi}_\alpha^{(0)} \delta\bar{a}_r \delta\bar{a}_s + \mathrm{O}(N^{-3/2}), \tag{6.228}$$

in which

$$\hat{\phi}_\alpha^{(k)} = \phi_\alpha^{(k)}(\langle A_{r_1} \rangle, \langle A_{r_2} \rangle, \dots), \qquad \partial_r = \frac{\partial}{\partial\langle A_r \rangle}. \tag{6.229}$$

Evidently, since the leading term $\hat{\phi}_\alpha^{(0)}$ coincides with Q_α, the deviation $\delta\phi_\alpha$ is given by the sum of all other terms on the right of Eq. (6.228).

It is important to realize that the only stochastic variables in the expansion (6.228) are the deviations $\delta\bar{a}_r$. The averages (6.225) and (6.226) over repeated simulations can therefore be computed straightforwardly using the results obtained in Section 6.4.1. In a few lines one then obtains the result

$$B_\alpha = \frac{1}{N}\left\{ \hat{\phi}_\alpha^{(1)} + \frac{1}{2}\sum_{r,s} \partial_r \partial_s \hat{\phi}_\alpha^{(0)} \langle A_r A_s \rangle_\mathrm{c} \right\} + \mathrm{O}(N^{-2}), \tag{6.230}$$

$$C_{\alpha\beta} = \frac{1}{N}\sum_{r,s} \partial_r \hat{\phi}_\alpha^{(0)} \partial_s \hat{\phi}_\beta^{(0)} \langle A_r A_s \rangle_\mathrm{c} + \mathrm{O}(N^{-2}). \tag{6.231}$$

In particular, the bias of the stochastic estimators is of order N^{-1}, while their statistical fluctuations are of order $N^{-1/2}$.

6.4.3 Jackknife error estimation

The bias B_α and the covariance matrix $C_{\alpha\beta}$ usually need to be estimated from the data. Such estimates can in principle be obtained via Eqs. (6.230) and (6.231) by calculating the expectation values and two-point functions of the relevant primary observables. The coefficient functions $\phi^{(0)}$ and $\phi^{(1)}$ must be known explicitly in this case.

The jackknife method allows the bias and covariance matrix to be estimated even if the coefficient functions are inaccessible or too complicated to be used directly. More precisely, the method constructs stochastic estimators for the large-N limits of NB_α and $NC_{\alpha\beta}$ (which are functions of the expectation values of the primary observables and therefore physical quantities).

6.4.3.1 Jackknife samples

A jackknife sample of the measured values $a_{r,1}, \dots, a_{r,N}$ of a primary observable A_r is obtained by omitting one measurement from the full series. If, say, the ith measurement is omitted, the corresponding jackknife sample consists of the measurements

$$a_{r,1}, \dots, a_{r,i-1}, a_{r,i+1}, \dots a_{r,N}. \tag{6.232}$$

Evidently, there are N distinct jackknife samples of $N-1$ measurements, labelled by the number i of the omitted measurement.

The average of the measurements included in the jackknife sample number i is denoted by

$$\bar{a}^J_{r,i} = \frac{1}{N-1} \sum_{j=1,j\neq i}^{N} a_{r,j}. \tag{6.233}$$

More generally, a stochastic estimator ϕ assumes some value ϕ^J_i if the ith measurement of the primary observables is discarded. Note that the jackknife samples are treated like any other measurement series of length $N-1$ in this context. From the expansion (6.224) one then infers that

$$\phi^J_i \underset{N\to\infty}{\sim} \sum_{k=0}^{\infty} (N-1)^{-k} \phi^{(k)}(\bar{a}^J_{r_1,i}, \bar{a}^J_{r_2,i}, \dots), \tag{6.234}$$

where the coefficient functions $\phi^{(k)}$ are the same as before.

6.4.3.2 Estimators for the bias and the covariance matrix

The jackknife estimators for the bias and the covariance matrix are now given by the elegant formulae

$$B^J_\alpha = \sum_{i=1}^{N} (\phi^J_{\alpha,i} - \phi_\alpha), \tag{6.235}$$

$$C^J_{\alpha\beta} = \sum_{i=1}^{N} (\phi^J_{\alpha,i} - \phi_\alpha)(\phi^J_{\beta,i} - \phi_\beta). \tag{6.236}$$

An important result of the discussion below is going to be that the scaled expressions NB_α^J and $NC_{\alpha\beta}^J$ are stochastic estimators in the sense of Section 6.4.2.2, the associated physical quantities being

$$\lim_{N\to\infty} NB_\alpha = \hat\phi_\alpha^{(1)} + \frac{1}{2}\sum_{r,s}\partial_r\partial_s\hat\phi_\alpha^{(0)}\langle A_r A_s\rangle_c, \tag{6.237}$$

$$\lim_{N\to\infty} NC_{\alpha\beta} = \sum_{r,s}\partial_r\hat\phi_\alpha^{(0)}\partial_s\hat\phi_\beta^{(0)}\langle A_r A_s\rangle_c. \tag{6.238}$$

The bias B_α and the covariance matrix $C_{\alpha\beta}$ are therefore approximated by the estimators B_α^J and $C_{\alpha\beta}^J$ up to statistical fluctuations of order $N^{-3/2}$ and further terms of order N^{-2}.

The expansion of the jackknife estimators (6.235) and (6.236) for $N\to\infty$ is obtained straightforwardly by substituting

$$\bar a_{r,i}^J = \bar a_r + \frac{1}{N-1}(\bar a_r - a_{r,i}) \tag{6.239}$$

in Eq. (6.234) and by systematically expanding all terms in powers of N^{-1}. In the case of the covariance matrix, for example, this leads to the expression

$$C_{\alpha\beta}^J = \frac{1}{N}\sum_{r,s}\bar\partial_r\phi_\alpha^{(0)}(\bar a_{r_1},\dots)\bar\partial_s\phi_\beta^{(0)}(\bar a_{r_1},\dots)(\bar a_{rs} - \bar a_r\bar a_s) + \mathrm{O}(N^{-2}), \tag{6.240}$$

where the notation has been simplified by setting

$$\bar a_{rs} = \frac{1}{N}\sum_{i=1}^{N} a_{r,i}a_{s,i}, \qquad \bar\partial_r = \frac{\partial}{\partial\bar a_r}. \tag{6.241}$$

The higher-order terms in Eq. (6.240) involve averages of increasingly longer products of the measured values $a_{r,i}$, but are otherwise of the same structure as the leading one. Since the corresponding products of the primary observables are primary observables too, the expansion shows that $NC_{\alpha\beta}^J$ is a stochastic estimator of the kind specified in Section 6.4.2.2. Moreover, proceeding as in Section 6.4.2.3, the physical quantity approximated by $NC_{\alpha\beta}^J$ is seen to coincide with the expression on the right of Eq. (6.238), as asserted above.

An interesting feature of the jackknife method is the fact that it does not require the derivatives of the coefficient function $\phi^{(0)}$ in the asymptotic expressions (6.237) and (6.238) to be calculated. As is evident from the derivation of Eq. (6.240), the method effectively performs a numerical differentiation by probing the stochastic estimators ϕ_α through the jackknife samples. The estimators are thus implicitly assumed to vary smoothly on the scale of the deviations $\bar a_{r,i}^J - \bar a_r = \mathrm{O}(N^{-1})$.

6.4.3.3 Error propagation

The jackknife method is easy to apply, because it requires no more than the evaluation of the stochastic estimators of interest for the full sample and the jackknife samples of the measured values of the primary observables. Note that $\psi_i^J = \psi(\phi_{1,i}^J, \ldots, \phi_{m,i}^J)$ if ψ is a function of the stochastic estimators $\phi_1, \ldots, \phi_m$. The calculation can therefore often be organized in a hierarchical manner, where one proceeds from the primary observables to more and more complicated stochastic estimators.

In the case of the computation of the pion mass sketched in Section 6.4.2.1, for example, the estimation of the statistical errors starts from the jackknife averages $\overline{\mathcal{O}}_\pi(x_0, y)_i^J$ of the primary observables $\mathcal{O}_\pi(x_0, y)$. The bias and covariance matrix of the effective mass are then obtained from the jackknife sample values

$$m_{\text{eff}}(x_0)_i^J = -a^{-1} \ln \frac{\overline{\mathcal{O}}_\pi(x_0 + a, y)_i^J}{\overline{\mathcal{O}}_\pi(x_0, y)_i^J} \tag{6.242}$$

using Eqs. (6.235) and (6.236). In particular, the jackknife estimator of the covariance matrix is given by

$$C_{x_0 x_0'}^J = \sum_{i=1}^N \{m_{\text{eff}}(x_0)_i^J - m_{\text{eff}}(x_0)\}\{m_{\text{eff}}(x_0')_i^J - m_{\text{eff}}(x_0')\}. \tag{6.243}$$

The statistical error of $m_{\text{eff}}(x_0)$, for example, is estimated to be $(C_{x_0 x_0}^J)^{1/2}$.

In the next step, the pion mass m_π is computed by minimizing the χ^2-statistic

$$\chi^2 = \sum_{x_0 = t_0}^{t_1} \sum_{x_0' = t_0}^{t_1} \{m_\pi - m_{\text{eff}}(x_0)\}[(C^J)^{-1}]_{x_0 x_0'} \{m_\pi - m_{\text{eff}}(x_0')\} \tag{6.244}$$

in some range $[t_0, t_1]$ of time. The pion mass

$$m_\pi = \frac{\sum_{x_0, x_0' = t_0}^{t_1} [(C^J)^{-1}]_{x_0 x_0'} m_{\text{eff}}(x_0')}{\sum_{x_0, x_0' = t_0}^{t_1} [(C^J)^{-1}]_{x_0 x_0'}} \tag{6.245}$$

thus calculated is algebraically expressed through the effective mass and the scaled jackknife estimator of the associated covariance matrix. In particular, since all of these are stochastic estimators of the kind specified in Section 6.4.2.2, the same is the case for the pion mass determined in this way. Its statistical error can therefore be computed using the jackknife method.

Following the general rules, the application of the jackknife method requires the pion mass to be recomputed for each jackknife sample of the primary data. In particular, the covariance matrix of the effective mass must be recomputed, which requires the jackknife samples of the jackknife samples to be considered, i.e. samples where a second measurement is discarded from the full data series. The computational effort thus grows like N^2, but the calculation is otherwise entirely straightforward.

In practice, a simplified procedure is often adopted, where the statistical error of the covariance matrix is ignored. The matrix is calculated for the full sample in this case and is held fixed when the errors of the effective mass are propagated to those of the pion mass. An advantage of this procedure is that the computational effort increases like N rather than N^2, but the correctness of the error estimation can only be shown for asymptotically large values of N and if the systematic deviation of the effective mass from being constant in time is much smaller than the statistical errors. Moreover, the bias of the pion mass is no longer correctly given by the jackknife formula (6.235). It is therefore advisable to pass to the simplified procedure only after having checked its consistency with the results of the full procedure.

Acknowledgments

I am indebted to the organizers for inviting me to lecture at this Summer School and for providing a very pleasant and stimulating atmosphere. During my lectures many questions were asked that helped to clarify some subtle points. I have rarely had such an attentive audience and would like to thank the students for patiently going with me through the rather technical material covered in the course. Finally, I wish to thank the University of Grenoble, the director of the school and the support staff for running this wonderful place in a perfect manner.

Notes

1. A rigorous mathematical result, known as Zarantello's lemma, asserts that it is not possible to obtain a more stringent bound by choosing a different polynomial, i.e. the polynomial (6.18) is the optimal one.
2. The deflation subspace can alternatively be generated "on the fly" while solving the Dirac equation, exploiting the fact that the residue of an approximate solution tends to align to the low modes of the Dirac operator (Brannick *et al.*, 2008; Clark *et al.*, 2008*a*).
3. The force $F(x,\mu)$ informally coincides with $\partial S(U)/\partial U(x,\mu)$. Derivatives with respect to the link variables, however, need to be properly defined. According to Eq. (6.117), the force field is obtained by substituting $U(x,\mu) \to \exp\{\omega^a(x,\mu)T^a\}U(x,\mu)$, differentiating with respect to the real variables $\omega^a(x,\mu)$ and setting $\omega^a(x,\mu) = 0$ at the end of the calculation.
4. An efficient ISO C program implementing the DD-HMC algorithm for two-flavor QCD can be downloaded from `http://cern.ch/luscher/DDHMC/index.html` under the GNU Public License (GPL). Many of the acceleration techniques discussed here are included in this package.
5. Zolotarev obtained two different optimal rational approximations $R(y)$ to the function $1/\sqrt{y}$, one of degree $[n,n]$ and the other of degree $[n-1,n]$. Both are used in lattice QCD and are commonly referred to as the Zolotarev rational approximation.

References

Abramowitz, M. and Stegun, I. A. (1972). *Handbook of mathematical functions.* Dover Publications, New York.

Achiezer, N. I. (1992). *Theory of approximation.* Dover Publications, New York.

Aoki, S. *et al.* (2002). *Phys. Rev.*, **D65**, 094507.

Bali, G. S. *et al.* (2005). *Phys. Rev.*, **D71**, 114513.

Banks, T. and Casher, A. (1980). *Nucl. Phys.*, **B169**, 103.

Brannick, J. *et al.* (2008). *Phys. Rev. Lett.*, **100**, 041601.

Brower, R. C., Ivanenko, T., Levi, A. R., and Orginos, K. N. (1997). *Nucl. Phys.*, **B484**, 353.

Brown, F. R. and Woch, T. J. (1987). *Phys. Rev. Lett.*, **58**, 2394.

Cabibbo, N. and Marinari, E. (1982). *Phys. Lett.*, **119B**, 387.

Clark, M. A. *et al.* (2008*a*). *PoS(LATTICE 2008)*, 035.

Clark, M. A. and Kennedy, A. D. (2007). *Phys. Rev. Lett.*, **98**, 051601.

Clark, M. A., Kennedy, A. D., and Silva, P. J. (2008*b*). *PoS(LATTICE 2008)*, 041.

Creutz, M. (1980). *Phys. Rev.*, **D21**, 2308.

Creutz, M. (1987). *Phys. Rev.*, **D36**, 515.

DeGrand, T. A. and Schaefer, S. (2004). *Comput. Phys. Commun.*, 185.

Del Debbio, L. *et al.* (2006). *J. High Energy Phys.*, **0602**, 011.

Del Debbio, L. *et al.* (2007). *J. High Energy Phys.*, **0702**, 056.

Della Morte, M. and Giusti, L. (2009). *Comput. Phys. Commun.*, **180**, 819.

Duane, S., Kennedy, A. D., Pendleton, B. J., and Roweth, D. (1987). *Phys. Lett.*, **B195**, 216.

Efron, B. and Tibshirani, R. J. (1993). *An introduction to the bootstrap.* Chapman & Hall, CRC Press, Boca Raton.

Fabricius, K. and Haan, O. (1984). *Phys. Lett.*, **143B**, 459.

Foley, J. *et al.* (2005). *Comput. Phys. Commun.*, **172**, 145.

Frezzotti, R. and Jansen, K. (1997). *Phys. Lett.*, **B402**, 328.

Frezzotti, R. and Jansen, K. (1999). *Nucl. Phys.*, **B555**, 395 and 432.

Frommer, A. *et al.* (1994). *Int. J. Mod. Phys.*, **C5**, 1073.

Giusti, L. *et al.* (2004). *J. High Energy Phys.*, **0404**, 013.

Giusti, L., Hoelbling, C., Lüscher, M., and Wittig, H. (2003). *Comput. Phys. Commun.*, **153**, 31.

Giusti, L. and Lüscher, M. (2009). *J. High Energy Phys.*, **0903**, 013.

Golub, G. H. and van Loan, C. F. (1989). *Matrix computations* (2nd edn). The Johns Hopkins University Press, Baltimore.

Hairer, E., Lubich, C., and Wanner, G. (2006). *Geometric numerical integration: structure-preserving algorithms for ordinary differential equations* (2nd edn). Springer, Berlin.

Hasenbusch, M. (2001). *Phys. Lett.*, **B519**, 177.

Hasenbusch, M. and Jansen, K. (2003). *Nucl. Phys.*, **B659**, 299.

Horvath, I., Kennedy, A. D., and Sint, S. (1999). *Nucl. Phys. (Proc. Suppl.)*, **73**, 834.

James, F. (1994). *Comput. Phys. Commun.*, **79**, 111. [E: *ibid.* **97** (1996) 357].

Jegerlehner, B. (1996). arXiv:hep-lat/9612014 (unpublished).

Kennedy, A. D. and Pendleton, B. J. (1985). *Phys. Lett.*, **156B**, 393.

Knuth, D. E. (1997). *The art of computer programming* (3rd edn). Volume 2. Addison–Wesley, Reading MA.

Kratochvila, S. and de Forcrand, Ph. (2003). *Nucl. Phys.*, **B671**, 103.

Leimkuhler, B. and Reich, S. (2004). *Simulating Hamiltonian dynamics.* Cambridge University Press, Cambridge.

Lüscher, M. (1994). *Comput. Phys. Commun.*, **79**, 100.

Lüscher, M. (2004). *Comput. Phys. Commun.*, **156**, 209.

Lüscher, M. (2005). *Comput. Phys. Commun.*, **165**, 199.

Lüscher, M. (2007*a*). *J. High Energy Phys.*, **0707**, 081.

Lüscher, M. (2007*b*). *J. High Energy Phys.*, **0712**, 011.

Lüscher, M. and Weisz, P. (2001). *J. High Energy Phys.*, **09**, 010.

Lüscher, M. and Weisz, P. (2002). *J. High Energy Phys.*, **07**, 049.

Madras, N. and Sokal, A. D. (1988). *J. Stat. Phys.*, **50**, 109.

Metropolis, N. *et al.* (1953). *J. Chem. Phys.*, **21**, 1087.

Meyer, H. B. (2003). *J. High Energy Phys.*, **0301**, 048.

Meyer, H. B. (2004). *J. High Energy Phys.*, **0401**, 030.

Meyer, H. B. (2008). *Phys. Rev. Lett.*, **100**, 162001.

Michael, C. and Peisa, J. (1998). *Phys. Rev.*, **D58**, 034506.

Neff, H. *et al.* (2001). *Phys. Rev.*, **D64**, 114509.

Omelyan, I. P., Mryglod, I. M., and Folk, R. (2002). *Phys. Rev.*, **E65**, 056706.

Omelyan, I. P., Mryglod, I. M., and Folk, R. (2003). *Comput. Phys. Commun.*, **151**, 272.

Parisi, G., Petronzio, R., and Rapuano, F. (1983). *Phys. Lett.*, **B128**, 418.

Pepe, M. and Wiese, U.-J. (2009). *Phys. Rev. Lett.*, **102**, 191601.

Saad, Y. (2003). *Iterative methods for sparse linear systems* (2nd edn). SIAM, Philadelphia. See also http://www-users.cs.umn.edu/~saad/.

Sexton, J. C. and Weingarten, D. H. (1992). *Nucl. Phys.*, **B380**, 665.

Takaishi, T. and de Forcrand, Ph. (2002). *Int. J. Mod. Phys.*, **C13**, 343.

Takaishi, T. and de Forcrand, Ph. (2006). *Phys. Rev.*, **E73**, 036706.

Tierney, L. (1994). *Ann. Statist.*, **22**, 1701.

Ukawa, A. (2002). *Nucl. Phys. B (Proc. Suppl.)*, **106–107**, 195.

Urbach, C., Jansen, K., Shindler, A., and Wenger, U. (2006). *Comput. Phys. Commun.*, **174**, 87.

van der Vorst, H. A. (1992). *SIAM J. Sci. Stat. Comput.*, **13**, 631.

Wilcox, W. (2007). *PoS(LATTICE 2007)*, 025.

Wolff, U. (2004). *Comput. Phys. Commun.*, **156**, 143. [E: *ibid.* **176** (2007) 383].

7

Simulations with the Hybrid Monte Carlo algorithm: implementation and data analysis

Stefan SCHAEFER

Humboldt Universität zu Berlin, Institut für Physik, Newtonstr. 15, 12489 Berlin, Germany

7.1 Introduction

The Hybrid Monte Carlo (HMC) algorithm (Duane *et al.*, 1987), together with its variants, is currently the most popular method for simulations of full QCD on the lattice. It is an exact algorithm, which can be used for virtually any theory with continuous variables. It is a method of choice when many degrees of freedom are coupled and single-variable updates are not feasible.

Because modern QCD packages are rather involved, the idea behind this chapter, was to write an HMC code for ϕ^4 theory in D dimensions from scratch and then perform some simulations. This was targeted at students who do not work daily on the numerical aspects of lattice field theory and want to gain familiarity with numerical computations, data analysis and the language used to discuss it. The mathematical background of Markov chain Monte Carlo and the HMC were introduced in the lectures by M. Lüscher at this school (Lüscher, 2011), however, for consistency, the basic definitions are given below. Whereas writing a full code for SU(3) gauge theory would have been beyond the scope of this chapter, a simple theory like ϕ^4 provides a feasible object for the basic ideas behind such a simulation and, apart from the complications concerning the group variables, the ϕ^4 code proceeds exactly as one for the gauge theory.

Studying ϕ^4 theory numerically has a long tradition in field theory. It is believed to be in the Ising universality class and for the $d = 3$ dimensions, to which we restrict ourselves here, it has been used to extract the corresponding critical exponents to high accuracy, see, e.g., Refs. (Hasenbusch *et al.*, 1999; Hasenbusch, 1999). Obviously, the HMC algorithm is not the most efficient choice for these computations. In the literature, local heatbath and overrelaxation sweeps are combined with cluster updates.

The course resembled more a recitation class than a lecture, in the sense that after a short introduction the students were working on a problem sheet. The presentation here goes along the line of these problem sheets, essentially solving them. This might explain some repetitions and also the fact that many issues are not covered.

During the school, virtually all the code was written from scratch, with two exceptions: one is a routine that fills the neighbor field used to navigate the D-dimensional lattice. The other is a routine that computes the action on a given field configuration. This essentially served the purpose of exemplifying the use of the hopping field. The starting code and also the final HMC are available under

`http://nic.desy.de/leshouches`

or directly from the author.

The course proceeded as follows: in the first lesson, the model was defined and to gain some familiarity with the code, the measurement routines and the momentum refreshment were implemented. With the second sheet, the molecular dynamics was implemented and tested. The third class had the full HMC algorithm with the acceptance test, again checks of the code and first running studying thermalization. The fourth and final lesson was devoted to autocorrelation analysis and measurement of critical exponents. This text follows loosely this structure, keeping the style and also large parts of the text of the problem sheets. The actual tasks are given at the end of each section.

7.1.1 Definition of the model

We immediately define the model on a D-dimensional L^D lattice with lattice spacing a and size $L = Na$. Thus, the lattice points are given by

$$x = a(n_0, \ldots, n_{D-1}); \quad n_i \in \mathbb{N}_0, \quad 0 \leq n_i < N.$$

Since this writeup is about numerical techniques, from now on a will be set to one. The real-valued fields ϕ are living on the sites, $\phi_x \in \mathbb{R}$ and we adopt periodic boundary conditions: if $\hat{\mu}$ is the unit vector in the μ-direction, then $\phi_{x+L\hat{\mu}} = \phi_x$. The action S is given by

$$S(\phi) = \sum_x \left[-2\kappa \sum_{\mu=0}^{D-1} \phi_x \phi_{x+\hat{\mu}} + \phi_x^2 + \lambda(\phi_x^2 - 1)^2 \right] \tag{7.1}$$

and expectation values can be computed using the path-integral method

$$\langle A \rangle = \frac{1}{Z} \int \prod_x \mathrm{d}\phi_x \exp(-S(\phi)) A(\phi), \tag{7.2}$$

with the partition function $Z = \int \prod_x \mathrm{d}\phi_x \exp(-S(\phi))$.

The model has two well-known limits: $\lambda = 0$ gives a Gaussian model, in the limit $\lambda \to \infty$ the Ising model is recovered. We will only do computations in three dimensions, where it has a second-order critical line in the space spanned by the two coupling constants λ and κ.

7.1.1.1 Observables

We will be looking just at a few observables. The most important ones are powers of the magnetization m

$$m = \sum_x \phi_x. \tag{7.3}$$

Note that on a finite lattice $\langle m \rangle = 0$, because the action is invariant under $\phi_x \to -\phi_x$ and the fact that spontaneous symmetry breaking only occurs in an infinite volume. So, interesting quantities are the magnetic susceptibility

$$\chi = \frac{1}{V} \langle m^2 \rangle$$

and the Binder cumulant U

$$U = \frac{\langle m^4 \rangle}{(\langle m^2 \rangle)^2}. \tag{7.4}$$

U is dimensionless and therefore can be used as a phenomenological coupling.

Other interesting quantities to look at are derivatives of observables with respect to κ. If A does not depend on κ, we can get them from correlations with the interaction term W

$$\frac{\partial}{\partial \kappa}\langle A \rangle = \langle W A \rangle - \langle W \rangle \langle A \rangle \qquad \text{with} \qquad W = 2 \sum_x \sum_{\mu=0}^{D-1} \phi_x \phi_{x+\hat{\mu}}. \qquad (7.5)$$

7.1.2   Implementation

The starting point for the implementation are routines for the layout of the fields and the navigation of the D-dimensional lattice, along with an example provided by a function to compute the action of a given field ϕ. They are available along with the code for the solutions under the URL given above.

Most of it is in file `phi4.c`. It contains the main program, which at the moment does read in the basic parameters of the action, initializes the hopping field, fills the field ϕ with random numbers and computes the action S on a given field configuration ϕ.

7.1.2.1   Layout of the lattice

The ϕ field and the hopping field are global in `lattice.h`, where also the lattice size L, the dimension D and the volume V are defined. The lattice is ordered lexicographically, each point with coordinate $(n_0, n_1, \ldots, n_{D-1})$ gets assigned a unique index j. It can be computed by

$$j = \sum_{i=0}^{D-1} n_i L^i.$$

In this way, the field ϕ_x of the Lagrangian is realized as the one-dimensional field `phi[j]` with length $V = \prod_i L_i$ of the computer program.

Fortunately, the specific ordering of the points rarely needs to be understood. However, the hopping field `hop[V][2*D]` is important to understand: it is used to navigate the D-dimensional lattice. The index of the neighbor of the point with index i in direction μ is `hop[i][mu]` in the forward direction $0 \leq \mu < D$ and `hop[i][D+mu]` in the backward direction. If this point has coordinates (n_0, n_1, n_2), then the point with coordinates $(n_0, n_1 + 1, n_2)$ has index `jp=hop[i][1]`. Analogously, if we want to increase n_2 by one, then `jp=hop[i][2]`. The field also takes care of the periodic boundary conditions. If you want to go into the negative direction, the corresponding indices are in the upper D entries of the hop field. So $(n_0, n_1 - 1, n_2)$ has index `jp=hop[i][D+1]`.

As an example, the code for the computation of the action corresponding to a field `phi[]` is given in Figure 7.1. In line 11 we perform a loop over all sites and for each of them sum over the values of the field in the D positive directions. This corresponds to the μ sum in Eq. 7.1. From this quantity and the value of the field at the site, the action density can be computed and accumulated for the total value of the action.

```
 1 double action(void)
 2 {
 3   int i;
 4   double J;
 5
 6   S=0;
 7   for (i=0;i<V;i++) /* loop over all sites */
 8   {
 9       /* sum over neighbors in positive direction */
10       J=0.0;
11       for (mu=0;mu<D;mu++) J+=phi[hop[i][mu]];
12
13       phi2=phi[i]*phi[i];
14       S+=-2*kappa*J*phi[i]+phi2+lambda*(phi2-1.0)*(phi2-1.0);
15   }
16   return S;
17 }
```

Fig. 7.1 Function for the computation of the action.

The "tasks" given below ask for measurement routines to be implemented. They are simple modifications of the action routine with the magnetization just the sum over the field $\sum_x$ phi[x]. For the derivatives, the interaction term is needed, which is just the sum over J*phi[x], as in the subroutine for the action.

To get familiar with the code here are a few simple exercises

Exercise 7.1 Get the code, type make phi4 to compile and then run it ./phi4 infile. infile is the input file from which the parameters are read.

Exercise 7.2 Understand the phi4.c file. First main() then action() that computes the action S for the global field phi[], in particular how the hopping field is used.

Exercise 7.3 Write a routine to measure the magnetization m defined in Eq. 7.3. Test this routine on field configurations ϕ of which you know the result.

Exercise 7.4 Write a routine to measure the other quantities necessary for the κ derivative of $\langle m^2 \rangle$ and $\langle m^4 \rangle$ using Eq. 7.5.

7.2 Hybrid Monte Carlo

Now that the fields are laid out and the action has been defined, we set out to implement the algorithm that generates a series of field configurations ϕ_i such that the expectation value in Eq. 7.2 can be estimated by a simple average over the measurements on those configurations

$$\langle A \rangle = \frac{1}{N} \sum_i A[\phi_i](1 + \mathcal{O}(1/\sqrt{N})),$$

where the error is purely statistical and is reduced with the square root of the number of measurements N. For this, an "exact" algorithm is needed, i.e. one that does not introduce a systematic error into this estimate, see Lüscher's lectures for more details.

HMC is an exact algorithm. It starts with enlarging the partition function by momenta π

$$Z = \int \prod_x \mathrm{d}\pi_x \prod_x \mathrm{d}\phi_x e^{-H(\pi,\phi)},$$

where for each site x, a momentum π_x is associated to the ϕ_x. The "Hamiltonian" H is given by

$$H(\pi, \phi) = \frac{1}{2} \sum_x \pi_x^2 + S(\phi).$$

Obviously, expectation values of observables $A[\phi]$, which are functions of ϕ only, are unaltered. At first sight, the gain from this augmented partition function is unclear, however, a particular kind of field update becomes possible.

It is based on molecular dynamics (MD) evolution, which means that we treat this system in close analogy to classical mechanics, where the fields play the role of the (generalized) position and the potential energy is given by the action of the model. The system obeys Hamilton's equations of motion,

$$\frac{\mathrm{d}\phi}{\mathrm{d}\tau} = \frac{\partial H}{\partial \pi} \quad \text{and} \quad \frac{\mathrm{d}\pi}{\mathrm{d}\tau} = -\frac{\partial H}{\partial \phi},$$

where a "Monte Carlo" time τ is introduced. (To avoid any misunderstanding: these are not the equations of motion of the underlying theory, but those associated to the artificial Hamiltonian H.) From the solution of these equations of motion, a valid update algorithm can be derived, because from classical mechanics we know that they conserve the Hamiltonian H and the phase-space volume, the latter by Liouville's theorem. Loosely speaking, a configuration of (π, ϕ) is equally likely to be (π', ϕ'), which one gets by using (π, ϕ) as initial condition and solving the equations of motion for some time τ. In the spirit of the analogy to classical systems, moving the fields in this way is called a trajectory of length τ. Since the energy H is conserved during this evolution, this is a microcanonical update.

But this is only one component of the HMC. The "hybrid" in the name of the algorithm comes from the fact that different algorithms are used for different parts of the partition function. For the momenta, heatbath updates are employed. This is easy, because the π follow a simple Gaussian distribution independent of the ϕ. The fields ϕ are updated with (quasi-) microcanonical molecular dynamics evolution, solving the equations of motion. An algorithm that alternates between these two steps is the "Hybrid Molecular Dynamics" algorithm. The classic QCD citation is Ref. (Gottlieb *et al.*, 1987).

The HMD algorithm would be exact if we could solve the equations of motion exactly. However, in general we have to use a numerical method to solve these equations and therefore integration errors occur. Given some properties of the integration procedure, namely the time-reversibility and the conservation of the phase-space volume, this algorithm can be made exact with a final Metropolis step at the end of each trajectory. One considers the fields (π', ϕ') at the end of the trajectory as a proposal. The new field configuration is accepted with probability $\exp(-\Delta H)$, where ΔH is the difference of the Hamiltonian at the beginning and the end of the trajectory. If the proposal is rejected, we go back to the field ϕ and start again at the heatbath for the momenta π. Note that when the solution of the equations of motion is exact, we know that $\dot{H} = 0$, which implies ΔH is zero and the proposed change is always accepted.

Before starting the implementation, let us summarize the HMC algorithm for a single trajectory. This is then repeated N_{tr} times, where N_{tr} depends on the accuracy required and the amount of computer time allocated for the project.

1. Momentum heatbath: Choose new random momenta according to the distribution $P(\pi_i) \propto \exp(-\pi_i^2/2)$.
2. Molecular dynamics evolution: Numerically solve the Hamiltonian equations of motion

$$
\begin{aligned}
\frac{\mathrm{d}}{\mathrm{d}\tau}\phi_x(\tau) &= \frac{\partial}{\partial \pi_x} H(\pi(\tau), \phi(\tau)) \\
\frac{\mathrm{d}}{\mathrm{d}\tau}\pi_x(\tau) &= -\frac{\partial}{\partial \phi_x} H(\pi(\tau), \phi(\tau))
\end{aligned}
\tag{7.6}
$$

for some interval of the fictitious time τ. This moves the fields from some initial (π, ϕ) to the proposed new fields (π', ϕ').
3. Acceptance step: Calculate the change in the Hamiltonian ΔH and accept the proposed new field ϕ' with probability

$$
P_{\text{acc}} = \min[1, \exp(-\Delta H)]; \quad \Delta H = H(\pi', \phi') - H(\pi, \phi).
$$

The implementation of these three steps will be discussed in the rest of the section, along with hints for the debugging of such code.

7.2.1 Momentum heatbath

The first step of the HMC algorithm is the momentum refreshment, i.e. the conjugate momenta π are filled with Gaussian random numbers. Random number generators by themselves are a complicated subject, which goes beyond the scope of this chapter. We therefore take as input a library routine that generates pseudorandom numbers, equally distributed in the range $[0, 1)$. A popular choice is `ranlux`, a high-quality random number generator (Lüscher, 1994) for which a C implementation is freely available.

These have then to be transformed such that they follow a normal distribution, e.g. by the Box–Muller procedure: Given x_1 and x_2 that are drawn from a flat distribution $x_i \in [0, 1)$, then y_1 and y_2 with

$$y_1 = \sqrt{-2\ln(1 - x_1)} \cos(2\pi(1 - x_2))$$

$$y_2 = \sqrt{-2\ln(1 - x_1)} \sin(2\pi(1 - x_2))$$

are distributed according to $P(y_i) \propto \exp(-y_i^2/2)$. We always take $(1 - x_i)$ in order to exclude the zero in the argument of the logarithm.

Once such a routine has been implemented, it needs testing. The most obvious requirement is that the generated random numbers have the expected distribution. For this, it is convenient to histogram the output and compare with the expected form. To give such a comparison a statistical significance, the Kolmogorov–Smirnov test should be applied. From a practical point of view, this is sufficient in our context, *if* we assume that the underlying random number generator is of sufficiently high quality for our purposes.

Exercise 7.5 Write a routine that fills a vector of length n with double precision Gaussian random numbers distributed according to $P(x) \propto \exp(-x^2/2)$. Test this routine by filling many such vectors and histogramming. Does the distribution match your expectation? (Do this only if time allows, otherwise get the routine from me.)

Exercise 7.6 Now introduce the global momentum field `mom[V]` that will hold the momenta π conjugate to the field variables `phi[V]` and fill it with normally distributed random numbers. Write a routine that, given the `phi` and `mom` fields, computes the molecular dynamics Hamiltonian H.

7.2.2 Molecular dynamics

The second step in the HMC algorithm, the molecular dynamics evolution, is the part in which the fields ϕ are actually changed. In general, it is by far the most difficult to implement part of the whole algorithm. On the other hand, it is relatively easy to debug: it consists of the numerical solution of the classical MD equations of motion, for which we know that the Hamiltonian is conserved. By making the numerical integration more and more precise, we can therefore check the setup by observing an improvement in the energy conservation in accordance with the expected scaling from the integrator.

The approximate numerical solution of differential equations like Eqs. 7.6 is a vast field of current research, the coverage of which is beyond the scope of these exercises. For more information, see again Lüscher's lectures at this school and the book by Reich and Leimkuhler (Leimkuhler and Reich, 2005).

In many applications, the integration algorithm is constructed by a so-called splitting method, based on a decomposition of the Hamiltonian in exactly integrable pieces

$$H(\phi, \pi) = H_1(\pi) + H_2(\phi)$$

with $H_1(\pi) = \sum_x \pi_x^2/2$ and $H_2(\phi) = S(\phi)$. Also, the splitting into more terms is possible and exploited in modern QCD codes. The algorithm consists of repeated application of the two elementary steps

$$\mathcal{I}_1(\epsilon) : (\pi, \phi) \to (\pi, \phi + \epsilon \nabla_\pi H_1(\pi)) \tag{7.7}$$

$$\mathcal{I}_2(\epsilon) : (\pi, \phi) \to (\pi - \epsilon \nabla_\phi S(\phi), \phi). \tag{7.8}$$

It turns out that basically any combination of steps $\mathcal{I}_1(x)$ and $\mathcal{I}_2(x)$ leads to a legal integrator as long as the time shifts x add up to τ during a trajectory. A simple algorithm is the Störmer–Verlet method, in the QCD literature mostly referred to as the leap-frog, which corresponds to the application of

$$\mathcal{J}_\epsilon(\tau) = [\mathcal{I}_1(\epsilon/2)\mathcal{I}_2(\epsilon)\mathcal{I}_1(\epsilon/2)]^{N_s} \tag{7.9}$$

with $\tau = N_s \epsilon$ the length of the trajectory. Because it is symmetric under time-reversal, a potential deviation from the exact solution $\mathcal{O}(\epsilon)$ drops out and the leading violation due to the finite step size ϵ is $\mathcal{O}(\epsilon^2)$. Observing this scaling behavior with varying step size serves as the check of the correctness of the code discussed in the beginning.

The leap-frog integrator is still the main work-horse of molecular dynamics simulations, however, in recent years it has been realized that modified schemes can lead to significant improvement at little extra cost. One example is the so-called Omelyan integrator (Omelyan *et al.*, 2003) that reduces the coefficient of the ϵ^2 term and for one particular criterion an optimal scheme is given by

$$[\mathcal{I}_1(\xi\epsilon)\mathcal{I}_2(\epsilon/2)\mathcal{I}_1((1-2\xi)\epsilon)\mathcal{I}_2(\epsilon/2)\mathcal{I}_1(\xi\epsilon)]^{N_s}$$

with a tunable parameter ξ. The canonical value is $\xi \approx 0.1931833$.

Both integration methods, leap-frog and Omelyan's prescription, satisfy the requirements, which are needed for the final HMC to be exact: it is time reversible and the phase-space measure is conserved, both properties of the exact solution. Reversibility means that if we perform one trajectory $(\pi^0, \phi^0) \to (\pi^1, \phi^1)$, then flip the momentum $\pi^1 \to -\pi^1$ and now use the same algorithm to run the trajectory back, we end up in exactly the same spot $(\pi^2, \phi^2) = (\pi^0, \phi^0)$, where we started. This is true at least in exact algebra. On a real computer with fixed-precision arithmetic, this is spoiled by rounding errors, which come from the violation of associativity of the summation. Though the effect of this violation is in general tiny, the influence on the final measurements is unclear, but in most cases negligible. (In QCD simulations, however, the issue is serious, because chronological inversion techniques cause coupling from previous MC time steps.) Reversibility is a standard check of the code.

The second important property is that the map in phase space, which is defined by the integration algorithm like the one given in Eq. 7.9,

$$\mathcal{J}_\epsilon(\tau) : (\pi, \phi) \to (\pi', \phi')$$

conserves the phase-space measure. As a consequence of this, the exponential of the change in the Hamiltonian $\Delta H = H(\pi', \phi') - H(\pi, \phi)$ is one on average

$$\langle \exp(-\Delta H) \rangle = \frac{1}{Z} \int [d\pi][d\phi] e^{-H(\pi,\phi)} e^{-\Delta H(\pi,\phi)} = \frac{1}{Z} \int [d\pi'][d\phi'] e^{-H(\pi',\phi')} = 1,$$

at least if the algorithm is exact, which is the case for the full HMC. This observable is therefore also routinely checked in simulations to provide yet another check for the code and the choice of parameters.

Here, a remark on the language used in the field is in place. In line with the language of classical mechanics, $F = -\nabla_\phi S(\phi)$ in Eq. 7.8 is called the "force". With a suitable norm, one can then speak of the size of such forces. Large forces typically require smaller step sizes for the integration and therefore measuring them serves as a common guide to tuning the various parameters of modern HMC simulations.

7.2.2.1   *Implementation*

For the ϕ^4 Hamiltonian, the derivatives are easily computed. The derivative of the Hamiltonian with respect to the momentum π_x in Eq. 7.8 evaluates independently of the model to

$$\nabla_{\pi_x} H_1(\pi) = \pi_x. \tag{7.10}$$

The corresponding code can be found in Fig. 7.2. The expression for the force in Eq. 7.7, however, will depend on the particular choice of the discrete action. For the one given in Eq. 7.1, the forces read

$$\nabla_{\phi_x} S(\phi) = -2\kappa \sum_{\mu=0}^{D-1} (\phi_{x+\hat{\mu}} + \phi_{x-\hat{\mu}}) + 2\phi_x + 4\lambda(\phi_x^2 - 1)\phi_x \tag{7.11}$$

$$= -2\kappa J_x + 2\phi_x + 4\lambda(\phi_x^2 - 1)\phi_x.$$

Using $J_x = \sum_{\mu=0}^{D-1} (\phi_{x+\hat{\mu}} + \phi_{x-\hat{\mu}})$, the sum of the field ϕ on all neighboring sites. For this purpose, the hopping field hop, introduced in Section 7.1.2.1, also contains the indices of the sites in negative direction. For the site x, the sum can thus easily be computed by adding up phi[y] over all sites with indices y=hop[x][mu], with mu running from 0 to $(2D-1)$. This is implemented in the code reproduced in Fig. 7.2.

Once this code is written, the routines have to be verified. Given a correct implementation of the computation of the action, a very stringent test is the behavior of the violation of energy along a trajectory. As already discussed, the change in $H(\pi, \phi)$ during the trajectory is expected to scale with ϵ^2, the square of the step size of the integration algorithm. The result of the test at the parameters suggested below is displayed in Fig. 7.3. We observe exactly the expected behavior. The Omelyan integrator performs roughly a factor of two better than the leap-frog, even after normalization by the number of force evaluations (typically the most expensive part of the simulation) per step. However, one should stress that the fact that $\Delta H \propto \epsilon^2$ tests only whether force and action match, matching errors in both routines will remain undetected.

Exercise 7.7 Write routines that perform the two elementary updates in Eq. 7.8 and Eq. 7.7.

Exercise 7.8 Write a routine that repeats the sequence of these three steps N_s times. This moves the fields to time $\tau = N_s \epsilon$.

```
 1 void move_phi(double eps)
 2 {
 3   int i;
 4   for (i=0;i<V;i++) phi[i]+=mom[i]*eps;
 5 }
 6
 7
 8 void move_mom(double eps)
 9 {
10   int i,mu;
11   double J, force;
12
13   for (i=0;i<V;i++)
14   {
15     J=0;
16     for (mu=0;mu<2*D;mu++) J+=phi[hop[i][mu]];
17
18     force=2*kappa*J-2*phi[i]-lambda*4*(phi[i]*phi[i]-1)*phi[i];
19     mom[i]+=force*eps;
20   }
21 }
```

Fig. 7.2 The two elementary updates corresponding to Eq. 7.7 (top) and Eq. 7.8 (bottom) for the action given in Eq. 7.1, implementing Eq. 7.10 and Eq. 7.11 respectively.

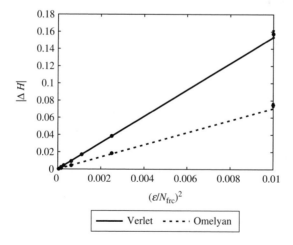

Fig. 7.3 Energy violation as a function of step size showing the expected quadratic convergence. Comparison between Verlet and Omelyan integrator.

Exercise 7.9 Test this routine by measuring the Hamiltonian after each application of the three steps. Since they are supposed to solve the Hamiltonian equations, the energy H should be conserved up to $\mathcal{O}(\epsilon^2)$. Use trajectories of length 1 and test for various ϵ that this is indeed the case. Suggestion: use a 4^4 lattice, $\kappa = 0.18169$ and $\lambda = 1.3282$, starting from a random ϕ field. Do 1000 trajectories, measure the energy violation $|\Delta H|$ after each.

Exercise 7.10 An important property of the integrator is that it is reversible. So, if we do a trajectory $(\pi, \phi) \to (\pi', \phi')$ and flip the momentum $\pi' \to -\pi'$, then the trajectory with $(-\pi', \phi')$ as initial values should lead to (π, ϕ). Check that this is the case. What spoils this?

Exercise 7.11 If you are ambitious: try the Omelyan integrator. Compare to the leap-frog.

7.2.3 Hybrid Monte Carlo

So far we have implemented the Hybrid Molecular Dynamics algorithm, which is inexact due to the integration errors of $\mathcal{O}(\epsilon^2)$. In a seminal paper on the Hybrid Monte Carlo algorithm (Duane *et al.*, 1987), Duane, Kennedy, Pendleton and Roweth realized that it can easily be made exact by a Metropolis acceptance step. Basically, one measures the value of the Hamiltonian $H_i = H(\pi, \phi)$ at the beginning of the trajectory and saves the field ϕ. Then, one performs the molecular dynamics evolution and again measures the Hamiltonian $H_f = H(\pi', \phi')$. This "proposed" new ϕ' is accepted with probability $\min[1, \exp(-\Delta H)]$, where $\Delta H = H_f - H_i$, else the new field is set to the initial field ϕ.

In a practical implementation, this acceptance step is realized by the following scheme: If $\Delta H < 0$, accept, else throw a random number r and accept if $\exp(-\Delta H) > r$, else reject the configuration. In case of rejection, copy back the ϕ at the beginning of the trajectory to the phi[] array.

More from the point of view of the implementation, a trajectory of the HMC algorithm therefore can be summarized in the following way

1. Momentum heatbath on field mom[].
2. Keep a copy of the current field phi[] in phiold[].
3. Measure the current value of the Hamiltonian H_i.
4. Molecular dynamics with initial values mom[] and phi[]. These fields get overwritten by the values after the evolution.
5. Measure the value of the Hamiltonian H_f.
6. If the Metropolis test rejects the new configuration, copy phiold[] to phi[].

This concludes the programming effort to implement the HMC algorithm for the ϕ^4 theory. An example for the implementation can be obtained at the URL given in Section 7.1. Using this code, we can now simulate the model and measure expectation values of the physics observables already implemented. The measurement would be step number seven in the list. It has to take place after the Metropolis step and after a potentially rejected configuration has been replaced. Typically, one does not measure after each trajectory. As will be discussed below, the configurations that are a few trajectories apart are correlated and the new measurement does not give much new information. The measurement frequency is therefore a balance between the cost of one measurement, the cost of a trajectory, the correlations between successive configurations and the amount of data to be stored. In our case, measurements are very cheap and since we are interested in the correlations, we can afford to measure after each trajectory. In QCD, measurements frequently cost a significant amount of computer time and consequently it is worth separating them by a couple of trajectories.

Exercise 7.12 Extend your program to the HMC algorithm: measure H and save the `phi[]` field at the beginning; perform the Metropolis step at the end.

Exercise 7.13 We also want to do physics measurements, so measure the magnetization m and the action S after each trajectory.

Exercise 7.14 Detour, if time allows: Computing the energy difference from H_i and H_f is susceptible to round-off errors. It is better to subtract first the energy densities and then perform the sum. Also, doing the sums hierarchically might be advised, first summing over sublattices and then accumulating these results. Implement these improvements in your code.

7.2.4 Simulation

Given a correct algorithm, i.e. one that is ergodic, stable and without errors in the implementation, the Markov chain can be started from any configuration, or distribution of configurations. Applying the algorithm for a sufficiently large number of iterations will deplete the wrong contributions of this distribution and in the end, one is left with the correct distribution given by the theory one wants to simulate. This decay is exponential but the corresponding decay rates can be very small. The process is called "thermalization" and once it is over, the simulation is in "equilibrium".

In practical simulations, one measures several quantities in regular intervals. During the thermalization, their values show a systematic drift; once equilibrium is reached, they fluctuate around their "true" average. This can look rather differently for various quantities, but as long as there are these systematic movements in any quantity, equilibrium has not been reached.

If one is uncertain about equilibration, a way to proceed is to start from very different starting configurations and observe whether or not common values for the observables are reached.

Once in equilibrium, there is a further test of the correctness of the code: we know that $\langle \exp(-\Delta H) \rangle = 1$. Of course, as will be discussed in the last section, measurements are only meaningful with a well-determined error. For this we have to measure the autocorrelations in this particular observable using the methods discussed in that section. For a run on a lattice with $L = 6$, $\kappa = 0.185825$ and $\lambda = 1.1689$ with trajectories of length 1 using 10 steps of the leap-frog algorithm, we get a rate of acceptance of 87%. The exponential of the energy violation averages beautifully to 1, $\langle \exp(-\Delta H) \rangle = 1.0002(6)$, despite fluctuations of ΔH with $\langle \Delta H^2 \rangle = 0.1097(8)$.

Another demonstration that the acceptance step makes the HMC algorithm exact is to look at the dependence of observables with changing step size. In Fig. 7.4 we observe that the extracted value of $\langle m^2 \rangle$ is independent of it, at least within error-bars. The

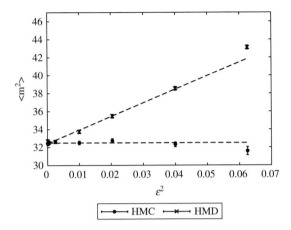

Fig. 7.4 Dependence of the magnetic susceptibility measured on the ensemble on the step size of the integration. The step-size error evident in the HMD algorithm is removed by the Metropolis step in the HMC. The parameters of the runs are given in the text.

same algorithm without the Metropolis step, the Hybrid Molecular Dynamics algorithm, however, shows a clear deviation from this value that scales consistent with ϵ^2.

In principle, the HMC can be run at any step size, which leads to the question of its optimal choice. For most simulations, acceptance rates between 70% and 90% are good target values. Higher rates typically mean that too much effort is spent on the integration. Low values of the acceptance rate, however, frequently can be improved by relatively little additional effort. Also, they might be an indicator of a problem in the numerical integration. The corresponding data for this simulation is plotted in Fig. 7.5. We show the acceptance rate as a function of the square of the step size and see a linear deviation from one. Ultimately, what counts is the cost of producing an independent configuration, as will be explained in the next section. To first approximation this is the cost per accepted trajectory, which is shown in the right plot. We observe a relatively shallow minimum in the region between 50% and 90% acceptance.

7.2.5   Physics

Using this code, one is in the position to compute estimates for the observables one might be interested in. For this, of course, a method to compute their errors is needed, since averages without giving their uncertainty are of no use. This will be explained in the next section. The comparison of such expectation values to data from the literature is as yet another test for the correctness of the code, and frequently a very important one. In our case, Refs. (Hasenbusch, 1999; Hasenbusch *et al.*, 1999) contain a large quantity of high-precision data to serve this purpose.

In order to actually do some physics with the program, one might follow the lines of these references and study critical phenomena of the Ising universality class. In principle, the extraction of the critical exponents is possible, even though the

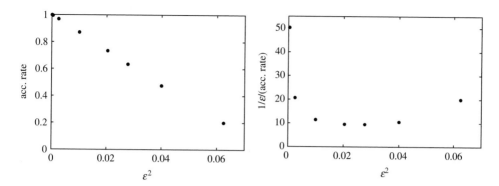

Fig. 7.5 Dependence of the acceptance rate on the step size (left) and the cost per accepted trajectory also as a function of the step size (right). The cost has a wide minimum that corresponds to acceptance rates between roughly 50% and 90%.

algorithm is definitely not the fastest one for this particular model. Had this course been one session longer, the computation of some leading critical exponents could have been attempted. The reader is referred to the above-mentioned publications for strategies. Figure 7.6 has to serve as a demonstration that the physics comes out about right. What is shown is the average of $|m|$ as it crosses the phase transition. The absolute value is taken, because in finite volume $\langle m \rangle$ is zero. In infinite volume, the magnetization is zero above the critical temperature, and from the critical point

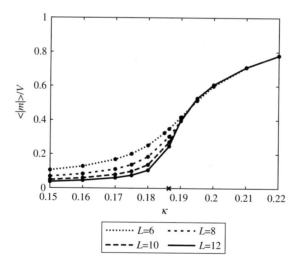

Fig. 7.6 Behavior of the magnetization $\langle m \rangle / V$ across the phase transition for different lattice sizes. In infinite volume, this should be zero for $\kappa < \kappa_{\mathrm{cr}}$ marked by the "x" on the x-axis. This behavior is slowly approximated for growing L.

on, it rises towards saturation. As we can see in the figure, this behavior is approached for L growing towards infinity.

Exercise 7.15 We do some initial test using $L = 6$, $\kappa = 0.185825$ and $\lambda = 1.1689$. This is pretty close to the critical line. Perform 5000 trajectories of length 1, $\epsilon = 0.05$, starting from a random initial configuration. Observe how the system thermalizes. Compare with runs at $\kappa = 0.1$ and $\kappa = 0.2$. How long (in MC time) do we have to wait until we can say that we are in equilibrium?

Exercise 7.16 Now make a longer run, e.g. 10^5 trajectories. Start measuring after the thermalization has been completed. What are the values of the Binder cumulant, the magnetization and the action? To avoid problems with autocorrelations, combine the results from $N_{\text{av}} = 1000$ consecutive measurements and then do a naive error analysis on these values. What is the acceptance rate?

Exercise 7.17 Verify that $\langle e^{-\Delta H} \rangle = 1$.

Exercise 7.18 Convince yourself that you actually have a correct algorithm. The results should not depend on the step size. If the steps are too large, only the acceptance rates goes down, the average values of the observables should not be affected. Again, block 1000 measurements.

Exercise 7.19 Now change to $\lambda = 1.145$ where the critical point is according to Ref. (Hasenbusch, 1999) at $\kappa_c = 0.1864463(4)$. How does $\langle |m| \rangle / V$ change when you cross the phase transition. How does the picture depend on L?

7.3 Error analysis

Most Monte Carlo algorithms produce a series of field configurations that are correlated among each other. These correlations die out with increasing Monte Carlo time separation, however, it still poses a problem to the estimation of the statistical errors of measured observables. Here, a brief summary of the method to deal with these correlations is given, for more detail see Refs. (Sokal, 1996; Wolff, 2004).

7.3.1 Theory

To be specific, our algorithm produces configurations $\phi^1 \to \phi^2 \to \cdots \to \phi^N$ on which observables A^α (labelled by an integer α) are measured. So, we get a set of measurements $\{A_i^\alpha : i = 1, \ldots, N\}$. Let us assume that the thermalization process is already completed and the ϕ^i are in equilibrium. To quantify the correlations between the successive configurations one looks at the *autocorrelation function* $\Gamma_{\alpha\beta}$

$$\Gamma_{\alpha\beta}(t) = \langle (A_i^\alpha - \langle A^\alpha \rangle)(A_{i+t}^\beta - \langle A^\beta \rangle) \rangle. \tag{7.12}$$

Autocorrelations let this be non-zero for $t > 0$. The brackets $\langle \cdots \rangle$ mean an average over repeated "experiments", i.e. independent sets of N configurations. Of course, we almost never do this in real life, but use an estimator from one Markov chain

$$\bar{\Gamma}_{\alpha\beta}(t) = \frac{1}{N-t} \sum_{i=1}^{N-t} (A_i^\alpha - \bar{A}^\alpha)(A_{i+t}^\beta - \bar{A}^\beta) + \mathcal{O}(1/N) \tag{7.13}$$

with $\bar{A}^\alpha = \frac{1}{N}\sum_i A_i^\alpha$.

In most lattice simulations, we compute observables, which are functions of such averages of primary observables

$$F = F(A^1, \ldots, A^n),$$

from which we also need the derivatives $f_\alpha = \partial F/\partial A^\alpha$. For example, the Binder cumulant Eq. 7.4 is constructed from two primary observables $U = A_2/A_1^2$ with $A_1 = \langle m^2 \rangle$ and $A_2 = \langle m^4 \rangle$. Accordingly, the derivatives are $f_1 = -2A_2/A_1^3$ and $f_2 = 1/A_1^2$.

The obvious estimator of F is $\bar{F} = F(\bar{A}^1, \ldots, \bar{A}^n)$ and its error σ_F is given by (Wolff, 2004)

$$\sigma_F^2 = \frac{2\tau_{\text{int}}}{N} v_F \tag{7.14}$$

with the variance v_F given by

$$v_F = \sum_{\alpha\beta} f_\alpha f_\beta \Gamma_{\alpha\beta}(0)$$

and the integrated autocorrelation time for the observable F

$$\tau_{\text{int},F} = \frac{1}{2} + \frac{1}{v_F} \sum_{t=1}^{\infty} \sum_{\alpha\beta} f_\alpha f_\beta \Gamma_{\alpha\beta}(t)$$
$$\equiv \frac{1}{2} + \sum_{t=1}^{\infty} \rho_F(t). \tag{7.15}$$

Since Eq. 7.14 is exactly the standard formula for the one sigma error, but where the number of measurements N is divided by $2\tau_{\text{int}}$, one colloquially speaks of twice τ_{int} as the time to produce an independent configuration. It has to be stressed, however, that this statement can strongly depend on the particular observable F. The error of the normalized autocorrelation function ρ_F can also be given

$$[\delta\rho_F(t)]^2 = \frac{1}{N} \sum_{k=1}^{\infty} (\rho_F(k+t) + \rho_F(k-t) - 2\rho_F(t)\rho_F(k))^2.$$

Note that $\rho_F(t) = \rho_F(-t)$.

In a practical implementation, it is frequently tedious to compute first $\Gamma_{\alpha\beta}(t)$ via Eq. 7.12 and then the matrix element $f_\alpha \Gamma_{\alpha\beta}(t) f_\beta$ according to Eq. 7.15. The easier path is to first get $B_i = \sum_\alpha f_\alpha (A_i^\alpha - \bar{A}^\alpha)$ and from this immediately the relevant correlation function $\Gamma_F(t)$

$$\Gamma_F(t) = \frac{1}{N-t} \sum_{i=1}^{N-t} (B_i B_{i+t}) + \mathcal{O}(1/N).$$

Because $\Gamma_{\alpha\beta}(t)$ is only poorly determined for large t, the sum for τ_{int} has to be truncated at some finite "window" W. Otherwise one would essentially sum up noise and increase the error of the computed $\tau_{\text{int},F}$

$$\bar{\tau}_{\text{int},F} = \frac{1}{2} + \sum_{k=1}^{W} \rho_F(t). \tag{7.16}$$

By neglecting the contribution from $k > W$, one introduces a bias in the estimator. According to Madras and Sokal (Madras and Sokal, 1988), the variance of this estimator of τ_{int} is given by

$$(\delta\tau_{\text{int},F})^2 \approx \frac{4W+2}{N} \tau_{\text{int},F}^2,$$

which shows that without a finite W the variance would be infinite. The right choice of W amounts to a balance between bias and variance of the estimator of τ_{int}. Madras and Sokal (Madras and Sokal, 1988) use as W the first point where $W \geq c\tau_{\text{int}}(W)$, e.g. with $c = 4$. For a single exponential decay, this leads to a 2% error on τ_{int}. Lüscher in Ref. (Lüscher, 2005) suggests to sum up to the smallest W such that $\sqrt{\langle \delta\rho(W)^2 \rangle} \geq \rho(W)$, i.e. the point in ρ where the noise starts to overwhelm the signal.

7.3.2 Examples

An example of this procedure is given in Fig. 7.7. This is data from a run on a $L/a = 6$ lattice with $\lambda = 1.145$ and $\kappa = 0.18$ from one million trajectories. This is more statistics than many runs in QCD. One can observe a nice exponential fall-off in the normalized autocorrelation function $\rho(t)$. In the lower plot the partial sum of τ_{int} as in Eq. 7.16 up to a window W is done. This saturates at around $W = 800$, a point where the measured $\rho(t)$ is still measured well enough to be different from zero.

The measurement of autocorrelation times can also serve as a rational of the choice of the trajectory length τ. So far it has been a free parameter of the algorithm. The basic idea is that for very short trajectory lengths (maybe one short elementary leap-frog step only), the momenta are refreshed very frequently and one moves in a new, arbitrary direction. This means the system performs a random walk, which is known to be inefficient as the distance from the origin after N_s steps scales with $\tau\sqrt{N_s}$; the amount of work, however, goes with τN_s. Scaling very short τ by a factor of x should therefore result in an improvement by $\sqrt{x}$, keeping the cost constant. At some point, this argument breaks down because the random-walk regime is left. Then longer trajectories just cost more without any further improvement in the autocorrelations. In Fig. 7.8, this behavior is shown for the autocorrelation time in units of molecular dynamics time, i.e. cost. One observes a clear minimum around $\tau = 2$. However, in a range between $\tau = 1$ and 4 remains within 25% of the optimum.

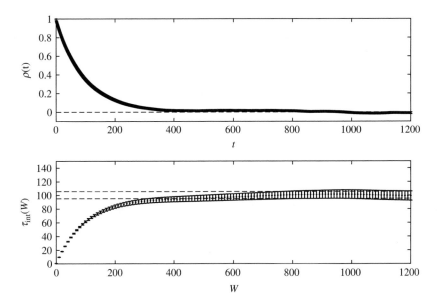

Fig. 7.7 Upper plot: Normalized autocorrelation function $\rho(t)$ of the magnetization m. The lower plot shows the accumulated sum for the computation of τ_{int} as function of the summation window W. The dashed lines indicate the 1σ range of its estimate.

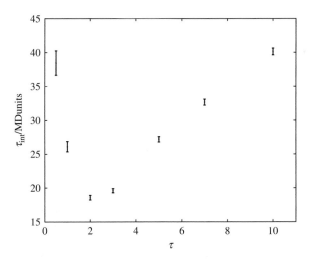

Fig. 7.8 Autocorrelation time of m^2 in units of Monte Carlo time as a function of trajectory length. In this simulation, trajectory length is almost directly proportional to the cost. The optimal trajectory length for this observable is therefore around $\tau = 2$.

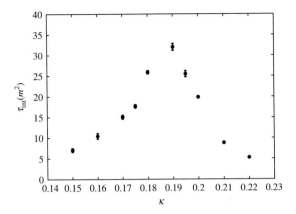

Fig. 7.9 Autocorrelation time of m^2 in units of Monte Carlo time as a function of κ for a $L = 6$ lattice and $\lambda = 1.145$. The critical point is at $\kappa_c = 0.1864463(4)$.

Autocorrelation times depend strongly on the physical system, the observable and the algorithm. They are therefore extremely difficult to predict. What virtually all algorithms have in common, though, is the fact that towards the critical point, they perform worse. This phenomenon is called critical slowing down, i.e. the integrated autocorrelation times of the observable in question increases as the critical point is approached. An example of this is shown in Fig. 7.9 with the critical point $\kappa_{\text{cr}} = 0.1864463(4)$ for $\lambda = 1.145$ taken from Ref. (Hasenbusch, 1999). In general, one expects a behavior similar to critical phenomena, i.e. a scaling with the correlation length according to a power law with a *dynamical* critical exponent z: $\tau_{\text{int}} \propto \xi^z$.

To summarize, the determination of integrated autocorrelation times is a central part of the planning and analysis of all Monte Carlo simulations. Before even starting the runs, one needs to have an idea of the magnitude of the τ_{int} of the observables to be measured in order to decide whether the computation is feasible at all. Once the measurements are done, not all autocorrelation functions will look as in Fig. 7.7. The errors are typically larger and also long tails can make a precise determination of τ_{int} difficult. However, without the autocorrelation time, we cannot give an error and without an error, the measurement is without meaning.

Exercise 7.20 Write a program that reads in the MC time history of the observables A^i and computes the expectation value of a derived quantity $\bar{F}$ and its error.

Exercise 7.21 Take a sample equilibrium history of m and m^2 and compute the normalized autocorrelation functions ρ. Plot them and look at the exponential fall-off. Then compute $\tau_{\text{int}}(W)$ and do some experiments regarding the summation window W. Suggestion: Use $L/a = 6$, $\lambda = 1.145$ and $\kappa = 0.18$ with 10^6 measurements.

Exercise 7.22 Study the dependence of the auto-correlation time of m^2 on the trajectory length and the step size. What are the issues in choosing it for production running?

Exercise 7.23 Now extend the program to analyze the Binder cumulant. Compute the necessary derivatives analytically and modify the `observable()` and `derivative()` routines.

Exercise 7.24 In the vicinity of phase transitions autocorrelation times tend to increase for most algorithms. This is called critical slowing down. Study τ_{int} of m and the Binder cumulant as κ crosses the critical point of $\kappa_c = 0.1864463(4)$ at $L/a = 6$, $\lambda = 1.145$.

Exercise 7.25 Try to get the exponential autocorrelation time τ_{exp}, for which should hold $\rho(t) = C\exp(-t/\tau_{\text{exp}})$ for $t \to \infty$. (To be more precise, it is the supremum over all possible observables.)

7.4 Summary

This chapter covers the implementation of an HMC for ϕ^4 theory, the measurement of simple observables and the data analysis. For the four sessions available, in which most of the programming was done, this was already a dense program. The next step could be the actual analysis of critical properties of this theory. However, one would like to use a better algorithm for this to be a satisfying experience.

Many other interesting topics are not covered as well. For example, one could have split the action and introduced multiple timescales for the integration, have gone more into the analysis of round-off errors or tried more advanced integrators. Also, a comparison to single-variable update algorithms would be interesting. However, I still hope that to those who have never done a complete simulation, it gives an impression of how a simulation with an HMC algorithm is set up and how data is analyzed.

Probably my biggest regret is that these exercises are very conservative by focusing on HMC instead of broadening the knowledge of alternative methods. The strength of the algorithm is that it can be used to simulate any theory with continuous variables, also when many degrees of freedom are coupled; a situation where single-variable updates typically fail. However, at least in the traditional form presented here, virtually no information about the theory is injected in the setup. This can be considered an advantage, because there is no risk of a bias in the result. Still, recent improvements of the HMC algorithms have shown that using knowledge about the physics of the system can lead to significant improvements in the performance. Hasenbusch's mass preconditioned HMC (Hasenbusch, 2001) or Lüscher's DD-HMC (Lüscher, 2005) are prime examples of this. Hopefully, the future will bring methods, that are even more adapted to the problem in question and are even more powerful for the simulation of QCD or other field theories.

Acknowledgments

I want to thank the organizers of this school for inviting me to give this course, Martin Hasenbusch for interesting discussions during its preparation and Rainer Sommer for useful remarks on the manuscript.

References

S. Duane, A.D. Kennedy, B.J. Pendleton and D. Roweth, "Hybrid Monte Carlo," Phys. Lett. B **195** (1987) 216.

S. A. Gottlieb, W. Liu, D. Toussaint, R. L. Renken and R. L. Sugar, "Hybrid Molecular Dynamics algorithms for the numerical simulation of Quantum Chromodynamics Phys. Rev. D **35** (1987) 2531.

M. Hasenbusch, "A Monte Carlo study of leading order scaling corrections of phi**4 theory on a three dimensional lattice," J. Phys. A **32** (1999) 4851 [arXiv:hep-lat/9902026].

M. Hasenbusch, K. Pinn and S. Vinti, "Critical exponents of the three-dimensional Ising universality class from finite-size scaling with standard and improved actions," Phys. Rev. B **59** (1999) 11471.

M. Hasenbusch, "Speeding up the hybrid Monte Carlo algorithm for dynamical fermions," Phys. Lett. **B519** (2001) 177-182. [hep-lat/0107019].

Leimkuhler, B. and Reich, S., Simulating Hamiltonian dynamics. 2005. Cambridge University Press, Cambridge.

M. Lüscher, "A portable high quality random number generator for lattice field theory simulations," Comput. Phys. Commun. **79** (1994) 100 [arXiv:hep-lat/9309020] http://luscher.web.cern.ch/luscher/ranlux/index.html.

M. Lüscher, "Schwarz-preconditioned HMC algorithm for two-flavour lattice QCD," Comput. Phys. Commun. **165** (2005) 199 [arXiv:hep-lat/0409106].

M. Lüscher,(2011) "Computational strategies in lattice QCD," arXiv:1002.4232 [hep-lat].

N. Madras and A. D. Sokal, "The Pivot algorithm: a highly efficient Monte Carlo method for selfavoiding walk," J. Stat. Phys. **50** (1988) 109.

I.P. Omelyan, I.M. Mryglod and R. Folk, Comput. Phys. Commun. **151** (2003) 272.

A. D. Sokal, Monte Carlo Methods in Statistical Mechanics: Foundations and New Algorithms, Lectures at the Cargèse Summer School, 1996.

U. Wolff [ALPHA collaboration], "Monte Carlo errors with less errors," Comput. Phys. Commun. **156** (2004) 143 [Erratum-ibid. **176** (2007) 383].

8

Applications of chiral perturbation theory to lattice QCD

Maarten GOLTERMAN

Department of Physics and Astronomy, San Francisco State University,
San Francisco, CA 94132, USA

Overview

This chapter contains the written version of lectures given at the 2009 Les Houches Summer School "Modern perspectives in lattice QCD: Quantum field theory and high performance computing." The goal is to provide a pedagogical introduction to the subject, and not a comprehensive review. Topics covered include a general introduction, the inclusion of scaling violations in chiral perturbation theory, partial quenching and mixed actions, chiral perturbation theory with heavy kaons, and the effects of finite volume, both in the p- and ϵ-regimes.

8.1 Why chiral perturbation theory for lattice QCD?

It is often claimed that lattice QCD provides a tool for computing hadronic quantities numerically, with fully controlled systematic errors, from first principles. This assertion is of course based on the fact that lattice QCD provides a non-perturbative definition of QCD (in fact, the only one to date). So, why not choose the parameters (in particular, the quark masses) at, or very near, their physical values, and compute all quantities of interest?

There are at least two major obstacles to this. The first obstacle is that lattice QCD is formulated in Euclidean space, rather than in Minkowski space. This is a necessary restriction if one wants to use Monte Carlo methods in order to evaluate expectation values of operators from which one extracts physical quantities. In Euclidean space correlation functions do not give direct access to physical scattering amplitudes; they first need to be continued to Minkowski space. If one is interested in physics at some scale for which an effective field theory (EFT) is available, one can use this EFT in order to match to euclidean lattice correlation functions in the regime of validity of the EFT. In the case of the chiral EFT for QCD, the form of the correlation functions is predicted in terms of a finite number of coupling constants to a given finite order in a momentum expansion. Once the value of these is known from lattice QCD computations, the EFT can then be used to continue to Minkowski space. Chiral perturbation theory (ChPT) provides this EFT framework for the low-energy physics of the (pseudo-) Nambu–Goldstone bosons of QCD.

A second obstacle is that in practice lattice QCD computations are carried out at values of the up-and down-quark mass larger than those observed in nature,[1] because of limits on the size of the physical volumes that can be handled with presently available computers, and because of the rapid increase in algorithmic cost with decreasing quark masses. If the quark masses are nevertheless still small enough for ChPT to be applicable, it can be used to extrapolate to the physical values for these light quark masses, again using ChPT to connect an "unphysical" lattice computation (namely, at values of the light quark masses larger than their real-world values) to physical quantities (those observed in nature).

It should be said that lattice QCD is moving toward the physical point, i.e. that numerical computations are being done with the light (up and down)-quark masses at or very near the physical values. If lattice-quark masses are very near the physical values, simple smooth extrapolations are in principle enough to obtain hadronic quantities of interest at the physical point, and the use of ChPT may become less important in this respect. I will return to this point below.

Both of these obstacles described above are examples of how ChPT (and, in general, EFTs) can be used to connect "unphysical" computations with physical quantities. This is particularly helpful in the case of lattice QCD, where often computations are easier, or even only possible, in some unphysical regime. The use of ChPT is by no means limited to extrapolations in quark masses, or continuation from Euclidean space. Other examples of the use of ChPT to bridge the gap between the lattice and the real world are, in increasing order of boldness:

1. Non-zero lattice spacing. While in state-of-the-art lattice computations the lattice spacing is small (in units of Λ_{QCD}), scaling violations are still significant. It is possible to extend ChPT to parametrize scaling violations afflicting the physics of Nambu–Goldstone bosons. In particular, depending on what lattice fermions one uses, chiral symmetry (and rotational invariance!) are broken on the lattice, and such breakings can be parametrized in ChPT through new coupling constants that vanish in the continuum limit. This will be the topic of Section 8.3. Another example in this category is the use of a finite volume in numerical computations. The dominant effects of this can also be studied in ChPT, as we will see in Section 8.6.

2. Partial quenching and the use of mixed actions. As we will see, on the lattice one is free to choose the valence-quark masses to be different from sea-quark masses. This is of course completely unphysical (it violates unitarity!), but it turns out to be very practical – it is (often much) cheaper to vary only valence-quark masses than to vary the sea (or "dynamical")-quark mass, which affects the ensemble of gauge configurations. Moreover, as we will see, varying valence- and sea-quark masses independently from each other gives us extra handles on the theory, making it easier to determine the couplings of ChPT than it would be without this possibility. This trick is known as "partial quenching."[2] One may even go further: one can use different discretizations for sea quarks (typically a computationally less-expensive method, such as staggered or Wilson quarks) and for valence quarks (a discretization using lattice fermions with very good chiral and flavor symmetries – operator mixing depends primarily on the symmetries of those operators, which are constructed out of the valence quarks). This generalization of partially quenched QCD (PQQCD) is now often referred to as "mixed-action" QCD. We will discuss these methods in Section 8.4.

3. Other unphysical constructions used in lattice QCD; a prominent example is the use of "fourth-rooted" staggered fermions. Again, EFT techniques provide much insight into the nature of this method, and the comparison of EFT calculations with numerical results can be used to test the validity of such methods.

Even if one is only interested in physical quantities that can directly be computed on the lattice, such as (stable) hadron masses at the physical values of the quark mass for a very small lattice spacing, ChPT may still be used to test the results. By varying the quark mass away from the physical value, one might discover that the numerical results do not match the predictions of continuum ChPT, for instance because the lattice spacing is still too large, the volume too small, or because of some other systematic effect. In other words, ChPT provides a useful tool for the validation of results obtained with lattice QCD.

Extending this latter argument, it can also be very interesting to use lattice QCD as a laboratory to explore what strong-interaction physics would look like at parameter values different from those in the real world. Again, interpreting the results of numerical computations in lattice QCD through the help of EFT techniques can be very useful in this respect. An example of this is the question about whether it is possible that the up-quark mass m_u could be zero.[3] This question has now been answered with the help of lattice QCD computations at *non-zero* m_u (the answer is that $m_u = 0$ is now exluded with a very high confidence level). This is only possible with the assistance of ChPT, which makes it possible to extrapolate to what physics would have looked like at $m_u = 0$, from computations done at non-zero values of m_u.

Finally, with limited computational resources, there is the important question as to how to use them most judiciously. Does one use all resources to do a few, very expensive, computations at physical quark masses, in very large volume? While this sounds very attractive, restrictions on available computational resources may force one to choose only larger lattice spacings, and one thus pays with the increase of one systematic error for the removal of other ones. Or does one try to strike a balance between the various systematic errors that afflict lattice computations, by using larger-than-physical quark masses, and less-expensive fermions with symmetry properties that are not as nice? What choices to make depends very much on the physics one is interested in, and they are to some extent subjective. However, ChPT and its extension to unphysical situations helps us with making such choices, and it therefore is, and will likely remain, a very important tool for lattice QCD practitioners.

The aim of this chapter is not to give a review of all the work that has been done in ChPT and its applications to lattice QCD. Rather, the aim is to provide a basic introduction to ChPT, with emphasis on its applications to lattice QCD, geared to readers with a general knowledge of quantum field theory and the basics of lattice QCD. The list of references included is heavily biased toward the papers from which I learned the subject. It will not be possible to refer to all papers that have been written on this and closely related topics and applications; there are probably at least a few hundred *more* than the references I did include. The references I did include are primarily those in which new concepts or methods are introduced and explained, and I strongly recommend them for delving deeper into the many aspects and applications of ChPT in its relation to QCD on the lattice.

It is not even possible to cover all the relevant and important applications that have been considered in the literature. Notable examples of topics left out are applications to baryons (including EFTs for two-nucleon systems), and to hadrons containing heavy quarks, although some aspects of these topics are shared by "heavy-kaon" ChPT, which we consider in Section 8.5. The topics I have chosen to include are a general introduction to ChPT in Section 8.2, the incorporation of scaling violations in ChPT in Section 8.3, partially quenched and mixed-action ChPT in Section 8.4, two-flavor ChPT with a heavy kaon in Section 8.5, and ChPT in finite volume in Section 8.6.

Good reviews of continuum chiral perturbation theory can, for instance, be found in Leutwyler (1994b), and, at the level of a textbook, Donoghue *et al.* (1992) and Georgi (2009). For a more general introduction to EFT methods, including ChPT, see Kaplan (2005). For applications of EFT methods to lattice QCD, see Kronfeld (2002); for a more detailed review of ChPT in the context of lattice QCD, see Sharpe (2006). Another lattice-oriented introduction, with applications of ChPT to lattice calculations of weak matrix elements, is Bernard (1990). For a recent review of applications of ChPT to hadron phenomenology, see Ecker (2008).

8.2 Continuum chiral perturbation theory

We begin with a review of ChPT in the continuum, in infinite volume. This is a vast subject in itself, with many important applications to phenomenology. I will not review any of those here; the idea is to get an understanding of the basics, which we will need for applications to lattice QCD.

8.2.1 Chiral symmetry

Let us consider QCD with $N_f = 3$ flavors, the up, down and strange quarks. (We will also have reason to consider the $N_f = 2$ theory, with only up and down quarks, in situations where the strange quark can be considered heavy.) Of course, heavier quarks exist, but we will be interested in physics at energy scales well below the charm mass, so we may consider the theory in which the charm, bottom and top quarks have been decoupled by integrating them out. In deference to the lattice, we will work in Euclidean space throughout these lectures, continuing back to Minkowski space when necessary. The fermion part of the QCD Lagrangian is

$$\mathcal{L} = \bar{q}_L \slashed{D} q_L + \bar{q}_R \slashed{D} q_R + \bar{q}_L M q_R + \bar{q}_R M^\dagger q_L, \tag{8.1}$$

in which q_i, $i = u, d, s$ is the three-flavor quark field. The subscripts L and R denote left- and right-handed projectors:

$$q_L = \frac{1}{2}(1 - \gamma_5)\, q, \qquad q_R = \frac{1}{2}(1 + \gamma_5)\, q, \tag{8.2}$$

$$\bar{q}_L = \bar{q}\,\frac{1}{2}(1 + \gamma_5), \qquad \bar{q}_R = \bar{q}\,\frac{1}{2}(1 - \gamma_5).$$

Note that we may define $\bar{q}_{L,R}$ as we like, since the Grassmann fields q and $\bar{q}$ are independent in Euclidean space; we have chosen them to be consistent with $\bar{q} = q^\dagger \gamma^0$ in the operator formalism. M is the quark-mass matrix, equal to

$$M = \begin{pmatrix} m_u & 0 & 0 \\ 0 & m_d & 0 \\ 0 & 0 & m_s \end{pmatrix}. \tag{8.3}$$

For massless QCD, with $M = 0$, the Lagrangian is invariant under the symmetry group $U(3)_L \times U(3)_R$:

$$q_L \to U_L q_L, \qquad q_R \to U_R q_R, \tag{8.4}$$

$$\bar{q}_L \to \bar{q}_L U_L^\dagger, \qquad \bar{q}_R \to \bar{q}_R U_R^\dagger,$$

with $U_{L,R} \in U(3)_{L,R}$. We observe that the full QCD Lagrangian is invariant if we also let M transform as

$$M \to U_L M U_R^\dagger. \tag{8.5}$$

Of course, the quark masses do not transform, but this "spurion" trick, of letting some parameters transform under the symmetry group, will be useful later, for instance in Section 8.2.5. Note that in Euclidean space, $\mathcal{L}$ (with Eq. (8.5)) is invariant under a larger group, $GL(3, \mathbb{C})_L \times GL(3, \mathbb{C})_R$. (If we consider this larger group, $U_L^\dagger$ and $U_R^\dagger$ in Eqs. (8.4) and (8.5) have to be replaced by U_L^{-1} and U_R^{-1}, respectively.) But if we identify $\bar{q}$ with $q^\dagger \gamma^0$, this reduces this larger group to $U(3)_L \times U(3)_R$, from which it can be shown that the larger symmetry group in Euclidean space has no additional physical consequences.

It is well known, of course, that "axial $U(1)$," i.e. transformations with $U_L = U_R^\dagger = \exp(i\theta)\mathbf{1}$, are not preserved when QCD is quantized, even when $M = 0$. The actual symmetry group is therefore $SU(3)_L \times SU(3)_R \times U(1)$, where the $U(1)$ factor is the quark number. On mesons, the topic of these lectures, quark-number $U(1)$ is trivially realized, and we may drop it from consideration.

In the real world, we find that there exists an octet of pseudoscalar mesons with masses smaller than any other hadron masses. Moreover, one observes that relations between their masses and interactions are reasonably well described by assuming an approximate $SU(3)$ flavor symmetry.[4] Under this $SU(3)$, the pseudoscalar mesons form an octet

$$\phi = \begin{pmatrix} \frac{\pi_0}{\sqrt{2}} + \frac{\eta}{\sqrt{6}} & \pi^+ & K^+ \\ \pi^- & -\frac{\pi_0}{\sqrt{2}} + \frac{\eta}{\sqrt{6}} & K^0 \\ K^- & \mathbf{K}^0 & -\frac{2\eta}{\sqrt{6}} \end{pmatrix} \sim \begin{pmatrix} u\bar{u} & u\bar{d} & u\bar{s} \\ d\bar{u} & d\bar{d} & d\bar{s} \\ s\bar{u} & s\bar{d} & s\bar{s} \end{pmatrix}, \tag{8.6}$$

where we also indicated the (valence) quark content.[5] (In the theory with only two light flavors, ϕ reduces to the two-by-two upper left-hand block, with the η omitted.) Under $SU(3)$ this octet transforms as

$$\phi \to U \phi U^\dagger, \tag{8.7}$$

and it is clear that this $SU(3)$ can be identified with the diagonal subgroup $SU(3)_V$ of $SU(3)_L \times SU(3)_R$, for which $U_L = U_R = U$ in Eq. (8.4).

If the full symmetry group $SU(3)_L \times SU(3)_R$ were realized "manifestly" in nature, we would observe larger hadronic multiplets, and in particular, we would observe "parity partners," i.e. pairs of hadrons with the same mass but opposite parity, since parity takes left-handed quarks into right-handed ones and vice versa.[6] Such parity partners are in general not observed, and specifically, there are no scalar mesons with masses near those of the pseudoscalar multiplet of Eq. (8.6). Instead, it is universally believed (with very strong evidence from both the real world and lattice computations) that, in a (hypothetical!) world with massless quarks, the chiral group is spontaneously broken to the diagonal subgroup, $SU(3)_L \times SU(3)_R \to SU(3)_V$. This requires eight Nambu–Goldstone bosons (NGBs), which are identified with the pions, eta and kaons of Eq. (8.6). Since the broken generators distinguish between left- and right-handed quarks, these NGBs have to be pseudoscalars. In the real world, quarks are not massless, but if the quark masses $m_{u,d,s}$ are small compared to Λ_{QCD}, they can be treated as a perturbation (as we will see below), giving the NGBs a small mass. Indeed, the pions have a very small mass of about 140 MeV, while the kaons and eta have a mass of about 500 MeV, which is at least smallish compared to other hadron masses. For a proof that $SU(3)_V$ does not undergo spontaneous symmetry breaking in continuum QCD, see Vafa and Witten (1984).

Our interest in these lectures will be in the physics of these NGBs, at energies below those at which any other hadrons can be produced. We therefore expect that it should be possible to find an EFT for the physics of NGBs. What that means is that it should be possible to write down a local Lagrangian in terms of the field ϕ that, in some systematic approximation, reproduces correlation functions involving only NGBs, with restrictions imposed by the principles of quantum field theory. The chiral Lagrangian is precisely this EFT.

8.2.2 The chiral Lagrangian

Let us therefore start with a closer look at spontaneous symmetry breaking of chiral symmetry, for the case that $M = 0$. An order parameter for this breaking is the renormalized condensate

$$\langle \bar{q}_{Ri} q_{Lj} \rangle = \langle \bar{q}_{Li} q_{Rj} \rangle \propto \Lambda_{QCD}^3 \delta_{ij}, \tag{8.8}$$

where the power of Λ_{QCD} estimates the magnitude of the condensate. As is always the case with spontaneous symmetry breaking, there is a manifold of equivalent vacua, and indeed, one may rotate this condensate as

$$\Omega_{ij} = \langle q_{Li} \bar{q}_{Rj} \rangle \to U_L \Omega U_R^\dagger, \tag{8.9}$$

where I indicated how Ω transforms under the chiral group (here I sum over spin and color indices, so that Ω is a color and spin singlet). These vacua are all equivalent, and they are rotated into each other by elements of the coset $SU(3)_L \times SU(3)_R / SU(3)_V$, which happens to be isomorphic to the group $SU(3)$. The standard choice of Eq. (8.8) leaves the diagonal subgroup $SU(3)_V$ invariant, but this does not mean that with another choice of condensate there would be no $SU(3)$ invariance: the unbroken $SU(3)$ would simply be differently embedded in $SU(3)_L \times SU(3)_R$.[7]

The color-singlet operators $\mathrm{tr}(\Gamma q_{Li} \bar{q}_{Rj})$ and their parity conjugates, when acting on the vacuum, create mesons with flavor quantum numbers corresponding to the flavor indices i and j, and spin parity corresponding to the matrix Γ, which is some product of Dirac gamma matrices. Since we are interested here in scalar and pseudoscalar mesons, we choose $\Gamma = 1$. Introducing a complex scalar field H_{ij} with the same quantum numbers, we can write down an effective Lagrangian of the form

$$\mathcal{L}_{eff} = \text{tr}(\partial_\mu H^\dagger \partial_\mu H) + \mathcal{L}_{int}(H, H^\dagger) + \mathcal{L}_{other}(H, H^\dagger, \text{other hadrons}). \qquad (8.10)$$

This Lagrangian should obey the same symmmetries as the QCD Lagrangian, and in particular, it should be invariant under $SU(3)_L \times SU(3)_R$, with $H \rightarrow U_L H U_R^\dagger$. We can now decompose $H = R\Sigma$ with R Hermitian and positive, and Σ unitary. Symmetry breaking with the pattern $SU(3)_L \times SU(3)_R \rightarrow SU(3)_V$ implies that R picks up an expectation value, which again we can take to be proportional to the unit matrix, as in Eq. (8.8). Σ parametrizes the vacuum manifold, and thus describes the NGBs predicted by chiral-symmetry breaking. If all other hadrons are massive, and we are interested only in the physics of NGBs below the typical hadronic scale, we can integrate out all non-Goldstone fields, leaving us with a low-energy effective Lagrangian in terms of only Σ. This Lagrangian is local at scales below the hadronic scale. Because of the anomalous axial $U(1)$, the meson associated with this symmetry is also heavy, and we make take Σ in $SU(3)$ rather than $U(3)$, and parametrize it as

$$\Sigma = \exp\left(2i\phi/f\right) \rightarrow U_L \Sigma U_R^\dagger, \qquad (8.11)$$

with ϕ the field of Eq. (8.6). The field Σ inherits its transformation under the chiral group from the chiral transformation of H, as indicated above. We inserted a dimensionful parameter $f \propto \Lambda_{\text{QCD}}$ so that the field ϕ has the canonical mass dimension of a scalar field. Observe that Σ precisely parametrizes the vacuum manifold as in Eq. (8.9) for the choice $\Omega_{ij} = -\langle \bar{q}q \rangle \delta_{ij}/(2N_f)$. The unbroken group $SU(3)_V$ is linearly realized: Eq. (8.11) implies indeed that ϕ transforms as in Eq. (8.7) under $SU(3)_V$. This is not true for the full chiral group: the field ϕ transforms nonlinearly under all symmetries outside the subgroup $SU(3)_V$. For more discussion, see Section 8.5.

Our Lagrangian thus simplifies to

$$\mathcal{L}_{eff} = \frac{1}{8}f^2 \text{tr}(\partial_\mu \Sigma^\dagger \partial_\mu \Sigma) + \mathcal{L}_{int}(\Sigma, \Sigma^\dagger), \qquad (8.12)$$

where I have normalized the first term such that, upon expanding Σ in terms of ϕ, the meson fields of Eq. (8.6) have properly normalized kinetic terms. While this is progress, we clearly need more input to turn this into any practical use.

The EFT does not have to be renormalizable. In fact, one expects the presence of a cutoff of order 1 GeV, because we integrated out all other hadrons, with masses of order 1 GeV and above. But we do expect that pion[8] interactions obey all fundamental properties of quantum field theory: unitarity and causality, crossing symmetry, clustering and Lorentz invariance (Weinberg 1979), of course all to the precision with which the EFT reproduces the physics predicted by QCD. We will assume[9] that all these properties are satisfied if we take $\mathcal{L}_{eff}$ to be the most general local, Lorentz (or Euclidean) invariant function of the field Σ and its derivatives (Weinberg 1979).[10] Moreover, since the EFT should obey all symmetries of the underlying theory, it should be invariant under $SU(3)_L \times SU(3)_R$.

Our task is therefore to construct local invariants out of the field Σ. Because of the transformation rule for Σ, cf. Eq. (8.11), such invariants can only be formed by alternating Σ with $\Sigma^\dagger$, and taking a trace of such a product. One may also multiply such traces together to form new invariants. However, all such invariants collapse to a constant, because $\Sigma\Sigma^\dagger = \mathbf{1}$! The only way out is to allow for derivatives on the fields, as in the first term of Eq. (8.12). Because of Lorentz invariance, we need at least two such derivatives, and we can organize terms in our Lagrangian into groups of terms with the same number of derivatives. The unique term with just two derivatives is the one shown as the first term in Eq. (8.12).

This construction leads to several fundamental consequences. First, since we need at least two derivatives in all terms, pion interactions vanish when their momenta vanish, and thus they become weak for small momenta (and masses, as we will see below). That, in turn, implies that we may organize our EFT in terms of a derivative expansion: pion correlation functions with small momenta on the external legs can be expanded in terms of a small parameter p/Λ_χ, where p is a typical pion momentum, and Λ_χ is the typical hadronic scale, of order 1 GeV.[11]

These observations give us a very powerful method for constructing the chiral Lagrangian. If our aim is to work only to leading order in the derivative expansion, the chiral Lagrangian is simply given by the first term in Eq. (8.12):

$$\mathcal{L}^{(2)} = \frac{1}{8}f^2\text{tr}(\partial_\mu\Sigma^\dagger\partial_\mu\Sigma). \tag{8.13}$$

Note that this is an interacting theory, because Σ is non-linear in the pion fields ϕ. It therefore describes all pion physics to leading order in pion momenta. For example, writing ϕ as $\phi = \phi_a T_a$, with T_a the generators of $SU(3)$ obeying

$$\text{tr}(T_a T_b) = \delta_{ab}, \qquad [T_a, T_b] = \sqrt{2}i f_{abc} T_c, \tag{8.14}$$

one finds the leading-order prediction for the pion scattering amplitude from a tree-level calculation using Eq. (8.13):

$$\mathcal{A}(ab \to cd) = \frac{1}{3f^2}\left\{(t-u)f_{abe}f_{cde} + (s-u)f_{ace}f_{bde} + (s-t)f_{ade}f_{bce}\right\}, \tag{8.15}$$

in which s, t and u are the Mandelstam variables. This is our first prediction from ChPT.

Beyond leading order, we need to worry about two different (but, as we will see, intricately related) issues. We need to construct the most general term of order p^4, i.e. with four derivatives, and we also need to worry about unitarity (up to a given order). A nice way of building terms with higher derivatives starts from the observation that we can define an object with one derivative that transforms only under $SU(3)_L$, or, analogously, an object that transforms only under $SU(3)_R$:

$$L_\mu = \Sigma\partial_\mu\Sigma^\dagger = -\partial_\mu\Sigma\Sigma^\dagger, \tag{8.16}$$
$$R_\mu = \Sigma^\dagger\partial_\mu\Sigma = -\partial_\mu\Sigma^\dagger\Sigma = -\Sigma^\dagger L_\mu\Sigma.$$

The most general Lagrangian with four derivatives is (Gasser and Leutwyler 1985) (for the two-flavor case, see Gasser and Leutwyler (1984))

$$\mathcal{L}^{(4)} = -L_1 \left(\mathrm{tr}(L_\mu L_\mu)\right)^2 - L_2 \,\mathrm{tr}(L_\mu L_\nu)\,\mathrm{tr}(L_\mu L_\nu) - L_3 \,\mathrm{tr}(L_\mu L_\mu L_\nu L_\nu), \qquad (8.17)$$

in which $L_{1,2,3}$ are new (dimensionless) coupling constants. For general N_f, there is an additional term of the form $\mathrm{tr}(L_\mu L_\nu L_\mu L_\nu)$, but for $N_f = 3$ this can be written in terms of the three terms appearing in Eq. (8.17) (Gasser and Leutwyler 1985). Even with only three light quarks, this is not true in the partially quenched case, see Section 8.4.

We see that the new terms in $\mathcal{L}^{(4)}$ will contribute to pion scattering at tree level: each of these terms starts off with a $(\partial\phi/f)^4$ term when we expand Σ in terms of ϕ. Thus, indeed, such contributions will be of order $(p/f)^4$, i.e. of order $(p/f)^2$ relative to Eq. (8.15) – apparently f plays the role of the typical hadronic scale ~ 1 GeV, and $(p/f)^2$ is the expansion parameter of ChPT – see below.

In order to preserve unitarity, we should also calculate loop contributions, insofar as they contribute to order p^4. They arise when we consider the one-loop scattering diagrams with two four-pion vertices from $\mathcal{L}^{(2)}$. This leads to contributions of the form $(s/4\pi f^2)^2 \log(-s/\Lambda^2)$ (and more such terms also involving t and u, as dictated by crossing symmetry), where Λ is the cutoff needed in order to define the theory, and the factor $1/(4\pi)^2$ comes from the loop integral. The appearance of the logarithm is interesting for two reasons: it shows that, as mandated by unitarity, there is a two-particle cut in the pion scattering amplitude. Secondly, it is clear that a cutoff is needed in order to define theory.

Physical quantities, such as the scattering amplitude, cannot depend on this cutoff, which has only been introduced in order to define the effective theory. But indeed, since also polynomial terms from $\mathcal{L}^{(4)}$ appear at this order (order p^4), the cutoff dependence can be removed by taking $L_{1,2,3}$ to be dependent on Λ, such that the scattering amplitudes (and other physical quantities) are not. We see that in order to renormalize the theory defined by $\mathcal{L}^{(2)}$, we need to introduce $\mathcal{L}^{(4)}$. The new constants $L_{1,2,3}$ represent the effect of the underlying physics that has been integrated out in the EFT. This is the beginning of a general pattern, as we will argue in Section 8.2.3 below.

Before we discuss higher orders, let us consider order p^4 in some more detail. First, note that the one-loop contribution is in fact of order $p^2/(4\pi f)^2$ relative to Eq. (8.15). Any EFT involves an expansion in a ratio of scales, and in ChPT the higher scale in this ratio appears in the form $4\pi f$. The size of f can be estimated from pion scattering (using Eq. (8.15)); for another method, see Section 8.2.4. At the level of ChPT, the "low-energy" constants (LECs) f and L_i are free parameters, which can only be determined either by comparison with experiment, or by matching ChPT to a lattice QCD computation. We can say something about the "generically expected" values of the L_i. Since the L_i absorb the scale dependence, we expect that their magnitude changes by an amount of order $1/(4\pi)^2 \log(\Lambda/\Lambda')$ when we change the scale $\Lambda \to \Lambda'$. Since the cutoff Λ appears because we integrated out all heavier hadrons, a physically sensible choice for Λ is the typical hadronic scale of 1 GeV. With this interpretation, varying the cutoff within an order of magnitude is reasonable. This shifts the L_i by an amount of order $1/(4\pi)^2$, and gives us a sense of what one expects the values of these LECs to be.

8.2.3 Power counting

Now, let us consider the derivative expansion more systematically, and show that indeed the chiral theory is a proper EFT. What this means is that there exists a systematic power counting, and that to any order in the expansion we need only a finite number of coupling constants to define the theory.

Consider an amputated connected diagram with V_d vertices with d derivatives. Collectively denoting the external momenta by p, this diagram is of order $\sum_d dV_d - 2I + 4L$ in p, where I is the number of internal lines, and L is the number of loops. This follows from simply counting powers of p. Using, as usual, that the number of loops $L = I - \sum_d V_d + 1$, we can rewrite this as $\sum_d (d - 2)V_d + 2L + 2$. Our diagram is thus of order

$$f^2 p^2 \left(\frac{p^2}{f^2}\right)^N \left(\frac{1}{f}\right)^E, \tag{8.18}$$

with

$$N = \sum_d \frac{1}{2}(d - 2)V_d + L, \tag{8.19}$$

and E the number of external legs. The powers of f in this result follow from dimensional analysis. Ignoring the cutoff for now, this is the only other scale in the problem, if we express all other LECs as products of dimensionless constants times the appropriate power of f.

We see that all contributions to a certain amplitude of a fixed order in external momenta correspond to a fixed value of N. For instance, for $N = 0$, only tree-level diagrams, calculated from $\mathcal{L}^{(2)}$ contribute, because $N = 0$ requires that $L = 0$ and $V_d = 0$ for $d > 2$. For $N = 1$, one-loop diagrams coming from $\mathcal{L}^{(2)}$ combine with tree-level diagrams coming from $\mathcal{L}^{(4)}$, consistent with our discussion in the previous section. In general, if we calculate to some fixed order in N, we need the chiral Lagrangian only up to $\mathcal{L}^{(2N+2)}$. In other words, while our EFT is not renormalizable, it is nevertheless predictive if we work to a fixed order in the derivative expansion, since to that order only a finite number of LECs occur in the chiral Lagrangian (Weinberg 1979).[12]

Each loop provides a factor $1/(4\pi)^2$. If we assume (cf. Section 8.2.2) that this number also sets the natural size of all dimensionless LECs, that turns the momentum expansion into an expansion in powers of $p^2/(4\pi f)^2$. Then, so far we have ignored the fact that loops lead to divergences. That means that there is another scale, the cutoff, that can appear in our result. Putting everything together, we therefore amend our power-counting result: our diagram takes the schematic form (restoring a delta function for momentum conservation)

$$(2\pi)^4 \delta \left(\sum_{i=1}^{E} p_i \right) f^2 p^2 \left(\frac{p^2}{(4\pi f)^2} \right)^N \left(\frac{1}{f} \right)^E F(p^2/\Lambda^2), \tag{8.20}$$

with F a dimensionless function. Contributions to F can be of three types. First, any positive powers of p^2/Λ^2 can be ignored, as they correspond to higher-order contributions. Then, depending on the regulator, negative powers may occur (i.e. positive powers of the cutoff); those correspond to power divergences. Since all divergences have to be local (see for instance Collins (1984)), such divergences can be absorbed into lower-order LECs. Finally, there can be logarithmic divergences, which, as we have seen, can be renormalized by LECs in $\mathcal{L}^{(2N+2)}$ if they occur at N loops. In dimensional regularization, power divergences do not occur, making this regulator the most practical one for calculations in ChPT.[13]

The conclusion of this section is that a well-defined power-counting scheme exists, turning our EFT, i.e. ChPT, into a systematic and practically useful tool. This observation will play an important role when we start using ChPT in applications to lattice QCD.

8.2.4 Conserved currents

It follows from Noether's theorem that there are sixteen conserved currents, associated with the generators of the group $SU(3)_L \times SU(3)_R$. At the level of QCD, these currents are fixed by the Lagrangian and the symmetry group, and the same should thus be true at the level of the chiral Lagrangian. In both cases, these currents can be calculated following Noether's procedure. Equivalently, one may couple the QCD Lagrangian to external gauge fields, ℓ_μ and r_μ (taken to be Hermitian), which transform under local chiral transformations as

$$\ell_\mu \to U_L \ell_\mu U_L^\dagger - i\partial_\mu U_L U_L^\dagger, \tag{8.21}$$

$$r_\mu \to U_R r_\mu U_R^\dagger - i\partial_\mu U_R U_R^\dagger,$$

with the covariant derivatives in Eq. (8.1) turning into $D_\mu \to D_\mu - i\ell_\mu$ when acting on left-handed quarks, and $D_\mu \to D_\mu - ir_\mu$ when acting on right-handed quarks. This substitution makes the QCD Lagrangian (we still are considering the case that $M = 0$!) invariant under local chiral transformations, and thus the chiral Lagrangian should also be made invariant under *local* chiral transformations (in addition to Lorentz (or Euclidean) invariance and parity). This can be accomplished by replacing

$$\partial_\mu \Sigma \to D_\mu \Sigma = \partial_\mu \Sigma - i\ell_\mu \Sigma + i\Sigma r_\mu \to U_L D_\mu \Sigma U_R^\dagger \tag{8.22}$$

everywhere in the chiral Lagrangian, given to order p^4 by the sum of Eqs. (8.13) and (8.17). The second arrow in Eq. (8.22) indicates how the covariant derivative of Σ transforms; $U_{L,R}$ are now local transformations in $SU(3)_{L,R}$. In addition, more invariant terms can be constructed, if we also use the building blocks

$$L_{\mu\nu} = \partial_\mu \ell_\nu - \partial_\nu \ell_\mu - i[\ell_\mu, \ell_\nu], \tag{8.23}$$
$$R_{\mu\nu} = \partial_\mu r_\nu - \partial_\nu r_\mu - i[r_\mu, r_\nu].$$

At order p^2 there are no new terms (because these field strengths have two Lorentz indices), but at order p^4 several new terms arise:

$$iL_9 \operatorname{tr}\left(L_{\mu\nu} D_\mu \Sigma (D_\nu \Sigma)^\dagger + R_{\mu\nu}(D_\mu \Sigma)^\dagger D_\nu \Sigma\right) - L_{10}\operatorname{tr}(L_{\mu\nu}\Sigma R_{\mu\nu}\Sigma^\dagger) \tag{8.24}$$
$$-H_1 \operatorname{tr}(L_{\mu\nu}L_{\mu\nu} + R_{\mu\nu}R_{\mu\nu}).$$

With the chiral Lagrangian as constructed above, we have that

$$\log Z_{ChPT}(\ell_\mu, r_\mu) = \log Z_{QCD}(\ell_\mu, r_\mu) + \text{constant}, \tag{8.25}$$

up to the order at which we work in the chiral expansion. Since ℓ_μ and r_μ couple to the left- and right-handed Noether currents, (connected) correlation functions of these currents are generated by taking derivatives with respect to ℓ_μ and r_μ. In words, Eq. (8.25) states that these correlation functions, as calculated in ChPT, equal those of QCD, to some given order in the derivative expansion.

By taking one derivative of the chiral Lagrangian with respect to ℓ_μ or r_μ, and setting these sources equal to zero, we find the conserved left- and right-handed currents. Showing only the lowest-order explicitly:

$$J_\mu^L = \frac{i}{4}f^2 \partial_\mu \Sigma \Sigma^\dagger + \cdots = -\frac{1}{2}f \partial_\mu \phi + \ldots, \tag{8.26}$$
$$J_\mu^R = -\frac{i}{4}f^2 \Sigma^\dagger \partial_\mu \Sigma + \cdots = \frac{1}{2}f \partial_\mu \phi + \ldots.$$

Therefore, to lowest order in the derivative expansion, and writing

$$\phi = \phi_a T_a, \tag{8.27}$$
$$J_\mu^{L,R} = J_{a\mu}^{L,R} T_a,$$

with T^a the $SU(3)$ generators normalized, as before, through $\operatorname{tr}(T_a T_b) = \delta_{ab}$, we get for the pion to vacuum matrix element of the axial current

$$\langle 0|J_{a\mu}^R(x) - J_{a\mu}^L(x)|\phi_b(p)\rangle = -ip_\mu f \delta_{ab}\, e^{-ipx}, \tag{8.28}$$

and we conclude that f is equal to the pion decay constant, f_π, in the chiral limit.[14] We also find another prediction: to leading order in the chiral expansion $f_\pi = f_K = f_\eta$.
  We end this section with a number of remarks:

1. The observation that we can obtain conserved currents in the effective theory just as well from the Noether procedure or from the "source method" that I described above is correct, but both methods do not in general lead to the same current. At order p^2 the currents obtained using either of these methods are the same, but at order p^4 they differ by a term proportional to L_9. Clearly, if we do not introduce the sources ℓ_μ and r_μ at all, the L_9 term in Eq. (8.24) never appears,

and therefore the Noether current has no term proportional to L_9. But if we use the source method described above Eq. (8.26) and apply it to the terms in Eq. (8.24), we find additional terms

$$\Delta J_\mu^L = iL_9\, \partial_\nu[L_\mu, L_\nu], \tag{8.29}$$

$$\Delta J_\mu^R = iL_9\, \partial_\nu[R_\mu, R_\nu].$$

However, these extra terms are automatically conserved, $\partial_\mu \Delta J_\mu^{L,R} = 0$ identically. Therefore, the most general form of the conserved current is that provided by the source method.

2. When we gauge the group $SU(3)_L \times SU(3)_R$, as we did above, it is in fact anomalous. In order to reproduce the "non-abelian" anomaly in the chiral theory, we need to add the gauged Wess–Zumino–Witten term (Wess and Zumino 1971, Witten 1983). Since this part of the chiral Lagrangian will play no role in the rest of these lectures, we will not discuss this any further.

3. While (apart from the anomaly), the derivation given above looks very straightforward, in fact the freedom to perform field redefinitions and to add total-derivative terms is needed to complete the proof that the correct recipe is to just choose the chiral Lagrangian to be locally invariant (Leutwyler 1994a).

4. A new term proportional to H_1, not containing the pion fields Σ, shows up. This term can only contribute contact terms to correlation functions and is thus not physical. For instance, if one considers the two-point function $\langle J_\mu^L(x) J_\nu^L(y) \rangle$ at one loop, this has a logarithmic divergence, which can be absorbed into H_1. Such parameters in the chiral Lagrangian are sometimes referred to as "high-energy constants."

5. The coupling of the chiral Lagrangian to the sources ℓ_μ and r_μ tells us how to couple pions to the photon and to (virtual) W and Z mesons. For instance, one obtains the photon coupling by taking

$$\ell_\mu = r_\mu = -\frac{1}{3} e a_\mu \begin{pmatrix} 2 & 0 & 0 \\ 0 & -1 & 0 \\ 0 & 0 & -1 \end{pmatrix}, \tag{8.30}$$

with a_μ the photon field and e the charge of the electron.

Generally, electromagnetic effects in hadronic quantities are of approximately the same size as the isospin breaking coming from the fact that in nature $m_u \neq m_d$. This implies that ultimately electromagnetic effects will have to be taken into account in lattice QCD computations. For explorations in this direction, see Blum *et al.* (2007) and Basak *et al.* (2008).

8.2.5 Quark masses

It is time to remember that in the real world the masses of the up, down and strange quark do not vanish. Generalizing Eq. (8.5), the mass terms in Eq. (8.1) can be replaced by source terms, with Hermitian scalar and pseudoscalar sources $s(x)$ and $p(x)$:

$$\bar{q}_L(s+ip)q_R + \bar{q}_R(s-ip)q_L,\tag{8.31}$$

with $SU(3)_L \times SU(3)_R$ transformation rules

$$s+ip \to U_L(s+ip)U_R^\dagger, \qquad s-ip \to U_R(s-ip)U_L^\dagger.\tag{8.32}$$

We recover the quark-mass terms by setting $s=M$ with M as in Eq. (8.3) and $p=0$.

This gives us a new building block for constructing terms in the chiral Lagrangian. To lowest order in s and p, using parity symmetry, there is a unique operator that can be added to $\mathcal{L}^{(2)}$, which now becomes

$$\mathcal{L}^{(2)} = \frac{1}{8}f^2 \operatorname{tr}\left((D_\mu \Sigma)^\dagger D_\mu \Sigma\right) - \frac{1}{8}f^2 \operatorname{tr}\left(\chi^\dagger \Sigma + \Sigma^\dagger \chi\right),\tag{8.33}$$

$$\chi \equiv 2B_0(s+ip),$$

with B_0 a new LEC. Setting $\chi = \chi^\dagger = 2B_0 M$, and expanding Σ to quadratic order in ϕ, we can read off the (tree-level) pion masses in terms of the quark masses:

$$m_{\pi+}^2 = B_0(m_u + m_d),\tag{8.34}$$

$$m_{K+}^2 = B_0(m_u + m_s),$$

$$m_{K^0}^2 = B_0(m_d + m_s),$$

$$m_{\pi^0}^2 = B_0\left(m_u + m_d + O\left(\frac{(m_u - m_d)^2}{m_s}\right)\right),$$

$$m_\eta^2 = \frac{1}{3}B_0\left(m_u + m_d + 4m_s + O\left(\frac{(m_u - m_d)^2}{m_s}\right)\right).$$

This gives our third prediction from chiral symmetry: five meson masses are expressed in terms of three parameters. If we ignore isospin breaking (i.e. set $m_\ell \equiv m_u = m_d$), we find that $m_{\pi^0} = m_{\pi+}$ and $m_\eta^2 = (2(m_{K+}^2 + m_{K^0}^2) - m_{\pi+}^2)/3$, relations that agree with experiment at the few per cent level. The latter relation is the well-known Gell-Mann–Okubo relation.

Since the masses on the left-hand side are physical quantities, also the expressions on the right-hand side should be physical; in particular, they should not depend on the renormalization scale of the underlying theory. Indeed, by taking a derivative with respect to s and then setting all sources equal to zero, we find that in the chiral limit

$$\langle \bar{u}u \rangle = \langle \bar{d}d \rangle = \langle \bar{s}s \rangle = -f^2 B_0,\tag{8.35}$$

and we then use that $m_u \langle \bar{u}u \rangle$, etc, are scale independent. The quark masses appear in the chiral Lagrangian only in the scale-independent combinations

$$\chi_u \equiv 2B_0 m_u, \qquad \chi_d \equiv 2B_0 m_d, \qquad \chi_s \equiv 2B_0 m_s.\tag{8.36}$$

From this tree-level exercise, we see that for on-shell momenta, one power of the quark mass should be counted as order p^2, because for an on-shell pion we have that

$p^2 = -m_{\pi^+}^2$, etc. With this power counting for the quark masses, the whole discussion of Section 8.2.3 applies, with the proviso that p^2 in formulas like Eq. (8.20) can now also stand for any of the squared meson masses. With $f = 130$ MeV, we note that the expansion parameters of the chiral expansion are

$$\frac{m_\pi^2}{(4\pi f)^2} \approx 0.007, \qquad \frac{m_K^2}{(4\pi f)^2} \approx 0.09, \tag{8.37}$$

which gives one hope that even for kaons the chiral expansion may be well behaved. We will return to this issue in Section 8.5. As an aside, we also see why the heavy-quark masses (charm, etc.) cannot be accounted for in ChPT: The corresponding "expansion" parameters would not be small. In fact, the situation is "upside down": one may instead consider expansions in $\Lambda_{\mathrm{QCD}}/m_{\mathrm{heavy}}$, leading to heavy-quark EFT.[15]

In order to find the meson masses, we expanded the nonlinear Σ around $\mathbf{1}$. Indeed, when all masses in Eq. (8.3) are positive, $\Sigma = \mathbf{1}$ is the minimum of the potential, which is given by the second term in Eq. (8.33). When we allow two of the masses to be negative, the situation is essentially the same. For instance, if $m_{u,d} < 0$ while $m_s > 0$, the vacuum is

$$\Sigma_{\mathrm{vac}} = \mathrm{diag}(-1, -1, 1), \tag{8.38}$$

which is equivalent to the trivial vacuum by an $SU(3)$ rotation. When, however, an odd number of quark masses are negative, there is no $SU(3)$ rotation relating the vacuum to the trivial vacuum, and the theory can be in a different phase, in which CP is broken because of the appearance of a θ-term with $\theta = \pi$ (Dashen 1971); for a nice analysis using the ChPT framework, see Witten (1980). Since in nature all quark masses are positive (or equivalently, $\theta = 0$), we will only consider the case of positive quark masses throughout this Section.[16] However, we will see in Section 8.3.3 that a non-trivial vacuum structure is nevertheless possible for some discretizations of QCD at non-zero lattice spacing.

Of course, with the scalar source χ as a new building block, more terms can appear in $\mathcal{L}^{(4)}$ as well. Gauging Eq. (8.17) and including Eq. (8.24), the most general form becomes[17]

$$\mathcal{L}^{(4)} = -L_1 \left(\mathrm{tr} \left(D_\mu \Sigma (D_\mu \Sigma)^\dagger \right) \right)^2 - L_2 \, \mathrm{tr} \left(D_\mu \Sigma (D_\nu \Sigma)^\dagger \right) \, \mathrm{tr} \left(D_\mu \Sigma (D_\nu \Sigma)^\dagger \right) \tag{8.39}$$

$$- L_3 \, \mathrm{tr} \left(D_\mu \Sigma (D_\mu \Sigma)^\dagger D_\nu \Sigma (D_\nu \Sigma)^\dagger \right)$$

$$+ L_4 \, \mathrm{tr} \left(D_\mu \Sigma (D_\mu \Sigma)^\dagger \right) \, \mathrm{tr}(\chi^\dagger \Sigma + \Sigma^\dagger \chi) + L_5 \, \mathrm{tr} \left(D_\mu \Sigma (D_\mu \Sigma)^\dagger (\chi^\dagger \Sigma + \Sigma^\dagger \chi) \right)$$

$$- L_6 \left(\mathrm{tr}(\chi^\dagger \Sigma + \Sigma^\dagger \chi) \right)^2 - L_7 \left(\mathrm{tr}(\chi^\dagger \Sigma - \Sigma^\dagger \chi) \right)^2 - L_8 \, \mathrm{tr}(\chi^\dagger \Sigma \chi^\dagger \Sigma + \Sigma^\dagger \chi \Sigma^\dagger \chi)$$

$$+ i L_9 \, \mathrm{tr} \left(L_{\mu\nu} D_\mu \Sigma (D_\nu \Sigma)^\dagger + R_{\mu\nu} (D_\mu \Sigma)^\dagger D_\nu \Sigma \right) - L_{10} \, \mathrm{tr}(L_{\mu\nu} \Sigma R_{\mu\nu} \Sigma^\dagger)$$

$$- H_1 \mathrm{tr}(L_{\mu\nu} L_{\mu\nu} + R_{\mu\nu} R_{\mu\nu}) - H_2 \, \mathrm{tr}(\chi^\dagger \chi).$$

The full Lagrangian $\mathcal{L}^{(2)} + \mathcal{L}^{(4)}$ of Eqs. (8.33) and (8.39) is invariant under the local group $SU(3)_L \times SU(3)_R$. We note that five new LECs and one new "high-energy" constant have been added. The term proportional to H_2 contributes to the mass dependence of the chiral condensate, and is needed for instance in order to match lattice QCD, where the condensate includes a term proportional to m/a^2 (with m the relevant quark mass).

As an application, we quote some results of the quark-mass dependence of various physical quantities to order p^4 (Gasser and Leutwyler 1985), in the isospin limit $m_u = m_d$:

$$f_\pi = f\left\{1 + \frac{8L_5}{f^2}\hat{m}_\ell + \frac{8L_4}{f^2}(2\hat{m}_\ell + \hat{m}_s) - 2L(m_\pi^2) - L(m_K^2)\right\}, \tag{8.40}$$

$$\frac{f_K}{f_\pi} = 1 + \frac{4L_5}{f^2}(\hat{m}_s - \hat{m}_\ell) + \frac{5}{4}L(m_\pi^2) - \frac{1}{2}L(m_K^2) - \frac{3}{4}L(m_\eta^2),$$

$$\frac{f_\eta}{f_\pi} = \left(\frac{f_K}{f_\pi}\right)^{4/3}\left\{1 + \frac{1}{48\pi^2 f^2}\left(3m_\eta^2 \log\frac{m_\eta^2}{m_K^2} + m_\pi^2 \log\frac{m_\pi^2}{m_K^2}\right)\right\},$$

$$\frac{m_\pi^2}{2\hat{m}_\ell} = 1 + \frac{8(2L_8 - L_5)}{f^2}2\hat{m}_\ell + \frac{16(2L_6 - L_4)}{f^2}(2\hat{m}_\ell + \hat{m}_s) + L(m_\pi^2) - \frac{1}{3}L(m_\eta^2),$$

$$\frac{m_K^2}{\hat{m}_\ell + \hat{m}_s} = 1 + \frac{8(2L_8 - L_5)}{f^2}(\hat{m}_\ell + \hat{m}_s) + \frac{16(2L_6 - L_4)}{f^2}(2\hat{m}_\ell + \hat{m}_s) + \frac{2}{3}L(m_\eta^2),$$

in which

$$\hat{m}_\ell = 2B_0 m_u = 2B_0 m_d, \qquad \hat{m}_s = 2B_0 m_s, \tag{8.41}$$

$$L(m^2) = \frac{m^2}{(4\pi f)^2}\log\left(\frac{m^2}{\Lambda^2}\right),$$

and where we can use the tree-level expressions (8.34) inside the logarithms. Because all meson interactions that follow from the chiral Lagrangian are even in the meson fields,[18] and the calculation of both masses and decay constants involves two-point functions, the only diagrams that appear at one loop are tadpole diagrams.

Again, ChPT makes a prediction: if we use $f_K/f_\pi = 1.2$ as input, one finds that (to this order in ChPT) $f_\eta/f_\pi = 1.3$. With these values for f_K/f_π and f_η/f_π we see that ChPT to order p^4 works reasonably well for quantities involving the strange quark, with order-p^4 corrections of about 20–30%.[19] For a discussion of meson masses and decay constants to order p^6 in ChPT, see Amoros *et al.* (2000). For a recent review of the phenomenological status of ChPT, and many references, see Ecker (2008).

An important observation is that all LECs, f, B_0 and the L_is, are independent of the quark mass; they only depend on the number of flavors, N_f.[20] All quark-mass dependence in Eq. (8.40) is explicit. It follows from this that the values of the LECs can, in principle, be obtained from lattice QCD with unphysical values of the quark masses (as long as they are small enough for ChPT to be valid – an important restriction). Thus, if one finds a set of quantities through which all LECs can be determined from the lattice, one can then use these values to calculate other quantities that are less easily accessible on the lattice. A simple example of this is that one can

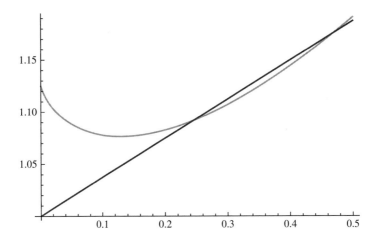

Fig. 8.1 The functions $1.125 + L(m_\pi^2)$ (curve) and $1 + m_\pi^2/(4\pi f)^2$ (straight line), as a function of m_π^2 in GeV^2. I took $\Lambda = m_\rho = 770$ MeV and $f = 130$ MeV.

extrapolate physical quantities to the physical values of the light quark masses, from lattice computations at larger values of m_ℓ in the isospin limit.[21] Unfortunately, it is not possible to extract all the L_i from two-point functions, since the constants $L_{1,2,3}$ do not show up in masses and decay constants.

We conclude this section with an important lesson about the use of ChPT for fitting the quark-mass dependence of a hadronic quantity on the lattice. As we see in Eq. (8.40), non-analytic terms show up, here in the form of so-called "chiral logarithms." Figure 8.1 shows how important it can be to include such non-analytic terms in chiral fits, by comparing the functions $1.125 + L(m_\pi^2)$ with $1 + m_\pi^2/(4\pi f)^2$. The logarithm has a dramatic effect: it looks nothing like the linear curve in the region around the physical pion mass, $m_\pi^2 = 0.02$ GeV^2. If one would only have lattice data points in the region $m_\pi^2 \gtrsim 0.2$ GeV^2 with error bars that would more or less overlap with both curves, and one would perform a linear fit,[22] we see that this might lead to errors of order 10% in the value at the physical pion mass. In order to confirm the existence of the chiral logarithm in the data, clearly data points in the region of the curvature of the logarithm are needed. In the cartoon example of Fig. 8.1 this means that we need data points in the region down to $m_\pi \sim 200$ MeV. In addition, data points with very good statistics help.

For a realistic example, see Fig. 8.2, which shows m_π^2/m_ℓ as a function of the light-quark mass, m_ℓ, for fixed (physical) strange mass, from Aoki *et al.* (2008b) (to which I refer for details). If one were to only consider the large m_π points (the black points), one would not be able to reliably fit the chiral logarithm in Eq. (8.40), and thus also not the linear combination of LECs that accompany the logarithm. Only with the gray points, which were obtained for small values of m_π, or equivalently m_ℓ, can one hope to perform sensible chiral fits. For a discussion of such fits, see Aoki *et al.* (2008b), only note that both $O(a^2)$ effects and finite volume can in principle

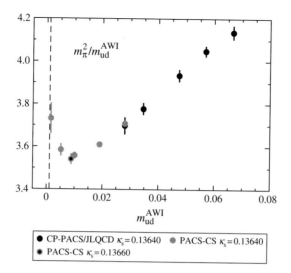

Fig. 8.2 Comparison of the PACS-CS (gray) and the CP-PACS/JLQCD (black) results for m_π^2/m_{ud}^{AWI} as a function of the light quark mass m_{ud}^{AWI} (the so-called "axial Ward identity" definition of the quark mass). The vertical line denotes the physical point. From Aoki *et al.* (2008b).

modify the chiral logarithms. These issues can be systematically studied with the help of ChPT, as we will see in Sections 8.3 and 8.6.

8.3 ChPT at non-zero lattice spacing

If we compute hadronic quantities on the lattice, the values we obtain will differ from their continuum values by scaling violations – terms of order a^n with some positive integer power n (possibly modified by logarithms), where a is the lattice spacing. For example, for the mass of some hadron evaluated on the lattice at some non-zero a we expect that

$$M_{\text{lattice}} = M_{\text{continuum}}(m_{\text{quark}}) + c_1 a \Lambda_{\text{QCD}}^2 + O(a^2, a m_{\text{quark}}), \qquad (8.42)$$

with c_1 at most logarithmically dependent on a.[23] In Eq. (8.42) I indicated schematically that the hadron mass also depends on the quark masses in the theory.

Evidently, at non-zero lattice spacing, there is a "new" infrared scale $a\Lambda_{\text{QCD}}^2$ that emerges from combining the physical infrared scale Λ_{QCD} with the ultraviolet scale a. Physical quantities thus depend not only on the quark masses, but also on this new scale, and one expects that a combined expansion in terms of m_{quark} and $a\Lambda_{\text{QCD}}^2$ should be possible when these parameters are both small enough. The goal of this section is to see how ChPT can be extended to provide this combined expansion.

Before we begin doing that, there are two important observations. First, ChPT is constructed by making use of the chiral symmetry of the underlying theory – it is nothing more or less than a very efficient way of implementing the Ward identities for chiral symmetry. This implies that if we use a chirally invariant regulator on the lattice, such as domain-wall fermions[24] or overlap fermions, there will be almost no change in the form of the chiral Lagrangian. Almost no change, because for any lattice regulator continuous rotational invariance is broken, so one expects that new operators will appear in the chiral Lagrangian that break continuous rotational invariance, but respect the lattice subgroup (usually the group of hypercubic rotations). Quite generally, in the construction of the chiral Lagrangian for Nambu–Goldstone bosons, the only quantity transforming as a four-vector is the derivative, so such terms will have to have at least four derivatives. Combined with the fact that they need to carry a positive power of the lattice spacing, they are usually of quite high order in the chiral expansion, as we will see. The only other difference is that the continuum LECs themselves are afflicted by scaling violations, which can in principle be removed by an extrapolation to $a = 0$.

It is only for regulators that break some of the continuum chiral symmetries that the form of the chiral Lagrangian will be different. The two most important examples are Wilson-like fermions (including also clover fermions and twisted-mass QCD), for which vector-like flavor symmetries are preserved but chiral symmetries are broken, and staggered fermions, for which flavor symmetries are broken, and only one chiral symmetry is preserved. Here, I will primarily discuss Wilson fermions, since they are conceptually the simplest. Section 8.3.6 is devoted to a brief overview of ChPT for QCD with staggered quarks.

A second point is that we will be dealing with an expansion in two small parameters, m_{quark} and $a\Lambda_{\mathrm{QCD}}^2$, and we thus have to define a power counting – we have to compare their relative size in order to know to what relative order in each parameter we have to expand. Typical quark masses in lattice computations range between 10 and 100 MeV, and typical lattice spacings vary between $a^{-1} \sim 2 - 3$ GeV. With $\Lambda_{\mathrm{QCD}} \sim 300$ MeV, this means that m_{quark} and $a\Lambda_{\mathrm{QCD}}^2$ are of the same order, and this is what I will assume to get the discussion started, until Section 8.3.3.

QCD with Wilson fermions has no chiral symmetry, and it thus seems that our guiding principle for constructing the chiral effective theory is completely lost: we should allow for all kinds of operators in the chiral Lagrangian that are only invariant under $SU(3)_V$, and the only thing we know is that any such operator that is forbidden in the continuum theory is multiplied by some positive power of the lattice spacing. It thus seems that, while an EFT at low energies should exist for small enough lattice spacing, it will not be of much practical use. It turns out, however, that this is not true.

The solution to this apparent roadblock consists of first considering a "low-energy" continuum EFT of quarks and gluons, where "low energy" means that we want to consider correlation functions of quarks and gluons at momenta $\Lambda_{\mathrm{QCD}} \ll p \ll 1/a$. Since at such momenta the degrees of freedom in this EFT are the same as in the lattice theory, no powers of Λ_{QCD} can appear in this EFT, and any operator of order a^n $(n \geq 0)$ thus has to be multiplied by an operator of mass dimension $4 + n$, so that the full Lagrangian always has dimension four. This means that, for small values of n the operators that can appear in this EFT are very constrained, and, as we will see, this allows us to also restrict the form of the operators for small values of n in the chiral theory. The idea of considering the quark–gluon continuum EFT for the lattice theory is due to Symanzik (Symanzik 1983a), and we will refer to this EFT as the Symanzik effective theory (SET). The insight that this can be used to develop ChPT including scaling violations systematically is due to Sharpe and Singleton (Sharpe and Singleton 1998).

8.3.1 Symanzik effective theory

Let us begin with a brief review of the construction of the SET for lattice QCD with Wilson fermions.[25] The Symanzik expansion is an expansion in powers of a,

$$\mathcal{L}_{SET} = \mathcal{L}_S^{(4)} + a\mathcal{L}_S^{(5)} + a^2\mathcal{L}_S^{(6)} + \dots . \tag{8.43}$$

The first term, $\mathcal{L}_S^{(4)}$, is the continuum theory,

$$\mathcal{L}_S^{(4)} = \mathcal{L}_{\text{gluons}} + \bar{q}\slashed{D}q + \bar{q}Mq, \tag{8.44}$$

with M the quark-mass matrix given in Eq. (8.3).[26] In fact, because of the lack of chiral symmetry of Wilson fermions, the first term in the Symanzik expansion is a term of dimension three of the form $(c/a)\bar{q}q$ with c a numerical constant. The quark masses in M thus have to be defined as the difference between the lattice-quark masses and the power divergence c/a:

$$m_i = Z_S^{-1}(m_{0,i} - c)/a, \quad i = u, d, s, \tag{8.45}$$

where $m_{0,i}$ is the bare lattice mass for flavor i in lattice units, and Z_S is the multiplicative renormalization factor needed to relate lattice masses to the continuum regulator that we use to define the SET. It is conceptually easiest to think of the continuum regulator as dimensional regularization (in some minimal subtraction scheme). This brings us to an important point: the fact that, for momenta $\Lambda_{\text{QCD}} \ll p \ll 1/a$, an effective theory of the form (8.43) can be used to represent the lattice theory was only argued to exist using perturbation theory.[27] It is an assumption, essentially based on locality, that the SET is also valid non-perturbatively: the claim is that discretization effects in correlation functions with all momenta $\Lambda_{\text{QCD}} \ll p \ll 1/a$ are given by insertions of local operators, to any finite order in an expansion in ap. We note that this is simply the basic assumption underlying the construction of all EFTs, if we replace ap by the ratio of scales that provides the small parameter for the EFT.

Of course, in order to "think" about the SET outside perturbation theory, one would have to use a non-perturbative regulator – such as the lattice! One can imagine using a lattice regulator to define the SET with a lattice spacing a' much smaller than a, so that $p \ll 1/a \ll 1/a'$, and since $a' \ll a$ one does not have to include powers of a' in the expansion (8.43). This ignores the possibility of power divergences in the smaller lattice spacing of the form $1/a'$ or $1/a'^2$, which one might have to subtract in order to define the SET. Since I will only use the form of the SET, without ever doing any calculations with it, I will sidestep this issue, and assume that we can work with the expansion (8.43).[28]

Next, let us consider $\mathcal{L}_S^{(5)}$. This consists of all dimension-five operators consistent with the exact symmetries of the underlying lattice theory. For Wilson fermions

$$\mathcal{L}_S^{(5)} = b_1\,\bar{q}is_{\mu\nu}G_{\mu\nu}q + b_2\,\bar{q}D_\mu D_\mu q + b_3\,\bar{q}M\!\!\!\!/\,\,q + b_4\,\bar{q}M^2 q + b_5\,\text{tr}(M)\mathcal{L}_{\text{gluons}},$$

$$(8.46)$$

in which the b_i are real dimensionless coefficients, independent of quark masses, that can be calculated in perturbation theory, or, non-perturbatively, on the lattice. $G_{\mu\nu}$ is the gluon field-strength tensor. Note that the b_i are not really constant: they depend on logarithms of the ratio of the scales used to define the lattice theory and the SET. If dimensional regularization is used for the SET, $b_i = b_i(g^2(\mu), \log a\mu)$. We will ignore this additional logarithmic dependence on a, as it should be much milder than the explicit power dependence on a in Eq. (8.43).

Since the b_3 and b_5 terms are of the form quark mass times terms that already occur in $\mathcal{L}_S^{(4)}$, they will translate into $O(p^6)$ terms in ChPT. The counting here works as follows. Terms in $\mathcal{L}^{(4)}$ translate into $O(p^2)$ terms in ChPT, cf. Eq. (8.13). The explicit factors of quark mass and a add two powers of p^2, since we chose a power-counting scheme in which $p^2 \sim m_{\text{quark}}\Lambda_{\text{QCD}} \sim a\Lambda_{\text{QCD}}^3$. Since here we will work to order p^4, we may drop these terms from consideration. The same is true for the b_4 term, which is of order $a(m_{\text{quark}})^2$.

At the next order in the Symanzik expansion, the number of terms of dimension six, consistent with all exact lattice symmetries, proliferates, and I will not list them all (Sheikholeslami and Wohlert 1985). There are purely gluonic operators, that do not break chiral symmetry, but they may break Lorentz (or, rather, Euclidean) invariance, such as

$$\sum_{\mu,\kappa}\text{tr}(D_\mu G_{\mu\kappa}D_\mu G_{\mu\kappa}),$$

$$(8.47)$$

where for clarity I showed index sums explicitly. While this term would thus lead to new terms in the chiral Lagrangian, it is easy to see that these are higher order than p^4. As we already noted before, in order to construct a chiral operator that breaks Euclidean invariance, one needs at least four derivatives. With the additional two powers of a that multiply terms in $\mathcal{L}_S^{(6)}$, this makes such terms $O(p^8)$ in our power counting.[29] Note that this observation hinges on the fact that in pion ChPT the derivative is the only object with a Lorentz index. For instance, in baryon ChPT this is not true (there is also the four-velocity of the heavy baryon, which is order one in power counting), and the effects of the breaking of continuous rotational symmetry already shows up at order p^4 (Tiburzi 2005).

Operators involving quarks are of two types: bilinear and quartic in the quark fields. In order to generate new terms in the chiral Lagrangian, they have to break one of the symmetries of the continuum theory. There are bilinears that break Euclidean invariance, which, as we argued above, we do not have to consider at order p^4. Dimension-six operators that break chiral symmetry (and are not of the form $\text{tr}(M)\mathcal{L}_S^{(5)}$) are, in the basis of Bär *et al.* (2004),

$$O_5^{(6)} = (\bar{q}q)^2, O_{10}^{(6)} = (\bar{q}t_r q)^2, \tag{8.48}$$
$$O_6^{(6)} = (\bar{q}\gamma_5 q)^2, O_{11}^{(6)} = (\bar{q}t_r \gamma_5 q)^2,$$
$$O_9^{(6)} = (\bar{q}s_{\mu\nu}q)^2, O_{14}^{(6)} = (\bar{q}t_r s_{\mu\nu}q)^2,$$

where t_r are the $SU(3)_{\text{color}}$ generators.

8.3.2 Transition to the chiral theory

Now we are ready to make the transition to the chiral theory. First, consider terms of order a, corresponding to $\mathcal{L}_S^{(5)}$ in Eq. (8.43). Using a trick similar to the one we used translating quark-mass terms to ChPT, we introduce a spurion field A, and write the first term in Eq. (8.46) as

$$ab_1 \, \bar{q}is_{\mu\nu}G_{\mu\nu}q \to b_1 \left(\bar{q}_L is_{\mu\nu}G_{\mu\nu}Aq_R + \bar{q}_R is_{\mu\nu}G_{\mu\nu}A^\dagger q_L \right), \tag{8.49}$$

which is invariant under $SU(3)_L \times SU(3)_R$ if we let A transform as (cf. Eqs. (8.5) and (8.32))

$$A \to U_L A U_R^\dagger. \tag{8.50}$$

Note that we recover the term on the left of Eq. (8.49) by setting $A = a\mathbf{1}$.

We see that the b_1 term, from the point of view of chiral symmetry, has exactly the same structure as the quark-mass term. At lowest order in ChPT, we thus find a new term at leading order in the chiral Lagrangian of the form

$$-\frac{1}{4}f^2 W_0 \, \text{tr}(A^\dagger \Sigma + \Sigma^\dagger A), \tag{8.51}$$

in which W_0 is a new LEC analogous to B_0. The b_2 term in Eq. (8.46) has exactly the same chiral structure, and thus does not lead to any new operator in ChPT. This is an example of the fact that there is in general not a one-to-one correspondence between operators in the SET and in ChPT.

Once we set $A = a\mathbf{1}$, the new term can actually be completely absorbed into the mass term in Eq. (8.33), if we shift the quark-mass matrix by

$$M \to M' \equiv M + \frac{W_0}{B_0}a\mathbf{1}, \tag{8.52}$$

$$\text{or} \quad m_i \to m_i' = m_i + \frac{W_0}{B_0}a, \quad i = u, d, s.$$

If we determine, non-perturbatively, the lattice-quark masses by requiring the pion mass to vanish,[30] then this determines M', rather than M – in other words, the shift of order a in Eq. (8.52) is automatically taken into account in the determination of the critical quark mass for Wilson fermions. Note that this does not remove all $O(a)$ effects from the theory when we extend it to include other hadrons,[31] or operators other than the action, cf. (Setion 8.3.5).

Next, we consider terms of order $ap^2 \sim am_{\text{quark}} \sim a^2$, all of order p^4 in our power counting (appropriate powers of Λ_{QCD} are implicit). These come from two sources: higher-order terms in the spurion A, as well as terms arising from $\mathcal{L}_S^{(6)}$. However, both of these lead to the same new $O(p^4)$ terms in the chiral Lagrangian. The point is that the same spurion field can be used to also make the operators in Eq. (8.48) invariant under $SU(3)_L \times SU(3)_R$:

$$a^2(\bar{q}q)^2 \to (\bar{q}_L A q_R + \bar{q}_R A^\dagger q_L)^2, \tag{8.53}$$

and similar for the other operators in Eq. (8.48). All these operators can thus be made invariant using the "composite" spurions

$$A \otimes A \to U_L A U_R^\dagger \otimes U_L A U_R^\dagger, \tag{8.54}$$

$$A \otimes A^\dagger \to U_L A U_R^\dagger \otimes U_R A^\dagger U_L^\dagger,$$

$$A^\dagger \otimes A \to U_R A^\dagger U_L^\dagger \otimes U_L A U_R^\dagger,$$

$$A^\dagger \otimes A^\dagger \to U_R A^\dagger U_L^\dagger \otimes U_R A^\dagger U_L^\dagger.$$

Using these spurions, we find the following new terms in the chiral Lagrangian

$$\text{tr}(A\Sigma^\dagger)\,\text{tr}(A\Sigma^\dagger) + \text{h.c.} \to a^2\left(\text{tr}(\Sigma^\dagger)\right)^2 + \text{h.c.}, \tag{8.55}$$

$$\text{tr}(A\Sigma^\dagger A\Sigma^\dagger) + \text{h.c.} \to a^2\,\text{tr}(\Sigma^\dagger \Sigma^\dagger) + \text{h.c.},$$

$$\text{tr}(A\Sigma^\dagger)\,\text{tr}(A^\dagger\Sigma) \to a^2\,\text{tr}(\Sigma^\dagger)\,\text{tr}(\Sigma).$$

Setting sources ℓ_μ and r_μ equal to zero,[32] the complete set of additional terms in the chiral Lagrangian at order p^4, including new LECs multiplying each of the new operators is (Bär *et al.* 2004)

$$\Delta\mathcal{L}^{(4)} = \hat{a}W_4\,\text{tr}(\partial_\mu\Sigma\partial_\mu\Sigma^\dagger)\,\text{tr}(\Sigma + \Sigma^\dagger) + \hat{a}W_5\,\text{tr}\left(\partial_\mu\Sigma\partial_\mu\Sigma^\dagger(\Sigma + \Sigma^\dagger)\right) \tag{8.56}$$

$$-\hat{a}W_6\,\text{tr}(\hat{M}\Sigma^\dagger + \Sigma\hat{M})\,\text{tr}(\Sigma + \Sigma^\dagger) - \hat{a}W_7\,\text{tr}(\hat{M}\Sigma^\dagger - \Sigma\hat{M})\,\text{tr}(\Sigma - \Sigma^\dagger)$$

$$-\hat{a}W_8\,\text{tr}(\hat{M}\Sigma^\dagger\Sigma^\dagger + \Sigma\Sigma\hat{M})$$

$$-\hat{a}^2 W_6'\left(\text{tr}(\Sigma + \Sigma^\dagger)\right)^2 - \hat{a}^2 W_7'\left(\text{tr}(\Sigma - \Sigma^\dagger)\right)^2 - \hat{a}^2 W_8'\,\text{tr}(\Sigma\Sigma + \Sigma^\dagger\Sigma^\dagger),$$

in which

$$\hat{M} \equiv 2B_0 M, \qquad \hat{a} \equiv 2W_0 a. \tag{8.57}$$

If one considers (non-perturbatively) $O(a)$ improved QCD with Wilson fermions, one can use the same chiral Lagrangian. It simply means that $W_0 = 0$ in Eq. (8.51). Note that this does *not* mean that we should set all terms of order a^2 equal to zero, despite our convention to include a factor of W_0 in the definition of $\hat{a}$! The reason is the following. If the underlying theory is $O(a)$ improved, all terms of order a in the SET (cf. Eq. (8.46)) vanish, and there is no need to introduce the spurion A at all at this order. However, this does not mean that there are no $O(a^2)$ terms in the SET, so we do need to introduce $O(a^2)$ spurions as in Eq. (8.53). It just so happens that we can "recycle" the same spurion A at order a^2 by taking tensor products as in Eq. (8.54). If underlying lattice theory is not improved, there are also contributions at order a^2 proportional to the $O(a)$ coefficients of Eq. (8.46), in addition to those proportional to appearing in the SET at order a^2. At the level of $\Delta\mathcal{L}^{(4)}$ both types of contribution at order a^2 take the same form, given in Eq. (8.56).

As an application, let us consider the modifications to the $O(p^4)$ result for the pion mass, cf. Eq. (8.40) (Bär *et al.* 2004). First, the overall factor $\hat{m}_\ell$ gets replaced by $\hat{m}_\ell + \hat{a} = 2B_0 m'_\ell$, with m'_ℓ the shifted light-quark masses of Eq. (8.52), and likewise, the logarithms get replaced by

$$L(m_\pi^2) \rightarrow L(2B_0 m_\ell + 2W_0 a) = L(2B_0 m'_\ell), \qquad (8.58)$$

etc. Both of these modifications are a consequence of the fact that the $O(a)$ term (8.51) can be absorbed into the shifted quark masses of Eq. (8.52). In this case, as we already saw above, the quark masses that one determines in a numerical computation are the shifted m' masses, so the $O(a)$ shift in Eq. (8.58) is not observable.

Secondly, the terms proportional to the L_is in Eq. (8.40) also change. Not surprisingly, instead of only terms proportional to $L_i m_{\text{quark}}^2$, one now also finds terms of the form $W_i a m_{\text{quark}}$ and $W'_i a^2$. At order a, the chiral Lagrangian only "knows" about the shifted quark masses m'_i of Eq. (8.52), as the $O(a)$ operator (8.51) is absorbed into the quark masses before tuning them to be near the critical value. The same mechanism does not work at order a^2. One may thus imagine working with quark masses so small that the shifted light-quark mass m'_ℓ becomes of order $a^2\Lambda_{\text{QCD}}^2$. This would imply that the power counting has to be modified; for instance, the $W'_{6,7,8}$ terms in Eq. (8.56) now become leading order. This is the so-called LCE (large cutoff effects) regime; the regime in which $m_\ell \sim m'_\ell \sim a\Lambda_{\text{QCD}}^2$ is often referred to as the GSM (generic small (quark) mass) regime. So far, our discussion has been in the GSM regime. Before we discuss some of the ChPT results in the LCE regime, however, we consider the phase diagram of the theory, as it turns out that the competition between the $O(m'_\ell)$ terms and $O(a^2)$ terms in the chiral Lagrangian can lead to various nontrivial phase structures, depending on the sign of an $O(a^2)$ LEC.

8.3.3 Phase diagram

So far, we have been expanding Σ around the trivial vacuum in order to calculate various physical quantities, such as in Eq. (8.40). Indeed, with $\chi = 2B_0 M$ with M as in Eq. (8.3) with all quark masses positive, $\Sigma = \mathbf{1}$ corresponds to the minimum of the classical potential energy, given by the second term in Eq. (8.33). For $a = 0$, higher-order terms in the chiral expansion of the potential are smaller, and thus do not change this observation.

When we include scaling violations, there are two small parameters, m_{quark} and $a\Lambda_{\text{QCD}}^2$. In the previous section, we took these two parameters to be of the same order, and again the leading-order non-derivative terms in Eqs. (8.33) and (8.51) constitute the leading-order classical potential. As long as the shifted quark mass of Eq. (8.52) stays positive and in the GSM regime, the vacuum remains trivial. However, if we now make the shifted quark mass smaller, so that at some point $m'_{\text{quark}} \sim a^2 \Lambda^3_{\text{QCD}}$, the terms with LECs $W'_{6,7,8}$ in Eq. (8.56) become comparable in size, and should be taken as part of the leading-order classical potential that determines the vacuum structure of the theory. This is the regime that we denoted as the LCE regime at the end of Section 8.3.2 (Sharpe and Singleton 1998) (see also Creutz (1995)).

The strange-quark mass is much larger than the up-or down-quark masses, so it is interesting to consider the case that $m_\ell = m_u = m_d \ll m_s$, with $m_\ell \sim a^2 \Lambda^2_{\text{QCD}}$, while m_s stays in the GSM regime.[33] We then expect the vacuum value of Σ to take the form

$$\Sigma_{vacuum} = \begin{pmatrix} \Sigma_2 & 0 \\ 0 & 1 \end{pmatrix}, \tag{8.59}$$

in which Σ_2 denotes an $SU(2)$ valued matrix. The interesting phase structure is thus essentially that of a two-flavor theory, and in the rest of this subsection we will analyze the phase structure for $N_f = 2$. For $N_f = 2$, the classical potential, which consists of the second term in Eq. (8.33), Eq. (8.51), and the $W'_{6,7,8}$ terms in Eq. (8.56), becomes

$$V = -\frac{1}{4} f^2 \hat{m}_\ell \, \text{tr}(\Sigma_2 + \Sigma_2^\dagger) - \hat{a}^2 (W'_6 + \frac{1}{2} W'_8) \left(\text{tr}(\Sigma_2 + \Sigma_2^\dagger) \right)^2 + \text{constant}, \tag{8.60}$$

where we have used the $SU(2)$ relation

$$\text{tr}(\Sigma_2 \Sigma_2 + \Sigma_2^\dagger \Sigma_2^\dagger) = \frac{1}{2} \left(\text{tr}(\Sigma_2 + \Sigma_2^\dagger) \right)^2 - 4. \tag{8.61}$$

All other terms in the chiral potential are of higher order.

This potential exhibits an interesting phase structure. First, consider the case that $W' \equiv 32(W'_6 + \frac{1}{2} W'_8)(2W_0)^2/f^2 > 0$. In that case, the potential is minimized for

$$\Sigma_2 = \begin{cases} +1, \, \hat{m}_\ell > 0 \\ -1, \, \hat{m}_\ell < 0 \end{cases} \quad (W' > 0). \tag{8.62}$$

The pion mass is given (to leading order) by

$$m^2_{\pi,LO} = 2|\hat{m}_\ell| + 2a^2 W' \tag{8.63}$$

(where LO remind us that this is the pion mass to leading order), and attains a non-zero minimum value at $\hat{m}_\ell = 0$. The phase transition is first order, with the vacuum expectation value Σ_2 exhibiting a discontinuous jump across $\hat{m}_\ell = 0$.

Equation (8.63) suggests what will happen for $W' < 0$.[34] The pion mass-squared turns negative when $|\hat{m}_\ell| < a^2|W'|$, which signals the spontaneous breakdown of a symmetry. Indeed, writing

$$\Sigma_2 = s + i\vec{\tau} \cdot \vec{\pi}, \quad s^2 + \vec{\pi}^2 = 1 \tag{8.64}$$

(with $\vec{\tau}$ the Pauli matrices), we find that the potential is minimized for

$$s = \begin{cases} +1, & \hat{m}_\ell \geq a^2|W'| \\ \dfrac{\hat{m}_\ell}{a^2|W'|}, & |\hat{m}_\ell| < a^2|W'| \\ -1, & \hat{m}_\ell \leq -a^2|W'| \end{cases} \qquad (W' < 0). \qquad (8.65)$$

A phase of width a^3 in lattice units opens up, in which isospin and parity are spontaneously broken, because of the formation of a pion condensate (Sharpe and Singleton 1998). We can take the condensate in the $\pi_3 = \pi_0$ direction in isospin space, in which case the charged pions become massless, becoming the exact Goldstone bosons for the breakdown of isospin. Since the condensate now varies continuously across the phase transition, this is a second-order transition. The possibility of such a phase was first noticed by Aoki (Aoki 1984), and provides a possible mechanism for pions to become exactly massless with Wilson fermions at non-zero lattice spacing by tuning the quark mass. At the phase transition the charged pions are still massless, and because of isospin symmetry at that point, the neutral pion also has to be massless at the transition. Note, however, that if $W' > 0$, pions will not be massless at the phase transition for non-zero a.

There is a lot more than can be said about the phase structure. We will not do so in these lectures, but end this section with a few comments:

1. An important observation is that the chiral Lagrangian does not only provide us with a systematic expansion in the small infrared scales of the theory (the quark masses and $a\Lambda^2_{\rm QCD}$), but it also provides non-perturbative information about the phase structure of the theory.

2. In two-flavor QCD with Wilson fermions, one may consider a more general quark-mass matrix

$$M = m_\ell \mathbf{1} + i\mu\tau_3. \qquad (8.66)$$

This leads to tmQCD, Wilson-fermion QCD with a twisted mass. In the continuum, one can perform a chiral rotation to make the mass matrix equal to $M = \sqrt{m^2 + \mu^2}\mathbf{1}$, but with Wilson fermions there is no chiral symmetry, and the theory for $\mu \neq 0$ differs from the one with $\mu = 0$. We note that for $\mu \neq 0$ this generalized mass matrix breaks isospin symmetry explicitly, and μ thus serves as a "magnetic field" pointing the condensate in the τ_3 direction in the limit $\mu \to 0$ (after the thermodynamic limit has been taken). It is straightforward to develop ChPT for this case; one simply substitutes this mass matrix into the (two-flavor) chiral Lagrangian. For detailed discussions of tmQCD, see for instance the reviews in Sharpe (2006) and Sint (2007).

8.3.4   The LCE regime

We have seen that $O(a)$ effects, while present for (unimproved) Wilson fermions, can be completely absorbed into a shifted quark mass in the pion chiral Lagrangian. While we already considered $O(a^2)$ effects in the pion masses in the GSM regime toward the end of Section 8.3.2, it is interesting to consider effects from $O(a^2)$ scaling violations in the LCE regime, where $m'_\ell \sim a^2 \Lambda^3_{QCD}$. We will always work with m'_ℓ, cf. Eq. (8.52), and in this section we will again restrict ourselves to $N_f = 2$, thinking of the strange quark as heavy. This makes sense, as we may expect that numerical computations with very small light-quark masses, which may then indeed turn out to be of order $a^2 \Lambda^3_{QCD}$, will become more prominent in the near future. The results we consider in this section have been obtained in Aohi *et al.* (2008).[35] They are valid in the phase with unbroken isospin and parity. In other words, we will choose m'_ℓ such that $m^2_{\pi,LO}$ of Eq. (8.63) is always non-negative.

The first thing to note is that, since our power counting changes relative to the GSM regime, new terms need to be added to the chiral Lagrangian if one wishes to work to order p^4:

$$\text{LCE regime: } O(p^2) : p^2, m'_\ell, a^2, \tag{8.67}$$

$$O(p^3) : ap^2, am'_\ell, a^3,$$

$$O(p^4) : p^4, p^2 m'_\ell, (m'_\ell)^2, a^2 p^2, a^2 m'_\ell, a^4.$$

Since we will be interested in the non-analytic terms at one loop, which follow from the lowest-order chiral Lagrangian, I just refer to Bär *et al.* (2008) for the explicit form for all new terms not present in Eq. (8.56), which does already include all $O(p^2)$ terms.

One finds for the pion mass to order p^4 (Bär *et al.* 2008):

$$m^2_\pi = \tilde{m}^2_{\pi,LO} \left\{ 1 + \frac{1}{16\pi^2 f^2} \left((\tilde{m}^2_{\pi,LO} - 10a^2 W') \log \left(\frac{\tilde{m}^2_{\pi,LO}}{\Lambda^2} \right) \right) \right. \tag{8.68}$$

$$\left. + \text{ analytic terms proportional to } \tilde{m}^2_{\pi,LO} \text{ and } a \right\}.$$

Here, $\tilde{m}^2_{\pi,LO}$ is equal to $m^2_{\pi,LO}$ (which is given in Eq. (8.63)) up to $O(a^3)$ and $O(a^4)$ terms that have been absorbed into the definition of the quark mass. This corresponds to defining the critical quark mass as the value where the pion mass vanishes, but one should keep in mind that these scaling violations are not universal.

This result is interesting because it shows that the continuum value of the coefficient of the chiral logarithm, which is a prediction of ChPT, can be significantly modified by scaling violations in the LCE regime. It also shows that at non-zero lattice spacing the chiral limit is unphysical: for $\tilde{m}^2_{\pi,LO} \to 0$ the chiral logarithm diverges, unless we take the lattice spacing to zero as well, at a rate not slower than the square root of the quark mass. Note, however, that if $W' > 0$, there is a non-vanishing minimum value of the pion mass that is of order a^2, cf. Eq. (8.63), so that this divergence cannot occur. In the case $W' < 0$ the divergence can happen, and the only way to make mathematical sense of the result (8.68) is to resum higher powers of $a^2 \log(\tilde{m}^2_{\pi,LO})$.[36] Similar effects also show up in pion scattering lengths (Bär *et al.*

2008).[37] Because of the absence of chiral symmetry at non-zero a, pion interactions do not vanish for vanishing external momenta; the leading term is of order a^2, and divergent chiral logarithms occur at order a^4. Note that the results of Bär *et al.* (2008) are obtained approaching the phase transition from the phase in which isospin is unbroken. If, for $W' < 0$, one calculates scattering lengths approaching the phase transition from the other side, one would expect them to vanish. The reason is that on that side isospin is broken, and two of the three pions are thus genuine Goldstone bosons at non-zero a, with vanishing scattering lengths. Since isospin is restored at the phase transition, all pion scattering lengths should thus vanish approaching the phase transition from the phase with broken isospin. This is not a contradiction: scattering lengths are discontinuous across the phase transition between a symmetric and a broken phase.

An important general lesson of this section is that Wilson ChPT allows us to choose quark masses so small that $m_\pi^2 \sim a^2 \Lambda_{\text{QCD}}^4$, i.e. in a regime where physical and unphysical effects compete. In contrast, if we only had continuum ChPT as a tool, we would have to first extrapolate to the continuum before doing any chiral fits. However, the examples we have discussed make it clear that, in general, we would not want to take the chiral and continuum limits "in the wrong order": if one first takes the physical infrared scale m_π to zero at fixed non-zero value of the unphysical infrared scale $a\Lambda_{\text{QCD}}^2$, ChPT tells us that infrared divergences may occur, as we have seen above.

We end this section by observing that the result for the GSM regime to $O(p^4)$ can be recovered from Eq. (8.68) by expanding in a^2 and dropping all terms of order a^3, $a^2 m'_\ell$ and a^4.

8.3.5   Axial current

Before leaving Wilson fermions for staggered fermions, let us briefly look at currents in the presence of scaling violations; as an example I will consider the non-singlet axial current, and we will work to order a, i.e., we are back in the GSM regime. I will highlight some of the arguments presented in Aoki *et al.* (2009), which contains references to earlier work, and that also discusses the vector current.

The issue is that for $a \neq 0$, the axial current for QCD with Wilson fermions is not conserved. Since there is no Noether current at the lattice QCD level, there isn't one at the SET level, nor in ChPT. Since there is no conserved current, the lattice current that is usually employed is the local current[38]

$$A_\mu^a(x) = \bar{q}(x)\gamma_\mu\gamma_5 T^a q(x), \tag{8.69}$$

in which the T^a are $SU(3)$ or $SU(2)$ flavor generators. First, this current needs a finite renormalization, in order to match to the (partially) conserved axial current in the continuum (Karsten and Smit 1981, Bochicchio *et al.* 1985),

$$A_{\mu,\text{ren}}^a = Z_A A_\mu^a, \tag{8.70}$$

with Z_A a renormalization constant determined non-perturbatively by enforcing Ward identities. Since Z_A is determined non-perturbatively, it includes scaling violations.

Then, since A_μ^a is not a Noether current following from the lattice action, the effective current that represents the axial current in the SET also does not follow from the SET action; instead, it is just an "external" operator for which we have to find the corresponding expression in the SET. To order a, the "Symanzik" current that represents the lattice current A_μ^a in the SET is[39]

$$A_{\mu,\text{SET}}^a = \frac{1}{Z_A^0}\left(1 + \bar{b}_A am\right)\left(A_{\mu,\text{cont}}^a + a\bar{c}_A\partial_\mu P_{\text{cont}}^a\right), \tag{8.71}$$

$$A_{\mu,\text{cont}}^a = \bar{q}\gamma_\mu T^a q, \qquad P_{\text{cont}}^a = \bar{q}\gamma_5 T^a q,$$

in which $\bar{b}_A$ and $\bar{c}_A$ are Symanzik coefficients that depend on the underlying lattice action; "cont" indicates continuum operators. (In a fully $O(a)$ improved theory both $\bar{b}_A$ and $\bar{c}_A$ vanish.) By combining Eqs. (8.70) and (8.71), we see that the overall multiplicative renormalization factor is $(1 + \bar{b}_A am)Z_A/Z_A^0$. Z_A^0 is the all-orders perturbative matching factor needed to convert a lattice current into the properly normalized (partially conserved) continuum current; we have that $Z_A = Z_A^0 + O(a)$.[40]

At this stage, two steps need to be carried out to translate this current to ChPT. First, the operator $A_{\mu,\text{cont}}^a + a\bar{c}_A\partial_\mu P_{\text{cont}}^a$ needs to be mapped into ChPT. Then, the renormalization factor following from Eqs. (8.70) and (8.71) has to be determined in exactly the same way as the lattice current is matched to the renormalized continuum current in an actual lattice computation. This matching has to be done to order a as well, if one consistently wants to include all $O(a)$ corrections in the chiral theory. Since this involves the matching of scaling violations, one does not expect the outcome of this second step to be universal, and indeed, Aoki *et al.* (2009) find that already at order a the matching depends on the precise correlation functions used for the matching, with a non-universal result for the ratio Z_A/Z_A^0; I refer to Aoki *et al.* (2009) for further discussion of this second step. Here, I will address the first step: the mapping to operators in the chiral theory, to order a.[41]

In order to find the ChPT expression for the axial current, Aoki *et al.* (2009) first observe that, in the case that $\bar{c}_A = 0$, $\bar{b}_A = 0$ and $Z_A^0 = 1$, the Symanzik current is just the continuum axial current, which can be rotated into the vector continuum current by a chiral transformation. This fixes the $\bar{c}_A = 0$ part of the axial current in ChPT in terms of the vector current. It turns out that the vector current is given by the Noether current of the chiral theory, because the $O(a)$ terms in the vector current are of the form $a\partial_\nu(\bar{q}s_{\mu\nu}T^a q)$, which is automatically conserved.[42] The Noether current following from $\mathcal{L}^{(2)} + \Delta\mathcal{L}^{(4)}$ (cf. Eqs. (8.33) and (8.56)) is

$$V_\mu = (J_\mu^R + J_\mu^L)\left(1 + \frac{8W_4}{f^2}\,\text{tr}(\hat{A}^\dagger\Sigma + \Sigma^\dagger\hat{A})\right) + \frac{4W_5}{f^2}\left\{J_\mu^R + J_\mu^L, \hat{A}^\dagger\Sigma + \Sigma^\dagger\hat{A}\right\}, \tag{8.72}$$

with $J_\mu^{R,L}$ defined to lowest order in Eq. (8.26); we omitted terms coming from $\mathcal{L}^{(4)}$ of Eq. (8.17). Here, $\hat{A} \equiv 2W_0 A$ with A the spurion of Eq. (8.50). We should set $\hat{A} = \hat{a}$ of course, but only *after* doing the chiral rotation to find the $\bar{c}_A = 0$ part of the axial current. We thus find

$$A_\mu = (J_\mu^R - J_\mu^L) \left(1 + \frac{8\hat{a}W_4}{f^2} \operatorname{tr}(\Sigma + \Sigma^\dagger)\right) + \frac{4\hat{a}W_5}{f^2} \left\{ J_\mu^R - J_\mu^L, \Sigma + \Sigma^\dagger \right\} \quad (\bar{c}_A = 0),$$

(8.73)

where now we have set $\hat{A} = \hat{a}$.

Finally, we have to worry about the $\bar{c}_A$ term in Eq. (8.71). This is rather simple, because for this we need the lowest-order ChPT operator for the pseudoscalar density, which is $-i(\Sigma - \Sigma^\dagger)$. We need to introduce a new LEC W_A for this contribution, since, unlike the contribution in Eq. (8.73), this term does not relate in any way to the Symanzik effective action, and the same is thus true in ChPT. We thus find for the complete expression for the axial current to order a:

$$A_\mu = (J_\mu^R - J_\mu^L) \left(1 + \frac{8\hat{a}W_4}{f^2} \operatorname{tr}(\Sigma + \Sigma^\dagger)\right) + \frac{4\hat{a}W_5}{f^2} \left\{ J_\mu^R - J_\mu^L, \Sigma + \Sigma^\dagger \right\} \qquad (8.74)$$

$$- 4i\hat{a}W_A \bar{c}_A \partial_\mu(\Sigma - \Sigma^\dagger) \quad (\bar{c}_A \neq 0),$$

where, following Aoki *et al.* (2009), I kept $\bar{c}_A$ outside W_A.[43] In closing this section, I should emphasize that the derivation given here is the "quick and dirty" derivation–for a careful analysis of both vector and axial currents, I refer to Aoki *et al.* (2009).

8.3.6 Staggered fermions

Before developing ChPT for staggered QCD, I give a brief review of the definition and symmetry properties of lattice QCD with staggered fermions, to make these lectures more self-contained.[44]

Brief review. The staggered action is (for the time being I will set the lattice spacing $a = 1$, but I will restore it when we get to the SET and ChPT)

$$S = \sum_{x,\mu,i} \frac{1}{2} \eta_\mu(x) \overline{\chi}_i(x) \left[U_\mu(x)\chi_i(x+\mu) - U_\mu^\dagger(x-\mu)\chi_i(x-\mu) \right] + \sum_{x,i} m_i \overline{\chi}_i(x)\chi_i(x),$$

$$\eta_\mu(x) = (-1)^{x_1 + \cdots + x_{\mu-1}}. \qquad (8.75)$$

Here, χ_i is a lattice fermion field with an explicit flavor index $i = 1, \ldots, N_f$, an implicit color index, and *no* Dirac index; $U_\mu(x)$ are the gauge-field links, with values in $SU(3)_{\text{color}}$.

The action (8.75) has species doubling, and because of the hypercubic structure of the theory, this doubling is sixteen-fold. The sixteen components that emerge in the continuum limit can be accounted for by an index pair α, a, with each index running over the values $1, \ldots, 4$. It turns out that the first index can be interpreted as a Dirac index, while the second index constitutes an additional flavor label; the continuum limit looks like (Sharatchandra *et al.* 1981, Golterman and Smit 1984, Kluberg-Stern *et al.* 1983)

$$S_{cont} = \int d^4x \left(\bar{q}_{i\alpha a} \gamma_{\mu,\alpha\beta} D_\mu q_{i\beta a} + m_i \bar{q}_{i\alpha a} q_{i\alpha a} \right), \tag{8.76}$$

in which repeated indices are summed. In brief, this structure emerges as follows. Species doublers live near the corners

$$\pi_A \in \{(0,0,0,0),(\pi,0,0,0),\ldots,(\pi,\pi,\pi,\pi)\}, \quad A = 1,\ldots,16, \tag{8.77}$$

in the Brillouin zone. It thus makes sense to split the lattice momenta p into

$$p = \pi_A + k, \quad -\frac{\pi}{2} \le k_\mu \le \frac{\pi}{2}. \tag{8.78}$$

We thus identify sixteen different fermion fields in momentum space, $\chi(p) = \chi(\pi_A + k) \equiv \chi_A(k)$, which represent the sixteen doublers. Since the phases $\eta_\mu(x)$ insert momenta with values in the set (8.77), it follows that the operation

$$T_{\pm\mu} : \chi_i(x) \to \eta_\mu(x)\chi_i(x \pm \mu) \tag{8.79}$$

mixes the sixteen doublers. In momentum space, after a basis transformation, the operations T_μ can be represented by

$$T_{\pm\mu} : q_{i\alpha a}(k) \to \gamma_{\mu,\alpha\beta} e^{\pm ik_\mu} q_{i\beta a}(k), \tag{8.80}$$

with the γ_μ a set of Dirac matrices. That these four matrices satisfy a Dirac algebra follows from the fact that the T_μ anticommute:

$$T_\mu T_\nu = -T_\nu T_\mu, \quad \mu \ne \nu, \tag{8.81}$$

and that

$$T_{\pm\mu}^2 : \chi_i(x) \to \chi_i(x \pm 2\mu). \tag{8.82}$$

Using these ingredients, we can see how Eq. (8.76) follows from Eq. (8.75). Using Eqs. (8.79) and (8.80), the free kinetic term in Eq. (8.75) can be written as (for one flavor)

$$\sum_{x,\mu} \frac{1}{2} \bar{\chi}(x) \left(T_\mu - T_{-\mu} \right) \chi(x) = \int \frac{d^4k}{(2\pi)^4} \sum_\mu \bar{q}(k) \gamma_\mu \, i \sin(k_\mu) q(k). \tag{8.83}$$

In the continuum limit, $\sin(k_\mu) \to k_\mu$; note that k is restricted to the reduced Brillouin zone (cf. Eq. (8.78)), on which no species doublers reside. Finally, since Eq. (8.75) is gauge invariant, the continuum limit (8.76) has to be gauge invariant as well. It follows that, at least classically, Eq. (8.76) is the continuum limit of Eq. (8.75).

The observation that the continuum limit of staggered QCD is given by Eq. (8.76) has been proven to one loop in perturbation theory, and this result has been understood in terms of the exact lattice symmetries, making it likely that this proof can be extended to all orders (Golterman and Smit 1984). There is also extensive numerical evidence (Bazavov *et al.* 2009). We conclude that the lattice theory defined by Eq. (8.75) constitutes a regularization of QCD with $4N_f$ flavors. In modern parlance, "flavor" is now usually used to denote the explicit multiplicity associated with the flavor index i, while "taste" is used to denote the implicit multiplicity associated with species doubling. If indeed the continuum limit is given by Eq. (8.76), this implies that a full $U(4)_{\text{taste}}$ symmetry emerges in the continuum limit, for each staggered flavor. This taste symmetry is not present in the lattice theory – we will return to this shortly.

We note that the taste degeneracy is somewhat of an embarrassment, if one wishes to use the flavor index i as physical flavor, as is usually done in practice. While this is not a problem for valence quarks (one simply picks out, say, taste $a = 1$ in order to construct operators, for example), this means that there are too many sea quarks – too many quarks on the internal fermion loops. This problem is solved in practice by taking the (positive) fourth root of the staggered determinant for each physical flavor. While this makes sense intuitively, it is a serious modification of the theory, which entails a violation of locality and unitarity at non-zero lattice spacing (Bernard *et al.* 2006). Much work has recently gone into attempts to show that the procedure is nevertheless correct in the sense that it yields the desired continuum limit, in which EFT techniques in fact play an important role. There is no space here to discuss this very interesting (and important!) topic; I refer to the recent reviews of Golterman (2008) and Bazavov *et al.* (2009). Here, I will focus on the construction of staggered ChPT for unrooted staggered fermions.

To construct the SET for staggered QCD it is important to use all exact symmetries of the lattice theory, in order to restrict the form of the operators that will appear in the Symanzik expansion, Eq. (8.43). The action (8.75) has an exact vector-like $SU(N_f)$ symmetry, broken only if the quark masses m_i are non-degenerate. Furthermore, there are space-time symmetries. In addition to hypercubic rotations and parity, the action is invariant under shift symmetry (Golterman and Smit 1984, van den Doel and Smit 1983):

$$S_{\pm\mu} : \chi_i(x) \to \zeta_\mu(x)\chi_i(x \pm \mu), \qquad \zeta_\mu(x) = (-1)^{x_{\mu+1}+\cdots+x_4}, \qquad (8.84)$$
$$S_{\pm\mu} : U_\nu(x) \to U_\nu(x \pm \mu),$$

because of the fact that

$$\zeta_\nu(x)\eta_\mu(x)\zeta_\nu(x + \mu) = \eta_\mu(x + \nu). \qquad (8.85)$$

On the basis $q_{i\alpha a}$ shift symmetry is represented by

$$S_{\pm\mu} : q_{i\alpha a}(k) \to \xi_{\mu,ab}e^{\pm ik_\mu}q_{i\alpha b}(k), \qquad (8.86)$$

with ξ_μ a different set of Dirac matrices (like Eq. (8.81) we have that $S_\mu S_\nu = -S_\nu S_\mu$ for $\mu \neq \nu$) acting on the taste index, commuting with the γ_ν of Eq. (8.80), because of Eq. (8.85).

Finally, if $m_i = 0$ for a certain flavor, there is a chiral symmetry ("$U(1)_\epsilon$ symmetry") (Kawamoto and Smit 1981)

$$\chi_i(x) \to e^{i\theta_i \epsilon(x)} \chi_i(x), \qquad \overline{\chi}_i(x) \to e^{i\theta_i \epsilon(x)} \overline{\chi}_i(x), \tag{8.87}$$

$$\epsilon(x) = (-1)^{x_1 + x_2 + x_3 + x_4}.$$

Since this symmetry is broken by a non-zero m_i, it should be interpreted as an axial symmetry of the theory. In addition, since it is an exact symmetry of the lattice theory (for $m_i = 0$) it has to correspond to a non-singlet axial symmetry in the continuum limit, because the continuum singlet axial symmetry is anomalous. If all quarks are massless, it is not difficult to show that this enlarges to an exact $U(N_f)_L \times U(N_f)_R$ symmetry.[45] The axial symmetries in this group are non-singlet symmetries on the basis $q_{i\alpha a}$, because the phase $\epsilon(x)$ inserts momentum, and thus acts non-trivially on the index pair α, a: Since

$$\epsilon(x) \chi_i(x) = \prod_\mu T_{-\mu} S_\mu \chi_i(x), \tag{8.88}$$

it follows that $U(1)_\epsilon$ symmetry is represented on the $q_{\alpha a}$ basis as

$$q_{i\alpha a} \to \left(e^{i\theta_i \gamma_5 \xi_5} \right)_{\alpha a, b} q_{ib}. \tag{8.89}$$

In staggered QCD there is extensive evidence that these axial symmetries are spontaneously broken, and for each such broken symmetry there is thus an exact Nambu–Goldstone boson. When m_i is turned on, this NG boson picks up a mass-squared proportional to m_i, as we will see below. Contrary to the case of Wilson fermions, we thus know the critical values of the quark masses m_i – the massless limit is at $m_i = 0$. Interpolating fields for these Nambu–Goldstone bosons are $\epsilon(x) \overline{\chi}_i(x) \chi_j(x)$ in position space, or $\overline{q}_i \gamma_5 \xi_5 q_j$ in momentum space.

As can be seen from Eq. (8.76), these lattice symmetries enlarge to $SU(4N_f)_L \times SU(4N_f)_R$ (and the usual Euclidean space-time symmetries) in the continuum limit. In the continuum limit, one thus expects $(4N_f)^2 - 1$ pions, associated with the spontaneous breakdown $SU(4N_f)_L \times SU(4N_f)_R \to SU(4N_f)_V$. At non-zero a, there are only N_f^2 exact Nambu–Goldstone bosons; the rest will pick up additional contributions to their mass squared of order a^2, as we will see in more detail below. These lattice-artifact mass splittings are usually referred to as "taste splittings," since they occur because of the breakdown of $U(4)_\text{taste}$ on the lattice. Indeed, this taste breaking is seen on the lattice very clearly in the meson spectrum, which is shown in Fig. 8.3. Clearly, we would like to understand these results with help of a ChPT framework.

Staggered ChPT. The SET is a continuum theory. This means that it is invariant under translations over an arbirary distance; in particular, it is invariant under translations over a distance a in any direction.[46] It is also invariant under any of the exact lattice symmetries. We may thus combine a shift S_μ in the positive μ direction by a translation over a distance a in the negative μ direction. Combining these two,

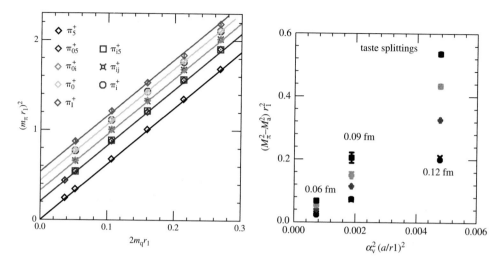

Fig. 8.3 (*a*) Taste splittings between pions made out of an up and (anti)down staggered quarks, in different representations of the staggered symmetry group, at fixed lattice spacing, $a = 0.125$ fm. The black points correspond to the exact Nambu–Goldstone boson, lines are chiral fits using staggered ChPT. (*b*) Taste splittings between the various non-exact Nambu–Goldstone bosons and the exact Namubu–Goldstone boson as a function of lattice spacing. These figures show the behavior predicted by Eq. (8.104). From Aubin *et al.* (2004), to which I refer for a detailed explanation.

we see that the SET is invariant under (Bernard *et al.* 2008, Lee and Sharpe 1999)

$$q_i(k) \rightarrow \xi_\mu q_i(k). \tag{8.90}$$

At the level of the SET, the "translation" and "taste" parts of shift symmetry thus decouple, and we find that the SET thus has to be invariant under the 32-element group Γ_4 of discrete taste transformations generated by ξ_μ, which is a subgroup of $U(4)_{\text{taste}}$.[47] We can thus use this group in order to restrict operators at each order in the Symanzik expansion, along with the other exact lattice symmetries.

The first term in the Symanzik expansion, $\mathcal{L}_S^{(4)}$, is given by (the integrand of) Eq. (8.76). There are no dimension-five operators that can be constructed from the quark fields q and $\bar{q}$ that respect the lattice symmetries (Sharpe 1994). Ignoring $U(1)_\epsilon$ symmetry, the possible dimension-five operators are those of Eq. (8.46). The first two operators in this list are immediately excluded by Eq. (8.89). The three other terms involve powers of the quark mass, and in order to make use of $U(1)_\epsilon$ symmetry, we thus have to make the quark mass a spurion, as in Eqs. (8.1) and (8.5), but we now do it with L and R projectors on the quark fields defined with $\gamma_5 \xi_5$, instead of γ_5. It is then not difficult to see that also the three last operators in Eq. (8.46) are excluded.

This implies that scaling violations for staggered QCD start at order a^2. In fact, before we start the discussion of $\mathcal{L}^{(6)}$, we should rethink power counting. If we use GSM power counting, all terms of order a^2 would be $O(p^4)$. However, it turns out that scaling violations with staggered fermions, even though formally $O(a^2)$, are numerically rather large, even with improved staggered actions, at current values of the lattice spacing.[48] From the taste splittings in pseudoscalar mesons, it seems that the contribution of the quark masses and $O(a^2)$ scaling violations to their masses is roughly equal, and we will therefore use a power counting in which $p^2 \sim m_{\text{quark}} \Lambda_{\text{QCD}} \sim a^2 \Lambda^4_{\text{QCD}}$. This means that $O(a^2)$ effects are of leading order in ChPT, and we would like to construct at least the $O(a^2)$ part of the chiral Lagrangian, because from the $O(p^2)$ Lagrangian we can obtain the $O(p^2)$ and non-analytic $O(p^4)$ parts of physical quantities in terms of quark masses and the lattice spacing. We thus set out to find the $O(a^2)$ part of $\mathcal{L}^{(2)}$, cf. Eq. (8.33). This new part contains no quark masses, and no derivatives, each of which would make such terms higher order in the chiral expansion.

The $O(a^2)$ part of the SET, $\mathcal{L}^{(6)}$, contains a large number of terms (Lee and Sharpe 1999, Aubin and Bernard 2003a), and I will only give some examples here. First, there are the purely gluonic terms that we already discussed in the context of Wilson ChPT, and thus do not have to revisit again. Similarly, fermion bilinears in $\mathcal{L}_S^{(6)}$ do not lead to any new terms at the desired order in ChPT (Lee and Sharpe 1999). The remaining terms are all four-fermion operators.

Taste-breaking, short-distance four-fermion operators occur in the theory because of diagram as shown in Fig. 8.4. The gluon carries a momentum near one of the values in Eq. (8.77), with $n > 0$ components near π/a, thus changing the taste of the staggered quarks at both vertices. For such gluon momenta, we have that the gluon propagator

$$\frac{1}{\sum_\mu \frac{4}{a^2} \sin^2\left(\frac{1}{2}ap_\mu\right)} \approx \frac{a^2}{4n}. \tag{8.91}$$

Effectively, this generates a four-fermion operator in the SET (in which gluons with momenta of order $1/a$ have been integrated out), for instance of the form[49]

$$\mathcal{O}_1 = (\bar{q}_i \xi_{5\nu} q_i)(\bar{q}_j \xi_{\nu 5} q_j), \tag{8.92}$$

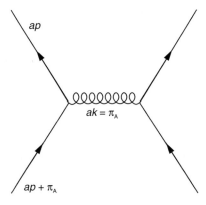

Fig. 8.4 Generation of taste-breaking four-fermion vertices through short-distance gluon exchange.

in which $-\xi_{\nu 5} = \xi_{5\nu} = \xi_5 \xi_\nu$, and all repeated indices are summed. This operator is invariant under the group Γ_4, but not under the full continuum taste group $U(4)_{\text{taste}}$ (it is, in fact, invariant under $SO(4)_{\text{taste}} \subset U(4)_{\text{taste}}$, accidentally). Since the gluon carries no flavor, the flavor-index structure has to be as indicated. Note that the spin $\otimes$ taste matrix appearing in each quark bilinear has to correspond to an odd number of applications of T_μ or S_μ, because gluons couple only to the kinetic term in Eq. (8.75), which is invariant under $U(1)_\epsilon$.

Note that the appearance of four-fermion operators due to the exchange of "heavy" gluons is analogous to the appearance of four-fermion terms with coefficients of order g_{weak}^2/M_W^2 (the Fermi EFT) in the Standard Model at low energy because of the exchange of the heavy W and Z bosons. Likewise, in the case at hand, the four-fermion operators in the SET are of order α_s, the strong "fine-structure" constant. Light gluons, with momenta $p \ll 1/a$ are still present in the SET, i.e., they have not been integrated out.

The spin matrices γ_μ can also appear in these four-fermion operators. This can happen in two ways, here is an example of each:

$$\mathcal{O}_2 = \sum_\mu (\bar{q}_i \gamma_\mu \xi_5 q_i)(\bar{q}_j \gamma_\mu \xi_5 q_j), \tag{8.93}$$

$$\mathcal{O}_3 = \sum_{\mu < \nu} (\bar{q}_i \gamma_\nu \xi_{\mu\nu} q_i)(\bar{q}_j \gamma_\nu \xi_{\nu\mu} q_j), \tag{8.94}$$

in which $\xi_{\mu\nu} = \xi_\mu \xi_\nu$ $(\mu \neq \nu)$, and where we now showed the sums over Lorentz indices explicitly. As before, neither of these operators is invariant under $U(4)_{\text{taste}}$, and both are invariant under Γ_4 (as well as all other lattice symmetries, in particular hypercubic rotations). The first operator is separately invariant under $SO(4)_{\text{Euclidean}} \times SO(4)_{\text{taste}}$, the second is not. Operators of the second type can only be represented in ChPT with at least two derivatives (and one factor a^2) (Sharpe and Van der Water 2005), which are of order p^4.[50] Only operators of the first type have $O(p^2)$ representations in ChPT. An interesting observation is thus that at lowest order staggered ChPT therefore has $SO(4)_{\text{taste}}$ symmetry, and not just the required Γ_4 symmetry (Lee and Sharpe 1999).

Finally, we need to make the transition from the SET to ChPT, for operators such as (8.92) and (8.93). For $N_f = 3$, our theory is now an $SU(12)$ theory, because of the three flavors times four tastes. The non-linear field

$$\Sigma_{ia,jb} \sim q_{iaL}\bar{q}_{jbR}, \tag{8.95}$$

is thus an $SU(12)$-valued field, with an index structure as shown. In order to streamline our notation, we introduce matrices (Aubin and Bernard 2003a)

$$\xi_\mu^{(3)} = \begin{pmatrix} \xi_\mu & 0 & 0 \\ 0 & \xi_\mu & 0 \\ 0 & 0 & \xi_\mu \end{pmatrix}, \tag{8.96}$$

so that we can write for example Eq. (8.92) as $(\bar{q}\xi_{5\nu}^{(3)}q)(\bar{q}\xi_{\nu 5}^{(3)}q)$. We now rewrite

$$-(\bar{q}\xi_{5\nu}^{(3)}q)(\bar{q}\xi_{\nu 5}^{(3)}q) = (\bar{q}_R\xi_{5\nu}^{(3)}q_L + \bar{q}_L\xi_{5\nu}^{(3)}q_R)^2 \tag{8.97}$$

$$\to (\bar{q}_R X_R q_L + \bar{q}_L X_L q_R)^2,$$

where in the last line we introduced two spurions transforming as

$$X_L \to U_L X_L U_R^\dagger, \qquad X_R \to U_R X_R U_L^\dagger, \tag{8.98}$$

while they are interchanged by parity. With the field Σ we can construct three types of terms in ChPT with these spurion fields that do not contain the quark-mass matrix or any derivatives:

$$(\bar{q}\xi_{5\nu}^{(3)}q)(\bar{q}\xi_{\nu 5}^{(3)}q) \to \begin{cases} \text{tr}(X_R\Sigma)\,\text{tr}(X_L\Sigma^\dagger), \\ \left(\text{tr}(X_R\Sigma)\right)^2 + \left(\text{tr}(X_L\Sigma^\dagger)\right)^2, \\ \text{tr}(X_R\Sigma X_R\Sigma) + \text{tr}(X_L\Sigma^\dagger X_L\Sigma^\dagger), \end{cases} \tag{8.99}$$

which, setting the spurions equal to the values that reproduce $\mathcal{O}_1$, gives the three ChPT operators

$$\text{tr}(\xi_{5\nu}^{(3)}\Sigma)\,\text{tr}(\xi_{5\nu}^{(3)}\Sigma^\dagger), \ \left(\text{tr}(\xi_{5\nu}^{(3)}\Sigma)\right)^2 + \left(\text{tr}(\xi_{5\nu}^{(3)}\Sigma^\dagger)\right)^2, \ \text{tr}(\xi_{5\nu}^{(3)}\Sigma\xi_{5\nu}^{(3)}\Sigma) + \text{tr}(\xi_{5\nu}^{(3)}\Sigma^\dagger\xi_{5\nu}^{(3)}\Sigma^\dagger),$$

$$\tag{8.100}$$

in which sums over the index ν are understood.

Introducing spurions $Y_{L,R}$ for the taste matrices $\xi_5^{(3)}$ in $\mathcal{O}_2$, which transform as $Y_{L,R} \to U_{L,R} Y_{L,R} U_{L,R}^\dagger$, and working through a similar exercise, one finds the (unique) ChPT operator representing $\mathcal{O}_2$:

$$\sum_\mu (\bar{q}_i\gamma_\mu\xi_5^{(3)}q_i)(\bar{q}_j\gamma_\mu\xi_5^{(3)}q_j) \to \text{tr}(\xi_5^{(3)}\Sigma\xi_5^{(3)}\Sigma^\dagger). \tag{8.101}$$

Referring to Aubin and Bernard (2003a) for further details, we quote the $O(a^2)$ contribution to the chiral Lagrangian, to be added to Eq. (8.33). Writing this part as $a^2 c_V = a^2(\mathcal{U} + \mathcal{U}')$, one finds that

$$-\mathcal{U} = C_1 \text{tr}\left(\xi_5^{(3)}\Sigma\xi_5^{(3)}\Sigma^\dagger\right) + \frac{1}{2}C_3 \sum_\nu \left[\text{tr}\left(\xi_\nu^{(3)}\Sigma\xi_\nu^{(3)}\Sigma\right) + \text{h.c.}\right] \tag{8.102}$$

$$+ \frac{1}{2}C_4 \sum_\nu \left[\text{tr}\left(\xi_{5\nu}^{(3)}\Sigma\xi_{\nu 5}^{(3)}\Sigma\right) + \text{h.c.}\right] + C_6 \sum_{\mu<\nu} \text{tr}\left(\xi_{\mu\nu}^{(3)}\Sigma\xi_{\nu\mu}^{(3)}\Sigma^\dagger\right),$$

$$-\mathcal{U}' = \frac{1}{4}C_{2V} \sum_\nu \left[\text{tr}(\xi_\nu^{(3)}\Sigma)\,\text{tr}(\xi_\nu^{(3)}\Sigma) + \text{h.c.}\right] + \frac{1}{4}C_{2A} \sum_\nu \left[\text{tr}(\xi_{5\nu}^{(3)}\Sigma)\,\text{tr}(\xi_{\nu 5}^{(3)}\Sigma) + \text{h.c.}\right]$$

$$+ \frac{1}{2}C_{5V} \sum_\nu \text{tr}(\xi_\nu^{(3)}\Sigma)\,\text{tr}(\xi_\nu^{(3)}\Sigma^\dagger) + \frac{1}{2}C_{5A} \sum_\nu \text{tr}(\xi_{5\nu}^{(3)}\Sigma)\,\text{tr}(\xi_{\nu 5}^{(3)}\Sigma^\dagger).$$

Let us briefly consider some of the physics results that can be obtained from the $O(p^2)$ chiral Lagrangian, which is the sum of Eqs. (8.33) (for a twelve-flavor continuum theory) and (8.102). First, consider meson masses at tree level. Here, I will only quote results for flavored (i.e., off-diagonal in Eq. (8.6)) masses. The flavor-neutral sector is more complicated, and I refer to Aubin and Bernard (2003a) for details.

Each of the entries in Eq. (8.6) is now a four-by-four taste matrix, and we can thus expand each entry on a set of $U(4)_{\text{taste}}$ generators:

$$\phi_{ij} = \frac{1}{2} \sum_F \Xi_F \phi_{ij}^F, \quad i, j = u, d, s, \tag{8.103}$$

$$\Xi_F \in \{\xi_5, i\xi_{5\mu}, i\xi_{\mu\nu}, \xi_\mu, \mathbf{1}\}.$$

By expanding Σ to quadratic order, we can then read off the meson masses from Eqs. (8.33) and (8.102). In the flavored sector, only $\mathcal{U}$ contributes, and we find

$$m_{Fij}^2 = B_0(m_i + m_j) + a^2 \Delta(\Xi_F), \quad i, j = u, d, s, \quad i \neq j, \tag{8.104}$$

in which

$$\Delta(\xi_5) = 0, \tag{8.105}$$

$$\Delta(i\xi_{5\mu}) = \frac{16}{f^2} (C_1 + 3C_3 + C_4 + 3C_6),$$

$$\Delta(i\xi_{\mu\nu}) = \frac{16}{f^2} (2C_3 + 2C_4 + 4C_6),$$

$$\Delta(\xi_\mu) = \frac{16}{f^2} (C_1 + C_3 + 3C_4 + 3C_6),$$

$$\Delta(\xi_1) = \frac{16}{f^2} (4C_3 + 4C_4).$$

As we expect, there is one exact flavor multiplet of Nambu–Goldstone bosons, because of $U(1)_\epsilon$ symmetry, with a mass-squared proportional to the quark masses, as in the continuum. All other meson masses are shifted by terms of order a^2, and Fig. 8.3 is in good agreement with the behavior predicted by Eq. (8.104). The numerical results show that C_4 must be dominant among $C_{1,3,4,6}$. This is consistent with the fact that, as previously mentioned, $U(4)_{\text{taste}}$ is broken at this order in ChPT to $SO(4)$, not all the way down to Γ_4. Thus, there are only five non-degenerate meson masses (Lee and Sharpe 1999), whereas the irreducible representations of the staggered symmetry group would allow eight (Golterman 1986).

We end this section by quoting a one-loop result that can be obtained from our lowest-order staggered chiral Lagrangian. But before we do this, we have to address the fact that the $SU(12)$ theory has too many sea quarks, which on the lattice side is fixed by taking the fourth root of each staggered fermion determinant. The question is how to carry over this procedure to staggered ChPT. The way this can be done is through the "replica trick." One starts with *adding more* fermions to the theory by using n_r instead of one staggered fermions for each physical flavor. For each of the n_r up quarks, the quark mass is kept degenerate – all up quarks have mass m_u, and likewise for down and strange quarks. In the lattice QCD path integral, this results in raising the staggered determinant for each flavor to the n_rth power. The idea is now to do all calculations keeping the n_r dependence, and then at the end set $n_r = 1/4$, which corresponds precisely to taking the fourth root of the determinants.

As we already mentioned, it is by no means obvious that the fourth-root procedure is field-theoretically legitimate. However, if it is, it is possible to show that the replica trick is the correct way of incorporating the effects of the fourth root in staggered ChPT (and EFTs for staggered QCD in general) (Bernard *et al.* 2008, Bernard 2006). Assuming that this all works, we quote the $O(p^4)$ result for the two-flavor ξ_5-pion decay constant (which is the real Nambu–Goldstone boson for the breakdown of $U(1)_\epsilon$ symmetry), with equal light-quark masses, $m_u = m_d = m_\ell$:[51]

$$\frac{f_\pi^{\xi_5}}{f} = 1 - \frac{1}{8} \sum_B L(2B_0 m_\ell + a^2 \Delta(\Xi_B)) \tag{8.106}$$

$$- 4 \left(L(2B_0 m + a^2 \delta'_V) + L(2B_0 m + a^2 \delta'_A) - 2L(2B_0 m) \right)$$

$$+ \frac{16 B_0 m}{f^2} (2L_4 + L_5) + a^2 F,$$

in which

$$\delta'_V = \frac{16}{f^2} (C_{2V} - C_{5V}), \tag{8.107}$$

$$\delta'_A = \frac{16}{f^2} (C_{2A} - C_{5A}),$$

and F is a linear combination of $O(p^4)$ LECs.[52] Note that now the LECs are low-energy constants in the $N_f = 2$ theory. The LECs $C_{2V,A}$ and $C_{5V,A}$ only appear in flavor-neutral meson masses at tree level (as can be seen by expanding out $\mathcal{U}'$), and their appearance reflects the fact that flavor-neutral mesons appear in the one-loop corrections. Flavored mesons also appear in the loops; producing the logarithms in the sum over B.

We note that the masses as they appear in chiral logarithms get $O(a^2)$ corrections. This is of course natural, because all lattice mesons can in principle appear on loops, and most of these masses get $O(a^2)$ corrections, already at lowest order in ChPT. This change in the logarithms points at a reason as to why it is important to incorporate scaling violations in ChPT systematically: as long as $m_{\text{quark}} \sim a^2 \Lambda_{\text{QCD}}^3$, one cannot expand the logarithm in Eq. (8.106) in powers of a^2. If we would have naively assumed that all scaling violations can be understood in terms of simple powers of a, we would have missed this non-analytic dependence on the lattice spacing. The $O(a^2)$ term inside the logarithm reduces the curvature for small quark masses. Knowing the field-theoretical form of this non-analytic behavior makes it thus feasible to explore smaller quark masses at a given lattice spacing than otherwise would be possible. In

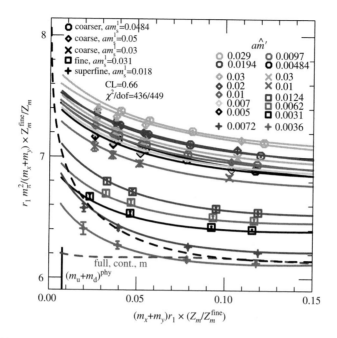

Fig. 8.5 $m_\pi^2/(m_x + m_y)$ as a function of $m_x + m_y$. The continuous curves are (partially quenched) chiral fits at fixed lattice spacing; the dashed line is the (partially quenched) continuum line (with sea-quark masses corresponding to those of the "superfine" 0.018/0.0072 ensemble), which has a much stronger curvature. Note how the curvature is reduced by non-vanishing lattice spacing. The lower dashed curve is the fully unquenched continuum curve. m_s' is the strange sea-quark mass and $\hat{m}'$ is the light sea-quark mass. For more explanation of the meaning of all symbols, see Aubin *et al* (2004). Figure courtesy C. Bernard.

Fig. 8.5, I show a plot of the staggered ChPT fits of the meson masses from the MILC collaboration (which, of course, use the full non-degenerate 2 + 1-flavor and partially quenched formulas of Aubin and Bernard (2003b). The indices x and y denote the flavors of the valence quarks, which for most of the data points in this figure have masses different from the sea quarks, a generalization we will consider in the next section. In this figure, we see that indeed the curvature on the lattice is significantly different from that of the continuum curve.

Finally, we observe that the form of the tree-level masses, Eq. (8.104), is very similar to Eq. (8.63). That raises the question whether, since we are working in the LCE regime in which $m_{\text{quark}} \sim a^2 \Lambda_{\text{QCD}}^3$, Aoki-like phases could occur for small enough quark mass. Using techniques similar to those used in QCD inequalities[53] it can be argued that all $\Delta(\Xi_B) \geq 0$, excluding this route to a possible Aoki-like phase. However, this is not true for the $O(a^2)$ corrections in the flavor-neutral sector, which are governed by the LECs $\delta_{V,A}'$. While no sign of an Aoki-like phase has been observed in numerical computations, this remains an interesting theoretical possibility (Aubin and Bernard 2003a, Aubin and Wang 2004).

8.4 Choosing valence and sea quarks to be different

Consider the two-point function for a charged pion in Euclidean QCD,

$$\langle \pi^+(x)\pi^-(y)\rangle = \langle (\overline{u}(x)\gamma_5 d(x))(\overline{d}(y)\gamma_5 u(y))\rangle \tag{8.108}$$

$$= \frac{1}{Z}\int \prod_\mu [dU_\mu] \prod_i [d\overline{q}_i][dq_i]\ \exp\left(-S_{QCD}\right)\overline{u}(x)\gamma_5 d(x)\overline{d}(y)\gamma_5 u(y)$$

$$= -\frac{1}{Z}\int \prod_\mu [dU_\mu]\ \mathrm{Det}\left(\slashed{D} + M\right)\mathrm{tr}\left[\gamma_5\left(\slashed{D} + m_u\right)^{-1}(y,x)\gamma_5\left(\slashed{D} + m_d\right)^{-1}(x,y)\right].$$

In an actual lattice computation, one usually does not use point sources and sinks; instead one sums with some weight over the timeslices at $t = x_4$ and $t' = y_4$, but this will not affect the observation we want to make.

In the third line of Eq. (8.108), the quarks enter in two clearly distinct ways: through the determinant, and through the quark propagators that follow from carrying out the Wick contractions. We refer to quarks of the first type as sea quarks, and those of the latter type as valence quarks. Of course, in nature these two types of quarks are the same, and they have to be in order to preserve unitarity of the theory. But nothing stops us on the lattice from taking the quark masses in the propagators unequal to those in the determinant, or, even more drastically, from choosing a completely different discretization for $\slashed{D}$ in the propagators and in the determinant! Let us first consider whether this seemingly sacrilegious idea might have any advantages, and if so, whether we can make any field-theoretical sense out of it.

First, consider only taking valence-quark masses unequal to sea-quark masses. This is a situation commonly referred to as partial quenching. This name has its origin in the fact that as we send the sea-quark masses to infinity, we effectively remove the determinant from the path integral, thus reverting to the quenched approximation. Unlike the quenched approximation, however, as long as we keep the sea-quark masses in the theory, we can always recover the physical theory by choosing the valence-and sea-quark masses equal (for each flavor) (Bernard and Golterman 1994). In other words, real or "full" QCD is a "special case" of partially quenched QCD (PQQCD).[54]

This observation should then carry over to any EFT for PQQCD, and thus, in particular, to partially quenched ChPT (PQChPT). (Of course, we will have to reconsider the arguments for the existence of ChPT, a point to which we will return below.) What we will see is that, if both sea-and valence-quark masses are light enough for the chiral expansion to apply, being able to vary the valence and sea-quark masses independently is very useful: It gives us an extra tool for matching lattice QCD computations to ChPT, and thus determine the numerical values of the LECs at some scale – which pin down the EFT quantitatively. The key observation here is that those LECs are the *same* in the full and partially quenched theories, because they are independent of the quark masses, and thus also of the distinction that we make between valence-and sea-quark masses in the partially quenched generalization of QCD (Sharpe and Shoresh 2000). In addition, a very practical consideration is that in most cases it is much less expensive in terms of computational cost to vary the valence-quark masses (which only show up in the propagators coming from the external operators) than the sea-quark masses, which are part of the effective gauge action used to generate an ensemble of gauge-field configurations. We will consider partial quenching in detail in Secstion 8.4.2 and 8.4.3.

The motivation for choosing even the lattice Dirac operators different for valence and sea quarks is different. Of course, the basic assumption will be that in the continuum limit the difference in discretization disappears if the bare valence-and sea-quark masses are tuned to yield the same renormalized quark mass – an extension of the notion of universality. But, the symmetry properties of a correlation function such as Eq. (8.108) are determined by the symmetry properties of the valence Dirac operator, and in many cases it is a great advantage to be able to use a discretization with symmetries (in particular chiral symmetry) as close as possible to those of the continuum. Often, because of the smaller amount of symmetry on the lattice, operator mixing is more severe on the lattice, making computations of, for instance, weak matrix elements much harder, if not impossible. More symmetry, in particular more chiral and flavor symmetry, means less-mixing. However, lattice Dirac operators with good symmetry properties, such as domain-wall and overlap operators, are numerically very expensive. Since a large part of the computational effort goes into evaluating the determinant, it makes sense to choose a less-expensive lattice version for the sea-quark Dirac operator, such as Wilson or staggered operators. This "mixed-action" option will be discussed in Section 8.4.4.

Having extolled the possible virtues, we should address the question whether these generalizations of QCD make any field-theoretical sense.

8.4.1 Path integral

Of course, operationally, the procedure of choosing different Dirac operators for the valence and sea quarks outlined in the previous section is well defined. But, does it make field-theoretical sense? For instance, can we hope to develop EFTs for these generalized, but – at this stage – suspect versions of QCD? Clearly, in order to address these questions, it would help if we can write down a path-integral expression that reflects the distinction between valence and sea quarks. This can be done using a trick proposed by Morel (Morel 1987). Obviously, separate fermionic fields are needed for sea and valence quarks, but doing just this would lead to fermion determinants for both. The trick is to cancel the valence determinant by introducing yet another set of quark fields, with Dirac operator and mass matrix identical to those of the valence quarks, but with opposite (bosonic) statistics. We will refer to these fields as ghosts.[55] The fermionic part of the QCD Lagrangian generalizes to[56]

$$\mathcal{L} = \bar{q}_s(D_s + M_s)q_s + \bar{q}_v(D_v + M_v)q_v + \tilde{q}^\dagger(D_v + M_v)\tilde{q} \qquad (8.109)$$

in which $\tilde{q}$ denotes the bosonic ghost-quark fields; D_s is the lattice Dirac operator for sea quarks, and D_v that for valence quarks. In general, we will label sea quantities with subscript s, valence quantities with subscript v, and put tildes on entities referring to ghosts. Note that the ghost and antighost fields, $\tilde{q}$ and $\tilde{q}^\dagger$, are not independent. The path integral for this theory is

$$Z = \int \prod_\mu [dU_\mu] \prod_i [d\bar{q}_{s,i}][dq_{s,i}] \prod_j [d\bar{q}_{v,j}][dq_{v,j}][d\tilde{q}_j^\dagger][d\tilde{q}_j] \tag{8.110}$$

$$\exp\left(-S_{gauge} - \sum_x [\bar{q}_s(D_s + M_s)q_s + \bar{q}_v(D_v + M_v)q_v + \tilde{q}^\dagger(D_v + M_v)\tilde{q}]\right)$$

$$= \int \prod_\mu [dU_\mu] \exp(-S_{gauge}) \, \text{Det} \, (D_s + M_s),$$

which follows because of the exact cancellation of the valence and ghost determinants. Intuitively, at the level of diagrams, valence and ghost loops cancel, because they always come in pairs with opposite signs between them. Clearly, it is possible to couple the path integral to sources for all quark fields, and we can thus generate correlation functions for operators made out of any of these fields. In numerical computations, of course the ghost fields are never introduced in the first place. But one still has access to correlation functions that include ghost quarks, because after doing the Wick contractions one simply finds that each ghost–antighost contraction is replaced by a valence propagator. For example, if one considers the pion two-point function of Eq. (8.108) for pions made entirely out of ghost quarks, one finds exactly the same expression as on the last line of Eq. (8.108), except without the minus sign, which would be absent because ghost fields commute, rather than anticommute.

In order for definition (8.110) to make sense, we have to make sure that the integral over ghost fields converges. This is the case if D_v is anti-Hermitian, as it is in the continuum and for staggered fermions, for instance. In addition, we have to require that the mass matrix, M_v, has strictly positive eigenvalues. For Wilson or domain-wall fermions the Dirac operator is not anti-Hermitian (or Hermitian), and the real part of its eigenvalues can be of both signs. For a method of defining PQ or mixed-action QCD with valence Wilson fermions, I refer to Golterman *et al.* (2005).[57] For overlap fermions, the eigenvalues are also complex, but one can prove that if the overlap operator satisfies $D^\dagger = \gamma_5 D \gamma_5$, all eigenvalues have a non-negative real part. Thus, the ghost integral converges again as long as M_v has only positive eigenvalues.

It is clear that, while a Euclidean path-integral definition can be given, this theory is sick. The ghost quarks have the wrong statistics for a spin-1/2 field. Remember that if you try to quantize a spin-1/2 field with the wrong statistics, one of the problems that occurs is that the Hamiltonian of the theory is unbounded from below (see for instance Peskin and Schroeder (1995))! This and other (Bernard and Golterman 1996) aspects indicate that it may be problematic to continue the theory to a theory in Minkowski space with a positive Hamiltonian. But even if we could continue to Minkowski space, there is a problem with unitarity if one considers correlation functions with valence quarks on the external lines, while *different* sea quarks run on the loops.[58]

Fortunately, lattice QCD, as applied in modern numerical computations, does not want to be in Minkowski space. It is perfectly fine to match Euclidean lattice computations to a Euclidean EFT. Once the EFT has been completely defined by obtaining the values of all its LECs, we can take the continuum limit, set valence-quark masses equal to sea-quark masses, and continue to Minkowski space. The fact that PQQCD and mixed-action QCD can only be defined in Euclidean space is not a problem.

However, for all this to work, we do need to develop ChPT for partially quenched and mixed-action theories, and it is not obvious that this can be done. The generalized theory does not satisfy some of the cherished properties of a healthy quantum field theory, in particular unitarity,[59] which are usually invoked to argue that a local, unitary EFT such as ChPT has to exist, and, is in fact the most general possibility (Weinberg 1979). We will say more about this in Section 8.4.2 for PQQCD, and in Section 8.4.4 for mixed-action QCD.

Before we delve into the specifics, let us discuss some generalities considering the symmetries of Eq. (8.109). If we have N_s sea quarks (we will always take $N_s = 3$ or 2, as in previous sections) and N_v valence quarks (we can always pick as many valence quarks as we need, so N_v is arbitrary), it looks like the chiral symmetry group of Eq. (8.109) (for $M_s = 0$ and $M_v = 0$) is $[U(N_s)_L \times U(N_s)_R] \times [U(N_v|N_v)_L \times U(N_v|N_v)_R]$.[60] Here, $U(N|N)$ is a "graded" group, which we will discuss in more detail in the next section;[61] for now, all we need is that the symmetry group of the ghost part of the Lagrangian Eq. (8.109) alone would be the subgroup $U(N_v)_L \times U(N_v)_R$. It turns out that this is not quite correct (Damgaard *et al.* 1999, Zirnbauer 1996, Golterman *et al.* 2005), as I will now explain.

Since $\tilde{q}^\dagger$ is not independent of $\tilde{q}$, it follows that the ghost part of Eq. (8.109) is invariant under

$$\tilde{q}_L \to V\tilde{q}_L, \qquad \tilde{q}_R \to V^{\dagger -1}\tilde{q}_R, \tag{8.111}$$

with $V \in GL(N_v)$. Here

$$\tilde{q}_L = \frac{1}{2}(1 - \gamma_5)\tilde{q}, \qquad \tilde{q}_R = \frac{1}{2}(1 + \gamma_5)\tilde{q}, \tag{8.112}$$

from which it follows that (unlike for quark fields $\bar{q}$ and q, for which the chiral projections can be chosen independently, cf. Eq. (8.2))

$$\tilde{q}_L^\dagger = \tilde{q}^\dagger \frac{1}{2}(1 - \gamma_5), \qquad \tilde{q}_R = \tilde{q}^\dagger \frac{1}{2}(1 + \gamma_5). \tag{8.113}$$

Note that the definition of the projected $q^\dagger$s is opposite to that of the projected $\bar{q}$s. Equation (8.111) then follows from the fact that

$$\tilde{q}^\dagger D_v \tilde{q} = \tilde{q}_L^\dagger D_v \tilde{q}_R + \tilde{q}_R^\dagger D_v \tilde{q}_L. \tag{8.114}$$

Here, I assumed that $\{D_v, \gamma_5\} = 0$, which holds in the continuum. The lattice equivalent of the decomposition (8.114) will depend on the type of lattice fermion employed, as already noted above.[62] But in all cases, the fact that $\tilde{q}^\dagger$ is not independent of $\tilde{q}$

determines the precise form of the chiral symmetry group, and in all cases it has to coincide with $GL(N_v)$ in the continuum limit.

The vector subgroup, defined as the group that leaves the ghost condensate $\tilde{q}_L^\dagger \tilde{q}_L + \tilde{q}_R^\dagger \tilde{q}_R$ invariant, is the unitary subgroup, $U(N_v) \subset GL(N_v)$. If we restrict ourselves momentarily to the ghost sector only, this implies that Σ has to be an element of the coset $GL(N_v)/U(N_v)$. If we want to write Σ in terms of a meson field $\tilde{\phi}$ analogous to ϕ in Eq. (8.6), but with its meson component fields "made out of ghost quarks and antiquarks," we should write $\Sigma = \exp(2\tilde{\phi}/f)$, with $\tilde{\phi}$ a *Hermitian* matrix. The number of mesonic degrees of freedom is the same as it would be for the coset $[U(N_v)_L \times U(N_v)_R]/U(N)_V$, but the expression for Σ in terms of the meson fields is different.[63]

For non-perturbative issues, it is crucial to work with the correct symmetry group. This can already be seen from Eq. (8.109): If we introduce spurion fields M and $\bar{\text{M}}$ to write the ghost-mass term as

$$\tilde{q}^\dagger M_v \tilde{q} \to \tilde{q}_L^\dagger M \tilde{q}_L + \tilde{q}_R^\dagger \bar{\text{M}} \tilde{q}_R, \tag{8.115}$$

transforming as

$$M \to V^{\dagger-1} M V^{-1}, \qquad \bar{\text{M}} \to V \bar{\text{M}} V^\dagger, \tag{8.116}$$

we see that both M and $\bar{\text{M}}$ remain strictly positive under $GL(N_v)$ transformations, consistent with the convergence of the ghost path integral. This would not be true if we replace $GL(N_v)$ by $U(N_v)_L \times U(N_v)_R$.

One may want to use ChPT to find the phase structure of a quenched or partially quenched theory, as we did for $N_f = 2$ Wilson fermions in Section 8.3.3 (Golterman *et al.* 2005). In that case, one will have to be careful to start from the correct symmetries in constructing the chiral Lagrangian. However, if one is in a phase in which the correct vacuum has been determined (and usually in applications, is given by $\Sigma = 1$), and one is only interested in calculating the chiral expansion of physical quantities, it can be shown that instead the symmetry group $U(N_v|N_v)_L \times U(N_v|N_v)_R$ can be used for constructing the chiral Lagrangian (Golterman *et al.* 2005, Sharpe and Shoresh 2001). This is what we will do in the next section. The reason is that *in perturbation theory*, which is defined by expanding Σ in terms of $\tilde{\phi}$, we can perform a field redefinition $\tilde{\phi} \to i\tilde{\phi}$ that brings Σ into the "standard form," $\Sigma = \exp(2i\tilde{\phi}/f)$, i.e. the form that would have followed from starting with the ghost-sector symmetry group $U(N_v)_L \times U(N_v)_R$.

As an aside for the interested reader, there is another way to define PQQCD and mixed-action QCD, through the replica trick (Damgaard 2000, Damgaard and Spittorff 2000). Instead of introducing ghost quarks, one only introduces separate valence quarks, n_i for each flavor. One then takes the limit $n_i \to 0$ at the end of a calculation.[64] Formally, this sends the valence determinant to a constant, thus removing it from the QCD partition function. For all positive integer values of the n_i, this defines a physically sensible theory, in which we just happen to discretize different quarks in different ways. The disadvantage is that not much is known about the properties of the limit $n_i \to 0$. In this case, one bases the construction of the chiral Lagrangian on the symmetry group $[U(N_s)_L \times U(N_s)_R] \times [U(N_v)_L \times U(N_v)_R]$ (with $N_v = \sum_i n_i$), and considers ChPT in the $n_i \to 0$ limit. In all cases where explicit calculations have been done, the results agree from those obtained with Morel's trick, using ghost quarks.

8.4.2 Partially quenched ChPT

In this section, we will consider ChPT for PQQCD, the generalization of QCD in which we only choose the masses of valence and sea quarks to be independent, while using the same discretization of the Dirac operator. The first question is whether indeed EFTs such as ChPT can be constructed for PQQCD. We will assume that it can, as a Euclidean EFT for the underlying Euclidean theory. There exists no proof of this assertion. Some observations in support can be made, however.

1. One expects the SET to exist, with almost the same level of rigor as in the physical case. This expectation is based on two observations: First, I expect that the perturbative construction of the SET goes through just the same for the partially quenched case. Then, the fact that the SET also is valid beyond perturbation theory amounts basically to the assumption that the coefficients in the Symanzik expansion have a well-defined meaning beyond perturbation theory,[65] and it does not look like a drastic step to expect that this assumption is also valid in the partially quenched case. While this observation does not directly address the validity of partially quenched ChPT (PQChPT), the SET is a "stepping stone," and its validity is important for the construction of the appropriate chiral theory.

2. PQQCD contains full QCD as a subset: setting a valence-quark mass equal to a sea-quark mass makes that valence quark the same as a sea quark, and correlation functions with only sea quarks on the external lines are exactly those of full QCD. The collection of all correlation functions of PQQCD thus contains all correlation functions of full QCD as a subset. This allows us to make two observations: First, chiral-symmetry breaking in the sea sector (i.e. through a sea-quark condensate) takes place as usual. In PQQCD, one can use vector-like symmetries to rotate sea quarks into valence quarks, thus producing also a valence condensate (and, by extension, a ghost condensate equal to the valence condensate).[66] Secondly, for full QCD the SET and ChPT exist, and it seems not unreasonable to assume that the match between PQQCD and PQChPT can be extended away from $m_v = m_s$, at least as long as the difference $m_v - m_s$ does not become too large, relative to m_s and m_v. Below, we will have more to say about what "too large" means.

3. The assumption that EFTs such as SET and ChPT exist for purely Euclidean theories appears rather natural if one would imagine constructing the EFTs through the renormalization group. In this framework one would expect the EFT to be local below the scale above which one has integrated out all modes through some type of renormalization group blocking procedure. To be sure, there is no proof that the EFT obtained if one actually knew how to carry out the renormalization group blocking is local. But it is a "folklore" that seems equally applicable to PQQCD as to the standard case of full QCD.

Based on these observations, we will assume that, as before, the Lagrangian of the EFT of interest should be taken to be the most general local function of the fields consistent with the symmetries of the theory. In fact, it is interesting to follow this route, and see where it will lead us. Hopefully, PQChPT will exhibit, in a more concrete form, what the diseases of the theory are, and, hopefully, these will be the only diseases! If this is true, this makes it possible to use PQChPT in the interpretation of lattice results (read: use the predictions of PQChPT to fit data). If this works to a high degree of precision, that constitutes a non-trivial and important test of PQChPT, and thus of our understanding of PQQCD. As we will see, the tests become non-trivial once numerical computations "can see" the non-analytic terms predicted by PQChPT. Numerical evidence for such non-analytic behavior exists, and I expect that it will fairly rapidly become more extensive, at least in the Nambu–Goldstone-boson sector.

PQQCD has a much larger symmetry than Eq. (8.109), because now sea, valence, and ghost quarks can all be rotated into each other. (Only the mass matrices cause a soft breaking of this symmetry.) Following our earlier discussion, we will now think of $\tilde{q}^\dagger$ as independent of $\tilde{q}$, just as $\bar{q}$ is independent of q. The full chiral symmetry group (ignoring the mass matrix) is then $SU(N_s + N_v|N_v)_L \times SU(N_s + N_v|N_v)_R$, where we now omit an anomalous axial $U(1)$, as well as the $U(1)$ for quark number, which is trivially represented on mesons. The non-linear field $\Sigma = \exp{(2i\Phi/f)}$ now lives in the coset $[SU(N_s + N_v|N_v)_L \times SU(N_s + N_v|N_v)_R]/SU(N_s + N_v|N_v)_V$, and correspondingly, the field Φ describes $(N_s + 2N_v)^2 - 1$ mesons, counting precisely the number of pseudoscalar meson fields one can construct from the sea, valence, and ghost quarks and antiquarks when we leave out the singlet field that corresponds to the axial $U(1)$ (see below).

Before we continue, let us have a brief look at the graded group $SU(N_s + N_v|N_v)$ (for properties of graded groups, see for example deWitt (1984), Freund (1986)). An element $U \in SU(N_s + N_v|N_v)$ can be written in block form as

$$U = \begin{pmatrix} A & B \\ C & D \end{pmatrix}, \tag{8.117}$$

in which A is an $(N_s + N_v) \times (N_s + N_v)$ matrix of commuting numbers, D an $N_v \times N_v$ matrix of commuting numbers, while B and C are $(N_s + N_v) \times N_v$ and $N_v \times (N_s + N_v)$ matrices of anticommuting numbers, respectively. It is straightforward to check that if Σ has this structure, also Φ has to have this structure. The quark fields

$$Q = \begin{pmatrix} q_s \\ q_v \\ \tilde{q} \end{pmatrix}, \qquad Q = \begin{pmatrix} \bar{q}_s & \bar{q}_v & \tilde{q}^\dagger \end{pmatrix} \tag{8.118}$$

transform in the fundamental and antifundamental representations of $SU(N_s + N_v|N_v)$,[67] and in terms of these "super" quark fields, the PQQCD Lagrangian takes the simple form

$$\mathcal{L}_{PQQCD} = \mathcal{L}_{\text{gauge}} + Q(D + \mathcal{M})Q, \tag{8.119}$$

in which $\mathcal{M}$ is the mass matrix

$$\mathcal{M} = \begin{pmatrix} M_s & 0 & 0 \\ 0 & M_v & 0 \\ 0 & 0 & M_v \end{pmatrix}. \tag{8.120}$$

In the chiral theory, invariants are constructed using traces and determinants, and we need the generalization of these to graded groups. The "supertrace" of a matrix U as in Eq. (8.117) is defined as

$$\mathrm{str}(U) = \mathrm{tr}(A) - \mathrm{tr}(D). \tag{8.121}$$

The minus sign maintains the cyclic property, $\mathrm{str}(U_1 U_2) = \mathrm{str}(U_2 U_1)$. The "superdeterminant" is then defined through

$$\mathrm{sdet}(U) = \exp(\mathrm{str}\log(U)) = \det(A - BD^{-1}C)/\det(D), \tag{8.122}$$

for U of the form Eq. (8.117); this definition implies that $\mathrm{sdet}(U_1 U_2) = \mathrm{sdet}(U_1)\mathrm{sdet}(U_2)$. To understand this expression, first decompose U as

$$U = \begin{pmatrix} A & B \\ C & D \end{pmatrix} = \begin{pmatrix} 1 & BD^{-1} \\ 0 & 1 \end{pmatrix} \begin{pmatrix} A - BD^{-1}C & 0 \\ 0 & D \end{pmatrix} \begin{pmatrix} 1 & 0 \\ D^{-1}C & 1 \end{pmatrix}. \tag{8.123}$$

Now, using the definition $\mathrm{sdet}(U) = \exp(\mathrm{str}\log(U))$, the result follows, with the superdeterminant of the first and last factors on the right-hand side being equal to one.

Hermitian conjugation is defined as usual, with the proviso that complex conjugation reorders: $(ab)^* = b^* a^*$,[68] from which it follows that $(U_1 U_2)^\dagger = U_2^\dagger U_1^\dagger$. The group $SU(N_s + N_v | N_v)$ is now defined as the group of all unitary graded $(N_s + 2N_v) \times (N_s + 2N_v)$ matrices, with grading as in Eq. (8.117), and superdeterminant equal to one. Clearly, if $\Sigma = \exp(2i\Phi/f)$ has $\mathrm{sdet}(\Sigma) = 1$, this is equivalent with $\mathrm{str}(\Phi) = 0$.

The matrix Φ can be written in block form

$$\Phi = \begin{pmatrix} \phi & \eta \\ \bar{\eta} & \tilde{\phi} \end{pmatrix}, \tag{8.124}$$

in which ϕ contains all meson fields made out of fermionic quarks, $\tilde{\phi}$ contains all fields made out of ghost quarks, η contains those made out of a fermionic quark and a ghost antiquark, whereas $\bar{\eta}$ contains meson fields made out of a ghost quark and a fermionic antiquark. Note that ϕ is not the same as Eq. (8.6), because it also contains meson fields made out of valence and sea quarks ("mixed" pions) and meson fields made out of valence quarks only ("valence" pions). The upper left-hand $N_s \times N_s$ block inside ϕ can be identified with Eq. (8.6), for $N_s = 3$.

In practical applications, it turns out to be useful to relax the restriction that $\mathrm{str}(\Phi) = 0$, which makes $\Sigma \in U(N_s + N_v | N_v)$ instead of $SU(N_s + N_v | N_v)$. In conjunction, it is also useful to parametrize the diagonal of Φ not in terms of physical fields (i.e., those found by diagonalizing the mass matrix), but in terms of "single-flavor" fields:

$$\mathrm{diag}(\Phi) = (U, D, S, X, Y, \ldots, \tilde{X}, \tilde{Y}, \ldots). \tag{8.125}$$

Here, we have introduced a now commonly used notation for the valence quarks $x, y, \ldots$, with $U \sim u\bar{u}$, $X \sim x\bar{x}$, etc, instead of $u_v, d_v, \ldots$, to save writing indices. One can have as many valence quarks as one needs – since they do not contribute to the dynamics, their number need not be fixed. This explains the dots in Eq. (8.125); of course, the number of ghost quarks needs to be equal to the number of valence quarks.[69] The "super"-singlet field (the "super-η'"),

$$\Phi_0 \equiv \mathrm{str}(\Phi) = U + D + S + X + Y + \cdots - (\tilde{X} + \tilde{Y} + \ldots), \tag{8.126}$$

is not a Goldstone meson, because of the axial anomaly. This can easily be seen from the contribution of all quarks to the triangle diagram: in order for the ghost-quark contributions to *add* to the anomaly, one should include their contributions with an explicit minus sign, because they do not get a sign from the loop. This means that we should really remove the field Φ_0 from the chiral Lagrangian (as we have been doing thus far). We will do this by giving it a large mass, that we will eventually send to infinity, thus decoupling the super-η'.[70]

It is time to get to the partially quenched chiral Lagrangian! The lowest-order form is

$$\mathcal{L}^{(2)} = \frac{1}{8} f^2 \, \mathrm{str}(\partial_\mu \Sigma^\dagger \partial_\mu \Sigma) - \frac{1}{8} f^2 \, \mathrm{str}(\chi^\dagger \Sigma + \Sigma^\dagger \chi) + \frac{1}{6} m_0^2 \, (\mathrm{str}(\Phi))^2, \tag{8.127}$$

in which χ is a spurion for the quark masses, that should be set equal to $2B_0\mathcal{M}$. We note that the field $\Phi_0 = \mathrm{str}(\Phi)$ is not constrained by symmetry (except parity), and thus we should really multiply every term in Eq. (8.127) by an arbitrary (even) function of Φ_0.[71] However, we will be decoupling the super-η', and that removes the dependence on all parameters contained in these potentials (Sharpe and Shoresh 2001).

From this Lagrangian, we can read off the lowest-order expression for all meson masses in the theory. This is straightforward for the off-diagonal fields in Φ. By expanding Eq. (8.127) to quadratic order, one finds that the mass of a meson with flavors i and j is given by

$$m_{ij}^2 = B_0(m_i + m_j). \tag{8.128}$$

The only thing "different" is that the ghost-meson propagators get an extra minus sign because of the supertrace in Eq. (8.127),

$$\langle \tilde{\phi}_{ij}(p)\tilde{\phi}_{ji}(q) \rangle = \frac{-1}{p^2 + m_{ij}^2} \, \delta(p - q). \tag{8.129}$$

This is a first example of how PQChPT makes the "diseases" of PQQCD visible. For fermionic mesons, one has to remember that the ordering of η and $\bar{\eta}$ matters. The fermionic nature of these fields will provide minus signs when calculating loops, thus providing a mechanism through which the cancellation between valence-quark and ghost-quark loops is built into PQChPT.

The flavor-neutral sector is a little more complicated, because of the m_0^2 term (or, equivalently, because of mixing between the U, D, etc fields). Let us calculate the propagator $\langle \Phi_{ii} \Phi_{jj} \rangle$. From the first two terms in Eq. (8.127), there is a contribution

$$G_{ij} \equiv \langle \Phi_{ii} \Phi_{jj} \rangle = \frac{\epsilon_i \delta_{ij}}{p^2 + m_{ii}^2}, \tag{8.130}$$

$$\epsilon_i = \begin{cases} +1, \ i \ \text{sea or valence} \\ -1, \ i \ \text{ghost} \end{cases},$$

from the first two terms in Eq. (8.127), with a minus sign as in Eq. (8.129) if i refers to a ghost. There are additional contributions coming from the m_0^2 term, that can be found by treating this term as a two-point vertex, and doing the geometric sums. In detail, we can rewrite

$$\frac{1}{6} m_0^2 \left(\text{str}(\Phi) \right)^2 = \frac{1}{6} m_0^2 \Phi_{ii} K_{ij} \Phi_{jj}, \tag{8.131}$$

with

$$K_{ij} = \epsilon_i \epsilon_j. \tag{8.132}$$

One finds, using that

$$(KGK)_{ij} = K_{ij} \sum_k G_{kk}, \tag{8.133}$$

for the contribution proportional to m_0^2 (Bernard and Golterman 1994, Sharpe and Shoresh 2000):

$$-m_0^2/3 \, (GKG)_{ij} + \left(-m_0^2/3 \right)^2 (GKGKG)_{ij} + \ldots \tag{8.134}$$

$$= \frac{-m_0^2/3}{(p^2 + m_{ii}^2)(p^2 + m_{jj}^2)} \left(\frac{1}{1 + \sum_{k=u,d,s} \frac{m_0^2/3}{p^2 + m_{kk}^2}} \right)$$

$$= \frac{-m_0^2/3}{(p^2 + m_{ii}^2)(p^2 + m_{jj}^2)} \frac{(p^2 + m_U^2)(p^2 + m_D^2)(p^2 + m_S^2)}{(p^2 + m_{\pi^0}^2)(p^2 + m_\eta^2)(p^2 + m_{\eta'}^2)}$$

$$\rightarrow \frac{-1/3}{(p^2 + m_{ii}^2)(p^2 + m_{jj}^2)} \frac{(p^2 + m_U^2)(p^2 + m_D^2)(p^2 + m_S^2)}{(p^2 + m_{\pi^0}^2)(p^2 + m_\eta^2)}.$$

Let us step through this equation line by line. In the first expression on the right-hand side, the sum in the denominator really extends over all flavors, but the ghost terms cancel the valence terms, reducing the sum to run over the sea flavors u, d and s only. The second expression is just a rewriting of the first line. It is clear that the denominator of the second factor is a third-order polynomial in p^2, and that it has to have poles at the physical masses (this factor only refers to the sea sector). These are of course the masses of the π^0, η and η'. The η' mass-squared contains a term proportional to m_0^2,

$$m_{\eta'}^2 = \frac{N_s}{3} m_0^2 + \text{terms proportional to sea-quark masses}, \tag{8.135}$$

whereas the π_0 and η become independent of m_0^2 when we take it to infinity. This observation explains the last line in Eq. (8.134). In Eq. (8.135) I made the dependence on the number of sea quarks, N_s, explicit – of course, in our explicit example above $N_s = 3$.

This result for the neutral propagators exhibits a new property of the partially quenched theory. The easiest way to see this is to note that, if we take $i = j$ to be a valence flavor, there is a double pole in Eq. (8.134). This is a clear sickness of the partially quenched theory, reflecting the fact that it does not correspond to a standard field theory in Minkowski space. We should expect, however, that if we take both i and j to correspond to sea flavors, that everything works as in the full theory: by design the sea sector does not know about valence and ghost quarks! Indeed, if we pick both i and j one of u, d, s, we see that at least one of these poles cancel against a factor in the numerator, leaving us with an expression that can be written as a sum over single poles. (This also works if there is a degeneracy in the sea sector.) It is a nice exercise to work out the propagators for the sea π^0 and η, and find that indeed they agree with the expressions one finds in the full theory. A very simple example is the case of degenerate sea masses, $m_U = m_D = m_S \equiv m_{\text{sea}}$, for which we obtain for the neutral sea propagator the expression

$$\langle \Phi_{\text{sea}} \Phi_{\text{sea}} \rangle = \left(1 - \frac{1}{3}\right) \frac{1}{p^2 + m_{\text{sea}}^2}, \tag{8.136}$$

where the "1" term comes from the first two terms in Eq. (8.127), and the "1/3" term comes from Eq. (8.134), and projects out the η'.

So, while the sea sector is healthy, as it should be, the neutral valence sector is sick. So what? Maybe this just tells us that we should not consider flavor-neutral valence pions.[72] However, even if we make this choice, the sickness permeates to the flavored sector as well, as I will now argue.

We should be able to calculate chiral logarithms in meson masses, decay constants, etc, from Eq. (8.127) as usual. The four-point vertices following from $\mathcal{L}^{(2)}$ have a flavor structure $\Phi_{ij}\Phi_{jk}\Phi_{kl}\Phi_{li}$, dressed up with derivatives or quark-mass factors. Consider a one-loop contribution to $\langle \Phi_{ij}\Phi_{ji} \rangle$, with $i \neq j$ so as to stay away from the dangerous neutral sector. Because we only have four-point vertices, all one-loop diagrams are tadpoles, with Φ_{ij} and Φ_{ji} connected to the same vertex. One of the Wick contractions is the one in which the index structure on the vertex is $\Phi_{ij}\Phi_{jj}\Phi_{ji}\Phi_{ii}$, where Φ_{ii} and Φ_{jj} have to contract and form the loop! We see that the neutral propagator terms of Eq. (8.134) unavoidably show up, if we want to make use of partial quenching at all. For simplicity, let us work out the degenerate case $m_U = m_D = m_S = m_{\text{sea}}$, $m_{ii} = m_{jj} \equiv m_{\text{val}}$. Simplifying Eq. (8.134) accordingly, and integrating over p, as one would in the one-loop diagram, we obtain

$$-\frac{1}{3} \int \frac{d^4p}{(2\pi)^4} \frac{p^2 + m_{\text{sea}}^2}{(p^2 + m_{\text{val}}^2)^2} = \frac{1}{3} \frac{d}{dm_{\text{val}}^2} \int \frac{d^4p}{(2\pi)^4} \frac{p^2 + m_{\text{sea}}^2}{p^2 + m_{\text{val}}^2} \tag{8.137}$$

$$\to \frac{1}{48\pi^2}(m_{\text{sea}}^2 - 2m_{\text{val}}^2)\log\left(\frac{m_{\text{val}}^2}{\Lambda^2}\right).$$

The arrow in the second line indicates that I only kept the chiral logarithm. This chiral logarithm is not of the structure we encounter in full QCD, where no infrared divergences can occur. Here, there is an infrared divergence: Eq. (8.137) diverges when we take the valence mass to zero at fixed sea mass. Only when we take these masses in a fixed ratio can we define the limit. This means that the chiral expansion in the partially quenched theory will not converge if we take the sea and valence masses too different.[73] We should not only keep m_{val}^2 and m_{sea}^2 small in the sense of the chiral expansion, but we should also keep the unphysical infrared scale $m_{\text{sea}}^2 - m_{\text{val}}^2$ small compared to either of those, $|m_{\text{sea}}^2 - m_{\text{val}}^2| \lesssim \min(m_{\text{sea}}^2, m_{\text{val}}^2)$. For current numerical simulations, these conditions are well satisfied, so that the infrared divergences of the partially quenched theory do not constitute a problem in practice.

There are many examples of the unphysical infrared behavior of the partially quenched theory – the phenomenon is generic. Here, I briefly mention two of them; for detailed explanations, see the original papers quoted below.

The first example occurs in scalar-isoscalar and scalar-isovector propagators (Bardeen *et al.* 2001, Prelovsek *et al.* 2004). In that case, there are one-loop contributions with $\pi - \pi$ or $\pi - \eta$ intermediate states. In the isopin limit, Eq. (8.134) contributes to the η internal line. With the additional pion internal line, this leads to an integral like Eq. (8.137), but with three powers of $p^2 + m_{\text{val}}^2$ in the denominator.[74] This leads to an infrared divergence (at fixed m_{sea}^2) that goes like $(m_{\text{sea}}^2 - m_{\text{val}}^2)/m_{\text{val}}^2$, worse than the logarithmic divergence of Eq. (8.137).

A second example where the double pole has a dramatic effect is in the contribution from one-pion exchange to the nucleon–nucleon potential. In full QCD one-pion exchange leads to a Yukawa potential, $\exp(-m_{\text{val}}r)/r$, whereas double-pole terms in the exchange lead, instead, to a potential of the form $(m_{\text{sea}}^2 - m_{\text{val}}^2)\exp(-m_{\text{val}}r)/m_{\text{val}}$, as can be seen by differentiating with respect to m_{val}^2, like we did in Eq. (8.137) (Beane and Savage 2002). The double pole thus leads to an unphysical interaction that dominates the physical term at large distance.

In all cases, the unphysical infrared effects are, of course, proportional to $m_{\text{sea}}^2 - m_{\text{val}}^2$. In general, the message is that partial quenching can be a useful tool, but in order to make use of PQChPT, we do not only need m_{sea}^2 and m_{val}^2 to be small enough, but also their difference, $m_{\text{sea}}^2 - m_{\text{val}}^2$.

Having done all the preparatory work, let us consider the $O(p^4)$ result for the valence pion mass, in the case of degenerate sea quarks, and for $m_x = m_y$ (Sharpe 1997):[75]

$$m_X^2 = 2B_0 m_x \left\{ 1 + \frac{2B_0}{24\pi^2 f^2}\left((2m_x - m_{\text{sea}})\log\left(\frac{2B_0 m_x}{\Lambda^2}\right) + m_x - m_{\text{sea}}\right) \right.$$

$$\left. + \frac{32B_0}{f^2}\left((2L_8 - L_5)m_x + 3(2L_6 - L_4)m_{\text{sea}}\right) \right\}. \qquad (8.138)$$

The chiral logarithms in this expression precisely come from the "double-pole" part of the neutral propagator, Eq. (8.134), i.e. they are of the type Eq. (8.137).[76]

This result, finally, allows us to return to the point as to why partial quenching is useful. As we have seen, full QCD is contained in PQQCD – the only difference is that we choose different masses for the valence and the sea quarks. Since the LECs are, by construction, independent of the quark masses, the LECs in PQChPT and full ChPT are identical. In other words, partially quenched lattice computations give us access to the real-world values of the LECs of ChPT (Sharpe and Shoresh 2000); the fact that we can vary the valence-quark masses independently simply gives us another "knob to turn." This can be very useful: it is clear that by varying the valence-quark masses independently, one can extract $2L_8 - L_5$ from the pion mass, without varying the sea-quark masses. This stands in contrast to the full theory: as can be seen from Eq. (8.40), in order to separate $2L_8 - L_5$ from $2L_6 - L_4$, one would have to vary the sea-quark mass. It is important to note that the LECs *do* depend on the number of "dynamical" flavors, N_s – one still has to do the numerical lattice computation with the correct number of sea quarks.

While it is true that the *values* of the LECs in PQChPT and full ChPT are the same, there can be *more* LECs in the partially quenched case. This already shows up at order p^4: the operator $\mathrm{tr}(L_\mu L_\nu L_\mu L_\nu)$ is not independent from those in Eq. (8.17) for $SU(3)$ cf. Section 8.2.2, but $\mathrm{str}(L_\mu L_\nu L_\mu L_\nu)$ is for $SU(3 + N_v|N_v)$ (Sharpe and Van der Water 2004).[77] One thus obtains the most general $O(p^4)$ Lagrangian for PQChPT by replacing traces with supertraces in Eq. (8.39), and by adding

$$L_{PQ}\left(\mathrm{str}(L_\mu L_\nu L_\mu L_\nu) + 2\,\mathrm{str}(L_\mu L_\mu L_\nu L_\nu) + \tfrac{1}{2}\left(\mathrm{str}(L_\mu L_\mu)\right)^2 - \right.$$

$$\mathrm{str}(L_\mu L_\nu)\,\mathrm{str}(L_\mu L_\nu)), \tag{8.139}$$

where the new operator is written in this way because this combination vanishes for $SU(3)$. This new operator does not contribute at tree level, because at tree level there is no distinction with full QCD, for which $SU(3)$ is the relevant symmetry group. It can contribute to one-loop diagrams, and does so in the case of pion scattering, for instance (Sharpe and Van der Water 2004), where, of course, this only happens if valence masses are chosen different from sea masses.

Finally, we remark that quenched QCD, and thus quenched ChPT corresponds to the special case that $N_s = 0$ (Bernard and Golterman 1992, Sharpe 1992). While we are not pursuing quenched QCD in these lectures, it is worth noting that this special case has, in fact, some special properties. For example, one cannot take the limit $m_0^2 \to \infty$, as inspection of Eq. (8.134) reveals (the factor in parentheses on the first line gets replaced by one). Of course, quenched QCD does not have the correct number of light sea quarks, and is thus not only sick, but terminally ill. No physical result with controlled errors can ever be obtained from the quenched approximation.

8.4.3   Quark flow

In the previous section, we have seen that PQChPT can be systematically developed for PQQCD. Once we assume that this EFT exists just as in the full-QCD case, the construction is quite straightforward. But, one may like to have a more pictorial understanding of how it all works, and this is provided by the quark-flow picture (which, in a sense, is formalized through the replica trick, but here I will not attempt to present it in that way).

We have two types of quarks: valence and sea quarks. Valence quarks are represented by propagators, as in Eq. (8.108), and in a diagrammatic language, can be represented by valence-quark lines. Each valence-quark line has to have a beginning and an end at some external source or sink. Valence-quark loops can be formed for instance if more than one valence line connects the same source and sink, as in the example of Eq. (8.108). The "valence quark part" of a contribution to a correlation function can thus be precisely defined, and the topology of valence lines represents the various possible Wick contractions that contribute to a particular correlation function.

In perturbation theory, the (logarithm of the) fermion determinant, which comes only from sea quarks, can be represented by an infinite sum over one-loop diagrams with any number of gluon lines attached to the sea-quark loop. One thus imagines the diagram coming from a certain Wick contraction in the valence sector to be dressed up with gluons and sea quark loops in all possible ways. In this way one can also represent sea-quark contributions through loops, but clearly this picture is not rigorous outside perturbation theory. In this picture, however, the reason that only internal (i.e. not connected to a source or sink) sea-quark loops occur is because valence and ghost loops, which in principle also are present,[78] cancel each other. It is instructive to consider PQChPT in terms of quark loops.

Figure 8.6 shows quark-flow representations of the meson propagators at tree level, as well as the four-point meson vertex. The flavored propagators that follow from Eq. (8.127) are represented in "double-line" notation, with each line denoting a quark of a certain flavor, Fig. 8.6(a). Note that we only draw valence lines – one is to imagine these diagrams arbitrarily dressed up with gluons and sea-quark loops. In the flavor-neutral sector, we have seen that there are additional contributions, which follow from iterating the two-point vertices proportional to m_0^2. Flavor-neutral mesons get a contribution as in Fig. 8.6(a) (just set $i = j$), but also get a contribution from Fig. 8.6(b), where the "double hairpin" represents the m_0^2 two-point vertices. This is what we usually call a "disconnected" diagram in lattice QCD, because it represents a disconnected valence Wick contraction, but again, it is not really disconnected in terms of gluons. Note that the double hairpin stands for the full geometric sum (8.134) of all insertions of the m_0^2 vertices, where each additional insertion of a double-hairpin vertex corresponds to another sea-quark loop inserted between the two hairpins.

Figure 8.6(c) represents a vertex with flavor structure $\Phi_{ij}\Phi_{jk}\Phi_{kn}\Phi_{ni}$, which follows from the single-trace terms in $\mathcal{L}^{(2)}$, whereas Fig. 8.6(d) represents the flavor structure $\Phi_{ij}\Phi_{ji}\Phi_{kn}\Phi_{nk}$, which is a double-trace term. The latter do not occur in $\mathcal{L}^{(2)}$, but they do occur in $\mathcal{L}^{(4)}$. Various one-loop contributions to a pion propagator are shown in Fig. 8.7. Each internal quark loop has to be summed over only the sea quarks, because valence and ghost internal loops cancel each other. It so happens that only diagrams of type 8.7(c) contribute at order p^4 to meson masses; more types contribute to decay constants.

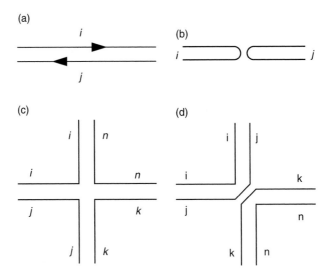

Fig. 8.6 Quark-flow diagrams for ChPT Feynman rules. (*a*): flavored propagator, (*b*): m_0^2 part of the flavor-neutral propagator, cf. Eq. (8.134), (*c*): single-trace four-point vertex, (*d*): double-trace four-point vertex. Arrows on quark lines are shown in (*a*), and understood in the other figures. See text.

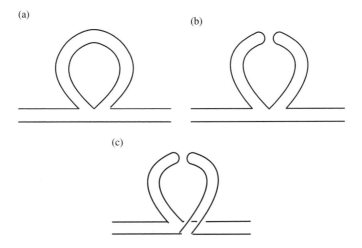

Fig. 8.7 Quark-flow diagrams for various one-loop contributions to the pion self-energy.

From these examples, it is clear that there is no one-to-one correspondence between PQChPT diagrams and quark-flow diagrams. The PQChPT diagram representing all diagrams of Fig. 8.7 is just the tadpole shown in Fig. 8.8(a). One can refine this a little by representing the hairpin vertex by a cross, as in Fig. 8.8(b). Of course, one can also use the quark-flow picture for correlation functions in full ChPT, and one

(a)   (b)

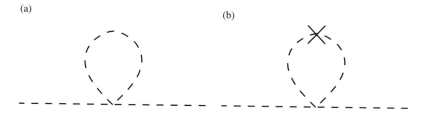

Fig. 8.8 ChPT diagrams for the pion one-loop self energy. A dashed line stands for one of the meson fields in Φ. The cross stand for an insertion of Eq. (8.134).

can use those as a starting point to develop PQChPT intuitively, as was originally done for quenched ChPT (Sharpe 1992). It can be very helpful in seeing how ChPT should work for some "modified" version of QCD. For an example of applying this to staggered ChPT for staggered QCD with the fourth-root procedure, see Aubin and Bernard (2003a). However, it is important to have a path-integral representation for each modified version of QCD,[79] as this gives us the opportunity to use many of the standard tools of field theory. Examples of these tools are the precise definition and role of the symmetries of the modified theory, and the availability of field redefinitions (Bernard and Golterman 1994, 1992).

Double-hairpin diagrams as in Fig. 8.6(b) contribute only to flavor-neutral propagators (at tree level). The quarks on each of the hairpin lines are valence quarks, and the quark–antiquark pair that each of these hairpins represents gets bound into a valence meson – we thus expect each of these hairpins to represent a single meson pole with the mass of a valence pion. Any other quark loops in this diagram are not explicitly shown, and can only come from the fermion determinant. In other words, any such quark-loop contributions can only involve the sea quarks, and, if sea-quark masses do not equal valence-quark masses, nothing in the diagram can remove the two valence-pion poles. We conclude, at this heuristic level, that double-pole terms have to occur in neutral pion propagators in PQQCD, and this is, of course, confirmed in PQChPT.[80] Because double poles cannot occur in full QCD, we also conclude that the residue of these double poles has to be proportional to $m_{\text{sea}} - m_{\text{val}}$.

Another observation that is easily made in the quark-flow picture is how PQQCD gives access to Wick contractions not available in full QCD. Suppose we do take all valence masses equal to the sea masses, so that the theory is not really partially quenched. The only effect of partial quenching is that we have more flavors available to put on the external lines than run around on the internal loops that come from the fermion determinant. We can use this, for example, to separate the disconnected part, Fig. 8.6(b), from the connected part, Fig. 8.6(a), of any flavor-neutral meson propagator. In other words, partial quenching separates the contribution of different Wick contractions to the same correlation function. At the level of quark-flow diagrams, this is obvious. But since all that happens at the quark-flow level is encoded in PQChPT, we can use this to obtain systematic chiral expansions for these separate Wick contractions. A simple example follows from Eqs. (8.130) and (8.134). In QCD with one flavor, these two contributions cannot be separated. But, one can add valence quarks to the theory (keeping the number of sea quarks at just one), and use this to separate the connected and disconnected parts (8.130) and (8.134) even in QCD with one flavor. This can for instance be used to define what it means to set the quark mass to zero, by requiring that the pion mass obtained from the connected part vanishes (Farchioni *et al.* 2007).

Another example of this is given in Sharpe and Shoresh (2001), in which it is shown how this can be used for a determination of L_7. For other work using the "extra" quarks available in the partially quenched theory to separate connected and disconnected Wick contractions, see Chen and Savage (2002) for an application to electromagnetic properties of baryons, Golterman and Pallante (2004), Aubin *et al.* (2006), for an application to quenched and partially quenched nonleptonic kaon decays, and Juttner and Della Morte (2009) in the context of the hadronic contribution to $g - 2$.

8.4.4 ChPT for mixed actions

QCD with a mixed action is a generalization of PQQCD, because not only do we treat the valence- and sea-quark masses as independent, but also their lattice Dirac operators. An important consequence is that mixed-action QCD has fewer flavor symmetries than its partially quenched cousin (Bär *et al.* 2003). Revisiting the general observations in support of the validity of PQQCD in Section 8.4.2, we note that observations 1 and 3 still apply, but that observation 2 has to be modified. It is still true that full QCD is a subset of mixed-action QCD, but it is no longer true that one can rotate sea quarks into valence quarks.

If we consider a theory with two sets of sea quarks, with each set discretized with a different lattice Dirac operator, there would be little doubt that this constitutes a valid discretization of QCD, because of universality. One would expect that the framework of Section 8.3 applies to such a theory as well, with the necessary technical modifications. In mixed-action QCD, however, one of these sets is quenched, making it less obvious that a universal continuum limit exists. But if it exists, it will clearly have to be the corresponding partially quenched version of QCD. Hence, if one combines the assumption that PQChPT is the correct EFT for PQQCD with the notion of universality of different discretizations, it appears reasonable that we can combine the techniques of Sections 8.3 and 8.4.2 to construct a chiral EFT for mixed-action QCD.

The relevant symmetry group clearly depends on what discretization we use for the valence and sea quarks. For definiteness, let us consider the case that we choose Wilson fermions for the sea quarks, and overlap fermions[81] for the valence quarks.[82] The lattice symmetry group (ignoring the soft breaking by quark masses, and removing "overall" $U(1)$ factors as usual) is then $SU(N_s)_V \times [SU(N_v|N_v)_L \times SU(N_v|N_v)_R]$. With overlap fermions, the symmetry group in the valence sector is the same as in the continuum, but Wilson fermions have no chiral symmetry, and the symmetry group is thus the smaller group of vector-like transformations only.

We can now construct the chiral Lagrangian (valid in the GSM regime to order p^4, and in the LCE regime to order p^2) by writing down the SET for the mixed-action theory, introducing spurions, and making the transition to the chiral Lagrangian as we did in Section 8.3, and this has been done systematically in Bär *et al.* (2004). However, it is rather simple to understand the result, so, rather than repeat the whole analysis, I will first quote the result, and then comment on the various terms. The result is

$$\mathcal{L} = \mathcal{L}_{\text{cont}}^{PQQCD}(2B_0 M_s \rightarrow 2B_0 M_s' \equiv 2B_0 M_s + \hat{a}\mathbf{1}) \tag{8.140}$$

$$+\Delta\mathcal{L}^{(4)}(\hat{a} \rightarrow \hat{a}P_s) - \hat{a}^2 W_M \, \text{str}(P_s \Sigma P_s \Sigma^\dagger),$$

with $\Delta\mathcal{L}^{(4)}$ given by Eq. (8.56), and the subscript cont stands for "continuum." Here, P_s is the projector on the sea sector, defined by $P_s q_s = q_s$ and $P_s q_v = 0$, $P_s \tilde{q} = 0$. While this equation contains a lot of information, it is easy to understand this result. First, since the valence quarks have exactly the same chiral symmetries as in the continuum, there are no explicit $O(a^2)$ terms in the valence sector of the chiral Lagrangian. The only such terms come from the Wilson sea sector, and this explains the appearance of the projector P_s in Eq. (8.140), and the fact that only the sea-quark masses get shifted by an $O(a)$ term, as in Eq. (8.52). Of course, there are scaling violations in the valence sector as well, but they can only occur as $O(a^2)$ corrections to the continuum LECs, f, B_0 and L_i; but in the GSM regime these are of order p^6.[83] Rotational symmetry is always broken on the lattice, but, as we have seen before, the effect of that only shows up at higher orders in the chiral Lagrangian. Essentially, the form of the Lagrangian (8.140) is thus that of the continuum Lagrangian, plus the symmetry-breaking lattice part for the sea sector. However, one completely new term does appear in Eq. (8.140), and that is the term with the new LEC W_M. This term is allowed, because there are no symmetries that connect the valence and sea sectors. The new term does not break (continuum) chiral symmetry restricted to the sea sector, so a similar term also appears if we choose different types of sea quarks, such as staggered or domain-wall quarks. Since this new term does not break chiral symmetry, a similar term also arises in the valence sector. But since we have that $P_v = 1 - P_s$ and $\Sigma\Sigma^\dagger = 1$, we can write

$$\text{str}(P_v \Sigma P_v \Sigma^\dagger) = \text{str}(P_s \Sigma P_s \Sigma^\dagger) + \text{str}(\mathbf{1}) - 2\,\text{str}(P_s), \tag{8.141}$$

so only one of these terms is independent. If we take the valence quarks to be the same as the sea quarks, the projector P_s is replaced by the unit matrix, and the new term is an irrelevant constant. But with a mixed action this is not the case, and the new term should be taken into account.

Let us consider the $O(p^2)$ expressions for meson masses in the LCE regime, beginning with the flavored ones. By expanding the Lagrangian to quadratic order in Φ, we find, using now Latin indices for the valence flavors and Greek indices for sea flavors, in order to distinguish them more clearly:

$$m_{ij}^2 = B_0(m_i + m_j), \tag{8.142}$$

$$m_{\alpha\beta}^2 = B_0(m_\alpha' + m_\beta') + \frac{32\hat{a}^2}{f^2}(N_s W_6' + W_8'),$$

$$m_{i\alpha}^2 = B_0(m_i + m_\alpha') + \frac{4\hat{a}^2}{f^2}(4N_s W_6' + 2W_8' + W_M).$$

The factors N_s in these expressions come from factors $\mathrm{str}(P_s)$ that occur working out the expansion in Φ. The new LEC W_M only shows up in the "mixed" pion mass, i.e. the pion made of a valence and a sea quark, hence there is no $O(p^2)$ relation between the pure valence, pure sea, and mixed pion masses.[84] In any application of QCD with a mixed action, the mixed pion mass gives thus an important indication whether one should use GSM or LCE power counting.

In the flavor-neutral sector, all the complications of PQQCD of course are inherited by the mixed-action case, including the appearance of double-pole terms, but a new issue also arises. In the sea sector, to quadratic order, there is an additional contribution to the m_0^2 term in Eq. (8.127), coming from the W_7' term in Eq. (8.56) shifting the value of m_0^2 *only in the sea sector* (because of the P_s projectors, cf. Eq. (8.140)) effectively to

$$m_{0,\mathrm{eff}}^2 = m_0^2 + \frac{96\hat{a}^2}{f^2}W_7'. \tag{8.143}$$

For a theory with degenerate valence and degenerate sea quarks, one finds for the flavor-neutral propagator in the valence sector (Golterman *et al.* 2005)

$$\langle\phi_{ii}\phi_{jj}\rangle = \frac{\delta_{ij}}{p^2 + m_{\mathrm{val}}^2} - \frac{1}{3}\frac{m_0^2(p^2 + m_{\mathrm{sea}}^2) + (N_s m_0^2/3)(m_{0,\mathrm{eff}}^2 - m_0^2)}{(p^2 + m_{\mathrm{val}}^2)^2(p^2 + m_{\mathrm{sea}}^2 + N_s m_{0,\mathrm{eff}}^2/3)} \tag{8.144}$$

$$\rightarrow \left(\delta_{ij} - \frac{1}{N_s}\right)\frac{1}{p^2 + m_{\mathrm{val}}^2} - \frac{(m_{\mathrm{sea}}^2 - m_{\mathrm{val}}^2)/N_s + 32\hat{a}^2 W_7'/f^2}{(p^2 + m_{\mathrm{val}}^2)^2},$$

where the second line is obtained by taking $m_0 \to \infty$ in the first line of this equation.

The importance of this result lies in the fact that scaling violations are not universal. Naively, one might think that the best way to conduct a mixed-action lattice computation is to tune the quark masses such that the valence-pion mass is equal to the sea-pion mass. However, comparison between Eqs. (8.142) and (8.144) shows that this is not the same as tuning quark masses such that the residue of the double pole vanishes. If one tunes quark masses to make sea- and valence-pion masses equal, there will still be a double pole in the theory, with a residue of order a^2. Since the double pole can lead to infrared "enhancement" effects (such as the enhanced chiral logarithms and other such examples we encountered in Section 8.4.2), such scaling violations could be larger than one would expect based on a simple estimate of the size of $(a\Lambda_{\mathrm{QCD}})^2$. This phenomenon does not occur at this order in the case that one uses overlap valence quarks on a staggered sea; because both valence and staggered quarks have exact chiral symmetries, in that case the effect can occur at the earliest at order $a^2 m_{\mathrm{quark}}^2$ (Golterman *et al.* 2005, Bär *et al.* 2005).

One-loop universality. As soon as the lattice spacing is large enough that we need to include it as a variable in chiral fits, the version of ChPT we need to use will depend in some detail on what kind of lattice action is being used. For instance, ChPT for Wilson fermions is quite different from ChPT for staggered fermions. This is in general also true when a mixed action is used. This is not a problem of principle, but for each new type of lattice action, the necessary ChPT calculations need to be repeated.

It would be very nice if some general prescription existed that allowed one to take a calculation carried out in continuum PQQCD, and modify it using some simple rules into the appropriate mixed-action ChPT expressions. While this is clearly not possible in general, it is possible under some limitations, which, however, cover many cases of practical interest.

As we noted before, the use of a mixed action is particularly useful when the valence fermions have exact chiral symmetry (broken only by the physical quark masses), while the sea sector has much less symmetry. If we allow only valence quarks on the external lines of any correlation function of interest, in order to access what we will refer to as "valence quantities," and we work to leading order (i.e. tree level) in ChPT, one clearly needs just that part of the $O(p^2)$ chiral Lagrangian that refers only to the valence sector – one can simply set all sea-meson and mixed-meson fields equal to zero. This is simply the observation that sea quarks (and thus sea and mixed mesons) can only appear on loops. If the lattice valence quarks have exact chiral symmetry, the $O(p^2)$ chiral Lagrangian is the same as in the continuum, i.e. it takes the form of Eq. (8.33).

What is remarkable is that a similar argument holds at order p^4, i.e. to one loop (Chen *et al.* 2007, 2009b). For one-loop contributions we need only the $O(p^2)$ Lagrangian, and, in one-loop diagrams with *only valence quarks on the external legs*, only one sea-quark loop can occur, as can easily be seen in the quark-flow picture.[85] Let us see how this works in the case that the sea quarks are Wilson-like, and that we are in the LCE regime. To order p^2, the chiral Lagrangian is then

$$\mathcal{L}_{LO} = \frac{1}{8}f^2 \, \text{str} \left((D_\mu \Sigma)^\dagger D_\mu \Sigma \right) - \frac{1}{4} B_0 f^2 \, \text{str} \left(\mathcal{M} \Sigma + \Sigma^\dagger \mathcal{M} \right) \tag{8.145}$$

$$-\hat{a}^2 W_6' \left(\text{str}(P_s \Sigma + \Sigma^\dagger P_s) \right)^2 - \hat{a}^2 W_7' \left(\text{str}(P_s \Sigma - \Sigma^\dagger P_s) \right)^2$$

$$-\hat{a}^2 W_8' \, \text{str}(P_s \Sigma P_s \Sigma + \Sigma^\dagger P_s \Sigma^\dagger P_s) - \hat{a}^2 W_M \, \text{str}(P_s \Sigma P_s \Sigma^\dagger),$$

from Eqs. (8.33) and (8.56), where I have inserted the projectors P_s, and in which the $O(a)$ term has been absorbed into the sea-quark mass matrix, as in Eq. (8.140). As noted, we can have only one sea quark on the loop. This implies that if we expand out Σ in terms of Φ, we will only use those terms in the expansion of any of the $O(a^2)$ operators in which two of the indices on the Φ fields correspond to sea quarks (these two indices correspond to the "beginning" and the "end" of the sea-quark loop), with the rest of the indices referring to valence quarks. Because of the P_s projectors in the $O(a^2)$ terms, such terms only arise when we set one Σ (or one $\Sigma^\dagger$) equal to one, in all possible ways in which this can be done. Applying this "1-loop valence" rule to Eq. (8.145), it simplifies to

$$\mathcal{L}_{LO} \bigg|_{\text{1−loop valence rule}} = \frac{1}{8}f^2 \, \text{str} \left((D_\mu \Sigma)^\dagger D_\mu \Sigma \right) - \frac{1}{4} B_0 f^2 \, \text{str} \left(\mathcal{M} \Sigma + \Sigma^\dagger \mathcal{M} \right)$$

$$-\hat{a}^2 \left(4 N_s W_6' + 2 W_8' + W_M \right) \, \text{str} \left(P_s \Sigma + \Sigma^\dagger P_s \right). \tag{8.146}$$

In other words, all dependence (to one loop) on the LECs $W'_{6,8}$ and W_M can be absorbed completely into a shift of the sea-quark mass matrix of the form (Chen *et al.* 2009a)

$$1\text{-loop valence rule:} \quad B_0 M_s \to B_0 M_s + \frac{4\hat{a}^2}{f^2}\left(4N_s W'_6 + 2W'_8 + W_M\right)\mathbf{1}. \qquad (8.147)$$

Note, however, that this simplified form *cannot* be used to calculate any quantity involving sea quarks! For instance, the leading-order meson mass for a meson made only out of sea quarks is given by the middle equation of Eq. (8.142), which does *not* correspond to shifting the sea-quark masses as in Eq. (8.147); in particular, the LEC W_M does not occur. It also does not apply to the disconnected term in Eq. (8.144), in which again the sea-meson masses are given by Eq. (8.142). However, since this is the only place where the sea-meson mass shows up at one loop in valence quantities, the LEC W'_7 can be absorbed into the sea-meson mass-squared, for the one-loop contributions to valence quantities.

In summary, the one-loop expressions of continuum PQChPT for valence quantities, such as meson masses, decay constants and scattering amplitudes (Chen *et al.* 2006), carry over to the mixed case, if one takes the mixed-meson mass to be given by the last line of Eq. (8.142),

$$m^2_{i\alpha} = B_0(m_i + m'_\alpha) + \frac{4\hat{a}^2}{f^2}\left(4N_s W'_6 + 2W'_8 + W_M\right), \qquad (8.148)$$

and the sea-meson mass by

$$m^2_{\alpha\beta} = B_0(m'_\alpha + m'_\beta) + \frac{32\hat{a}^2}{f^2}\left(N_s W'_6 + W'_8 + N_s W'_7\right), \qquad (8.149)$$

by combining Eqs. (8.142) and (8.144). It is only through these masses that the LECs $W'_{6,7,8}$ and W_M occur.

In a calculation to order p^4, there are also tree-level contributions originating from terms in the chiral Lagrangian of order p^3 and p^4, cf. Eq. (8.67). Single-trace terms containing a P_s thus do not contribute. In double-trace terms any Σ multiplied by a P_s should be set equal to one. That means that such terms do contribute, but again they can be absorbed into a redefinition of other LECs. For instance, the W_4 and W_6 terms in Eq. (8.56), reduce to

$$\hat{a} W_4 \, \text{str}\left(\partial_\mu \Sigma \partial_\mu \Sigma^\dagger\right) \, \text{str}\left(P_s \Sigma + \Sigma^\dagger P_s\right) \qquad (8.150)$$

$$-2\hat{a} B_0 W_6 \, \text{str}\left(\mathcal{M}\Sigma^\dagger + \Sigma\mathcal{M}\right) \, \text{str}\left(P_s\Sigma + \Sigma^\dagger P_s\right)$$

$$\to \; 2N_s \hat{a} W_4 \, \text{str}\left(\partial_\mu \Sigma \partial_\mu \Sigma^\dagger\right) - 4N_s \hat{a} B_0 W_6 \, \text{str}\left(\mathcal{M}\Sigma^\dagger + \Sigma\mathcal{M}\right),$$

or

$$f^2 \to f^2 + \frac{16\hat{a}}{f^2} N_s W_4, \qquad B_0 \to B_0\left(1 + \frac{16\hat{a}}{f^2} N_s W_6\right). \qquad (8.151)$$

All double-trace terms in the chiral Lagrangian need to be treated this way, leading to more redefinitions of the $O(p^2)$ LECs B_0 and f^2. In this way the redefined LECs B_0 and f also pick up dependence on the sea-quark masses, from the L_4 and L_6 terms in Eq. (8.39). Whether this is convenient or not depends on the application. Note that if one trades sea-quark masses for sea-meson masses, one has to account for the leading-order $O(a^2)$ shifts of Eq. (8.142). For more discussion, including the example of $I = 2$ $\pi\pi$ scattering to order p^4, see Chen *et al.* (2007, 2009b).

In summary, what we find is that, to order p^4, the formulas of continuum PQChPT can be used for correlation functions with only valence quarks on the external legs, with a suitable redefinition of the mixed- and sea-meson masses and the $O(p^2)$ LECs. In other words, once these redefinitions have been made, the theory does not remember the fact that the sea quarks are Wilson-like. That means that we uncovered something universal (Chen *et al.* 2007, 2009b): The same argument should also apply when the sea quarks are staggered! And indeed, it does: For example, it is straightforward to check from Eq. (8.102) that the shift in M_s corresponding to Eq. (8.147) is

$$\text{1-loop valence rule:} \quad B_0 M_s \to B_0 M_s + \frac{4a^2}{f^2}\left(C_1 + 4C_3 + 4C_4 + 6C_6 + C_{\text{mix}}\right),$$

$$(8.152)$$

where C_{mix} is the staggered equivalent of W_M (Bär *et al.* 2005).

8.5 *SU*(2) **versus** *SU*(3): **physics with heavy kaons**

So far, we have mostly discussed ChPT for QCD with three flavors to order p^4. An important question is whether this is sufficiently high order for practical applications to lattice QCD, and the answer is, unfortunately, that in general it is not. It is a fact that the kaon mass is not so small in the real world, and even if ChPT applies to the strange-quark mass as well, the expansion to order p^4 may not suffice. Indeed, $m_K^2/m_\pi^2 \approx 13$, suggesting that an expansion that works very well for m_π^2 may not work so well when m_K^2 is the expansion parameter.

One option for dealing with this problem is to go to higher order in the chiral expansion. In applications to phenomenology, there has been much work to extend ChPT calculations to order p^6 (see for instance Bijnens (2007a)). For applications to the lattice, $O(p^6)$ expressions, which involve two-loop calculations (as can be seen from Eq. (8.18)), need to be extended to the partially quenched case, and ideally, also to include scaling-violation effects. Meson masses and decay constants have been calculated in partially quenched, three-flavor ChPT to order p^6 (Bijnens *et al.* 2006); see also (Bijnens 2007b), and it will be interesting to see what happens if the two-loop chiral logarithms are included in fits to lattice data. Scaling violation effects have not been calculated in ChPT to the same order.[86] While this means that in principle lattice results would have to be first extrapolated to the continuum, before chiral fits are performed, it may be interesting to include "what we know," i.e., the continuum part at order p^6, because the continuum $O(p^6)$ terms are expected to be important at larger quark masses, where scaling violations are less important. To date, something similar has been done by MILC, which included all the analytic terms to order p^8 in their mass and decay constant fits, but no logarithms beyond order p^4 (yet).[87] For an early attempt to fit $N_f = 2 + 1$ full-QCD overlap results to $O(p^6)$ ChPT including two-loop non-analytic terms, see Noaki *et al.* (2008). It has recently been suggested that kaon loops are not a reliable part of ChPT (Donoghue 2009), because the relevant scale is $2m_K \sim 1$ GeV, rather than m_K itself. Diagrams can be reconstructed from their cuts and poles, and, for a meson of mass m, these cuts typically start at $4m^2$, and not m^2. For instance, the contribution of a kaon pair to the imaginary part of the $\pi\pi$ scattering amplitude starts at $s = 4m_K^2$; for other examples (F_π and the pion electromagnetic form factor), see Donoghue (2009). Of course, $2m_K \sim 1$ GeV is too large to trust a chiral expansion in that scale. However, this appears to be a quantity-dependent observation. It applies to purely pionic quantities, since (virtual) kaons always have to contribute in pairs to these quantities. But if we consider for instance πK scattering, the cut starts at $(m_\pi + m_K)^2 \approx m_K^2$, and the chiral expansion for this scattering amplitude as a function of m_K may have better convergence properties. A similar argument applies to F_K.

For lattice computations relevant for phenomenology, there is a different option. It is not difficult to do lattice computations with the strange quark mass adjusted to (close to) its physical value, so that only extrapolations in terms of the light-quark masses (m_u and m_d) are needed. In this case, we do not need three-flavor ChPT, but, rather, we can work with two-flavor ChPT, treating the strange quark as "heavy." Note, however, that we do not integrate out the strange quark, which we cannot do, because m_s is not large compared to $\Lambda_{\rm QCD}$. The actual lattice computations still have to be done with three flavors, up, down and strange.

In the pion sector it is straightforward to develop ChPT again, one simply starts from the group $SU(2)_L \times SU(2)_R$, instead of $SU(3)_L \times SU(3)_R$.[88] But the properties of kaons (its mass, decay constant, the kaon B parameter, etc) will depend on m_u and m_d through interactions of kaons with pions, and since lattice computations will still typically use unphysical values for the up- and down-quark masses, we will still need to know how to extrapolate kaon properties in those light-quark masses. We thus need to extend two-flavor ChPT systematically to include the interactions of kaons with pions. How to do this will be the topic of this section. Studies using this EFT in application to three-flavor lattice data were recently carried out in Allton *et al.* (2008) and Kadoh *et al.* (2008). While the case of kaons is particularly simple, the basic setup also applies to EFTs for the coupling of baryons and heavy-light mesons to pions.

While this alternative approach may turn out to be a useful approach, it is a more limited approach, because the kaons are not treated as Nambu–Goldstone bosons. For instance, while the dependence of the kaon mass on m_u and m_d is predicted by this EFT, it will not go to zero in the chiral limit. Instead, the chiral-limit value of the kaon mass is a new parameter in the EFT. Another example is that in this approach, one cannot ask certain questions, such as the value of the three-flavor condensate in the chiral limit. This particular question is interesting, because it has been argued that if this condensate were very small, that would lead to the need to redefine the power counting of ChPT, and thus rearrange the chiral Lagrangian order by order (see for instance Knecht *et al.* (1995)). To answer this question, one needs to work with the three-flavor theory, and vary the strange mass. In the setup we will investigate in this section, one only gets to ask about the two-flavor condensate, at the physical value of the strange mass.[89]

8.5.1   Including a kaon in two-flavor ChPT

We need to develop two-flavor ChPT for the three pions, coupled to the isospin-1/2 kaon. In the special case of the kaon, it should be possible to deduce the results from ChPT with three flavors. For instance, one may take the result for f_K/f_π in Eq. (8.40), and expand it in the light quark mass for fixed strange quark mass. One then absorbs the strange quark mass dependence into the LECs, and this parameterizes the dependence of f_K/f_π on the light-quark mass, while giving up information about the dependence on the strange-quark mass. However, this method is not available for other hadrons such as baryons, and it is useful and instructive to develop a method that does not rely on three-flavor ChPT. After developing the method, we will return to the matching with three-flavor ChPT in some examples.

The systematic method for coupling pions to a non-Goldstone hadron was worked out a long time ago in Coleman *et al.* (1969) and Callan *et al.* (1969), which shows how to couple any field with a given isospin to pions in the most general way. In fact, the method generalizes to other groups as well: The general case is that of some continuous group G that is spontaneously broken to some subgroup H. Since only H is realized in Wigner mode, all fields need to transform in representations of H, and not G, as we will see in more detail below. Here, we will restrict ourselves to $G = SU(2)_L \times SU(2)_R$, and $H = SU(2)_V$, i.e., isospin.

We will present the construction as explained in Roessl (1999). As we have seen in Section 8.2.2, the pion fields parametrize the coset $[SU(2)_L \times SU(2)_R]/SU(2)_V$ (G/H in the general case), because $SU(2)_V$ transformations leave the vacuum invariant. An element of the coset can be represented by picking an element (u_L, u_R) of the full symmetry group $SU(2)_L \times SU(2)_R$, and defining an equivalence class by making any two such elements equivalent when they differ by a transformation in $SU(2)_V$. Each equivalence class is an element of the coset. We can pick a "standard" representative of each equivalence class by imposing a condition on u_L and u_R; we pick the condition that $u_L = u_R^\dagger \equiv u$. We can do this: by definition, if we multiply (u_L, u_R) on the right by an element $(h^\dagger, h^\dagger) \in SU(2)_V$, obtaining $(u_L h^\dagger, u_R h^\dagger)$, this group element is in the same equivalence class, and thus represents the same coset element. If we pick h such that it solves $u_R = h u_L^\dagger h$, this brings the coset element into standard form. If we now multiply the group element (u_L, u_R) by another group element, (v_L, v_R):

$$(u_L, u_R) \rightarrow (v_L u_L, v_R u_R), \tag{8.153}$$

the new group element in general corresponds to a different element of the coset. But again we can choose a $(h^\dagger, h^\dagger) \in SU(2)_V$ such that

$$(u_L, u_R) \rightarrow (v_L u_L h^\dagger, v_R u_R h^\dagger) \tag{8.154}$$

keeps the transformed element in standard form, where, of course, $h = h(u, v_{L,R})$ now depends both on u and (v_L, v_R). If we choose $(v_L, v_R) \in SU(2)_V$, i.e. $v_L = v_R = v$, we can choose the "compensator" field h equal to v, and it does not depend on u. It follows that, as expected, an $SU(2)_V$ transformation stays in the same equivalence class.

We now may define Σ in terms of u by[90]

$$\Sigma = u_L u_R^\dagger = u^2 \rightarrow v_L \Sigma v_R^\dagger, \tag{8.155}$$

$$u = \exp(i\phi/f),$$

where now

$$\phi = \begin{pmatrix} \frac{\pi^0}{\sqrt{2}} & \pi^+ \\ \pi^- & -\frac{\pi^0}{\sqrt{2}} \end{pmatrix}, \tag{8.156}$$

because then Σ also parametrizes the coset, and has the correct transformation properties under the chiral group if we identify $v_{L,R} = U_{L,R}$, cf. Eq. (8.11).

This construction provides a straightforward way to include the kaon. Introducing a field K in the isospin-1/2 representation of $SU(2)_V$, we can extend the non-linear represention of the chiral group to

$$u \rightarrow v_L u h^\dagger(u, v_{L,R}) = h(u, v_{L,R}) u v_R^\dagger, \tag{8.157}$$

$$K \rightarrow h(u, v_{L,R}) K.$$

These two fields, their (covariant, see below) derivatives and the sources ℓ_μ, r_μ, χ and $\chi^\dagger$ now form the set of building blocks from which to construct the chiral Lagrangian.

Note that we only have to specify the $SU(2)_V$ representation of the kaon field K, and not the representation of the full chiral group $SU(2)_L \times SU(2)_R$. There is a simple intuitive reason for that (Kaplan 2005): Suppose, for example, we specify that a kaon field K' transforms in the $(1/2, 0)$ representation of $SU(2)_L \times SU(2)_R$, $K' \rightarrow v_L K'$. Then, performing the field redefinition $K'' = \Sigma^\dagger K' = (1 + O(\pi)) K'$ changes the representation to $(0, 1/2)$, since $K'' \rightarrow v_R K''$. We see that the presence of pions (or, in general, the Goldstone bosons associated with G/H) gives us the freedom to choose any representation of $SU(2)_L \times SU(2)_R$ we want! The field K used in Eq. (8.157) is obtained by defining $K = u^\dagger K'$.[91] We note that the field K' transforms under parity to

$$K' \rightarrow K'^{(P)} = -\Sigma^\dagger K', \tag{8.158}$$

because the parity transform of K' should transform as $K'^{(P)} \rightarrow v_R K'^{(P)}$ under the chiral group; the minus sign indicates that the kaon is a pseudoscalar. The field K transforms into $-K$ under parity, consistent with its transformation rule Eq. (8.157) under the chiral group ($h \in SU(2)_V$).

In order to construct the chiral Lagrangian, we need to construct invariants out of u and K, which transform as in Eq. (8.157). Since h is local, we will need a gauge connection for $SU(2)_V$, constructed from u. We thus introduce

$$u_\mu^L = u^\dagger \partial_\mu u, \tag{8.159}$$

$$u_\mu^R = u \partial_\mu u^\dagger,$$

which transform into each other under parity. Under Eq. (8.157), the parity odd and even combinations transform as

$$\Delta_\mu \equiv \frac{1}{2} \left(u_\mu^R - u_\mu^L \right) \to \frac{1}{2} h \left(u_\mu^R - u_\mu^L \right) h^\dagger, \tag{8.160}$$

$$\frac{1}{2} \left(u_\mu^R + u_\mu^L \right) \to \frac{1}{2} h \left(u_\mu^R + u_\mu^L \right) h^\dagger - \partial_\mu h h^\dagger.$$

We see that Δ_μ transforms homogeneously, while the other combination transforms as an $SU(2)_V$ gauge connection. The latter can thus be used to define a covariant derivative for K:

$$D_\mu K = \partial_\mu K + \frac{1}{2} \left(u^\dagger \partial_\mu u + u \partial_\mu u^\dagger \right) K. \tag{8.161}$$

If we wish to include the sources ℓ_μ and r_μ, this can be done by replacing $\partial_\mu \to \partial_\mu - i\ell_\mu$ in u_μ^L and $\partial_\mu \to \partial_\mu - ir_\mu$ in u_μ^R.

The chiral Lagrangian consists of three parts. First, there is the purely pionic part of the Lagrangian, which is given by $\mathcal{L}^{(2)} + \mathcal{L}^{(4)}$ of Eqs. (8.33) and (8.39). For two flavors, a number of simplifications occur because of special properties of $SU(2)$. Then, there is the quadratic term in the kaon fields that defines the kaon propagator in ChPT,

$$\mathcal{L}_K^{(0)} = (D_\mu K)^\dagger D_\mu K + M^2 K^\dagger K. \tag{8.162}$$

Here, M is a free parameter equal to the kaon mass in the chiral limit.

As already noted, even with $\ell_\mu = r_\mu = 0$ the covariant derivative (8.161) is required, because h in Eq. (8.157) is a local, field-dependent transformation. Interactions between pions and kaons are therefore necessarily present in our EFT. In this respect, it is instructive to see how this same conclusion would have been reached if we started from the field $K' = uK$ discussed above. It looks like we can write down a kaon kinetic term without any interactions in terms of this field: $\partial_\mu K'^\dagger \partial_\mu K'$ is invariant under the chiral group. However, this is not invariant under parity. Using the parity transform $K'^{(P)}$ introduced in Eq. (8.158), a Lagrangian invariant under the chiral group and parity is

$$\frac{1}{2} \left(\partial_\mu K'^\dagger \partial_\mu K' + \partial_\mu (K'^\dagger \Sigma) \partial_\mu (\Sigma^\dagger K') \right), \tag{8.163}$$

which, using $K' = uK$, is equal to the first term in Eq. (8.162), modulo a term proportional to $K^\dagger L_\mu L_\mu K = 2 \operatorname{tr}(\Delta_\mu \Delta_\mu) K^\dagger K$ that can be absorbed into the A_1 term in Eq. (8.166) below.

Many more terms involving pion–kaon couplings can be constructed. For a more extensive discussion of building blocks and possible terms in the chiral Lagrangian, I refer to Roessl (1999).[92] Here, I give only one more building block:

$$\chi_+ = u^\dagger \chi u^\dagger + u \chi^\dagger u, \tag{8.164}$$

which is even under parity (the odd combination, with a minus sign instead of a plus sign, appears in the chiral Lagrangian at order p^4).

With these definitions, noting that $\Delta_\mu = \frac{1}{2} u^\dagger L_\mu u$, with $L_\mu = \Sigma \partial_\mu \Sigma^\dagger$ (cf. Eq. (8.16)), Eq. (8.33) can be written as

$$\mathcal{L}^{(2)} = -\frac{1}{2} f^2 \operatorname{tr}(\Delta_\mu \Delta_\mu) - \frac{1}{8} f^2 \operatorname{tr}(\chi_+), \tag{8.165}$$

and more terms involving the kaon field are given by

$$\begin{aligned}
\mathcal{L}_K^{(2)} = &A_1 \operatorname{tr}(\Delta_\mu \Delta_\mu)\, K^\dagger K + A_2 \operatorname{tr}(\Delta_\mu \Delta_\nu)\, (D_\mu K)^\dagger D_\nu K \\
&- A_3 K^\dagger \chi_+ K - A_4 \operatorname{tr}(\chi_+) K^\dagger K \\
&+ A_2' \operatorname{tr}(\Delta_\nu \Delta_\nu)\, (D_\mu K)^\dagger D_\mu K + A_3'(D_\mu K)^\dagger \chi_+ D_\mu K + A_4' \operatorname{tr}(\chi_+)(D_\mu K)^\dagger D_\mu K,
\end{aligned} \tag{8.166}$$

where $A_{1,2,3,4}$ and $A'_{2,3,4}$ are seven new LECs.[93] Formally, we see that $\mathcal{L}_K^{(0)}$ and $\mathcal{L}_K^{(2)}$ are the first two terms in a small-derivative/light-quark mass expansion, because only derivatives on the pion field u should be counted as small, and not those acting on the kaon field K. We note that some of the terms in $\mathcal{L}_K^{(2)}$ would be counted as $O(p^4)$ if the kaon is treated as a Nambu–Goldstone boson, cf. Section 8.2.

8.5.2   Power counting

At this point, we need to revisit power counting. First, there are now three scales in the problem, m_π, m_K and $4\pi f$, cf. Section 8.2.3, and we are now *not* assuming that $m_K^2/(4\pi f)^2$ is small. Therefore, it is not obvious that a systematic chiral expansion in $m_\pi^2/(4\pi f)^2$ and/or m_π^2/m_K^2 can be set up. Secondly, our new Lagrangian also contains interactions contributing for example to the scattering process $\pi^+ \pi^- \to K^+ K^-$. But, clearly, with the kaon mass not treated as a small parameter, the initial-state pions in this process also do not have small momenta in the sense of a chiral expansion. The prediction for this process from our new Lagrangian can thus not be argued to be the leading order in a systematic expansion.

We conclude that "kaon ChPT" cannot be used for any process near or above the two-kaon threshold. However, it can be used to calculate correlation functions in which a kaon goes in and comes out, so that the large energy residing in the kaon mass does not get converted to pions. This means that one can use the new Lagrangian to calculate the dependence of kaonic quantities, such as its mass and decay constant, on the light-quark masses m_u and m_d. Note that in the leptonic decay of a kaon, its mass does get converted into the energy of the outgoing leptons, but that does not affect the calculation of its dependence on light-quark masses.[94]

While this claim turns out to be correct, we still need to deal with the first problem mentioned above, i.e. the appearance of a new "large" scale m_K, in order to show its validity. From Eq. (8.162) it follows that the parameter M^2 only shows up in the kaon propagator. It can therefore only get promoted to the numerator of some contribution to a physical quantity by appearing in a loop. The simplest case is that of a closed kaon loop; for instance, the kaon tadpole diagram contribution to the pion self-energy that would follow from $\mathcal{L}_K^{(0)} + \mathcal{L}_K^{(2)}$. In dimensional regularization this would give a contribution of order $m_K^2 \log\left(m_K^2/\Lambda^2\right)$, which is not small. However, such contributions can be absorbed into the low-energy constants of the pion Lagrangian, because they are independent of the pion mass.[95] We conclude from this example that we can thus *omit* closed kaon loops from the calculation of correlation functions with kaon ChPT.[96] With this rule that we omit closed kaon loops, we conclude that correlation functions with only pions on the external legs should be calculated with the pion chiral Lagrangian only.

In effect, because the kaon is heavy compared to the pion, we have integrated it out, and including closed kaon loops as discussed above would amount to double counting. This is of course the reason that we can use two-flavor ChPT for pion physics well below the kaon mass. The new element is that we can also consider processes with a kaon going in and coming out – as long as no kaons get annihilated or produced, the energy stored in the kaon mass does not play a dynamical role. An example is $\pi K \to \pi K$ scattering: As long as the energy of the incoming pion is small, the energy of the outgoing pion also has to be small, and two-flavor ChPT (in the presence of a "heavy" kaon) should apply. For a detailed analysis of this scattering process, see Roessl (1999).

There are also loops with internal kaon lines where the kaon does not form a closed loop, for instance in the just-mentioned example of kaon–pion scattering. To one-loop order, they do not occur in loop corrections to pion and kaon masses and decay constants, because all one-loop diagrams for these quantities are tadpole diagrams, which follows from the fact that there are no three-meson vertices in our EFT. However, such vertices do occur in the kaon theory if we also include the K^*, described by a vector field K_μ also in the isospin-1/2 representation of $SU(2)_V$. In this extended version of kaon ChPT, a possible invariant is

$$-igfK_\mu^\dagger(u^\dagger\partial_\mu u)K = gK_\mu^\dagger K\partial_\mu\phi + \dots, \qquad (8.167)$$

i.e., now the theory does contain a $K^*K\pi$ three-meson vertex.

Let us consider an example. The vertex (8.167) leads to a correction to the kaon self-energy through the diagram of Fig. 8.9. The amputated part of this diagram stands for an integral of the form (Allton *et al.* 2008) (using dimensional regularization)

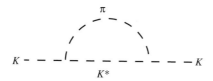

Fig. 8.9 One-loop contribution to the kaon self-energy with a $K^*\pi$ intermediate state.

$$I = \int \frac{d^d q}{(2\pi)^d} \frac{q_\mu q_\nu \left(\delta_{\mu\nu} + (p-q)_\mu (p-q)_\nu / M_*^2 \right)}{(q^2 + m_\pi^2)((p-q)^2 + M_*^2)}, \tag{8.168}$$

in which M_* is the mass of the K^*. We need this integral for an on-shell kaon, so we need to set $p = Mv$, with v a four-vector with $v^2 = -1$. The denominator of the K^* propagator then becomes

$$q^2 - 2Mv \cdot q + M_*^2 - M^2. \tag{8.169}$$

Non-analytic dependence on m_π comes from the region $q \sim m_\pi$. The behavior of the integral now depends on the various other scales. If M_* and M were the masses of spin-zero and spin-one heavy-light mesons (instead of the K^* and the kaon), the term linear in M would be the dominant term, because $M_*^2 - M^2 \sim \Lambda_{\rm QCD}^2$ and $M_* \sim M \gg \Lambda_{\rm QCD}$.[97] Instead, for the KK^* system both masses as well as their difference are of order $\Lambda_{\rm QCD}$, and the difference $M_*^2 - M^2$ is the dominant term. This leads to a different infrared behavior for each of these cases. For the heavy-light case, the denominator (8.169) can be of order m_π (times M),[98] while for the kaonic case, the denominator does not become small. In the latter case, we can thus effectively replace this denominator by $\Lambda_{\rm QCD}^2$, and read off the infrared behavior from the rest of the integral, which gives rise to a contribution of order $m_\pi^4 \log(m_\pi^2/\Lambda^2)$ (Allton *et al.* 2008). The power m_π^4 has to occur because of dimensional analysis, and the fact that the kaon propagator, even though it appears inside a loop integral, can be "power-counted" as order Λ_{QCD}^{-2}. I would expect this type of argument to extend to higher-loop contributions, thus explaining how higher-loop contributions lead to non-analytic terms at higher orders in m_π^2. For an extensive discussion of all this in the context of baryon ChPT, I refer to Jenkins and Manohar (1991) and Becher and Leutwyler (1999).

It is straightforward to show that there are no $m_\pi^2 \log(m_\pi^2/\Lambda^2)$ terms coming from $\mathcal{L}_K^{(0)} + \mathcal{L}_K^{(2)}$, and thus (Roessl 1999)

$$m_K^2 = M^2 - 2(A_3 + 2A_4)\hat{m}_\ell + O(\hat{m}_\ell^2 \log(\hat{m}_\ell)). \tag{8.170}$$

This is consistent with the result for m_K^2 shown in Eq. (8.40), which does not contain a term of the form $m_\pi^2 \log(m_\pi^2/\Lambda^2)$.[99] It should be noted that Roessl (1999) obtains this result in the theory without the K^*. Indeed, since the K^* is significantly heavier than the kaon, we should be able to integrate it out. Doing this leads to an effective coupling of the form

$$g^2 \frac{f^2}{M_*^2} K^\dagger \Delta_\mu \Delta_\mu K = g^2 \frac{f^2}{2M_*^2} \, {\rm tr}(\Delta_\mu \Delta_\mu) K^\dagger K \tag{8.171}$$

(where we used that $\Delta_\mu = -u^\dagger \partial_\mu u$ is an element of the Lie algebra for $SU(2)$), and we see that this is of the form of the first term in Eq. (8.166).

8.5.3 An application: decay constants

The pion decay constant follows simply from Eq. (8.40), by expanding out $L(m_K^2)$ in $\hat{m}_\ell$:

$$f_\pi = f' \left(1 + \frac{4l_4}{f^2} B_0 m_\ell - 2L(2B_0 m_\ell) + O(\hat{m}_\ell^2) \right), \tag{8.172}$$

in which we can match the LECs f' and l_4 to their three-flavor counterparts (Gasser and Leutwyler 1985),

$$f' = f \left(1 - L(B_0 m_s) + \frac{16L_4}{f^2} B_0 m_s \right), \tag{8.173}$$

$$l_4 = 4(L_5 + 2L_4) - \frac{1}{(4\pi)^2} \left(\log \left(\frac{B_0 m_s}{\Lambda^2} \right) + 1 \right).$$

The matching formulas (8.173) only make sense if m_s is small enough for three-flavor ChPT to apply; otherwise f' and l_4 should be treated as two-flavor LECs that depend in an unknown way on the strange quark mass. The expressions (8.173) give the first few orders of the expansion of f' and l_4 in m_s. It is possible that the strange-quark mass is so large that this expansion simply does not converge. But it is also possible that this expansion does (asymptotically) converge, but too slowly for the expressions (8.173) to be of practical use. In that case, one should either use kaon ChPT, or three-flavor ChPT to higher order than p^4. Whether the expansion in the strange-quark mass converges to some order in ChPT or not is quantity dependent.

In order to obtain the expansion of f_K in terms of m_ℓ, we can follow two routes. One is to simply use the three-flavor result for f_K in Eq. (8.40), and expand in m_ℓ, as we did above for f_π. However, it is instructive to see how one obtains this expansion in kaon ChPT independently of the three-flavor ChPT calculation (Allton *et al.* 2008). In order to do this, we need to translate the currents

$$J_\mu^{L,R} = \bar{q}_{L,R} \gamma_\mu s_{L,R}, \tag{8.174}$$

$$\bar{q} = \begin{pmatrix} \bar{u} & \bar{d} \end{pmatrix},$$

into the appropriate operators in kaon ChPT. These currents now do not correspond to the Noether currents of (softly broken) symmetries, because we are not assuming that the strange quark is small. Therefore, their translations into kaon ChPT do not follow from the Lagrangian $\mathcal{L}_K^{(0)} + \mathcal{L}_K^{(2)}$. We thus introduce spurions $H_{L,R}$ into these currents, with transformation rules such as to make these currents invariant under $SU(2)_L \times SU(2)_R$:

$$J_\mu^{L,R} = \bar{q}_{L,R} H_{L,R} \gamma_\mu s_{L,R}, \quad \text{with} \quad H_{L,R} \to v_{L,R} H_{L,R}, \tag{8.175}$$

in which $v_{L,R} \in SU(2)_{L,R}$. Making the transition to kaon ChPT, the first two terms in the expansion in terms of derivatives of the pion fields are

$$J_\mu^R = C_1 (D_\mu K)^\dagger u H_R + i C_2 K^\dagger \Delta_\mu u H_R, \tag{8.176}$$

$$J_\mu^L = -C_1 (D_\mu K)^\dagger u^\dagger H_L + i C_2 K^\dagger \Delta_\mu u^\dagger H_L,$$

where $C_{1,2}$ are new LECs (not present in the Lagrangian). The left- and right-handed currents are related by parity. In these ChPT expressions for the currents, one can now set $H_{L,R} = \begin{pmatrix} 1 \\ 0 \end{pmatrix}$, or $H_{L,R} = \begin{pmatrix} 0 \\ 1 \end{pmatrix}$ in order to pick out the charged or neutral kaon, corresponding to picking out the up or down quark in Eq. (8.175), respectively. Taking the matrix element of the axial current $J_\mu^R - J_\mu^L$ between a kaon state and the vacuum, one finds that

$$f_K = 2C_1 \left(1 + \frac{C}{f^2} \hat{m}_\ell - \frac{3}{4} L(2\hat{m}_\ell) + O(\hat{m}_\ell^2) \right), \tag{8.177}$$

where C is a linear combination of LECs in the kaon EFT Lagrangian.[100] Note that the LEC C_1 is unrelated to f' in Eq. (8.172), because there is no $SU(3)$ symmetry in kaon ChPT. If we do use $SU(3)$, we may again match the LECs in Eq. (8.177) to the three-flavor ChPT result, from which we find

$$C_1 = \frac{1}{2} f \left(1 - 3L(\hat{m}_s) + \frac{8(2L_4 + L_5)}{f^2} \hat{m}_s \right), \tag{8.178}$$

$$C = 8(4L_4 + L_5) - \frac{5}{2} \frac{1}{(4\pi)^2} \left(\log\left(\frac{\hat{m}_s}{\Lambda^2} \right) + 1 \right).$$

As it should, the coefficient of $L(m_\pi^2)$ in Eq. (8.177) matches that following from Eq. (8.40).

We close this section with another simple example of the translation of a weak operator into kaon ChPT (Allton *et al.* 2008). The $\Delta S = 2$ operator $(\bar{s}_L \gamma_\mu d_L)(\bar{s}_L \gamma_\mu d_L)$ can be written in $SU(2)_L$-invariant form by again using the spurion H_L as in

$$(\bar{s}_L \gamma_\mu H_L^\dagger q_L)(\bar{s}_L \gamma_\mu H_L^\dagger q_L) \rightarrow B(H_L^\dagger u K)(H_L^\dagger u K), \tag{8.179}$$

where in the second step I gave the leading-order translation into kaon ChPT, introducing another weak LEC B. Finally, one sets $H_L^\dagger = \begin{pmatrix} 0 & 1 \end{pmatrix}$, because for this choice $H_L^\dagger q_L = d_L$. This technique for finding the ChPT representations of electroweak operators is of course not special to kaon ChPT, and the same method can also be used in three-flavor ChPT (Bernard 1990).

It turns out that $SU(2)$ ChPT can also be applied to semileptonic and non-leptonic kaon decays, despite the fact that the final-state pions can have an energy up to half the kaon mass. This is not at all obvious, because in the $SU(2)$ framework the kaon mass is not a small parameter. For more on this topic, see Flynn and Sachrajda (2009), and Bijnens and Celis (2009).

8.6 Finite volume

As a final topic, I would like to touch, rather briefly, on the fact that all lattice QCD computations are, necessarily, performed in a finite spatial volume, and with a finite maximal extent L_4 in the (Euclidean) time direction.

Usually, the time dimension of the four-dimensional Euclidean volume is chosen larger than the spatial directions, large enough so that only the lowest (or lowest few) intermediate states contribute to the correlation functions of interest at large times. In order to mimimize, or eliminate, systematic effects from the finite extent in spatial directions, the conceptually easiest method would be to choose the spatial volume of a cubic box $V_3 = L^3$ (where L is the linear size of the spatial box, and we will assume periodic boundary conditions for all fields), sufficiently large that one may ignore the effects of L being finite. However, the cost of numerical computations grows rapidly with L, and one would thus like to choose the volume just large enough that finite-volume effects can be neglected, but not much larger. The question then arises whether we have any theoretical insight into what "just large enough" means.

In fact, as we will see, it is possible to develop systematic expansions that provide quantitative information on finite-volume effects, without the introduction of any new parameters in the chiral Lagrangian. This allows us to quantitatively estimate the effects of working in a finite volume, and even to make use of finite-volume effects as a probe of hadronic quantities such as the pion decay constant and the chiral condensate.

The reason that again ChPT gives us the tool to develop these systematic expansions is that pions[101] are the lightest hadrons. Thus, if we imagine reducing L from some very large value, the pions feel the effects of being in a finite volume first. Therefore, ChPT should give us access to the dominant finite-volume effects on many hadronic quantities, if we know how to apply it in this setting.

It already follows from this qualitative discussion that the "figure of merit" for what "large" or "small" volume means is the linear size of the box in terms of the wavelength of the pion, $z = m_\pi L$. Clearly, large volume means $z \gg 1$. But only a more quantitative analysis can tell us whether this means, say, $z \gtrsim 3$, or whether maybe $z \gtrsim 10$ is needed.

The key theorem in setting up finite-volume ChPT states that the Lagrangian to be used is the infinite-volume Lagrangian (Gasser and Leutwyler 1987a). The argument goes as follows. First, assume that the spatial volume is infinite, but that the Euclidean time extent L_4 is finite, and that the pion fields (8.156) obey periodic boundary conditions in that direction. The pion effective theory with these boundary conditions describes the theory at a finite temperature $T = 1/L_4$. The important observation here is that to calculate pion correlation functions at finite temperature, one uses the *same* chiral Lagrangian as at zero temperature (i.e. infinite L_4). The temperature, and thus the dependence on L_4 enters only through the boundary conditions.[102] But, if this is the case, this then also has to be true for the dependence on the three spatial dimensions of the box, if we use periodic boundary conditions also in those directions (Gasser and Leutwyler 1988). In other words, the LECs in a finite-volume calculation should be taken the same as in infinite volume.

This result makes sense from the EFT point of view. If we remember that the LECs originated from integrating out high-energy degrees of freedom, they can pick up exponentially suppressed finite-volume corrections of order $\exp(-EL)$, where E is the energy of the integrated mode, relative to their infinite-volume values. If we take for example $E = m_\rho$ and $L = 2$ fm, $\exp(-EL) = 4.5 \times 10^{-4}$, which is very small in comparison with the present precision of our knowledge of LECs. We conclude that we can safely ignore such effects, and thus take the LECs to be those of the infinite-volume theory.[103] This more intuitive argument also shows that the theorem should hold for more general boundary conditions, such as antiperiodic ones.

It follows that the way in which the volume dependence enters the Feynman rules is through the propagator, which now is periodic, since it solves, for a particle with mass m_π,

$$(-\Box + m_\pi^2)G_L(x - y) = \bar{\delta}(x - y) = \prod_\mu \sum_{n_\mu} \delta(x_\mu - y_\mu + n_\mu L_\mu), \qquad (8.180)$$

where L_μ is the linear extent of the volume in the μ direction, and $\bar{\delta}$ is the periodic delta function, with period L_μ in the μ direction. The solution is, of course,

$$G_L(x - y) = \sum_{n_\mu} G_\infty(x - y + nL), \qquad (8.181)$$

with G_∞ the infinite-volume propagator, and in which nL is the four-vector with components $n_\mu L_\mu$. In words, finite-volume effects occur because pions can "travel around the world," multiple times in each direction, in a world that is a box with periodic boundary conditions. Because of the periodicity, momenta are quantized, $p_\mu = 2\pi n_\mu/L_\mu$, with n_μ integer, and the propagator can also be represented by

$$G_L(x - y) = \frac{1}{L_1 L_2 L_3 L_4} \sum_p \frac{e^{ip(x-y)}}{p^2 + m_\pi^2}, \qquad (8.182)$$

where the sum is over the discrete momenta. Feynman rules for vertices are the same as in the infinite-volume theory. With these rules, one may extend the calculations of meson masses, decay constants, and other quantities to include finite-volume (and/or finite-temperature) corrections. We will return to this in Section 8.6.1.

However, before we extend ChPT calculations to finite volume in this way, we need to address a serious physics issue: it is well known that in the chiral limit, the chiral condensate vanishes in a finite volume. In order to obtain a non-vanishing condensate, one should take the chiral limit after taking the infinite-volume limit, and not before; these two limits do not commute. But, if there is no chiral-symmetry breaking, that brings into question the whole framework of ChPT!

Let us see what happens in the chiral theory at zero quark mass in finite volume. The condensate can be found from the logarithmic derivative of the partition function with respect to $\chi^\dagger$:

$$\langle \bar{q}_R q_L \rangle = -\frac{2B_0}{Z} \operatorname{tr}\left(\frac{\partial Z}{\partial \chi^\dagger} \right)\bigg|_{\chi=0} = -\frac{1}{4} f^2 B_0 \langle \operatorname{tr}(\Sigma) \rangle, \qquad (8.183)$$

where in the second step we made the transition to the chiral theory. We can decompose

$$\Sigma(x) = U \, \exp(2i\phi(x)/f), \qquad (8.184)$$

with U a constant $SU(N_f)$ matrix, and in which ϕ does not contain any constant mode, i.e. $\int d^4x \, \phi(x) = 0$.[104] The matrix U represents an element of the coset $[SU(N_f)_L \times SU(N_f)_R]/SU(N_f)$, and varying U over the group $SU(N_f)$ covers the vacuum manifold, cf. Section 8.2.2. Since U represents an $SU(N_f)_L$ symmetry transformation, it follows that for vanishing quark masses, the chiral Lagrangian does not depend on U. This, in turn, implies that the condensate (8.183) vanishes, because

$$\int dU \, U = 0, \qquad (8.185)$$

with dU the Haar measure on the group $SU(N_f)$.

What we see is that ChPT reproduces the correct behavior in finite volume. The intuitive insight is that as long as the volume is large compared to the typical hadronic scale, $L \gg 1/\Lambda_{\mathrm{QCD}}$, QCD is non-perturbative, and hadrons form. The lightest hadrons are the pions, and all that happens if we lower the value of $z = m_\pi L$ is that the pions get very distorted. However, if this is the correct picture, ChPT should still be the correct effective theory. Only when we make $\Lambda_{\mathrm{QCD}} L$ too small does QCD become perturbative, and ChPT is no longer the appropriate effective theory. As long as $\Lambda_{\mathrm{QCD}} L \gg 1$, ChPT is the appropriate EFT for the low-energy physics of QCD, even if L is small in units of m_π.

When we turn on quark masses, the chiral condensate no longer vanishes. For a degenerate quark mass m and a zero mode $\Sigma_0 \equiv (1/V) \int d^4x \operatorname{tr}(\Sigma)$, the size of the leading-order mass term in the chiral Lagrangian is of order

$$N_f f^2 B_0 m V \Sigma_0, \qquad (8.186)$$

with $V = L^4$ the four-dimensional finite volume, if we take our volume to be a four-dimensional box with equal size in each (Euclidean) direction. We may now consider different regimes. If we keep Eq. (8.186) fixed, we choose $m \sim 1/L^4$, or, equivalently, $m_\pi = \sqrt{2B_0 m} \sim 1/L^2$; we will refer to this regime as the "ϵ-regime" (Gasser and Leutwyler 1987a).[105] This is a different regime from the large-volume regime we considered in Section 8.2, in which we found that $m_\pi \sim p \sim 1/L$, and thus $m \sim 1/L^2$; this regime is usually referred to as the "p-regime." In the ϵ-regime, with $\epsilon \equiv 1/L$, we have that $p \sim \epsilon$, while $m_\pi \sim \epsilon^2 \ll p$, while in the p-regime they are of the same order.

The kinetic term of the chiral Lagrangian, $\int d^4x \, \frac{1}{2} \operatorname{tr}(\partial_\mu \phi \partial_\mu \phi)$, suppresses the non-zero modes ϕ, and limits the fields ϕ to be of order $1/L$. If we expand

$$\Sigma = U \, \exp(2i\phi/f) = U \left(1 + \frac{2i}{f}\phi - \frac{2}{f^2}\phi^2 + \dots \right), \qquad (8.187)$$

it is easy to see that, with $m_u = m_d \equiv m$, the first term in

$$\int d^4 x \, \frac{1}{4} f^2 m B_0 \, \mathrm{tr}(\Sigma + \Sigma^\dagger) = \frac{1}{4} f^2 B_0 m V \, \mathrm{tr}(U + U^\dagger) \qquad (8.188)$$

$$- \int d^4 x \, \frac{1}{2} m B_0 \left(\mathrm{tr}(U + U^\dagger)\phi^2 \right) + O(\phi^4)$$

is of order L^0 in the ϵ-regime, while the second term is of order $1/L^2$. To leading order, U is non-perturbative: The zero modes in Σ do not have a Gaussian damping, and can thus not be taken into account through standard perturbation theory. When we increase the quark mass from $m \sim 1/L^4$ to $m \sim 1/L^2$, the first term in Eq. (8.188) grows to be of order L^2, still dominating the second term, which is now of order L^0. This forces U to approach $\mathbf{1}$, making also the zero mode perturbative. We thus make the transition from the ϵ-regime to the p-regime. This is also reflected by the behavior of the $p = 0$ mode contribution to the pion propagator (8.182) in finite volume, which, taking all $L_\mu = L$, is equal to $1/(L^4 m_\pi^2)$. For $m_\pi^2 = 2B_0 m \sim 1/L^4$, this is of order one, whereas all other momentum modes contribute terms of order $1/L^2$. When m_π is of order $1/L$, also the zero-momentum contribution is of order $1/L^2$, and it does not dominate the propagator.

There is thus a region in which the chiral expansion in the ϵ-regime coincides with that in the p-regime. Both representations have to agree for $m \sim 1/L^2$, making it possible to match correlation functions between the two regimes (Gasser and Leutwyler 1987a). It is only in the p-regime that the Feynman rules in infinite volume, where only the propagator (8.181) knows about the finite volume, apply.

Next, we will consider each of these two regimes separately. I will only consider the finite-volume physics of pions, because in the real world they are much lighter than kaons, and this is now also the case in state-of-the-art lattice QCD computations. Therefore, finite-volume effects for kaons and other hadrons will be dominated by their interactions with pions. To put it differently, the kaons are always in the p-regime ($m_K \gg 1/L$; for example, for $L = 2$ fm, $m_K L \approx 5$), and in the p-regime finite-volume effects due to kaons travelling around the world are much suppressed compared to those due to pions.[106]

8.6.1   *p*-regime

In the p-regime, it is quite straightforward to adapt the infinite-volume expressions of ChPT to finite volume. Here, for definiteness, we will take a spatial volume $V_3 = L^3$, with periodic boundary conditions, while, for simplicity, we will assume that the (Euclidean) time extent is large enough to be taken as infinity. We will also assume that only finite-volume effects due to pions are significant.

Finite-volume effects start occurring at one loop, because in the loops the pion propagator gets replaced by its finite-volume version, Eq. (8.181). For quantities like meson masses and decay constants, cf. Eq. (8.40), all one-loop integrals are tadpole integrals, for instance of the form

$$G_\infty(0) = \int \frac{d^4p}{(2\pi)^4} \frac{1}{p^2 + m_\pi^2} \doteq \frac{m_\pi^2}{16\pi^2}\left(\frac{2}{d-4} + \log\left(\frac{m_\pi^2}{\Lambda^2}\right) + \text{finite}\right), \qquad (8.189)$$

where the second equation holds using dimensional regularization, and, of course, we have to stay away from the pole at $d = 4$. In finite volume, this gets replaced by

$$G_L(0) = G_\infty(0) + (G_L(0) - G_\infty(0)). \qquad (8.190)$$

Using that

$$G_\infty(x) = \frac{1}{4\pi^2} \frac{m_\pi}{\sqrt{x^2}} K_1(m_\pi\sqrt{x^2}), \qquad (8.191)$$

in which K_1 is the modified Bessel function of the second kind of order one, we find that

$$G_L(0) - G_\infty(0) = \sum_{\vec{n}} G_\infty(x = (\vec{n}L, 0)) - G_\infty(0) \qquad (8.192)$$

$$= \sum_{\vec{n}\neq 0} G_\infty(x = (\vec{n}L, 0))$$

$$= \sum_{n=1}^{\infty} k(n) \frac{1}{4\pi^2} \frac{m_\pi}{\sqrt{n}L} K_1(\sqrt{n}m_\pi L).$$

Here, $k(n)$ counts the number of vectors $\vec{n}$ with integer-valued components of length $\sqrt{n}$. For instance, $k(1) = 6$, $k(2) = 12$, $k(3) = 8$, $k(4) = 6$, etc. Note that, as one would expect, the UV divergence of $G_L(0)$ is the same as in infinite volume, because finite-volume effects come from pions traveling a "long" distance. Of course, in the one-loop calculations leading to Eq. (8.40), we also encounter integrals such as Eq. (8.189) with an extra p^2 in the numerator. Here we can use that, for $x \neq 0$,

$$\int \frac{d^4p}{(2\pi)^4} \frac{p^2 e^{ipx}}{p^2 + m_\pi^2} = -m_\pi^2 \int \frac{d^4p}{(2\pi)^4} \frac{e^{ipx}}{p^2 + m_\pi^2} = -\frac{1}{4\pi^2} \frac{m_\pi^3}{\sqrt{x^2}} K_1(m_\pi\sqrt{x^2}). \qquad (8.193)$$

We thus find, for example, for the finite-volume pion and kaon masses and decay constants to one loop (Gasser and Leutwyler 1987a)

$$m_\pi(L) = m_\pi \left(1 + \frac{m_\pi^2}{(4\pi f_\pi)^2} \sum_{n=1}^{\infty} \frac{4k(n)}{\sqrt{n}m_\pi L} K_1(\sqrt{n}m_\pi L)\right) \qquad (8.194)$$

$$= m_\pi \left(1 + \frac{m_\pi^2}{(4\pi f_\pi)^2} \frac{24}{(m_\pi L)^{3/2}} \sqrt{\frac{\pi}{2}} e^{-m_\pi L} + O\left(e^{-\sqrt{2}m_\pi L}\right)\right),$$

$$f_\pi(L) = f_\pi \left(1 - 2\frac{m_\pi^2}{(4\pi f_\pi)^2} \sum_{n=1}^{\infty} \frac{4k(n)}{\sqrt{n}m_\pi L} K_1(\sqrt{n}m_\pi L)\right)$$

$$= f_\pi \left(1 - \frac{m_\pi^2}{(4\pi f_\pi)^2} \frac{48}{(m_\pi L)^{3/2}} \sqrt{\frac{\pi}{2}} \, e^{-m_\pi L} + O\left(e^{-\sqrt{2} m_\pi L}\right) \right),$$

$$f_K(L) = f_K \left(1 - \frac{3}{4} \frac{m_\pi^2}{(4\pi f_\pi)^2} \sum_{n=1}^{\infty} \frac{4k(n)}{\sqrt{n} m_\pi L} K_1(\sqrt{n} m_\pi L) \right)$$

$$= f_K \left(1 - \frac{m_\pi^2}{(4\pi f_\pi)^2} \frac{18}{(m_\pi L)^{3/2}} \sqrt{\frac{\pi}{2}} \, e^{-m_\pi L} + O\left(e^{-\sqrt{2} m_\pi L}\right) \right),$$

where m_π and f_π are the pion mass and decay constant in infinite volume. The kaon mass does not receive any finite-volume corrections due to pions, as we can infer from the fact that the infinite-volume one-loop contribution has no contribution from pion loops (but only from an η' loop, cf. Eq. (8.40)). In the second of each of these equations, we used the asymptotic expansion $K_1(z) \sim \sqrt{\pi/(2z)} \exp(-z)$. The dominant corrections do not come from higher orders in the asymptotic expansion of K_1, but from the term $n = 2$ in the sum in Eq. (8.194). For $m_\pi \approx 300$ MeV and $m_\pi L \approx 3$, the size of the corrections in Eq. (8.194) is of order one per cent; making L smaller the corrections grow rapidly in size. Detailed studies of finite-volume effects for the pion and kaon masses and decay constants as well as the η mass have been made in Colangelo *et al.* (2005) using a partial resummation of higher loops based on a different approach (Lüscher 1986a), and, for the pion mass to two loops in ChPT (Colangelo and Haefeli 2006).[107]

Many investigations of finite-volume effects in the p-regime have been carried out, using both ChPT and other techniques, and it is beyond the scope of these lectures to cover this topic in any more detail. However, there is one application in which ChPT can play an uniquely important role, and that is the effect of partial quenching on the study of two-particle correlation functions in a finite box.

In a finite volume, the energy spectrum of QCD is discrete, and that also holds for two-particle (and many-particle) states. In fact, the energy levels of two-particle states in finite volume give information on scattering phase shifts in infinite volume (Lüscher 1986b, 1991).[108] The intuitive idea is the following. The normalized wave function of a particle in a box of dimension L^3 is of order $L^{-3/2}$. If two such particles are put inside the box, and they interact, one thus expects a shift in the two-particle energy of order the product of the interaction strength time overlap of the two one-particle wave functions, which is of order $1/L^3$. Therefore, the dependence of the two-particle energy shifts on the volume gives us information about the interactions between the two particles. Unitarity plays an important role in the systematic analysis (Lüscher 1986b, 1991), making it not straightforward to generalize these ideas to PQQCD or QCD with mixed actions. However, ChPT gives us a handle on this problem, since, as we have seen, it can be extended (in Euclidean space) to the partially quenched and mixed-action cases. The two-particle correlation functions, from which in the unitary case the two-particle energy levels are determined, can thus be calculated in PQChPT (or mixed-action ChPT). One finds that the double-hairpin contribution to flavor-neutral propagators gives rise to unphysical phenomena such as enhanced finite-volume effects that can be estimated using ChPT (Bernard and Golterman 1994,

Golterman and Pallante 2000, Beane and Savage 2002, Lin *et al.* 2004, Golterman *et al.* 2005). More study of these intricacies would be interesting.

8.6.2 ϵ-regime

In this section, we will restrict ourselves to two flavors, under the assumption that the strange quark is not in the ϵ-regime. It can thus be taken into account for instance by the method described in Section 8.5. Alternatively, one can start with $N_f = 3$ ChPT, but keep all fields describing mesons containing strange quarks in the p-regime.

Since in this section we will restrict ourselves to two flavors, $\Sigma \in SU(2)$, and instead of the parametrization (8.187) it is easier to use

$$\Sigma = U \left(s + i \frac{\sqrt{2}}{f} \, \vec{\tau} \cdot \vec{\pi} \right), \tag{8.195}$$

$$s = \sqrt{1 - \frac{2\vec{\pi} \cdot \vec{\pi}}{f^2}},$$

with $\vec{\tau}$ the Pauli matrices, and $\int d^4x \, \pi_i(x) = 0$. With $m_u = m_d = m$, $V = L^4$, and using Eq. (8.195), we expand the chiral Lagrangian to second order in the pion field:

$$\int d^4x \, \mathcal{L} = -\frac{1}{4} f^2 B_0 m V \, \mathrm{tr}(U + U^\dagger) \tag{8.196}$$

$$+ \frac{1}{2} \int d^4x \left[\partial_\mu \vec{\pi} \cdot \partial_\mu \vec{\pi} + \frac{1}{2} B_0 m \, \vec{\pi} \cdot \vec{\pi} \, \mathrm{tr}(U + U^\dagger) \right] + O(\pi^3).$$

In the ϵ-regime, as we have seen already, the first term and the kinetic term are of order one. This makes the mass-dependent $O(\pi^2)$ term a term of order $1/L^2$, and the whole $O(p^4)$ Lagrangian (8.39) of order $1/L^4$. At this point, we observe several differences between the p-regime and the ϵ-regime. First, in the p-regime, finite-volume effects are exponentially suppressed, while in the ϵ-regime, they are suppressed by powers of $1/(fL)$. Then, we also see that the chiral expansion in the ϵ-regime rearranges itself: the LECs L_i only show up at next-to-next-to-leading order in the expansion. To leading order, we thus find for the partition function, as a function of the quark mass:

$$Z(m) = \int [d\Sigma] \, \exp \left(\frac{1}{4} f^2 B_0 m V \, \mathrm{tr}(U + U^\dagger) - \frac{1}{2} \int d^4x \, \partial_\mu \vec{\pi} \cdot \partial_\mu \vec{\pi} \right). \tag{8.197}$$

Higher-order terms in Eq. (8.196) can be expanded out from the exponent in the integrand, and thus be taken into account systematically. The notation $[d\Sigma]$ indicates the $SU(2)$-invariant Haar measure for each $\Sigma(x)$.

In order to proceed, we need to split the measure $[d\Sigma]$ into an integration over U and an integration over $\vec{\pi}$. Following Hasenfratz and Leutwyler (1990), we insert

$$1 = \int dH_0 dH_1 dH_2 dH_3 \, \delta \left(H - \frac{1}{V} \int d^4x \, \Sigma(x) \right) \tag{8.198}$$

$$= \int h^3 dh \, dU \, \delta \left(hU - \frac{1}{V} \int d^4x \, \Sigma(x) \right),$$

$$H = H_0 + i\vec{\tau} \cdot \vec{H},$$

where, with $h = |H| \equiv \sqrt{\det(H)}$, we can write $H = hU$ with $U \in SU(2)$, and with dU the Haar measure on $SU(2)$. Now, we transform variables $\Sigma = U\tilde{\Sigma}$, so that

$$\tilde{\Sigma} = s + i \frac{\sqrt{2}}{f} \, \vec{\tau} \cdot \vec{\pi}, \tag{8.199}$$

and use the fact that $d\Sigma = d\tilde{\Sigma}$ to obtain

$$Z(m) = \int dU \int [d\tilde{\Sigma}] \prod_{i=1}^{3} \delta \left(\frac{\sqrt{2}}{fV} \int d^4x \, \pi_i(x) \right) \tag{8.200}$$

$$\times \exp \left(\frac{1}{4} f^2 B_0 mV \, \mathrm{tr}(U + U^\dagger) - \frac{1}{2} \int d^4x \, \partial_\mu \vec{\pi} \cdot \partial_\mu \vec{\pi} + 3 \log \left(\frac{1}{V} \int d^4x \, s(x) \right) \right)$$

$$= \mathcal{N} \int dU \, \exp \left(\frac{1}{4} f^2 B_0 mV \, \mathrm{tr}(U + U^\dagger) \right),$$

where $\mathcal{N}$ is independent of the quark mass m. Carrying out the integral over U with

$$U = \cos\theta \, \mathbf{1} + i \, \sin\theta \, \vec{n} \cdot \vec{\tau}, \quad |\vec{n}| = 1, \quad 0 \le \theta \le \pi, \tag{8.201}$$

$$dU = \frac{1}{4\pi^2} \, d\Omega(\vec{n}) \sin^2(\theta) \, d\theta,$$

we find

$$Z(m) = \mathcal{N} \frac{2}{f^2 B_0 mV} \, I_1 \left(f^2 B_0 mV \right), \tag{8.202}$$

where I_1 is the modified Bessel function of order one. From this, we find for the chiral condensate

$$\langle \bar{q}q \rangle = -\frac{1}{ZV} \frac{dZ}{dm} = -f^2 B_0 \left(\frac{I_1'(f^2 B_0 mV)}{I_1(f^2 B_0 mV)} - \frac{1}{f^2 B_0 mV} \right) \tag{8.203}$$

$$\overset{m \to 0}{=} -\frac{1}{4} f^4 B_0^2 mV.$$

Consistent with our previous discussion, the condensate vanishes for $m \to 0$. For $m \ne 0$ fixed and $V \to \infty$, Eq. (8.203) reproduces the infinite-volume result $\langle \bar{q}q \rangle = -f^2 B_0$, cf. Eq. (8.35). Note that the dimensionless variable $f^2 B_0 mV$ can take any value; to be in the ϵ-regime, the requirements are only that $f^2 L^2 \gg 1$ and $B_0 m L^2 \ll 1$. This does not restrict the size of their product.

Equation (8.203) provides an example of a ChPT calculation in the ϵ-regime to lowest order. Starting from Eq. (8.196), it is in principle straightforward to go to higher orders in $1/(fL)$ (Hasenfratz and Leutwyler 1990, Hansen and Leutwyler 1991, Hansen 1990a, 1990b). For recent work discussing the ChPT calculation of the chiral condensate, see Damgaard and Fukaya (2009).

For instance, it is rather easy to see that the $m\,\vec{\pi}\cdot\vec{\pi}$ term in Eq. (8.196) leads to a renormalization of the first term in that Lagrangian by (Hasenfratz and Leutwyler 1990)

$$r \equiv 1 - \frac{1}{f^2 V}\int d^4 x\langle\vec{\pi}(x)\cdot\vec{\pi}(x)\rangle = 1 - \frac{3}{f^2 V}\sum_{p\neq 0}\frac{1}{p^2} = 1 - \frac{3\beta_1}{f^2 L^2}, \tag{8.204}$$

where β_1 is a numerical constant (Hasenfratz and Leutwyler 1990), and we used that $p_\mu = 2\pi n_\mu/L$ with n_μ integer.[109]

As we have noted before, the $O(p^4)$ LECs L_i only come into play at order $1/L^4$, i.e. at next-to-next-to-leading order in an expansion in $1/(f^2 L^2)$. This is to be contrasted with the chiral expansion in the p-regime, where they appear already at next-to-leading order. This might be helpful for a precise determination of the leading-order LECs (f and B_0), because only these two parameters appear in fits using next-to-leading order expressions in the ϵ-regime. This is another example in which we can extract physical quantities (f and B_0) from unphysical computations – clearly, a small, Euclidean, four-dimensional, finite volume with periodic boundary conditions is unphysical, but accessible to lattice QCD.

This concludes my brief overview of the application of ChPT to the study of volume dependence. For applications to the distribution of topological charge, see Leutwyler and Smilga (1992). I have already mentioned the extension of the chiral expansion to the partially quenched case with some quark masses are in the ϵ-regime, with others in the p-regime (with the obvious application to QCD with $2+1$ flavors with only the strange quark mass in the p-regime), see Bernardoni *et al.* (2008). For an extension including lattice-spacing artifacts for Wilson fermions, see Bär *et al.* (2009) and Shindler (2009); for a recent higher-order study in a partially quenched setting, see Lehner and Wettig (2009). These references contain fairly complete pointers to earlier work in the ϵ-regime.

8.7   Concluding remarks

Clearly, ChPT is a very useful, and, in practice, indispensable tool for extracting hadronic physical quantities from lattice QCD, which, in almost all cases can only obtain correlation functions in some unphysical regime. We have seen that this includes non-zero lattice spacing and finite volume, necessarily, but also "much more" unphysical situations, such as independent choices for valence and sea quarks on the lattice. The chiral Lagrangian, as well as other EFTs, such as baryon ChPT, heavy-light ChPT and two-nucleon EFT, are local theories, defined in terms of a number of LECs. Once lattice QCD determines the values of these LECs, even if this is by matching unphysical correlation functions, they can be used to evaluate any hadronic quantity accessible to the various EFTs. While we have restricted ourselves in this chapter to the physics of Nambu–Goldstone bosons, the conceptual points all carry over to EFTs that include the interactions of other hadrons with Nambu–Goldstone bosons.

There are many applications of ChPT, relevant to lattice QCD, that were left out of this chapter. As already mentioned in this chapter, one can extend ChPT to include the interactions of pions with heavier hadrons. While we briefly discussed this in the context of "heavy" kaons in Section 8.5, the methods described there can be extended to baryons (Jenkins and Manohar 1991, Becher and Leutwyler 1999) and hadrons containing heavy quarks (Manohar and Wise 2000, Sharpe and Zhang 1996, Savage 2002). For a recent review of hadron interactions, including π–π and baryon–baryon scattering, see Beane *et al.* (2008).

Some exercises

Most explicit equations in these lecture notes can be derived with relative ease. None of them require more than a one-loop calculation, and in all cases except Eq. (8.168) the corresponding diagrams are tadpole diagrams. Here are a few more exercises:

1. Derive Eq. (8.15), which is, in fact, valid for an arbitrary number of flavors. Simplify for $N_f = 2$.
2. Show that $\mathrm{tr}(L_\mu L_\nu L_\mu L_\nu)$ can be written as a linear combination of the terms in Eq. (8.17), using the Cayley–Hamilton theorem, which, for a three-by-three matrix A says that

$$A^3 - \mathrm{tr}(A)\, A^2 + \frac{1}{2}\left(\mathrm{tr}(A)^2 - \mathrm{tr}(A^2)\right) A - \det(A)\, \mathbf{1} = \mathbf{0}. \tag{8.205}$$

   [Hint: one way to proceed is to consider $\mathrm{tr}\left((L_\mu + L_\nu)^4\right)$.] Simplify Eq. (8.17) for $N_f = 2$.
3. Show that the scattering lengths for pion scattering in the $O(N)$ linear sigma model are discontinuous across the phase transition between the symmetric phase and the phase in which $O(N)$ breaks down to $O(N-1)$ (cf. discussion after Eq. (8.68)).
4. Consider quenched ChPT (i.e. no sea quarks at all), with one valence quark and ignore the anomaly. Show that using the naive (but wrong!) symmetry group $SU(1|1)_L \otimes SU(1|1)_R$ for constructing the chiral Lagrangian leads to nonsense for the vacuum. See Golterman *et al.* (2005) for more discussion.
5. Verify cyclicity of the supertrace on 2×2 graded matrices.
6. Derive Eq. (8.142). Show that in the LCE regime all other terms in the chiral Lagrangian contribute at next-to-leading order.
7. Investigate possible higher-order terms, to order p^4, in the LCE regime for Wilson ChPT (see Aoki *et al.* (2008a)).

8. Show that a solution of the equation $u_R = h u_L^\dagger h$ exists, for given u_R and u_L. Here, $u_{R,L}$ and h are unitary matrices with determinant one.
9. Show that Eq. (8.163) is equivalent to Eq. (8.162) plus a term that can be absorbed into the A_1 term of Eq. (8.166).
10. Calculate the integral in Eq. (8.168) and convince yourself of the correctness of the discussion following Eq. (8.169).
11. Extend the discussion of Section 8.3.3 to $N_f = 3$ degenerate light quarks (in the LCE regime). I do not know the answer – as far as I know, you can publish the result!

Acknowledgments

First, I would like to thank the organizers for inviting me to present these lectures at this School. I thank the students for the many questions they asked, and all participants, students, lecturers and organizers alike, for a lively and stimulating experience.

I would like to thank Oliver Bär, Claude Bernard, Santi Peris, Yigal Shamir, Steve Sharpe, and André Walker-Loud for helping me understand many of the topics and concepts covered in these lectures, as well as for useful comments on the manuscript. I also would like to thank Claude Bernard for providing me with a copy of his unpublished notes of the lectures on ChPT he presented at the 2007 INT Summer School on "Lattice QCD and its applications." I thank IFAE at the Universitat Autònoma de Barcelona and the KITPC in Beijing, where some of these lectures were written, for hospitality. This work is supported in part by the US Department of Energy.

Notes

1. The strange-quark mass can be taken at its physical value, although it is theoretically interesting to see what happens if one varies the strange-quark mass as well.
2. "Quenched QCD" corresponds to the case in which the sea-quark masses are taken to infinity, making the fermion determinant constant. This is of course unphysical "beyond repair": all effects of quark loops are irrevocably gone.
3. That would solve the strong CP problem.
4. We will return to the issue that the strange-quark mass is much larger than the up-and down-quark masses, see Section 8.5.
5. We ignore isospin breaking, which causes the fields π^0 and η of Eq. (8.6) to mix, in most of this chapter.
6. In mathematical terms, parity is an automorphism of the group $SU(3)_L \times SU(3)_R$.
7. If we choose some other value Ω for the condensate, the subgroup leaving this invariant is the subgroup for which $U_L = \Omega U_R \Omega^{-1}$, which is isomorphic to

$SU(3)_V$. Note that for a different value of Ω also the definition of parity would need to be modified.

8. I will often use "pion" to refer to all (pseudo-) Nambu–Goldstone bosons, kaons and eta included.

9. I am not aware of any general proof from the underlying theory, QCD.

10. A more constructive argument, in which clustering plays a central role, was given in Leutwyler (1994a).

11. This assumes that there are no "accidentally" massless (or very light) hadrons other than the pions.

12. The number of LECs grows rapidly with N, rendering our EFT practically useless beyond $N = 2$.

13. For a different chirally invariant regularization using a lattice cutoff, see Lewis and Ouimet (2001).

14. Our normalization is such that in the real world $f_\pi = 130.4$ MeV. Another common convention uses a value smaller by a factor $\sqrt{2}$.

15. For an introduction to heavy quarks, see the lectures by Rainer Sommer at this school.

16. As mentioned already in the first section, the scenario with $m_u = 0$ is ruled out (Aubin *et al.* 2004, Bernard *et al.* 2007).

17. Some terms in Eq. (8.39) have been removed by field redefinitions.

18. This is not true for the Wess–Zumino–Witten term, which, however, at this order does not contribute to the quantities of Eq. (8.40) to order p^4.

19. For a recent discussion of the experimental value of f_η, including mixing because of $SU(3)$ breaking, see Escribano and Frère (2005).

20. For the relation between $N_f = 3$ LECs and $N_f = 2$ LECs, see Gasser and Leutwyler (1985).

21. Since also EM interactions break isospin, they also will have to be taken into account if one reaches a precision at which isospin breaking becomes significant.

22. Both functions can be thought of as two-parameter fits. In the case of the logarithm, the two parameters are the constant and Λ, in the case of the linear function, they are the constant and the slope.

23. For lattice regulators with an exact chiral symmetry, one expects scaling violations of order a^2 instead of order a. See for example Section 8.3.6.

24. At very small m_{residual}; see David Kaplan's lectures at this school.

25. For a more detailed discussion of Symanzik effective theories in the context of improvement, see the lectures of Peter Weisz at this school.

26. For the application of these ideas to twisted-mass QCD, see the review of Sharpe (2006).

27. And, in the original work by Symanzik, only for ϕ^4 theory (Symanzik 1983a) and the two-dimensional non-linear sigma model (Symanzik 1983b), although there is little doubt that the ideas carry over to gauge theories as well.

28. Since we are considering a renormalizable theory, there should only be a finite number of power divergences in a'. For more discussion of related issues in the context of staggered fermions, see Bernard *et al.* (2008).

29. Such terms would be $O(p^6)$ in a power counting in which $m_{\text{quark}} \sim a^2 \Lambda_{\text{QCD}}^3$.

30. Or the PCAC quark mass to go to zero.

31. For instance, the nucleon mass has $O(a)$ corrections that are not removed by the shift (8.52) (Beane and Savage 2003, Tiburzi 2005).

32. For the case including all sources and new "high-energy" constants, see Sharpe and Wu (2004).

33. I will drop the primes on the shifted quark masses in the rest of this section.

34. For $W' = 0$ higher-order terms in the potential would have to be considered (Sharpe 2005).

35. For earlier work, see Aoki (2003).

36. It is not clear to me that this would lead to a less-divergent result after resummation for all physical quantities in which such enhanced logarithms may occur, as turns out to be the case for the pion mass (Aoki 2003).

37. This reference contains an interesting idea on how to determine the value of W' from the $I = 2$ pion scattering length.

38. Or an improved version of the local current.

39. See the lectures by Peter Weisz at this school.

40. Only the ratio Z_A/Z_A^0 appears when we combine Eqs. (8.70) and (8.71).

41. This step was already carried out in Sharpe and Wu (2005) using the source method; see also Sharpe (2006) and Aoki *et al.* (2009).

42. There is a vector Noether current on the lattice (Karsten and Smit 1981, Bochicchio *et al.* 1985), but often the local vector current is used instead.

43. This result can now be multiplied again by the factor $(1 + \bar{b}_A a)/Z_A^0$.

44. For more introduction, see the lectures by Pilar Hernández at this school, as well as the recent review by the MILC collaboration (Bazavov *et al.* 2009).

45. $U(N_f)$ instead of $SU(N_f)$, because, as we just noted, the axial symmetries in this group are all non-singlet!

46. In contrast, staggered QCD is only invariant under ordinary translations over a distance $2a$ in any of the principal lattice directions.

47. Note that this group does not become larger due to the presence of N_f flavors, because the gluon field also transforms non-trivially under shifts; cf. Eq. (8.84), from which it follows that the analog of Eq. (8.90) is trivial for the gluon field. Thus, all quark flavors must be shifted together. But because U_μ shifts by a pure translation, the analog of Eq. (8.90) for gluons is trivial: $U_\mu(p) \to U_\mu(p)$.

48. This state of affairs may change in the relatively near future, with more highly improved lattice actions, and even smaller lattice spacings.

49. Four-fermion operators can be color mixed or unmixed. For instance, Eq. (8.92) can take the form $(\bar{q}_{ia}\xi_{5\nu}q_{ia})(\bar{q}_{jb}\xi_{\nu5}q_{jb})$ or $(\bar{q}_{ia}\xi_{5\nu}q_{ib})(\bar{q}_{jb}\xi_{\nu5}q_{ja})$, where a and b are color indices. We see that color indices can be contracted in two different ways. They will be omitted here, because it makes no difference for our analysis.

50. Now both derivatives and ξ_μ matrices can provide Lorentz indices, so the argument following Eq. (8.47) does not apply.

51. This result can be derived from Eq. (27) of Aubin and Bernard (2003b), by making the strange quark heavy (a situation we will discuss more in Section 8.5). For non-degenerate results, see Aubin and Bernard (2003b), for $O(p^4)$ meson masses, see Aubin and Bernard (2003a).

52. The complete $O(p^4)$ staggered chiral Lagrangian has been worked out in Sharpe and Van der Water (2005).
53. For a review, see Nussinov and Lampert (2002).
54. If one chooses the number of light sea quarks to be equal to three! In that case, the name "partially quenched" is slightly unfortunate. However, plenty of lattice computations are still really partially quenched, in that they include only two sea quarks.
55. These ghosts have nothing to do with Faddeev–Popov ghosts! They are called ghosts because they share the similarly "incorrect" spin-statistics properties.
56. I write the Dirac operators D_s and D_v without a slash, because not all lattice Dirac operators are of the form $\slashed{D} = \gamma_\mu D_\mu$.
57. A similar trick as described in Golterman *et al.* (2005) exists for domain-wall fermions.
58. It is clear that this sickness already occurs in weak-coupling perturbation theory.
59. And causality, if these generalized theories only live in Euclidean space. However, since PQQCD and mixed-action QCD are local theories, we expect that the cluster property is realized on correlation functions.
60. In the partially quenched case, for which $D_s = D_v$ in Eq. (8.109), the symmetry group is larger, as we will discuss in Section 8.4.2.
61. See the text starting around Eq. (8.117).
62. For instance, for staggered fermions, one can define $\tilde{q}_R$ and $\tilde{q}_L$ using the generator $\epsilon(x)$ of $U(1)_\epsilon$ transformations instead of γ_5.
63. I am skipping a number of further technical details about the precise symmetry group of Eq. (8.109). Also, when all is said and done, the anomaly will have to be taken into account, as usual. For the latter, see Section 8.4.2.
64. The idea is similar to the usage of the replica trick in defining the fourth root with staggered fermions, hence the same name. Note, however, that the aim is different: here, we take $n_i \to 0$, rather than continuing it to $1/4$.
65. An assumption that is used in non-perturbative improvement.
66. The argument that vector symmetries are not spontaneously broken (Vafa and Witten 1984) extends to continuum PQQCD. However, not all discretizations obey the conditions for this theorem, as the possible existence of the Aoki phase in QCD with Wilson fermions shows (Sharpe and Singleton 1998).
67. Not all products of irreducible representations of graded groups are fully reducible. This plays a (minor) role for weak matrix elements in the partially quenched theory with $N_s = 2$, see Golterman and Pallante (2001, 2006), and references therein.
68. Of course, this only makes a difference when both a and b are Grassmann variables.
69. Within this chapter, I'll need only two valence quarks, x and y. But for instance for non-leptonic kaon decays, one needs three.
70. For a proof that this is a correct procedure, see Sharpe and Shoresh (2001).
71. And add more terms, see Gasser and Leutwyler (1985) and Bernard and Golterman (1992).
72. They are numerically hard, because of disconnected diagrams, anyway.

73. For a heuristic argument that such infrared divergences are a property of QCD, and not an artifact of ChPT, in the quenched case, see Bernard and Golterman (1993).

74. Using the quark-flow picture of the next section, it is straightforward to see that all propagators in this contribution are valence propagators.

75. See also Golterman and Leung (1998), which also discusses the case in which χ_{sea} is not very small compared to m_0^2; this can lead to a reduction of the coefficients of chiral logarithms.

76. All contributions to the self-energy with mesons containing sea quarks on the loop get absorbed by the wave-function renormalization.

77. For another example relevant for weak matrix elements, see Laiho and Soni (2005). For a discussion of the implications of graded group representation theory for "external operators" such as electroweak operators, see Golterman and Pallante (2001, 2006).

78. For instance, they would be present if one takes the ghost masses different from the valence masses.

79. This is the problem with staggered QCD with the fourth-root trick, see Golterman (2008).

80. For a much more detailed argument along these lines, see Section 6 of Sharpe and Shoresh (2001).

81. Or domain-wall fermions with a very small residual mass.

82. For the case of overlap fermions on a staggered sea, see Bär *et al.* (2005).

83. To the best of my knowledge, the $O(p^4)$ chiral Lagrangian with LCE power counting has not been constructed.

84. In the GSM regime there is, because the $O(a^2)$ terms are of higher order in that regime. In the GSM regime we do thus have that $m_{\text{mixed}}^2 = (m_{\text{sea}}^2 + m_{\text{val}}^2)/2$ to lowest order.

85. For disconnected contributions, cf. the second term in Eq. (8.144), see below.

86. In the GSM regime, $m_{\text{quark}} \sim a\Lambda_{\text{QCD}}^2$, this would be easier to carry out than in the LCE regime, $m_{\text{quark}} \sim a^2\Lambda_{\text{QCD}}^3$.

87. This means that the fitted LECs at order p^6 from the lattice cannot be compared to any continuum results. While this "hybrid" fitting method might also affect lower-order LECs, they are expected to be much less sensitive, because their values are predominantly determined by lattice results at lower quark masses, where the $O(p^6)$ effects are less important (Aubin *et al.* 2004).

88. In this section we will use the word "pion" to refer to $\pi^\pm$ and π^0, because we will not think of the kaon as an approximate Nambu–Goldstone boson.

89. Recent results for both condensates in the three-flavor theory can be found in Bernard *et al.* (2007). It is found that while the three-flavor condensate is smaller than the two-flavor condensate, it is not very small.

90. Whether with the symbol "u" I refer to the up quark or the non-linear pion field should always be clear from the context.

91. It can be shown that any other non-linear representation of the group $SU(2)_L \times SU(2)_R$ with the same field content can be brought into the form (8.157) by a field redefinition (Coleman *et al.* 1969, Callan *et al.* 1969).

92. Note that conventions in Roessl (1999) differ from those used here.

93. The primed As are not present in $\mathcal{L}_K^{(2)}$ of Roessl (1999), and can probably be traded for higher-order terms by field redefinitions. However, this is not mandatory, and I choose to keep them in $\mathcal{L}_K^{(2)}$.

94. Using kaon ChPT for semileptonic decays only works when the momentum of the outgoing pion is small.

95. Similar to the way that the physics of r mesons and other heavy hadrons appears through the values of the LECs in the pion Lagrangian.

96. The same argument was used for baryon ChPT already in Gasser *et al.* (1988).

97. Of course, $Mv \cdot q$ can be smaller than Λ_{QCD}^2, but the q integral gets cut off in the infrared by m_π. The situation is more complicated if there are also other infrared scales present, such as momentum transfer in a form factor or scattering amplitude. For detailed discussions, see for instance Becher and Leutwyler (1999) and Walker-Loud (2008).

98. This leads to the terms of order m_π^3 that appear in the heavy-light and baryon cases.

99. Expanding $L(m_\eta^2)$ in m_π^2/m_K^2 gives rise to analytic terms in m_π^2 only.

100. It turns out that C_2 does not contribute to this order.

101. In this section, I will use the word "pions" to refer only to $\pi^\pm$ and π^0, and assume that their masses are equal, i.e. that isospin is conserved.

102. For field theory at finite temperature, see the lectures by Owe Philipsen at this school.

103. A corollary is that the chiral theory at non-zero temperature only makes sense for small T, so that $\exp(-E/T)$ is small.

104. Here, I have arbitrarily placed U on the left of $\exp(2i\phi(x)/f)$. More generally, we can write $\Sigma(x) = U_L \exp(2i\phi(x)/f)U_R$, but if we set $U_L = UU_R^\dagger$, we have that $\Sigma(x) = U\exp(2iU_R^\dagger\phi(x)U_R/f)$, and $U_R^\dagger\phi U_R$ does not contain a constant mode if ϕ does not.

105. In this small four-dimensional volume, this basically constitutes our definition of the pion mass, as we can clearly not define it as the parameter characterizing the large-t behavior of the Euclidean pion two-point function.

106. For an investigation including both pions in the ϵ-regime and kaons in the p-regime, see Bernardoni *et al.* (2008).

107. For earlier work, see these references.

108. For an application to non-leptonic kaon decays, see Lellouch and Lüscher (2001).

109. The sum over p is of course infinite, but we are interested here only in the $1/L^2$ part, which is finite. The volume-independent infinite part can be absorbed into a multiplicative renormalization of m. In dimensional regularization, the "infinite" part actually vanishes.

References

Allton C. *et al.* [RBC-UKQCD Collaboration], *Phys. Rev. D* **78**, 114509 (2008) [arXiv:0804.0473 [hep-lat]].

Amoros G., Bijnens J. and Talavera P., *Nucl. Phys. B* **568**, 319 (2000) [arXiv:hep-ph/9907264].

Aoki S., *Phys. Rev. D* **30**, 2653 (1984).

Aoki S., *Phys. Rev. D* **68**, 054508 (2003) [arXiv:hep-lat/0306027].

Aoki S., Bär O. and Biedermann B., *Phys. Rev. D* **78**, 114501 (2008a) [arXiv:0806.4863 [hep-lat]].

Aoki S., Bär O. and Sharpe S. R., *Phys. Rev. D* **80**, 014506 (2009) [arXiv:0905.0804 [hep-lat]].

Aoki S. *et al.* [PACS-CS Collaboration], arXiv:0807.1661 [hep-lat] (2008b).

Aubin C. and Bernard C., *Phys. Rev. D* **68**, 034014 (2003a) [arXiv:hep-lat/0304014].

Aubin C. and Bernard C., *Phys. Rev. D* **68**, 074011 (2003b) [arXiv:hep-lat/0306026].

Aubin C. and Wang Q., *Phys. Rev. D* **70**, 114504 (2004) [arXiv:hep-lat/0410020].

Aubin C. *et al.* [MILC Collaboration], *Phys. Rev. D* **70**, 114501 (2004) [arXiv:hep-lat/0407028].

Aubin C., Christ N. H., Dawson C., Laiho J. W., Noaki J., Li S. and Soni A., *Phys. Rev. D* **74**, 034510 (2006) [arXiv:hep-lat/0603025].

Bär O., Bernard C., Rupak G. and Shoresh N., *Phys. Rev. D* **72**, 054502 (2005) [arXiv:hep-lat/0503009].

Bär O., Necco S. and Schaefer S., *JHEP* **0903**, 006 (2009) [arXiv:0812.2403 [hep-lat]].

Bär O., Rupak G. and Shoresh N., *Phys. Rev. D* **67**, 114505 (2003) [arXiv:hep-lat/0210050].

Bär O., Rupak G. and Shoresh N., *Phys. Rev. D* **70**, 034508 (2004) [arXiv:hep-lat/0306021].

Bardeen W. A., Duncan A., Eichten E., Isgur N. and Thacker H., *Phys. Rev. D* **65**, 014509 (2001) [arXiv:hep-lat/0106008].

Basak S. *et al.* [MILC Collaboration], arXiv:0812.4486 [hep-lat] (2008).

Bazavov A. *et al.*, arXiv:0903.3598 [hep-lat] (2009).

Beane S. R., Orginos K. and Savage M. J., *Int. J. Mod. Phys. E* **17**, 1157 (2008) [arXiv:0805.4629 [hep-lat]].

Beane S. R. and Savage M. J., *Phys. Lett. B* **535**, 177 (2002) [arXiv:hep-lat/0202013].

Beane S. R. and Savage M. J., *Phys. Rev. D* **68**, 114502 (2003) [arXiv:hep-lat/0306036].

Becher T. and Leutwyler H., *Eur. Phys. J. C* **9**, 643 (1999) [arXiv:hep-ph/9901384].

Bernard C., in *"From actions to answers,"* proceedings of TASI'89, eds. DeGrand T. and Toussaint D. (World Scientific, Singapore, 1990).

Bernard C., *Phys. Rev. D* **73**, 114503 (2006) [arXiv:hep-lat/0603011].

Bernard C. and Golterman M., *Phys. Rev. D* **46**, 853 (1992) [arXiv:hep-lat/9204007].

Bernard C. and Golterman M., *Nucl. Phys. Proc. Suppl.* **30**, 217 (1993) [arXiv:hep-lat/9211017].

Bernard C. and Golterman M., *Phys. Rev. D* **49**, 486 (1994) [arXiv:hep-lat/9306005].

Bernard C. and Golterman M., *Phys. Rev. D* **53**, 476 (1996) [arXiv:hep-lat/9507004].

Bernard C., Golterman M. and Shamir Y., *Phys. Rev. D* **73**, 114511 (2006) [arXiv:hep-lat/0604017].

Bernard C., Golterman M. and Shamir Y., *Phys. Rev. D* **77**, 074505 (2008) [arXiv:0712.2560 [hep-lat]].

Bernard C. *et al.*, *PoS* **LAT2007**, 090 (2007) [arXiv:0710.1118 [hep-lat]].

Bernardoni F., Damgaard P. H., Fukaya H. and Hernández P., *JHEP* **0810**, 008 (2008) [arXiv:0808.1986 [hep-lat]].

Bijnens J., *Prog. Part. Nucl. Phys.* **58**, 521 (2007a) [arXiv:hep-ph/0604043].

Bijnens J., *PoS* **LAT2007**, 004 (2007b) [arXiv:0708.1377 [hep-lat]].

Bijnens J. and Celis A., *Phys. Lett. B* **680**, 466 (2009) [arXiv:0906.0302 [hep-ph]].

Bijnens J., Danielsson N. and Lahde T. A., *Phys. Rev. D* **73**, 074509 (2006) [arXiv:hep-lat/0602003].

Blum T., Doi T., Hayakawa M., Izubuchi T. and Yamada N., *Phys. Rev. D* **76**, 114508 (2007) [arXiv:0708.0484 [hep-lat]].

Bochicchio M., Maiani L., Martinelli G., Rossi G. C. and Testa M., *Nucl. Phys. B* **262**, 331 (1985).

Callan C. G., Coleman S. R., Wess J. and Zumino B., *Phys. Rev.* **177**, 2247 (1969).

Chen J. W., Golterman M., O'Connell D. and Walker-Loud A., *Phys. Rev. D* **79**, 117502 (2009a) [arXiv:0905.2566 [hep-lat]].

Chen J. W., O'Connell D., Van de Water R. S. and Walker-Loud A., *Phys. Rev. D* **73**, 074510 (2006) [arXiv:hep-lat/0510024].

Chen J. W., O'Connell D. and Walker-Loud A., *Phys. Rev. D* **75**, 054501 (2007) [arXiv:hep-lat/0611003];

Chen J. W., O'Connell D. and Walker-Loud A., *JHEP* **0904**, 090 (2009b) [arXiv:0706.0035 [hep-lat]].

Chen J. W. and Savage M. J., *Phys. Rev. D* **65**, 094001 (2002) [arXiv:hep-lat/0111050].

Colangelo G. and Haefeli C., *Nucl. Phys. B* **744**, 14 (2006) [arXiv:hep-lat/0602017].

Colangelo G., Dürr S. and Haefeli C., *Nucl. Phys. B* **721**, 136 (2005) [arXiv:hep-lat/0503014].

Coleman S. R., Wess J. and Zumino B., *Phys. Rev.* **177**, 2239 (1969).

Collins J. C., *Renormalization*, Cambridge University Press, Cambridge, 1984.

Creutz M., *Phys. Rev. D* **52**, 2951 (1995) [arXiv:hep-th/9505112].

Damgaard P. H., *Phys. Lett. B* **476**, 465 (2000) [arXiv:hep-lat/0001002].

Damgaard P. H. and Fukaya H., *JHEP* **0901**, 052 (2009) [arXiv:0812.2797 [hep-lat]].

Damgaard P. H. and Splittorff K., *Phys. Rev. D* **62**, 054509 (2000) [arXiv:hep-lat/0003017].

Damgaard P., Osborn J.C., Toublan D. and Verbaarschot J.J.M., *Nucl. Phys. B* **547**, 305 (1999) [arXiv:hep-th/9811212].

Dashen R. F., *Phys. Rev. D* **3**, 1879 (1971).

DeWitt B., *Supermanifolds*, Cambridge University Press, Cambridge, 1984.

Donoghue J. F., arXiv:0909.0021 [hep-ph] (2009).

Donoghue J. F., Golowich E. and Holstein B. R., *Dynamics of the standard model*, Cambridge University Press, Cambridge, 1992.

Ecker G., *PoS* **CONFINEMENT8**, 025 (2008) [arXiv:0812.4196 [hep-ph]].

Escribano R. and Frère J. M., *JHEP* **0506**, 029 (2005) [arXiv:hep-ph/0501072].

Farchioni F., Münster G., Sudmann T., Wuilloud J., Montvay I. and Scholz E. E., *PoS* **LAT2007**, 135 (2007) [arXiv:0710.4454 [hep-lat]].

Flynn J. M. and Sachrajda C. T. [RBC Collaboration and UKQCD Collaboration], *Nucl. Phys. B* **812**, 64 (2009) [arXiv:0809.1229 [hep-ph]].

Freund P. G. O., *Introduction to supersymmetry*, Cambridge University Press, 1986.

Gasser J. and Leutwyler H., *Annals Phys.* **158**, 142 (1984).

Gasser J. and Leutwyler H., *Nucl. Phys.* B **250**, 465 (1985).

Gasser J. and Leutwyler H., *Phys. Lett.* B **184**, 83 (1987a).

Gasser J. and Leutwyler H., *Phys. Lett.* B **188**, 477 (1987b).

Gasser J. and Leutwyler H., *Nucl. Phys.* B **307**, 763 (1988).

Gasser J., Sainio M. E. and Švarc A., *Nucl. Phys.* B **307**, 779 (1988).

Georgi H., *Weak interactions and modern particle theory*, Dover, Minèola, NY 2009.

Golterman M., *Nucl. Phys.* B **273**, 663 (1986).

Golterman M., *PoS* **CONFINEMENT8**, 014 (2008) [arXiv:0812.3110 [hep-ph]].

Golterman M., Izubuchi T. and Shamir Y., *Phys. Rev.* D **71**, 114508 (2005) [arXiv: hep-lat/0504013].

Golterman M. and Leung K. C. L., *Phys. Rev.* D **57**, 5703 (1998) [arXiv: hep-lat/9711033].

Golterman M. and Pallante E., *Nucl. Phys. Proc. Suppl.* **83**, 250 (2000) [arXiv: hep-lat/9909069].

Golterman M. and Pallante E., *JHEP* **0110**, 037 (2001) [arXiv:hep-lat/0108010].

Golterman M. and Pallante E., *Phys. Rev.* D **69**, 074503 (2004) [arXiv:hep-lat/0212008].

Golterman M. and Pallante E., *Phys. Rev.* D **74**, 014509 (2006) [arXiv:hep-lat/0602025].

Golterman M., Sharpe S. R. and Singleton R. L., Jr., *Phys. Rev.* D **71**, 094503 (2005) [arXiv:hep-lat/0501015].

Golterman M. and Smit J., *Nucl. Phys.* B **245**, 61 (1984).

Hansen F. C., *Nucl. Phys.* B **345**, 685 (1990a).

Hansen F. C., BUTP-90-42-BERN (1990b).

Hansen F. C. and Leutwyler H., *Nucl. Phys.* B **350**, 201 (1991).

Hasenfratz P. and Leutwyler H., *Nucl. Phys.* B **343**, 241 (1990).

Jenkins E. E. and Manohar A. V., *Phys. Lett.* B **255**, 558 (1991).

Juttner A. and Della Morte M., *PoS* **LAT2009**, 143 (2009) [arXiv:0910.3755 [hep-lat]].

Kadoh D. *et al.* [PACS-CS Collaboration], *PoS* **LAT2008**, 092 (2008) arXiv:0810.0351 [hep-lat].

Kaplan D. B., *Five lectures on effective field theory*, arXiv:nucl-th/0510023 (2005).

Karsten L. H. and Smit J., *Nucl. Phys.* B **183**, 103 (1981).

Kawamoto N. and Smit J., *Nucl. Phys.* B **192**, 100 (1981).

Kluberg-Stern H., Morel A., Napoly O. and Petersson B., *Nucl. Phys.* B **220**, 447 (1983).

Knecht M., Moussallam B., Stern J. and Fuchs N. H., *Nucl. Phys.* B **457**, 513 (1995) [arXiv:hep-ph/9507319].

Kronfeld A. S., *Uses of effective field theory in lattice QCD*, arXiv:hep-lat/0205021 (2002).

Laiho J. and Soni A., *Phys. Rev.* D **71**, 014021 (2005) [arXiv:hep-lat/0306035].

Lee W. J. and Sharpe S. R., *Phys. Rev.* D **60**, 114503 (1999) [arXiv:hep-lat/9905023].

Lehner C. and Wettig T., *JHEP* **0911**, 005 (2009) [arXiv:0909.1489 [hep-lat]].

Lellouch L. and Lüscher M., *Commun. Math. Phys.* **219**, 31 (2001) [arXiv:hep-lat/0003023].

Leutwyler H., *Annals Phys.* **235**, 165 (1994a) [arXiv:hep-ph/9311274].

Leutwyler H., *Principles of chiral perturbation theory*, arXiv:hep-ph/9406283 (1994b).

Leutwyler H. and Smilga A. V., *Phys. Rev. D* **46**, 5607 (1992).

Lewis R. and Ouimet P. P., *Phys. Rev. D* **64**, 034005 (2001) [arXiv:hep-ph/0010043].

Lin C. J. D., Martinelli G., Pallante E., Sachrajda C. T. and Villadoro G., *Phys. Lett. B* **581**, 207 (2004) [arXiv:hep-lat/0308014].

Lüscher M., *Commun. Math. Phys.* **104**, 177 (1986a).

Lüscher M., *Commun. Math. Phys.* **105** (1986b) 153.

Lüscher M., *Nucl. Phys. B* **364**, 237 (1991).

Manohar A. V. and Wise M. B., *Heavy quark physics*, Cambridge University Press, Cambridge, 2000.

Morel A., *J. Phys. (France)* **48**, 1111 (1987).

Noaki J. *et al.*, *PoS* **LAT2008**, 107 (2008) [arXiv:0810.1360 [hep-lat]].

Nussinov S. and Lampert M. A., *Phys. Rept.* **362**, 193 (2002) [arXiv:hep-ph/9911532].

Peskin M. and Schroeder D., *An introduction to quantum field theory*, Perseus Books, Reading, Massachusetts, 1995.

Prelovsek S., Dawson C., Izubuchi T., Orginos K. and Soni A., *Phys. Rev. D* **70**, 094503 (2004) [arXiv:hep-lat/0407037].

Roessl A., *Nucl. Phys. B* **555**, 507 (1999) [arXiv:hep-ph/9904230].

Savage M. J., *Phys. Rev. D* **65**, 034014 (2002) [arXiv:hep-ph/0109190].

Sharatchandra H. S., Thun H. J. and Weisz P., *Nucl. Phys. B* **192**, 205 (1981).

Sharpe S. R., *Phys. Rev. D* **46**, 3146 (1992) [arXiv:hep-lat/9205020].

Sharpe S. R., *Nucl. Phys. Proc. Suppl.* **34**, 403 (1994) [arXiv:hep-lat/9312009].

Sharpe S. R., *Phys. Rev. D* **56**, 7052 (1997) [Erratum-ibid. D **62**, 099901 (2000)] [arXiv:hep-lat/9707018].

Sharpe S. R., arXiv:hep-lat/0607016 (2006).

Sharpe S. R., *Phys. Rev. D* **72**, 074510 (2005) [arXiv:hep-lat/0509009].

Sharpe S. R. and Shoresh N., *Phys. Rev. D* **62**, 094503 (2000) [arXiv:hep-lat/0006017].

Sharpe S. R. and Shoresh N., *Phys. Rev. D* **64**, 114510 (2001) [arXiv:hep-lat/0108003].

Sharpe S. R. and Singleton R. L., *Phys. Rev. D* **58**, 074501 (1998) [arXiv:hep-lat/9804028];

Sharpe S. R. and Van de Water R. S., *Phys. Rev. D* **69**, 054027 (2004) [arXiv:hep-lat/0310012].

Sharpe S. R. and Van de Water R. S., *Phys. Rev. D* **71**, 114505 (2005) [arXiv:hep-lat/0409018].

Sharpe S. R. and Wu J. M. S., *Phys. Rev. D* **70**, 094029 (2004) [arXiv:hep-lat/0407025].

Sharpe S. R. and Wu J. M. S., *Phys. Rev. D* **71**, 074501 (2005) [arXiv:hep-lat/0411021].

Sharpe S. R. and Zhang Y., *Phys. Rev. D* **53**, 5125 (1996) [arXiv:hep-lat/9510037].

Sheikholeslami B. and Wohlert R., *Nucl. Phys. B* **259**, 572 (1985).

Shindler A., *Phys. Lett. B* **672**, 82 (2009) [arXiv:0812.2251 [hep-lat]].

Sint S., arXiv:hep-lat/0702008 (2007).

Symanzik K., *Nucl. Phys. B* **226**, 187 (1983a).

Symanzik K., *Nucl. Phys. B* **226**, 205 (1983b).

Tiburzi B. C., *Nucl. Phys. A* **761**, 232 (2005) [arXiv:hep-lat/0501020].

Vafa C. and Witten E., *Nucl. Phys. B* **234**, 173 (1984).

Van den Doel C. and Smit J., *Nucl. Phys. B* **228**, 122 (1983).

Walker-Loud A., *PoS* Lattice 2008, 005 (2008) [arXiv:0810.0663[hep-lat]]. [arXiv: 0812.2723 [nucl-th]].

Weinberg S., *Physica A* **96**, 327 (1979).

Wess J. and Zumino B., *Phys. Lett. B* **37**, 95 (1971).

Witten E., *Annals Phys.* **128**, 363 (1980).

Witten E., *Nucl. Phys. B* **223**, 422 (1983).

Zirnbauer M., *J. Math. Phys. (N.Y.)* **37**, 4986 (1996).

9

Non-perturbative heavy quark effective theory

Rainer SOMMER

NIC, DESY, Platanenallee 6, 15738 Zeuthen, Germany

Overview

This chapter on the effective field theory for heavy quarks, an expansion around the static limit, concentrates on the motivation and formulation of HQET, its renormalization and discretization. This provides the basis for understanding that and how this effective theory can be formulated fully non-perturbatively in the QCD coupling, while by the very nature of an effective field theory, it is perturbative in the expansion parameter $1/m$. After the couplings in the effective theory have been determined, the result at a certain order in $1/m$ is unique up to higher-order terms in $1/m$. In particular, the continuum limit of the lattice-regularized theory exists and leaves no trace of how it was regularized. In other words, the theory yields an asymptotic expansion of the QCD observables in $1/m$ – as usual in a quantum field theory modified by powers of logarithms. None of these properties has been shown rigorously (e.g. to all orders in perturbation theory) but perturbative computations and recently also non-perturbative lattice results give strong support to this "standard wisdom".

A subtle issue is that a theoretically consistent formulation of the theory is only possible through a non-perturbative matching of its parameters with QCD at finite values of $1/m$ (Section 9.4.4). As a consequence, one finds immediately that the splitting of a result for a certain observable into, for example, lowest order and first order is ambiguous. Depending on how the matching between effective theory and QCD is done, a first-order contribution may vanish and appear instead in the lowest order. For example, the often-cited phenomenological HQET parameters $\bar{\Lambda}$ and λ_1 lack a unique non-perturbative definition. But this does not affect the precision of the asymptotic expansion in $1/m$. The final result for an observable is correct up to order $(1/m)^{n+1}$ if the theory was treated including $(1/m)^n$ terms.

Clearly, the weakest point of HQET is that it intrinsically is an expansion. In practice, carrying it out non-perturbatively beyond the order $1/m$ will be very difficult. In this context two observations are relevant. First, the expansion parameter for HQET applied to B-physics is $\Lambda_{\rm QCD}/m_{\rm b} \sim 1/(r_0 m_{\rm b}) = 1/10$ and indeed recent computations of $1/m_{\rm b}$ corrections showed them to be very small. Secondly, since HQET yields the asymptotic expansion of QCD, it becomes more and more accurate the larger the mass is. It can therefore be used to constrain the large-mass behavior of QCD computations done at finite, varying, quark masses. At some point, computers and computational strategies will be sufficient to simulate with lattice spacings that are small enough for a relativistic b-quark. One would then like to understand the full mass-behavior of observables and a combination of HQET and relativistic QCD will again be most useful. Already now, there is a strategy (De Divitiis *et al.*, 2003*a*;*b*; Guazzini *et al.*, 2008), which is related to the one discussed in Section 9.5.3 and which, in its final version combines HQET and QCD in such a manner. For a short review of this aspect I refer to (Tantalo, 2008).

9.1   Introduction

9.1.1   Conventions

Our conventions for gauge fields, lattice derivatives, etc. are summarized in the appendix.

9.1.2 The rôle of HQET

This book focuses on lattice gauge theories. How does heavy-quark effective theory (HQET) fit into it? The first part of the answer is that HQET is expected to provide the true asymptotic expansion of quantities in powers (accompanied by logarithms) of $1/m$, the mass of the heavy quark, with all other scales held fixed. The accessible quantities are energies, matrix elements and Euclidean correlation functions with a single heavy (valence) quark, while all other quarks are light. A full understanding of QCD should contain this kinematical region.

The second part of the answer has to do with the challenge we are facing when we perform a Monte Carlo (MC) evaluation of the QCD path integral. This becomes apparent by considering the scales that are relevant for QCD. For low-energy QCD and flavor physics excluding the top quark, they range from

$$m_\pi \approx 140\,\text{MeV over } m_\text{D} = 2\,\text{GeV to } m_\text{B} = 5\,\text{GeV}.$$

In addition, the ultraviolet cutoff of $\Lambda_\text{UV} = a^{-1}$ of the discretized theory has to be large compared to all physical energy scales if the theory discretized with a lattice spacing a is to be an approximation to a continuum. Finally, the linear extent of space-time has to be restricted to a finite value L in a numerical treatment: there is an infrared cutoff L^{-1}. Together, the following constraints have to be satisfied.

$$\Lambda_\text{IR} = L^{-1} \ll m_\pi, \ldots, m_\text{D}, m_\text{B} \ll a^{-1} = \Lambda_\text{UV}. \tag{9.1}$$

The infrared and the ultraviolet effects are systematic errors that have to be controlled. Infrared effects behave as (Lüscher, 1986) $\text{O}(e^{-Lm_\pi})$ and are known from chiral perturbation theory (Colangelo *et al.*, 2005) to be at the per cent level when $L \gtrsim 4/m_\pi \approx 6\,\text{fm}$, while the UV, discretization, errors are $\text{O}((a\,m_\text{quark})^2)$ in $\text{O}(a)$-improved theories.[1] With a charm quark mass of around $1\,\text{GeV}$ we have a requirement of $a \lesssim 1/(2m_\text{c}) \ldots 1/(4m_\text{c}) \approx 0.1 \ldots 0.05\,\text{fm}$ (Kurth and Sommer, 2002) and thus

$$L/a \approx 60 \ldots 120. \tag{9.2}$$

Including b-quarks would increase the already rather intimidating estimate of L/a by a factor 4. It is thus mandatory to resort to an effective theory where degrees of freedom with energy scales around the b-quark mass and higher are summarized in the coefficients of terms in the effective Lagrangian. A *precise* treatment of this theory has become very relevant because the search for physics beyond the Standard Model in the impressive first generation of B-physics flavor experiments has been unsuccessful so far. New physics contributions are very small and even higher precision is needed both in experiment and in theory to possibly reveal them. HQET is a very important ingredient in this effort.

Before we focus on our topic let us note that a factor two or so in L/a may be saved by working at somewhat higher pion mass and extrapolating with chiral perturbation theory, see M. Golterman's chapter.

9.1.3 On continuum HQET

9.1.3.1 Idea

We consider hadrons with a single very heavy quark, e.g. a B-meson. Physical intuition tells us that these will be similar to a hydrogen atom with the analogy

hydrogen atom	:	heavy proton	+	light electron
B-meson	:	heavy b-quark	+	light antiquark
b-baryons	:	heavy b-quark	+	two light quarks

and so on.

When we take the limit $m = m_b \rightarrow \infty$ ("static") the b-quark is at rest in the *rest-frame of the b-hadron* (B, Λ_b, ...). In this situation, we should be able to find an effective Lagrangian describing the dynamics of the light quarks and glue with the heavy quark just representing a color source. Corrections in $1/m_b$ should be systematically included in a series expansion in that variable. The Lagrangian is then expected to be given as a series in D_k/m where the covariant derivatives act on the heavy-quark field and correspond to its spatial momenta in the rest-frame of the heavy hadron.

Before proceeding to a heuristic derivation of the effective field theory, let us note some general properties of what we are actually seeking, comparing to other familiar effective field theories. In contrast to the low-energy effective field theory for electroweak interactions, where the heavy particles (W- and Z-boson, top quark) are removed completely from the Lagrangian we here want to consider processes with b-quarks in initial and/or final states. The b-quark field is thus contained in the Lagrangian and we have to find its relevant modes to be kept.[2]

Another important effective field theory to compare to is the chiral effective theory, covered here by Maarten Golterman. The main differences are that this is a fully relativistic theory with loops of the (pseudo-) Goldstone bosons and that the interaction of the fields in the effective Lagrangian disappears for zero momentum. The theory can therefore be evaluated perturbatively. It is also called chiral perturbation theory. In contrast, the b-quarks in HQET still interact non-perturbatively with the light quarks and gluons. This effective-field theory therefore needs a lattice implementation in order to come to predictions beyond those that can be read off from its symmetries.

9.1.3.2 Derivation of the form of the effective field theory: FTW trafo

Strategy. Our strategy is to carry out the following steps, which we discuss in more detail below.

- We start from a Euclidean action.
- We identify the dominant degrees of freedom for the kinematical situation we are interested in: the "large" components of the b-quark field for the quark and the "small" components for the antiquark.

- We decouple large components and small components, order by order in D_k/m [$\overline{\psi}_{\rm h} D_k/m\, \psi_{\rm h} \ll \overline{\psi}_{\rm h}\psi_{\rm h}$]. This assumes smooth gauge (and other) fields. It is thus essentially a classical derivation. The decoupling is achieved by a sequence of Fouldy Wouthuysen-Tani (FTW) transformations (see e.g. (Itzykson and Zuber, 1980)), following essentially (Körner and Thompson, 1991).

- The irrelevant modes are dropped from the theory (often it is said they are integrated out). Their effects are *not* expected to change the form of the local Lagrangian, but just to renormalize its parameters. Still, it could be that local terms allowed by the symmetries happen to vanish in the classical theory. Thus, the symmetries have to be considered and all terms of the proper dimension compatible with the symmetries have to be taken into account.

- At tree level the values of the parameters in the effective Lagrangian are given by the FTW transformation. In general (i.e. for any value of the QCD coupling) they have to be determined by matching to QCD: one expands QCD correlation functions in $1/m_{\rm b}$ and compares to HQET. This part of the strategy will be discussed in detail in later sections.

Identifying the degrees of freedom. We consider the *free* propagator of a Dirac-fermion in Euclidean space, in the time / space-momentum representation:[3]

$$S(x_0; \mathbf{k}) = \int {\rm d}^3\mathbf{x}\, {\rm e}^{-i\mathbf{k}\mathbf{x}} \langle \psi(x)\overline{\psi}(0)\rangle = \int \frac{{\rm d}k_0}{(2\pi)} {\rm e}^{ik_0 x_0} \left[ik_\mu\gamma_\mu + m\right]^{-1} \tag{9.3}$$

$$= S_+(x_0; \mathbf{k}) + S_-(x_0; \mathbf{k}),$$

with

$$S_+(x_0; \mathbf{p}) = \theta(x_0)\frac{m}{E(\mathbf{p})}{\rm e}^{-E(\mathbf{p})x_0}P_+(u), \qquad P_+(u) = \frac{1 - iu_\mu\gamma_\mu}{2}, \quad u_\mu = p_\mu/m, \tag{9.4}$$

$$S_-(x_0; \mathbf{p}) = \theta(-x_0)\frac{m}{E(\mathbf{p})}{\rm e}^{E(\mathbf{p})x_0}P_-(u), \qquad P_-(u) = \frac{1 + iu_\mu\gamma_\mu}{2},$$

where p_μ is the on-shell momentum, i.e.

$$p_0 = iE(\mathbf{p}) = i\sqrt{m^2 + \mathbf{p}^2}. \tag{9.5}$$

Here, $S_+(x_0; \mathbf{p})$ describes the propagation of a quark from time $t = 0$ to $t = x_0$ and $S_-(x_0; \mathbf{p})$ describes the propagation of an antiquark from $t = -x_0$ to $t = 0$. Since the Euclidean 4-velocity vector u satisfies $u^2 = u_\mu u_\mu = -1$, the matrices $P \in \{P_+, P_-\}$ are projection operators,

$$[P(u)]^2 = P(u), \; P_+(u)P_-(u) = 0, \; P_+(u) + P_-(u) = 1. \tag{9.6}$$

They allow us to project onto the on-shell components of a quark with velocity **u**.

The "large" field components corresponding to the quark are given by the projection

$$\psi_{h,u}(x) = P(u)\psi(x), \ \overline{\psi}_{h,u}(x) = \overline{\psi}(x)P(u) \tag{9.7}$$

and the "small" ones, the antiquark field, are

$$\psi_{\bar{h},u}(x) = P(-u)\psi(x), \ \overline{\psi}_{\bar{h},u}(x) = \overline{\psi}(x)P(-u), \tag{9.8}$$

such that for free quarks

$$\int d^3\mathbf{x}\, e^{-i\mathbf{p}\mathbf{x}} \langle \psi_{h,u}(x)\overline{\psi}_{h,u}(0)\rangle = S_+(x_0; \mathbf{p}) \tag{9.9}$$

and similarly for the antiquark.[4]

For a b-hadron with velocity $\mathbf{u}$, the fields $\psi_{h,u}(x)$, $\overline{\psi}_{h,u}(x)$ are expected to be the relevant ones with the other field-components giving subdominant contributions in the path-integral representation of correlation functions (or scattering amplitudes in Minkowski space), while for a $\bar{b}$-hadron $\psi_{\bar{h},u}(x)$, $\overline{\psi}_{\bar{h},u}(x)$ are expected to dominate.

In the presence of a gauge field. When a gauge field is present, we therefore expect an effective Lagrangian for the b-hadrons in terms of $\psi_{h,u}, \overline{\psi}_{h,u}$ plus a term for the antiquark. When we rewrite the Dirac Lagrangian in terms of these fields,

$$\mathcal{L} = \overline{\psi}(m + \mathcal{D})\psi \tag{9.10}$$
$$= \overline{\psi}_{h,u}(m + \mathcal{D}_\parallel)\psi_{h,u} + \overline{\psi}_{\bar{h},u}(m + \mathcal{D}_\parallel)\psi_{\bar{h},u} + \overline{\psi}_{h,u}\mathcal{D}_\perp\psi_{\bar{h},u} + \overline{\psi}_{\bar{h},u}\mathcal{D}_\perp\psi_{h,u},$$

there are mixed contributions that involve

$$\mathcal{D}_\perp = \gamma_\mu D_\mu^\perp, \quad D_\mu^\perp = (\delta_{\mu\nu} + u_\mu u_\nu)D_\nu, \tag{9.11}$$

where the derivative is projected orthogonal to u_μ. Analogously, we have

$$\mathcal{D}_\parallel = \gamma_\mu D_\mu^\parallel, \quad D_\mu^\parallel = -u_\mu D_\nu u_\nu. \tag{9.12}$$

From our general consideration of the kinematical situation that we want to describe, $D_\mu^\perp$ acting on the heavy quark field is to be considered small (compared to m). In contrast, $D_\mu^\parallel$ applied to the field will yield approximately $p_\mu = u_\mu m$. We therefore carry out an expansion with

$$\mathcal{D}_{\|}\psi = \mathrm{O}(m)\,\psi,$$

$$\mathcal{D}_{\perp}\psi = \mathrm{O}(1)\,\psi \tag{9.13}$$

and all other fields, such as $F_{\mu\nu}$, treated as order one. This is often called the power-counting scheme.

FTW trafo and Lagrangian at zero velocity. Having identified the expansion, we perform a field rotation (FTW transformation) to decouple large and small components order by order in $1/m$. First, we consider the special case of zero velocity,

$$u_k = 0: \quad \mathcal{D}_{\|} = D_0\gamma_0, \quad \mathcal{D}_{\perp} = D_k\gamma_k,$$

$$P(u) = P_+ = \frac{1+\gamma_0}{2}, \quad P(-u) = P_- = \frac{1-\gamma_0}{2}. \tag{9.14}$$

The FTW transformation is

$$\psi \to \psi' = \mathrm{e}^S\psi, \quad S = \frac{1}{2m}D_k\gamma_k = -S^\dagger,$$

$$\overline{\psi} \to \overline{\psi}' = \overline{\psi}\mathrm{e}^{-\overleftarrow{S}} = \overline{\psi}\mathrm{e}^{-\overleftarrow{D}_k\gamma_k/(2m)}. \tag{9.15}$$

Its Jacobian is one. The Lagrangian written in terms of the transformed fields,

$$\mathscr{L} = \overline{\psi}'(\mathcal{D}'+m)\psi', \tag{9.16}$$

yields a Dirac operator (note that S acts to the right everywhere)

$$\mathcal{D}'+m = \mathrm{e}^{-S}(\mathcal{D}+m)\mathrm{e}^{-S}. \tag{9.17}$$

Expanding $\mathrm{e}^{-S} = 1 - S + \frac{1}{2}S^2 - \dots$ in $S = \mathrm{O}(1/m)$ yields

$$\mathcal{D}'+m = \underbrace{\mathcal{D}+m}_{\mathrm{O}(m)} + \underbrace{\{-S,\mathcal{D}+m\}}_{\mathrm{O}(1)} + \underbrace{\frac{1}{2}\{-S,\{-S,\mathcal{D}+m\}\}}_{\mathrm{O}(1/m)} + \dots \tag{9.18}$$

In the evaluation of the different terms we count all fields and derivatives of fields (e.g. $F_{\mu\nu}$) as order one except for D_0 *acting onto the heavy quark-field*. We work out the expansion up to order $1/m$. A little algebra yields

$$\mathcal{D}+m+\{-S,\mathcal{D}+m\} = D_0\gamma_0 - \frac{1}{2m}[\gamma_k\gamma_0 F_{k0} + \frac{1}{i}\sigma_{kl}F_{kl} + 2D_k D_k] \tag{9.19}$$

with $\sigma_{\mu\nu} = \frac{i}{2}[\gamma_\mu,\gamma_\nu]$, $F_{kl} = [D_k, D_l]$ and

$$\underbrace{\frac{1}{2}\{-S,\{-S,\mathcal{D}+m\}\}}_{-D_k\gamma_k + \mathrm{O}(1/m)} = \frac{1}{4m}[\frac{1}{i}\sigma_{kl}F_{kl} + 2D_k D_k], \tag{9.20}$$

such that

$$\mathcal{D}' = D_0\gamma_0 - \frac{1}{2m}[\underbrace{\gamma_k\gamma_0 F_{k0}}_{\text{off-diagonal}} + \frac{1}{2i}\sigma_{kl}\,F_{kl} + D_kD_k] + \mathrm{O}(1/m^2). \qquad (9.21)$$

In the static part, $D_0\gamma_0$, the large and small components are decoupled, but one of the $1/m$ terms, $\gamma_k\gamma_0 F_{k0}$, is off-diagonal with respect to this split. We therefore seek a second transformation $\psi'' = e^{S'}\psi'$ to cancel also that term, namely we want

$$\{-S', \mathcal{D}' + m\} = \frac{1}{2m}\gamma_k\gamma_0 F_{k0} + \mathrm{O}(1/m^2). \qquad (9.22)$$

The simple choice $S' = \frac{1}{4m^2}\gamma_0\gamma_k F_{k0}$ does the job. Now we have the classical HQET Lagrangian

$$\mathcal{L} = \mathcal{L}_{\mathrm{h}}^{\mathrm{stat}} + \frac{1}{2m}\mathcal{L}_{\mathrm{h}}^{(1)} + \mathcal{L}_{\bar{\mathrm{h}}}^{\mathrm{stat}} + \frac{1}{2m}\mathcal{L}_{\bar{\mathrm{h}}}^{(1)} + \mathrm{O}(\frac{1}{m^2}) \qquad (9.23)$$

$$\mathcal{L}_{\mathrm{h}}^{\mathrm{stat}} = \bar{\psi}_{\mathrm{h}}(m + D_0)\psi_{\mathrm{h}}, \quad P_+\psi_{\mathrm{h}} = \psi_{\mathrm{h}}, \quad \bar{\psi}_{\mathrm{h}}P_+ = \bar{\psi}_{\mathrm{h}}, \quad P_\pm = \frac{1\pm\gamma_0}{2} \qquad (9.24)$$

$$\mathcal{L}_{\bar{\mathrm{h}}}^{\mathrm{stat}} = \bar{\psi}_{\bar{\mathrm{h}}}(m - D_0)\psi_{\bar{\mathrm{h}}}, \quad P_-\psi_{\bar{\mathrm{h}}} = \psi_{\bar{\mathrm{h}}}, \quad \bar{\psi}_{\bar{\mathrm{h}}}P_- = \bar{\psi}_{\bar{\mathrm{h}}}, \qquad (9.25)$$

$$\mathcal{L}_{\mathrm{h}}^{(1)} = -(\mathcal{O}_{\mathrm{kin}} + \mathcal{O}_{\mathrm{spin}}), \quad \mathcal{L}_{\bar{\mathrm{h}}}^{(1)} = -(\bar{\mathcal{O}}_{\mathrm{kin}} + \bar{\mathcal{O}}_{\mathrm{spin}}), \qquad (9.26)$$

correct up to terms of order $1/m^2$. We introduced

$$\mathcal{O}_{\mathrm{kin}}(x) = \bar{\psi}_{\mathrm{h}}(x)\,\mathbf{D}^2\,\psi_{\mathrm{h}}(x), \; \mathcal{O}_{\mathrm{spin}}(x) = \bar{\psi}_{\mathrm{h}}(x)\,\boldsymbol{\sigma}\cdot\mathbf{B}(x)\,\psi_{\mathrm{h}}(x), \qquad (9.27)$$

$$\bar{\mathcal{O}}_{\mathrm{kin}}(x) = \bar{\psi}_{\bar{\mathrm{h}}}(x)\,\mathbf{D}^2\,\psi_{\bar{\mathrm{h}}}(x), \; \bar{\mathcal{O}}_{\mathrm{spin}}(x) = \bar{\psi}_{\bar{\mathrm{h}}}(x)\,\boldsymbol{\sigma}\cdot\mathbf{B}(x)\,\psi_{\bar{\mathrm{h}}}(x), \qquad (9.28)$$

$$\sigma_k = \frac{1}{2}\epsilon_{ijk}\sigma_{ij}, \quad B_k = i\frac{1}{2}\epsilon_{ijk}F_{ij}, \qquad (9.29)$$

and the heavy-quark fields are the transformed ones, i.e. we renamed $\psi_{\mathrm{h}}'' \to \psi_{\mathrm{h}}$, etc.

Depending on the process/correlation function, just the heavy-quark part or just the heavy-antiquark part of the Lagrangian will contribute, but there are also processes such as $\mathrm{B} - \bar{\mathrm{B}}$ oscillations where both are needed.

It is worth summarizing some issues that arose in this formal derivation.

- Assuming $D_k = \mathrm{O}(1)$ means that this is a classical derivation: in the quantum field theory path integral we integrate over rough fields, i.e. there are arbitrarily large derivatives.

  As emphasized before we therefore take this as a classical Lagrangian. Its renormalization will be discussed later, guided by dimensional counting.

- The derivation is perturbative in $1/m$, order by order. This is all that we want. In this way we expect to obtain the *asymptotic* expansion in powers of $1/m$.

- We note that there are alternative ways to derive the form of the Lagrangian. One may integrate out the components $\bar{\psi}_{\bar{\mathrm{h}}}, \psi_{\bar{\mathrm{h}}}$ in a path integral and then perform a formal expansion of the resulting non-local action for the remaining fields in terms of a series of local operators (Mannel *et al.*, 1992). Another option is to perform a hopping-parameter expansion of the Wilson–Dirac lattice propagator. The leading term gives the propagator of the static action; see Exercise 9.1.

FTW transformation and Lagrangian at finite velocity. At finite velocity the transformation is given again by Eq. (9.15) but with $S = D_\mu^\perp \gamma_\mu / (2m)$. For the lowest-order (static) approximation, just the anticommutator

$$\{\mathcal{D}^\perp, \mathcal{D}^\|\} = \frac{1}{2}\{D_\mu^\perp, D_\nu^\|\} 2\delta_{\mu\nu} + \frac{1}{2}[D_\mu^\perp, D_\nu^\|][\gamma_\mu, \gamma_\nu]$$

is needed. Since $D_\mu^\perp D_\mu^\| = 0 = D_\mu^\| D_\mu^\perp$ and the second term just involves a commutator of derivatives, we see that $\{\mathcal{D}^\perp, \mathcal{D}^\|\} = O(1)$. Consequently, we find

$$\mathcal{L} = \overline{\psi}_{h,u}(m + \mathcal{D}^\|)\psi_{h,u} + \overline{\psi}_{\bar{h},u}(m + \mathcal{D}^\|)\psi_{\bar{h},u} + O(1/m)$$

$$= \overline{\psi}_{h,u}(m - iu_\mu D_\mu)\psi_{h,u} + \overline{\psi}_{\bar{h},u}(m + iu_\mu D_\mu)\psi_{\bar{h},u} + O(1/m) \qquad (9.30)$$

with the projected fields Eq. (9.7) and Eq. (9.8).[5]

Let us add a few *comments on the finite-velocity theory*, since we will not discuss it further.

- O(4) (or Lorentz) invariance is broken. One therefore has to expect a different renormalization of D_0 and D_k (or as is usually said, a renormalization of **u** (Christensen *et al.*, 2000; Mandula and Ogilvie, 1998)).
- The operator $-iD_k u_k$ is unbounded from below. Since it enters the Hamiltonian the theory seems to contain states with arbitrarily large negative energies. Resulting problems in the Euclidean formulation of the theory have been discussed in the literature (Aglietti *et al.*, 1992; Aglietti, 1994), but a compelling formulation of the theory seems not to have been found. There are also no modern applications of the finite-velocity theory on the lattice. We will therefore concentrate entirely on zero-velocity HQET from now on.

9.1.3.3 *Propagator and symmetries*

The continuum propagator. We consider the static approximation at zero velocity and the latter always from now on. The static Dirac operator for the quark is just $D_0 + m$ so its Green function, G_h, (the propagator) in a gauge field $A_\mu(x)$ then satisfies

$$(\partial_{x_0} + A_0(x) + m)G_h(x, y) = \delta(x - y)\, P_+. \qquad (9.31)$$

The solution of this equation is simply

$$G_h(x, y) = \theta(x_0 - y_0)\, \exp(-m\,(x_0 - y_0))\, \mathcal{P}\exp\left\{-\int_{y_0}^{x_0} \mathrm{d}z_0\, A_0(z_0, \mathbf{x})\right\} \delta(\mathbf{x} - \mathbf{y})\, P_+,$$

$$(9.32)$$

were $\mathcal{P}$ denotes path ordering (fields at the end of the integration path to the left). In the same way the propagator for the antiquark is[6]

$$G_{\bar{h}}(x,y) = \theta(y_0 - x_0)\, \exp(-m\,(y_0 - x_0))\, \mathcal{P} \exp\left\{ -\int_{y_0}^{x_0} dz_0\, A_0(z_0, \mathbf{x}) \right\} \delta(\mathbf{x} - \mathbf{y})\, P_-,$$

$$(-\partial_{x_0} - A_0(x) + m)G_{\bar{h}}(x,y) = \delta(x - y)\, P_-. \tag{9.33}$$

The mass appears in a trivial way, with an explicit factor $\exp(-m\,|x_0 - y_0|)$ for any gauge field A_μ. This exponential decay is then present also after path integration over the gauge fields in any 2-point function with a heavy quark,

$$C_{\mathrm{h}}(x,y;m) = C_{\mathrm{h}}(x,y;0)\, \exp(-m\,(x_0 - y_0)). \tag{9.34}$$

An explicit example is

$$C_{\mathrm{h}}^{\mathrm{PP}}(x,y;m) = \langle \bar{\psi}_l(x)\gamma_5 \psi_{\mathrm{h}}(x)\, \bar{\psi}_{\mathrm{h}}(y)\gamma_5 \psi_l(y) \rangle, \tag{9.35}$$

with $\psi_l(x)$ a light-quark fermion field. Equation (9.34) means that m shifts *all* energies in the sector of the Hilbert space with a single heavy quark (or antiquark). We may remove m from the effective Lagrangian and add it to the energies later. We only have to be careful that $m \geq 0$ in Eq. (9.31), Eq. (9.33) selects the forward/backward propagation. Therefore, we set

$$\mathcal{L}_{\mathrm{h}}^{\mathrm{stat}} = \bar{\psi}_{\mathrm{h}}(D_0 + \epsilon)\psi_{\mathrm{h}}, \quad \mathcal{L}_{\bar{\mathrm{h}}}^{\mathrm{stat}} = \bar{\psi}_{\bar{\mathrm{h}}}(-D_0 + \epsilon)\psi_{\bar{\mathrm{h}}}, \quad E_{\mathrm{h}/\bar{\mathrm{h}}}^{\mathrm{QCD}} = E_{\mathrm{h}/\bar{\mathrm{h}}}^{\mathrm{stat}} + m, \tag{9.36}$$

where the limit $\epsilon \to 0_+$ is to be understood.

We note that after performing this shift of the energies, there is no difference in the Lagrangian of a charm or a b-quark if both are treated at the lowest order in this expansion. We turn to discussing this as well as other symmetries of the static theory.

Symmetries

1. Flavor

   If there are F heavy quarks, we just add a corresponding flavor index and use a notation

   $$\psi_{\mathrm{h}} \to \psi_{\mathrm{h}} = (\psi_{\mathrm{h}1}, \ldots, \psi_{\mathrm{h}F})^T, \quad \bar{\psi}_{\mathrm{h}} \to \bar{\psi}_{\mathrm{h}} = (\bar{\psi}_{\mathrm{h}1}, \ldots, \bar{\psi}_{\mathrm{h}F}) \tag{9.37}$$

   $$\mathcal{L}_{\mathrm{h}}^{\mathrm{stat}} = \bar{\psi}_{\mathrm{h}}(D_0 + \epsilon)\psi_{\mathrm{h}}. \tag{9.38}$$

   Then, we obviously have the symmetry

   $$\psi_{\mathrm{h}}(x) \to V\,\psi_{\mathrm{h}}(x), \quad \bar{\psi}_{\mathrm{h}}(x) \to \bar{\psi}_{\mathrm{h}}(x)V^\dagger, \quad V \in \mathrm{SU}(F) \tag{9.39}$$

   and the same for the antiquarks. Note that this symmetry emerges in the large-mass limit irrespective of how the limit is taken. For example, we may take ($F = 2$ with the first heavy flavor identified with charm and the second with beauty)

   $$m_{\mathrm{b}} - m_{\mathrm{c}} = c \times \Lambda_{\mathrm{QCD}}, \quad \text{or} \quad m_{\mathrm{b}}/m_{\mathrm{c}} = c', \quad m_{\mathrm{b}} \to \infty \tag{9.40}$$

   with either c or c' fixed when taking $m_{\mathrm{b}} \to \infty$.

2. Spin

We further note that for each field there are also the two spin components but the Lagrangian contains no spin-dependent interaction. The associated SU(2) rotations are generated by the spin matrices Eq. (9.29) (remember that ψ_h, $\overline{\psi}_\mathrm{h}$ are kept as 4-component fields with 2 components vanishing)

$$\sigma_k = \frac{1}{2}\epsilon_{ijk}\sigma_{ij} \equiv \begin{pmatrix} \sigma_k & 0 \\ 0 & \sigma_k \end{pmatrix}, \tag{9.41}$$

where the symbol σ_k is used at the same time for the Pauli matrices and the 4×4 matrix. We here are in the Dirac representation where

$$\gamma_0 = \begin{pmatrix} 1 & 0 \\ 0 & -1 \end{pmatrix}, \ P_+ = \begin{pmatrix} 1 & 0 \\ 0 & 0 \end{pmatrix}, \ P_- = \begin{pmatrix} 0 & 0 \\ 0 & 1 \end{pmatrix}. \tag{9.42}$$

The spin rotation is then

$$\psi_\mathrm{h}(x) \rightarrow \mathrm{e}^{i\alpha_k\sigma_k}\,\psi_\mathrm{h}(x), \qquad \overline{\psi}_\mathrm{h}(x) \rightarrow \overline{\psi}_\mathrm{h}(x)\mathrm{e}^{-i\alpha_k\sigma_k}, \tag{9.43}$$

with arbitrary real parameters α_k. It acts on each flavor component of the field. Obviously, the symmetry is even bigger. We can take $V \in \mathrm{SU}(2F)$ in Eq. (9.39). This plays a rôle in heavy meson ChPT (Wise, 1992; Grinstein *et al.*, 1992; Burdman and Donoghue, 1992).

3. Local flavor-number

The static Lagrangian contains no space derivative. The transformation

$$\psi_\mathrm{h}(x) \rightarrow \mathrm{e}^{i\eta(\mathbf{x})}\,\psi_\mathrm{h}(x), \qquad \overline{\psi}_\mathrm{h}(x) \rightarrow \overline{\psi}_\mathrm{h}(x)\mathrm{e}^{-i\eta(\mathbf{x})}, \tag{9.44}$$

is therefore a symmetry for any local phase $\eta(\mathbf{x})$. For every point $\mathbf{x}$ there is a corresponding Noether charge

$$Q_\mathrm{h}(x) = \overline{\psi}_\mathrm{h}(x)\psi_\mathrm{h}(x)\,[\,= \overline{\psi}_\mathrm{h}(x)\gamma_0\psi_\mathrm{h}(x)\,], \tag{9.45}$$

which we call local quark number. It is conserved,

$$\partial_0 Q_\mathrm{h}(x) = 0 \ \forall x. \tag{9.46}$$

9.1.3.4 *Renormalizability of the static theory*

Our effective-field theory is in the category of local field theories with a Lagrangian made up from local fields. In d space-time dimensions, standard wisdom says that such theories are renormalizable if the mass-dimension of the fields in the Lagrangian does not exceed d. Ultraviolet divergences can then be absorbed by adding a complete set of (composite) local fields with mass dimension smaller than or equal to d to the Lagrangian.

According to this (unproven[7]) rule, the static theory is renormalizable. The possible counterterms have to share the symmetries of the bare Lagrangian. They are easily found. From the kinetic term in the Lagrangian Eq. (9.36) we see that the dimension of the fields is $[\psi_\mathrm{h}] = 3/2$. Only 2-fermion terms with up to one derivative are then possible. Space derivatives are excluded by the local phase invariance Eq. (9.44). We then have the total quantum Lagrangian

$$\mathscr{L}_\mathrm{h}(x) = c_1 \mathcal{O}_1(x) + c_2 \mathcal{O}_2(x) \tag{9.47}$$

$$\mathcal{O}_1(x) = \overline{\psi}_\mathrm{h}(x)\psi_\mathrm{h}(x), \quad \mathcal{O}_2(x) = \overline{\psi}_\mathrm{h}(x)D_0\psi_\mathrm{h}(x), \tag{9.48}$$

where the convention $c_2 = 1$ can be chosen since it only fixes the unphysical field normalization, and $c_1 = \delta m$ has mass dimension $[\delta m] = 1$ and corresponds to an additive mass renormalization. From dimensional analysis and neglecting for simplicity the masses of the light quarks, it can be written as $\delta m = (e_1 g_0^2 + e_2 g_0^4 + \ldots) \Lambda_\mathrm{cut}$ in terms of the bare gauge coupling g_0 and a cutoff Λ_cut, which in lattice regularization is $\Lambda_\mathrm{cut} = 1/a$. For a static quark there is of course no chiral symmetry to forbid additive mass renormalization.

This is the complete static Lagrangian. After the standard QCD renormalization of coupling and light-quark masses, all divergences can be absorbed in δm, i.e. an energy shift. Flavor symmetry tells us that with several heavy flavors, δm is proportional to the unit matrix in flavor space. Energies of *any state* are then

$$E^\mathrm{QCD}_{\mathrm{h/\bar{h}}} = \left. E^\mathrm{stat}_{\mathrm{h/\bar{h}}} \right|_{\delta m=0} + \delta m + m = \left. E^\mathrm{stat}_{\mathrm{h/\bar{h}}} \right|_{\delta m=0} + m_\mathrm{bare}. \tag{9.49}$$

Here, m_bare and δm compensate the linear divergence (self-energy) of the static theory, while m is finite. Note that there is no symmetry that would suggest a natural way of splitting m_bare into δm and m. This split is arbitrary and convention dependent. The quantity δm is often called the residual mass.

A rigorous proof of *renormalizability* to all orders in perturbation theory has not been given but we note the following.

- Perturbative computations have confirmed the standard wisdom. These computations reach up to three loops in dimensional regularization (Chetyrkin and Grozin, 2003; Grozin *et al.*, 2008), while in various different lattice regularizations 1-loop computations have been carried out (Eichten and Hill, 1990*a;c;b*; Boucaud *et al.*, 1989; 1993; Flynn *et al.*, 1991; Borrelli and Pittori, 1992; Kurth and Sommer, 2001; 2002; Della Morte *et al.*, 2005; Palombi, 2008; Guazzini *et al.*, 2007; Grimbach *et al.*, 2008; Palombi *et al.*, 2006; Blossier *et al.*, 2006)
- We will see non-perturbative results that again yield a rather strong confirmation.
- Nevertheless, a proof of renormalizability would be very desirable.

Exercise 9.1 Static quarks from the hopping parameter expansion

Consider a Wilson quark propagator in a gauge background field. Evaluate the leading non-vanishing term in the hopping parameter expansion (with non-zero time separation). Check that it is the continuum HQET propagator (restricted to the lattice points) up to an energy shift. Even if this is a nice piece of confirmation, note that one here takes the limit $\kappa \to 0$ corresponding to $ma \to \infty$, while the true limit for relating QCD observables Φ^QCD to those of HQET is

$$\Phi^{\text{HQET}} \sim \lim_{m \to \infty} \lim_{a \to 0} \Phi^{\text{QCD}},$$

in that order!

9.1.3.5 Normalization of states, scaling of decay constants

For the discussion of the mass dependence of matrix elements we have to think about the normalization of states. Standard, relativistic invariant, normalization of bosonic one-particle states is

$$\langle \mathbf{p} | \mathbf{p}' \rangle_{\text{rel}} = (2\pi)^3 \, 2E(\mathbf{p}) \, \delta(\mathbf{p} - \mathbf{p}'). \tag{9.50}$$

The states have a mass dimension $[\,|\mathbf{p}\rangle_{\text{rel}}\,] = -1$. The factor $E(\mathbf{p})$ introduces a spurious mass dependence. In the large-mass limit, relativistic invariance plays no rôle and we should choose a mass-independent normalization instead. The standard convention for such a non-relativistic normalization is

$$\langle \mathbf{p} | \mathbf{p}' \rangle_{\text{NR}} \equiv \langle \mathbf{p} | \mathbf{p}' \rangle = 2 \, (2\pi)^3 \, \delta(\mathbf{p} - \mathbf{p}') \tag{9.51}$$

with $[\,|\mathbf{p}\rangle\,] = -3/2$ and

$$|\mathbf{p}\rangle_{\text{rel}} = \sqrt{E(\mathbf{p})} \, |\mathbf{p}\rangle. \tag{9.52}$$

Consider as an example where the normalization of states plays a role, the leptonic decay of a B-meson, $B^- \to \tau^- \bar{\nu}_\tau$. The transition amplitude $\mathcal{A}$ for this decay is given to a good approximation in terms of the effective weak Hamiltonian. It factorizes into a leptonic and a hadronic part as

$$\mathcal{A} \propto \langle \tau \, \bar{\nu} | \tau(x) \gamma_\mu (1 - \gamma_5) \bar{\nu}_\tau(x) | 0 \rangle \, \langle 0 | \bar{u}(x) \gamma_\mu (1 - \gamma_5) b(x) | B^- \rangle. \tag{9.53}$$

Using parity and Lorentz invariance, the hadronic part is

$$\langle 0 | \bar{u}(x) \gamma_\mu (1 - \gamma_5) b(x) | B^-(\mathbf{p}) \rangle = \langle 0 | A_\mu(x) | B^-(\mathbf{p}) \rangle = p_\mu f_{\text{B}} e^{ipx} \tag{9.54}$$

in terms of the flavored axial current

$$A_\mu(x) = \bar{u}(x) \gamma_\mu \gamma_5 b(x). \tag{9.55}$$

There is a single hadronic parameter f_{B} (matrix element) parameterizing the bound-state dynamics in this decay. We note that it is very relevant for the phenomenological analysis of the CKM matrix (Antonelli *et al.*, 2009).

We may now use HQET to find the asymptotic mass-dependence of f_{B} for large $m = m_{\text{b}}$. Since to lowest order in $1/m$ the FTW transformation is trivial, the HQET current is just

$$A_0^{\text{HQET}}(x) = A_0^{\text{stat}}(x) + \mathrm{O}(1/m), \quad A_0^{\text{stat}}(x) = \bar{u}(x) \gamma_0 \gamma_5 \psi_{\text{h}}(x). \tag{9.56}$$

The static current A_0^{stat} has no explicit mass dependence. In static approximation we then have

$$\langle 0 | A_0^{\text{stat}}(0) | B^-(\mathbf{p}=0)\rangle = \Phi^{\text{stat}}, \tag{9.57}$$

with a mass-independent Φ^{stat}. Its relation to f_{B},

$$\Phi^{\text{stat}} = m_{\text{B}}^{-1/2} p_0 f_{\text{B}} = m_{\text{B}}^{1/2} f_{\text{B}}, \tag{9.58}$$

takes Eq. (9.52) into account ($p_0 = E(\mathbf{0}) = m_{\text{B}}$). We arrive at the prediction

$$f_{\text{B}} = \frac{\Phi^{\text{stat}}}{\sqrt{m_{\text{B}}}} + \mathrm{O}(1/m_{\text{b}}), \quad \frac{f_{\text{B}}}{f_{\text{D}}} = \frac{\sqrt{m_{\text{D}}}}{\sqrt{m_{\text{B}}}} + \mathrm{O}(1/m_{\text{c}}). \tag{9.59}$$

The latter use of course assumes $\Lambda_{\text{QCD}}/m_{\text{c}} \ll 1$. We will see later that these predictions are modified by the renormalization of the effective theory.

9.1.3.6   *HQET and phenomenology*

Heavy-quark spin/flavor symmetry is very useful to classify the spectrum in terms of a few non-perturbative parameters or predict relations between different masses, e.g.

$$m_{\text{B}^*}^2 - m_{\text{B}}^2 \approx m_{\text{D}^*}^2 - m_{\text{D}}^2, \tag{9.60}$$

$$m_{\text{B}'} - m_{\text{B}} \approx m_{\text{D}'} - m_{\text{D}}, \tag{9.61}$$

where m_{B^*}, m_{D^*} are the vector meson masses and with $m_{\text{B}'}$, $m_{\text{D}'}$ we indicate the first excitation in the pseudoscalar sector. The first of these relations has been seen to be approximately realized in nature.

More detailed statements about semileptonic transitions $B \to Dl\nu$, $B^* \to D^*l\nu$ are possible. In the heavy-quark limit for both beauty and charm these are described by a single form factor, the Isgur–Wise function, instead of several (Isgur and Wise, 1989; 1990). These topics and many others are discussed in many reviews, e.g. (Neubert, 1994). We here concentrate on lattice HQET and where HQET helps to understand lattice results for states with a b-quark.

9.2   Lattice formulation

We start with the static approximation. The $1/m$ terms will be added after a discussion of the renormalization of the static theory.

9.2.1   Lattice action

For a static quark there is no chiral symmetry. Since we want to avoid doublers, we discretize à la Wilson (with $r = 1$). The continuum $D_0 \psi_{\text{h}}(x)$ is transcribed to the lattice as

$$D_0 \gamma_0 \to \frac{1}{2}\{(\nabla_0 + \nabla_0^*)\gamma_0 - a\nabla_0^*\nabla_0\}, \tag{9.62}$$

and with $P_+\psi_h = \psi_h$, $P_-\psi_{\bar{h}} = \psi_{\bar{h}}$, we have the lattice identities

$$D_0\,\psi_h(x) = \nabla_0^*\psi_h(x), \quad D_0\,\psi_{\bar{h}}(x) = \nabla_0\psi_{\bar{h}}(x). \tag{9.63}$$

For later convenience we insert a specific normalization factor, defining the static lattice Lagrangians

$$\mathscr{L}_h = \frac{1}{1+a\delta m}\overline{\psi}_h(x)[\nabla_0^* + \delta m]\psi_h(x), \tag{9.64}$$

$$\mathscr{L}_{\bar{h}} = \frac{1}{1+a\delta m}\overline{\psi}_{\bar{h}}(x)[-\nabla_0 + \delta m]\psi_{\bar{h}}(x). \tag{9.65}$$

The following points are worth noting.

- Formally, this is just a one-dimensional Wilson fermion replicated for all space points $\mathbf{x}$, see also Exercise 9.1.

- As a consequence there are no doubler modes.

- The construction of a positive Hermitian transfer matrix for Wilson fermions (Lüscher, 1977, Montvay and Münster, 1994) can just be taken over.

- The choice of the backward derivative for the quark and the forward derivative for the antiquark is selected by the Wilson term. We will see that this selects forward/backward propagation and an ϵ-prescription as in Eq. (9.36) is not needed.

- The form of this Lagrangian was first written down by Eichten and Hill (Eichten and Hill, 1990a).

- The lattice action preserves all the continuum heavy-quark symmetries discussed in the previous section.

9.2.2 Propagator

From the Lagrangian Eq. (9.64) we have the defining equation for the propagator

$$\frac{1}{1+a\,\delta m}(\nabla_0^* + \delta m)G_h(x,y) = \delta(x-y)P_+ \equiv a^{-4}\prod_\mu \delta_{\frac{x_\mu}{a}\frac{y_\mu}{a}}P_+. \tag{9.66}$$

Obviously $G_h(x,y)$ is proportional to $\delta(\mathbf{x}-\mathbf{y})$. Writing $G_h(x,y) = g(n_0, k_0; \mathbf{x})\delta(\mathbf{x}-\mathbf{y})P_+$ with $x_0 = an_0$, $y_0 = ak_0$, the above equation yields a simple recursion for $g(n_0 + 1, k_0; \mathbf{x})$ in terms of $g(n_0, k_0; \mathbf{x})$ that is solved by

$$g(n_0, k_0; \mathbf{x}) = \theta(n_0 - k_0)(1+a\delta m)^{-(n_0-k_0)}\mathcal{P}(y, x; 0)^\dagger, \tag{9.67}$$

$$\mathcal{P}(x, x; 0) = 1, \quad \mathcal{P}(x, y+a\hat{0}; 0) = \mathcal{P}(x, y; 0)U(y, 0), \tag{9.68}$$

where

$$\theta(n_0 - k_0) = \begin{cases} 0 & n_0 < k_0 \\ 1 & n_0 \geq k_0. \end{cases} \tag{9.69}$$

The static propagator reads

$$G_{\mathrm{h}}(x,y) = \theta(x_0 - y_0)\,\delta(\mathbf{x} - \mathbf{y})\,\exp\left(-\widehat{\delta m}\,(x_0 - y_0)\right)\mathcal{P}(y,x;0)^\dagger\,P_+, \quad (9.70)$$

$$\widehat{\delta m} = \frac{1}{a}\ln(1 + a\delta m). \tag{9.71}$$

The object $\mathcal{P}(x,y;0)$ parallel transports fields in the fundamental representation from y to x along a time-like path. Note that the derivation fixes $\theta(0) = 1$ for the lattice θ-function. As in the continuum, the mass counterterm δm just yields an energy shift; now, on the lattice, the shift is

$$E_{\mathrm{h/\bar{h}}}^{\mathrm{QCD}} = E_{\mathrm{h/\bar{h}}}^{\mathrm{stat}}\Big|_{\delta m=0} + m_{\mathrm{bare}}, \quad m_{\mathrm{bare}} = \widehat{\delta m} + m. \tag{9.72}$$

It is valid for all energies of states with a single heavy quark or antiquark. As in the continuum the split between δm and the finite m is convention dependent.

In complete analogy the antiquark propagator is given by

$$G_{\bar{\mathrm{h}}}(x,y) = \theta(y_0 - x_0)\,\delta(\mathbf{x} - \mathbf{y})\,\exp\left(-\widehat{\delta m}\,(y_0 - x_0)\right)\mathcal{P}(x,y;0)\,P_-\,. \tag{9.73}$$

9.2.3   Symmetries

All HQET symmetries are preserved on the lattice, in particular the $U(2F)$ spin-flavor symmetry and the local flavor-number conservation. The symmetry transformations can literally be carried over from the continuum, e.g. Eq. (9.44). One just replaces the continuum fields by the lattice ones.

Note that these HQET symmetries are defined in terms of transformations of the heavy-quark fields, while the light-quark fields do not change (unlike, e.g., standard parity). Integrating out just the quark fields in the path integral while leaving the integral over the gauge fields, they thus yield identities for the integrand or one may say for "correlation functions in any fixed gauge background field".

9.2.4   Symanzik analysis of cutoff effects

According to the – by now well tested[8] – Symanzik conjecture, the cutoff effects of a lattice theory can be described in terms of an effective *continuum* theory (Symanzik, 1983a;b; Lüscher *et al.*, 1996). Once the terms in Symanzik's effective Lagrangian are known, the cutoff effects can be cancelled by adding terms of the same form to the lattice action, resulting in an improved action.

For a static quark, Symanzik's effective action is (Kurth and Sommer, 2001)

$$S_{\mathrm{eff}} = S_0 + aS_1 + \dots, \quad S_i = \int \mathrm{d}^4x\,\mathscr{L}_i(x), \tag{9.74}$$

where $\mathscr{L}_0(x) = \mathscr{L}_{\mathrm{h}}^{\mathrm{stat}}(x)$ is the continuum static Lagrangian of Eq. (9.47) and

$$\mathscr{L}_1(x) = \sum_{i=3}^{5} c_i \, \mathcal{O}_i(x), \tag{9.75}$$

is given in terms of local fields with mass dimension $[\mathcal{O}_i(x)] = 5$. Their coefficients c_i are functions of the bare gauge coupling. Assuming for simplicity mass-degenerate light quarks with a mass m_{l}, the set of possible dimension five fields, which share the symmetries of the lattice theory, is

$$\mathcal{O}_3 = \overline{\psi}_{\mathrm{h}} D_0 D_0 \psi_{\mathrm{h}}, \quad \mathcal{O}_4 = m_{\mathrm{l}} \, \overline{\psi}_{\mathrm{h}} D_0 \psi_{\mathrm{h}}, \quad \mathcal{O}_5 = m_{\mathrm{l}}^2 \, \overline{\psi}_{\mathrm{h}} \psi_{\mathrm{h}}. \tag{9.76}$$

Note that $P_+ \sigma_{0j} P_+ = 0$ means there is no term $\overline{\psi}_{\mathrm{h}} \sigma_{0j} F_{0j} \psi_{\mathrm{h}}$, and $\overline{\psi}_{\mathrm{h}} D_j D_j \psi_{\mathrm{h}}$ can't occur because it violates the local phase invariance Eq. (9.44). Finally, $\overline{\psi}_{\mathrm{h}} \sigma_{jk} F_{jk} \psi_{\mathrm{h}}$ is not invariant under the spin rotations Eq. (9.43).

Furthermore, we are only interested in on-shell correlation functions and energies. For this class of observables $\mathcal{O}_3$, $\mathcal{O}_4$ do not contribute (Lüscher and Weiss, 1985; Lüscher *et al.*, 1996) because they vanish by the equation of motion,[9]

$$D_0 \psi_{\mathrm{h}} = 0. \tag{9.77}$$

The only remaining term, $\mathcal{O}_5$, induces a redefinition of the mass counterterm δm that therefore depends explicitly on the light-quark mass.

We note that for almost all applications, δm is explicitly cancelled in the relation between physical observables and one thus has automatic on-shell O(a) improvement for the static action. No parameter has to be tuned to guarantee this property. Still, the improvement of matrix elements and correlation functions requires us to also consider composite fields in the effective theory.

Exercise 9.2 The static quark antiquark potential.

A (time-local) field

$$O(t, \mathbf{x}, \mathbf{y}) = \overline{\psi}_{\mathrm{h}}(x) \, \mathcal{P}(x, y) \gamma_5 \, \psi_{\bar{\mathrm{h}}}(y), \quad x_0 = y_0 = t$$

with $\mathcal{P}(x, y)$ being a parallel transporter from y to x in $x_0 = t$ plane, can be used to annihilate a quark–antiquark pair at a separation $\mathbf{x} - \mathbf{y}$, while

$$\overline{O}(t, \mathbf{x}, \mathbf{y}) = -\overline{\psi}_{\bar{\mathrm{h}}}(y) \, \mathcal{P}(y, x) \gamma_5 \, \psi_{\mathrm{h}}(x), \quad x_0 = y_0 = t \tag{9.78}$$

will create a quark–antiquark pair at a separation $\mathbf{x} - \mathbf{y}$.

Show that for $t > 0$

$$\langle \overline{O}(t, \mathbf{x}, \mathbf{y}) \, O(0, \mathbf{x}, \mathbf{y}) \rangle = \mathrm{const.} \; \mathrm{e}^{-2t \, \widehat{\delta m}} W(t, \mathbf{x} - \mathbf{y}), \tag{9.79}$$

where W is the Wilson loop introduced in the chapter of P. Hernandez. Since the energy levels of HQET are finite (after inclusion of a suitable δm), one can conclude that

$$V_{\mathrm{R}}(\mathbf{x} - \mathbf{y}) = - \lim_{t \to \infty} \partial_t \ln(W(t, \mathbf{x} - \mathbf{y})) + 2\widehat{\delta m} \tag{9.80}$$

is a finite quantity: the divergent constant in the bare potential is absorbed by $\widehat{\delta m}$, i.e. by a renormalization of the heavy-quark mass.

Furthermore, from the $O(a)$ improvement of HQET, one concludes (Necco and Sommer, 2002)

$$V_R(\mathbf{x} - \mathbf{y}) = V_R^{\text{cont}}(r = |\mathbf{x} - \mathbf{y}|) + O(a^2) \tag{9.81}$$

if the action for the light fields is $O(a)$ improved.

9.2.4.1   Renormalized and improved axial current

We now also have to specify the discretization of the light-quark field ψ. We will generically think of a standard $O(a)$-improved Wilson discretization (Sheikholeslami and Wohlert, 1985; Lüscher *et al.*, 1996) but occasionally mention changes that occur when one has an action with exact chiral symmetry (Neuberger, 1998; Hasenfratz *et al.*, 1998; Lüscher, 1998)[10] or a Wilson regularization with a twisted-mass term (Frezzotti *et al.*, 2001*a*;*b*; Frezzotti and Rossi, 2004). As an example, we study the time component of the axial current. In Symanzik's effective theory it is represented by

$$(A_0^{\text{stat}})_{\text{eff}} = A_0^{\text{stat}} + a \sum_{k=1}^{4} \omega_k (\delta A_0^{\text{stat}})_k, \quad A_0^{\text{stat}} = \overline{\psi}\gamma_0\gamma_5\psi_{\text{h}} \tag{9.82}$$

with some coefficients ω_k. Here, the flavor index of the field $\overline{\psi}$ is suppressed. It is considered to have some fixed but arbitrary value for our discussion, except where we indicate this explicitly. A basis for the dimension-four fields $\{(\delta A_0^{\text{stat}})_k\}$ is

$$(\delta A_0^{\text{stat}})_1 = \overline{\psi}\overleftarrow{D}_j\gamma_j\gamma_5\psi_{\text{h}}, \quad (\delta A_0^{\text{stat}})_2 = \overline{\psi}\gamma_5 D_0\psi_{\text{h}},$$

$$(\delta A_0^{\text{stat}})_3 = \overline{\psi}\overleftarrow{D}_0\gamma_5\psi_{\text{h}}, \quad (\delta A_0^{\text{stat}})_4 = m_{\text{l}}\overline{\psi}\gamma_0\gamma_5\psi_{\text{h}}. \tag{9.83}$$

From Eq. (9.77) we see that $k = 2$ does not contribute, while the equation of motion for $\overline{\psi}$ relates $(\delta A_0^{\text{stat}})_3$, $(\delta A_0^{\text{stat}})_4$ and $(\delta A_0^{\text{stat}})_1$. We choose to remain with $k = 1$ (and in principle $k = 4$), but for simplicity assume[11] $a\, m_{\text{l}} \ll 1$; we can then drop $(\delta A_0^{\text{stat}})_4$. So, for on-shell quantities the effective theory representation is

$$(A_0^{\text{stat}})_{\text{eff}} = A_0^{\text{stat}} + a\tilde{\omega}_1(\delta A_0^{\text{stat}})_1. \tag{9.84}$$

In order to achieve a cancellation of the $O(a)$ lattice spacing effects, we add a corresponding combination of correction terms to the axial current in the lattice theory and write the improved and renormalized current in the form

$$(A_R^{\text{stat}})_0 = Z_A^{\text{stat}}(g_0, a\mu)\,(A_I^{\text{stat}})_0, \tag{9.85}$$

$$(A_I^{\text{stat}})_0 = A_0^{\text{stat}} + ac_A^{\text{stat}}(g_0)\,\overline{\psi}\gamma_j\gamma_5\frac{1}{2}(\overleftarrow{\nabla}_j + \overleftarrow{\nabla}_j^*)\psi_{\text{h}}, \tag{9.86}$$

with a mass-independent renormalization constant $Z_\mathrm{A}^\mathrm{stat}$ and a dimensionless improvement coefficient, $c_\mathrm{A}^\mathrm{stat}$, depending again on g_0 but not on the light-quark mass.

The improvement coefficients can be determined such that for this (time component of the) *improved axial current* we have the representation

$$(A_0^\mathrm{stat})_\mathrm{eff} = \overline{\psi}\gamma_0\gamma_5\psi_\mathrm{h} + \mathrm{O}(a^2), \tag{9.87}$$

in the Symanzik effective theory. In other words $\tilde{\omega}_1$ is then $\mathrm{O}(a)$ and cutoff effects are $\mathrm{O}(a^2)$.

The symmetries of the static theory are strong enough to improve all components of the flavor currents in terms of just $c_\mathrm{A}^\mathrm{stat}$ and to renormalize them by $Z_\mathrm{A}^\mathrm{stat}$. Let us discuss how this works.

9.2.5 The full set of flavor currents

The previous discussion literally carries over to the time component of the vector current,

$$V_0^\mathrm{stat} = \overline{\psi}\gamma_0\psi_\mathrm{h}. \tag{9.88}$$

Its improved and renormalized lattice version may be chosen as

$$(V_\mathrm{R}^\mathrm{stat})_0 = Z_\mathrm{V}^\mathrm{stat}(V_\mathrm{I}^\mathrm{stat})_0 \tag{9.89}$$

$$(V_\mathrm{I}^\mathrm{stat})_0 = \overline{\psi}\gamma_0\psi_\mathrm{h} + ac_\mathrm{V}^\mathrm{stat}\overline{\psi}\gamma_j\frac{1}{2}(\overleftarrow{\nabla}_j + \overleftarrow{\nabla}_j^*)\psi_\mathrm{h}. \tag{9.90}$$

The chiral symmetry of the continuum limit can be used to relate $Z_\mathrm{V}^\mathrm{stat}$, $c_\mathrm{V}^\mathrm{stat}$ to $Z_\mathrm{A}^\mathrm{stat}$, $c_\mathrm{A}^\mathrm{stat}$ in the following way. We assume $N_\mathrm{f} \geq 2$ massless light quarks. Then, the infinitesimal transformation

$$\delta_\mathrm{A}^a\psi(x) = \frac{1}{2}\tau^a\gamma_5\psi(x), \qquad \delta_\mathrm{A}^a\overline{\psi}(x) = \overline{\psi}(x)\gamma_5\frac{1}{2}\tau^a, \tag{9.91}$$

with the Pauli matrices τ^a acting on two of the flavor components of the light-quark fields ψ, $\overline{\psi}$, is a (non-anomalous) symmetry of the theory. Identifying $V_0^\mathrm{stat} = \overline{\psi}_1\gamma_0\psi_\mathrm{h}$, where $\overline{\psi}_1$ is the first flavor component of $\overline{\psi}$, the vector current transforms as $\delta_\mathrm{A}^3 V_0^\mathrm{stat} = -\frac{1}{2}A_0^\mathrm{stat}$. The same property can then be required for the renormalized and improved lattice fields,

$$\delta_\mathrm{A}^3(V_\mathrm{R}^\mathrm{stat})_0 = -\frac{1}{2}(A_\mathrm{R}^\mathrm{stat})_0 + \mathrm{O}(a^2). \tag{9.92}$$

This condition can be implemented in the form of Ward identities relating different correlation functions, in particular in the Schrödinger functional. We refer to A. Vladikas' chapter and (Lüscher, 1998) for the principle; practical implementations have been studied in (Hashimoto *et al.*, 2002; Palombi, 2008). Such Ward identities determine $Z_\mathrm{V}^\mathrm{stat}$, $c_\mathrm{V}^\mathrm{stat}$ in terms of $Z_\mathrm{A}^\mathrm{stat}$, $c_\mathrm{A}^\mathrm{stat}$.

Furthermore, by a finite spin-symmetry transformation (with σ_k of Eq. (9.41))

$$\psi_\mathrm{h} \to \psi_\mathrm{h}' = \mathrm{e}^{-i\pi\sigma_k/2}\psi_\mathrm{h} = -i\sigma_k\psi_\mathrm{h}, \qquad \overline{\psi}_\mathrm{h}' = \overline{\psi}_\mathrm{h}i\sigma_k, \tag{9.93}$$

we have

$$V_0^{\text{stat}} \to \left[V_0^{\text{stat}}\right]' = A_k^{\text{stat}} \equiv \overline{\psi}\gamma_k\gamma_5\psi_{\text{h}}, \quad \left[A_0^{\text{stat}}\right]' = V_k^{\text{stat}} \equiv \overline{\psi}\gamma_k\psi_{\text{h}}, \tag{9.94}$$

and we can require the same for the correction terms,

$$\left[\delta V_0^{\text{stat}}\right]' = \delta A_k^{\text{stat}}, \quad \left[\delta A_0^{\text{stat}}\right]' = \delta V_k^{\text{stat}}. \tag{9.95}$$

We leave it as an exercise to determine the form of δA_k^{stat}, δV_k^{stat}. The discussed transformations are valid for the bare lattice fields at any lattice spacing. Thus, renormalization and improvement of the spatial components is given completely in terms of the time components once we define the renormalized fields to transform in the same way as the bare fields. A last property to note before writing down the renormalized and improved fields is that we have

$$Z_{\text{V}}^{\text{stat}}(g_0, a\mu) = Z_{\text{V/A}}^{\text{stat}}(g_0)\, Z_{\text{A}}^{\text{stat}}(g_0, a\mu) \tag{9.96}$$

with a μ-independent function $Z_{\text{V/A}}^{\text{stat}}(g_0)$ and up to $\text{O}(a^2)$, as soon as we require Eq. (9.92).[12]

Let us disregard the $\text{O}(a)$ improvement terms for simplicity. We can then summarize what we have learnt about the renormalization of the static-light bilinears as

$$(A_{\text{R}}^{\text{stat}})_0 = Z_{\text{A}}^{\text{stat}}(g_0, a\mu)\, A_0^{\text{stat}}, \tag{9.97}$$

$$(V_{\text{R}}^{\text{stat}})_0 = Z_{\text{A}}^{\text{stat}}(g_0, a\mu)\, Z_{\text{V/A}}^{\text{stat}}(g_0)\, V_0^{\text{stat}}, \tag{9.98}$$

$$(V_{\text{R}}^{\text{stat}})_k = Z_{\text{A}}^{\text{stat}}(g_0, a\mu)\, V_k^{\text{stat}}, \tag{9.99}$$

$$(A_{\text{R}}^{\text{stat}})_k = Z_{\text{A}}^{\text{stat}}(g_0, a\mu)\, Z_{\text{V/A}}^{\text{stat}}(g_0)\, A_k^{\text{stat}}, \tag{9.100}$$

where $Z_{\text{V/A}}^{\text{stat}}(g_0)$ can be determined from a chiral Ward identity (Hashimoto *et al.*, 2002; Palombi, 2008). Note that we denote the flavor currents in HQET in complete analogy to QCD. Still, they do not form 4-vectors, as 4-dimensional rotation invariance is broken in HQET. For example, $(A_{\text{R}}^{\text{stat}})_0$ cannot be rotated into $(A_{\text{R}}^{\text{stat}})_k$ by a 90-degree lattice rotation.

The only bilinears that are missing here are scalar, pseudoscalar densities (and the tensor). These are equivalent to A_0^{stat} and V_0^{stat} in the static approximation, for example

$$\overline{\psi}\gamma_5\psi_{\text{h}} = \overline{\psi}\gamma_5\gamma_0\psi_{\text{h}} = -A_0^{\text{stat}}, \quad \overline{\psi}\psi_{\text{h}} = \overline{\psi}\gamma_0\psi_{\text{h}} = V_0^{\text{stat}}. \tag{9.101}$$

At this stage it is therefore unnecessary to introduce renormalized scalar and pseudoscalar densities.

We have so far written down expressions for the relevant renormalized heavy-light quark bilinears. The Z-factors can be chosen such that correlation functions of these fields have a continuum limit (with δm, gauge coupling and light-quark masses properly determined). Beyond this requirement, however, also the finite parts need to be fixed by renormalization conditions. We have fixed some of them such that the renormalized fields satisfy chiral symmetry and heavy-quark spin symmetry.

Only one finite part (in $Z_{\mathrm{A}}^{\mathrm{stat}}$) then remains free. Preserving these symmetries by the renormalization is natural, but not absolutely required; e.g. Eq. (9.94) could be violated in terms of the renormalized fields. As long as one just remains inside the effective-field theory these ambiguities are not fixed. The proper conditions for the finite parts, valid for HQET as an effective theory of QCD, have to be determined from QCD with finite heavy-quark masses. We will return to this later.

We may, however, already note that for renormalization-group-invariant fields, these ambiguities are not present. The renormalization-group invariants are thus very appropriate. Still, relating the bare lattice fields to the renormalization-group invariant ones is a non-trivial task in practice (Lüscher *et al.*, 1991; Capitani *et al.*, 1999). We will briefly discuss how it can be done (and has been done) for the static-light bilinears (Kurth and Sommer, 2001; Heitger *et al.*, 2003; Della Morte *et al.*, 2007*b*). For this and other purposes we need the Schrödinger functional. In the following, we just give a simplified review of it and describe how static quarks are incorporated. Some more details are discussed by Peter Weisz.

9.2.6   HQET and Schrödinger functional

The Schrödinger functional (Symanzik, 1981; Lüscher *et al.*, 1992; Sint, 1994; 1995) can just be seen as QCD in a finite Euclidean space-time of size $T \times L^3$, with specific boundary conditions. It is useful as a renormalizable probe of QCD, providing a definition of correlation functions that are accessible at all distances, short or long: gauge invariance is manifest and even at short distances (large momenta) cutoff effects can be kept small. It will help us to perform the non-perturbative renormalization of HQET and its matching to QCD. In all these applications it is advantageous to have a variety of kinematics at one's disposal. One element is to have access to finite but small momenta of the quarks (think of the free theory, a relevant starting point for the short-distance regime).

To this end, the spatial boundary conditions were chosen to be $\psi(x + L\hat{k}) = e^{i\theta_k}\psi(x)$, $\overline{\psi}(x + L\hat{k}) = e^{-i\theta_k}\overline{\psi}(x)$ in (Sint and Sommer, 1996), which allows momenta

$$p_k = \frac{2\pi l_k}{L} + \frac{\theta_k}{L}, \ l_k \in \mathbb{Z}, \tag{9.102}$$

in particular small ones when $l_k = 0$. Performing a variable transformation $\psi(x) \to e^{i\theta_k x_k/L}\psi(x)$, $\overline{\psi}(x) \to e^{-i\theta_k x_k/L}\psi(x)$, for $0 \le x_k \le L - a$, we see that this boundary condition is equivalent to periodic boundary conditions (without a phase) for the new fields, while the spatial covariant derivatives contain an additional phase, for example

$$\nabla_k\psi(x) = \frac{1}{a}\left[e^{i\theta_k a/L}U(x,\mu)\psi(x + a\hat{k}) - \psi(x)\right], \tag{9.103}$$

see also Section 9.6.1. The phase $\theta_k a/L$ can be seen as a constant abelian gauge potential and the above variable transformation as a gauge transformation. Of course, the angles θ_k that we will set all equal from now on ($\theta_k = \theta$), are not specific to the Schrödinger functional ; they were just first used in this context.

The standard Schrödinger functional boundary conditions in time are (Sint, 1994, Sint, 1995)

$$P_+\psi(x)|_{x_0=0} = 0, \quad P_-\psi(x)|_{x_0=T} = 0, \tag{9.104}$$

and

$$\overline{\psi}(x)P_-|_{x_0=0} = 0, \quad \overline{\psi}(x)P_+|_{x_0=T} = 0. \tag{9.105}$$

The gauge fields are taken periodic in space and the space components of the continuum gauge fields are set to zero at $x_0 = 0$ and $x_0 = T$ (on the lattice the boundary links $U(x,k)$ are set to unity).[13]

For the static quark the components projected by P_- vanish anyway, so there is just

$$P_+\psi_{\rm h}(x)|_{x_0=0} = 0, \quad \overline{\psi}_{\rm h}(x)P_+|_{x_0=T} = 0. \tag{9.106}$$

Defining

$$\psi_{\rm h}(x) = 0 \quad \text{if } x_0 < 0 \text{ or } x_0 \geq T, \tag{9.107}$$

the lattice action for the static quark with Schrödinger functional boundary conditions can be written as

$$S_{\rm h} = \frac{1}{1+a\delta m} a^4 \sum_x \overline{\psi}_{\rm h}(x)[\nabla_0^* + \delta m]\psi_{\rm h}(x) \tag{9.108}$$

as before. In general, the improvement of the Schrödinger functional requires to add boundary terms to the action as a straightforward generalization of Symanzik improvement. These terms are dimension-four composite fields located on or at the boundaries, summed over space (Lüscher *et al.*, 1992; 1996). Since they are not so important here and are also known sufficiently well, we do not discuss them. We just note that no boundary improvement terms involving static fields are needed (Kurth and Sommer, 2001), since the dimension four fields vanish either due to the equation of motion or the heavy-quark symmetries.

We take the same periodicity in space as for relativistic quarks,

$$\psi_{\rm h}(x + L\hat{k}) = \psi_{\rm h}(x), \quad \overline{\psi}_{\rm h}(x + L\hat{k}) = \overline{\psi}_{\rm h}(x). \tag{9.109}$$

In the static theory this has no effect, since quarks at different **x** are not coupled, but it plays a rôle at order $1/m$, where θ is a useful kinematical variable.

An important feature of the Schrödinger functional is that one can form gauge-invariant correlation functions of boundary quark fields. In particular, one can project those quark fields to small spatial momentum, e.g. $\mathbf{p} = 1/L \times (\theta, \theta, \theta)$ for the quarks and $-\mathbf{p}$ for the antiquarks. For the precise definition of the boundary quark fields we refer to (Lüscher *et al.*, 1996) or for an alternative view we refer to (Lüscher, 2006). The details are here not so important. We only need to know that these boundary fields, i.e. fermion fields localized at the boundaries, exist. Those at $x_0 = 0$ are denoted by

$$\zeta_l(\mathbf{x}),\ \overline\zeta_l(\mathbf{x}),\ \zeta_{\bar h}(\mathbf{x}),\ \overline\zeta_h(\mathbf{x}),$$

and those at $x_0 = T$ by

$$\zeta_l{}'(\mathbf{x}),\ \overline\zeta_l{}'(\mathbf{x}),\ \zeta_h{}'(\mathbf{x}),\ \overline\zeta_{\bar h}{}'(\mathbf{x}).$$

9.2.6.1 Renormalization

These boundary quark fields are multiplicatively renormalized with factors Z_ζ, Z_{ζ_h}, such that $(\zeta_l(\mathbf{x}))_{\mathrm R} = Z_\zeta \zeta_l(\mathbf{x})$, etc.

To illustrate a first use of the Schrödinger functional and the boundary fields we introduce three correlation functions

$$f_{\mathrm A}^{\mathrm{stat}}(x_0,\theta) = -\frac{a^6}{2}\sum_{\mathbf{y},\mathbf{z}}\big\langle (A_{\mathrm I}^{\mathrm{stat}})_0(x)\,\overline\zeta_h(\mathbf{y})\gamma_5\zeta_l(\mathbf{z})\big\rangle: \tag{9.110}$$

$$f_1^{\mathrm{stat}}(\theta) = -\frac{a^{12}}{2L^6}\sum_{\mathbf{u},\mathbf{v},\mathbf{y},\mathbf{z}}\big\langle \overline\zeta_l{}'(\mathbf{u})\gamma_5\zeta_h{}'(\mathbf{v})\,\overline\zeta_h(\mathbf{y})\gamma_5\zeta_l(\mathbf{z})\big\rangle: \tag{9.111}$$

$$f_1^{\mathrm{hh}}(x_3,\theta) = -\frac{a^8}{2L^2}\sum_{x_1,x_2,\mathbf{y},\mathbf{z}}\big\langle \overline\zeta_{\bar h}{}'(\mathbf{x})\gamma_5\zeta_h{}'(\mathbf{0})\,\overline\zeta_h(\mathbf{y})\gamma_5\zeta_{\bar h}(\mathbf{z})\big\rangle: \tag{9.112}$$

In the graphs, double lines are static quark propagators. Note that the sum in Eq. (9.112) runs on x_1 and x_2 and therefore yields an x_3-dependent correlation function. We further point out that $\sum_{\mathbf{y}}$, etc. project the boundary quark fields onto zero (space) momentum, but together with the abelian gauge field, this is equivalent to a physical momentum $p_k = \theta/L$. For example, the time decay of a free massless quark propagator projected this way contains an energy $E(\theta/L,\theta/L,\theta/L) = \sqrt{3}\theta/L$, cf. Eq. (9.4).

The above functions are renormalized as

$$\big[f_{\mathrm A}^{\mathrm{stat}}\big]_{\mathrm R} = Z_{\mathrm A}^{\mathrm{stat}}\,Z_{\zeta_h}Z_\zeta\,f_{\mathrm A}^{\mathrm{stat}},\quad \big[f_1^{\mathrm{stat}}\big]_{\mathrm R} = Z_{\zeta_h}^2 Z_\zeta^2\,f_1^{\mathrm{stat}},\quad \big[f_1^{\mathrm{hh}}\big]_{\mathrm R} = Z_{\zeta_h}^4\,f_1^{\mathrm{hh}}. \tag{9.113}$$

We remind the reader that an additional renormalization is the mass counterterm of the static action.

The ratio

$$\left[\frac{f_{\mathrm A}^{\mathrm{stat}}(T/2,\theta)}{\sqrt{f_1^{\mathrm{stat}}(\theta)}}\right]_{\mathrm R} = Z_{\mathrm A}^{\mathrm{stat}}(g_0,a\mu)\,\frac{f_{\mathrm A}^{\mathrm{stat}}(T/2,\theta)}{\sqrt{f_1^{\mathrm{stat}}(\theta)}} \tag{9.114}$$

renormalizes in a simple way and also needs no knowledge of δm, since it cancels out due to Eq. (9.70). It is hence an attractive possibility to *define* the renormalization constant $Z_{\mathrm A}^{\mathrm{stat}}$ through this ratio. Explicitly, we may choose

$$Z_A^{\text{stat}}(g_0, a\mu) \equiv \frac{\sqrt{f_1^{\text{stat}}(\theta)}}{f_A^{\text{stat}}(L/2, \theta)} \left[\frac{f_A^{\text{stat}}(L/2, \theta)}{\sqrt{f_1^{\text{stat}}(\theta)}} \right]_{g_0=0} \qquad \mu = 1/L, \quad T = L, \quad \theta = \frac{1}{2},$$

$$(9.115)$$

which defines the finite part of Z_A^{stat} in a so-called Schrödinger functional scheme. As usual the factor $\left[f_A^{\text{stat}}(L/2, \theta)/\sqrt{f_1^{\text{stat}}(\theta)} \right]_{g_0=0}$ is inserted to ensure $Z_A^{\text{stat}} = 1 + O(g_0^2)$. The name Schrödinger functional scheme just refers to the fact that the renormalization factor is defined in terms of correlation functions with Schrödinger functional boundary conditions. While Z_A^{stat} refers to a specific regularization, the renormalization scheme is independent of that and can in principle be applied in a continuum regularization. Many similar Schrödinger functional schemes can be defined (e.g. (Heitger *et al.*, 2003)), but by the choice $T = L, \theta = 0.5$ we have made Eq. (9.115) unique. It is implied that the light-quark masses are set to zero. We will soon come back to the μ dependence of the renormalized current and its relation to the RGI current. First, let us show some numerical results that provide a non-perturbative test of the renormalizability of the static theory.

9.2.7   Numerical test of the renormalizability

The above-listed renormalization structure of the Schrödinger functional correlation functions is just deduced from a simple dimensional analysis. A number of 1-loop calculations of the correlation functions defined above as well as of others (Kurth and Sommer, 2001; 2002; Della Morte *et al.*, 2005; Palombi, 2008) confirm the structure Eq. (9.113) and more generally the renormalizability of the theory (by local counterterms).

Also, non-perturbative tests exist. A stringent and precise one (Della Morte *et al.*, 2005) is based on the ratios

$$\xi_A(\theta, \theta') = \frac{f_A^{\text{stat}}(T/2, \theta)}{f_A^{\text{stat}}(T/2, \theta')}, \quad \xi_1(\theta, \theta') = \frac{f_1^{\text{stat}}(\theta)}{f_1^{\text{stat}}(\theta')}, \quad h(d/L, \theta) = \frac{f_1^{\text{hh}}(d, \theta)}{f_1^{\text{hh}}(L/2, \theta)}. \quad (9.116)$$

The additional dependence on L and the lattice resolution a/L of these ratios is not indicated explicitly. With Eq. (9.113), we see that all renormalization factors cancel in these ratios. They should have a finite limit $a/L \to 0$, approached asymptotically with a rate $(a/L)^2$. This is tested in Fig. 9.1, where L is kept fixed in units of the reference length scale r_0 (Sommer, 1994) to $L/r_0 = 1.436$. This choice corresponds to about $L \approx 0.7$ fm. The same continuum limit has to be reached for different lattice discretizations. Also, this universality is tested in the graphs, where four different choices of the covariant derivative D_0 in the static action are used. All actions defined by the different choices of D_0 have the symmetries discussed earlier.

9.2.8   Scale dependence of the axial current and the RGI current

Let us first recapitulate the scale dependence in perturbation theory. At one-loop order one has

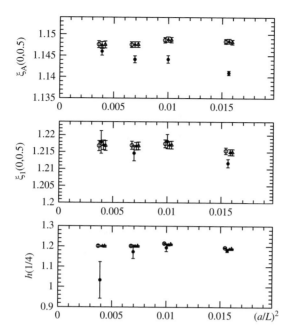

Fig. 9.1 Lattice-spacing dependence of various ratios of correlation functions for which Z-factors cancel. Different symbols correspond to different actions. Computation and figure from (Della Morte *et al.*, 2005).

$$(A_{\mathrm{R}}^{\mathrm{stat}})_0(x) = Z_{\mathrm{A}}^{\mathrm{stat}}(g_0, \mu a) A_0^{\mathrm{stat}}(x), \qquad (9.117)$$

$$Z_{\mathrm{A}}^{\mathrm{stat}}(g_0, \mu a) = 1 + g_0^2 \left[B_0 - \gamma_0 \ln(a\,\mu) \right] + \dots, \quad \gamma_0 = -\frac{1}{4\pi^2}. \qquad (9.118)$$

In the lattice minimal subtraction scheme the Z factors are polynomials in $\ln(a\mu)$ without constant part; thus $B_0 = 0$. Instead, when the renormalization scheme is defined by Eq. (9.115) a one-loop computation of $f_{\mathrm{A}}^{\mathrm{stat}}$, f_1^{stat} yields (Kurth and Sommer, 2001)[14] $B_0 = -0.08458$. As usual there is the renormalization group equation (RGE) (remember $g_0^2 = \bar{g}^2 + \mathrm{O}(\bar{g}^4)$)

$$\mu \frac{\partial}{\partial \mu} (A_{\mathrm{R}}^{\mathrm{stat}})_0 = \gamma(\bar{g})(A_{\mathrm{R}}^{\mathrm{stat}})_0, \quad \gamma(\bar{g}) = -\bar{g}^2 \left\{ \gamma_0 + \bar{g}^2 \gamma_1 + \dots \right\}. \qquad (9.119)$$

Combining it with the RGE for the coupling Eq. (9.281) it is easily integrated to (see Eq. (9.284) for the definition of the beta-function coefficients b_i)

$$(A_{\mathrm{R}}^{\mathrm{stat}})_0(\mu) = (A^{\mathrm{RGI}})_0 \, \exp \left\{ \int^{\bar{g}(\mu)} \mathrm{d}x \, \frac{\gamma(x)}{\beta(x)} \right\} \qquad (9.120)$$

$$\equiv (A^{\mathrm{RGI}})_0 \left[2b_0 \bar{g}^2 \right]^{\gamma_0/2b_0} \exp \left\{ \int_0^{\bar{g}} \mathrm{d}x \left[\frac{\gamma(x)}{\beta(x)} - \frac{\gamma_0}{b_0 x} \right] \right\}, \qquad (9.121)$$

where Eq. (9.121) provides the definition of the lax notation for the second factor in Eq. (9.120). The integration "constant" is the renormalization-group-invariant field. It can also be written as

$$(A^{\mathrm{RGI}})_0 = \lim_{\mu \to \infty} \left[2b_0 \bar{g}^2(\mu) \right]^{-\gamma_0/2b_0} (A^{\mathrm{stat}}_{\mathrm{R}})_0(\mu) \, ,$$

since the last factor in Eq. (9.121) converges to one as $\mu \to \infty$. Using also that γ_0, b_0 are independent of the renormalization scheme, as well as $O_S(\mu) = O_{S'}(\mu)(1 + \mathrm{O}(\bar{g}^2(\mu))$ (valid for any operator O and standard schemes S, S'), this representation also shows that the renormalization-group-invariant operator $(A^{\mathrm{RGI}})_0$ is independent of scale and scheme.[15]

Let us now go beyond perturbation theory and start from a non-perturbative definition of the renormalized current, such as Eq. (9.115), together with a non-perturbative definition of a renormalized coupling (Lüscher *et al.*, 1991; 1992; 1994). With the step-scaling method discussed in more detail by Peter Weisz, one can then determine the change

$$(A^{\mathrm{stat}}_{\mathrm{R}})_0(\mu) = \sigma^{\mathrm{stat}}_{\mathrm{A}}(\bar{g}^2(2\mu)) \, (A^{\mathrm{stat}}_{\mathrm{R}})_0(2\mu), \quad \mu = 1/L \tag{9.122}$$

of the renormalized field $(A^{\mathrm{stat}}_{\mathrm{R}})_0(\mu)$ when the renormalization scale μ is changed by a factor of two. The so-called step-scaling function $\sigma^{\mathrm{stat}}_{\mathrm{A}}$ is parameterized in terms of the running coupling $\bar{g}(\mu)$. Its argument is $\mu = 1/L$ in terms of the linear extent, $L = T$, of a Schrödinger functional.

Instead of the scale dependence of $(A^{\mathrm{stat}}_{\mathrm{R}})_0(\mu)$ we will often discuss a generic matrix element

$$\Phi(\mu) = \langle \alpha | (\widehat{A^{\mathrm{stat}}_{\mathrm{R}}})_0(\mu) | \beta \rangle \tag{9.123}$$

of the associated operator $\widehat{A^{\mathrm{stat}}_0}$ in Hilbert space.

In a non-perturbative calculation, the continuum $\sigma^{\mathrm{stat}}_{\mathrm{A}}$ is obtained through a numerical extrapolation

$$\sigma^{\mathrm{stat}}_{\mathrm{A}}(u) = \lim_{a/L \to 0} \Sigma^{\mathrm{stat}}_{\mathrm{A}}(u, a/L) \tag{9.124}$$

of the lattice step-scaling functions

$$\Sigma^{\mathrm{stat}}_{\mathrm{A}}(u, a/L) = \frac{Z^{\mathrm{stat}}_{\mathrm{A}}(g_0, a/2L)}{Z^{\mathrm{stat}}_{\mathrm{A}}(g_0, a/L)} \bigg|_{\bar{g}^2(1/L)=u} \tag{9.125}$$

obtained directly from simulations. Here, $\bar{g}^2(1/L)$ is kept fixed to remain at constant L, while L/a is varied in the continuum extrapolation.

The μ dependence of Φ can then be constructed iteratively via

$$u_0 = \bar{g}^2(1/L_0), \quad L_n = 2^n L_0, : \Phi(1/L_{n+1}) = \sigma^{\mathrm{stat}}_{\mathrm{A}}(u_n) \, \Phi(1/L_n)$$

$$u_{n+1} = \sigma(u_n),$$

where the step-scaling function σ of the running coupling enters. The length scale L_0 is chosen deep in the perturbative domain, typically $L_0 \approx 1/100\,\mathrm{GeV}$ and therefore the μ dependence can be completed perturbatively to infinite μ, i.e. to the RGI using Eq. (9.121).

For $N_\mathrm{f} = 2$ the analysis has been done for $\mu \approx 300\,\mathrm{MeV} - 80\,\mathrm{GeV}$. After it was verified that the steps at smallest L ($L \le L_2$) are accurately described by perturbation theory (see Fig. 9.2), the two-loop anomalous dimension was used in Eq. (9.121) with $\mu = 1/L_p$ to connect to the RGI current. The result (Kurth and Sommer, 2001; Heitger *et al.*, 2003; Della Morte *et al.*, 2007*b*) is conveniently written as

$$Z_{\mathrm{A,RGI}}^{\mathrm{stat}}(g_0) = \frac{\Phi_{\mathrm{RGI}}}{\Phi(\mu)} \times Z_{\mathrm{A}}^{\mathrm{stat}}(g_0, a\mu)\Big|_{\mu = 1/(2L_{\mathrm{max}})}, \qquad (9.126)$$

where only the second factor depends on the lattice action, and $\bar{g}^2(1/L_{\mathrm{max}}) = u_{\mathrm{max}}$ is a convenient value covered by the non-perturbative results for the above recursion.

We show the result for the first factor in Fig. 9.2 for a series of μ with $N_\mathrm{f} = 2$ dynamical quarks. The different points in the graph correspond to different n in the recursion. Note that the two-loop running becomes accurate only at rather small L. There is an about 5% difference in $\frac{\Phi_{\mathrm{RGI}}}{\Phi(\mu)}$ between a two-loop result and the non-perturbative one at the smallest μ.

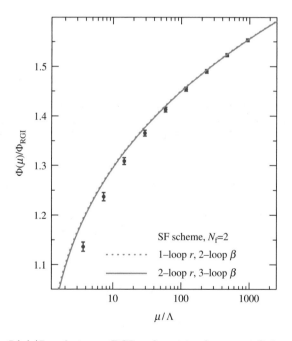

Fig. 9.2 Relation $\Phi(\mu)/\Phi_{\mathrm{RGI}}$ between RGI and matrix element at finite μ in a Schrödinger functional scheme and for $N_\mathrm{f} = 2$. The Λ-parameter in the SF-scheme is around $100\,\mathrm{MeV}$. Everything was computed non-perturbatively from continuum extrapolated step-scaling functions (Della Morte *et al.*, 2007*b*).

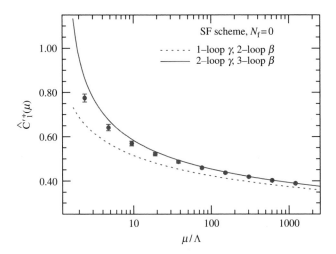

Fig. 9.3 Relation $\hat{c} = \Phi_{\mathrm{RGI}}/\Phi(\mu)$ of the RGI matrix element Φ_{RGI} and the matrix element at finite μ of the "$VA + AV$" four-fermion operator in a Schrödinger functional scheme. It was computed non-perturbatively from continuum extrapolated step-scaling functions (Dimopoulos *et al.*, 2008).

The details of this calculation and strategy are not that important for the following. We have mainly discussed it since

– first, the RGI matrix elements play a prominent role in HQET in static approximation and it is relevant to understand that they can be obtained completely non-perturbatively; and

– secondly, we also want to later emphasize the difference between the here used – by now more classic – renormalization of the static theory and the strategy discussed in Section 9.5.3.

Let us further note that the strategy above has been extended to four-fermion operators relevant for $B - \bar{B}$ oscillations in (Palombi *et al.*, 2006; Dimopoulos *et al.*, 2008). Since more than one operator is involved and twisted-mass QCD is used in order to avoid mixing with operators of wrong chirality, the strategy and computation are somewhat more involved. In particular, two different four-fermion operators contribute at leading order in $1/m$. As an example, we just show in Fig. 9.3 the result for the four-fermion operator that dominates in the physical process.

9.2.9 Eigenstates of the Hamiltonian

The eigenstates of the static Hamiltonian can be diagonalized simultaneously with the local heavy-flavor number operator (remember $Q_{\mathrm{h}}(x) = \bar{\psi}_{\mathrm{h}}(x)\psi_{\mathrm{h}}(x)$). We consider a finite volume with periodic boundary conditions. Since the theory is translation invariant, there is a $k \times L^3/a^3$ - fold degeneracy of states with a single heavy quark,[16] where k arises from degeneracies on top of the translation invariance discussed here. For the lowest-energy level one can choose a basis of eigenstates of the Hamiltonian as

$$|\tilde{B}(\mathbf{x})\rangle, \quad \langle \tilde{B}(\mathbf{x})|\tilde{B}(\mathbf{y})\rangle = 2\delta(\mathbf{x} - \mathbf{y}), \quad \hat{Q}_\text{h}(\mathbf{y})|\tilde{B}(\mathbf{x})\rangle = \delta(\mathbf{x} - \mathbf{y}) \quad (9.127)$$

or their Fourier transforms

$$|B(\mathbf{p})\rangle = a^3 \sum_\mathbf{x} \text{e}^{-i\mathbf{p}\mathbf{x}}|\tilde{B}(\mathbf{x})\rangle, \quad \langle \tilde{B}(\mathbf{p}')|\tilde{B}(\mathbf{p})\rangle = 2(2\pi)^3\delta(\mathbf{p} - \mathbf{p}') \quad (9.128)$$

$$\delta(\mathbf{p} - \mathbf{p}') = (L/(2\pi))^3 \prod_i \delta_{l_i l_i'}, \quad k_i = \frac{2\pi l_i}{L}, l_i \in \mathbf{Z}. \quad (9.129)$$

(Here we set $\theta = 0$.) We usually work with the zero-momentum eigenstate, denoted for short by $|B\rangle = |B(\mathbf{p} = 0)\rangle$, as this is related to an eigenstate of the finite-mass QCD Hamiltonian, which in finite volume has normalization $\langle B|B\rangle = 2L^3$.

9.3 Mass dependence at leading order in $1/m$: matching

We now discuss the "matching" of HQET to QCD using the example of a simple correlation function. As mentioned before, the issue is to fix the finite parts of renormalization constants such that the effective theory describes the underlying theory QCD. Throughout this section we remain in the static approximation. Matching including $1/m$ terms will be discussed in the next section.

9.3.1 A correlation function in QCD

We start from a simple QCD correlation function, which we write down in the lattice regularization,

$$C_\text{AA,R}^\text{QCD}(x_0) = Z_\text{A}^2\, a^3 \sum_\mathbf{x} \left\langle A_0(x)A_0^\dagger(0) \right\rangle_\text{QCD} \quad (9.130)$$

with the bare heavy-light axial current in QCD, $A_\mu = \overline{\psi}\gamma_\mu\gamma_5\psi_\text{b}$, and $A_\mu^\dagger = \overline{\psi}_\text{b}\gamma_\mu\gamma_5\psi$. The current is formed with the relativistic b-quark field ψ_b. In QCD, the renormalization factor, $Z_\text{A}(g_0)$, is fixed by chiral Ward identities (Bochicchio *et al.*, 1985; Lüscher *et al.*, 1997). It therefore does not depend on a renormalization scale.

One reason to consider this correlation function is that at large time the B-meson state dominates its spectral representation via

$$C_\text{AA,R}^\text{QCD}(x_0) = Z_\text{A}^2 a^3 \sum_\mathbf{x} \langle 0|A_0^\dagger(x)|B\rangle \frac{1}{2L^3}\langle B|A_0(0)|0\rangle \text{e}^{-x_0 m_\text{B}}\left[1 + \text{O}(\text{e}^{-x_0\Delta})\right]$$

$$= Z_\text{A}^2 \frac{1}{2}\langle 0|A_0^\dagger(0)|B\rangle\langle B|A_0(0)|0\rangle \text{e}^{-x_0 m_\text{B}}\left[1 + \text{O}(\text{e}^{-x_0\Delta})\right] \quad (9.131)$$

and the B-meson mass and its decay constant can be obtained from[17]

$$\Gamma_\text{AA}^\text{QCD}(x_0) = -\tilde{\partial}_0 \ln(C_\text{AA}^\text{QCD}(x_0)) = m_\text{B} + \text{O}(\text{e}^{-x_0\Delta}) \quad (9.132)$$

$$\left[\Phi^\text{QCD}\right]^2 \equiv f_\text{B}^2\, m_\text{B} \quad (9.133)$$

$$= \left|\langle B|Z_\text{A}A_0|0\rangle\right|^2 = 2 \lim_{x_0 \to \infty} \exp(x_0\,\Gamma_\text{AA}^\text{QCD}(x_0))C_\text{AA}^\text{QCD}(x_0).$$

Note that we use the normalization Equation (9.52) (in finite volume $\langle B|B\rangle = 2L^3$) for the zero-momentum state $|B\rangle$. The gap Δ is the energy difference between the second energy level and the first energy level in the zero-momentum (flavored) sector of the Hilbert space of the finite-volume lattice theory.

9.3.2 The correlation function in static approximation

In the static approximation we replace $Z_A A_0 \rightarrow Z_A^{\text{stat}}(g_0, \mu a) A_0^{\text{stat}}$ and define

$$C_{\text{AA,R}}^{\text{stat}}(x_0) = (Z_A^{\text{stat}})^2 \, C_{\text{AA}}^{\text{stat}}(x_0) = (Z_A^{\text{stat}})^2 \, a^3 \sum_{\mathbf{x}} \Big\langle A_0^{\text{stat}}(x)(A_0^{\text{stat}})^\dagger(0) \Big\rangle_{\text{stat}} \quad (9.134)$$

$$\Gamma_{\text{AA}}^{\text{stat}}(x_0) = -\widetilde{\partial}_0 \ln(C_{\text{AA}}^{\text{stat}}(x_0)), \quad (9.135)$$

$$\left[\Phi(\mu)\right]^2 \equiv \left|\langle B|Z_A^{\text{stat}} A_0^{\text{stat}}|0\rangle_{\text{stat}}\right|^2 \quad (9.136)$$

$$= 2 \lim_{x_0 \rightarrow \infty} \exp(x_0 \, \Gamma_{\text{AA}}^{\text{stat}}(x_0))(Z_A^{\text{stat}})^2 \, C_{\text{AA}}^{\text{stat}}(x_0).$$

The μ dependence of Φ results from the renormalization of the current in the effective theory,

$$Z_A^{\text{stat}}(g_0, \mu a) = 1 + g_0^2 \left[B_0 - \gamma_0 \ln(a\,\mu)\right] + \text{O}(g_0^4). \quad (9.137)$$

Different renormalization schemes have different constants B_0. Alternatively one uses the renormalization-group-invariant operator $(A^{\text{RGI}})_0$. We will come to that shortly.

9.3.3 Matching

The correlation function $C_{\text{AA}}^{\text{QCD}}$ and the matrix element Φ^{QCD}, Eq. (9.133), are independent of any renormalization scale, due to the chiral symmetry of QCD in the massless limit. But of course they depend on the mass of the b-quark.

In the effective theory we first renormalize in an arbitrary scheme, which we do not need to specify for the following, resulting in a scale-dependent $\Phi(\mu)$. The two quantities are then related through the matching equation (without explicit superscripts "QCD" we refer to HQET quantities, here static),

$$\Phi^{\text{QCD}}(m) = \widetilde{C}_{\text{match}}(m, \mu) \times \Phi(\mu) + \text{O}(1/m). \quad (9.138)$$

Somewhat symbolically the same equation could be written for the current instead of its matrix element; we write "symbolically" since the two currents belong to theories with different field contents. However, thinking in terms of the currents, it is clear that Eq. (9.138) can be thought of as a change of renormalization scheme in the effective theory, where the new renormalization scale is $m = m_{\text{b}}$ and the finite part is exactly fixed by Eq. (9.138). In fact, since at tree level we have constructed the effective theory such that $\Phi = \Phi^{\text{QCD}}$, the tree-level value for $\widetilde{C}_{\text{match}}$ is one and we have a perturbative expansion

$$\widetilde{C}_{\text{match}}(m, \mu) = 1 + c_1(m/\mu)\bar{g}^2(\mu) + \dots. \quad (9.139)$$

The finite renormalization factor $\widetilde{C}_{\text{match}}$ may be determined such that Eq. (9.138) holds for some particular matrix element of the current and will then be valid *for all matrix elements* or correlation functions. Some aspects of the above equation still need explanation. The μ dependence is not present on the left-hand side and this should be made explicit also on the right-hand side; further, one may wonder which definition of the quark mass and coupling constant one is to choose.

9.3.3.1 *One loop*

Before coming to these issues, it is illustrative to write down explicitly what Eq. (9.138) looks like at 1-loop order. Ignore for now how we renormalized the current in Eq. (9.115) and use instead lattice minimal subtraction,

$$Z_{\text{A}}^{\text{stat}}(g_0, \mu a) = 1 - \gamma_0 \ln(a\,\mu)\, g_0^2 + \dots, \tag{9.140}$$

in the static theory.

Instead of the decay constant, use as an observable a perturbatively accessible quantity. We take[18]

$$\Phi^{\text{QCD}} = Y_{\text{R}}^{\text{QCD}}(\theta, m_{\text{R}}, L) \equiv \lim_{a \to 0} Z_{\text{A}}(g_0) \frac{f_{\text{A}}(L/2, \theta, m_{\text{R}})}{\sqrt{f_1(\theta, m_{\text{R}})}}, \tag{9.141}$$

where for our one-loop discussion we do not need to specify the normalization condition for the renormalized heavy quark mass $\bar{m} = m_{\text{R}}$ and coupling $\bar{g} = g_{\text{R}}$. The one-loop expansion of these functions has been computed (Kurth and Sommer, 2002), and the result can be summarized as

$$\Phi = Y_{\text{R}}^{\text{stat}}(\theta, \mu, L) = \lim_{a \to 0} Z_{\text{A}}^{\text{stat}}(g_0, \mu a) \frac{f_{\text{A}}^{\text{stat}}(L/2, \theta)}{\sqrt{f_1^{\text{stat}}(\theta)}} \tag{9.142}$$

$$= A(\theta)[1 - \gamma_0 \ln(\mu L)\, g_{\text{R}}^2] + D(\theta) g_{\text{R}}^2 + \text{O}(g_{\text{R}}^4)$$

in static approximation and

$$\Phi^{\text{QCD}} = A(\theta)[1 + (D' - \gamma_0 \ln(m_{\text{R}} L))g_{\text{R}}^2] + D(\theta) g_{\text{R}}^2 + \text{O}(1/(m_{\text{R}} L)) + \text{O}(g_{\text{R}}^4)$$

in QCD. From these expressions we can read off

$$c_1(m_{\text{R}}/\mu) = \gamma_0 \ln(\mu/m_{\text{R}}) + D'. \tag{9.143}$$

Furthermore, the fact that the same functions $A(\theta)$, $D(\theta)$ appear in the static theory and in QCD is a (partial) confirmation that the static approximation is the effective theory for QCD. In particular, the logarithmic L dependence in QCD matches the one in the static theory. With Eq. (9.143), the matching of QCD and static theory holds for all θ, and also for other matrix elements of A_0.

9.3.3.2 *Renormalization-group invariants*

Having seen how QCD and effective theory match at one-loop order, we now proceed to a general discussion of Eq. (9.138), beyond one loop. Obviously, the μ dependence in Eq. (9.138) is artificial, since we have a scale-independent quantity in QCD. Only the mass dependence is for real. We may then choose any value for μ. For convenience, we set all renormalization scales equal to the mass itself,[19]

$$\mu = m_\star = \bar{m}(m_\star), \quad g_\star = \bar{g}(m_\star), \tag{9.144}$$

where $\bar{m}(\mu), \bar{g}(\mu)$ are running mass and coupling in an unspecified massless renormalization scheme.[20] This simplifies the matching function to

$$\widetilde{C}_{\text{match}}(m_\star, m_\star) = C_{\text{match}}(g_\star) = 1 + c_1(1)\, g_\star^2 + \dots. \tag{9.145}$$

Further, we want to eliminate the dependence on the renormalization scheme for $\bar{m}, \bar{g}, (A_{\text{R}}^{\text{stat}})_0$. As a first step we change from $\Phi(\mu)$ to the RGI matrix element

$$\Phi_{\text{RGI}} = \exp\left\{ -\int^{\bar{g}(\mu)} \mathrm{d}x\, \frac{\gamma(x)}{\beta(x)} \right\} \Phi(\mu), \tag{9.146}$$

and arrive at the form

$$\Phi^{\text{QCD}} = C_{\text{match}}(g_\star) \times \Phi(\mu) = C_{\text{match}}(g_\star) \exp\left\{ \int^{g_\star} \mathrm{d}x\, \frac{\gamma(x)}{\beta(x)} \right\} \Phi_{\text{RGI}} \tag{9.147}$$

$$\equiv \exp\left\{ \int^{g_\star} \mathrm{d}x\, \frac{\gamma_{\text{match}}(x)}{\beta(x)} \right\} \Phi_{\text{RGI}}. \tag{9.148}$$

Everywhere terms of order $1/m$ are dropped, since we are working to static order. Equation (9.148) defines γ_{match}, which describes the physical mass dependence via,

$$\frac{m_\star}{\Phi^{\text{QCD}}} \frac{\partial \Phi^{\text{QCD}}}{\partial m_\star} = \gamma_{\text{match}}(g_\star), \tag{9.149}$$

but it still depends on the chosen renormalization scheme through the choice of $\bar{m}$ (the scheme, not the scale). We eliminate also this scheme dependence by switching to the RGI mass, M, and the Λ parameter,

$$\frac{\Lambda}{\mu} = \exp\left\{ -\int^{\bar{g}(\mu)} \mathrm{d}x\, \frac{1}{\beta(x)} \right\}, \tag{9.150}$$

$$\frac{M}{\bar{m}(\mu)} = \exp\left\{ -\int^{\bar{g}(\mu)} \mathrm{d}x\, \frac{\tau(x)}{\beta(x)} \right\}. \tag{9.151}$$

Exact expressions, defining the constant parts in these equation, are given in the appendix.

Just based on dimensional analysis, we expect a relation

$$\Phi^{\text{QCD}} = C_{\text{PS}}(M/\Lambda) \times \Phi_{\text{RGI}} \tag{9.152}$$

to hold. Indeed, remembering Eq. (9.144), $\mu = m_\star = \bar{m}$, we can combine Eq. (9.150) and Eq. (9.151) to

$$\frac{\Lambda}{M} = \exp\left\{ - \int^{g_\star(M/\Lambda)} dx\, \frac{1 - \tau(x)}{\beta(x)} \right\}, \tag{9.153}$$

from which $g_\star$ can be determined for any value of M/Λ; we write $g_\star = g_\star(M/\Lambda)$. It follows that

$$M \frac{\partial g_\star(m_\star(M/\Lambda))}{\partial M} = \frac{\beta(g_\star)}{1 - \tau(g_\star)}, \tag{9.154}$$

and the matching function is

$$C_{\mathrm{PS}}(M/\Lambda) = \exp\left\{ \int^{g_\star(M/\Lambda)} dx\, \frac{\gamma_{\mathrm{match}}(x)}{\beta(x)} \right\}. \tag{9.155}$$

We note that the dependence on M is described by a function[21]

$$\left. \frac{M}{\Phi} \frac{\partial \Phi}{\partial M} \right|_\Lambda = \left. \frac{M}{C_{\mathrm{PS}}} \frac{\partial C_{\mathrm{PS}}}{\partial M} \right|_\Lambda = \frac{\gamma_{\mathrm{match}}(g_\star)}{1 - \tau(g_\star)}, \quad g_\star = g_\star(M/\Lambda). \tag{9.156}$$

With

$$\gamma_{\mathrm{match}}(g_\star) \overset{g_\star \to 0}{\sim} - \gamma_0 g_\star^2 - \gamma_1^{\mathrm{match}} g_\star^4 + \dots, \qquad \beta(\bar{g}) \overset{\bar{g} \to 0}{\sim} - b_0 \bar{g}^3 + \dots \tag{9.157}$$

we can now give the leading large-mass behavior

$$C_{\mathrm{PS}} \overset{M \to \infty}{\sim} (2b_0 g_\star^2)^{-\gamma_0/2b_0} \sim [\ln(M/\Lambda)]^{\gamma_0/2b_0}. \tag{9.158}$$

Functions such as C_{PS} convert from the static RGI matrix elements to the QCD matrix element; we call them conversion functions.

An interesting application is the asymptotics of the decay constant of a heavy-light pseudoscalar (e.g. B):[22]

$$F_{\mathrm{PS}} \overset{M \to \infty}{\sim} \frac{[\ln(M/\Lambda)]^{\gamma_0/2b_0}}{\sqrt{m_{\mathrm{PS}}}} \Phi_{\mathrm{RGI}} \times [1 + \mathrm{O}([\ln(M/\Lambda)]^{-1})]. \tag{9.163}$$

At leading order in $1/m$ the conversion function C_{PS} contains the full (logarithmic) mass dependence. The non-perturbative effective theory matrix elements, Φ_{RGI}, are mass-independent numbers. Conversion functions such as C_{PS} are universal for all (low-energy) matrix elements of their associated operator. For example,

$$C_{\mathrm{AA,R}}^{\mathrm{QCD}}(x_0) \overset{x_0 \gg 1/m}{\sim} [C_{\mathrm{PS}}(\frac{M}{\Lambda_{\overline{\mathrm{MS}}}})\, Z_{\mathrm{A,RGI}}^{\mathrm{stat}}]^2 \underbrace{\langle A_0^{\mathrm{stat}}(x)^\dagger A_0^{\mathrm{stat}}(0)\rangle}_{C_{\mathrm{AA}}^{\mathrm{stat}}(x_0)\ (\mathrm{bare})} + \mathrm{O}(\frac{1}{m}), \tag{9.164}$$

is a straightforward generalization of Eq. (9.138).

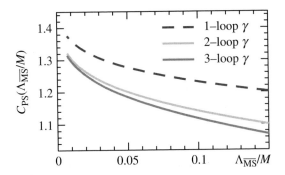

Fig. 9.4 C_{PS} estimated in perturbation theory. For B-physics we have $\Lambda_{\overline{\mathrm{MS}}}/M_{\mathrm{b}} \approx 0.04$. Figure from (Heitger *et al.*, 2004).

Analogous expressions for the conversion functions are valid for the time component of the axial current replaced by other composite fields, for example the space components of the vector current. Based on the work of (Broadhurst and Grozin, 1991; Shifman and Voloshin, 1987; Politzer and Wise, 1988) and recent efforts their perturbative expansion is known including the 3-loop anomalous dimension γ_{match} obtained from the 3-loop anomalous dimension γ (Chetyrkin and Grozin, 2003) in the $\overline{\mathrm{MS}}$-scheme and the 2-loop matching function C_{match} (Ji and Musolf, 1991; Broadhurst and Grozin, 1995; Gimenez, 1992).

Figure 9.4 seems to indicate that the remaining $O(\bar{g}^6(m_{\mathrm{b}}))$ errors in C_{PS} are relatively small. However, as discussed in more detail in Section 9.6.2, such a conclusion is premature. By now, *ratios* of conversion functions for different currents are known to even one more order in perturbation theory (Bekavac *et al.*, 2010). We show an example in the first column of Fig. 9.5, where the x-axis is approximately proportional to $g_\star^2(M/\Lambda)$ and for B-physics one needs $1/\ln(\Lambda_{\overline{\mathrm{MS}}}/M_{\mathrm{b}}) \approx 0.3$. For a quark mass around the mass of the b-quark and lower, the higher-order contributions in perturbation theory do not decrease significantly and perturbation theory is not trustworthy. It seems impossible to estimate a realistic error of the perturbative expansion. Only for somewhat higher masses does the expansion look reasonable.

Moreover, using the freedom to choose the scale μ in Eq. (9.139), the lth order coefficients (as far as they are known) can be brought down in magnitude below about $(4\pi)^{-l}$, which means there is a fast decrease of terms in the perturbative series once $\alpha(\mu) \lesssim 1/3$. This is shown in columns two and three of the figure. Unfortunately, the required scale μ is around a factor 4 or more below the mass of the quark. For the b-quark, α is rather large at that scale and the series is again unreliable. Only for even larger masses, say $m_\star > 15\,\mathrm{GeV}$, is the asymptotic convergence of the series noticeably better after adjusting the scale. More details are found in Section 9.6.2. Unfortunately, we see no way out of the conclusion that for B-physics with a trustworthy error budget aiming at the few per cent level, one needs a non-perturbative matching, *even in the static approximation.*

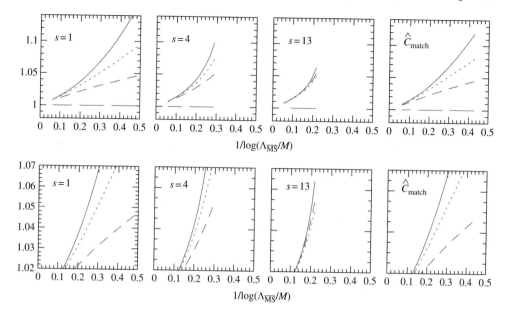

Fig. 9.5 The ratio $C_{\text{PS}}/C_{\text{V}}$, evaluated in the first column as described here. In columns two and three the expansion in $g_\star$ is generalized to an expansion in $\bar{g}(m_\star/s)$, see Section 9.6.2.2. The last column contains the conventionally used $\hat{C}_{\text{match}}^{\text{PS}}(m_{\text{Q}}, m_{\text{Q}}, m_{\text{Q}})/\hat{C}_{\text{match}}^{\text{V}}(m_{\text{Q}}, m_{\text{Q}}, m_{\text{Q}})$, see Section 9.6.2. For B-physics we have $\Lambda_{\overline{\text{MS}}}/M_{\text{b}} \approx 0.04$ and $1/\ln(\Lambda_{\overline{\text{MS}}}/M_{\text{b}}) \approx 0.3$. The loop order changes from one-loop (long dashes) up to 4-loop (full line) anomalous dimension.

We return to the full set of heavy-light flavor currents of Section 9.2.5. The bare fields satisfy the symmetry relations Eq. (9.94). The same is then true for the RGI fields in static approximation. It follows that *in static approximation* the effective currents are given by

$$A_0^{\text{HQET}} = C_{\text{PS}}(M_{\text{b}}/\Lambda_{\overline{\text{MS}}})\, Z_{\text{A,RGI}}^{\text{stat}}(g_0)\, A_0^{\text{stat}}, \tag{9.165}$$

$$V_k^{\text{HQET}} = C_{\text{V}}(M_{\text{b}}/\Lambda_{\overline{\text{MS}}})\, Z_{\text{A,RGI}}^{\text{stat}}(g_0)\, V_k^{\text{stat}}, \tag{9.166}$$

$$V_0^{\text{HQET}} = C_{\text{PS}}(M_{\text{b}}/\Lambda_{\overline{\text{MS}}})\, Z_{\text{A,RGI}}^{\text{stat}}(g_0)\, Z_{\text{V/A}}^{\text{stat}}(g_0)\, V_0^{\text{stat}}, \tag{9.167}$$

$$A_k^{\text{HQET}} = C_{\text{V}}(M_{\text{b}}/\Lambda_{\overline{\text{MS}}})\, Z_{\text{A,RGI}}^{\text{stat}}(g_0) Z_{\text{V/A}}^{\text{stat}}(g_0)\, A_k^{\text{stat}}. \tag{9.168}$$

The factor $Z_{\text{A,RGI}}^{\text{stat}}(g_0)$ is known as discussed in the previous lecture. Note that $Z_{\text{A,RGI}}^{\text{stat}}(g_0)$ is common to all (components of the) currents. Due to the HQET symmetries, there is one single anomalous dimension. A dependence on the different fields comes in only through matching, i.e. through the QCD matrix elements. In the above equations, chiral symmetry (of the continuum theory), Eq. (9.158), has been used to relate conversion functions of axial and vector currents.

Exercise 9.3 Pseudoscalar and scalar densities

Start from the PCAC, PCVC relations in QCD

$$\partial_\mu (A_R)_\mu = (\overline{m}_b(\mu) + \overline{m}_l(\mu)) P_R(\mu), \tag{9.169}$$

$$\partial_\mu (V_R)_\mu = (\overline{m}_b(\mu) - \overline{m}_l(\mu)) S_R(\mu). \tag{9.170}$$

Replace all quantities by their RGIs. Take the matrix elements between vacuum and a suitable B-meson state to show that

$$P^{\text{HQET}} = -C_{\text{PS}}(M_b/\Lambda_{\overline{\text{MS}}}) \frac{m_B}{M_b} Z_{\text{A,RGI}}^{\text{stat}}(g_0) A_0^{\text{stat}}, \tag{9.171}$$

$$S^{\text{HQET}} = C_V(M_b/\Lambda_{\overline{\text{MS}}}) \frac{m_B}{M_b} Z_{V/A}^{\text{stat}}(g_0) Z_{\text{A,RGI}}^{\text{stat}}(g_0) V_0^{\text{stat}}, \tag{9.172}$$

is valid up to terms of order $1/m$. What happens if you choose a different matrix element?

9.3.3.3 Applications

As an application, we can now modify the scaling law for the decay constant to include renormalization and matching effects

$$\frac{f_B \sqrt{m_B}}{C_{\text{PS}}(M_b/\Lambda_{\overline{\text{MS}}})} = \Phi_{\text{RGI}} + O(1/m) \tag{9.173}$$

$$\frac{f_B}{f_D} \approx \frac{\sqrt{m_D}}{\sqrt{m_B}} \frac{C_{\text{PS}}(M_b/\Lambda_{\overline{\text{MS}}})}{C_{\text{PS}}(M_c/\Lambda_{\overline{\text{MS}}})}, \tag{9.174}$$

where the latter equation is maybe stretching the applicability domain of HQET.

Despite the discussion above, let us assume that the conversion functions C are known with reasonably small errors from perturbation theory. In this case, the knowledge of the leading term in expansions such as Eq. (9.173) is very useful to constrain the large-mass behavior of QCD observables, computed on the lattice with unphysical quark masses $m_h < m_b$, typically $m_h \approx m_{\text{charm}}$. (Such a calculation is done with a relativistic (Wilson, tmQCD, ...) formulation, extrapolating $am_h \to 0$ at fixed m_h.) As illustrated in Fig. 9.6, one can then, with a reasonable smoothness assumption, interpolate to the physical point.

Given the unclear precision of the perturbative predictions, the above interpolation method has to be taken with care. The inherent perturbative error remains to be estimated.

The relation between the RGI fields and the bare fields has also been obtained for the two parity-violating $\Delta B = 2$ four-fermion operators (Palombi *et al.*, 2006; 2007) for $N_f = 0$ and $N_f = 2$ (Dimopoulos *et al.*, 2008). Their matrix elements, evaluated in twisted-mass QCD will give the standard model B-parameter for B–B̄ mixing.

We now turn to the natural question whether one can directly compute the $1/m$ corrections in HQET, which will lead us again to the necessity of performing a non-perturbative matching between HQET and QCD.

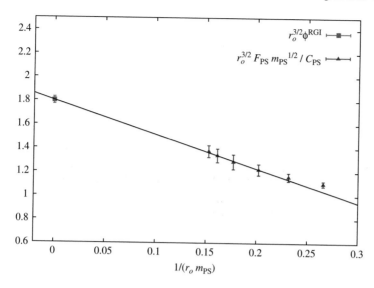

Fig. 9.6 Example of an interpolation between a static result and results with $m_{\mathrm{h}} < m_{\mathrm{b}}$. The function C_{PS} is estimated at three-loop order. Continuum extrapolations are done before the interpolation (Blossier *et al.*, 2010c). The point at $1/r_0 m_{\mathrm{PS}} = 0$ is given by $r_0^{3/2}\Phi_{\mathrm{RGI}}$. This quenched computation is done for validating and demonstrating the applicability of HQET.

Exercise 9.4 Anomalous dimension γ_{match}
Show that

$$\gamma_{\mathrm{match}} = -\gamma_0 g_\star^2 - [\gamma_1 + 2b_0 c_1(1)]g_\star^4 + \dots . \tag{9.175}$$

where c_1 is the 1-loop matching coefficient in the same scheme as γ_1.

9.4   Renormalization and matching at order $1/m$

9.4.1   Including $1/m$ corrections

We here work directly in lattice regularization. The continuum formulae are completely analogous. The expressions for $\mathcal{O}_{\mathrm{kin}}, \mathcal{O}_{\mathrm{spin}}$ are discretized in a straightforward way,

$$D_k D_k \to \nabla_k^* \nabla_k, \qquad F_{kl} \to \widehat{F}_{kl} \tag{9.176}$$

with the latter given by the clover-leaf representation, defined, e.g., in (Lüscher *et al.*, 1996). Of course other discretizations of these composite fields are possible.

Apart from the terms in the classical Lagrangian, renormalization can in principle introduce new local fields compatible with the symmetries (but not necessarily the heavy-quark symmetries that are broken by $\mathcal{O}_{\mathrm{spin}}, \mathcal{O}_{\mathrm{kin}}$) and with dimension $d_{\mathrm{op}} \le 5$. Also, the field equations can be used to eliminate terms. With these rules one finds that no new terms are needed and it suffices to treat the coefficients of $\mathcal{O}_{\text{spin}}, \mathcal{O}_{\text{kin}}$ as free parameters that depend on the bare coupling of the theory and on m.

The $1/m$ Lagrangian then reads

$$\mathscr{L}_{\text{h}}^{(1)}(x) = -(\omega_{\text{kin}}\,\mathcal{O}_{\text{kin}}(x) + \omega_{\text{spin}}\,\mathcal{O}_{\text{spin}}(x)). \tag{9.177}$$

Since these terms are composite fields of dimension five, the theory defined with a path-integral weight ($\mathscr{L}_{\text{light}}$ collects all contributions of QCD with the heavy quark(s) dropped)

$$W_{\text{NRQCD}} \propto \exp(-a^4 \sum_x [\mathscr{L}_{\text{light}}(x) + \mathscr{L}_{\text{h}}^{\text{stat}}(x) + \mathscr{L}_{\text{h}}^{(1)}(x)]) \tag{9.178}$$

is *not* renormalizable. In perturbation theory, new divergences will occur at each order in the loop expansion, which necessitate introduction of new counterterms. The continuum limit of the lattice theory will then not exist (Thacker and Lepage, 1991). Since the effective theory is "only" supposed to reproduce the $1/m$ expansion of the observables order by order in $1/m$, we instead expand the weight W in $1/m$, counting $\omega_{\text{kin}} = \text{O}(1/m) = \omega_{\text{spin}}$,

$$W_{\text{NRQCD}} \to W_{\text{HQET}} \equiv \exp(-a^4 \sum_x [\mathscr{L}_{\text{light}}(x) + \mathscr{L}_{\text{h}}^{\text{stat}}(x)]) \left\{ 1 - a^4 \sum_x \mathscr{L}_{\text{h}}^{(1)}(x) \right\}.$$

This rule is part of the definition of HQET, just like the same step is part of Symanzik's effective theory discussed by Peter Weisz.

Let us remark here on the difference from chiral perturbation theory. In chiral perturbation theory one computes the asymptotic expansion in powers of p^2. Each term in the expansion requires a finite number of counterterms, since there are only a finite number of (pion) loops. The theory is thus renormalizable order by order in the expansion. In NRQCD and HQET one expands in $1/m$. At each order of the expansion an arbitrary number of loops remain, coming from the gluons and light quarks. In fact, we are even interested in more than an arbitrary number of loops: in non-perturbative results in α.

NRQCD can then only be formulated with a cutoff and results depend on how the cutoff is introduced and on its value. On the lattice, the cutoff is identified with the one present for the other fields, $\Lambda_{\text{cut}} \sim 1/a$. Instead of taking a continuum limit, one then relies on physics results not depending on the lattice spacing (the cutoff) within a window (Thacker and Lepage, 1991)

$$1/m \ll a \ll \Lambda_{\text{QCD}} \qquad \text{[in NRQCD]}. \tag{9.179}$$

In HQET the discussion is rather simple, since the static theory is (believed to be) renormalizable; we will come to the renormalization of the insertion of $\mathscr{L}_{\text{h}}^{(1)}$ shortly.

Up to and including $\text{O}(1/m)$, expectation values in HQET are therefore defined as

$$\langle \mathcal{O} \rangle = \langle \mathcal{O} \rangle_{\text{stat}} + \omega_{\text{kin}} a^4 \sum_x \langle \mathcal{O}\mathcal{O}_{\text{kin}}(x) \rangle_{\text{stat}} + \omega_{\text{spin}} a^4 \sum_x \langle \mathcal{O}\mathcal{O}_{\text{spin}}(x) \rangle_{\text{stat}}$$

$$\equiv \langle \mathcal{O} \rangle_{\text{stat}} + \omega_{\text{kin}} \langle \mathcal{O} \rangle_{\text{kin}} + \omega_{\text{spin}} \langle \mathcal{O} \rangle_{\text{spin}}, \tag{9.180}$$

where the path-integral average

$$\langle \mathcal{O} \rangle_{\text{stat}} = \frac{1}{\mathcal{Z}} \int_{\text{fields}} \mathcal{O} \exp(-a^4 \sum_x [\mathscr{L}_{\text{light}}(x) + \mathscr{L}_{\text{h}}^{\text{stat}}(x)]) \tag{9.181}$$

is taken with respect to the lowest-order action. The integral extends over all fields and the normalization $\mathcal{Z}$ is fixed by $\langle 1 \rangle_{\text{stat}} = 1$.[23]

In order to compute matrix elements or correlation functions in the effective theory, we also need the effective composite fields. At the classical level they can again be obtained from the Fouldy–Wouthuysen rotation. In the quantum theory one adds all local fields with the proper quantum numbers and dimensions. For example, the effective axial current (time component) is given by

$$A_0^{\text{HQET}}(x) = Z_A^{\text{HQET}} [A_0^{\text{stat}}(x) + \sum_{i=1}^{2} c_A^{(i)} A_0^{(i)}(x)], \tag{9.182}$$

$$A_0^{(1)}(x) = \overline{\psi}(x) \frac{1}{2} \gamma_5 \gamma_i (\nabla_i^{\text{S}} - \overleftarrow{\nabla}_i^{\text{S}}) \psi_{\text{h}}(x), \tag{9.183}$$

$$A_0^{(2)}(x) = -\frac{1}{2} \widetilde{\partial}_i A_i^{\text{stat}}, \tag{9.184}$$

where all derivatives are symmetric,

$$\widetilde{\partial}_i = \frac{1}{2}(\partial_i + \partial_i^*), \quad \overleftarrow{\nabla}_i^{\text{S}} = \frac{1}{2}(\overleftarrow{\nabla}_i + \overleftarrow{\nabla}_i^*), \quad \nabla_i^{\text{S}} = \frac{1}{2}(\nabla_i + \nabla_i^*), \tag{9.185}$$

and we recall $A_i^{\text{stat}}(x) = \overline{\psi}(x) \gamma_i \gamma_5 \psi_{\text{h}}(x)$. One arrives at these currents, writing down all dimension-four operators with the right flavor structure and transformation under spatial lattice rotations and parity. The equations of motion of the light and static quarks are used to eliminate terms but heavy-quark symmetries (spin and local flavor) can't be used since they are broken at order $1/m$.[24]

For completeness, let us write down the other HQET currents:

$$A_k^{\text{HQET}}(x) = Z_A^{\text{HQET}} [A_k^{\text{stat}}(x) + \sum_{i=3}^{6} c_A^{(i)} A_k^{(i)}(x)], \tag{9.186}$$

$$A_k^{(3)}(x) = \overline{\psi}(x) \frac{1}{2} \gamma_k \gamma_5 \gamma_i (\nabla_i^{\text{S}} - \overleftarrow{\nabla}_i^{\text{S}}) \psi_{\text{h}}(x), \quad A_k^{(4)}(x) = \overline{\psi}(x) \frac{1}{2} (\nabla_k^{\text{S}} - \overleftarrow{\nabla}_k^{\text{S}}) \gamma_5 \psi_{\text{h}}(x),$$

$$A_k^{(5)}(x) = \frac{1}{2} \widetilde{\partial}_i \left(\overline{\psi}(x) \gamma_k \gamma_5 \gamma_i \psi_{\text{h}}(x) \right), \quad A_k^{(6)}(x) = \frac{1}{2} \widetilde{\partial}_k A_0^{\text{stat}}.$$

The vector current components are just obtained by dropping γ_5 in these expressions and changing $c_A^{(i)} \to c_V^{(i)}$. The classical values of the coefficients are $c_A^{(1)} = c_A^{(2)} = c_A^{(3)} = c_A^{(5)} = -\frac{1}{2m}$, while $c_A^{(4)} = c_A^{(6)} = 0$. We note that with periodic boundary conditions in space we have

$$a^3 \sum_x A_0^{(1)}(x) = a^3 \sum_x \overline{\psi}(x) \overleftarrow{\nabla}_i^{\text{S}} \gamma_i \gamma_5 \psi_{\text{h}}(x), \quad a^3 \sum_x A_0^{(2)}(x) = 0, \tag{9.187}$$

which for instance may be used in the determination of the B-decay constant.

Before entering into details of the renormalization, we show some examples of how the $1/m$-expansion works.

9.4.2 $1/m$-expansion of correlation functions and matrix elements

For now we assume that the coefficients

$$O(1) : \delta m, \ Z_A^{\text{HQET}},$$

$$O(1/m) : \omega_{\text{kin}}, \ \omega_{\text{spin}}, \ c_A^{(1)},$$

(9.188)

are known as a function of the bare coupling g_0 and the quark mass m. Their non-perturbative determination will be discussed later.

The rules of the $1/m$-expansion are illustrated on the example of $C_{\text{AA,R}}^{\text{QCD}}(x_0)$, Eq. (9.130). One uses Eq. (9.180) and the HQET representation of the composite field Eq. (9.182). Then, the expectation value is expanded consistently in $1/m$, counting powers of $1/m$ as in Eq. (9.188). At order $1/m$, terms proportional to $\omega_{\text{kin}} \times c_A^{(1)}$, etc. are to be dropped. As a last step, we have to take the energy shift between HQET and QCD into account. Therefore, correlation functions with a time separation x_0 obtain an extra factor $\exp(-x_0 \, m)$, where the scheme dependence of m is compensated by a corresponding one in δm. Dropping all terms $O(1/m^2)$ without further notice, one arrives at the expansion

$$C_{\text{AA}}^{\text{QCD}}(x_0) = e^{-mx_0} (Z_A^{\text{HQET}})^2 \left[C_{\text{AA}}^{\text{stat}}(x_0) + c_A^{(1)} \, C_{\delta \text{AA}}^{\text{stat}}(x_0) \right.$$

(9.189)

$$\left. + \omega_{\text{kin}} C_{\text{AA}}^{\text{kin}}(x_0) + \omega_{\text{spin}} C_{\text{AA}}^{\text{spin}}(x_0) \right]$$

$$\equiv e^{-mx_0} (Z_A^{\text{HQET}})^2 \, C_{\text{AA}}^{\text{stat}}(x_0) \left[1 + c_A^{(1)} \, R_{\delta A}^{\text{stat}}(x_0) \right.$$

(9.190)

$$\left. + \omega_{\text{kin}} R_{\text{AA}}^{\text{kin}}(x_0) + \omega_{\text{spin}} R_{\text{AA}}^{\text{spin}}(x_0) \right]$$

with (remember the definitions in Eq. (9.180))

$$C_{\delta \text{AA}}^{\text{stat}}(x_0) = a^3 \sum_{\mathbf{x}} \langle A_0^{\text{stat}}(x)(A_0^{(1)}(0))^\dagger \rangle_{\text{stat}} + a^3 \sum_{\mathbf{x}} \langle A_0^{(1)}(x)(A_0^{\text{stat}}(0))^\dagger \rangle_{\text{stat}},$$

$$C_{\text{AA}}^{\text{kin}}(x_0) = a^3 \sum_{\mathbf{x}} \langle A_0^{\text{stat}}(x)(A_0^{\text{stat}}(0))^\dagger \rangle_{\text{kin}}$$

$$C_{\text{AA}}^{\text{spin}}(x_0) = a^3 \sum_{\mathbf{x}} \langle A_0^{\text{stat}}(x)(A_0^{\text{stat}}(0))^\dagger \rangle_{\text{spin}}.$$

The contribution of $A_0^{(2)}$ vanishes due to Eq. (9.187). It is now a straightforward exercise to obtain the expansion of the B-meson mass[25]

$$m_{\rm B} = - \lim_{x_0 \to \infty} \widetilde{\partial_0} \ln C_{\rm AA}^{\rm QCD}(x_0) \tag{9.191}$$

$$= m_{\rm bare} - \lim_{x_0 \to \infty} \widetilde{\partial_0} \big[\ln C_{\rm AA}^{\rm stat}(x_0) + c_{\rm A}^{(1)} R_{\delta A}^{\rm stat}(x_0) \tag{9.192}$$

$$+ \omega_{\rm kin} R_{\rm AA}^{\rm kin}(x_0) + \omega_{\rm spin} R_{\rm AA}^{\rm spin}(x_0) \big]_{\delta m = 0}$$

$$= m_{\rm bare} + E^{\rm stat} + \omega_{\rm kin} E^{\rm kin} + \omega_{\rm spin} E^{\rm spin}, \tag{9.193}$$

$$E^{\rm stat} = - \lim_{x_0 \to \infty} \widetilde{\partial_0} \ln C_{\rm AA}^{\rm stat}(x_0) \Big|_{\delta m = 0}, \tag{9.194}$$

$$E^{\rm kin} = - \lim_{x_0 \to \infty} \widetilde{\partial_0} R_{\rm AA}^{\rm kin}(x_0), \quad E^{\rm spin} = - \lim_{x_0 \to \infty} \widetilde{\partial_0} R_{\rm AA}^{\rm spin}(x_0). \tag{9.195}$$

Again, we have made the dependence on δm explicit through $m_{\rm bare} = m_{\rm b} + \widehat{\delta m}$ and then quantities in the theory with $\delta m = 0$ appear. Note that the ratios $R_{\rm AA}^{x}$ (and therefore $E^{\rm kin}, E^{\rm spin}$) do not depend on δm; the quantities $E^{\rm kin}, E^{\rm spin}$ have mass dimension two and we have already anticipated Eq. (9.200).

The expansion for the decay constant is

$$f_{\rm B} \sqrt{m_{\rm B}} = \lim_{x_0 \to \infty} \big\{ 2 \exp(m_{\rm B} x_0) C_{\rm AA}^{\rm QCD}(x_0) \big\}^{1/2} \tag{9.196}$$

$$= Z_{\rm A}^{\rm HQET} \, \Phi^{\rm stat} \lim_{x_0 \to \infty} \big\{ 1 + \tfrac{1}{2} x_0 \big[\omega_{\rm kin} E^{\rm kin} + \omega_{\rm spin} E^{\rm spin} \big]$$

$$+ \tfrac{1}{2} c_{\rm A}^{(1)} R_{\delta A}^{\rm stat}(x_0) + \tfrac{1}{2} \omega_{\rm kin} R_{\rm AA}^{\rm kin}(x_0) + \tfrac{1}{2} \omega_{\rm spin} R_{\rm AA}^{\rm spin}(x_0) \big\}, \tag{9.197}$$

$$\Phi^{\rm stat} = \lim_{x_0 \to \infty} \big\{ 2 \exp(E^{\rm stat} x_0) C_{\rm AA}^{\rm stat}(x_0) \big\}^{1/2}. $$

Using the transfer-matrix formalism (with normalization $\langle B|B \rangle = 2L^3$), one further observes that (do it as an exercise)

$$E^{\rm kin} = - \frac{1}{2L^3} \langle B| a^3 \sum_{\mathbf{z}} \mathcal{O}_{\rm kin}(0, \mathbf{z}) | B \rangle_{\rm stat} = - \frac{1}{2} \langle B| \mathcal{O}_{\rm kin}(0) | B \rangle_{\rm stat} \tag{9.198}$$

$$E^{\rm spin} = - \frac{1}{2} \langle B| \mathcal{O}_{\rm spin}(0) | B \rangle_{\rm stat}, \tag{9.199}$$

$$0 = \lim_{x_0 \to \infty} \widetilde{\partial_0} R_{\delta A}^{\rm stat}(x_0). \tag{9.200}$$

As expected, only the parameters of the action are relevant in the expansion of hadron masses.

A correct split of the terms in Eq. (9.193) and Eq. (9.197) into leading order and next to leading order pieces that are separately renormalized and that hence *separately have a continuum limit* requires more thought on the renormalization of the $1/m$-expansion. We turn to this now.

9.4.3 Renormalization beyond leading order

For illustration we check the self-consistency of Eq. (9.189). The relevant question concerns renormalization: are the "free" parameters $\delta m \ldots c_A^{(1)}$ sufficient to absorb all divergences on the r.h.s.? We consider the term $\propto C_{AA}^{kin}(x_0)$ since its renormalization displays all subtleties. As a first step, we rewrite $\omega_{kin}\mathcal{O}_{kin} = \frac{1}{2m_R}(\mathcal{O}_{kin})_R$ in terms of a renormalized mass and the renormalized operator

$$(\mathcal{O}_{kin})_R(z) = Z_{\mathcal{O}_{kin}}\left(\mathcal{O}_{kin}(z) + \frac{c_1}{a}\overline{\psi}_h(z)D_0\psi_h(z) + \frac{c_2}{a^2}\overline{\psi}_h(z)\psi_h(z)\right). \quad (9.201)$$

The latter involves a subtraction of lower-dimensional ones with dimensionless coefficients $c_i(g_0)$. The renormalization scheme for m_R is irrelevant, as any change of scheme can be compensated by $Z_{\mathcal{O}_{kin}}, c_i$ whose finite parts need to be fixed by matching to QCD. We further expand

$$(Z_A^{HQET})^2 = (Z_A^{stat})^2 + 2Z_A^{stat}Z_A^{(1/m)} + O(1/m^2), \quad (9.202)$$

which we will discuss more below. With these rules we then have

$$\left(Z_A^{stat}\right)^2 \omega_{kin}C_{AA}^{kin}(x_0) = \frac{1}{2m_R}a^7\sum_{x,z}G(x,z) + \text{subtraction terms}, \quad (9.203)$$

where

$$G(x,z) = \left\langle[A_0^{stat}]_R(x)\,([A_0^{stat}]_R(0))^\dagger\,(\mathcal{O}_{kin})_R(z)\right\rangle_{stat}. \quad (9.204)$$

The subtraction terms are due to the lower-dimensional operators with coefficients c_1 and c_2. Since we are interested in on-shell observables ($x_0 > 0$ in Eq. (9.189)), we may use the equation of motion $D_0\psi_h(z) = 0$ to see that the c_1-term does not contribute, while $\frac{c_2}{a^2}\overline{\psi}_h(z)\psi_h(z)$, is equivalent to a mass shift. In the full correlation function Eq. (9.189) it hence contributes to δm that becomes quadratically divergent when the $1/m$ terms are included.

While $G(x,z)$ is a renormalized correlation function for all physical separations, its integral over z (or on the lattice the continuum limit of the sum over z) does not exist due to singularities at $z \to 0$ and as $z \to x$. These contact-term singularities can be analyzed by the operator product expansion. We discuss them first in the continuum and regulate the short-distance region by just integrating for $z^2 \geq r^2$ with some small r. The operator product expansion then yields

$$\int_{z^2 \geq r^2} d^4z\, G(x,z) \quad (9.205)$$

$$\overset{r\to 0}{\sim}\left\langle[A_0^{stat}]_R(x)\,[d_1''\frac{1}{r}\,(A_0^{stat}(0))^\dagger + d_2''(A_0^{(1)}(0))^\dagger + d_3''(A_0^{(2)}(0))^\dagger]\right\rangle_{stat}$$

up to terms that are finite as $r \to 0$. The coefficients d_i'' in the operator product expansion have a further logarithmic dependence on r.[26] For (the continuum version of) Eq. (9.204) we need $r \to 0$. In this limit short-distance divergences emerge that have to be subtracted by counterterms. In the lattice regularization, short-distance singularities are regulated by the lattice spacing a and we have in full analogy

$$\left\langle [A_0^{\text{stat}}]_{\text{R}}(x)\left[a^4 \sum_z ([A_0^{\text{stat}}]_{\text{R}}(0))^\dagger \left(\mathcal{O}_{\text{kin}} \right)_{\text{R}}(z) \right] \right\rangle_{\text{stat}} \qquad (9.206)$$

$$\overset{a\to 0}{\sim}\left\langle [A_0^{\text{stat}}]_{\text{R}}(x)\left[d_1' \frac{1}{a}\left(A_0^{\text{stat}}(0) \right)^\dagger + d_2'(A_0^{(1)}(0))^\dagger + d_3'(A_0^{(2)}(0))^\dagger \right] \right\rangle_{\text{stat}}$$

up to terms that have a continuum limit $a \to 0$ and up to the singular terms originating from $z \approx x$. The coefficients d_i contain a logarithmic dependence on a. Treating the singular terms at $z \approx x$ in the same way and noting that the term with $A_0^{(2)}(0)$ vanishes upon summation over $\mathbf{x}$ we find

$$Z_{\text{A}}^{\text{stat}}\left[d_1 \frac{1}{a} C_{\text{AA}}^{\text{stat}}(x_0) + d_2 C_{\delta\text{A}}^{\text{stat}}(x_0) \right] \qquad (9.207)$$

for the contact-term singularities in Eq. (9.203). These are absorbed in Eq. (9.189) through counterterms contained in $Z_{\text{A}}^{\text{HQET}}$ and $c_{\text{A}}^{(1)}$,

$$2Z_{\text{A}}^{(1/m)} = -\frac{d_1}{2am_{\text{R}}} + \dots, \qquad c_{\text{A}}^{(1)} = -\frac{d_2}{2m_{\text{R}} Z_{\text{A}}^{\text{stat}}} + \dots. \qquad (9.208)$$

The change from d_i' to d_i is due to the use of the equation of motion above. This step is valid only up to contact terms, resulting in the shift $d' \to d$. The ellipses contain the physical, finite $1/m$ terms.

We now comment further on the expansion Eq. (9.202). Our discussion shows that the quadratic term $(Z_{\text{A}}^{(1/m)})^2$ in Eq. (9.202) *must be dropped*; otherwise an uncancelled $1/(a^2 m^2)$ divergence remains. As we have seen, there is no $1/(a^2 m^2)$ in $C_{\text{AA}}^{\text{kin}}(x_0)$ and the other pieces in Eq. (9.189) are less singular. This is just a manifestation of the general rule of an effective-field theory that all quantities are to be expanded in $1/m$ whether they are divergent or not. With this rule the various HQET parameters can be determined such that they absorb all divergences.[27]

The lesson of our discussion is that counterterms with the correct structure are automatically present because in the effective theory all the relevant local composite fields are included with free coefficients. These free parameters may thus be chosen such that the continuum limit of the HQET correlation functions exists. Finally, their finite parts are to be determined such that the effective theory yields the $1/m$ expansion of the QCD observables.

9.4.4 The need for non-perturbative conversion functions

An important step remains to be explained: the determination of the HQET parameters. As discussed in Section 9.3 at the leading order in $1/m$, this can be done with the help of perturbation theory for conversion functions such as C_{PS}. However, as soon as a $1/m$ correction is to be included, the leading-order conversion functions have to be known non-perturbatively. This general feature in the determination of power corrections in QCD is seen in the following way. Consider the error made in

Eq. (9.138), when the anomalous dimension has been computed at l loops and C_{match} at $l-1$ loop order. The conversion function

$$C_{\text{PS}} = \exp\left\{-\int^{g_*} dx \frac{\gamma_0 x^2 + \ldots + \gamma_{l-1}^{\text{match}} x^{2l}}{\beta(x)}\right\} + \Delta(C_{\text{PS}}) \qquad (9.209)$$

is then known up to a relative *error*

$$\frac{\Delta(C_{\text{PS}})}{C_{\text{PS}}} \propto [\bar{g}^2(m)]^l \sim \left\{\frac{1}{2b_0 \ln(m/\Lambda_{\text{QCD}})}\right\}^l \overset{m \to \infty}{\gg} \frac{\Lambda_{\text{QCD}}}{m}. \qquad (9.210)$$

As m is made large, this perturbative error becomes dominant over the power correction one wants to determine. Taking a perturbative conversion function and adding power corrections to the leading-order effective theory is thus a phenomenological approach, where one assumes that for example at the b-quark mass, the coefficient of the $[\bar{g}^2(m_b)]^l$ term (as well as higher-order ones) is small, such that the Λ/m_b corrections dominate. In such a phenomenological determination of a power correction, its size depends on the order of perturbation theory considered. A theoretically consistent evaluation of power corrections requires a fully non-perturbative formulation of the theory including a non-perturbative matching to QCD. Note that the essential point of Eq. (9.210) is not the expected factorial growth of the coefficients of the perturbative expansion. Rather, it is due to the truncation of perturbation theory as such. Of course, a renormalon-like growth of the coefficients does not help.

The foregoing discussion is completely generic, applying to any regularization. When we define the theory on the lattice, there are in addition power divergences, e.g. in Eq. (9.206). It is well known that they have to be subtracted non-perturbatively if one wants the continuum limit to exist.

9.4.5 Splitting leading order (LO) and next-to-leading order (NLO)

We just learned that the very definition of a NLO correction to f_B means to take Eq. (9.197) with all coefficients $Z_A^{\text{HQET}} \ldots c_A^{(1)}$ determined non-perturbatively. We want to briefly explain that, as a consequence, the split between LO and NLO is not unique. This is fully analogous to the case of standard perturbation theory in α, where the split between different orders depends on the renormalization scheme used, and on the experimental observable used to determine α in the first place.

Consider the lowest order. The only coefficient needed in Eq. (9.197) is then $Z_A^{\text{HQET}} = C_{\text{PS}} Z_{A,\text{RGI}}^{\text{stat}}$. It has to be fixed by matching some matrix element of A_0^{stat} to the matrix element of A_0 in QCD. For example, one may choose $\langle B'|A_0^\dagger|0\rangle$, with $|B'\rangle$ denoting some other state such as an excited pseudoscalar state. Or one may take a finite-volume matrix element defined through the Schrödinger functional as we will do later. Since the matching involves the QCD matrix element, there are higher order in $1/m$ "pieces" in these equations. There is no reason for them to be independent of the particular matrix element. So, from matching condition to matching condition, $C_{\text{PS}} Z_{A,\text{RGI}}^{\text{stat}}$ determined at the leading order in $1/m$ differs by $O(\Lambda_{\text{QCD}}/m_b)$ terms.

The matrix element f_B in static approximation inherits this $O(\Lambda_{QCD}/m_b)$ ambiguity. These corrections are hence not unique. Fixing a matching condition, the leading order f_B as well as the one including the corrections can be computed and have a continuum limit. Their difference can be defined as the $1/m$ correction. However, what matters is not the ambiguous NLO term, but the fact that the uncertainty is reduced from $O(\Lambda_{QCD}/m_b)$ in the LO term to $O(\Lambda_{QCD}^2/m_b^2)$ in the sum.

The following table illustrates the point explicitly.

Observables	$\langle B|A_0^\dagger|0\rangle$	$\langle B'|A_0^\dagger|0\rangle$	$\langle B''|A_0^\dagger|0\rangle$
matching condition	*		
error in HQET result	0	$O(\Lambda/m_b)$	$O(\Lambda/m_b)$
matching condition		*	
error in HQET result	$O(\Lambda/m_b)$	0	$O(\Lambda/m_b)$
matching condition			*
error in HQET result	$O(\Lambda/m_b)$	$O(\Lambda/m_b)$	0

As a consequence, there is no strict meaning to the statement "*the $1/m$ correction to f_B is 10%*".

9.4.6 Mass formulae

Often-cited mass formulae are

$$m_B^{\text{av}} \equiv \frac{1}{4}[m_B + 3m_{B^*}] = m_b + \overline{\Lambda} + \frac{1}{2m_b}\lambda_1 + O(1/m_b^2) \qquad (9.211)$$

$$\Delta m_B \equiv m_{B^*} - m_B = -\frac{2}{m_b}\lambda_2 + O(1/m_b^2) \qquad (9.212)$$

with (ignoring renormalization)

$$\lambda_1 = \langle B|\mathcal{O}_{\text{kin}}|B\rangle, \quad \lambda_2 = \frac{1}{3}\langle B|\mathcal{O}_{\text{spin}}|B\rangle. \qquad (9.213)$$

The quantity $\overline{\Lambda}$ is termed the "static binding energy". Also here, depending on how one formulates the matching condition that determines m_b, one changes $\overline{\Lambda}$ by a term of order Λ_{QCD}, e.g. one may define $\overline{\Lambda} = 0$. Similarly, the kinetic term $\lambda_1/(2m_b)$ has a non-perturbative matching-scheme dependence of order Λ_{QCD} and thus λ_1 itself has a matching-scheme dependence of order m_b. The situation for $\overline{\Lambda}$ is similar to the gluon "condensate". The non-perturbative scheme dependence has the same size as the gluon "condensate" itself. In contrast, λ_2 is the leading term in the $1/m$ expansion and does not have such an ambiguity. We refer also to the more detailed discussion in (Sommer, 2006).

9.4.7 Non-perturbative determination of HQET parameters

We close our theoretical discussion of HQET by stating the correct procedure to determine the N_{HQET} parameters in the effective theory at a certain order in $1/m$. One requires

$$\Phi_i^{\text{QCD}}(m) = \Phi_i^{\text{HQET}}(m, a), \quad i = 1 \ldots N_{\text{HQET}}, \tag{9.214}$$

where the m dependence on the r.h.s. is entirely inside the HQET parameters. On the l.h.s. the continuum limit in QCD is assumed to have been taken, but the r.h.s. refers to a given lattice spacing where it defines the bare parameters of the theory at that value of a. We emphasize that as this matching has to be invoked by numerical data, it is done at a given finite value of $1/m$. Carrying it out with just the static parameters defines the static approximation, etc.

As simple as it is written down, it is non-trivial to implement Eq. (9.214) in practice such that

1) the HQET expansion is accurate and one may thus truncate at a given order;
2) the numerical precision is sufficient;
3) lattice spacings are available for which large-volume computations of physical matrix elements can be performed.

In the following section we explain how these criteria can be satisfied using Schrödinger functional correlation functions and a step-scaling method. The first part will be a test of HQET on some selected correlation functions. This establishes how 1) and 2) can be met. We can then explain the complete strategy that also achieves 3).

9.4.8 Relation to RGI matrix elements and conversion functions

The matching equations Eq. (9.214) provide a definition of all HQET parameters, in principle at any given order in the expansion. If considered at the static order, it also provides the renormalization of the static axial current, which we discussed at length in Section 9.3. The relation between the two ways of parametrizing the current in the static approximation are

$$Z_A^{\text{HQET}} = Z_{A,\text{RGI}}^{\text{stat}} \, C_{\text{PS}}(M/\Lambda) + \text{O}(1/m, a). \tag{9.215}$$

While Eq. (9.214) is a matching equation determining directly the product $Z_{A,\text{RGI}}^{\text{stat}} C_{\text{PS}}$, the r.h.s. separates the problem into a pure HQET problem, the determination of the RGI operator, and a pure QCD problem, the "anomalous dimension" γ_{match}, see Eq. (9.149). Note that in this simple form, such a separation is only possible at the lowest order in $1/m$.

Since the breaking of spin symmetry is due to a single operator at order $1/m$, there is also an analogous representation of ω_{spin}. We refer the interested reader to (Guazzini *et al.*, 2007).

9.5 Non-perturbative HQET

After our long discussion of the theoretical issues in the renormalization of HQET, we turn to a complete strategy for the non-perturbative implementation. To this end the three criteria in Section 9.4.7 have to be fulfilled. Establishing 1) is equivalent to testing HQET. We therefore start with such a test. Item 2) has to do with finding matching conditions sensitive to the $1/m$-suppressed contributions. For this purpose

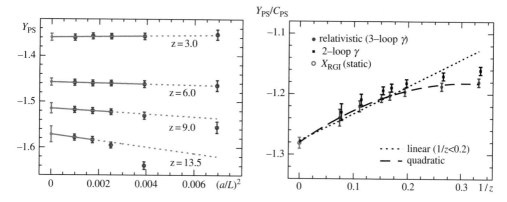

Fig. 9.7 Testing Eq. (9.221) through numerical simulations in the quenched approximation and for $L \approx 0.2$ fm (Heitger *et al.*, 2004). (The graph uses notation $Y_{\mathrm{R}}^{\mathrm{QCD}} \equiv Y_{\mathrm{PS}}$). The physical mass of the b-quark corresponds to $z \approx 6$. Two different orders of perturbation theory for C_{PS} are shown.

we then expand a little on correlation functions in the Schrödinger functional before coming to a full description of the matching strategy.

9.5.1 Non-perturbative tests of HQET

Although it is generally accepted that HQET is an effective theory of QCD, tests of this equivalence are rare and mostly based on phenomenological analysis of experimental results. A pure theory test can be performed if QCD including a heavy-enough quark can be simulated on the lattice at lattice spacings that are small enough to be able to take the continuum limit. This has been done in the last few years (Heitger *et al.*, 2004; Della Morte *et al.*, 2008) and will be summarized below.

We start with the QCD side of such a test. Lattice spacings such that $am_{\mathrm{b}} \ll 1$ can be reached if one puts the theory in a finite volume, $L^3 \times T$ with L, T not too large. We shall use $T = L$. For various practical reasons, Schrödinger functional boundary conditions are chosen. Equivalent boundary conditions are imposed in the effective theory. As in Section 9.3.3 we consider the ratio $Y_{\mathrm{R}}^{\mathrm{QCD}}(\theta, m_{\mathrm{R}}, L)$ built from the correlation functions f_{A} and f_1.

It can be written as

$$Y_{\mathrm{R}}^{\mathrm{QCD}}(\theta, m_{\mathrm{R}}, L) = \frac{\langle \Omega(L)|A_0|B(L)\rangle}{||\,|\Omega(L)\rangle\,||\;||\,|B(L)\rangle\,||}, \tag{9.216}$$

$$|B(L)\rangle = \mathrm{e}^{-L\mathbb{H}/2}|\varphi_{\mathrm{B}}(L)\rangle, \; |\Omega(L)\rangle = \mathrm{e}^{-L\mathbb{H}/2}|\varphi_0(L)\rangle,$$

in terms of the boundary states $|\varphi_{\mathrm{B}}(L)\rangle$, $|\varphi_0(L)\rangle$. Expanded in energy eigenstates with energies $E_n \geq m_{\mathrm{B}}$ in the B-sector and energies $\tilde{E}_n$ in the vacuum sector, we have

$$|B(L)\rangle = \sum_n e^{-LE_n/2} \langle n, B | \varphi_{\rm B}(L) \rangle \, |n, B\rangle \tag{9.217}$$

$$\sim \sum_{n \,|\, E_n - m_{\rm B} < k/L} e^{-LE_n/2} \langle n, B | \varphi_{\rm B}(L) \rangle \, |n, B\rangle + {\rm O}(e^{-k/2}), \tag{9.218}$$

$$|\Omega(L)\rangle = \sum_n e^{-L\tilde{E}_n/2} \langle n, 0 | \varphi_0(L) \rangle \, |n, 0\rangle, \tag{9.219}$$

$$\sim \sum_{n \,|\, \tilde{E}_n < k/L} e^{-L\tilde{E}_n/2} \langle n, 0 | \varphi_0(L) \rangle \, |n, 0\rangle + {\rm O}(e^{-k/2}), \tag{9.220}$$

which shows that only energy eigenstates with $E_n - E_0 = {\rm O}(1/L)$ contribute significantly. For $z = LM_{\rm b} \gg 1$, HQET will thus describe the correlation functions and the ratio $Y_{\rm R}^{\rm QCD}$. We come to the conclusion that

$$Y_{\rm R}^{\rm QCD}(\theta, m_{\rm R}, L) = C_{\rm PS}(M_{\rm b}/\Lambda) \, X_{\rm RGI} + {\rm O}(1/z), \quad z = M_{\rm b}L, \tag{9.221}$$

$$X_{\rm RGI} = \lim_{a \to 0} Z_{\rm A, RGI}^{\rm stat}(g_0) \frac{f_{\rm A}^{\rm stat}(L/2, \theta)}{\sqrt{f_1^{\rm stat}(\theta)}}, \tag{9.222}$$

and similarly for other observables. Note that one could also just argue that the only relevant scales are $L, \Lambda, m_{\rm b}$. Therefore, with $L \approx 1/\Lambda$ there is a $\Lambda/m_{\rm b} \sim 1/z$ expansion.

Of course relations such as Eq. (9.221) are expected after the continuum limit of both sides has been taken separately. For the case of $Y_{\rm R}^{\rm QCD}$, this is done by the following steps:

- Fix a value u_0 for the renormalized coupling $\bar{g}^2(L)$ (in the Schrödinger functional scheme) at vanishing quark mass. In (Heitger *et al.*, 2004) u_0 was chosen such that $L \approx 0.2\,{\rm fm}$.
- For a given resolution L/a, determine the bare coupling from the condition $\bar{g}^2(L) = u_0$. This step is well known by now (Capitani *et al.*, 1999).
- Fix the bare-quark mass $m_{\rm q}$ of the heavy quark such that $LM = z$ using the known renormalization factors $Z_{\rm M}, Z$ in $M = Z_{\rm M} Z \, (1 + ab_{\rm m}m_{\rm q}) \, m_{\rm q}$, where $Z, Z_{\rm M}, b_{\rm m}$ are all known non-perturbatively (Guagnelli *et al.*, 2001; Della Morte *et al.*, 2007*a*).
- Evaluate $Y_{\rm R}^{\rm QCD}$ and repeat for better resolution a/L.
- Extrapolate to the continuum as shown in Fig. 9.7, left.

In the effective theory the same steps are followed. As a simplification, no quark mass needs to be fixed and the continuum extrapolation is much easier, as illustrated in Fig. 9.8.

The comparison of the static result and the relativistic theory, Fig. 9.7, looks rather convincing,[28] but we note that the b-quark mass point is $1/z = 1/z_{\rm b} \approx 0.17$, where $1/z^2$ terms are not completely negligible. The displayed fit has an 8% contribution by the $1/z$ term and a 2% $1/z^2$ piece.

For a precision application (Section 9.5.3) it is thus safer to have $L \gtrsim 0.4\,{\rm fm}$ instead of the $L = 0.2\,{\rm fm}$ chosen in the first test, reducing $1/z^2$ by a factor four. For $L \approx 0.5\,{\rm fm}$ we show two different examples, Fig. 9.9, and Fig. 9.10 that involve

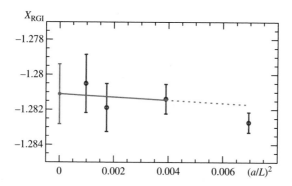

Fig. 9.8 Continuum extrapolation of X_{RGI} (Heitger *et al.*, 2004).

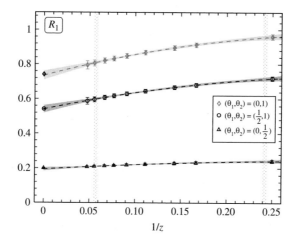

Fig. 9.9 The logarithmic ratio R_1 for different pairs (θ_1, θ_2) with $N_{\text{f}} = 2$ flavors and for $L = T \approx 0.5\,\text{fm}$ (Della Morte *et al.*, 2008) with $N_{\text{f}} = 2$. The value of $1/z$ for charm and bottom quarks are indicated by the vertical bands.

$$k_1(\theta) = -\frac{a^{12}}{6L^6} \sum_{\mathbf{u},\mathbf{v},\mathbf{y},\mathbf{z},k} \left\langle \overline{\zeta_1}{}'(\mathbf{u})\gamma_k\zeta_{\text{b}}'(\mathbf{v})\, \overline{\zeta}_{\text{b}}(\mathbf{y})\gamma_k\zeta_{\text{l}}(\mathbf{z}) \right\rangle \tag{9.223}$$

in addition to the previously introduced correlation functions. The considered combinations are

$$R_1 = \frac{1}{4}\left(\ln\left(\frac{f_1(\theta_1)k_1(\theta_1)^3}{f_1(\theta_2)k_1(\theta_2)^3} \right) \right) \tag{9.224}$$

$$\widetilde{R}_1 = \frac{3}{4}\ln\left(\frac{f_1}{k_1} \right). \tag{9.225}$$

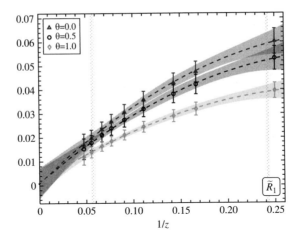

Fig. 9.10 The logarithmic ratio $\widetilde{R}_1$ for different values θ_0 with $N_{\mathrm{f}} = 2$ flavors and for $L = T \approx 0.5\,\mathrm{fm}$ (Della Morte *et al.*, 2008) with $N_{\mathrm{f}} = 2$.

Their HQET expansion contains no conversion functions at leading order and they are thus free of the associated perturbative uncertainty. While R_1 has a finite static limit, $\widetilde{R}_1$ vanishes as $z \to \infty$ due to the spin symmetry. The expected HQET behavior is confirmed with surprisingly small $1/z^2$ corrections for a charm quark. The quadratic fits in $1/z$ displayed in the figures are not constrained to pass through the separately displayed static limit.

9.5.2 HQET expansion of Schrödinger functional correlation functions

In complete analogy to the case of a manifold without boundary we can write down the expansions of the Schrödinger functional correlation functions to first order in $1/m$:

$$[f_{\mathrm{A}}]_{\mathrm{R}} = Z_{\mathrm{A}}^{\mathrm{HQET}} Z_{\zeta_{\mathrm{h}}} Z_{\zeta} e^{-m_{\mathrm{bare}}x_0} \left\{ f_{\mathrm{A}}^{\mathrm{stat}} + c_{\mathrm{A}}^{(1)} f_{\delta\mathrm{A}}^{\mathrm{stat}} + \omega_{\mathrm{kin}} f_{\mathrm{A}}^{\mathrm{kin}} + \omega_{\mathrm{spin}} f_{\mathrm{A}}^{\mathrm{spin}} \right\}, \quad (9.226)$$

$$[f_1]_{\mathrm{R}} = Z_{\zeta_{\mathrm{h}}}^2 Z_{\zeta}^2 e^{-m_{\mathrm{bare}}T} \left\{ f_1^{\mathrm{stat}} + \omega_{\mathrm{kin}} f_1^{\mathrm{kin}} + \omega_{\mathrm{spin}} f_1^{\mathrm{spin}} \right\}, \quad (9.227)$$

$$[k_1]_{\mathrm{R}} = Z_{\zeta_{\mathrm{h}}}^2 Z_{\zeta}^2 e^{-m_{\mathrm{bare}}T} \left\{ f_1^{\mathrm{stat}} + \omega_{\mathrm{kin}} f_1^{\mathrm{kin}} - \frac{1}{3}\omega_{\mathrm{spin}} f_1^{\mathrm{spin}} \right\}. \quad (9.228)$$

Apart from

$$f_{\delta\mathrm{A}}^{\mathrm{stat}}(x_0, \theta) = -\frac{a^6}{2} \sum_{\mathbf{y}, \mathbf{z}} \left\langle A_0^{(1)}(x) \, \overline{\zeta}_{\mathrm{h}}(\mathbf{y}) \gamma_5 \zeta_{\mathrm{l}}(\mathbf{z}) \right\rangle \quad (9.229)$$

the labelling of the different terms follows directly the one introduced in Eq. (9.180). The relation between the $1/m$ terms in f_1 and k_1 is a simple consequence of the spin symmetry of the static action, valid at any lattice spacing. A further simplicity is that no $1/m$ boundary corrections are present. Potentially, such terms have dimension four. After using the equations of motion, only one candidate remains, which, however, does not contribute to any correlation function.[29]

9.5.3 Strategy for non-perturbative matching

After the tests of HQET described above, it is clear how one can non-perturbatively match HQET to QCD. Consider the action as well as A_0 (just at $\mathbf{p} = 0$) and denote the free parameters of the effective theory by ω_i, $i = 1 \ldots N_{\mathrm{HQET}}$. In the static approximation we then have

$$\omega^{\mathrm{stat}} = (m_{\mathrm{bare}}^{\mathrm{stat}}, \, [\ln(Z_{\mathrm{A}})]^{\mathrm{stat}})^t, \quad N_{\mathrm{HQET}} = 2 \tag{9.230}$$

and including the first-order terms in $1/m$ together with the static ones, the HQET parameters are

$$\omega^{\mathrm{HQET}} = (m_{\mathrm{bare}}, \, \ln(Z_{\mathrm{A}}^{\mathrm{HQET}}), \, c_{\mathrm{A}}^{(1)}, \, \omega_{\mathrm{kin}}, \, \omega_{\mathrm{spin}})^t \quad N_{\mathrm{HQET}} = 5. \tag{9.231}$$

The pure $1/m$ parameters may be defined as $\omega^{(1/m)} = \omega^{\mathrm{HQET}} - \omega^{\mathrm{stat}}$, with all of them, e.g. also $m_{\mathrm{bare}}^{(1/m)}$, non-zero. In fact, our discussion of renormalization of the $1/m$ terms shows that $m_{\mathrm{bare}}^{(1/m)}$ diverges as $1/(a^2 m)$.

With suitable observables

$$\Phi_i(L_1, M, a), \, i = 1 \ldots N_{\mathrm{HQET}},$$

in a Schrödinger functional with $L = T = L_1 \approx 0.5\,\mathrm{fm}$, we then require matching[30]

$$\Phi_i(L_1, M, a) = \Phi_i^{\mathrm{QCD}}(L_1, M, 0), \, i = 1 \ldots N_{\mathrm{HQET}}. \tag{9.232}$$

Note that the continuum limit is taken in QCD, while in HQET we want to extract the bare parameters of the theory from the matching equation and thus have a finite value of a. It is convenient to pick observables with HQET expansions linear in ω_i,

$$\Phi(L, M, a) = \eta(L, a) + \phi(L, a)\,\omega(M, a), \tag{9.233}$$

in terms of a $N_{\mathrm{HQET}} \times N_{\mathrm{HQET}}$ coefficient matrix ϕ. A natural choice for the first two observables is

$$\Phi_1 = L\Gamma^{\mathrm{P}} \equiv -L\tilde{\partial}_0 \ln(-f_{\mathrm{A}}(x_0))_{x_0 = L/2} \overset{L \to \infty}{\sim} Lm_{\mathrm{B}} \tag{9.234}$$

$$\Phi_2 = \ln(Z_{\mathrm{A}} \frac{-f_{\mathrm{A}}}{\sqrt{f_1}}) \overset{L \to \infty}{\sim} L^{3/2} f_{\mathrm{B}} \sqrt{m_{\mathrm{B}}/2}, \tag{9.235}$$

since in the static approximation these determine directly ω_1 and ω_2. We will introduce the other Φ_i later. The explicit form of η, ϕ is

$$\eta = \begin{pmatrix} \Gamma^{\mathrm{stat}} \\ \zeta_{\mathrm{A}} \\ \cdots \end{pmatrix}, \quad \phi = \begin{pmatrix} L & 0 & \cdots \\ 0 & 1 & \cdots \\ \cdots & & \end{pmatrix} \tag{9.236}$$

with

$$\Gamma^{\text{stat}} = -L\widetilde{\partial_0}\ln(f_{\text{A}}^{\text{stat}}(x_0))_{x_0=L/2}, \quad \zeta_{\text{A}} = \ln(\frac{-f_{\text{A}}^{\text{stat}}}{\sqrt{f_1^{\text{stat}}}}). \tag{9.237}$$

In the static approximation, the structure of the matrix ϕ is perfect: one observable determines one parameter. This is possible since there is no (non-trivial) mixing at that order.

Having specified the matching conditions, the HQET parameters $\omega_i(M,a)$ can be obtained from Eqs.(9.232, and 9.233), but only for rather small lattice spacings since a reasonable suppression of lattice artifacts requires $L_1/a = \text{O}(10)$ and thus $a = \text{O}(0.05\,\text{fm})$.

Larger lattice spacings as needed in large volume, can be reached by adding a step-scaling strategy, illustrated in Fig. 9.11. Let us now go through the various steps of this strategy.

(1) Take the continuum limit

$$\Phi_i^{\text{QCD}}(L_1, M, 0) = \lim_{a/L_1 \to 0} \Phi_i^{\text{QCD}}(L_1, M, a). \tag{9.238}$$

This is similar to the HQET tests and as we saw there, it requires $L_1/a = 20\ldots40$, or $a = 0.025 - 0.012\,\text{fm}$.

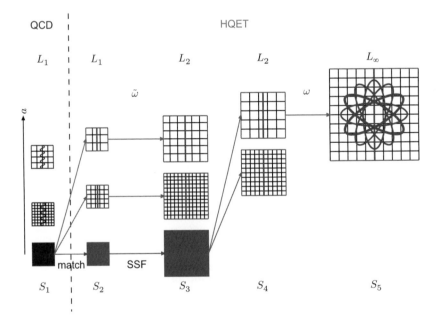

Fig. 9.11 Strategy for non-perturbative HQET (Blossier *et al.*, 2010*b*). Note that in the realistic implementation (Blossier *et al.*, 2010*b*) finer resolutions are used.

(2a) Set the HQET observables equal to the QCD ones, Eq. (9.232) and extract the parameters

$$\tilde{\omega}(M, a) \equiv \phi^{-1}(L_1, a) \left[\Phi(L_1, M, 0) - \eta(L_1, a) \right] \tag{9.239}$$

$$= \begin{pmatrix} L_1^{-1} \Phi_1(L_1, M, 0) - \Gamma^{\text{stat}}(L_1, a) \\ \Phi_2(L_1, M, 0) - \zeta_A(L_1, a) \\ \cdots \end{pmatrix}. \tag{9.240}$$

The only restriction here is $L_1/a \gg 1$, so one can use $L_1/a = 10 \ldots 20$, which means $a = 0.05 - 0.025 \, \text{fm}$.

(2b.) Insert $\tilde{\omega}$ into $\Phi(L_2, M, a)$:

$$\Phi(L_2, M, 0) = \lim_{a/L_2 \to 0} \left\{ \eta(L_2, a) + \phi(L_2, a) \, \tilde{\omega}(M, a) \right\} \tag{9.241}$$

$$= \lim_{a/L_2 \to 0} \begin{pmatrix} L_2 \Gamma^{\text{stat}}(L_2, a) + \frac{L_2}{L_1} \Phi_1(L_1, M, 0) - L_2 \Gamma^{\text{stat}}(L_1, a) \\ \zeta_A(L_2, a) + \Phi_2(L_1, M, 0) - \zeta_A(L_1, a) \\ \cdots \end{pmatrix}$$

$$= \lim_{a/L_2 \to 0} \underbrace{\begin{pmatrix} L_2 [\Gamma^{\text{stat}}(L_2, a) - \Gamma^{\text{stat}}(L_1, a)] \\ \zeta_A(L_2, a) - \zeta_A(L_1, a) \\ \cdots \end{pmatrix}}_{\text{finite HQET SSF's}} + \underbrace{\begin{pmatrix} \frac{L_2}{L_1} \Phi_1(L_1, M, 0) \\ \Phi_2(L_1, M, 0) \\ \cdots \end{pmatrix}}_{\text{QCD, mass dependence}}.$$

In the last line we have identified pieces that are separately finite. This step can be done as long as the lattice spacing is common to the $n_2 = L_2/a$ and $n_1 = L_1/a$ lattices and

$$s = L_2/L_1 = n_2/n_1 \tag{9.242}$$

is kept at a fixed, small, ratio.[31]

(3.) Repeat (2a.) for $L_1 \to L_2$:

$$\omega(M, a) \equiv \phi^{-1}(L_2, a) \left[\Phi(L_2, M, 0) - \eta(L_2, a) \right]. \tag{9.243}$$

With the same resolutions $L_2/a = 10 \ldots 20$ one has now reached $a = 0.1 - 0.05 \, \text{fm}$.

(4.) Finally, insert ω into the expansion of large-volume observables, e.g.

$$m_B = \omega_1 + E^{\text{stat}}. \tag{9.244}$$

In the chosen example the result is the relation between the RGI b-quark mass and the B-meson mass m_B. It is illustrative to put the different steps into one equation,

$$(9.245)$$

$$m_B =$$

$$\lim_{a \to 0} \left[E^{\text{stat}} - \Gamma^{\text{stat}}(L_2, a) \right] \qquad a = 0.1 - 0.05 \,\text{fm} \qquad [S_4, S_5]$$

$$+ \lim_{a \to 0} \left[\Gamma^{\text{stat}}(L_2, a) - \Gamma^{\text{stat}}(L_1, a) \right] \qquad a = 0.05 - 0.025 \,\text{fm} \qquad [S_2, S_3]$$

$$+ \frac{1}{L_1} \lim_{a \to 0} \Phi_1(L_1, M_b, a) \qquad a = 0.025 - 0.012 \,\text{fm}. \qquad [S_1].$$

We have indicated the lattices drawn in Fig. 9.11 and the typical lattice spacings of these lattices. The explicit expression for the decay constant in static approximation is even simpler; write it down as an exercise!

So far we have spelled out only those observables that are needed in the static approximation. The following heuristics helps to find observables suitable for the determination of the $1/m$-terms. Recall that $\theta \ne 0$ means $\frac{1}{2}(\nabla_j + \nabla_j^*) \sim i\theta/L$ (acting onto a quark field) when the gauge fields are weak, as is the case in small volume. Hence, expanding in $1/m$

$$\Phi_3(L, M, a) = \frac{f_A(\theta_1)}{f_A(\theta_2)} \sim \ldots + c_A^{(1)} \left[\theta_2 - \theta_1 \right]/L \qquad (9.246)$$

for weakly coupled quarks. In the same way the combination (recall Eq. (9.224))

$$\Phi_4(L, M, a) = R_1 = R_1^{\text{stat}} + \omega_{\text{kin}} R_1^{\text{kin}}$$

has a sensitivity to ω_{kin} of $R_1^{\text{kin}} \propto \theta_1^2 - \theta_2^2$, while in the specific linear combination of f_1 and k_1 that form R_1 the parameter ω_{spin} drops out. Finally, the choice

$$\Phi_5(L, M, a) = \tilde{R}_1 = \omega_{\text{spin}} R_1^{\text{spin}} \qquad (9.247)$$

allows for a direct determination of ω_{spin}. These choices leave relatively many zeros in the matrix ϕ, which has a block structure,

$$\phi = \begin{pmatrix} C & B \\ 0 & A \end{pmatrix}, \quad \phi^{-1} = \begin{pmatrix} C^{-1} & -C^{-1}BA^{-1} \\ 0 & A^{-1} \end{pmatrix}, \quad C = \begin{pmatrix} L & 0 \\ 0 & 1 \end{pmatrix}. \qquad (9.248)$$

The listed observables Φ_i have been shown to work in practice, i.e. in a numerical application (Blossier *et al.*, 2010*b*).

9.5.4 Numerical computations in the effective theory

Before showing some results, we should briefly mention that it is not entirely straight-forward to obtain precise numerical results in the effective theory. The reason is a generically rather strong growth of statistical errors as a function of the Euclidean time separation of the correlation functions. Two ideas help to overcome this problem. We sketch them here; more details are available in the cited literature.

9.5.4.1 The static action

Consider a typical two-point function, for example Eq. (9.134). At large time it decays exponentially and so does the variance. Setting $\delta m = 0$ the decay of the signal is

$$C(x_0) \sim \mathrm{e}^{-E_{\mathrm{stat}} x_0}, \tag{9.249}$$

while the variance decays with an exponential rate given by the pion mass. Thus, the noise-to-signal ratio for the B-meson correlation function behaves as

$$R_{\mathrm{NS}} \propto \mathrm{e}^{[E_{\mathrm{stat}} - m_\pi/2] x_0} . \tag{9.250}$$

The self-energy of a static quark is power divergent, in particular in perturbation theory

$$E^{\mathrm{stat}} \sim \left(\frac{1}{a} r^{(1)} + \mathrm{O}(a^0) \right) g_0^2 + \mathrm{O}(g_0^4). \tag{9.251}$$

This divergence yields the leading behavior of Eq. (9.250) for small a. It is potentially dangerous since we are interested in the continuum limit. The scale of the problem can be reduced considerably by the replacement

$$U(x,0) \rightarrow W_{\mathrm{HYPi}}(x,0), \tag{9.252}$$

in the covariant derivative ∇_0^* in the static action. Here, W_{HYPi} is a so-called HYP-smeared link. Table 9.1 shows how the self-energy is reduced for two choices of W_{HYPi}.

It is mandatory to check that such a change of action does not introduce large cutoff effects. This was done for single smearing in (Della Morte *et al.*, 2005): the points with smallest error bars in Fig. 9.1 are for these actions. We expect that large cutoff effects would, however, appear if smearing was repeated several times.

Table 9.1 One-loop coefficients $r^{(1)}$, Eq. (9.251) and non-perturbative values for aE_{stat} at $\beta = 6/g_0^2 = 6$ and a (quenched) light quark with the mass of the strange quark. "EH" refers to Eichten–Hill, i.e. $W(x,0) = U(x,0)$, while "HYP1,HYP2" are two versions of HYP-smearing (Hasenfratz and Knechtli, 2001; Della Morte *et al.*, 2005).

$S_{\mathrm{h}}^{\mathrm{W}}$	$r^{(1)}$	aE_{stat}
$S_{\mathrm{h}}^{\mathrm{EH}}$	0.16845(2)	0.68(9)
$S_{\mathrm{h}}^{\mathrm{HYP1}}$	0.04844(1)	0.44(2)
$S_{\mathrm{h}}^{\mathrm{HYP2}}$	0.03523(1)	0.41(1)

9.5.4.2   Generalized eigenvalue method

For the numerical evaluation of matrix elements such as Φ^{stat}, Eq. (9.57), or of energy levels it is advisable to use an improvement over the straightforward formula Eq. (9.136). The reason is as follows. Let us label the energies in the sector contributing to a given correlation function by E_n, $n = 1, 2, 3$. Then, there are corrections to the desired ground-state matrix element due to excited-state contaminations of order $e^{-x_0 \Delta}$ and $\Delta = E_2 - E_1$. From an investigation of the spectrum in the B-meson sector one finds numerically $\Delta \approx 600\,\text{MeV}$ and thus $\Delta\, x_0 \approx 3x_0/\text{fm}$. The suppression of excited-state contaminations is then not necessarily good enough for $x_0 \sim 1\,\text{fm}$ but using Eq. (9.136) beyond $x_0 \sim 1\,\text{fm}$ is very difficult because statistical errors grow quite rapidly with x_0.

A considerable improvement is achieved if one considers the generalized eigenvalue problem (GEVP) (Michael and Teasdale, 1983; Lüscher and Wolff, 1990; Blossier *et al.*, 2009). It uses additional information in the form of a matrix correlation function formed from N different interpolating fields on one timeslice and the same interpolating fields on another timeslice. When this matrix correlation function is analyzed in a specific way, described in (Blossier *et al.*, 2009), one can prove that a much larger gap, $\Delta = E_{N+1} - E_1$ appears for the dominating correction terms due to excited states. These then disappear much more quickly with growing time.

The GEVP is straightforwardly applicable to HQET, order by order in $1/m$. The precision of the numerical results that we show below is largely due to this method, together with the use of HYP1/2 actions.

9.5.5   **Examples of results**

We now discuss a few numerical results (Blossier *et al.*, 2010*b*;*a*;*c*) in order to give an indication of what can be done at present. The graphs and numbers are for the quenched approximation (the light quark is a strange quark) but these computations are also on the way for dynamical fermions. The statistics employed in the quenched approximation is rather modest: only 100 configurations were analyzed. One can easily use a larger number, even with dynamical fermions. We skip numerical details in the following discussion.

As a first step, one wants to fix the b-quark mass. This is done through Eq. (9.245) and its $1/m$ corrections. Its graphical solution is illustrated in Fig. 9.12 where all plotted numbers originate from prior continuum extrapolations. The resulting mass of the b-quark is displayed in Table 9.2. Observe that it depends very little on the matching condition, i.e. the choice of $\theta_0, \theta_1, \theta_2$ and moreover the $1/m$ corrections are small.

Next, we look at the lattice-spacing dependence of the decay constant. For the results including $1/m$ corrections no significant dependence on a is seen in Fig. 9.13 despite a good precision of about 2%. In the static approximation, discretization errors are visible but small. Table 9.3 lists the B_s decay constant using $r_0 = 0.5\,\text{fm}$ to convert to MeV for illustration. The actual number is affected by an unknown "quenching

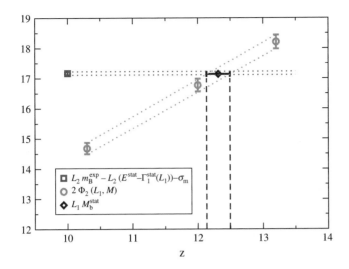

Fig. 9.12 Numerical solution of the equation for M_b (Blossier *et al.*, 2010*b*) made dimensionless by multiplication with L_2. The figure uses a notation $\sigma_m = \lim_{a \to 0} L_2 \left[\Gamma^{\text{stat}}(L_2, a) - \Gamma^{\text{stat}}(L_1, a) \right]$ and Φ_2 in the figure is Φ_1 in our notation.

effect" and thus not so important. It is more relevant to observe the precision that can be reached with just 100 configurations and how the spread in the numbers in the static approximation is reduced when the $1/m$ corrections are included.

Further, the comparison with results in the charm mass region, Fig. 9.14, seems to indicate that the $1/m$ expansion works very well even for charm quarks. This is a bit surprising and certainly requires further confirmation. Note also that this comparison makes use of the perturbatively evaluated C_{PS} whose intrinsic uncertainty due to perturbation theory is difficult to evaluate. Of course this uncertainty does neither affect the non-perturbatively computed static value at $1/(r_0 m_{\text{PS}}) = 0$, nor $f_{B_s} \sqrt{m_{B_s}}$ computed with $1/m$ corrections at the mass of the b-quark, corresponding to $1/(r_0 m_{\text{PS}}) \approx 0.07$. It only affects the comparison to the results for $1/(r_0 m_{\text{PS}}) \gtrsim 0.15$

Table 9.2 Dimensionless b-quark mass, $r_0 M_b$, obtained from the B_s meson mass, for different values of θ_i.

	LO (static)	NLO (static + $O(1/m)$)		
		$(\theta_1, \theta_2) = (0, 0.5)$	$(\theta_1, \theta_2) = (0.5, 1)$	$(\theta_1, \theta_2) = (0, 1)$
$\theta_0 = 0$	17.1 ± 0.2	17.1 ± 0.2	17.1 ± 0.2	17.1 ± 0.2
$\theta_0 = 0.5$	17.2 ± 0.2	17.2 ± 0.2	17.2 ± 0.2	17.1 ± 0.2
$\theta_0 = 1$	17.2 ± 0.2	17.3 ± 0.3	17.3 ± 0.3	17.3 ± 0.3

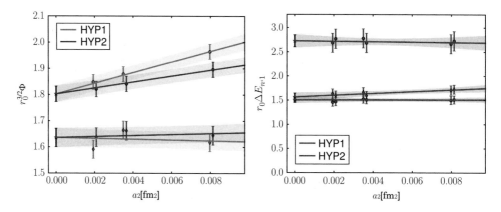

Fig. 9.13 Continuum extrapolations in HQET. Left: $\Phi^{\mathrm{HQET}} = f_{B_s}\sqrt{m_{B_s}}/C_{\mathrm{PS}}$ (diamonds) in HQET with $1/m$ corrections included (Blossier *et al.*, 2010c) and its static limit Φ_{RGI} (circles). The value of C_{PS} does not depend on the lattice spacing. It renders the two quantities directly comparable. Right: pseudoscalar energy levels (Blossier *et al.*, 2010a). From bottom to top: 2s − 1s splitting static, 2s − 1s splitting static + $1/m$, 3s − 1s splitting static.

Table 9.3 Pseudoscalar heavy-light decay constant f_{B_s} in MeV, for different values of θ_i.

	LO (static)	NLO (static + O(1/m))		
		$(\theta_1, \theta_2) = (0, 0.5)$	$(\theta_1, \theta_2) = (0.5, 1)$	$(\theta_1, \theta_2) = (0, 1)$
$\theta_0 = 0$	233 ± 6	220 ± 9	218 ± 9	218 ± 9
$\theta_0 = 0.5$	229 ± 7	221 ± 9	219 ± 8	219 ± 9
$\theta_0 = 1$	219 ± 6	223 ± 9	221 ± 8	222 ± 8

since only for the purpose of this comparison does the logarithmic mass dependence described by C_{PS} have to be divided out.

Finally, we show some results concerning the spectrum. The splitting between radial excitations in the pseudoscalar sector is displayed in the right part of Fig. 9.13. As throughout in our results, the $1/m$ correction is rather small.

9.5.6   Perspectives

Meanwhile, it has been established that HQET with non-perturbatively determined parameters is a precision tool. However, we are still at the beginning concerning applications. Results for the quantities shown here will be available for $N_{\mathrm{f}} = 2$ dynamical fermions rather soon. But there are many more applications that remain unexplored and open interesting avenues of research for the future.

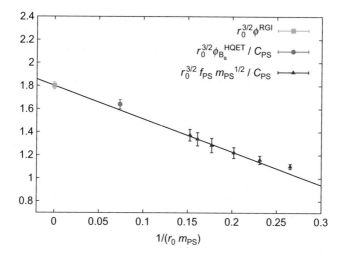

Fig. 9.14 Static results together with results with $m_{\rm h} < m_{\rm b}$ and an HQET computation with $1/m$ corrections included. Continuum extrapolations are done before the interpolation (Blossier *et al.*, 2010c). $C_{\rm PS}$ is evaluated with the three-loop approximation of $\gamma_{\rm match}$.

9.6 Appendix

9.6.1 Notation

9.6.1.1 *Index conventions*

Lorentz indices $\mu, \nu, \ldots$ are taken from the middle of the Greek alphabet and run from 0 to 3. Latin indices $k, l, \ldots$ run from 1 to 3 and are used to label the components of spatial vectors. For the Dirac indices capital letters $A, B, \ldots$ from the beginning of the alphabet are taken. They run from 1 to 4. Color vectors in the fundamental representation of $\mathrm{SU}(N)$ carry indices $\alpha, \beta, \ldots$ ranging from 1 to N, while for vectors in the adjoint representation, Latin indices $a, b, \ldots$ running from 1 to $N^2 - 1$ are employed.

Repeated indices are always summed over unless otherwise stated and scalar products are taken with Euclidean metric.

9.6.1.2 *Dirac matrices*

In the *chiral representation* for the Dirac matrices, we have

$$\gamma_\mu = \begin{pmatrix} 0 & e_\mu \\ e_\mu^\dagger & 0 \end{pmatrix}. \tag{9.253}$$

The 2×2 matrices e_μ are taken to be

$$e_0 = -1, \qquad e_k = -i\sigma_k, \tag{9.254}$$

with σ_k the Pauli matrices. It is then easy to check that

$$\gamma_\mu^\dagger = \gamma_\mu, \qquad \{\gamma_\mu, \gamma_\nu\} = 2\delta_{\mu\nu}. \tag{9.255}$$

Furthermore, if we define $\gamma_5 = \gamma_0\gamma_1\gamma_2\gamma_3$, we have

$$\gamma_5 = \begin{pmatrix} 1 & 0 \\ 0 & -1 \end{pmatrix}. \tag{9.256}$$

In particular, $\gamma_5 = \gamma_5^\dagger$ and $\gamma_5^2 = 1$. The Hermitian matrices

$$\sigma_{\mu\nu} = \frac{i}{2}[\gamma_\mu, \gamma_\nu] \tag{9.257}$$

are explicitly given by $(\sigma_i\sigma_j = i\epsilon_{ijk}\sigma_k)$

$$\sigma_{0k} = \begin{pmatrix} \sigma_k & 0 \\ 0 & -\sigma_k \end{pmatrix}, \qquad \sigma_{ij} = -\epsilon_{ijk}\begin{pmatrix} \sigma_k & 0 \\ 0 & \sigma_k \end{pmatrix} \equiv -\epsilon_{ijk}\sigma_k, \tag{9.258}$$

where ϵ_{ijk} is the totally antisymmetric tensor with $\epsilon_{123} = 1$.

In the *Dirac representation* we have

$$\gamma_k = \begin{pmatrix} 0 & -i\sigma_k \\ i\sigma_k & 0 \end{pmatrix}, \qquad \gamma_0 = \begin{pmatrix} 1 & 0 \\ 0 & -1 \end{pmatrix}, \tag{9.259}$$

$$\gamma_5 = \begin{pmatrix} 0 & 1 \\ 1 & 0 \end{pmatrix}, \qquad \sigma_{ij} = -\epsilon_{ijk}\begin{pmatrix} \sigma_k & 0 \\ 0 & \sigma_k \end{pmatrix} = \sigma_k. \tag{9.260}$$

9.6.1.3 *Lattice conventions*

Ordinary forward and backward lattice derivatives act on color singlet functions $f(x)$ and are defined through

$$\partial_\mu f(x) = \frac{1}{a}[f(x + a\hat{\mu}) - f(x)],$$

$$\partial_\mu^* f(x) = \frac{1}{a}[f(x) - f(x - a\hat{\mu})], \tag{9.261}$$

where $\hat{\mu}$ denotes the unit vector in direction μ. We also use the symmetric derivative

$$\tilde{\partial}_\mu = \frac{1}{2}(\partial_\mu + \partial_\mu^*). \tag{9.262}$$

The gauge covariant derivative operators, acting on a quark field $\psi(x)$, are given by

$$\nabla_\mu\psi(x) = \frac{1}{a}[\lambda_\mu U(x, \mu)\psi(x + a\hat{\mu}) - \psi(x)], \tag{9.263}$$

$$\nabla_\mu^*\psi(x) = \frac{1}{a}[\psi(x) - \lambda_\mu^{-1}U(x - a\hat{\mu}, \mu)^{-1}\psi(x - a\hat{\mu})], \tag{9.264}$$

with the constant phase factors

$$\lambda_\mu = e^{ia\theta_\mu/L}, \qquad \theta_0 = 0, \quad -\pi < \theta_k \leq \pi, \tag{9.265}$$

explained in Section 9.2.6. The left action of the lattice derivative operators is defined by

$$\overline{\psi}(x)\overleftarrow{\nabla}_\mu = \frac{1}{a}\left[\overline{\psi}(x + a\hat{\mu})U(x,\mu)^{-1}\lambda_\mu^{-1} - \overline{\psi}(x)\right], \tag{9.266}$$

$$\overline{\psi}(x)\overleftarrow{\nabla}_\mu^* = \frac{1}{a}\left[\overline{\psi}(x) - \overline{\psi}(x - a\hat{\mu})U(x - a\hat{\mu},\mu)\lambda_\mu\right]. \tag{9.267}$$

Our lattice versions of δ-functions are

$$\delta(x_\mu) = a^{-1}\delta_{x_\mu 0}, \quad \delta(\mathbf{x}) = \prod_{k=1}^3 \delta(x_k), \quad \delta(x) = \prod_{\mu=0}^3 \delta(x_\mu) \tag{9.268}$$

and we use

$$\theta(x_\mu) = 1 \text{ for } x_\mu \geq 0 \tag{9.269}$$

$$\theta(x_\mu) = 0 \text{ otherwise.}$$

Fields in momentum space are introduced by the Fourier transformation

$$\tilde{f}(p) = a^4 \sum_x e^{-ipx} f(x) \Leftrightarrow \begin{cases} f(x) = \frac{1}{L^3T}\sum_p e^{ipx}\tilde{f}(p) & \text{in a } T \times L^3 \text{ volume} \\ f(x) = \int_{-\pi/a}^{\pi/a}\frac{d^4p}{(2\pi)^4}e^{ipx}\tilde{f}(p) & \text{in infinite volume.} \end{cases} \tag{9.270}$$

9.6.1.4 *Continuum gauge fields*

An SU(N) gauge potential in the continuum theory is a vector field $A_\mu(x)$ with values in the Lie algebra su(N). It may thus be written as

$$A_\mu(x) = A_\mu^a(x)T^a \tag{9.271}$$

with real components $A_\mu^a(x)$ and

$$(T^a)^\dagger = -T^a, \quad \text{tr}\{T^aT^b\} = -\frac{1}{2}\delta^{ab}. \tag{9.272}$$

The associated field tensor,

$$F_{\mu\nu}(x) = \partial_\mu A_\nu(x) - \partial_\nu A_\mu(x) + [A_\mu(x), A_\nu(x)], \tag{9.273}$$

may be decomposed similarly and the right and left action of the covariant derivative D_μ is defined by

$$D_\mu\psi(x) = (\partial_\mu + A_\mu)\psi(x), \tag{9.274}$$

$$\overline{\psi}(x)\overleftarrow{D}_\mu = \overline{\psi}(x)(\overleftarrow{\partial}_\mu - A_\mu). \tag{9.275}$$

We note that periodic boundary conditions up to a phase θ_μ are equivalent to adding a constant abelian gauge field $i\theta_\mu/L$: in the above we replace $A_\mu \to A_\mu + i\theta_\mu/L$.

9.6.1.5 Lattice action

Let us first assume that the theory is defined on an infinite lattice. A gauge field U on the lattice is an assignment of a matrix $U(x,\mu) \in \mathrm{SU}(N)$ to every lattice point x and direction $\mu = 0,1,2,3$. Quark and antiquark fields, $\psi(x)$ and $\overline{\psi}(x)$, reside on the lattice sites and carry Dirac, color and flavor indices. The (unimproved) lattice action is of the form

$$S[U,\overline{\psi},\psi] = S_G[U] + S_F[U,\overline{\psi},\psi], \tag{9.276}$$

where S_G denotes the usual Wilson plaquette action and S_F the Wilson quark action. Explicitly, we have

$$S_G[U] = \frac{1}{g_0^2}\sum_p \mathrm{tr}\{1 - U(p)\} = \frac{1}{g_0^2}\sum_x\sum_{\mu,\nu} P_{\mu\nu}(x), \tag{9.277}$$

$$P_{\mu\nu}(x) = U(x,\mu)\,U(x+a\hat{\mu},\nu)\,U(x+a\hat{\nu},\mu)^{-1}\,U(x,\nu)^{-1} \tag{9.278}$$

with g_0 being the bare gauge coupling and $U(p)$ the parallel transporter around the plaquette p. The sum runs over all *oriented* plaquettes p on the lattice, i.e. independently over μ,ν. The quark action,

$$S_F[U,\overline{\psi},\psi] = a^4\sum_x \overline{\psi}(x)(D_W + m_0)\psi(x), \tag{9.279}$$

is defined in terms of the Wilson–Dirac operator

$$D_W = \frac{1}{2}\{\gamma_\mu(\nabla_\mu^* + \nabla_\mu) - a\nabla_\mu^*\nabla_\mu\}, \tag{9.280}$$

which involves the gauge covariant lattice derivatives ∇_μ and ∇_μ^*, Eq. (9.261), and the bare quark mass *matrix*, $m_0 = \mathrm{diag}(m_{0u}, m_{0d}, \ldots)$.

9.6.1.6 Renormalization-group functions and invariants

Our RG functions are defined through

$$\mu\frac{\partial\bar{g}}{\partial\mu} = \beta(\bar{g}), \tag{9.281}$$

$$\frac{\mu}{\overline{m}}\frac{\partial\overline{m}}{\partial\mu} = \tau(\bar{g}), \tag{9.282}$$

$$\frac{\mu}{\Phi}\frac{\partial\Phi}{\partial\mu} = \gamma(\bar{g}) \tag{9.283}$$

in terms of running coupling and running quark mass as well as some matrix element Φ of a (multiplicatively renormalizable) composite field. They have asymptotic expansions

$$\beta(\bar{g}) \overset{\bar{g}\to 0}{\sim} -\bar{g}^3 \left\{ b_0 + \bar{g}^2 b_1 + \ldots \right\}, \tag{9.284}$$

$$b_0 = \frac{1}{(4\pi)^2} \left(11 - \frac{2}{3} N_{\mathrm{f}} \right), \quad b_1 = \frac{1}{(4\pi)^4} \left(102 - \frac{38}{3} N_{\mathrm{f}} \right),$$

$$\tau(\bar{g}) \overset{\bar{g}\to 0}{\sim} -\bar{g}^2 \left\{ d_0 + \bar{g}^2 d_1 + \ldots \right\}, \qquad d_0 = 8/(4\pi)^2, \tag{9.285}$$

$$\gamma(\bar{g}) \overset{\bar{g}\to 0}{\sim} -\bar{g}^2 \left\{ \gamma_0 + \bar{g}^2 \gamma_1 + \ldots \right\}. \tag{9.286}$$

The integration constants of the solutions to the RGEs define the RG invariants

$$\Lambda = \mu \left(b_0 \bar{g}^2 \right)^{-b_1/(2b_0^2)} e^{-1/(2b_0 \bar{g}^2)} \exp \left\{ -\int_0^{\bar{g}} \mathrm{d}x \left[\frac{1}{\beta(x)} + \frac{1}{b_0 x^3} - \frac{b_1}{b_0^2 x} \right] \right\}, \tag{9.287}$$

$$M = \overline{m} \left(2 b_0 \bar{g}^2 \right)^{-d_0/2b_0} \exp \left\{ -\int_0^{\bar{g}} \mathrm{d}x \left[\frac{\tau(x)}{\beta(x)} - \frac{d_0}{b_0 g} \right] \right\} \tag{9.288}$$

$$\Phi_{\mathrm{RGI}} = \Phi \left[2 b_0 \bar{g}^2 \right]^{-\gamma_0/2b_0} \exp \left\{ -\int_0^{\bar{g}} \mathrm{d}x \left[\frac{\gamma(x)}{\beta(x)} - \frac{\gamma_0}{b_0 x} \right] \right\}, \tag{9.289}$$

where $\bar{g} \equiv \bar{g}(\mu) \ldots \Phi \equiv \Phi(\mu)$. We will also use the shorthand notation

$$\frac{\Lambda}{\mu} = \varphi_g(\bar{g}) = \exp \left\{ -\int^{\bar{g}} \mathrm{d}x \frac{1}{\beta(x)} \right\}, \tag{9.290}$$

$$\frac{M}{\overline{m}} = \varphi_m(\bar{g}) = \exp \left\{ -\int^{\bar{g}} \mathrm{d}x \frac{\tau(x)}{\beta(x)} \right\}, \tag{9.291}$$

$$\frac{\Phi_{\mathrm{RGI}}}{\Phi} = \varphi_\Phi(\bar{g}) = \exp \left\{ -\int^{\bar{g}} \mathrm{d}x \frac{\gamma(x)}{\beta(x)} \right\}, \tag{9.292}$$

with the constants exactly as defined above.

9.6.2 Conversion functions and anomalous dimensions

Conversion functions and the anomalous dimensions γ_{match} are not part of the standard phenomenology literature. For completeness, we give the explicit relations to the matching coefficients found directly in the literature and discuss the accuracy of their perturbative expansion.

9.6.2.1 Matching coefficients and anomalous dimension

We here describe the result (Bekavac *et al.*, 2010) and its relation to the anomalous dimension. We denote a matrix element of some heavy–light quark bilinear $\bar{\psi} \Gamma \psi_{\mathrm{h}}$ in the effective theory by $\Phi(\mu)$. The Dirac structure Γ is left implicit.[32]

All quantities are renormalized in the $\overline{\mathrm{MS}}$-scheme, with a scale μ_o for the QCD bilinear and a scale μ in HQET. Choosing the pole quark mass m_{Q},[33] the matrix element is then (without explicit superscripts "QCD" we refer to HQET quantities, in the static approximation),

$$\Phi^{\text{QCD}}(m_{\text{Q}}, \mu_o; \mathcal{V}_{\text{kin}}) = \widehat{C}_{\text{match}}(m_{\text{Q}}, \mu_o, \mu) \times \Phi(\mu; \mathcal{V}_{\text{kin}}) + O(1/m). \qquad (9.293)$$

The kinematical variables entering the matrix element Φ are denoted by $\mathcal{V}_{\text{kin}}$. For the (partially) conserved currents V_μ, A_μ there is no μ_o-dependence on the l.h.s. of Eq. (9.293), $(\partial_{\mu_o} \Phi^{\text{QCD}}(m_{\text{Q}}, \mu_o) = 0)$, while in general we have

$$\frac{\mu}{\Phi^{\text{QCD}}(m_{\text{Q}}, \mu_o)} \frac{\partial \Phi^{\text{QCD}}(m_{\text{Q}}, \mu_o)}{\partial \mu_o} = \frac{\partial \ln(\Phi^{\text{QCD}}(m_{\text{Q}}, \mu_o))}{\partial \ln(\mu_o)} \equiv \gamma_o(\bar{g}(\mu_o)). \qquad (9.294)$$

We pass to the RGI matrix element in QCD via $(O(1/m)$ is dropped without notice)

$$\Phi_{\text{RGI}}^{\text{QCD}} = \exp\left\{ -\int^{\bar{g}(\mu_o)} dx \frac{\gamma_o(x)}{\beta(x)} \right\} \Phi^{\text{QCD}}(m_{\text{Q}}, \mu_o; \mathcal{V}_{\text{kin}}) \qquad (9.295)$$

$$= \exp\left\{ -\int^{\bar{g}(\mu_o)} dx \frac{\gamma_o(x)}{\beta(x)} \right\} \widehat{C}_{\text{match}}(m_{\text{Q}}, \mu_o, \mu) \times \Phi(\mu; \mathcal{V}_{\text{kin}}) \qquad (9.296)$$

$(O(1/m)$ is dropped without notice). It depends on the quark mass but not on a renormalization scale. The physical anomalous dimension is given by

$$\gamma_{\text{match}}(g_\star) = \frac{d \ln(m_{\text{Q}})}{d \ln(m_\star)} \frac{\partial \ln(\widehat{C}_{\text{match}}(m_{\text{Q}}, \mu_o, \mu))}{\partial \ln(m_{\text{Q}})}, \qquad (9.297)$$

where the first factor is computed from the expansion (Gray *et al.*, 1990; Fleischer *et al.*, 1999; Melnikov and Ritbergen, 2000; Bekavac *et al.*, 2010)

$$m_{\text{Q}} = m_\star \left[1 + \sum_{l \geq 1} k_l [\bar{a}(m_\star)]^l \right], \quad \bar{a}(\mu) = \frac{\bar{g}^2(\mu)}{4\pi^2} \qquad (9.298)$$

$$k_1 = 4/3, \quad k_2 = -1.0414(N_{\text{f}} - 1) + 13.4434,$$

$$k_3 = 0.6527(N_{\text{f}} - 1)^2 - 26.655(N_{\text{f}} - 1) + 190.595.$$

The authors of Ref. (Bekavac *et al.*, 2010) set $\mu_o = \mu$. Building on (Ji and Musolf, 1991; Broadhurst and Grozin, 1995; Gimenez, 1992), they give the perturbative expansion

$$\widehat{C}_{\text{match}}(m_{\text{Q}}, \mu, \mu) = 1 + \sum_{l \geq 1} \sum_{k=0}^{l} L_{lk} [\ln(m_{\text{Q}}^2/\mu^2)]^k [\bar{a}(m_{\text{Q}})]^l, \qquad (9.299)$$

with coefficients L_{lk} depending on the Dirac structure, Γ.

Independence of the l.h.s. of Eq. (9.293) of μ yields

$$\frac{\partial \ln(\widehat{C}_{\text{match}}(m_{\text{Q}}, \mu_o, \mu))}{\partial \ln(\mu)} = -\frac{\partial \ln(\Phi(\mu))}{\partial \ln(\mu)} = -\gamma^{\text{stat}}(\bar{g}(\mu)), \qquad (9.300)$$

and with $\frac{\partial \ln(\widehat{C}_{\text{match}}(m_Q, \mu_o, \mu))}{\partial \ln(\mu_o)} = \gamma_o(\bar{g}(\mu_o))$ we have

$$\frac{\mathrm{d} \ln(\widehat{C}_{\text{match}}(m_Q, m_Q, m_Q))}{\mathrm{d} \ln(m_Q)} \tag{9.301}$$

$$= \left. \frac{\partial \ln((\widehat{C}_{\text{match}}(m_Q, \mu_o, \mu))}{\partial \ln(m_Q)} \right|_{\mu_o = \mu = m_Q} + \gamma_o(\bar{g}(m_Q)) - \gamma_{\text{stat}}(\bar{g}(m_Q)).$$

From these equations $\gamma_{\text{match}}(g_\star)$ can be determined up to three-loop order and the differences $\gamma_{\text{match}}^{\Gamma'}(g_\star) - \gamma_{\text{match}}^{\Gamma}(g_\star)$ up to four-loop order.

9.6.2.2 *Numerical results and the behavior of perturbation theory*

Let us now look at the numerical size of the perturbative coefficients of the RG functions. The following table lists results for $N_f = 3$. This is enough to understand the general picture since for smaller N_f the higher-order coefficients are generically somewhat larger, but not by much.

coefficient	$i = 1$	$i = 2$	$i = 3$	$i = 4$
$(4\pi)^i b_{i-1}$	0.71620	0.40529	0.32445	0.47367
$(4\pi)^i d_{i-1}$	0.63662	0.76835	0.80114	0.90881
$(4\pi)^i \gamma_{\text{stat},i-1}$	−0.31831	−0.26613	−0.25917	
$(4\pi)^i \gamma_{\text{match},i-1}^{\gamma_0 \gamma_5}$	−0.31831	−0.57010	−0.94645	
$(4\pi)^i \gamma_{\text{match},i-1}^{\gamma_k}$	−0.31831	−0.87406	−3.12585	
$(4\pi)^i [\gamma_{\text{match},i-1}^{\gamma_0\gamma_5} - \gamma_{\text{match},i-1}^{\gamma_k}]$	0	0.30396	2.17939	14.803
$(4\pi)^i [\gamma_{\text{match},i-1}^{\gamma_0\gamma_5} - \gamma_{\text{match},i-1}^{\gamma_5}]$				

The normalization $(4\pi)^i$ has been inserted such that the series is well behaved for $\alpha \lesssim 1/3$ if the coefficients are order one. Indeed this is the magnitude of the coefficients in the first three rows that show as a comparison the beta function, mass anomalous dimension and the anomalous dimension of the static–light bilinears (all in the $\overline{\text{MS}}$–scheme). In contrast in the physical anomalous dimension of the vector current $\gamma_{\text{match}}^{\gamma_k}$, the 3-loop coefficient is rather big and the difference $(4\pi)^i [\gamma_{\text{match},3}^{\gamma_0\gamma_5} - \gamma_{\text{match},3}^{\gamma_k}]$ is even above ten. Perturbation theory is then useful only at rather small α; in particular not really for the b-quark.

An attempt to improve the perturbative series is to re-expand γ_{match} in the coupling at a different scale, adjusting the scale to obtain smaller coefficients. In fact, since the effective theory is valid at energy scales below the mass of the quark, it is plausible that scales smaller than $m_\star$ are more suitable. So we choose a coupling $\hat{g}^2 = \bar{g}^2(s^{-1} m_\star) = \sigma(g_\star^2, s)$ and

$$\hat{\gamma}_{\text{match}}(\hat{g}) = \gamma_{\text{match}}([\sigma(\hat{g}^2, 1/s)]^{1/2}), \tag{9.302}$$

which is of course expanded order by order,

$$g_\star^2 = \sigma(\hat{g}^2, 1/s) = \hat{g}^2 - 2b_0 \ln(s)\, \hat{g}^4 + \ldots. \tag{9.303}$$

The conversion functions are then expressed as

$$C_{\mathrm{PS}}(M/\Lambda) = \exp\left\{ \int^{\hat{g}} \mathrm{d}x \frac{\hat{\gamma}_{\mathrm{match}}(x)}{\beta(x)} \right\}. \tag{9.304}$$

The difference comes from truncating Eq. (9.302) as a series in $\hat{g}^2$. The argument above suggests $s > 1$. The perturbative coefficients are listed in the following table for a few choices of s, for example the one that brings the two-loop coefficient γ_1 to zero.

coefficient	$i = 1$	$i = 2$	$i = 3$	$i = 4$	s
$(4\pi)^i\, \gamma_{\mathrm{match},i-1}^{\gamma_0\gamma_5}$	-0.31831	-0.57010	-0.94645		1
	-0.31831	0	0.39720		3.4916
$(4\pi)^i\, \gamma_{\mathrm{match},i-1}^{\gamma_k}$	-0.31831	-0.87406	-3.12585		1
	-0.31831	0	-0.231121		6.8007
$(4\pi)^i\, [\gamma_{\mathrm{match},i-1}^{\gamma_0\gamma_5} - \gamma_{\mathrm{match},i-1}^{\gamma_k}]$	0	0.30396	2.17939	14.803	1
	0	0.30396	0.972221	4.733	4
	0	0.30396	-0.05414	1.82678	13
	0	0.30396	-0.23495	1.85344	16

The higher-order coefficients can indeed be reduced significantly but $s \gtrsim 4$ is required. For B-physics $\alpha(m_{\star\mathrm{b}}/s)$ is then not small and there is no really useful improvement for phenomenology, see Fig. 9.5. We emphasize, however, that with $s \approx 4$ the series is much better behaved for masses that are a factor two or more higher than the b-quark mass. The pattern visible in the tables reflects itself in Fig. 9.5.

Let us finally mention that the same behavior is found for $\hat{C}_{\mathrm{match}}(m_{\mathrm{Q}}, m_{\mathrm{Q}}, m_{\mathrm{Q}})$ for all Dirac structures of the currents. Their perturbative expansion in a coupling $\bar{g}(m_{\mathrm{Q}}/s)$ is better behaved for $s \gtrsim 4$ than for $s = 1$.

Acknowledgments

I am thankful for the nice collaboration in the team of organizers, with the director of the school and the staff of the school. It is also a pleasure to thank the members of the LGT discussion seminar at Humboldt-University and DESY, in particular Hubert Simma and Ulli Wolff, for their valuable suggestions on a first version of this chapter. I am grateful for a fruitful collaboration with Benoit Blossier, Michele Della Morte, Patrick Fritzsch, Nicolas Garron, Jochen Heitger, Georg von Hippel, Tereza Mendes, Mauro Papinutto and Hubert Simma on several of the subjects of this chapter and thank Nicolas Garron for providing me with tables and figures. Most of all I would like to thank Dorothy for her patience with me spending much time on this chapter and the school.

Notes

1. See Peter Weisz' chapter for the general discussion of discretization errors and improvement of lattice gauge theories.
2. However, when one carries out the expansion to include $1/m_b^2$ terms, also a whole set of terms generated by b-quark loops in QCD that do not contain the b-quark field in the effective theory have to be taken into account. An example are 4-fermion operators made of the light quarks, just as they appear when one "integrates out" the W and Z-bosons in the Standard Model.
3. The expectation value $\langle . \rangle$ refers to the Euclidean path integral, here with the free Dirac action.
   We suggest to verify these formulae as an exercise.
4. The terms "large" and "small" components are commonly used when discussing the non-relativistic limit of the Dirac equation for bound states, see e.g. (Itzykson and Zuber, 1980).
5. One can also obtain this Lagrangian by performing a boost of the zero-velocity theory (Horgan *et al.*, 2009). In the quoted reference also the next-to-leading order terms are found.
6. Note $\mathcal{P} \exp\left\{ -\int_{y_0}^{x_0} \mathrm{d}z_0 A_0(z_0, \mathbf{x}) \right\} = \mathcal{P} \exp\left\{ -\int_{x_0}^{y_0} \mathrm{d}z_0 A_0(z_0, \mathbf{x}) \right\}^{\dagger}$.
7. Power counting as discussed by Peter Weisz in this book is not applicable here, since the propagator does not fall off with all momentum components.
8. See Peter Weisz' chapter for a theoretical discussion and Chapter I of (Sommer, 2006) for an overview of tests. Finally, (Balog *et al.*, 2009b;a) represents the most advanced understanding of the subject.
9. The equations of motion follow just from a change of variable in the path integral. Contact terms are reabsorbed into the free coefficients c_i. We refer to Peter Weisz' chapter or (Lüscher *et al.*, 1996) for a more detailed discussion.
10. When the chiral-symmetry realization of domain-wall fermions (Shamir, 1993) is good enough, these fermions can of course also be considered to have an, in practice, exact chiral symmetry.
11. If the light-quark action has an exact chiral symmetry or the light quarks are discretized with a twisted-mass term at full twist, this restriction is unnecessary, since the term is excluded by the symmetry. Note that $(\delta A_0^{\mathrm{stat}})_1$ is, however, not forbidden by chiral symmetry and c_A^{stat} is necessary for $O(a)$-improvement in any case.
12. A formal argument is as follows. Rewrite Eq. (9.92) in terms of the bare operators, $Z_V^{\mathrm{stat}}(g_0, a\mu)\delta_A^3 V_0^{\mathrm{stat}} = -\frac{1}{2} Z_A^{\mathrm{stat}}(g_0, a\mu) A_0^{\mathrm{stat}} + O(a^2)$. Since the bare, regularized, operators $V_0^{\mathrm{stat}}, A_0^{\mathrm{stat}}$ carry no μ dependence, we see that $Z_V^{\mathrm{stat}}(g_0, a\mu)/Z_A^{\mathrm{stat}}(g_0, a\mu)$ is a function of g_0 only, apart from $O(a^2)$ cutoff effects. To make the argument more rigorous one should rewrite the equation in the form of correlation functions that represent a Ward-identity equivalent to Eq. (9.92).

13. In (Lüscher *et al.*, 1992; 1994) the definition of a renormalized coupling uses more general boundary conditions for the gauge fields, but these are not needed here.
14. As usual in perturbation theory, terms of order $(a\mu)^n, n \geq 1$ are dropped.
15. Of course, a trivial definition dependence due to the choice of prefactors in Eq. (9.121) is present. Unfortunately, there is no uniform choice for those in the literature.
16. In certain types of quark smearing, this has to be properly taken into account (Christ *et al.*, 2007).
17. It is technically of advantage to consider so-called smeared-smeared and local-smeared correlation functions, but this is irrelevant in the present discussion.
18. The correlation functions f_A, f_1 are the relativistic versions of f_A^{stat}, f_1^{stat}.
19. Note that $m_\star$ is implicitly defined through $m_\star = \overline{m}(m_\star)$.
20. In a massless renormalization scheme, the renormalization factors do not depend on the masses. Consequently, the renormalization group functions do not depend on the masses.
21. This is seen from

$$\frac{M}{\Phi}\frac{\partial\Phi}{\partial M} = \underbrace{\frac{M}{m_\star}\frac{\partial m_\star}{\partial M}}_{\frac{1}{1-\tau(g_\star)}}\underbrace{\frac{m_\star}{\Phi}\frac{\partial\Phi}{\partial m_\star}}_{\gamma_{\text{match}}(g_\star)} = \frac{\gamma_{\text{match}}(g_\star)}{1-\tau(g_\star)}, \tag{9.159}$$

where we used

$$m_\star = M \exp\left\{\int^{g_\star} \mathrm{d}x \frac{\tau(x)}{\beta(x)}\right\} \tag{9.160}$$

$$\frac{\partial m_\star}{\partial M} = \frac{m_\star}{M} + \frac{\tau(g_\star)}{\beta(g_\star)}\frac{\partial g_\star}{\partial M}m_\star = \frac{m_\star}{M} + \frac{\tau(g_\star)}{\beta(g_\star)}\beta(g_\star)\frac{\partial m_\star}{\partial M}, \tag{9.161}$$

which shows that

$$\frac{M}{m_\star}\frac{\partial m_\star}{\partial M} = \frac{1}{1-\tau(g_\star)}. \tag{9.162}$$

22. Note the slow, logarithmic, decrease of the corrections in Eq. (9.163). We will see below, in the discussion of Figs. 9.4 and 9.5, that the perturbative evaluation of $C_{\text{PS}}(M_{\text{b}}/\Lambda)$ is somewhat problematic.
23. A straight expansion gives, e.g., $\omega_{\text{kin}}a^4 \sum_x \langle \mathcal{O}[\mathcal{O}_{\text{kin}}(x) - \langle\mathcal{O}_{\text{kin}}(x)\rangle_{\text{stat}}\rangle_{\text{stat}}$, but this just corresponds to an irrelevant shift of $\mathcal{O}_{\text{kin}}(x)$, etc. by a constant.
24. An operator $\frac{m_l}{m}A_0^{\text{stat}}$ is included as a corresponding mass dependence of Z_A^{HQET}. In practice, since $\frac{m_l}{m_b} \ll 1$, and this term appears only at one-loop order, this dependence on the light-quark mass can be neglected.
25. It follows from the simple form of the static propagator that there is no dependence on δm except for the explicitly shown energy shift $\widehat{\delta m}$.
26. We have written down the integrated version, since then a smaller number of operators can appear and we are ultimately interested in the integral.

27. It is convenient to avoid the multiplication of $1/m$ terms explicitly by a choice of observables, for example

$$\tilde{\Phi} = \ln(f_B\sqrt{m_B}) = \ln(Z_A^{\text{HQET}}) + \ln(\Phi^{\text{stat}}) + \lim_{x_0\to\infty}\left\{\frac{1}{2}x_0\omega_{\text{kin}}E^{\text{kin}} + \right.$$

$$\left.\frac{1}{2}\omega_{\text{kin}}R_{\text{AA}}^{\text{kin}}(x_0) + \dots\right\}, \ln(Z_A^{\text{HQET}}) = \ln(Z_A^{\text{stat}}) + \frac{Z_A^{(1/m)}}{Z_A^{\text{stat}}} \equiv \ln(Z_A^{\text{stat}}) + [\ln(Z_A)]^{1/m}.$$

In this convention all $1/m$ terms appear linearly.

28. Note that the comparison Fig. 9.7 has to be taken with a grain of salt due to the perturbative uncertainty in C_{PS} discussed in Section 9.3.3.2.

29. In the notation of (Lüscher *et al.*, 1996) it reads $\bar{\rho}_h(\mathbf{x})\gamma_k D_k\rho_h(\mathbf{x})$ at $x_0 = 0$. Such a term does not contribute to any correlation function due to the form of the static propagator.

30. Recall that observables without a superscript refer to HQET.

31. A fixed ratio s ensures that the cutoff effects are a smooth function of a/L_i.

32. The notation $C_{\tilde{\Gamma}}$ of (Broadhurst and Grozin, 1995) translates to our Γ as $\tilde{\Gamma} = (1, \gamma_0, \gamma_1, \gamma_0\gamma_1) \to \Gamma = (\gamma_5, \gamma_0\gamma_5, \gamma_k, \gamma_0\gamma_k)$ and (Bekavac *et al.*, 2010) uses the notation of (Broadhurst and Grozin, 1995) when one sets $v_\mu\gamma_\mu = \gamma_0$, $\gamma_\perp = \gamma_k$ as is the case in the rest-frame. We will also refer to the bilinears as $(\text{PS}, A_0, V_k, \text{T})$. In comparison to (Bekavac *et al.*, 2010) we add a subscript Q to the pole-quark mass and a bar to the running mass $(m \to m_Q, m(\mu) \to \overline{m}(\mu))$ for clarity.

33. While in the complete, non-perturbative theory, the pole mass is ill-defined, in perturbation theory it exists order by order in the expansion. We use it here, because the formulae in the literature are written in terms of it. It will be eliminated in the final formulae.

References

Aglietti, U. (1994). Consistency and lattice renormalization of the effective theory for heavy quarks. *Nucl. Phys.*, **B421**, 191–216.

Aglietti, U., Crisafulli, M., and Masetti, M. (1992). Problems with the euclidean formulation of heavy quark effective theories. *Phys. Lett.*, **B294**, 281–285.

Antonelli, Mario *et al.* (2009). Flavor Physics in the Quark Sector arXiv:0907.5386.

Balog, Janos, Niedermayer, Ferenc, and Weisz, Peter (2009*a*). Logarithmic corrections to O(a^2) lattice artifacts. *Phys. Lett.*, **B676**, 188–192.

Balog, Janos, Niedermayer, Ferenc, and Weisz, Peter (2009*b*). The puzzle of apparent linear lattice artifacts in the 2d non-linear sigma-model and Symanzik's solution. *Nucl. Phys.*, **B824**, 563–615.

Bekavac, S., Grozin, A.G., Marquard, P., Piclum, J.H., Seidel, D. et al. (2010). Matching QCD and HQET heavy-light currents at three loops. *Nucl. Phys.*, **B833**, 46–63.

Blossier, Benoit *et al.* (2010*a*). HQET at order $1/m$: II. Spectroscopy in the quenched approximation. *JHEP*, **1005**, 074.

Blossier, Benoit, Della Morte, Michele, Garron, Nicolas, and Sommer, Rainer (2010*b*). HQET at order $1/m$: I. Non-perturbative parameters in the quenched approximation. *JHEP*, **1006**, 002.

Blossier, Benoit, Della Morte, Michele, Garron, Nicolas, von Hippel, Georg, Mendes, Tereza *et al.* (2010*c*). HQET at order 1/m: III. Decay constants in the quenched approximation. *arXiv:1006.5816* JHEP, 1012, 039.

Blossier, Benoit, Della Morte, Michele, von Hippel, Georg, Mendes, Tereza, and Sommer, Rainer (2009). On the generalized eigenvalue method for energies and matrix elements in lattice field theory. *JHEP*, **04**, 094.

Blossier, B., Le Yaouanc, A., Morenas, V., and Pene, O. (2006). Lattice renormalization of the static quark derivative operator. *Phys. Lett.*, **B632**, 319–325.

Bochicchio, Marco, Maiani, Luciano, Martinelli, Guido, Rossi, Gian Carlo, and Testa, Massimo (1985). Chiral symmetry on the lattice with Wilson fermions. *Nucl. Phys.*, **B262**, 331.

Borrelli, A. and Pittori, C. (1992). Improved renormalization constants for B-decay and $B\overline{B}$ mixing. *Nucl. Phys.*, **B385**, 502–524.

Boucaud, Ph., Leroy, J. P., Micheli, J., Pène, O., and Rossi, G. C. (1993). Rigorous treatment of the lattice renormalization problem of f_b. *Phys. Rev.*, **D47**, 1206.

Boucaud, Ph., Lin, C. L., and Pène, O. (1989). B-meson decay constant on the lattice and renormalization. *Phys. Rev.*, **D40**, 1529–1545. Erratum Phys. Rev. D41 (1990) 3541.

Broadhurst, D. J. and Grozin, A. G. (1991). Two-loop renormalization of the effective field theory of a static quark. *Phys. Lett.*, **B267**, 105–110.

Broadhurst, D. J. and Grozin, A. G. (1995). Matching QCD and HQET heavy-light currents at two loops and beyond. *Phys. Rev.*, **D52**, 4082–4098.

Burdman, Gustavo and Donoghue, John F. (1992). Union of chiral and heavy quark symmetries. *Phys. Lett.*, **B280**, 287–291.

Capitani, Stefano, Lüscher, Martin, Sommer, Rainer, and Wittig, Hartmut (1999). Non-perturbative quark mass renormalization in quenched lattice QCD. *Nucl. Phys.*, **B544**, 669.

Chetyrkin, K. G. and Grozin, A. G. (2003). Three-loop anomalous dimension of the heavy-light quark current in HQET. *Nucl. Phys.*, **B666**, 289–302.

Christ, Norman H., Dumitrescu, Thomas T., Loktik, Oleg, and Izubuchi, Taku (2007). The static approximation to B meson mixing using light domain-wall fermions: perturbative renormalization and ground state degeneracies. *PoS*, **LAT2007**, 351.

Christensen, Joseph C., Draper, T., and McNeile, Craig (2000). Renormalization of the lattice HQET Isgur-Wise function. *Phys. Rev.*, **D62**, 114006.

Colangelo, Gilberto, Dürr, Stephan, and Haefeli, Christoph (2005). Finite volume effects for meson masses and decay constants. *Nucl. Phys.*, **B721**, 136–174.

de Divitiis, G. M., Guagnelli, M., Palombi, F., Petronzio, R., and Tantalo, N. (2003*a*). Heavy-light decay constants in the continuum limit of lattice QCD. *Nucl. Phys.*, **B672**, 372–386.

de Divitiis, Giulia Maria, Guagnelli, Marco, Petronzio, Roberto, Tantalo, Nazario, and Palombi, Filippo (2003*b*). Heavy quark masses in the continuum limit of lattice QCD. *Nucl. Phys.*, **B675**, 309–332.

Della Morte, Michele *et al.* (2007*a*). Towards a non-perturbative matching of HQET and QCD with dynamical light quarks. *PoS*, **LAT2007**, 246.

Della Morte, Michele, Fritzsch, Patrick, and Heitger, Jochen (2007*b*). Non-perturbative renormalization of the static axial current in two-flavour QCD. *JHEP*, **02**, 079.

Della Morte, Michele, Fritzsch, Patrick, Heitger, Jochen, and Sommer, Rainer (2008). Non-perturbative quark mass dependence in the heavy-light sector of two-flavour QCD. *PoS*, **LATTICE2008**, 226.

Della Morte, Michele, Shindler, Andrea, and Sommer, Rainer (2005). On lattice actions for static quarks. *JHEP*, **08**, 051.

Dimopoulos, P. *et al.* (2008). Non-perturbative renormalisation of Delta F=2 four-fermion operators in two-flavour QCD. *JHEP*, **05**, 065.

Eichten, Estia and Hill, Brian (1990*a*). An effective field theory for the calculation of matrix elements involving heavy quarks. *Phys. Lett.*, **B234**, 511.

Eichten, Estia and Hill, Brian (1990*b*). Renormalization of heavy - light bilinears and f_b for Wilson fermions. *Phys. Lett.*, **B240**, 193.

Eichten, Estia and Hill, Brian (1990*c*). Static effective field theory: 1/m corrections. *Phys. Lett.*, **B243**, 427–431.

Fleischer, J., Jegerlehner, F., Tarasov, O. V., and Veretin, O. L. (1999). Two-loop QCD corrections of the massive fermion propagator. *Nucl. Phys.*, **B539**, 671–690.

Flynn, Jonathan M., Hernandez, Oscar F., and Hill, Brian R. (1991). Renormalization of four fermion operators determining B anti-B mixing on the lattice. *Phys. Rev.*, **D43**, 3709–3714.

Frezzotti, Roberto, Grassi, Pietro Antonio, Sint, Stefan, and Weisz, Peter (2001*a*). Lattice QCD with a chirally twisted mass term. *JHEP*, **08**, 058.

Frezzotti, R. and Rossi, G. C. (2004). Chirally improving Wilson fermions. i: O(a) improvement. *JHEP*, **08**, 007.

Frezzotti, Roberto, Sint, Stefan, and Weisz, Peter (2001*b*). O(a) improved twisted mass lattice QCD. *JHEP*, **07**, 048.

Gimenez, V. (1992). Two loop calculation of the anomalous dimension of the axial current with static heavy quarks. *Nucl. Phys.*, **B375**, 582–624.

Gray, N., Broadhurst, David J., Grafe, W., and Schilcher, K. (1990). Three loop relation of quark (modified) ms and pole masses. *Z. Phys.*, **C48**, 673–680.

Grimbach, Alois, Guazzini, Damiano, Knechtli, Francesco, and Palombi, Filippo (2008). O(a) improvement of the HYP static axial and vector currents at one-loop order of perturbation theory. *JHEP*, **03**, 039.

Grinstein, Benjamin, Jenkins, Elizabeth, Manohar, Aneesh V., Savage, Martin J., and Wise, Mark B. (1992). Chiral perturbation theory for f_{D_s}/f_D and B_{B_s}/B_B Nucl. *Phys.*, **B380**, 369–376.

Grozin, A. G., Marquard, P., Piclum, J. H., and Steinhauser, M. (2008). Three-loop chromomagnetic interaction in HQET. *Nucl. Phys.*, **B789**, 277–293.

Guagnelli, Marco *et al.* (2001). Non-perturbative results for the coefficients b_m and $b_A - b_P$ in O(a) improved lattice QCD. *Nucl. Phys.*, **B595**, 44–62.

Guazzini, Damiano, Meyer, Harvey B., and Sommer, Rainer (2007). Non-perturbative renormalization of the chromo-magnetic operator in heavy quark effective theory and the B* - B mass splitting. *JHEP*, **10**, 081.

Guazzini, Damiano, Sommer, Rainer, and Tantalo, Nazario (2008). Precision for B-meson matrix elements. *JHEP*, **01**, 076.

Hasenfratz, Anna and Knechtli, Francesco (2001). Flavor symmetry and the static potential with hypercubic blocking. *Phys. Rev.*, **D64**, 034504.

Hasenfratz, Peter, Laliena, Victor, and Niedermayer, Ferenc (1998). The index theorem in QCD with a finite cut-off. *Phys. Lett.*, **B427**, 125–131.

Hashimoto, S., Ishikawa, T., and Onogi, T. (2002). Nonperturbative calculation of Z(A) / Z(V) for heavy light currents using Ward-Takahashi identity. *Nucl. Phys. Proc. Suppl.*, **106**, 352–354.

Heitger, Jochen, Jüttner, Andreas, Sommer, Rainer, and Wennekers, Jan (2004). Non-perturbative tests of heavy quark effective theory. *JHEP*, **11**, 048.

Heitger, Jochen, Kurth, Martin, and Sommer, Rainer (2003). Non-perturbative renormalization of the static axial current in quenched QCD Nucl. Phys., B669, 173–206.

Horgan, R. R. *et al.* (2009). Moving NRQCD for heavy-to-light form factors on the lattice. *Phys. Rev.*, **D80**, 074505.

Isgur, Nathan and Wise, Mark B. (1989). Weak decays in the static quark approximation. *Phys. Lett.*, **B232**, 113–117.

Isgur, Nathan and Wise, Mark B. (1990). Weak transition form factors between heavy mesons. *Phys. Lett.*, **B237**, 527–530.

Itzykson, C. and Zuber, J. B. (1980). *Quantum field theory.* McGraw-Hill Inc. New York.

Ji, X. and Musolf, M. J. (1991). Subleading logarithmic mass dependence in heavy meson form- factors. *Phys. Lett.*, **B257**, 409.

Körner, J. G. and Thompson, George (1991). The heavy mass limit in field theory and the heavy quark effective theory. *Phys. Lett.*, **B264**, 185–192.

Kurth, Martin and Sommer, Rainer (2001). Renormalization and $O(a)$-improvement of the static axial current. *Nucl. Phys.*, **B597**, 488–518.

Kurth, Martin and Sommer, Rainer (2002). Heavy quark effective theory at one-loop order: An explicit example. *Nucl. Phys.*, **B623**, 271–286.

Lüscher, M. (1977). Construction of a self-adjoint, strictly positive transfer matrix for euclidean lattice gauge theories. *Commun. Math. Phys.*, **54**, 283.

Lüscher, M. (1986). Volume dependence of the energy spectrum in massive quantum field theories. 1. stable particle states. *Commun. Math. Phys.*, **104**, 177.

Lüscher, Martin (1998a). Advanced lattice QCD arXiv:hep-lat/9802029.

Lüscher, Martin (1998b). Exact chiral symmetry on the lattice and the Ginsparg-Wilson relation. *Phys. Lett.*, **B428**, 342–345.

Lüscher, Martin (2006). The Schrödinger functional in lattice QCD with exact chiral symmetry. *JHEP*, **05**, 042.

Lüscher, Martin, Narayanan, Rajamani, Weisz, Peter, and Wolff, Ulli (1992). The Schrödinger functional: A renormalizable probe for nonabelian gauge theories. *Nucl. Phys.*, **B384**, 168–228.

Lüscher, Martin, Sint, Stefan, Sommer, Rainer, and Weisz, Peter (1996). Chiral symmetry and $O(a)$ improvement in lattice QCD. *Nucl. Phys.*, **B478**, 365–400.

Lüscher, Martin, Sint, Stefan, Sommer, Rainer, and Wittig, Hartmut (1997). Non-perturbative determination of the axial current normalization constant in $O(a)$ improved lattice QCD. *Nucl. Phys.*, **B491**, 344–364.

Lüscher, Martin, Sommer, Rainer, Weisz, Peter, and Wolff, Ulli (1994). A precise determination of the running coupling in the SU(3) Yang-Mills theory. *Nucl. Phys.*, **B413**, 481–502.

Lüscher, M. and Weisz, P. (1985). On-shell improved lattice gauge theories. *Commun. Math. Phys.*, **97**, 59.

Lüscher, Martin, Weisz, Peter, and Wolff, Ulli (1991). A numerical method to compute the running coupling in asymptotically free theories. *Nucl. Phys.*, **B359**, 221–243.

Lüscher, Martin and Wolff, Ulli (1990). How to calculate the elastic scattering matrix in two-dimensional quantum field theories by numerical simulation. *Nucl. Phys.*, **B339**, 222–252.

Mandula, Jeffrey E. and Ogilvie, Michael C. (1998). Nonperturbative evaluation of the physical classical velocity in the lattice heavy quark effective theory. *Phys. Rev.*, **D57**, 1397–1410.

Mannel, Thomas, Roberts, Winston, and Ryzak, Zbigniew (1992). A derivation of the heavy quark effective lagrangian from qcd. *Nucl. Phys.*, **B368**, 204–220.

Melnikov, Kirill and Ritbergen, Timo van (2000). The three-loop relation between the MS-bar and the pole quark masses. *Phys. Lett.*, **B482**, 99–108.

Michael, Christopher and Teasdale, I. (1983). Extracting glueball masses from lattice QCD. *Nucl. Phys.*, **B215**, 433.

Montvay, I. and Münster, G. (1994). *Quantum fields on a lattice*. Cambridge Monographs on Mathematical Physics, Cambridge University Press, Cambridge.

Necco, Silvia and Sommer, Rainer (2002). The $N_f = 0$ heavy quark potential from short to intermediate distances. *Nucl. Phys.*, **B622**, 328–346.

Neuberger, Herbert (1998). Exactly massless quarks on the lattice. *Phys. Lett.*, **B417**, 141–144.

Neubert, Matthias (1994). Heavy quark symmetry. *Phys. Rep.*, **245**, 259–396.

Palombi, Filippo (2008). Non-perturbative renormalization of the static vector current and its O(a)-improvement in quenched QCD. *JHEP*, **01**, 021.

Palombi, Filippo, Papinutto, Mauro, Pena, Carlos, and Wittig, Hartmut (2006). A strategy for implementing non-perturbative renormalisation of heavy-light four-quark operators in the static approximation. *JHEP*, **08**, 017.

Palombi, Filippo, Papinutto, Mauro, Pena, Carlos, and Wittig, Hartmut (2007). Non-perturbative renormalization of static-light four-fermion operators in quenched lattice QCD. *JHEP*, **0709**, 062.

Politzer, H. D. and Wise, M. B. (1988). *Phys. Lett.*, **B206**, 681.

Shamir, Yigal (1993). Chiral fermions from lattice boundaries. *Nucl. Phys.*, **B406**, 90–106.

Sheikholeslami, B. and Wohlert, R. (1985). Improved continuum limit lattice action for QCD with Wilson fermions. *Nucl. Phys.*, **B259**, 572.

Shifman, Mikhail A. and Voloshin, M. B. (1987). On annihilation of mesons built from heavy and light quark and anti-$B_0 \leftrightarrow B_0$ oscillations. *Sov. J. Nucl. Phys.*, **45**, 292.

Sint, Stefan (1994). On the Schrödinger functional in QCD. *Nucl. Phys.*, **B421**, 135–158.

Sint, Stefan (1995). One loop renormalization of the QCD Schrödinger functional. *Nucl. Phys.*, **B451**, 416–444.

Sint, Stefan and Sommer, Rainer (1996). The running coupling from the QCD Schrödinger functional: A one loop analysis. *Nucl. Phys.*, **B465**, 71–98.

Sommer, R. (1994). A new way to set the energy scale in lattice gauge theories and its applications to the static force and α_s in SU(2) Yang-Mills theory. *Nucl. Phys.*, **B411**, 839.

Sommer, Rainer (2006). Non-perturbative QCD: Renormalization, O(a)-improvement and matching to heavy quark effective theory. *In perspectives in lattice QCD*, edition Y. Kuramashi World Scientific 2008.

Symanzik, K. (1981). Schrodinger representation and Casimir effect in renormalizable quantum field theory. *Nucl. Phys.*, **B190**, 1.

Symanzik, K. (1983a). Continuum limit and improved action in lattice theories. 1. Principles and ϕ^4 theory. *Nucl. Phys.*, **B226**, 187.

Symanzik, K. (1983b). Continuum limit and improved action in lattice theories. 2. O(N) nonlinear sigma model in perturbation theory. *Nucl. Phys.*, **B226**, 205.

Tantalo, N. (2008). Heavy-light meson's physics in Lattice QCD arXiv:0810.3624.

Thacker, B. A. and Lepage, G. Peter (1991). Heavy quark bound states in lattice QCD. *Phys. Rev.*, **D43**, 196–208.

Wise, Mark B. (1992). Chiral perturbation theory for hadrons containing a heavy quark. *Phys. Rev.*, **D45**, 2188–2191.

10
Lattice QCD and nuclear physics

Sinya AOKI

Graduate School of Pure and Applied Sciences,
University of Tsukuba Tsukuba, Ibaraki 305-8571, Japan
Lectures delivered at the École d'Été de Physique Théorique, Les Houches

10.1   Introduction: nuclear forces

In 1935 Yukawa (Yukawa, 1935) introduced virtual particles—pions—to explain the nuclear force that binds protons and neutrons inside nuclei. Since then, enormous efforts have been devoted to understanding the nucleon–nucleon (NN) interaction at low energies both from theoretical and experimental points of view. The notion of an NN potential turns out to be very useful in describing elastic NN scattering at low energy, that is, below the pion-production threshold, as well as properties of the deuteron (Taketani *et al.*, 1967; Hoshizaki *et al.*, 1968; Brown and Jackson, 1976; Machleidt, 1989; Machleidt and Slaus, 2001). This potential can be constructed phenomenologically so as to reproduce the scattering phase shifts and bound-state properties. Once the potential is determined, it can be used to study systems with more than two nucleons by using various many-body techniques.

Phenomenological NN potentials that can fit the NN data precisely (e.g., more than 2000 data points with $\chi^2/\mathrm{dof} \simeq 1$) at scattering energies $T_{\mathrm{lab}} < 300$ MeV are called high-precision NN potentials. Viewed in coordinate space, as in Fig.10.1, they reflect some characteristic features of the NN interaction at different length scales (Taketani *et al.*, 1967; Hoshizaki *et al.*, 1968; Brown and Jackson, 1976; Machleidt, 1989; Machleidt and Slaus, 2001):

(i) The long-range part of the nuclear force (distance $r > 2$ fm) is dominated by the one-pion exchange introduced by Yukawa (Yukawa, 1935). Because of the pion's Nambu–Goldstone character, it couples to the spin–isospin density of

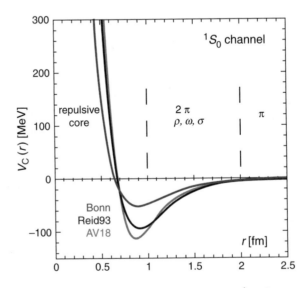

Fig. 10.1   Three examples of modern NN potentials in the $^1S_0$ (spin-singlet and s-wave) channel: CD–Bonn (Machleidt, 2001), Reid93 (Stoks *et al.*, 1994) and Argonne v_{18} (Wiringa *et al.*, 1995).

the nucleon and hence leads to a strong spin–isospin-dependent force, namely the tensor force.

(ii) The medium-range part (1 fm $< r <$ 2 fm) receives significant contributions from the exchange of two pions and of the heavy mesons ρ, ω, and σ. In particular, the spin–isospin-independent attraction of about 50–100 MeV in this region plays an essential role in the binding of atomic nuclei.

(iii) The short-range part ($r <$ 1 fm) is best described by a strong repulsive core as originally introduced by Jastrow (Jastrow, 1951). Such a short-range repulsion is important for the stability of atomic nuclei against collapse, for determining the maximum mass of neutron stars, and for igniting Type II supernova explosions (Tamagaki *et al.*, 1993; Heiselberg and Pandharipande, 2000; Lattimer and Prakash, 2000).

A repulsive core surrounded by an attractive well is in fact a common feature of the "effective potential" between composite particles. The Lennard-Jones potential between neutral atoms or molecules is a well-known example in atomic physics. The potential between ^{4}He nuclei is a typical example in nuclear physics. The origin of the repulsive cores in these examples is known to be Pauli exclusion among electrons or among nucleons. The same idea, however, is not directly applicable to the NN potential, because the quark has not only spin and flavor but also color, which allows six quarks to occupy the same state without violating the Pauli principle. To account for the repulsive core of the NN force, then, various other ideas have been proposed (Myhrer and Wroldsen, 1988; Oka *et al.*, 2000; Fujiwara *et al.*, 2007): the exchange of the neutral ω meson (Nambu, 1957), the exchange of a non-linear pion field (Jackson *et al.*, 1985; Yabu and Ando, 1985), and a combination of the Pauli principle with one-gluon exchange between quarks (Otsuki *et al.*, 1965; Machida and Namiki, 1965; Neudachin *et al.*, 1977; Liberman, 1977; DeTar, 1979; Oka and Yazaki, 1980; 1981*a*;*b*; Toki, 1980; Faessler *et al.*, 1982). Despite all these efforts, a convincing account of the nuclear force has yet to be produced.

In this situation, it is highly desirable to study the NN interaction with first-principle lattice QCD simulations. A theoretical framework suitable for this purpose was first proposed by Lüscher (Lüscher, 1991). For two hadrons in a finite box of size $L \times L \times L$, with periodic boundary conditions, an exact relation between the energy spectrum and the elastic-scattering phase shift can be derived. If the range R of the hadron interaction is much smaller than the size of the box, $R \ll L/2$, the behavior of the two-particle Bethe–Salpeter (BS) wave function $\psi(\mathbf{r})$ in the interval $R < |\mathbf{r}| < L/2$ contains sufficient information to relate the phase shift to the two-particle spectrum. Lüscher's method bypasses the difficulty of treating the real-time scattering process on the Euclidean lattice. Furthermore, it utilizes the finiteness of the lattice box effectively to extract the desired information about the on-shell scattering matrix and hence the phase shift.

An alternative but closely related approach to the NN interaction from lattice QCD has been proposed recently (Ishii *et al.*, 2007; Aoki *et al.*, 2008; 2010*b*). The starting point is the same BS wave function $\psi(\mathbf{r})$ as discussed in (Lüscher, 1991). Instead of looking at the wave function outside the range of the interaction, the authors consider the internal region $|\mathbf{r}| < R$ and define an energy-independent non-local potential $U(\mathbf{r}, \mathbf{r}')$ from $\psi(\mathbf{r})$ that obeys a Schrödinger-type equation in a finite box. For the NN interaction, $U(\mathbf{r}, \mathbf{r}')$ is localized in its spatial coordinates due to confinement of quarks and gluons, and hence the potential suffers from finite-volume effects only weakly in a large box. Thus, once U is determined and appropriately extrapolated to $L \to \infty$, one may simply use the Schrödinger equation in *infinite* space to calculate the scattering phase shifts and bound-state spectra to compare with experimental data. A further advantage of this potential is that it should be a smooth function of the quark masses, which makes it relatively easy to handle on the lattice. This is in sharp contrast to the scattering length, which shows singular behavior around a quark mass corresponding to the formation of the NN bound state.

This chapter is organized as follows. We first introduce Lüscher's method for the scattering phase shift in Section 10.2. Since the method is not only well established but also well explained in (Lüscher, 1991), we mainly describe the properties of the BS wave function, in terms of which the scattering phase shift can be related to the energy shift of the two-particle state in the finite box. These properties are also used to define the NN potential in Section 10.3, where the new method of (Ishii *et al.*, 2007; Aoki *et al.*, 2008; 2010*b*) is explained in detail. Finally, in Section 10.4 we consider a very recent attempt to understand the origin of the repulsive core in the NN potential. Using the operator product expansion and renormalization group analysis in QCD, the potential derived from the BS wave function in Section 10.3 is shown to have a repulsive core, whose functional form is also theoretically predicted (Aoki *et al.*, 2010*a*). Brief concluding remarks are given in Section 10.5.

10.2 Phase shift from lattice QCD: Lüscher's formula in finite volume

10.2.1 Preparation: The scattering phase shift in quantum mechanics

In this subsection, as a preparation for the later sections, we give some basics of scattering theory in quantum mechanics.

Let us consider the 3-dimensional Schrödinger equation,

$$[H_0 + V(\mathbf{r})]\varphi(\mathbf{r}) = E\varphi(\mathbf{r}), \tag{10.1}$$

where

$$H_0 = -\frac{\nabla^2}{2m}, \quad \nabla^2 = \partial_x^2 + \partial_y^2 + \partial_z^2. \tag{10.2}$$

Hereafter, we consider only a spherically symmetric potential, i.e. $V(\mathbf{r}) = V(r)$ with $r = |\mathbf{r}|$. In this case it is convenient to use polar coordinates, so that

$$\nabla^2 = \frac{1}{r^2}\frac{\partial}{\partial r}r^2\frac{\partial}{\partial r} - \frac{\hat{L}^2}{r^2}, \quad \hat{L}^2 = -\left[\frac{1}{\sin\theta}\frac{\partial}{\partial\theta}\left(\sin\theta\frac{\partial}{\partial\theta}\right) + \frac{1}{\sin^2\theta}\frac{\partial^2}{\partial\phi^2}\right]. \tag{10.3}$$

Using separation of variables, we consider a solution of the form

$$\varphi(\mathbf{r}) = \sum_{lm} A_{lm} R_l(r) Y_{lm}(\theta, \phi), \tag{10.4}$$

where the spherical harmonic function Y_{lm} satisfies

$$\hat{L}^2 Y_{lm}(\theta, \phi) = l(l+1) Y_{lm}(\theta, \phi) \tag{10.5}$$

$$\hat{L}_z Y_{lm}(\theta, \phi) = m Y_{lm}(\theta, \phi) \tag{10.6}$$

and is normalized as

$$\int_0^{2\pi} d\phi \int_0^{\pi} \sin\theta \, d\theta \, \overline{Y_{lm}(\theta, \phi)} \, Y_{l'm'}(\theta, \phi) = \delta_{ll'} \delta_{mm'}. \tag{10.7}$$

(Note that $\overline{X}$ means the complex conjugate of X while $X^\dagger$ is the Hermitan conjugate of X.) Explicitly, Y_{lm} is given by

$$Y_{lm}(\theta, \phi) = c\sqrt{\frac{2l+1}{4\pi} \cdot \frac{(l-|m|)!}{(l+|m|)!}} P_{lm}(\cos\theta) \, e^{im\phi}, \tag{10.8}$$

$$c = \begin{cases} (-1)^m & m > 0 \\ 1 & m \le 0, \end{cases} \tag{10.9}$$

where $P_{lm}(z)$ is the associated Legendre polynomial of degree l, defined by

$$P_{lm}(z) = (1 - z^2)^{m/2} \frac{d^{|m|} P_l(z)}{dz^{|m|}}, \quad |m| \le l, \tag{10.10}$$

$$P_l(z) = \frac{1}{2^l l!} \frac{d^l}{dz^l} (z^2 - 1)^l. \tag{10.11}$$

For small l and m, for example, we have

$$Y_{00}(\theta, \phi) = \frac{1}{\sqrt{4\pi}}, \quad Y_{10}(\theta, \phi) = \sqrt{\frac{3}{4\pi}} \cos\theta, \quad Y_{1,\pm1} = \mp\sqrt{\frac{3}{8\pi}} \sin\theta \, e^{\pm i\phi},$$

$$Y_{20}(\theta, \phi) = \sqrt{\frac{5}{4\pi}} \frac{1}{2} (3\cos^2\theta - 1), \quad Y_{2,\pm1}(\theta, \phi) = \mp\sqrt{\frac{15}{8\pi}} \sin\theta \cos\theta \, e^{\pm i\phi},$$

$$Y_{2,\pm2}(\theta, \phi) = \mp\sqrt{\frac{15}{32\pi}} \sin\theta \, e^{\pm 2i\phi}. \tag{10.12}$$

Using Eqs. (10.4) and (10.5), the 3-dimensional Schrödinger equation (10.1) is reduced to the 1-dimensional equation for R_l,

$$\frac{1}{r} \frac{d^2}{dr^2} [r R_l(r)] + \left\{ 2m [E - V(r)] - \frac{l(l+1)}{r^2} \right\} R_l(r) = 0. \tag{10.13}$$

Usually, we further assume the following properties for the potential $V(r)$:

$$\lim_{r \to 0} r^2 V(r) = 0, \tag{10.14}$$

$$\lim_{r \to \infty} r^n V(r) = 0 \qquad \text{for } \forall n \in \mathbf{Z}. \tag{10.15}$$

The first condition means

$$V(r) < O\left(\frac{1}{r^2}\right) \tag{10.16}$$

for small r, which leads to

$$R_l(r) = O(r^l) \tag{10.17}$$

for small r. The second condition means

$$V(r) = 0 \qquad \text{for } r > R, \tag{10.18}$$

for sufficiently large R, so that Eq. (10.13) becomes

$$R_l''(y) + \frac{2}{y} R_l'(y) + \left[1 - \frac{l(l+1)}{y^2}\right] R_l(y) = 0 \tag{10.19}$$

for $r > R$, where $y = kr$ and $k^2 = 2mE$. The general solution to this equation is

$$R_l(y) = A_l j_l(y) + B_l n_l(y), \tag{10.20}$$

where the spherical Bessel functions are given by

$$j_l(x) = (-x)^l \left(\frac{1}{x}\frac{d}{dx}\right)^l \left(\frac{\sin x}{x}\right) \simeq \begin{cases} \dfrac{x^l}{(2l+1)!!}, & x \to 0 \\[2mm] \dfrac{\sin(x - l\pi/2)}{x}, & x \to \infty \end{cases}, \tag{10.21}$$

$$n_l(x) = (-x)^l \left(\frac{1}{x}\frac{d}{dx}\right)^l \left(\frac{\cos x}{x}\right) \simeq \begin{cases} \dfrac{x^{-(l+1)}}{(2l-1)!!}, & x \to 0 \\[2mm] \dfrac{\cos(x - l\pi/2)}{x}, & x \to \infty \end{cases}. \tag{10.22}$$

In the $r \to \infty$ limit, then, the above solution becomes

$$R_l(r) \simeq A_l \frac{\sin\left(kr - \frac{l}{2}\pi\right)}{kr} + B_l \frac{\cos\left(kr - \frac{l}{2}\pi\right)}{kr} = C_l \frac{\sin\left(kr - \frac{l}{2}\pi + \delta_l(k)\right)}{kr}, \tag{10.23}$$

where the scattering phase shift $\delta_l(k)$ is given by

$$\tan \delta_l(k) = \frac{B_l}{A_l}, \qquad C_l = \sqrt{A_l^2 + B_l^2} \tag{10.24}$$

and $A_l = C_l \cos \delta_l(k)$, $B_l = C_l \sin \delta_l(k)$. If $\delta_l(k) > 0$ the interaction is attractive for this k, while it is repulsive if $\delta_l(k) < 0$.

For the particle–particle scattering in quantum mechanics, we consider the Lagrangian

$$L = \frac{1}{2} m_1 \dot{\mathbf{r}}_1^2 + \frac{1}{2} m_2 \dot{\mathbf{r}}_2^2 - V(|\mathbf{r}_1 - \mathbf{r}_2|). \tag{10.25}$$

By introducing the relative coordinate $\mathbf{r} = \mathbf{r}_1 - \mathbf{r}_2$ and the center of gravity $\mathbf{R} = (m_1 \mathbf{r}_1 + m_2 \mathbf{r}_2)/(m_1 + m_2)$, the above Lagrangian becomes

$$L = \frac{1}{2} M \dot{\mathbf{R}}^2 + \frac{1}{2} \mu \dot{\mathbf{r}}^2 - V(r), \tag{10.26}$$

where $M = m_1 + m_2$ is the total mass and $\mu = m_1 m_2/(m_1 + m_2)$ is the reduced mass. The corresponding Hamiltonian is given by

$$H = H_G + H_{\mathrm{rel}}, \quad H_G = \frac{1}{2M} \mathbf{P}^2, \ H_{\mathrm{rel}} = \frac{1}{2\mu} \mathbf{p}^2 + V(r), \tag{10.27}$$

where $\mathbf{P} = M \dot{\mathbf{R}}$ and $\mathbf{p} = \mu \dot{\mathbf{r}}$. While H_G is a Hamiltonian for a free particle, H_{rel} corresponds to the Hamiltonian for a particle in the potential $V(r)$, whose Schrödinger equation is identical to Eq. (10.1).

10.2.2 Bethe–Salpeter wave function and phase shift in quantum field theory

In this section, we construct a "scattering wave" in quantum field theory whose asymptotic behavior is identical to that of the scattering wave in quantum mechanics given in Eq. (10.23). Moreover, we show that the phase shift corresponds to the phase of the S matrix required by unitarity. For notational simplicity we consider $\pi\pi$ scattering in QCD.

The unitarity of the S-matrix $S^\dagger S = SS^\dagger = \mathbf{1}$ with $S = 1 + iT$ leads to

$$\langle f|T|i\rangle - \langle f|T^\dagger|i\rangle = i \sum_n \langle f|T^\dagger|n\rangle \langle n|T|i\rangle, \tag{10.28}$$

where $|n\rangle$ are asymptotic states. In the case of $\pi\pi$ scattering in the center of mass frame, $k_a + k_b \to k_c + k_d$, where $k_a = (E_k, \mathbf{k})$, $k_b = (E_k, -\mathbf{k})$ and $k_c = (E_p, \mathbf{p})$, $k_d = (E_p, -\mathbf{p})$ with $E_k = \sqrt{\mathbf{k}^2 + m_\pi^2}$ and $E_p = \sqrt{\mathbf{p}^2 + m_\pi^2}$, we write explicitly

$$\langle k_c, k_d|T|k_a, k_b\rangle = (2\pi)^4 \delta^{(4)}(k_a + k_b - k_c - k_d) \, T(\mathbf{p}, \mathbf{q}). \tag{10.29}$$

We consider elastic scattering, where the total energy is below the 4π threshold, $2\sqrt{k^2 + m_\pi^2} < 4m_\pi$; equivalently, $k^2 < 3m_\pi^2$ with $k = |\mathbf{k}| = |\mathbf{p}|$. In this case, due to energy–momentum conservation, the sum over intermediate state n in Eq. (10.28) is restricted to $\pi\pi$ states according to

$$\sum_n |n\rangle\langle n| = \int \frac{d^3\mathbf{p_1}}{(2\pi)^3 E_{p_1}} \frac{d^3\mathbf{p_2}}{(2\pi)^3 E_{p_2}} |p_1, p_2\rangle\langle p_1, p_2|. \tag{10.30}$$

Inserting this into Eq. (10.28), we have

$$T(\mathbf{p}, \mathbf{k}) - T^\dagger(\mathbf{p}, \mathbf{k}) = i\frac{k}{32\pi^2 E_k} \int d\Omega_q\, T^\dagger(\mathbf{p}, \mathbf{q})\, T(\mathbf{q}, \mathbf{k}), \tag{10.31}$$

where $|\mathbf{q}| = k$ and Ω_q is the solid angle of the vector $\mathbf{q}$. Using the partial wave decomposition,

$$T(\mathbf{p}, \mathbf{k}) = 4\pi \sum_{l=0}^\infty \sum_{m=-l}^l T_l(k) Y_{lm}(\Omega_p)\overline{Y_{lm}(\Omega_k)}, \tag{10.32}$$

and the orthogonality of the spherical harmonics $Y_{lm}(\theta, \phi) = Y_{lm}(\Omega_q)$,

$$\int d\Omega_q\, \overline{Y_{lm}(\Omega_q)}\, Y_{l'm'}(\Omega_q) = \delta_{ll'}\delta_{mm'}, \tag{10.33}$$

the unitarity relation (10.31) becomes

$$T_l(k) - \overline{T_l(k)} = i\frac{k}{8\pi E_k}\overline{T_l(k)}\, T_l(k). \tag{10.34}$$

A solution to this unitarity condition is easily obtained,

$$T_l(k) = \frac{16\pi E_k}{k} e^{i\delta_l(k)} \sin\delta_l(k), \tag{10.35}$$

where $\delta_l(k)$ is an arbitrary real function of k that can be interpreted as the scattering phase shift, as will be seen.

We now introduce the Bethe–Salpeter (BS) wave function for the $\pi\pi$ system, defined by

$$\varphi(\mathbf{r}) = \langle 0|T[\pi_a(\mathbf{x} + \mathbf{r}, t_a)\pi_b(\mathbf{x}, t_b)]|k_a, a, k_b, b; \text{in}\rangle, \tag{10.36}$$

where $|k_a, a, k_b, b; \text{in}\rangle$ is a $\pi\pi$ asymptotic in-state in the center-of-mass system such that $k_a = (E_k, \mathbf{k})$ and $k_b = (E_k, -\mathbf{k})$ with flavors a and b. The pion interpolating operator is given by

$$\pi_a(\mathbf{x}, x_0) = \bar{q}(x)i\gamma_5\tau_a q(x), \quad q(x) = \begin{pmatrix} u(x) \\ d(x) \end{pmatrix} \tag{10.37}$$

with the Pauli matrix τ_a. For simplicity we take $t_a = t_b + \epsilon$ with $\epsilon \geq 0$ and the $\epsilon \to 0$ limit. We then simply write

$$\varphi(\mathbf{r}) = \langle 0|\pi_a(\mathbf{x} + \mathbf{r}, t)\pi_b(\mathbf{x}, t)|k_a, a, k_b, b; \text{in}\rangle. \tag{10.38}$$

The name of the Bethe–Salpeter wave function comes from the fact that this quantity satisfies the Bethe–Salpeter equation (Bethe and Salpeter, 1951). Unlike field equations such as the Dyson–Schwinger equations, the BS equation is derived from the diagrammatic expansion. We here consider the BS wave function from a different point of view.

By inserting the complete set of out-states such that

$$1 = \sum_c \int \frac{d^3\mathbf{p}}{(2\pi)^3 2p_0} |\mathbf{p}, c; \text{out}\rangle \, \langle \mathbf{p}, c; \text{out}| + \sum_X \frac{|X; \text{out}\rangle \langle X; \text{out}|}{2E_X}, \qquad (10.39)$$

we have

$$\varphi(\mathbf{r}) = \varphi^{\text{elastic}}(\mathbf{r}) + \varphi^{\text{inelastic}}(\mathbf{r}), \qquad (10.40)$$

where

$$\varphi^{\text{elastic}}(\mathbf{r}) = \sqrt{Z_\pi} \int \frac{d^3\mathbf{p}}{(2\pi)^3 2p_0} e^{i\mathbf{p}\cdot(\mathbf{x}+\mathbf{r})-ip_0 t} \langle \mathbf{p}, a; \text{out}|\pi_b(\mathbf{x}, t)|\mathbf{k}, a, -\mathbf{k}, b; \text{in}\rangle,$$

$$\qquad (10.41)$$

$$\varphi^{\text{inelastic}}(\mathbf{r}) = \sum_X \langle 0|\pi_a(\mathbf{x}+\mathbf{r}, t)|X; \text{out}\rangle \frac{1}{2E_X} \langle X; \text{out}|\pi_b(\mathbf{x}, t)|\mathbf{k}, a, -\mathbf{k}, b; \text{in}\rangle. (10.42)$$

Here, $|\mathbf{p}, a, ; \text{out}\rangle$ is a one-pion out-state with momentum $\mathbf{p}$, which satisfies

$$\langle 0|\pi_a(\mathbf{x}, x_0)|\mathbf{p}, b; \text{out}\rangle = \delta_{ab}\sqrt{Z_\pi}\, e^{-ip\cdot x}, \quad p_0 = \sqrt{\mathbf{p}^2 + m_\pi^2}, \qquad (10.43)$$

while X represents general states other than one-pion states.

Using the reduction formula,

$$a_{\text{out}}(\mathbf{p})T(O) - T(O)a_{\text{in}}(\mathbf{p}) = (-p^2 + m_\pi^2)T(\pi(p)O) \qquad (10.44)$$

$$T(O)a_{\text{in}}^\dagger(\mathbf{p}) - a_{\text{out}}^\dagger(\mathbf{p})T(O) = (-p^2 + m_\pi^2)T(O\pi^\dagger(p)), \qquad (10.45)$$

where O is an arbitrary field operator and

$$\pi(p) = \int d^4x \frac{e^{ipx}}{\sqrt{Z_\pi}} \pi(x), \qquad (10.46)$$

we obtain

$$\langle \mathbf{p}, a; \text{out}|\pi_b(\mathbf{x}, t)|\mathbf{k}, a, -\mathbf{k}, b; \text{in}\rangle = \sqrt{Z_\pi}(2\pi)^3 2k_0 \, \delta^3(\mathbf{p} - \mathbf{k})\, e^{-ikx}$$

$$+ \sqrt{Z_\pi} \frac{e^{-iqx}}{m_\pi^2 - q^2 - i\varepsilon} \hat{T}(p, q, k_a, k_b), \qquad (10.47)$$

where the off-shell T-matrix $\hat{T}$ is defined by

$$\hat{T}(p, q, k_a, k_b) = (-p^2 + m_\pi^2)(-q^2 + m_\pi^2)G(p, q, k_a, k_b)(-k_a^2 + m_\pi^2)(-k_b^2 + m_\pi^2),$$
(10.48)

$$G(p, q, k_a.k_b)i(2\pi)^4\delta^{(4)}(p + q - k_a - k_b) = \langle 0|T\{\pi_a(p)\pi_b(q)\pi_a^\dagger(k_a)\pi_b^\dagger(k_b)\}|0\rangle.$$
(10.49)

Here, $p = (\mathbf{p}, p_0)$, $k_a = (\mathbf{k}, k_0)$, and $k_b = (-\mathbf{k}, k_0)$ are on-shell 4-momenta, while $q = (-\mathbf{p}, 2k_0 - p_0)$ is generally off-shell. Using this expression we obtain

$$\varphi^{\text{elastic}}(\mathbf{r}) = Z_\pi e^{-i2k_0 t}e^{i\mathbf{k}\cdot\mathbf{r}} + Z_\pi e^{-i2k_0 t}\int \frac{d^3\mathbf{p}}{(2\pi)^3 2p_0}\frac{e^{i\mathbf{p}\cdot\mathbf{r}}}{m_\pi^2 - q^2 - i\varepsilon}\hat{T}(p, q, k_a, k_b).$$
(10.50)

Similarly we have

$$\varphi^{\text{inelastic}}(\mathbf{r}) = e^{-i2k_0 t}\sum_X \frac{\sqrt{Z_\pi Z_X}}{2E_X}\frac{e^{i\mathbf{p}_X\mathbf{r}}}{m_\pi^2 - q^2 - i\varepsilon}\hat{T}_X(p_X, q, k_a, k_b),$$
(10.51)

where $q = (-\mathbf{p}_X, 2k_0 - (p_X)_0)$. For simplicity we hereafter set $t = 0$. We rescale φ^{elastic} as

$$\varphi(\mathbf{r}) = Z_\pi \varphi^{\text{elastic}}(\mathbf{r}) + \varphi^{\text{inelastic}}(\mathbf{r}),$$
(10.52)

where

$$\varphi^{\text{elastic}}(\mathbf{r}) = e^{i\mathbf{k}\cdot\mathbf{r}} + \int \frac{d^3\mathbf{p}}{(2\pi)^3}\frac{1}{p^2 - k^2 - i\varepsilon}H(\mathbf{p}, \mathbf{k})e^{i\mathbf{p}\cdot\mathbf{r}},$$
(10.53)

with

$$H(\mathbf{p}, \mathbf{k}) = \frac{p_0 + k_0}{8p_0 k_0}T(\mathbf{p}, \mathbf{q}),$$
(10.54)

and $k = |\mathbf{k}|$ and $p = |\mathbf{p}|$.

We now investigate the large $r = |\mathbf{r}|$ behavior of the BS wave function below the 4π inelastic threshold. We first consider φ^{elastic}. Using the following partial wave decomposition,

$$H(\mathbf{p}, \mathbf{k}) = 4\pi \sum_{lm} H_l(p, k)Y_{lm}(\Omega_\mathbf{p})\overline{Y_{lm}(\Omega_\mathbf{k})}$$
(10.55)

$$\varphi^{\text{elastic}}(\mathbf{r}) = 4\pi \sum_{lm} i^l \varphi_l^{\text{elastic}}(r, k)Y_{lm}(\Omega_\mathbf{r})\overline{Y_{lm}(\Omega_\mathbf{k})}$$
(10.56)

$$e^{i\mathbf{p}\cdot\mathbf{r}} = 4\pi \sum_{lm} i^l j_l(pr)Y_{lm}(\Omega_\mathbf{r})\overline{Y_{lm}(\Omega_\mathbf{p})},$$
(10.57)

we have

$$\varphi_l^{\text{elastic}}(r,k) = j_l(kr) + \int \frac{p^2 dp}{2\pi^2} \frac{1}{p^2 - k^2 - i\varepsilon} H_l(p,k) j_l(pr). \tag{10.58}$$

We assume that the interaction vanishes for large r:

$$-(\nabla^2 + k^2)\varphi^{\text{elastic}}(\mathbf{r}) = \int \frac{d^3 p}{(2\pi)^3} H(\mathbf{p},\mathbf{k}) e^{i\mathbf{p}\cdot\mathbf{r}} \xrightarrow[r\to\infty]{} 0, \tag{10.59}$$

which, in terms of the partial wave, gives

$$\int \frac{p^2 dp}{2\pi^2} H_l(p,k) j_l(pr) \xrightarrow[r\to\infty]{} 0. \tag{10.60}$$

We evaluate the second term of Eq. (10.58). Using the explicit form of the spherical Bessel function $j_l(x)$ in Eq. (10.21) we have

$$\int \frac{p^2 dp}{2\pi^2} \frac{1}{p^2 - k^2 - i\varepsilon} H_l(p,k) j_l(pr) = (-r)^l \left(\frac{1}{r}\frac{d}{dr}\right)^l \int_0^\infty \frac{p^{2-l} dp}{4\pi^2 ipr} \frac{e^{ipr} - e^{-ipr}}{p^2 - k^2 - i\varepsilon} H_l(p,k)$$

$$= (-r)^l \left(\frac{1}{r}\frac{d}{dr}\right)^l \int_{-\infty}^\infty \frac{p^{2-l} dp}{4\pi^2 ipr} \frac{e^{ipr}}{p^2 - k^2 - i\varepsilon} H_l(p,k). \tag{10.61}$$

Here, we used the property that $H_l(-p,k) = (-1)^l H_l(p,k)$, which is shown as

$$H(\mathbf{p},\mathbf{k}) = 4\pi \sum_{lm} H_l(p,k) Y_{lm}(\Omega_{\mathbf{p}}) \overline{Y_{lm}(\Omega_{\mathbf{k}})} = 4\pi \sum_{lm} H_l(-p,k) Y_{lm}(\Omega_{-\mathbf{p}}) \overline{Y_{lm}(\Omega_{\mathbf{k}})}$$

$$= 4\pi \sum_{lm} (-1)^l H_l(-p,k) Y_{lm}(\Omega_{\mathbf{p}}) \overline{Y_{lm}(\Omega_{\mathbf{k}})}, \tag{10.62}$$

where $\Omega_{-\mathbf{p}}$ means $\theta \to \pi - \theta$ and $\phi \to \pi + \phi$. To proceed, we assume $H_l(p,k) = O(p^l)$ for small p, so that no contribution around $p = 0$ appears in the above integral.

We know that the half-off-shell T-matrix $H_l(p,k)$ does not have any poles or cuts on the real axis, since k^2 is below the inelastic threshold. For simplicity, we assume that $H_l(p,k)$ has only poles in the complex p plane (this argument may also be generalized for cuts):

$$H_l(p,k) = \sum_{\substack{n>0,\\ \text{Im } k_n > 0}} \frac{Z_n}{p - k_n} + \sum_{\substack{n<0,\\ \text{Im } k_n < 0}} \frac{Z_n}{p - k_n} + \tilde{H}_l(p,k), \tag{10.63}$$

where $\tilde{H}_l(p,k)$ is analytic in p. Note that k_n and Z_n implicitly depend on k. This form of the assumption satisfies the condition (10.60), which becomes

$$(-r)^l \left(\frac{1}{r}\frac{d}{dr}\right)^l \sum_{\substack{n>0,\\ \text{Im } k_n > 0}} \frac{k_n^{2-l} Z_n}{2\pi k_n r} e^{i(\text{Re} k_n)r} e^{-(\text{Im } k_n)r} \simeq 0, \tag{10.64}$$

for $\bar{k}_0 r \gg 1$, where $\bar{k}_0 = \min_{n>0} \mathrm{Im}\, k_n > 0$. Using the assumption (10.63), we can evaluate Eq. (10.61) as

$$\frac{4}{4\pi} H_l(k,k)(-kr)^l \left(\frac{1}{kr}\frac{d}{d(kr)}\right)^l \frac{e^{ikr}}{kr} + (-r)^l \left(\frac{1}{r}\frac{d}{dr}\right)^l \sum_{n>0} \frac{k_n^{2-l} Z_n e^{ik_n r}}{2\pi k_n r(k_n^2 - k^2)}$$

$$= \frac{k}{4\pi} H_l(k,k)\{n_l(kr) + ij_l(kr)\} + (-r)^l \left(\frac{1}{r}\frac{d}{dr}\right)^l \sum_{n>0} \frac{k_n^{2-l} Z_n}{2\pi k_n r(k_n^2 - k^2)} e^{i(\mathrm{Re} k_n)r} e^{-(\mathrm{Im}\, k_n)r}.$$

$$(10.65)$$

Since $\mathrm{Im}\, k_n > 0$, the sum over $n > 0$ vanishes exponentially for large r, which satisfies $\bar{k}_0 r \gg 1$. Similarly, we can show that $\varphi^{\mathrm{inelastic}}(\mathbf{r})$ vanishes exponentially for large r as long as k^2 is below the inelastic threshold.

Therefore, for large r ($\bar{k}_0 r \gg 1$) we finally obtain

$$\varphi_l(r,k) \simeq Z_\pi \left\{ j_l(kr) + \frac{k}{4\pi} H_l(k,k)[n_l(kr) + ij_l(kr)] \right\}$$

$$\simeq Z_\pi \frac{e^{i\delta_l(k)}}{kr} \sin(kr + \delta_l(k) - l\pi/2), \qquad (10.66)$$

where we have used the form of the on-shell T-matrix determined from unitarity in Eq. (10.35),

$$H(k,k) = \frac{4\pi}{k} e^{i\delta_l(k)} \sin\delta_l(k), \qquad (10.67)$$

and the asymptotic behavior of $j_l(x)$ and $n_l(x)$,

$$j_l(x) \simeq \frac{\sin(x - l\pi/2)}{x}, \quad n_l(x) \simeq \frac{\cos(x - l\pi/2)}{x}. \qquad (10.68)$$

The derivation of the large-r behavior of the BS wave function in this section is similar to, but a little different from, that in (Lin *et al.*, 2002; Aoki *et al.*, 2005*b*) for $\pi\pi$ and in (Ishizuka, 2009; Aoki *et al.*, 2010*b*) for NN, though the final results are the same.

10.2.3 Lüscher's formula for the phase shift in finite volume

We now consider finite volume (Lüscher, 1991). We assume that there is no interaction (except for exponentially small contributions) for $r \geq R$, where R is sufficiently large. Therefore, if the box size L is larger than $2R$, there exists a region $R < r < L/2$ where

$$(\nabla^2 + k^2)\varphi_L(\mathbf{r}; k) = 0 \qquad (10.69)$$

is satisfied for the BS wave function $\varphi_L(\mathbf{r}; k)$ given by

$$\varphi_L(\mathbf{r}; k) = \langle 0|\pi_a(\mathbf{x}+\mathbf{r}, 0)\pi_b(\mathbf{y}, 0)|k_a, a, k_b, b\rangle_L, \qquad (10.70)$$

where the subscript L indicates that the state is constructed in the finite box. We expand this wave function in terms of the BS wave function in the infinite volume, introduced above, as

$$\varphi_L(\mathbf{r}; k) = 4\pi \sum_{lm} C_{lm}(k)\varphi_l(r, k)Y_{lm}(\Omega_\mathbf{r}), \tag{10.71}$$

where the coefficient $C_{lm}(k)$ is introduced to satisfy the periodic boundary conditions $\varphi_L(\mathbf{r} + \mathbf{n}L; k) = \varphi_L(\mathbf{r}; k)$ for $\mathbf{n} = (n_x, n_y, n_z) \in \mathbf{Z}^3$. Note that

$$\varphi_l(r, k) = n_l(kr)e^{i\delta_l(k)}\sin\delta_l(k) + j_l(kr)e^{i\delta_l(k)}\cos\delta_l(k) \tag{10.72}$$

for $r \geq R$.

On the other hand, we can construct the solution of the Helmholtz equation,

$$(\nabla^2 + k^2)\varphi_L(\mathbf{r}; k) = 0, \tag{10.73}$$

for $\mathbf{r} \neq 0$ with periodic boundary conditions as

$$\varphi_L(\mathbf{r}; k) = \sum_{lm} v_{lm}(k)G_{lm}(\mathbf{r}, k), \tag{10.74}$$

where

$$G_{lm}(\mathbf{r}, k) = \sqrt{4\pi}\mathbf{Y}_{lm}(\nabla)G(\mathbf{r}, k) \tag{10.75}$$

$$G(\mathbf{r}, k) = \frac{1}{L^3}\sum_{\mathbf{p}\in\Gamma}\frac{e^{i\mathbf{p}\cdot\mathbf{r}}}{\mathbf{p}^2 - k^2}, \quad \Gamma = \left\{\mathbf{p}\Big|\mathbf{p} = \mathbf{n}\frac{2\pi}{L}, \mathbf{n} \in \mathbf{Z}^3\right\}, \tag{10.76}$$

$$\mathbf{Y}_{lm}(\mathbf{p}) \equiv p^l Y_{lm}(\Omega_\mathbf{p}), \quad p = |\mathbf{p}|. \tag{10.77}$$

It is easy to see that $\varphi_L(\mathbf{r}; k)$ satisfies both the Helmholtz equation and the periodic boundary conditions for arbitrary $v_{lm}(k)$s since

$$(\nabla^2 + k^2)\varphi_L(\mathbf{r}; k) = \sum_{lm} v_{lm}(k)\sqrt{4\pi}\mathbf{Y}_{lm}(\nabla)(\nabla^2 + k^2)G(\mathbf{r}, k)$$

$$= \sum_{lm} v_{lm}(k)\sqrt{4\pi}\mathbf{Y}_{lm}(\nabla)\delta^3(\mathbf{r}) = 0 \tag{10.78}$$

for $\mathbf{r} \neq 0$ and

$$\varphi_L(\mathbf{r} + \mathbf{n}L; k) = \sum_{lm} v_{lm}(k)\sqrt{4\pi}\mathbf{Y}_{lm}(\nabla)G(\mathbf{r} + \mathbf{n}L, k)$$

$$= \sum_{lm} v_{lm}(k)\sqrt{4\pi}\mathbf{Y}_{lm}(\nabla)G(\mathbf{r}, k) = \varphi_L(\mathbf{r}; k). \tag{10.79}$$

The coefficients v_{lm} can be determined by comparing Eq. (10.74) with Eq. (10.71). We first rewrite

$$G(\mathbf{r}, k) = \frac{k}{4\pi} n_0(kr) + \sum_{lm} \sqrt{4\pi} Y_{lm}(\Omega_{\mathbf{r}}) g_{lm}(k) j_l(kr), \qquad (10.80)$$

where

$$g_{lm}(k) = \sqrt{4\pi} \frac{1}{L^3} \sum_{\mathbf{p} \in \Gamma} \frac{(ip/k)^l}{\mathbf{p}^2 - k^2} \overline{Y_{lm}(\Omega_{\mathbf{p}})}. \qquad (10.81)$$

This can be easily seen as follows. Since

$$(\nabla^2 + k^2) \frac{k}{4\pi} n_0(kr) = \delta(\mathbf{r}), \qquad (10.82)$$

the quantity

$$G(\mathbf{r}, k) - \frac{k}{4\pi} n_0(kr) \qquad (10.83)$$

satisfies the Helmholtz equation for all $\mathbf{r}$ and is smooth at $r \to 0$, so that it can be expanded in spherical harmonics as

$$G(\mathbf{r}, k) - \frac{k}{4\pi} n_0(kr) = \sum_{lm} \sqrt{4\pi} g_{lm}(k) j_l(kr) Y_{lm}(\Omega_{\mathbf{r}}). \qquad (10.84)$$

Using

$$e^{i\mathbf{p} \cdot \mathbf{r}} = 4\pi \sum_{lm} i^l j_l(pr) Y_{lm}(\Omega_{\mathbf{r}}) \overline{Y_{lm}(\Omega_{\mathbf{p}})} \qquad (10.85)$$

in Eq. (10.76), and considering the $r \to 0$ limit of both sides of Eq. (10.84), we obtain Eq. (10.81). (Note that $j_l(x) \simeq x^l/(2l+1)!!$ as $x \to 0$.)

We next observe (Lüscher, 1991) that

$$G_{lm}(\mathbf{r}, k) = \sqrt{4\pi} \mathbf{Y}_{lm}(\nabla) G(\mathbf{r}, k) = \frac{(-k)^l k}{4\pi}$$

$$\times \left[Y_{lm}(\Omega_{\mathbf{r}}) n_l(kr) + \sum_{l', m'} \mathbf{M}_{lm, l'm'} Y_{l'm'}(\Omega_{\mathbf{r}}) j_{l'}(kr) \right], \qquad (10.86)$$

where the non-zero elements of $\mathbf{M}_{lm, l'm'}$ are given by linear combinations of

$$M_{lm} = \frac{1}{i^l(2l+1)} \frac{4\pi}{k} g_{lm}(k). \qquad (10.87)$$

The following properties generally hold:

$$\mathbf{M}_{lm, l'm'} = \mathbf{M}_{l'm', lm} = \mathbf{M}_{l-m, l'-m'}. \qquad (10.88)$$

The non-zero elements for $l, l' \leq 3$ can be expressed as

$$\mathbf{M}_{lm,l'm'} = a M_{00} + b M_{40} + c M_{60}, \tag{10.89}$$

with a, b, c given in Table 10.1. See (Lüscher, 1991) for more details.

We now consider the cubic group $O(3, \mathbf{Z})$, which has 24 elements and is generated by following elements in the special cubic group $SO(3, \mathbf{Z})$

$$R_x = \begin{pmatrix} 1 & 0 & 0 \\ 0 & 0 & -1 \\ 0 & 1 & 0 \end{pmatrix}, \quad R_y = \begin{pmatrix} 0 & 0 & 1 \\ 0 & 1 & 0 \\ -1 & 0 & 0 \end{pmatrix}, \quad R_z = \begin{pmatrix} 0 & -1 & 0 \\ 1 & 0 & 0 \\ 0 & 0 & 1 \end{pmatrix}, \tag{10.90}$$

together with the parity transformation $P\mathbf{r} = -\mathbf{r}$. There are five irreducible representations of $SO(3, \mathbf{Z})$, denoted by A_1, A_2, E, T_1, and T_2, whose dimensions are 1, 1, 2, 3, and 3, respectively. The irreducible representations of $O(3, \mathbf{Z})$ are constructed from these five irreducible representations of $SO(3, \mathbf{Z})$ and the parity eigenvalue ± 1. The irreducible representations of the rotation group $O(3, \mathbf{R})$ can be decomposed in terms of these irreducible representations. For example,

$$\mathbf{0} = A_1^+, \ \mathbf{1} = T_1^-, \ \mathbf{2} = E^+ \oplus T_2^+,$$

$$\mathbf{3} = A_2^- \oplus T_1^- \oplus T_2^-, \ \mathbf{4} = A_1^+ \oplus E^+ \oplus T_1^+ \oplus T_2^+, \tag{10.91}$$

Table 10.1 Non-zero independent elements of $\mathbf{M}_{lm,l'm'}$.

$\mathbf{M}_{lm,l'm'}$	a (M_{00})	b (M_{40})	c (M_{60})
$\mathbf{M}_{00,00}$	1	0	0
$\mathbf{M}_{1m,1m}$	1	0	0
$\mathbf{M}_{20,20}$	1	$18/7$	0
$\mathbf{M}_{21,21}$	1	$-12/7$	0
$\mathbf{M}_{22,22}$	1	$3/7$	0
$\mathbf{M}_{22,2-2}$	0	$15/7$	0
$\mathbf{M}_{30,10}$	0	$-4\sqrt{21}/7$	0
$\mathbf{M}_{31,11}$	0	$3\sqrt{14}/7$	0
$\mathbf{M}_{33,1-1}$	0	$\sqrt{210}/7$	0
$\mathbf{M}_{30,30}$	1	$18/11$	$100/33$
$\mathbf{M}_{31,31}$	1	$3/11$	$-25/11$
$\mathbf{M}_{32,32}$	1	$-21/11$	$10/11$
$\mathbf{M}_{32,3-2}$	0	$15/11$	$-70/11$
$\mathbf{M}_{33,33}$	1	$9/11$	$-5/33$
$\mathbf{M}_{33,3-1}$	0	$3\sqrt{15}/11$	$35\sqrt{15}/33$

Table 10.2 Decomposition of the angular momentum into irreducible representations of the cubic group

l	rep.	basis polynomials	independent elements
0	A_1^+	1	
1	T_1^-	r_i	$i = 1, 2, 3$
2	E^+	$r_i^2 - r_j^2$	$(i, j) = (1, 2), (2, 3)$
2	T_2^+	$r_i r_j$	$i \neq j$
3	A_2^-	$r_1 r_2 r_3$	
3	T_1^-	$5r_i^3 - 3r^2 r_j$	$i = 1, 2, 3$
3	T_2^-	$r_i(r_j^2 - r_k^2)$	$(i, j, k) = (1, 2, 3), (2, 3, 1), (3, 1, 2)$
4	A_1^+	$5(r_1^4 + r_2^4 + r_3^4) - 3r^4$	
4	E^+	$7(r_i^4 - r_j^4) - 6r^2(r_i^2 - r_j^2)$	$(i, j) = (1, 2), (2, 3)$
4	T_1^+	$r_i r_j^3 - r_j r_i^3$	$i \neq j$
4	T_2^+	$7(r_i r_j^3 + r_j r_i^3) - 6r^2 r_i r_j$	$i \neq j$

where each number is the eigenvalue of the angular momentum l that defines the representation of $O(3, \mathbf{R})$. The basis polynomials for each cubic representation are given in Table 10.2.

We now compare the two expressions for $\varphi_L(\mathbf{r}; k)$ in some irreducible representations of the cubic group. First, we project the BS wave function to the A_1^+ representation, which contains the $l = 0$ partial wave as well as $l \geq 4$ contributions. Neglecting $l \geq 4$, Eq. (10.71) becomes

$$\varphi_L^{A_1^+}(\mathbf{r}; k) = \sqrt{4\pi} C_{00}(k) e^{i\delta_0(k)} \left[n_0(kr) \sin \delta_0(k) + j_0(kr) \cos \delta_0(k) \right] \quad (10.92)$$

for $r \geq R$. In order to match this expression, Eq. (10.74) must be

$$\varphi_L^{A_1^+}(\mathbf{r}; k) = \sqrt{4\pi} v_{00}(k) G_{00}(\mathbf{r}, k)$$

$$= \sqrt{4\pi} v_{00} \left[n_0(kr) + \sum_{lm} \mathbf{M}_{00, lm} Y_{lm}(\Omega_{\mathbf{r}}) j_l(kr) \right], \quad (10.93)$$

since $G_{lm}(\mathbf{r}, k)$ with $l \neq 0$, which contains $n_l(kr)$, can not appear in this equation. By comparing the two, we have

$$C_{00}(k) e^{i\delta_0(k)} \sin \delta_0(k) = v_{00} \quad (10.94)$$

$$C_{00}(k) e^{i\delta_0(k)} \cos \delta_0(k) = v_{00} \mathbf{M}_{00,00} = v_{00} M_{00}, \quad (10.95)$$

which leads to the famous Lüscher formula (Lüscher, 1991),

$$\cot(\delta_0(k)) = M_{00} = \frac{4\pi}{k} g_{00}(k) = \frac{4\pi}{k} \frac{1}{L^3} \sum_{\mathbf{p} \in \Gamma} \frac{1}{\mathbf{p}^2 - k^2}. \quad (10.96)$$

Note that the unmatched components proportional to $\mathbf{M}_{00,4m}$ ($m = 0, \pm 4$) give the $l = 4$ contributions.

Let us briefly explain how to use this formula. We first calculate the energy $E_2(L)$ of two pions in the center-of-mass frame on the finite L^3 box with periodic boundary conditions, where L is assumed to be larger than $2R$. We then determine k, the magnitude of the relative momentum of the two pions, from

$$E_2(L) = 2\sqrt{k^2 + m_\pi^2}, \qquad (10.97)$$

where m_π is the pion mass in the infinite volume limit. We finally determine $\delta_0(k)$ by solving Eq. (10.96). It should be noted that the momentum for one pion is quantized as $\mathbf{p} = 2\pi\mathbf{n}/L$ in the finite box with the periodic boundary conditions. If the interaction between the two pions were absent, we would have $E_2(L) = 2\sqrt{\mathbf{p}^2 + m_\pi^2}$, so that $k = |\mathbf{p}|$. The presence of the interaction makes k a little different from $|\mathbf{p}|$ in the finite box. The Lüscher formula relates this difference to the scattering phase shift. By applying the formula for two pions with a range of values of $\mathbf{p}$ and L, we can determine the scattering phase shift $\delta_0(k)$ at several values of k.

Similarly, we consider the T_1^- representation, which contains the $l = 1$ partial wave as well as $l \geq 3$ contributions. If the latter are neglected, we again obtain

$$\cot(\delta_1(k)) = M_{00} = \frac{4\pi}{k}g_{00}(k). \qquad (10.98)$$

On the other hand, if the contribution from $l = 3$ partial wave can not be neglected, in addition to the $l = 1$ component we have

$$\varphi_L^{T_1^-}(\mathbf{r}; k) = 4\pi \sum_m C_{1m}(k)Y_{1m}(\Omega_{\mathbf{rr}})e^{i\delta_1(k)}\left[n_1(kr)\sin\delta_1(k) + j_1(kr)\cos\delta_1(k)\right]$$

$$+ 4\pi \sum_m C_{3m}(k)Y_{3m}(\Omega_{\mathbf{r}})e^{i\delta_3(k)}\left[n_3(kr)\sin\delta_3(k) + j_3(kr)\cos\delta_3(k)\right],$$

$$(10.99)$$

which should be compared with

$$\varphi_L^{T_1^-}(\mathbf{r}; k) = \sum v_{1m}(k)G_{1m}(\mathbf{r}, k) + \sum v_{3m}(k)G_{3m}(\mathbf{r}, k). \qquad (10.100)$$

From the matching condition we obtain, after a little algebra,

$$\det \begin{pmatrix} M_{10,10} - \cot\delta_1(k), & M_{30,10} \\ M_{30,10}, & M_{30,30} - \cot\delta_3(k) \end{pmatrix} = 0, \qquad (10.101)$$

for $m = 0$, and

$$\det \begin{pmatrix} M_{11,11} - \cot\delta_1(k), & M_{31,11}, & M_{33,1-1} \\ M_{31,11}, & M_{31,31} - \cot\delta_3(k), & M_{33,3-1} \\ M_{33,1-1}, & M_{33,3-1}, & M_{33,33} - \cot\delta_3(k) \end{pmatrix} = 0, \quad (10.102)$$

for $m = 1, -3$ or $m = -1, 3$, while we have

$$\det \begin{pmatrix} \mathbf{M}_{32,32} - \cot \delta_3(k), & \mathbf{M}_{32,3-2} \\ \mathbf{M}_{32,3-2}, & \mathbf{M}_{32,32} - \cot \delta_3(k) \end{pmatrix} = 0, \qquad (10.103)$$

for $m = 2, -2$. From the last equation, we can determine $\delta_3(k)$ at some k and L. Putting this $\delta_3(k)$ into the first or the second equation, we can also extract $\delta_1(k)$.

10.2.4 Some references for the $\pi\pi$ phase shift from lattice QCD

The BS wave function for the $\pi\pi$ system in the isospin $I = 2$ channel has been investigated in quenched QCD at $k \simeq 0$ (Aoki *et al.*, 2005*a*) to extract the scattering length a_0 through the Lüscher formula in the center-of-mass system. The scattering length is related to the scattering phase shift $\delta_0(k)$ via

$$\frac{k}{\tan \delta_0(k)} = \frac{1}{a_0} + r_0 \frac{k^2}{2} + O(k^4), \qquad (10.104)$$

where r_0 is called the effective range. The calculation of the BS wave function for the $\pi\pi$ system in the $I = 2$ channel has been extended to the case of non-zero momentum in quenched QCD (Sasaki and Ishizuka, 2008) and the $I = 2$ $\pi\pi$ scattering phase shift can be extracted using the Lüscher formula in the laboratory system (Rummukainen and Gottlieb, 1995).

There are only a few calculations of the $\pi\pi$ scattering phase shift in unquenched theories. The $I = 2$ $\pi\pi$ scattering length and phase shift have been calculated through the Lüscher formula in 2-flavor lattice QCD with $O(a)$-improved Wilson fermions in both center-of-mass and laboratory systems (Yamazaki *et al.*, 2004). Both chiral and continuum extrapolations have been taken, though the pion masses in the simulation are rather heavy.

The $I = 2$ $\pi\pi$ scattering length has been calculated in 2+1-flavor mixed-action lattice QCD, using domain-wall valence quarks with asqtad-improved staggered sea quarks for $m_\pi^{\text{sea}} \simeq 294$, 348 and 484 MeV at $a \simeq 0.125$ fm (Beane *et al.*, 2006; 2008). The scattering phase shift has also been calculated at $k \simeq 544$ MeV and $m_\pi \simeq 484$ MeV.

Recently, the $I = 2$ $\pi\pi$ scattering length has been calculated in 2-flavor twisted-mass lattice QCD for pion masses ranging from 270 MeV to 485 MeV at $a \simeq 0.086$ fm (Feng *et al.*, 2010). The lattice spacing error is estimated at $a \simeq 0.067$ fm for one pion mass.

The P-wave scattering phase shift for the $I = 1$ $\pi\pi$ system has been calculated in 2-flavor lattice QCD with an improved Wilson fermion at $a = 0.22$ fm in the laboratory system (Aoki *et al.*, 2007). Since $m_\pi/m_\rho \simeq 0.41$ in this calculation, the decay width of ρ meson can be estimated from the scattering phase shift.

10.3 Nuclear potential from lattice QCD

In the previous section, we explained Lüscher's method for extracting the scattering phase shift from the two-particle energy in a finite box, considering the $\pi\pi$ case as an example. To show the relation between the phase shift and the energy, we used the fact that the BS wave function at large separation, $r \geq R$, where R is the interaction range in infinite volume, satisfies the free Schrödinger equation (the Helmholtz equation) with periodic boundary conditions.

In this section, instead of the large-distance behavior, we consider the short-distance properties of the BS wave function, from which we define the "potential" between two particles. We mainly consider the NN potential, though the method in this section can be applied to any two particles in principle.

10.3.1 Strategy to extract potentials in quantum field theory

In this subsection, we describe the strategy to extract the NN potentials in QCD (Ishii *et al.*, 2007; Aoki *et al.*, 2008; 2010*b*). As a preparation, we introduce the T-matrix for NN scattering below the $NN\pi$ inelastic threshold. The 4×4 T-matrix for a given total angular momentum J is decomposed into two 1×1 submatrices and one 2×2 submatrix (Ishizuka, 2009; Aoki *et al.*, 2010*b*),

$$T^J = \begin{pmatrix} T^J_{l=J,s=0} & 0 & 0_{1\times2} \\ 0 & T^J_{l=J,s=1} & 0_{1\times2} \\ 0_{2\times1} & 0_{2\times1} & T^J_{l=J\mp1,s=1} \end{pmatrix}, \tag{10.105}$$

where l is the orbital angular momentum between the two nucleons and s is the total spin. Unitarity tells us that

$$T^J_{l=J,s} = \hat{T}_{Js}, \quad T^J_{l=J\mp1,s=1} = O(k) \begin{pmatrix} \hat{T}_{J-1,1} & 0 \\ 0 & \hat{T}_{J+1,1} \end{pmatrix} O^{-1}(k), \tag{10.106}$$

with

$$\hat{T}_{ls} = \frac{16\pi E_K}{k} e^{i\delta_{ls}(k)} \sin\delta_{ls}(k), \quad O(k) = \begin{pmatrix} \cos\epsilon_J(k) & -\sin\epsilon_J(k) \\ \sin\epsilon_J(k) & \cos\epsilon_J(k) \end{pmatrix}, \tag{10.107}$$

where $\delta_{ls}(k)$ is the scattering phase shift, whereas $\epsilon_J(k)$ is the mixing angle between $l = J \pm 1$. Here, the total energy of the two nucleons is given by $2E_k = 2\sqrt{k^2 + m_N^2}$ in the center-of-mass frame.

Let us start describing the strategy to extract the potential in QCD. We first define the BS amplitude for two nucleons in the center-of-mass frame as

$$\varphi^E_{\alpha\beta}(\mathbf{r}) = \langle 0| T\left\{N_\alpha(\mathbf{y},0)N_\beta(\mathbf{x},0)\right\} |\mathbf{k},s_a,-\mathbf{k},s_b; \text{in}\rangle, \tag{10.108}$$

where the relative coordinate is denoted by $\mathbf{r} = \mathbf{x} - \mathbf{y}$, the spatial momenta and the helicities for the incoming nucleons are $(\mathbf{k}, s_a)$ and $(-\mathbf{k}, s_b)$, and the total energy is $E = 2E_k$ with $k = |\mathbf{k}|$. The local composite nucleon operator is given by

$$N^f_\alpha(x) = \epsilon^{abc} q^{a,f}_\alpha(x) q^{b,g}_\beta(x) (i\tau_2)_{gh} (C\gamma_5)^{\beta\gamma} q^{c,h}_\gamma(x)$$

$$= \epsilon^{abc} q^{a,f}_\alpha(x) \left[q^b(x) i\tau_2 C\gamma_5 q^c(x) \right], \tag{10.109}$$

where $q_\alpha^{a,f}$ is a quark field with the color index a, the flavor index f, and the spinor index α. Here, a repeated index assumes a sum, $C = \gamma_2\gamma_4$ is the charge conjugation matrix, and $i\tau_2$ acts on the flavor index. Unless necessary, the flavor indices are implicit.

Let us briefly consider the meaning of the BS wave function. By writing

$$\langle 0|T\left\{N(\mathbf{y},0)N(\mathbf{x},0)\right\} = \sum_{n,m=0}^{\infty} \int \frac{dE}{2E} \langle 2N, n(\bar{N}N), m\pi, E|f_{nm}(\mathbf{r}, E), \qquad (10.110)$$

where $|2N, n(\bar{N}N), m\pi, E\rangle$ is an in-state containing two nucleons, n nucleon–antinucleon pairs, and m pions, with total energy E, we see that $\varphi^E(\mathbf{r}) = f_{00}(\mathbf{r}, E)$. (Our normalization is $\langle 2N, n(\bar{N}N), m\pi, E||2N, n'(\bar{N}N), m'\pi, E'\rangle = 2E\,\delta(E - E')$ $\delta_{nn'}\delta_{mm'}$.) Therefore, the BS wave function $\varphi^E(\mathbf{r})$ is an amplitude to find the in-state $|2N, E\rangle$ in $T\left\{N(\mathbf{y},0)N(\mathbf{x},0)\right\}|0\rangle$.

As in the case of $\pi\pi$, the asymptotic behavior of the BS wave function at $r = |\mathbf{r}| > R$, where R is the interaction range of the two nucleons, agrees with that of the scattering wave in quantum mechanics (Ishizuka, 2009; Aoki *et al.*, 2010*b*). The BS wave function for a given total angular momentum J has 4 components. For example, the BS wave function for $(l = J, s = 0)$ at large r becomes

$$\varphi_{l=J,s=0}(\mathbf{r}; k) \xrightarrow[r>R]{} Z\, Y_{JJ_z}(\Omega_\mathbf{r})e^{i\delta_{J0}(k)}\left[j_J(kr)\cos\delta_{J0}(k) + n_J(kr)\sin\delta_{J0}(k)\right]$$

$$\simeq Z\, Y_{JJ_z}(\Omega_\mathbf{r})\frac{e^{i\delta_{J0}(k)}}{kr}\sin(kr + \delta_{J0}(k) - \pi J/2), \qquad (10.111)$$

where J_z is the z component of the total angular momentum. See (Ishizuka, 2009; Aoki *et al.*, 2010*b*) for more details. This shows that the BS wave function can be regarded as the NN scattering wave.

Now we define the non-local NN "potential" through $\varphi^E(\mathbf{r})$ (Ishii *et al.*, 2007; Aoki *et al.*, 2008; 2010*b*),

$$\left(\frac{k^2}{2\mu} - H_0\right)\varphi_{\alpha\beta}^E(\mathbf{x}) = \int d^3y\, U_{\alpha\beta,\gamma\delta}(\mathbf{x}, \mathbf{y})\varphi_{\gamma\delta}^E(\mathbf{y}), \qquad H_0 = \frac{-\nabla^2}{2\mu}, \qquad (10.112)$$

where $\mu = m_N/2$ is the reduced mass of the two nucleons. It is noted that $U(\mathbf{x}, \mathbf{y})$ is non-local but energy-independent and this potential is equivalent to the local but energy dependent potential $V(\mathbf{r}, \mathbf{k})$, which is defined by

$$K(\mathbf{r}, \mathbf{k}) \equiv \left(\frac{k^2}{2\mu} - H_0\right)\varphi(\mathbf{r}, \mathbf{k}) = V(\mathbf{r}, \mathbf{k})\varphi(\mathbf{r}, \mathbf{k}), \qquad (10.113)$$

where we write $\varphi(\mathbf{r}, \mathbf{k}) = \varphi^E(\mathbf{r})$. To see the equivalence, we construct the dual basis $\tilde{\varphi}(\mathbf{k}, \mathbf{r})$ via

$$\tilde{\varphi}(\mathbf{k}, \mathbf{r}) = \int d^3p\, \eta^{-1}(\mathbf{k}, \mathbf{p})\overline{\varphi(\mathbf{r}, \mathbf{p})}, \qquad (10.114)$$

where the metric is given by

$$\eta(\mathbf{k}, \mathbf{p}) = \int d^3 r \; \overline{\varphi(\mathbf{r}, \mathbf{k})} \varphi(\mathbf{r}, \mathbf{p}). \tag{10.115}$$

It is easy to see that the dual basis satisfies

$$\int d^3 y \; \tilde{\varphi}(\mathbf{k}, \mathbf{r}) \varphi(\mathbf{r}, \mathbf{p}) = \delta^3(\mathbf{k} - \mathbf{p}) \tag{10.116}$$

$$\int d^3 p \; \varphi(\mathbf{x}, \mathbf{p}) \tilde{\varphi}(\mathbf{p}, \mathbf{y}) = \delta^3(\mathbf{x} - \mathbf{y}). \tag{10.117}$$

Using the dual basis, we obtain the non-local potential from the local one via

$$U(\mathbf{x}, \mathbf{y}) = \int d^3 p \; K(\mathbf{x}, \mathbf{p}) \tilde{\varphi}(\mathbf{p}, \mathbf{y}) = \int d^3 p \; V(\mathbf{x}, \mathbf{p}) \varphi(\mathbf{x}, \mathbf{p}) \tilde{\varphi}(\mathbf{p}, \mathbf{y}). \tag{10.118}$$

This establishes an one-to-one correspondence between the non-local but energy-independent potential $U(\mathbf{x}, \mathbf{y})$ and the local but energy-dependent $V(\mathbf{r}, \mathbf{k})$.

The equivalence also tells us that we need to know the BS wave function at all energies to completely construct $U(\mathbf{x}, \mathbf{y})$. Although this is possible in principle, it is very difficult in practice. We therefore consider the following derivative expansion of $U(\mathbf{x}, \mathbf{y})$,

$$U(\mathbf{x}, \mathbf{y}) = V(\mathbf{x}, \nabla) \delta^3(\mathbf{x} - \mathbf{y}). \tag{10.119}$$

The structure of the $V(\mathbf{r}, \nabla)$ can be determined as follows (Okubo and Marshak, 1958). The most general (non-relativistic) NN potential is parameterized as

$$V(\mathbf{r}_1, \mathbf{r}_2, \mathbf{p}_1, \mathbf{p}_2, \vec{\sigma}_1, \vec{\sigma}_2, \vec{\tau}_1, \vec{\tau}_2, t), \tag{10.120}$$

where $\mathbf{r}_i$, $\mathbf{p}_i$, $\vec{\sigma}_i$, and $\vec{\tau}_i$ are the coordinate, the momentum, the spin, and the isospin of the ith nucleon, respectively, and t is the time. There are several conditions that this potential should satisfy.

1. Probability conservation implies the hermiticity of the potential, $V^\dagger = V$.
2. Energy conservation implies t independence while momentum conservation says that the potential depends on the combination $\mathbf{r} = \mathbf{r}_1 - \mathbf{r}_2$ only.
3. Galileian invariance tells us that the potential contains $\mathbf{p} = \mathbf{p}_1 - \mathbf{p}_2$ only. From these three conditions, we have $V = V(\mathbf{r}, \mathbf{p}, \vec{\sigma}_1, \vec{\sigma}_2, \vec{\tau}_1, \vec{\tau}_2)$.
4. Angular momentum conservation implies that V commutes with $\vec{J} = \vec{L} + \vec{S}$ with the orbital angular momentum $\vec{L} = \mathbf{r} \times \mathbf{p}$ and the total spin $\vec{S} = (\vec{\sigma}_1 + \vec{\sigma}_2)/2$.
5. The potential should be invariant under parity, $(\mathbf{r}, \mathbf{p}, \vec{\sigma}_i) \to (-\mathbf{r}, -\mathbf{p}, \vec{\sigma}_i)$.
6. The potential is invariant under time reversal, $(\mathbf{r}, \mathbf{p}, \vec{\sigma}_i) \to (\mathbf{r}, -\mathbf{p}, -\vec{\sigma}_i)$.
7. Quantum statistics of the exchange of two nucleons implies the invariance of the potential under $(\mathbf{r}, \mathbf{p}, \vec{\sigma}_1, \vec{\sigma}_2, \vec{\tau}_1, \vec{\tau}_2) \to (-\mathbf{r}, -\mathbf{p}, \vec{\sigma}_2, \vec{\sigma}_1, \vec{\tau}_2, \vec{\tau}_1)$.
8. From isospin invariance, V contains only $\mathbf{1} \cdot \mathbf{1}$ or $\vec{\tau}_1 \cdot \vec{\tau}_2$ in the isospin space.

9. The potential has only $\vec{\sigma}_1^n \vec{\sigma}_2^m$ terms with $(n, m) = (0, 0), (1, 0), (0, 1), (1, 1)$. The other higher-order terms can be reduced to these terms because of the property that $\sigma^i \sigma^j = \delta^{ij} + i\epsilon^{ijk}\sigma^k$.

The terms that contain Pauli matrices and satisfy the above conditions are constructed as

$$\vec{\sigma}_1 \cdot \vec{\sigma}_2, \; (\vec{\sigma}_1 + \vec{\sigma}_2) \cdot \vec{L}, \; (\vec{\sigma}_1 \cdot \mathbf{r})(\vec{\sigma}_2 \cdot \mathbf{r}), \; (\vec{\sigma}_1 \cdot \mathbf{p})(\vec{\sigma}_2 \cdot \mathbf{p}), \; (\vec{\sigma}_1 \cdot \vec{L})(\vec{\sigma}_2 \cdot \vec{L}), \tag{10.121}$$

which are customarily reorganized as

$$\vec{\sigma}_1 \cdot \vec{\sigma}_2, \quad S_{12} \equiv 3(\vec{\sigma}_1 \cdot \hat{\mathbf{r}})(\vec{\sigma}_2 \cdot \hat{\mathbf{r}}) - \vec{\sigma}_1 \cdot \vec{\sigma}_2, \quad \vec{L} \cdot \vec{S},$$

$$P_{12} \equiv (\vec{\sigma}_1 \cdot \mathbf{p})(\vec{\sigma}_2 \cdot \mathbf{p}), \quad W_{12} \equiv Q_{12} - \frac{1}{3}\vec{\sigma}_1 \cdot \vec{\sigma}_2 \vec{L}^2, \tag{10.122}$$

where S_{12} is called the tensor operator, and

$$Q_{12} \equiv \frac{1}{2}\left[(\vec{\sigma}_1 \cdot \vec{L})(\vec{\sigma}_2 \cdot \vec{L}) + (\vec{\sigma}_2 \cdot \vec{L})(\vec{\sigma}_1 \cdot \vec{L}).\right]. \tag{10.123}$$

Finally, we obtain

$$V = \sum_{I=1,2} V^I(\mathbf{r}, \mathbf{p}, \vec{\sigma}_1, \vec{\sigma}_2)P_I^\tau, \tag{10.124}$$

where

$$V^I = V_0^I + V_\sigma^I(\vec{\sigma}_1 \cdot \vec{\sigma}_2) + V_{LS}^I(\vec{L} \cdot \vec{S}) + \frac{1}{2}\{V_T^I, S_{12}\} + \frac{1}{2}\{V_P^I, P_{12}\} + \frac{1}{2}\{V_W^I, W_{12}\} \tag{10.125}$$

with coefficient functions $V_X^I = V_X^I(\mathbf{r}^2, \mathbf{p}^2, \vec{L}^2)$ for $I = 0, 1$ and $X = 0, \sigma, T, LS, P, W$. Here, P_I^τ is the projection operator to the state with the total isospin I, given by

$$P_{I=0}^\tau = \frac{1}{4} - \vec{\tau}_1 \cdot \vec{\tau}_2, \qquad P_{I=1}^\tau = \frac{3}{4} + \vec{\tau}_1 \cdot \vec{\tau}_2. \tag{10.126}$$

The anticommutators in Eq. (10.125) are necessary to make the potential Hermitian, since S_{12}, P_{12} and W_{12} do not commute with the scalar potentials $V_X^I(\mathbf{r}^2, \mathbf{p}^2, \vec{L}^2)$.

From the general considerations above, we have, for example at $O(\nabla)$,

$$V(\mathbf{r}, \nabla) = \sum_{I=0,1}\left[V_0^I(r) + V_\sigma^I(r)\vec{\sigma}_1 \cdot \vec{\sigma}_2 + V_T^I(r)S_{12} + V_{LS}^I(r)\vec{L} \cdot \vec{S}\right]P_I^\tau$$

$$+ O(\nabla^2). \tag{10.127}$$

This form of the potential has often been used in nuclear physics. Note that the first three terms are $O(1)$, while the LS potential is $O(\nabla)$.

This is the strategy to define and extract the NN potential in QCD. There are two important and related remarks.

The leading-order potential in the derivative expansion is simply the local potential $V(\mathbf{r}, \mathbf{k})$. By construction, this (local) potential reproduces the correct phase shift $\delta(k)$ at $k = |\mathbf{k}|$, while it is not guaranteed that this potential gives the correct phase shift at different $k' = |\mathbf{k}'|(\neq k)$. This means that the local potential $V(\mathbf{r}, \mathbf{k})$ and $V(\mathbf{r}, \mathbf{k}')$ may differ, and this energy dependence of the local potential gives a measure for the non-locality of the non-local but energy-independent potential $U(\mathbf{x}, \mathbf{y})$ because of the equivalence between V and U. If the first order in the derivative expansion is good at low energy, we expect that energy dependence of the potential V is small at small k.

Secondly, it should be mentioned that the potential U defined through the BS amplitude of course depends on the choice of the interpolating field operators $N(x)$. In principle, one may choose any (local) composite operators with the same quantum numbers as the nucleon to define the BS wave function. Different choices for the nucleon operator give different BS wave functions, which may lead to different NN potentials, though they all give the same scattering phase shift. While the potential is not a physical observable in this sense, it does not mean that it is useless, however. The strategy in this chapter gives one specific scheme for the NN potential in QCD, which is defined through the BS amplitude constructed from the local nucleon field without derivatives. This is quite analogous to the situation for the running coupling in QCD. Although the running coupling is scheme dependent, it is useful to understand and describe deep inelastic electron–proton scattering.

Let us make this analogy more concrete. In both cases, in deep inelastic scattering and in NN scattering, the physical observables are the scattering data. An example of a physical interpretation is the almost-free partons in the proton for the case of deep inelastic scattering, while it is the existence of the repulsive core for the case of NN scattering. A theoretical explanation for the phenomena is the asymptotic freedom of the QCD running coupling for the free partons, while no valid theoretical explanation exists so far for the repulsive core. In this section, we introduced one definite scheme for the potential based on QCD, in order to show the existence of the repulsive core. Although the choice of the scheme is irrelevant in principle, it is better to use a "good" scheme in practice. In the case of the running coupling, good convergence of the pertubative expansion may give one criterion, though the popularly used $\overline{\text{MS}}$ coupling may not be the best one for this criterion. In the case of the NN potential, on the other hand, good convergence of the derivative expansion may give a criterion for a "good" potential. In other words, the good potential is almost local and energy independent. The NN potential that is completely local and energy independent at all energies is therefore the best one. It is also unique if the inverse scattering method holds for the NN case.

10.3.2 Extraction of the BS wave function on the lattice

In this subsection, we explain how to extract the BS wave function from correlation functions on the lattice. For simplicity, we here consider the $l = 0$ state, namely the S-state.

The BS wave function on a lattice with lattice spacing a and spatial lattice volume L^3 is extracted from the 4-point correlation function by inserting the complete set of QCD eigenstates in the finite box,

$$G_{\alpha\beta}(\mathbf{x}, \mathbf{y}, t - t_0; J^P) = \langle 0 | n_\beta(\mathbf{y}, t) p_\alpha(\mathbf{x}, t) \overline{J}_{pn}(t_0; J^P) | 0 \rangle \qquad (10.128)$$

$$= \sum_{n=0}^{\infty} A_n \langle 0 | n_\beta(\mathbf{y}, t) p_\alpha(\mathbf{x}, t) | E_n \rangle e^{-E_n(t - t_0)} \qquad (10.129)$$

$$\xrightarrow[t \gg t_0]{} A_0 \, \varphi_{\alpha\beta}(\mathbf{r}; J^P), \qquad \mathbf{r} = \mathbf{x} - \mathbf{y} \qquad (10.130)$$

with the matrix element $A_n = \langle E_n | \overline{J}_{pn}(0) | 0 \rangle$, where p (n) is the proton (neutron) interpolating operator, $p = N^u$ ($n = N^d$), and $|E_n\rangle$ is the QCD eigenstate with baryon number 2 and total energy $E_n = 2\sqrt{k_n^2 + m_N^2}$. The state created by the source $\overline{J}_{pn}$ has the conserved quantum numbers (J, J_z) (total angular momentum and its z-component), I (total isospin) and P (parity). To study the NN potential in the $J^P = 0^+$ with $I = 1$ ($^1S_0$) channel and the $J^P = 1^+$ with $I = 0$ ($^3S_1$ and $^3D_1$) channel, a wall source located at $t = t_0$ with Coulomb gauge fixing only at $t = t_0$ is used,

$$J_{pn}(t_0, J^P) = P_{\beta\alpha}^s \left[p_\alpha^{\text{wall}}(t_0) n_\beta^{\text{wall}}(t_0) \right], \qquad (10.131)$$

where $p^{\text{wall}}(t_0)$ and $n^{\text{wall}}(t_0)$ are obtained by replacing the local quark fields $q(x)$ in $N(x)$ by the wall-quark fields,

$$q^{\text{wall}}(t_0) = \sum_{\mathbf{x}} q(\mathbf{x}, t_0). \qquad (10.132)$$

By construction, the source operator Eq. (10.131) has zero orbital angular momentum at $t = t_0$, so that states with fixed (J, J_z) are obtained by the spin projection with $(s, s_z) = (J, J_z)$, e.g., $P_{\beta\alpha}^{s=0} = (\sigma^2)_{\beta\alpha}$ and $P_{\beta\alpha}^{s=1, s_z=0} = (\sigma^1)_{\beta\alpha}$. Note that l and s are not separately conserved, so that the state created by the source $J_{pn}(t_0; 1^+)$ becomes a mixture of $l = 0$ (S state) and $l = 2$ (D state) at later time $t > t_0$.

The BS wave function in the orbital S state is then defined with the projection operator for the cubic group P^R with the irreducible representation R and that for the spin P^s as

$$\varphi(r; {}^1S_0) = P^{A_1^+} P^{s=0} \varphi(\mathbf{r}; 0^+) \equiv \frac{1}{24} \sum_{g \in O(3, \mathbf{Z})} P_{\alpha\beta}^{s=0} \varphi_{\alpha\beta}(g^{-1}\mathbf{r}; 0^+), \qquad (10.133)$$

$$\varphi(r; {}^3S_1) = P^{A_1^+} P^{s=1} \varphi(\mathbf{r}; 1^+) \equiv \frac{1}{24} \sum_{g \in O(3, \mathbf{Z})} P_{\alpha\beta}^{s=1} \varphi_{\alpha\beta}(g^{-1}\mathbf{r}; 1^+), \qquad (10.134)$$

where the summation over $g \in O(3, \mathbf{Z})$ is taken for the cubic transformation group with 24 elements to project out the $l = 0$ component in the A_1^+ representation; contributions from the higher orbital waves with $l \geq 4$ contained in the A_1^+ rep. are expected to be negligible at low energy.

From these BS wave functions, we can construct the local potentials at the leading order of the derivative expansion. For the $^1S_0$ channel, the central potential becomes

$$V_C(r; {}^1S_0) \equiv V_0^{I=1}(r) + V_\sigma^{I=1}(r) = \frac{k^2}{m_N} + \frac{1}{m_N} \frac{\nabla^2 \varphi(r; {}^1S_0)}{\varphi(r; {}^1S_0)}, \qquad (10.135)$$

while for the $^3S_1$ channel there are two independent terms, $V_C(r, {}^3S_1) = V_0^{I=0}(r) - 3V_\sigma^{I=0}(r)$ and $V_T^{I=0}(r)$, at the leading order. For a while we ignore V_T and define the effective central potential as

$$V_C^{\text{eff}}(r; {}^3S_1) = \frac{k^2}{m_N} + \frac{1}{m_N} \frac{\nabla^2 \varphi(r; {}^3S_1)}{\varphi(r; {}^3S_1)}, \qquad (10.136)$$

where "effective' means that the potential includes the effect of the tensor potential V_T as a second-order perturbation.

It is noted here that k^2 is determined from the total energy E_0 of the two nucleons as $E_0 = 2\sqrt{k^2 + m_N^2}$.

10.3.3 Tensor potential

While the central potential acts separately on the S and D components, the tensor potential provides a coupling between these two components. We therefore consider a coupled-channel Schrödinger equation in the $J^P = 1^+$ channel, in which the BS wave function has both S-wave and D-wave components:

$$\left[H_0 + V_C(r; 1^+) + V_T(r)S_{12} \right] \varphi(\mathbf{r}; 1^+) = \frac{k^2}{m_N} \varphi(\mathbf{r}; 1^+). \qquad (10.137)$$

The projections to the S-wave and D-wave components, similar to Eq. (10.134), are defined by

$$P\varphi_{\alpha\beta}(r) \equiv P^{A_1^+} \varphi_{\alpha\beta}(\mathbf{r}; 1^+), \qquad (10.138)$$

$$Q\varphi_{\alpha\beta}(r) \equiv (1 - P^{A_1^+})\varphi_{\alpha\beta}(\mathbf{r}; 1^+). \qquad (10.139)$$

Here, both $P\varphi_{\alpha\beta}$ and $Q\varphi_{\alpha\beta}$ contain additional components with $l \geq 4$ but they are expected to be small at low energy.

Multiplying Eq. (10.137) by P and Q from the left and using the properties that H_0, $V_C(r; 1^+)$, and $V_T(r)$ commute with P and Q, we obtain

$$H_0[P\varphi](r) + V_C(r; 1^+)[P\varphi](r) + V_T(r)[PS_{12}\varphi](r) = \frac{k^2}{m_N}[P\varphi](r), \quad (10.140)$$

$$H_0[Q\varphi](r) + V_C(r; 1^+)[Q\varphi](r) + V_T(r)[QS_{12}\varphi](r) = \frac{k^2}{m_N}[Q\varphi](r), \quad (10.141)$$

where, for simplicity, the spinor indices α and β are suppressed. Note that S_{12} does not commute with P or with Q.

By solving these equations for the $(\alpha, \beta) = (2, 1)$ component, we finally extract

$$V_C(r; 1^+) = \frac{k^2}{m_N} - \frac{1}{D(r)} \{[QS_{12}\varphi]_{21}(r)H_0[P\varphi]_{21}(r) - [PS_{12}\varphi]_{21}(r)H_0[Q\varphi]_{21}(r)\}$$

(10.142)

$$V_T(r) = \frac{1}{D(r)} \{[Q\varphi]_{21}(r)H_0[P\varphi]_{21}(r) - [P\varphi]_{21}(r)H_0[Q\varphi]_{21}(r)\},$$ (10.143)

where

$$D(r) \equiv [P\varphi]_{21}(r)H_0[QS_{12}\varphi]_{21}(r) - [Q\varphi]_{21}(r)H_0[PS_{12}\varphi]_{21}(r).$$ (10.144)

Note that the effective central potential is expressed as

$$V_C^{\text{eff}}(r; {}^3S_1) = \frac{k^2}{m_N} - \frac{H_0[P\varphi]_{21}(r)}{[P\varphi]_{21}(r)}$$ (10.145)

with $H_0 = -\nabla^2/m_N$.

10.3.4   Results in lattice QCD

The first result for the NN potential in lattice QCD based on the strategy in the previous subsections appeared in (Ishii *et al.*, 2007), where the (effective) central potential was calculated for the ${}^1S_0$ and ${}^3S_1$ channels in quenched QCD with lattice spacing $a \simeq 0.137$ fm and spatial extent $L \simeq 4.4$ fm. Details of the numerical simulations can be found in (Ishii *et al.*, 2007).

Figure 10.2 shows the BS wave function in the ${}^1S_0$ and ${}^3S_1$ channels at $m_\pi = 529$ MeV and $k^2 \simeq 0$, measured at $t - t_0 = 6a$. The wave functions are normalized to be 1 at the largest spatial coordinate $r = 2.192$ fm.

The reconstructed central and effective central potentials in the ${}^1S_0$ and ${}^3S_1$ channels at $m_\pi = 529$ MeV, calculated from the BS wave functions with the formulae (10.135) and (10.136) are shown in Fig. 10.3. The overall structure of the potentials is similar to that of the known phenomenological NN potentials discussed in Section 10.1, showing the repulsive core at short distance surrounded by the attractive well at medium and long distances. The figure also shows that the interaction between two nucleons is practically switched off for $r > 1.5$ fm, so that the condition $R < L/2 \simeq 2.2$ fm is satisfied.

To check the stability of these potentials against changing the timeslice t adopted to define the BS wave function, the t dependence of the ${}^1S_0$ potential for several different values of r is shown in Fig. 10.4 at $m_\pi = 529$ MeV. In this case, the choice $t - t_0 = 6a$ for the extraction of $V_C(r)$ is large enough to assure stability within statistical errors, which indicates the ground state's dominance at this t.

The NN potentials in the ${}^1S_0$ channel are compared among three different quark masses in Fig. 10.5. As the quark mass decreases, the repulsive core at short distance and the attractive well at medium distance become stronger.

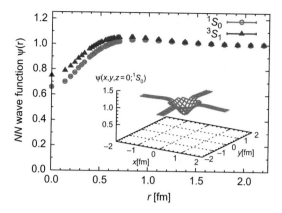

Fig. 10.2 The NN wave function in $^1S_0$ and $^3S_1$ channels at $m_\pi = 529$ MeV, measured at $(t - t_0) = 6a$. The inset is a three-dimensional plot of the wave function $\varphi(x, y, z = 0; ^1S_0)$.

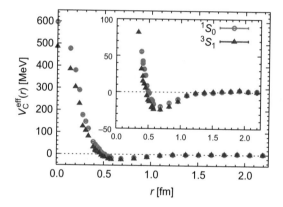

Fig. 10.3 The central potential in the $^1S_0$ channel and the effective central potential in the $^3S_1$ channel at $m_\pi = 529$ MeV.

The $^3S_1$ and $^3D_1$ components of the BS wave functions obtained from $J^P = 1^+$, $J_z = 0$ state at $m_\pi = 529$ MeV and $k^2 \simeq 0$ are plotted in Fig. 10.6(a), according to Eqs. (10.138) and (10.139). Note that the $^3D_1$ wave function becomes multivalued as a function of r due to its angular dependence. It is expected that $(\alpha, \beta) = (2, 1)$ spin component of the D-state wave function for $J^P = 1^+$ and $J_z = 0$ is proportional to the $Y_{20}(\theta, \phi) \propto 3\cos^2\theta - 1$. As shown in Fig. 10.6(b), the D-state wave function divided by $Y_{20}(\theta, \phi)$ becomes almost single-valued, so that the D-wave component is indeed dominant in $Q\varphi(r)$. The central potential $V_C(r; 1^+)$ and the tensor potential $V_T(r)$ together with the effective central potential $V_C^{\text{eff}}(r; ^3S_1)$ in the $^3S_1$ channel are plotted in Fig. 10.7. Note that $V_C^{\text{eff}}(r; ^3S_1)$ contains the effect of $V_T(r)$ implicitly as higher-order effects through processes such as $^3S_1 \to {}^3D_1 \to {}^3S_1$. In the real world, $V_C^{\text{eff}}(r; ^3S_1)$ is expected to acquire a large attraction from the tensor force, which is a

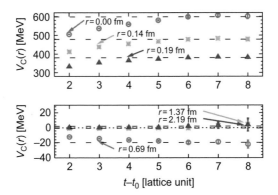

Fig. 10.4 The t dependence of the potential at $r = 0$, 0.14, 0.19, 1.37, 2.19, 0.69 fm, from top to bottom, in the $^1S_0$ channel at $m_\pi = 529$ MeV.

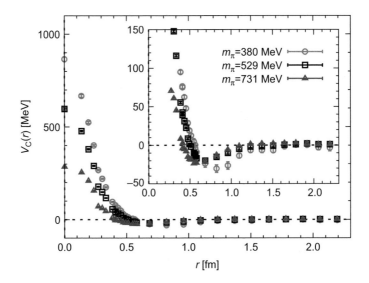

Fig. 10.5 The central potentials in the $^1S_0$ channel at three different quark masses.

reason why the deuteron bound state exists in the $^3S_1$ channel, while no bound states appears for a dineutron. As seen from Fig. 10.7, the difference between $V_C(r; 1^+)$ and $V_C^{\text{eff}}(r; {}^3S_1)$ is still small in this quenched simulation due to the relatively large quark masses.

The tensor potential $V_T(r)$ shown in Fig. 10.7 is negative for the whole range of r within statistical errors and has a minimum at short distance around 0.4 fm. If the tensor force receives a significant contribution from one-pion exchange as expected from the meson theory, V_T would be rather sensitive to the change of the quark mass. As shown in Fig. 10.8, indeed the attraction of $V_T(r)$ substantially increases as the quark mass decreases.

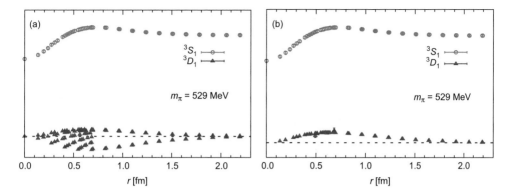

Fig. 10.6 (a) $(\alpha, \beta) = (2, 1)$ components of the S-state and the D-state wave functions projected out from a single state with $J^P = 1^+$, $J_z = 0$. (b) The D-state wave function divided by Y_{20}.

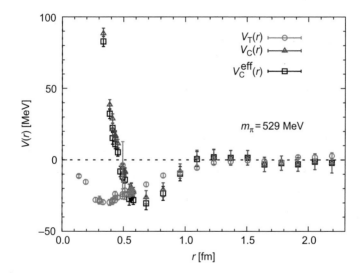

Fig. 10.7 The central potential $V_C(r; 1^+)$ and the tensor potential $V_T(r)$ obtained from the J^+ BS wave function at $m_\pi = 529$ MeV, together with $V_C^{\mathrm{eff}}(r; {}^3S_1)$.

At present, the potentials are determined at leading order of the derivative expansion, and the examples presented so far are extracted from lattice data taken at $k \simeq 0$. If the higher-order terms such as $V_{LS}(r)\vec{L} \cdot \vec{S}$ become important, the LO local potentials determined at $k > 0$ are expected to be different from those at $k \simeq 0$. From such k dependence of the LO local potentials, some of the higher-order terms can in principle be determined. A lattice QCD analysis of the k dependence has been recently carried out by changing the spatial boundary conditions of the quark field from periodic to antiperiodic, which corresponds to a change from $k \simeq 0$

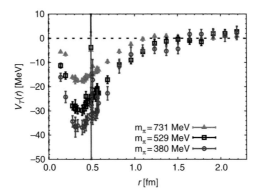

Fig. 10.8   Quark mass dependence of the tensor potential $V_T(r)$.

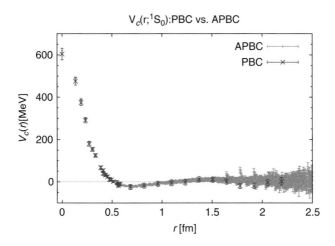

Fig. 10.9 A comparison of central potentials at $k \simeq 0$ (PBC, blue) and at $k \simeq 250$ MeV (APBC, red) for the $^1S_0$ state in quenched QCD at $a = 0.137$ fm and $m_\pi = 529$ MeV.

to $k = \sqrt{3(\pi/L)^2} \simeq 250$ MeV. In Fig. 10.9, the local potential for the $^1S_0$ channel obtained at $k \simeq 250$ MeV is compared with the one at $k \simeq 0$ in quenched QCD at $a = 0.137$ fm and $m_\pi = 529$ MeV. As seen from the figure, the k dependence of the local potential turns out to be very small for every r within statistical errors. This means that the non-locality of the potential, with the choice of the local interpolating operator for the nucleon, is small; hence the present local potential at the LO can be used to well describe physical observables such as the phase shift $\delta_0(k)$ from $k \simeq 0$ to $k \simeq 250$ MeV without significant modification—at least in quenched QCD at $a = 0.137$ fm and $m_\pi = 529$ MeV. This also indicates that the definition of the potential through the BS wave function with the local nucleon operator is a "good scheme."

10.4   Repulsive core and operator product expansion in QCD

As seen in the previous section, lattice QCD calculations show that the NN potential defined through the BS wave function has not only the attraction at medium to long distance that has long been well understood in terms of pion and other heavier meson exchanges, but also a characteristic repulsive core at short distance, whose origin is still theoretically unclear. A recent attempt (Aoki *et al.*, 2010*a*) to understand theoretically the short-distance behavior of the potential in terms of the operator product expansion (OPE) is explained in this section.

10.4.1   Basic idea

Let us first explain the basic idea. We consider the equal time BS wave function defined by

$$\varphi_{AB}^{E}(\mathbf{r}) = \langle 0|O_A(\mathbf{r}/2, 0)O_B(-\mathbf{r}/2, 0)|E\rangle, \tag{10.146}$$

where $|E\rangle$ is some eigenstate of a certain system with total energy E, and O_A, O_B are some operators of this system. (We suppress other quantum numbers of the state $|E\rangle$ for simplicity.) The OPE reads

$$O_A(\mathbf{r}/2, 0)O_B(-\mathbf{r}/2, 0) \simeq \sum_C D_{AB}^{C}(\mathbf{r})O_C(\mathbf{0}, 0), \tag{10.147}$$

Suppose that the coefficient function of the OPE behaves in the small $r = |\mathbf{r}|$ limit as

$$D_{AB}^{C}(\mathbf{r}) \simeq r^{\alpha_C}(-\log r)^{\beta_C} f_C(\theta, \phi), \tag{10.148}$$

where θ, ϕ are the angles of $\mathbf{r}$, the BS wave function becomes

$$\varphi_{AB}^{E}(\mathbf{r}) \simeq \sum_C r^{\alpha_C}(-\log r)^{\beta_C} f_C(\theta, \phi)D_C(E), \tag{10.149}$$

where

$$D_C(E) = \langle 0|O_C(\mathbf{0}, 0)|E\rangle. \tag{10.150}$$

The potential at short distances can be calculated from this expression. For example, in the case of the Ising field theory in two dimensions, the OPE for the spin field σ is given by

$$\sigma(x, 0)\sigma(0, 0) \simeq G(r)\mathbf{1} + c\,r^{3/4}O_1(0) + \cdots, \quad r = |x|, \tag{10.151}$$

where $O_1(x)$ $(=: \bar{\psi}\psi(x) :$ in terms of free fermion fields) is an operator of dimension one. This leads to

$$\varphi(r, E) \simeq r^{3/4}D(E) + O(r^{7/4}), \quad D(E) = c\langle 0|O_1(0)|E\rangle, \tag{10.152}$$

where $|E\rangle$ is a two-particle state with energy $E = 2\sqrt{k^2 + m^2}$. From this expression the potential becomes

$$V(r) = \frac{\varphi''(r, E) + k^2\varphi(r, E)}{m\varphi(r, E)} \simeq -\frac{3}{16}\frac{1}{mr^2} \tag{10.153}$$

in the $r \to 0$ limit. The OPE predicts not only the r^{-2} behavior of the potential at short distance but also its coefficient $-3/16$. Furthermore, the potential at short distance does not depend on the energy of the state in this example (Aoki *et al.*, 2009*b*).

In QCD the dominant terms at short distance have $\alpha_C = 0$. Among these terms, we assume that C has the largest contribution such that $\beta_C > \beta_{C'}$ for $\forall C' \neq C$. Since such dominant operators with $\alpha_C = 0$ mainly couple to the zero angular momentum $(L = 0)$ state, let us consider the BS wave function with $L = 0$. Applying ∇^2 to this wave function, we obtain the following classification of the short-distance behavior of the potential.

(1) $\beta_C \neq 0$: The potential at short distance is energy independent and becomes

$$V(r) \simeq -\frac{\beta_C}{mr^2(-\log r)}, \tag{10.154}$$

which is attractive for $\beta_C > 0$ and repulsive for $\beta_C < 0$.
(2) $\beta_C = 0$: In this case the potential becomes

$$V(r) \simeq \frac{D_{C'}(E)}{D_C(E)}\frac{-\beta_{C'}}{mr^2}(-\log r)^{\beta_{C'}-1}, \tag{10.155}$$

where $\beta_{C'} < 0$ is the second largest exponent. The sign of the potential at short distance depends on the sign of $D_{C'}(E)/D_C(E)$.

On the lattice, we do not expect divergence at $r = 0$ due to lattice artifacts at short distance. The above classification hold at $a \ll r \ll 1/\Lambda_{\rm QCD}$, while the potential becomes finite even at $r = 0$ on the lattice.

10.4.2   Renormalization-group analysis and operator-product expansion

Since QCD is the asymptotic free theory, the 1-loop calculation for anomalous dimensions becomes exact at short distance. The OPE in QCD is written as

$$O_A(y/2)O_B(-y/2) = \sum_C D^C_{AB}(r, g, m, \mu)O_C(0), \tag{10.156}$$

where $g\,(m)$ is the renormalized coupling constant (quark mass) at scale μ. In the limit that $r = |y| = e^{-t}R \to 0$ ($t \to \infty$ with fixed R), the renormalization-group analysis leads to

$$\lim_{r \to 0} D^C_{AB}(r, g, m, \mu) = (-2\beta^{(1)}g^2 \log r)^{\gamma^{C,(1)}_{AB}/(2\beta^{(1)})}D^C_{AB}(R, 0, 0, \mu), \tag{10.157}$$

where $\beta^{(1)} = \dfrac{1}{16\pi^2}\left(11 - \dfrac{2N_f}{3}\right)$ is the QCD beta function at 1-loop, and

$$\beta_{AB}^{C,(1)} = \gamma_C^{(1)} - \gamma_A^{(1)} - \gamma_B^{(1)}. \tag{10.158}$$

Here, $\gamma_X^{(1)}$ is the 1-loop anomalous dimension of the operator O_X. An appearance of $D_{AB}^C(R,0,0,\mu)$ in the right-hand side tells us that it is enough to know the OPE only at tree level. From the above expression, β_C in the previous subsection is given by $\beta_C = \dfrac{\gamma_{AB}^{C,(1)}}{2\beta^{(1)}}$. Therefore, our task is to calculate $\beta_X^{(1)}$ for 6 quark operators.

10.4.3 OPE and anomalous dimensions for two nucleons

10.4.3.1 6 quark operators

The OPE of two baryon operators at tree level is given by

$$B_1(x/2)B_2(-x/2) = B_1(0)B_2(0) + \frac{x^\mu}{2}\left\{\partial_\mu B_1(0) \cdot B_2(0) - B_1(0) \cdot \partial_\mu B_2(0)\right\} + \cdots, \tag{10.159}$$

where the first term corresponds to the $L = 0$ contribution, the second to the $L = 1$, and so on. In this report we consider $L = 0$ case (the first term) only. We denote the general form of a gauge-invariant 3-quark operator as $B_{\alpha\beta\gamma}^{fgh}(x) \equiv B_\Gamma^F(x) = \varepsilon^{abc}q_\alpha^{a,f}(x)q_\beta^{b,g}(x)q_\gamma^{c,h}(x)$, where α, β, γ are spinor, f, g, h are flavor, a, b, c are color indices of quark field q. The 6-quark operator is constructed from two 3-quark operators as $B_{\Gamma_1,\Gamma_2}^{F_1,F_2}(x) = B_{\Gamma_1}^{F_1}(x)B_{\Gamma_2}^{F_2}(x)$, where $\Gamma_i = \alpha_i\beta_i\gamma_i$ and $F_i = f_ig_ih_i$ $(i = 1, 2)$. Note, however, that such operators are not all linearly independent. Relations between them follow from a general identity satisfied by the totally antisymmetric epsilon symbol that for N labels reads

$$N\varepsilon^{a_1\ldots a_N}\varepsilon^{b_1\ldots b_N} = \sum_{j,k}\varepsilon^{a_1\ldots a_{j-1}b_k a_{j+1}\ldots a_N}\varepsilon^{b_1\ldots b_{k-1}a_j b_{k+1}\ldots b_N}. \tag{10.160}$$

For our special case, $N = 3$, this identity implies the following identities among the 6-quark operators

$$3B_{\Gamma_1,\Gamma_2}^{F_1,F_2} + \sum_{i,j=1}^{3}B_{(\Gamma_1,\Gamma_2)[i,j]}^{(F_1 F_2)[i,j]} = 0, \tag{10.161}$$

where the ith index of abc and the jth index of def are interchanged in $(abc, def)[i, j]$. For example, $(\Gamma_1,\Gamma_2)[1,1] = \alpha_2\beta_1\gamma_1, \alpha_1\beta_2\gamma_2$ or $(\Gamma_1,\Gamma_2)[2,1] = \alpha_1\alpha_2\gamma_1, \beta_1\beta_2\gamma_2$. Note that the interchange of indices occurs simultaneously for both Γ_1, Γ_2 and F_1, F_2 in the above formula. The plus sign in Eq.(10.161) appears because the quark fields are Grassmann.

As an example of identities, let us consider the case that $\Gamma_1, \Gamma_2 = \alpha\alpha\beta, \alpha\beta\beta$ $(\alpha \neq \beta$ and $F_1, F_2 = ffg, ffg$ $(f \neq g)$. The constraint gives

$$3B^{ffg,ffg}_{\alpha\alpha\beta,\alpha\beta\beta} + (3-2)B^{ffg,ffg}_{\alpha\alpha\beta,\alpha\beta\beta} + B^{fff,fgg}_{\alpha\alpha\alpha,\beta\beta\beta} + (2-1)B^{fgg,fff}_{\alpha\beta\beta,\alpha\alpha\beta}$$

$$= 4B^{ffg,ffg}_{\alpha\alpha\beta,\alpha\beta\beta} + B^{fff,fgg}_{\alpha\alpha\alpha,\beta\beta\beta} + B^{fgg,fff}_{\alpha\beta\beta,\alpha\alpha\beta} = 0, \tag{10.162}$$

where the minus signs in the first line come from the property that $B^{F_2,F_1}_{\Gamma_2,\Gamma_1} = -B^{F_1,F_2}_{\Gamma_1,\Gamma_2}$.

10.4.3.2 1-loop contributions

Only the divergent part of the gauge-invariant contribution at 1-loop is necessary to calculate the anomalous dimension of 6-quark operators at 1-loop. The building block of 1-loop calculations is the gluon exchange between two quark lines. If both quark lines belong to one operator, either $B^{F_1}_{\Gamma_1}$ or $B^{F_2}_{\Gamma_2}$, the contribution is cancelled by the renormalization factor of the 3-quark operator $B^{F_i}_{\Gamma_i}$, so that the divergent term does not contribute to $\gamma^{C,(1)}_{AB} = \gamma^{(1)}_C - \gamma^{(1)}_A - \gamma^{(1)}_B$. If one quark line comes from $B^{F_1}_{\Gamma_1}$ and the other from $B^{F_2}_{\Gamma_2}$, the divergent term contributes to $\gamma^{C,(1)}_{AB}$. Suppose that one quark line has indices (α_A, f_A) in one end and (α_1, f_1) in the other end and the other quark line has (α_B, f_B) and (α_2, f_2). The divergent contribution by the 1-gluon exchange can be expressed as

$$\frac{g^2}{32\pi^2}\frac{1}{\epsilon}\{\delta_{f_1 f_A}\delta_{f_2 f_B}[\delta_{\alpha_1\alpha_A}\delta_{\alpha_2\alpha_B} - 2\delta_{\alpha_2\alpha_A}\delta_{\alpha_1\alpha_B}]$$

$$+3\delta_{f_2 f_A}\delta_{f_1 f_B}[\delta_{\alpha_2\alpha_A}\delta_{\alpha_1\alpha_B} - 2\delta_{\alpha_1\alpha_A}\delta_{\alpha_2\alpha_B}]\} \tag{10.163}$$

for $(\alpha_1, \alpha_2) \in (R, R)$ or $(\alpha_1, \alpha_2) \in (L, L)$, where R and L means the right- and the left-handed component of the spinor indices. Other combinations, (R, L) or (L, R), vanish. This property comes from the fact that the gluon interaction to quarks is chirally symmetric and the chirally non-symmetric quark-mass term does not contribute to the divergence.

10.4.3.3 Chiral decomposition of 6-quark operators

The physical nucleon operators are constructed from general 3-quark operators as $B^f_\alpha = (P_{+4})_{\alpha\alpha'}B^{fgh}_{\alpha'\beta\gamma}(C\gamma_5)_{\beta\gamma}(i\tau_2)^{gh}$ where $P_{+4} = (1+\gamma_4)/2$, C is the charge conjugation matrix, f, g, h are u or d, and τ_2 is the Pauli matrix in the flavor space.

Let us consider $L = 0$ two nucleon states, which are $^1S_0$ and $^3S_1$. Here, we use the notation $^{2S+1}L_J$ where S is the total spin, L is the orbital angular momentum and J is the total angular momentum. The 6-quark operator for $^1S_0$, which is the spin-singlet and isospin-triplet state, and for $^3S_1$ (the spin-triplet and isospin-singlet state) are given by

$$BB(^1S_0) = (i\sigma_2)_{\alpha\beta}B^f_\alpha B^f_\beta, \qquad BB(^3S_1) = (i\tau_2)^{fg}B^f_\alpha B^g_\alpha, \tag{10.164}$$

where the summation is taken for the repeated index. Both 6-quark operators have the following chiral decomposition:

$$BB = B_{LL}B_{LL} + B_{LL}B_{LR} + B_{LL}B_{RL} + B_{LR}B_{LR} + B_{LL}B_{RR} + (L \leftrightarrow R),$$
$$(10.165)$$

where B_{XY} means $B_{\alpha,[\beta\gamma]}$ with $\alpha \in X$ and $[\beta,\gamma] \in Y$ for $X, Y = R$ or L.

10.4.3.4 *Anomalous dimension*

We now give our main results. We define

$$\gamma_{AB}^{C,(1)} = \gamma_C^{(1)} - \gamma_A^{(1)} - \gamma_B^{(1)} \equiv \frac{1}{48\pi^2}\gamma. \qquad (10.166)$$

The eigenoperators of the anomalous dimension matrix γ are found to correspond to chirally decomposed operators in the previous subsection. We give the eigenvalue of each operator in Table 10.3, which shows that the operator with zero anomalous dimension always exists and other anomalous dimensions are all negative for both $^1S_0$ and $^3S_1$ states. This corresponds to the case (2) of the argument in Section 10.4.1.

The appearance of zero eigenvalues in both $^1S_0$ and $^3S_1$ states can be understood as follows. As mentioned before, 1-loop contributions to the $\gamma_{AB}^{C,(1)}$ exist only if spinor indices of two quark lines, one from Γ_1 the other from Γ_2 in $B_{\Gamma_1}B_{\Gamma_2}$, belong to the same chirality (left or right). Since $B_{LL}B_{RR} + B_{RR}B_{LL}$ has no such combination, $\gamma_{AB}^{C,(1)}$ always zero for this type of operators. As pointed out before, this property is the consequence of the chiral symmetry in QCD interactions.

10.4.3.5 *Short-distance behavior of the potentials and the repulsive core*

The OPE and renormalization group analysis in QCD predicts the universal functional form of the nucleon-nucleon potential at short distance:

$$V(r) \simeq \frac{D_Y(E)}{D_X(E)} \frac{-\beta_Y(-\log r)^{\beta_Y - 1}}{m_N r^2}, \qquad r \to 0, \qquad (10.167)$$

which is a little weaker than the $1/r^2$ singularity. We obtain

$$\beta_Y(^1S_0) = -\frac{6}{33 - 2N_f}, \qquad \beta_Y(^3S_1) = -\frac{2}{33 - 2N_f}. \qquad (10.168)$$

The OPE, however, can not tell whether the potential at short distance is repulsive or attractive, which is determined by the sign of the coefficient. If we evaluate $D_X(E)$ and $D_Y(E)$ by the non-relativistic quark model at the leading order, we obtain

Table 10.3 The eigevalue γ for anomalous dimension of each eigenoperator in $^1S_0$ and $^3S_1$ states. In the table (XY, ZW) means $B_{XY}B_{ZW} + (R \leftrightarrow L)$.

	(LL, LL)	(LL, LR)	(LL, RL)	(LR, LR)	(LR, RL)	(LL, RR)
$\gamma(^1S_0)$	-36	-12	-24	-24	-18	0
$\gamma(^1S_0)$	-28	-4	-24	-16	-18	0

$$\frac{D_Y(E)}{D_X(E)}(^1S_0) \simeq \frac{D_Y(E)}{D_X(E)}(^3S_1) \simeq 2. \tag{10.169}$$

For both cases, the ratio has positive sign, which gives repulsion at short distance, the repulsive core.

10.5   Concluding remarks

In this chapter, we have considered two different but closely related topics in lattice QCD approaches to hadron interactions.

In the first part, we explained Lüscher's formula, which relates the scattering phase shift to the two-particle energy in finite volume. Although this formula is well established and has appeared in the well-written reference (Lüscher, 1991), a comprehensive (but less rigorous) derivation for the formula has been attempted in this chapter for the $\pi\pi$ system as an example with the emphasis on the BS wave function. It is important to stress that the BS wave function at large separation behaves as a free scattering wave with a phase shift that is determined by the unitarity of the S matrix in QCD. The Lüscher formula can be obtained from this asymptotic behavior of the BS wave function.

In the second part, on the other hand, we considered the BS wave function in a non-asymptotic region where the interaction between two particles exists, in order to define the potential in quantum field theory. We applied this method to two nucleons to calculate the NN potential in QCD. The first result from lattice QCD has a good shape, which reproduces both the repulsive core at short distance and the attractive well at medium and long distances.

In the last part, the origin of the repulsive core was theoretically investigated by the OPE and the renormalization group. The analysis predicts the r dependence of the potentials at short distance, though it can not tell the sign of the potential, positive (repulsive) or negative (attractive). A crude estimate by a non-relativistic quark model gives the positive sign, the repulsive core.

The method to investigate potentials through the BS wave function in lattice QCD is a quite new approach, which may be applied in the following directions.

(1) In order to extract realistic NN potentials from lattice QCD, it is necessary to carry out full QCD simulations near the physical u, d quark masses. Studies along these lines using (2+1)-flavor QCD configurations generated by the PACS–CS Collaboration (Aoki *et al.*, 2009*a*) are currently under way (Ishii *et al.*, 2008).

(2) The hyperon–nucleon (YN) and hyperon–hyperon (YY) potentials are essential to understand properties of hypernuclei and the hyperonic matter inside neutron stars. While experimental scattering data are very limited due to the short lifetime of hyperons, the NN, YN and YY interactions on the lattice can be investigated in the same manner just by changing the quark flavors. The ΞN potential in quenched QCD (Nemura *et al.*, 2009) and the ΛN potential in both quenched and full QCD (Nemura *et al.*, 2008) are being examined as a first step toward the systematic understanding of baryonic potentials. In this connection,

the OPE analysis should be extended to the $N_f = 3$ case, in order to reveal the nature of the repulsive core in baryon–baryon potentials. Since quark-mass terms can be neglected in the OPE at short distance, the analysis can be done in the exact SU(3)-symmetric limit.

(3) The three-nucleon force is thought to play important roles in nuclear structures and in the equation of state in high-density matter. Since experimental information is very limited, the extension of the method to three nucleons may lead to first-principle extractions of the three-nucleon potentials in QCD. It is also interesting to investigate the existence or the absence of the repulsive core in the three-nucleon potentials. The calculation of anomalous dimensions of 9-quark operators will be required at 2-loop level.

(4) More precise evaluations including numerical simulations of matrix elements $\langle 0|O_X|E \rangle$ will be needed to understand the nature of the core in the potential through the OPE analysis.

Acknowledgments

I would like to thank the organizers of this Les Houches school for giving me an opportunity to give these lectures at the school and all attendees of the school for stimulating discussions. I also thank members of the HAL QCD (Hadron to Atomic nuclei from Lattice QCD) Collaboration, T. Doi, T. Hatsuda, Y. Ikeda, T. Inoue, N. Ishii, K. Murano, K. Nemura and K. Sasaki, for providing me the latest data and useful discussions, and my collaborators and friends, J. Balog, N. Ishizuka, W. Weise and P. Weisz, for valuable discussions and comments. This work is supported in part by a Grant-in-Aid of the Japanese Ministry of Education, Sciences and Technology, Sports and Culture (No. 20340047) and by a Grant-in-Aid for Scientific Research on Innovative Areas (No. 2004: 20105001, 20105003).

References

Aoki, S. et al. (2005a). *Nucl. Phys. Proc. Suppl.*, **140**, 305–307.

Aoki, S. et al. (2005b). *Phys. Rev.*, **D71**, 094504.

Aoki, S. et al. (2007). *Phys. Rev.*, **D76**, 094506.

Aoki, S. et al. (2009a). *Phys. Rev.*, **D79**, 034503.

Aoki, S., Balog, J., and Weisz, P. (2009b). *Prog. Theor. Phys.*, **121**, 1003–1034.

Aoki, S., Balog, J., and Weisz, P. (2010a). *JHEP*, **05**, 008.

Aoki, S., Hatsuda, T., and Ishii, N. (2008). *Comput. Sci. Dis.*, **1**, 015009.

Aoki, S., Hatsuda, T., and Ishii, N. (2010b). *Prog. Theor. Phys.*, **123**, 89–128.

Beane, S. R., Bedaque, P. F., Orginos, K., and Savage, M. J. (2006). *Phys.Rev.*, **D73**, 054503.

Beane, S. R., Luu, T. C., Orginos, K., Parreno, A., Savage, M. J., Torok, A., and Walker-Loud, A. (2008). *Phys. Rev.*, **D77**, 014505.

Bethe, H. A. and Salpeter, E. E. (1951). *Phys. Rev.*, **84**, 1232.

Brown, G. E. and Jackson, A. D. (1976). *Nucleon-nucleon interaction.* North-Holland, Amsterdam.

DeTar, C. (1979). *Phys. Rev.*, **D19**, 1451.

Faessler, A., Fernandez, F., Lubeck, G., and Shimizu, K. (1982). *Phys. Lett.*, **B112**, 201.

Feng, X., Jansen, K., and Renner, D. B. (2010). *Phys. Lett.*, **B684**, 268–274.

Fujiwara, Y., Suzuki, Y., and Nakamoto, C. (2007). *Prog. Part. Nucl. Phys.*, **58**, 439.

Heiselberg, H. and Pandharipande, V. (2000). *Annu. Rev. Nucl. Part. Sci.*, **50**, 481.

Hoshizaki, N. et al. (1968). *Prog. Theor. Phys. Suppl.*, **42**, 1.

Ishii, N., Aoki, S., and Hatsuda, T. (2007). *Phys. Rev. Lett.*, **99**, 022001.

Ishii, N., Aoki, S., and Hatsuda, T. (2008). *PoS*, **LATTICE2008**, 155.

Ishizuka, N. (2009). *PoS*, **LATTICE2009**, 119.

Itzykson, C. and Zuber, J.B. (1980). *Quantum field theory*. MacGraw-Hill, New York.

Jackson, A. Jackson, A. D. and Pasquier, V. (1985). *Nucl. Phys.*, **A432**, 567.

Jastrow, R. (1951). *Phys. Rev.*, **81**, 165.

Lattimer, J. M. and Prakash, M. (2000). *Phys. Rep.*, **50**, 481.

Liberman, D. A. (1977). *Phys. Rev.*, **D16**, 1542.

Lin, C. J. D., Martinelli, G., Sachrajda, C. T., and Testa, M. (2002). *Nucl. Phys. Proc. Suppl.*, **109A**, 218–225.

Lüscher, M. (1991). *Nucl. Phys.*, **B354**, 531–578.

Machida, S. and Namiki, M. (1965). *Prog. Theor. Phys.*, **33**, 125.

Machleidt, R. (1989). *Adv. Nucl. Phys.*, **19**, 189.

Machleidt, R. (2001). *Phys. Rev.*, **C63**, 024001.

Machleidt, R. and Slaus, I. (2001). *J. Phys.*, **G27**, R69.

Myhrer, F. and Wroldsen, J. (1988). *Rev. Mod. Phys.*, **60**, 629.

Nambu, Y. (1957). *Phys. Rev.*, **106**, 1366.

Nemura, H., Ishii, N., Aoki, S., and Hatsuda, T. (2008). *PoS*, **LATTICE2008**, 156.

Nemura, H., Ishii, N., Aoki, S., and Hatsuda, T. (2009). *Phys. Lett.*, **B673**, 136–141.

Neudachin, V. G., Smirnov, Yu. F., and Tamagaki, R. (1977). *Prog. Theor. Phys.*, **58**, 1072.

Oka, M., Shimizu, K., and Yazaki, K. (2000). *Prog. Theor. Phys. Suppl.*, **137**, 1.

Oka, M. and Yazaki, K. (1980). *Phys. Lett.*, **B90**, 41.

Oka, M. and Yazaki, K. (1981*a*). *Prog. Theor. Phys*, **66**, 556.

Oka, M. and Yazaki, K. (1981*b*). *Prog. Theor. Phys.*, **66**, 572.

Okubo, S. and Marshak, R. E. (1958). *Ann. Phys.*, **4**, 166–179.

Otsuki, S., Tamagaki, R., and Yasuno, M. (1965). *Prog. Theor. Phys. Suppl.*, **Extra Number**, 578.

Rummukainen, K. and Gottlieb, S. (1995). *Nucl. Phys.*, **B450**, 397–436.

Sasaki, K and Ishizuka, N. (2008). *Phys. Rev.*, **D78**, 014511.

Stoks, V. G. J., Klomp, R. A. M., Terheggen, C. P. F., and de Swart, J. J. (1994). *Phys. Rev.*, **C49**, 2950–2962.

Taketani, M. et al. (1967). *Prog. Theor. Phys. Suppl.*, **39**, 1.

Tamagaki, R. et al. (1993). *Prog. Theor. Phys. Suppl.*, **112**, 1.

Toki, H. (1980). *Z. Phys.*, **A294**, 173.

Wiringa, R. B., Stoks, V. G. J., and Schiavilla, R. (1995). *Phys. Rev.*, **C51**, 38–51.

Yabu, H. and Ando, K. (1985). *Prog. Theor. Phys.*, **74**, 750.

Yamazaki, T. et al. (2004). *Phys. Rev.*, **D70**, 074513.

Yukawa, H. (1935). *Proc. Math.-Phys. Soc. Jpn.*, **17**, 48.

11

Flavor physics and lattice quantum chromodynamics

Laurent LELLOUCH

Centre de Physique Théorique[1]
Case 907
CNRS Luminy
Université de la Méditerranée (Aix-Marseille II)
F-13288 Marseille Cedex 9
France

Summer school on "Modern Perspectives in Lattice QCD"
École de Physique des Houches, August 3–28, 2009

Abstract

Quark flavor physics and lattice quantum chromodynamics (QCD) met many years ago, and together have given rise to a vast number of very fruitful studies. All of these studies certainly cannot be reviewed within the course of these lectures. Instead of attempting to do so, I discuss in some detail the fascinating theoretical and phenomenological context and background behind them, and use the rich phenomenology of non-leptonic weak kaon decays as a template to present some key techniques and to show how lattice QCD can effectively help shed light on these important phenomena. Even though the lattice study of $K \to \pi\pi$ decays originated in the mid-1980s, it is still highly relevant. In particular, testing the consistency of the Standard Model with the beautiful experimental measurements of direct CP violation in these decays remains an important goal for the upcoming generation of lattice QCD practitioners.

The course begins with an introduction to the Standard Model, viewed as an effective field theory. Experimental and theoretical limits on the energy scales at which New Physics can appear, as well as current constraints on quark flavor parameters, are reviewed. The role of lattice QCD in obtaining these constraints is described. A second section is devoted to explaining the Cabibbo–Kobayashi–Maskawa mechanism for quark flavor mixing and CP violation, and to detailing its most salient features. The third section is dedicated to the study of $K \to \pi\pi$ decays. It comprises discussions of indirect CP violation through K^0-$\bar{K}^0$ mixing, of the $\Delta I = 1/2$ rule, and of direct CP violation. It presents some of the lattice QCD tools required to describe these phenomena *ab initio*.

11.1 Introduction and motivation

11.1.1 The Standard Model as a low-energy effective field theory

If elementary particles were massless, their fundamental interactions would be well described by the most general perturbatively renormalizable[2] relativistic quantum field theory based on:

- the gauge group

$$SU(3)_c \times SU(2)_L \times U(1)_Y, \tag{11.1}$$

where the subscript c stands for "color", L for left-handed weak isospin and Y for hypercharge;

- three families of quarks and leptons

$$\begin{cases} u\ d\ e^-\ \nu_e \\ c\ s\ \mu^-\ \nu_\mu \\ t\ b\ \tau^-\ \nu_\tau, \end{cases} \tag{11.2}$$

with prescribed couplings to the gauge fields (i.e. in specific representations of the gauge groups);

- and the absence of anomalies.

In the presence of masses for the weak gauge bosons $W^\pm$ and Z^0, for the quarks, and for the leptons, the most economical way known to keep this construction perturbatively renormalizable is to implement the Higgs mechanism (Englert and Brout, 1964; Higgs, 1964), as done in the Standard Model (SM). However, this results in adding a yet unobserved degree of freedom to the model, the Higgs boson.

By calling a theory renormalizable we mean that it can be used to make *predictions of arbitrarily high accuracy* over a very large interval of energies, ranging from zero to possibly infinite energy, with only a finite number of coupling constants.[3] These couplings are associated with operators of mass dimension less than or equal to four in $3 + 1$ dimensions.

Renormalizable field theories are remarkable in many ways. Consider an arbitrary high-energy theory described by a Lagrangian $\mathcal{L}_{\text{UV}}$ (e.g. a GUT, a string theory,...) with given low-energy spectrum and symmetries. At sufficiently low energies this theory is described by the unique renormalizable theory with the given spectrum and symmetries, whose Lagrangian we will denote $\mathcal{L}_{\text{ren}}$. Moreover, the deviations between the predictions of the two theories can be parametrized through a local low-energy effective field theory (EFT)

$$\mathcal{L}_{\text{UV}} = \mathcal{L}_{\text{ren}} + \sum_{d \geq 4} \sum_i \frac{C_{d,i}}{\Lambda_i^{d-4}} O_i^{(d)}, \qquad (11.3)$$

where the $O_i^{(d)}$ are operators of mass dimension $d \geq 4$ built up from fields of $\mathcal{L}_{\text{ren}}$. The Λ_i are mass scales which are much larger than the masses in the spectrum of $\mathcal{L}_{\text{ren}}$ – there may be one or many of them depending on the number of distinct scales in $\mathcal{L}_{\text{UV}}$. The $C_{d,i}$ are dimensionless coefficients whose sizes depend on how the corresponding operators are generated in the UV theory, e.g. at tree or loop level.

Thus, very generally, we can write down the Lagrangian of particle physics as a low-energy EFT with the gauge group of Eq. (11.1), the matter content of Eq. (11.2) and a Higgs mechanism:

$$\mathcal{L}_{\text{SM}}^{\text{eff}} = \mathcal{L}_{\text{SM}} + \frac{1}{M} O_{\text{Maj}}^{(5)} + \sum_{d \geq 6} \sum_i \frac{C_{d,i}}{\Lambda_i^{d-4}} O_i^{(d)}, \qquad (11.4)$$

where the left-handed neutrino Majorana mass term, $O_{\text{Maj}}^{(5)}$, and the $O_i^{(d)}$ must be invariant under the Standard Model gauge group (11.1). In Eq. (11.4), $\mathcal{L}_{\text{SM}}$ is the renormalizable Standard Model Lagrangian

$$\mathcal{L}_{\text{SM}} = \mathcal{L}_{\text{g+f}} + \mathcal{L}_{\text{flavor}} + \mathcal{L}_{\text{EWSB}} + \mathcal{L}_\nu, \qquad (11.5)$$

where $\mathcal{L}_{\text{g+f}}$ contains the gauge and fermion kinetic and coupling terms, $\mathcal{L}_{\text{flavor}}$, the Higgs–Yukawa terms, $\mathcal{L}_{\text{EWSB}}$, the Higgs terms and $\mathcal{L}_\nu$, the possible renormalizable neutrino mass and right-handed neutrino kinetic terms. In that sense, the renormalizable Standard Model is a low-energy approximation of a more complete high-energy theory involving scales of New Physics much larger than M_W.

Schematically, the gauge and fermion Lagrangian reads

$$\mathcal{L}_{g+f} = \frac{1}{4} F^a_{\mu\nu} F^{\mu\nu}_a + \bar{\psi} \slashed{D} \psi. \tag{11.6}$$

It has three parameters, the gauge couplings (g_1, g_2, g_3), and is very well tested through experiments conducted at LEP, SLC, the Tevatron, etc. Its parameters are known to better than per mil accuracy.

The Higgs–Yukawa terms are given by

$$\mathcal{L}_{\text{flavor}} = -\bar{\psi}_R^{(-1/2)} Y_{(-1/2)} \phi^\dagger \psi_L - \bar{\psi}_R^{(1/2)} Y_{(1/2)} \tilde{\phi}^\dagger \psi_L + \text{h.c.}, \tag{11.7}$$

with ψ_L corresponding to the left-handed $SU(2)_L$ doublets and $\psi_R^{(\pm 1/2)}$ the right-handed $SU(2)_L$ singlets, associated with the $I_3 = \pm\frac{1}{2}$ component of the doublets. In this equation, ϕ is the Higgs field and $\tilde{\phi}$ its conjugate, $(\phi^0, -\phi^{+*})$. The flavor component of the Standard Model Lagrangian has many more couplings, 13 in fact. It gives rise to the three charged lepton masses, six quark masses, and the quark flavor mixing matrix which has three mixing angles and one phase.[4] The understanding of this quark mixing and its associated CP violation will be the main focus of the present course.

There is also the electroweak symmetry breaking (EWSB) contribution

$$\mathcal{L}_{\text{EWSB}} = (D_\mu \phi)^\dagger (D^\mu \phi) - \mu^2 \phi^\dagger \phi - \lambda (\phi^\dagger \phi)^2. \tag{11.8}$$

It has only two couplings, the Higgs mass and self-coupling (μ, λ), and is very poorly tested so far, a situation which will change radically with the LHC.

As for the neutrino Lagrangian, little is known from experiment about its form. There are theoretically two possible, non-exclusive scenarios:

1. There are no right-handed neutrinos in sight. Thus, we give our left-handed neutrinos a mass without introducing a right-handed partner. In that case, $\mathcal{L}_\nu = 0$ and we have a Majorana mass term for the left-handed neutrinos

$$O^{(5)}_{\text{Maj}} = -\frac{1}{2} \nu_L^T C \tilde{\phi}^T A_\nu^L \tilde{\phi} \nu_L + \text{h.c.}, \tag{11.9}$$

   where C is the charge conjugation matrix (see Eq. (11.47)). That is, after EWSB the neutrino acquires a Majorana mass through the introduction of a non-renormalizable dimension-5 operator. This implies that the Standard Model is an EFT and that we already have a signal for a new mass scale. Indeed, with $m_\nu \sim 0.1$ eV (a plausible value), eigenvalues of the coupling matrix A_ν^L of order 1 and $\langle \phi \rangle \sim 246$ GeV, one finds for the mass scale M of Eq. (11.4)

$$M \sim \frac{\langle \phi \rangle^2}{m_\nu} \sim 10^{15} \text{ GeV} \tag{11.10}$$

   which is tantalizingly close to a possible unification scale.

2. We choose to allow right-handed neutrinos, N_R. These neutrinos must be singlets under the Standard Model group. Thus, they themselves may have a Majorana

mass, but this time a renormalizable one, in addition to allowing the presence of a Dirac mass term:

$$\mathcal{L}_\nu = N_R i \partial \!\!\!/ N_R - (\bar{L}_L Y_\nu^\dagger \tilde{\phi} N_R + \frac{1}{2} N_R^T C M_\nu^R N_R + \text{h.c.}) \tag{11.11}$$

$$= N_R i \partial \!\!\!/ N_R - \frac{1}{2}(\nu_L^T, N_R^{cT}) C \begin{pmatrix} 0 & Y_\nu^\dagger \phi^0 \\ Y_\nu^* \phi^0 & M_\nu^R \end{pmatrix} \begin{pmatrix} \nu_L \\ N_R^c \end{pmatrix} + \cdots, \tag{11.12}$$

where L_L stands for the left-handed lepton doublets, and N_R^c for the charge conjugate of N_R (see Eq. (11.47)).

There are here three more possibilities:

(a) $M_\nu^R = 0$.

In that case, the three neutrinos have Dirac masses and lepton number is conserved.

(b) $M_\nu^R \gg Y_\nu \langle \phi \rangle$.

Here, the see-saw mechanism comes into play: there are no right-handed neutrinos in sight and all three left-handed neutrinos acquire a mass through $d = 5$ operators:

$$\mathcal{L}_{\text{eff}}^{m_\nu} = -\frac{1}{2} \nu_L^T C \tilde{\phi}^T \left(Y_\nu^\dagger (M_\nu^R)^{-1} Y_\nu^* \right) \tilde{\phi} \nu_L + O \left((M_\nu^R)^{-2} \right). \tag{11.13}$$

Thus, we have an explicit realization of scenario (1).

Taking $\text{Re} Y_\nu \langle \phi \rangle \sim 1 \, \text{GeV}$ in rough analogy with the τ and again, $m_\nu \sim 0.1 \, \text{eV}$, we obtain for the mass scale M of Eq. (11.4)

$$M \sim \frac{(\text{Re} Y_\nu \langle \phi \rangle)^2}{m_\nu} \sim 10^{10} \, \text{GeV}. \tag{11.14}$$

(c) Some eigenvalues of $M \sim$ some eigenvalues of $Y_\nu \langle \phi \rangle$.

Then, the sea-saw neutrino mass matrix will have more than three small distinct eigenvalues (actually up to six), leading to more than three light neutrinos. Such a possibility is constrained by phenomenology, but is not excluded.

Though the topic of neutrino masses and associated mixing and CP violation is fascinating, it is not the flavor physics which is of interest to us here. Thus, this is all that we will say about the subject and, for the remainder of the course, we can safely take $m_\nu = 0$, forgetting about $O_{\text{Maj}}^{(5)}$ and $\mathcal{L}_\nu$ altogether.

Having explored the neutrino mass Lagrangian and some of the constraints which neutrinos place on the scale of New Physics, we now do the same for the other components of $\mathcal{L}_{\text{SM}}^{\text{eff}}$, generically denoting the scale of New Physics by Λ:

1. *EWSB and naturalness:* besides possible right-handed neutrinos, the Higgs boson is the only Standard Model particle whose mass is not protected by a symmetry from the physics at energy scales much larger than M_W. To get a very rough estimate of what the contributions of New Physics to the Higgs mass could be,

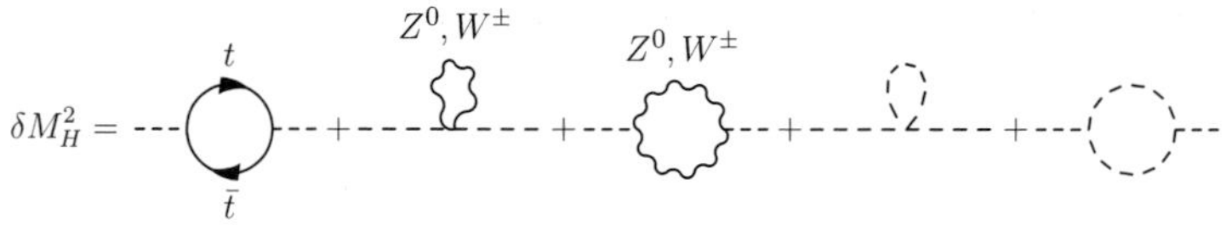

Fig. 11.1 Diagrams which contribute radiative corrections, δM_H^2, to the Higgs mass squared at one loop.

we assume that the effect of the new degrees of freedom can be approximated by computing Standard Model loop corrections to this mass, cut off at a scale $\Lambda \gg M_W$ that is characteristic of the new phenomena. Then, the contributions to the Higgs mass at one loop are given by the diagrams in Fig. 11.1 with a cutoff Λ. They yield

$$\delta M_H^2 = \frac{3\Lambda^2}{16\pi^2 \langle\phi\rangle^2}(4m_t^2 - 2M_W^2 - M_Z^2 - M_H^2) \tag{11.15}$$

which is dominated by the top contribution for $M_H \ll 350\,\mathrm{MeV}$. If, for naturalness reasons, we require that the physical squared Higgs mass is no less than a fraction f of the correction of Eq. (11.15), then we find that

$$\Lambda_{\mathrm{nat}} \leq \frac{4\pi\langle\phi\rangle}{\sqrt{3f}} \frac{M_H}{\sqrt{4m_t^2 - 2M_W^2 - M_Z^2 - M_H^2}}$$

$$\sim \frac{700\,\mathrm{GeV}}{\sqrt{f}} \times \frac{(M_H/115\,\mathrm{GeV})}{\sqrt{1 - \left(\frac{M_H - 115\,\mathrm{GeV}}{310\,\mathrm{GeV}}\right)^2}}. \tag{11.16}$$

Thus, if we allow at most 1% of fine tuning on the Higgs mass squared, Eq. (11.16) says that new physics must appear below $\Lambda_{\mathrm{nat}} \sim 7\,\mathrm{TeV}$.

2. *Gauge sector and flavor conserving $d = 6$ operators:* consider, for instance, $O_{WB} = g_1 g_2 (\phi^\dagger \sigma^a \phi) W_{\mu\nu}^a B_{\mu\nu}$, which couples the W bosons to the $U(1)_Y$ gauge boson B. Precision electroweak data, assuming that the Wilson coefficient of this operator is of order one, impose the following constraint on the scale of New Physics (Barbieri *et al.*, 2004):

$$\Lambda \gtrsim 5\,\mathrm{TeV} \qquad 95\%\,\mathrm{CL}. \tag{11.17}$$

3. *Flavor physics and, in particular, flavor changing neutral currents (FCNC):* consider, for instance, K^0-$\bar{K}^0$ mixing. In the absence of electroweak interactions, the long-lived K_L^0 is a CP odd state, whereas the short-lived K_S^0 is a CP even state. When these interactions are turned on, these two degenerate particles acquire a minuscule mass difference, which is measured experimentally to be:

$$\Delta M_K \equiv M_{K_L} - M_{K_S} \simeq 3.5 \times 10^{-12}\,\mathrm{MeV}. \tag{11.18}$$

Consider now the contribution to ΔM_K of an arbitrary $d = 6$, $\Delta S = 2$ operator schematically written as $(\bar{d}s)(\bar{d}s)$:

$$\Delta M_K = 2 \times \frac{1}{2M_K} \frac{\text{Re}C^*_{\Delta S=2}}{\Lambda^2} \langle \bar{K}^0|(\bar{s}d)(\bar{s}d)|K^0\rangle. \tag{11.19}$$

Now, assuming that $\text{Re}C^*_{\Delta S=2} \sim 1$ and that the matrix element is of order the fourth power of a typical QCD scale, e.g. $\sim M_\rho^4$, we get

$$\Lambda > \frac{M_\rho^2}{\sqrt{M_K \, \Delta M_K}} \sim 10^3 \text{ TeV} \tag{11.20}$$

which is orders of magnitudes larger than the lower bound imposed by gauge sector and flavor conserving transitions, as well as than the upper bound imposed by naturalness.

Thus, if we do not make any assumptions about how the New Physics breaks flavor symmetries, we are forced to push this physics to very high scales. Said differently, flavor physics is sensitive to very high energy scales if the New Physics is allowed to have a flavor structure which differs from that of the Standard Model. Therefore, one assumption commonly made is that the New Physics breaks the flavor symmetries with the same Yukawa couplings as in the Standard Model. This assumption is called Minimal Flavor Violation (MFV). For instance, we might have, in the case of K^0-$\bar{K}^0$ mixing, the following operators contributing: $\frac{1}{\Lambda^2}(\bar{s}_R Y_{sd} d_L)^2$, $\frac{1}{\Lambda^2}(\bar{s}_L Y^*_{sd} Y_{sd} \gamma_\mu d_L)^2$, etc. I will leave you to work out the corresponding scales, Λ, but they are certainly much lower and in line with those obtained from flavor-conserving physics.

11.1.2 Flavor physics phenomenology

As we shall see shortly in more detail, the Standard Model has a very rich and constrained flavor structure, which includes:

- mixing of quark flavors;
- CP violation by a unique invariant J (Jarlskog, 1985), discussed in Section 11.2.3;
- the absence of tree-level flavor changing neutral currents (FCNC).

All of these features are encapsulated in:

- the Cabibbo–Kobayashi–Maskawa (CKM) matrix (Cabibbo, 1963; Kobayashi and Maskawa, 1973)

$$V = \begin{pmatrix} V_{ud} & V_{us} & V_{ub} \\ V_{cd} & V_{cs} & V_{cb} \\ V_{td} & V_{ts} & V_{tb} \end{pmatrix}, \tag{11.21}$$

  which is unitary. It has three mixing angles and a single phase, which is responsible for CP violation.

- the quark masses: m_q, with $q = u, d, s, c, b, t$.

For their discovery, in 1973, that Nature's CP violation and rich flavor structure can be well described when a third generation is added to the $SU(2)_L \times U(1)_Y$ electroweak model (Kobayashi and Maskawa, 1973), Kobayashi and Maskawa were awarded part of the 2008 Physics Nobel Prize.

Because this flavor structure is so intriguing and most probably contains important information about physics at much higher energies than currently explored, particle physicists have invested a considerable amount of effort in exploring it theoretically and experimentally over the past five decades. This exploration has multiple goals:

(1) To determine from experiment the matrix elements of the CKM matrix V, which are important parameters of our fundamental theory.
(2) To verify that the CKM description of quark flavor mixing and CP violation is correct, e.g.:

- Can all of the observed CP violation in the quark sector be explained in terms of a single phase?
- Is the measured matrix V unitary?

The latter can be tested by verifying whether

$$\sum_{D=d,s,b} |V_{UD}|^2 = 1 \quad \text{and} \quad \sum_{U=u,c,t} |V_{UD}|^2 = 1. \tag{11.22}$$

If either of these sums turn out to be less than 1, that would signal an additional generation or family. On the other hand, if either one is larger than 1, completely new physics would have to be invoked. The unitarity of V also implies that the scalar product of any two distinct columns or rows of the matrix must vanish, i.e.

$$\sum_{U=u,c,t} V_{UD_1} V_{UD_2}^* = 0, \quad \text{for } D_1 \neq D_2, \tag{11.23}$$

$$\sum_{D=d,s,b} V_{U_1 D} V_{U_2 D}^* = 0, \quad \text{for } U_1 \neq U_2. \tag{11.24}$$

These relations can be represented as triangles in the complex plane, which are traditionally labeled by their unsummed flavors, i.e. (D_1, D_2) for those of Eq. (11.23) and (U_1, U_2) for those of Eq. (11.24). In the absence of CP violation, unitarity triangles would become degenerate. The (db) triangle is shown in Fig. 11.2. In Fig. 11.3, it is drawn to scale with two other triangles to give you a sense of the variety of unitarity triangles and the difficulties there may be in measuring some of their sides and angles.

The (db) triangle has been the focus of considerable experimental (LEP, B-factories, Tevatron, ...) and theoretical (QCD factorization, Soft Collinear Effective Theory (SCET), lattice QCD (LQCD), ...) effort in the last 10–15 years. With the arrival of the LHC and, in particular, the experiment LHCb, the focus is shifting from the study of the B_d towards the study of the B_s meson and thus toward investigations of the (sb) triangle. Here too, lattice QCD has

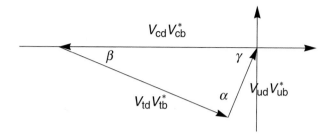

Fig. 11.2 The (db) unitarity triangle.

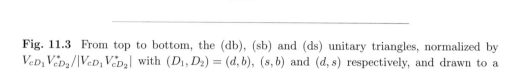

Fig. 11.3 From top to bottom, the (db), (sb) and (ds) unitary triangles, normalized by $V_{cD_1}V_{cD_2}^*/|V_{cD_1}V_{cD_2}^*|$ with $(D_1, D_2) = (d, b)$, (s, b) and (d, s) respectively, and drawn to a common scale.

a considerable role to play, most notably in the study of the B_s-$\bar{B}_s$ mass and width differences, of the leptonic decay $B_s \to \mu^+\mu^-$ or of the semileptonic decay $B_s \to \phi\mu^+\mu^-$.

The strategy here is to verify the unitarity of the CKM matrix by performing redundant measurements of:

- triangle sides with CP conserving decays;
- angles with CP violating processes;

and checking that the triangles indeed close.

(3) To determine in what processes there is still room for significant New Physics contributions. For instance, $O(40\%)$ effects are still possible in B^0-$\bar{B}^0$ mixing from New Physics with a generic weak phase (Lenz *et al.*, 2010).

(4) To constrain the flavor sectors of beyond the Standard Model (BSM) candidates. As we saw above, it is difficult to add new physics to the Standard Model without running into serious problems in the flavor sector.

(5) To actually find evidence for beyond the Standard Model physics. This is most likely to be found in processes which are highly suppressed in the Standard Model, such as FCNC.

(6) If new particles and interactions are discovered, it is important to investigate their quark and flavor structure.

All of these goals require being able to compute reliably and precisely flavor observables in the Standard Model or beyond. A high level of precision has been reached already on the magnitudes of individual CKM matrix elements (Charles *et al.*, 2005):

$$|V| = \begin{array}{c} \\ u \\ c \\ t \end{array} \begin{pmatrix} \overset{d}{0.97425^{+0.00018}_{-0.00018}} & \overset{s}{0.22543^{+0.00077}_{-0.00077}} & \overset{b}{0.00354^{+0.00016}_{-0.00014}} \\ 0.22529^{+0.00077}_{-0.00077} & 0.97342^{+0.00021}_{-0.00019} & 0.04128^{+0.00058}_{-0.00129} \\ 0.00858^{+0.00030}_{-0.00034} & 0.04054^{+0.00057}_{-0.00129} & 0.999141^{+0.000053}_{-0.000024} \end{pmatrix}, \qquad (11.25)$$

assuming the correctness of the Standard Model and, in particular, CKM unitarity. The most poorly known CKM matrix elements are $|V_{ub}|$ and $|V_{td}|$, both with an uncertainty around 4%. Then come $|V_{ts}|$ and $|V_{cb}|$, with an uncertainty of about 2%. Thus, to have an impact in testing the CKM paradigm of quark flavor mixing and CP violation, and to take full advantage of LHCb results, the precision of theoretical predictions must be of order a few percent (better in many cases). This is no small challenge when non-perturbative QCD dynamics is involved.

11.1.3 Flavor physics and lattice QCD

Lattice QCD plays and will continue to play a very important role in flavor physics, by providing reliable calculations of non-perturbative strong interaction corrections to weak processes involving quarks.

The processes for which LQCD gives the most reliable predictions are those which involve a single hadron that is stable against strong interaction decay in the initial state and, at most, one stable hadron in the final state. Resonances (i.e. unstable hadrons) are much more difficult to contend with, especially if many decay channels are open. Similarly, final states with more than a single stable hadron are much more difficult, especially if these hadrons can rescatter inelastically. This will be discussed in much more detail in Section 11.3.7.

Thus, the processes typically considered for determining the absolute values of the CKM matrix elements are the following

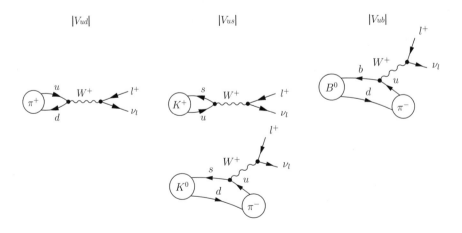

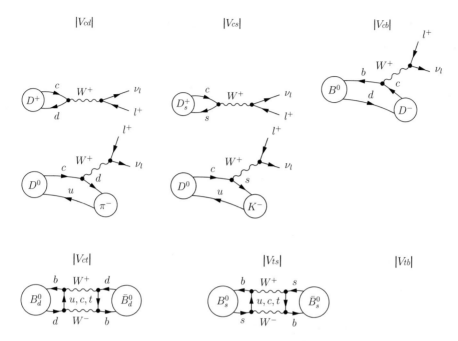

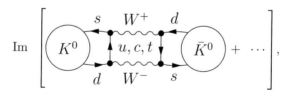

Now, to determine the unique CKM matrix phase or, more precisely, the CP violating parameter J, lattice QCD can have an important impact through the following processes:

- Indirect CP violation in $K \to \pi\pi$ decays. This occurs through the process of K^0-$\bar{K}^0$ mixing, which is given by the imaginary part

$$\mathrm{Im} \left[\raisebox{-1.5em}{\includegraphics{}} + \cdots \right],$$

where the ellipsis stands for the other box contribution. It is a $|\Delta S| = 2$ FCNC that will be discussed in detail below.

- Direct CP violation in $K \to \pi\pi$ decays. The processes which contribute are given by the following $|\Delta S| = 1$ amplitudes

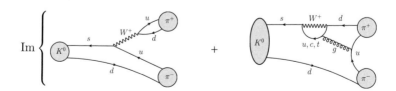

in the case of $K^0 \to \pi^+\pi^-$, where the ellipsis stands for missing diagrams similar to those drawn. The diagrams for $K^0 \to \pi^0\pi^0$ are analogous. Again, this particular phenomenon will be discussed in detail in the sequel.

11.1.4 Low-energy effective field theories of the Standard Model

With present knowledge and present computer resources, we cannot simulate the full Standard Model in lattice field theory calculations. In particular various degrees of freedom must be "eliminated" from the calculations for the following reasons:

- W, Z and t: there is no hope to be able to simulate these degrees of freedom whose masses are $M_{W,Z} \sim 80$–90 GeV and $m_t \sim 175$ GeV on lattices which must be large enough to accommodate 135 MeV pions, i.e. with sizes $L \gtrsim 4/M_\pi \sim 6$ fm.[5] Since we would also have to have $am_t \ll 1$, with a the lattice spacing, to guarantee controlled discretization errors, the number of points on the lattice would have to be $L/a \gg 4m_t/M_\pi \sim 5.2 \times 10^3$, which is beyond any foreseeable computing capabilities. Perhaps even more important, however, is the fact that we just do not know how to discretize non-abelian, chiral gauge theories (please see David's lecture notes in this volume (Kaplan, 2010)).

- b: even the b quark, with $m_b \sim 4.2$ GeV, would require lattices with $L/a \gg 120$, which is already too much for present technology.

- c: with $m_c \sim 1.3$ GeV, the charm is a borderline case, both on the lattice and in terms of QCD. On the lattice it can be included in simulations, but with $am_c \sim 0.35$ at best, discretization errors remain an important preoccupation. From the point of view of QCD, the charm is not quite a heavy quark–heavy quark effective theory is only marginally applicable since m_c is not much larger than typical QCD scales – and it is clearly not light – it is not in the regime of chiral perturbation theory. So its inclusion should be considered with care. In addition, its inclusion in weak processes can mean significantly more complicated correlation functions to compute. For instance, Fig. 11.4 illustrates the type of correlation functions required to determine the amplitude for K^0-$\bar{K}^0$ mixing, in the absence of charm (diagram on left) and in the presence of a dynamical charm quark (diagram on

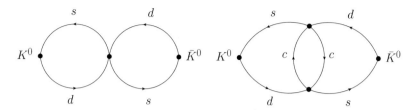

Fig. 11.4 Examples of correlation functions required for the lattice computation of the K^0-$\bar{K}^0$ mixing amplitude. The diagram on the left exhibits the type of three-point function with a four-quark operator insertion, required to obtain the K^0-$\bar{K}^0$ amplitude when the charm quark is integrated out. The diagram on the right shows a four-point function which is required when the charm quark is kept active.

right). In the absence of charm, it is a rather standard three-point function which must be computed while in its presence, it is a four-point function, with two four-quark operator insertions. However, in certain circumstances, the inclusion of a dynamical charm quark significantly simplifies the renormalization of the weak effective theory, as briefly discussed in Section 11.3.8 for the $\Delta I = 1/2$ rule.

Fortunately, with the precisions required at present and in any foreseeable future, it is not necessary to include virtual W, Z, t and b contributions. The situation with the charm is less clear, as $(M_\rho/m_c)^2$ sea corrections can, in principle, play a role when percent precisions are reached. So in considering processes involving these massive particles, we can turn to effective field theories (EFTs) in which the W, Z, t, b and possibly c and τ are no longer dynamical degrees of freedom. Thus, we are left with an $SU(3)_c \times U(1)_{\rm EM}$ gauge theory of color and electromagnetism. This theory includes the virtual effects of the following degrees of freedom:

- two massless gauge bosons: the qluon, g, and the photon, γ;
- three to four quarks: the u, d, s and possibly the c;
- two to three leptons: the e, μ and possibly the τ;
- three neutrinos: the ν_e, ν_μ and the ν_τ.

It also includes local operators of naive mass-dimension $d \geq 5$, which result from integrating out the heavy W, Z and t. Moreover, in this theory, the b – and possibly the c and the τ – are described by heavy fermion effective theories, in which the antiparticle of the fermion considered is integrated out. It is at this level that LQCD enters, to describe the non-perturbative effects of the strong interaction.

11.2   Standard Model and quark flavor mixing

As already discussed, the Standard Model has a highly constrained quark flavor structure, parametrized by the CKM matrix and the quark masses. We will now see more precisely how it arises and what its basic implications are.

11.2.1 On the origin of quark flavor mixing in the Standard Model

Let us look in more detail at the quark and gauge sectors of the Standard Model. With the notation (dim. rep. $SU(3)_c$, dim. rep. $SU(2)_L)_Y$, where "rep." stands for representation, the quark content of the Standard Model is given by

$$Q_L = \begin{pmatrix} U_L \\ D_L \end{pmatrix} \sim (3, 2)_{\frac{1}{2}}$$

$$U_R \sim (3, 1)_{2/3} \tag{11.26}$$

$$D_R \sim (3, 1)_{-1/3},$$

with $U = (u, c, t)^T$ and $D = (d, s, b)^T$. The coupling of these quarks with the gauge bosons is given by

$$\mathcal{L}_{g+q} = \cdots + \bar{Q}_L \slashed{D} Q_L + \bar{U}_R \slashed{D} U_R + \bar{D}_R \slashed{D} D_R, \tag{11.27}$$

where the ellipsis stands for the gauge kinetic and self-interaction terms. This Lagrangian has the following global symmetries: $U(3)_L$ on Q_L, $U(3)_{U_R}$ on U_R and $U(3)_{D_R}$ on D_R.

Now let us look at the quark Yukawa terms. After spontaneous symmetry breaking by the Higgs field, keeping only the terms proportional to the Higgs v.e.v. $\langle \phi \rangle$, we have:

$$\mathcal{L}_m^q \xrightarrow{\langle \phi \rangle \neq 0} -\bar{U}_R M_U U_L - \bar{D}_R M_D D_L + \text{h.c.}, \tag{11.28}$$

where M_U and M_D are arbitrary 3×3 complex matrices. On these terms we perform the following set of flavor transformations which leave $\mathcal{L}_{g+q}$ invariant:

- the $SU(3)_{U_R}$ transformation: $U_R \to V_R^U U_R$;
- the $SU(3)_L$ transformation: $U_L \to V_L^U U_L$ and $D_L \to V_L^U D_L$;
- and the $SU(3)_{D_R}$ transformation: $D_R \to V_R^D D_R$;

with V_R^U, V_L^U and V_R^D such that:

$$M_U^d = V_R^{U\dagger} M_U V_L^U, \tag{11.29}$$

and

$$V_R^{D\dagger} M_D = M_D^d V_L^{D\dagger} \tag{11.30}$$

where $M_{U,D}^d$ are diagonal matrices with real positive entries which are the quark masses. The second equation defines a fourth unitary matrix V_L^D. Under these rotations, the quark mass Lagrangian transforms as

$$\mathcal{L}_m^q \to -\bar{U}_R M_U^d U_L - \bar{D}_R M_D^d [V_L^{D\dagger} V_L^U D_L] + \text{h.c.} \tag{11.31}$$

The up quark mass matrix is diagonal in this basis, but not the down quark matrix. In addition, we have exhausted the flavor transformations allowed by $\mathcal{L}_{g+q}$. Thus, if we want to work in a mass basis (i.e. a basis in which all quark masses are diagonal), we have to perform the additional transformation:

- $D_L \to V_L^{U\dagger} V_L^D D_L$,

which is not a symmetry of $\mathcal{L}_{g+q}$. Clearly, the only terms in $\mathcal{L}_{g+q}$ which are affected are those which couple U_L and D_L. They are transformed in the following way:

$$\mathcal{L}_{CC}^q = \frac{g_2}{\sqrt{2}} \bar{U}_L W^{(+)} D_L + \text{h.c.}$$

$$\to \frac{g_2}{\sqrt{2}} \bar{U}_L W^{(+)} V D_L + \text{h.c.}, \tag{11.32}$$

where CC stands for charged current and $V \equiv V_L^{U\dagger} V_L^D$ is the CKM matrix. All other terms are left unchanged. In particular, the neutral currents:

$$\mathcal{L}_{NC}^q = \frac{g_2}{2} (\bar{U}_L, \bar{D}_L) W^3 \begin{pmatrix} 1 & 0 \\ 0 & -1 \end{pmatrix} \begin{pmatrix} U_L \\ D_L \end{pmatrix}$$

$$+ \frac{g_1}{6} (\bar{U}_L, \bar{D}_L) B \begin{pmatrix} U_L \\ D_L \end{pmatrix} \tag{11.33}$$

$$+ \frac{2}{3} g_1 \bar{U}_R B U_R - \frac{1}{3} g_1 \bar{D}_R B D_R,$$

which are flavor diagonal, remain diagonal in the mass basis. Moreover, the Higgs couples to fermions through their masses and therefore has a diagonal coupling to the quarks in the mass basis. The fact that all uncharged couplings remain diagonal in the mass basis implies that there are no tree level FCNC transitions in the Standard Model.

11.2.2 Properties of the CKM matrix

In this section, we look in more detail at what are the key properties of the CKM matrix.

Degrees of freedom of the CKM matrix

The flavor eigenstates, (d', s', b'), are related to the mass eigenstates, (d, s, b), through:

$$\begin{pmatrix} d' \\ s' \\ b' \end{pmatrix} = V \begin{pmatrix} d \\ s \\ b \end{pmatrix}, \tag{11.34}$$

where the CKM matrix is unitary, i.e. $V^\dagger V = V V^\dagger = 1$. Since V is a 3×3 complex matrix, it has nine phases and nine moduli. Unitarity imposes three real and three complex constraints. Thus, we are left with six phases and three moduli which, because of the normalization of the rows and columns to one, can be written as angles.

Now, aside from the CC interactions, all other terms involving quarks are diagonal in flavor, and have LL, LR, RL and RR chiral structures. Thus, we can perform vector (not axial) phase rotations on each flavor and leave all of these other terms invariant. However, under the phase rotations

$$U_{L,R} \to e^{i\theta_U} U_{L,R} \qquad \text{and} \qquad D_{L,R} \to e^{i\theta_D} D_{L,R}, \qquad (11.35)$$

with $U = u$, c or t and $D = d$, s or b, the UD component of the CKM matrix transforms as

$$V_{UD} \to V_{UD} \, e^{i(\theta_D - \theta_U)}. \qquad (11.36)$$

These transformations can be used to eliminate phases in V. Although there are six phases θ_U and θ_D, only five phase differences $\theta_D - \theta_U$ are independent. Thus, only five of the six phases can be eliminated.

We have now exhausted the field transformations that can be used to reduce the CKM matrix's degrees of freedom. Thus, V has three angles and one phase.

Standard parametrization of the CKM matrix

The idea behind this parametrization is to write V as a product of three rotations between pairs of generations, throwing the phase into the $1 \to 3$ rotation, so that it multiplies the smallest mixing coefficients. Thus,

$$V = R_{32} \, \text{diag}\{1, 1, e^{i\delta}\} R_{31} \{1, 1, e^{-i\delta}\} R_{21}, \qquad (11.37)$$

with the rotations

$$R_{21} = \begin{pmatrix} c_{12} & s_{12} & 0 \\ -s_{12} & c_{12} & 0 \\ 0 & 0 & 1 \end{pmatrix}, \qquad (11.38)$$

$c_{12} = \cos\theta_{12}$ and $s_{12} = \sin\theta_{12}$, and similarly for the other rotations. This yields the following expression for the CKM matrix:

$$V = \begin{pmatrix} c_{12}c_{13} & c_{13}s_{12} & s_{13}e^{-i\delta} \\ -s_{12}c_{23} - c_{12}s_{23}s_{13}e^{i\delta} & c_{13}c_{23} - s_{12}s_{23}s_{13}e^{i\delta} & c_{13}s_{23} \\ s_{12}s_{23} - c_{12}c_{23}c_{13}e^{i\delta} & -c_{12}s_{23} - c_{23}s_{13}s_{13}e^{i\delta} & c_{13}c_{23} \end{pmatrix}, \qquad (11.39)$$

and the angles are chosen to lie in the first quadrant. Note that this parametrization is not rephasing invariant.

Wolfenstein parametrization

Experimentally, it is found that $1 \gg s_{12} \gg s_{23} \gg s_{13}$, i.e. mixing gets smaller as one moves off the diagonal. It is convenient to exhibit this hierarchy by expanding V in powers of s_{12}, i.e. in the sine of the Cabibbo angle θ_{12} (Cabibbo, 1963; Gell-Mann and Levy, 1960). This yields the Wolfenstein parametrization (Wolfenstein, 1983). To implement this expansion, we define (Buras *et al.*, 1994)

$$\lambda \equiv s_{12} = \frac{|V_{us}|}{\sqrt{|V_{ud}|^2 + |V_{us}|^2}}$$

$$A\lambda^2 \equiv s_{23} = \frac{|V_{cb}|}{\sqrt{|V_{ud}|^2 + |V_{us}|^2}} \tag{11.40}$$

$$A\lambda^3(\rho + i\eta) \equiv s_{13}e^{i\delta} = V_{ub}^*.$$

and make the appropriate replacement in the standard parametrization of Eq. (11.39). Then,

$$V = \begin{pmatrix} 1 - \lambda^2/2 & \lambda & A\lambda^3(\rho - i\eta) \\ -\lambda & 1 - \lambda^2/2 & A\lambda^2 \\ A\lambda^3(1 - \rho - i\eta) & -A\lambda^2 & 1 \end{pmatrix} + O(\lambda^4), \tag{11.41}$$

which clearly exhibits the hierarchy of mixing.

CP violation

Let us now see how CP violation arises in the Standard Model. Under parity, the charged W-boson fields transform as

$$W_\mu^{(\pm)}(x) \xrightarrow{P} W^{(\pm)\mu}(x_P), \tag{11.42}$$

with $x_P = (x^0, -\vec{x})$. Similarly, under charge conjugation,

$$W_\mu^{(\pm)}(x) \xrightarrow{C} -W_\mu^{(\mp)}(x). \tag{11.43}$$

Thus, under CP, these bosons transform as

$$W_\mu^{(\pm)}(x) \xrightarrow{CP} -W^{(\mp)\mu}(x_P). \tag{11.44}$$

In my favorite Dirac spinor basis, the parity transform of a fermion field is given by:

$$\begin{pmatrix} \psi_L \\ \psi_R \end{pmatrix}(x) \xrightarrow{P} \begin{pmatrix} \psi_R \\ \psi_L \end{pmatrix}(x_P). \tag{11.45}$$

In this basis,

$$\gamma^\mu = \begin{pmatrix} 0 & \sigma^\mu \\ \bar{\sigma}^\mu & 0 \end{pmatrix} \quad \text{and} \quad \gamma^5 = \begin{pmatrix} -I & 0 \\ 0 & I \end{pmatrix}, \tag{11.46}$$

where I is the two-by-two unit matrix, $\sigma^\mu = (I, \vec{\sigma})$, $\bar{\sigma}^\mu = (I, -\vec{\sigma})$ and $\vec{\sigma}$ are the Pauli matrices. Clearly, parity is not a symmetry of the Standard Model since left and right fermions belong to different representations of the Standard Model group. Under charge conjugation, we have

$$\psi(x) = \begin{pmatrix} \psi_L \\ \psi_R \end{pmatrix}(x) \xrightarrow{C} \psi^c(x) = \begin{pmatrix} i\sigma^2\psi_R^* \\ -i\sigma^2\psi_L^* \end{pmatrix}(x) = i\gamma^2\gamma^0\bar{\psi}^T(x) = C\bar{\psi}^T(x). \tag{11.47}$$

Again, charge conjugation is clearly not a Standard Model symmetry. However, the CP operation,

$$\begin{pmatrix} \psi_L \\ \psi_R \end{pmatrix}(x) \xrightarrow{CP} \begin{pmatrix} -i\sigma^2\psi_L^* \\ i\sigma^2\psi_R^* \end{pmatrix}(x_P) = \gamma^0 C\bar{\psi}^T(x_P), \tag{11.48}$$

has a chance of being a symmetry transformation as it does not mix left and right-handed fields. Using the well-known Pauli matrix identity, $\sigma^2\sigma^i\sigma^2 = -\sigma^{i*}$ and the anticommutation of fermion fields, the CC quark term of the Standard Model Lagrangian transforms, under CP, as:

$$\frac{g_2}{\sqrt{2}}\left\{\bar{U}_L W^{(+)} V D_L + \bar{D}_L W^{(-)} V^\dagger U_L\right\} \xrightarrow{CP} \frac{g_2}{\sqrt{2}}\left\{\bar{U}_L W^{(+)} V^* D_L + \bar{D}_L W^{(-)} V^T U_L\right\}. \tag{11.49}$$

Since $V^* \neq V$ in the presence of a non-vanishing phase, δ, CP is potentially violated in the Standard Model. We will see below what the necessary conditions for CP to be violated are.

11.2.3 CP violation and rephasing invariants

The standard parametrization of the CKM matrix V, given in Eq. (11.39), corresponds to a particular choice of quark field phases. Observables cannot depend on such choices. Therefore, it is important to find rephasing invariant combinations of CKM matrix elements.

Quadratic invariants

The moduli

$$I_{UD}^{(2)} \equiv |V_{UD}|^2, \tag{11.50}$$

with $U = u, c, t$ and $D = d, s, b$, are clearly rephasing invariant. There are nine of these.

Now, unitarity requires that:

$$\begin{cases} \sigma_U = \sum_{D=d,s,b} I_{UD}^{(2)} = 1 \\ \sigma_D = \sum_{U=u,c,t} I_{UD}^{(2)} = 1, \end{cases} \tag{11.51}$$

which yields six constraints on the $I_{UD}^{(2)}$. However, we clearly have $\sum_U \sigma_U = \sum_D \sigma_D$, which means that there are only five independent constraints. In turn, this means that there are four independent quadratic invariants $I_{UD}^{(2)}$, which are obviously real.

Quartic invariants

We now define

$$I_{U_1 D_1 U_2 D_2}^{(4)} \equiv V_{U_1 D_1} V_{U_2 D_2} V_{U_1 D_2}^* V_{U_2 D_1}^*. \tag{11.52}$$

These products of CKM matrix elements are also clearly rephasing invariant, since for every field which one of the CKM factors in (11.52) multiplies, another factor multiplies its Dirac conjugate. In Eq. (11.52), U_1, U_2 (D_1, D_2) are chosen cyclically amongst u, c, t (d, s, b) so as to avoid $I^{(4)} = (I^{(2)})^2$ as well as to avoid obtaining complex conjugate invariants, e.g. $I^{(4)}_{U_1 D_2 U_2 D_1} = I^{(4)*}_{U_1 D_1 U_2 D_2}$. With these constraints, there are nine invariants.

However, not all of these invariants are independent. Indeed, unitarity yields

$$\begin{cases} \sum_{D=d,s,b} V_{U_1 D} V^*_{U_2 D} = 0 & U_1 \neq U_2 \\ \sum_{U=u,c,t} V_{U D_1} V^*_{U D_2} = 0 & D_1 \neq D_2. \end{cases} \tag{11.53}$$

This implies, in turn:

$$\begin{cases} V_{U_1 D_1} V^*_{U_2 D_1} = \sum_{D \neq D_1} V_{U_1 D} V^*_{U_2 D} & \text{(I)} \\ V_{U_1 D_1} V^*_{U_1 D_2} = \sum_{U \neq U_1} V_{U D_1} V^*_{U D_2} & \text{(II)}. \end{cases}$$

Multiplying both sides of (I) by $V_{U_2 D_2} V^*_{U_1 D_2}$ while maintaining the cyclicity of indices, yields

$$I^{(4)}_{U_1 D_1 U_2 D_2} = -|V_{U_1 D_1} V_{U_2 D_2}|^2 - I^{(4)}_{U_1 D_2 U_2 D_3} \qquad \text{(III)}.$$

Similarly, multiplying both sides of (II) by $V_{U_2 D_2} V^*_{U_2 D_1}$ gives

$$I^{(4)}_{U_1 D_1 U_2 D_2} = -|V_{U_2 D_1} V_{U_2 D_2}|^2 - I^{(4)}_{U_2 D_1 U_3 D_2} \qquad \text{(IV)}.$$

Thus, the nine $I^{(4)}$'s can all be written in terms of $I^{(4)}_{udcs}$, for instance, and the four independent $I^{(2)}_{UD}$. Moreover, (III) and (IV) imply that all nine $I^{(4)}$ have the same imaginary part.

Higher-order invariants

Higher-order invariants can, in fact, be written in terms of $I^{(2)}$s and $I^{(4)}$s. For instance, the sextic rephasing invariant $V_{U_1 D_1} V_{U_2 D_2} V_{U_3 D_3} V^*_{U_1 D_2} V^*_{U_2 D_3} V^*_{U_3 D_1}$ is equal to $I^{(4)}_{U_1 D_1 U_2 D_2} I^{(4)}_{U_2 D_1 U_3 D_3} / I^{(2)}_{U_2 D_1}$. This obviously fails in singular cases, but these will not be considered here because they are irrelevant in practice.

Jarlskog's invariant

This whole discussion of invariants implies that there is a unique, imaginary rephasing invariant combination of CKM matrix elements. This invariant must appear in all CP violating observables, because it is the imaginary component of the CKM matrix which is responsible for this violation. This invariant is known as the Jarlskog invariant (Jarlskog, 1985):

$$J \equiv \text{Im}\, I^{(4)} = c^2_{13} c_{12} c_{23} s_{12} s_{23} s_{13} s_\delta$$
$$= \lambda^6 A^2 \bar{\eta} + O(\lambda^{10}), \tag{11.54}$$

where $\bar{\eta} = \eta(1 - \lambda^2/2)$. This, in turn, means that to have CP violation in the Standard Model, θ_{12}, θ_{23}, θ_{13} must not be 0 or $\pi/2$ and $\delta \neq 0, \pi$. Moreover, CP violation is maximal for $s_\delta = 1$, $\theta_{12} = \theta_{23} = \pi/4$ and $s_{13} = 1/\sqrt{3}$. At that point, the Jarlskog invariant takes the value

$$J_{\text{max}} = \frac{1}{6\sqrt{3}} \simeq 0.1. \tag{11.55}$$

However, global CKM fits yield (Charles *et al.*, 2005)

$$J = 2.96^{+18}_{-17} \times 10^{-5} \ll J_{\text{max}}, \tag{11.56}$$

i.e. CP violation in Nature is very far from being as large as it could be.

Instead of looking at the CKM matrix for a rephasing invariant measure of CP violation, we can go back to the mass matrices M_U and M_D. In Section 11.2.1, after performing only flavor transformations which are symmetries of $\mathcal{L}_{g+q}$, we reached the point where the quark mass term was given by Eq. (11.31)

$$\mathcal{L}^q_m \to -\bar{U}_R M^d_U U_L - \bar{D}_R M^d_D V^\dagger D_L + \text{h.c.}, \tag{11.57}$$

where V is the CKM matrix. Now, by performing the flavor symmetry transformation

$$D_R \longrightarrow V^\dagger D_R, \tag{11.58}$$

we obtain

$$\mathcal{L}^q_m \to -\bar{U}_R M^d_U U_L - \bar{D}_R V M^d_D V^\dagger D_L + \text{h.c.} \tag{11.59}$$

In this equation, both M^d_U and $M^h_D \equiv V M^d_D V^\dagger$ are hermitian matrices. This means that the commutator $C_J \equiv [M^d_U, M^h_D]$ is pure imaginary. Since the only source of imaginary numbers in Eq. (11.59) is the phase of the CKM matrix, C_J carries information about CP violation in the Standard Model. However, C_J is not a rephasing invariant. Defining the matrices $P_U \equiv \text{diag}\{e^{-i\theta_u}, e^{-i\theta_c}, e^{-i\theta_t}\}$ and $P_D \equiv \text{diag}\{e^{-i\theta_d}, e^{-i\theta_s}, e^{-i\theta_b}\}$, under the rephasing operations of Eq. (11.35), C_J transforms as

$$C_J \to [M^d_U, P_D M^h_D P^\dagger_D], \tag{11.60}$$

where P_U cancels against $P^\dagger_U$ in the first term, because M^d_U and P_U are diagonal. Nonetheless, $\det C_J$ is rephasing invariant, because M^d_U and P_D are diagonal. Thus, following Jarlskog (1985), we consider

$$\det C_J = 2iJ \times \prod_{U_1 > U_2} (m_{U_1} - m_{U_2}) \prod_{D_1 > D_2} (m_{D_1} - m_{D_2}). \tag{11.61}$$

In light of what was discussed in the section on rephasing invariants, $\det C_J$ must be the only imaginary, rephasing-invariant quantity that can be obtained from the mass term of Eq. (11.59), which is the most general mass term that can be written for quarks in the Standard Model. This means that the presence of CP violation in the Standard Model is equivalent to $\det C_J \neq 0$. In turn this implies that there will be CP violation if and only if the conditions on the mixing angles and the phase of the CKM matrix given after Eq. (11.54) are obeyed, but also if and only if there are no mass degeneracies in the up and down quark sectors.

Unitarity triangle areas

As we saw earlier, the unitarity of the CKM matrix gives rise to six unitarity triangles, defined by Eqs. (11.23)–(11.24). The areas of the (D_1, D_2) triangles are given by

$$
\begin{aligned}
A_{D_1 D_2} &= \frac{1}{2} |V_{uD_1} V_{uD_2}^* \wedge V_{cD_1} V_{cD_2}^*| \\
&= \frac{1}{2} |-i\mathrm{Im}\left(V_{uD_1} V_{uD_2}^* V_{cD_1}^* V_{cD_2}\right)| \qquad\qquad (11.62) \\
&= \frac{1}{2} \mathrm{Im}\, I_{uD_1 cD_2}^{(4)} = \frac{1}{2} J,
\end{aligned}
$$

where the last line follows from Eq. (11.54). Similarly the areas of the (U_1, U_2) triangles are

$$
A_{U_1 U_2} = \frac{1}{2} \mathrm{Im}\, I_{U_1 dU_2 s}^{(4)} = \frac{1}{2} J.
$$

Thus, all six triangles have the same area, which is given by the Jarlskog invariant. Since CP violation can only arise if $J \neq 0$, none of the unitarity triangles can be degenerate if Standard Model CP violation is measured in Nature.

11.3   A lattice case study: $K \to \pi\pi$, CP violation and $\Delta I = 1/2$ rule

Having introduced the Standard Model and flavor physics, we now turn to an important set of processes that have been nagging theorists for over four decades: $K \to \pi\pi$ decays. These decays have been a rich source of information and of constraints on the weak interaction. In 1964 they provided the first evidence in Nature for indirect CP violation, which arises in the mixing of the neutral kaon with its antiparticle before the decay into two pions (Christenson *et al.*, 1964). Then, in 1999, CP violation which arises directly in the flavor-changing decay vertex was discovered in these same decays, after more than 20 years of experimental effort (Fanti *et al.*, 1999; Alavi-Harati *et al.*, 1999).

Currently these decays still give very important constraints on the CKM paradigm, through the measurement of indirect CP violation, parametrized by ϵ. And as far as we presently know, direct CP violation in these decays, parametrized by ϵ', may be harboring New Physics. Moreover, $K \to \pi\pi$ decays display what we believe are unusually large, and certainly poorly understood, non-perturbative QCD corrections, which go under the name of the $\Delta I = 1/2$ rule.

At first sight the study of these decays is a perfect problem for the lattice. Only u, d and s (valence) quarks are involved, so that one expects controllable discretization errors. Of course, pions are light, which makes them difficult to simulate, and there are also two hadrons in the final state. But $SU(3)$ chiral perturbation theory (χPT) at LO relates $K \to \pi\pi$ to $K \to \pi$ and $K \to 0$ amplitudes which are simpler to compute (Bernard *et al.*, 1985). Moreover, χPT at NLO relates $K \to \pi\pi$

amplitudes obtained with heavier pions to the same amplitudes with physically light pions (Kambor *et al.*, 1990, 1991). Given the typical size of chiral corrections, one would expect that we could at least be able to get an $O(20\text{–}30\%)$ estimate of the relevant amplitudes. In addition to which we might expect good signals since only pseudoscalar mesons are involved.

Despite all of the positive indications that the lattice should be able to provide valuable information about these decays, all attempts to account for non-perturbative strong interaction effects have failed, except in the study of indirect CP violation.[6] And though much progress has been made on many aspects of these decays over the years, providing a fully quantitative description still remains an open problem. Thus, I have chosen to focus on this particular topic in my discussion of the application of lattice methods to flavor phenomenology.

11.3.1 $K \to \pi\pi$ phenomenology[7]

Kaons have strong isospin 1/2, while pions have isospin 1. The weak decays of a kaon into two pions can thus occur through two channels in the isospin limit:

- the $\Delta I = 3/2$ channel, where the final two pions are in a state of isospin $I = 2$, a state which we label $(\pi\pi)_2$;
- the $\Delta I = 1/2$ channel, where the final two pions are in a state of isospin $I = 0$, a state which we label $(\pi\pi)_0$.

Decay into a two-pion state with isospin $I = 1$ is forbidden by Bose symmetry.

We denote the amplitudes for $K \to \pi\pi$ decays by:

$$T[K^0 \to (\pi\pi)_I] = iA_I e^{i\delta_I}, \tag{11.63}$$

where δ_I is the strong scattering phase of two pions in the isospin I, angular momentum $J = 0$ channel, defined through

$$T[(\pi\pi)_I \to (\pi\pi)_I] = 2e^{i\delta_I} \sin \delta_I. \tag{11.64}$$

In Eq. (11.63), K^0 is the flavor eigenstate with $I_3 = -1/2$ and strangeness $S = 1$: it is composed of a d and an $\bar{s}$ quark.

Using this notation, we have the following isospin decompositions for the $K \to \pi\pi$ amplitudes:

$$-iT[K^0 \to \pi^+\pi^-] = \frac{1}{\sqrt{6}}A_2 e^{i\delta_2} + \frac{1}{\sqrt{3}}A_0 e^{i\delta_0}$$

$$-iT[K^0 \to \pi^0\pi^0] = \sqrt{\frac{2}{3}}A_2 e^{i\delta_2} - \frac{1}{\sqrt{3}}A_0 e^{i\delta_0} \tag{11.65}$$

$$-iT[K^+ \to \pi^+\pi^0] = \frac{\sqrt{3}}{2}A_2 e^{i\delta_2},$$

where the coefficients of the various amplitudes are simply $SU(2)$ Clebsch-Gordan coefficients. If CP violation is present, then $A_I^* \neq A_I$.

Now, in the absence of CP violation, the two physical neutral kaon states K_S and K_L are also CP eigenstates:[8]

$$|K_{S/L}\rangle \simeq |K_\pm\rangle \equiv \frac{1}{\sqrt{2}} \left(|K^0\rangle \mp |\bar{K}^0\rangle\right), \tag{11.66}$$

with $CP|K_\pm\rangle = \pm|K_\pm\rangle$. The CP even K_S decays only into two pions, while the CP odd K_L decays into three. Because of the phase space available to the decay products, the former is much shorter lived than the latter, with a lifetime $\tau_S \sim 10^{-10}$ s versus $\tau_L \sim 5 \cdot 10^{-8}$ s. This explains the subscript S for short and L for long.

In Nature, the weak interaction breaks CP, and K_S and K_L are not pure CP eigenstates. As a result of K^0-$\bar{K}^0$ mixing through the weak interaction, we have

$$|K_{L/S}\rangle = \frac{1}{\sqrt{1 + |\tilde{\epsilon}|^2}} \left(|K_\mp\rangle + \tilde{\epsilon}|K_\pm\rangle\right) \tag{11.67}$$

$$= \frac{1}{\sqrt{2}\sqrt{1 + |\tilde{\epsilon}|^2}} \left((1 + \tilde{\epsilon})|K^0\rangle \pm (1 - \tilde{\epsilon})|\bar{K}^0\rangle\right), \tag{11.68}$$

where $\tilde{\epsilon}$ is a small complex parameter.

The neutral kaons form a two-state quantum mechanical system which can be described by a non-Hermitian, 2-by-2 Hamiltonian:

$$H_{ij} = M_{ij} - \frac{i}{2}\Gamma_{ij}, \tag{11.69}$$

where $i, j = 1$ corresponds to K^0 and $i, j = 2$ to $\bar{K}^0$. CPT implies that $H_{11} = H_{22}$ and $H_{21} = H_{12}^*$. To determine the elements of this matrix, we first decompose the effective Hamiltonian for the Standard Model into a QCD+QED part, $H_{\text{QCD+QED}}$, and a weak part, H_W, and work to second order in the weak interaction. Then,

$$H_{ij} = M_{K^0}\delta_{ij} + \frac{\langle i|H_W|j\rangle}{2M_{K^0}} + \frac{1}{2M_{K^0}} \sum_n \frac{\langle i|H_W|n\rangle\langle n|H_W|j\rangle}{M_{K^0} - E_n + i\epsilon} + \cdots, \tag{11.70}$$

where M_{K^0} is the mass common to K^0 and $\bar{K}^0$, as given in QCD and QED, and E_n is the energy of the intermediate state $|n\rangle$.

Now, Cauchy's theorem implies (with P the principal part)

$$\frac{1}{\omega - E + i\epsilon} = P\left(\frac{1}{\omega - E}\right) - i\pi\delta(E - \omega), \tag{11.71}$$

where the first term on the RHS of this equation will yield the dispersive contribution to the Hamiltonian of Eq. (11.69) (i.e. the mass term) and the second term, the absorptive part (i.e. the width term). Then, the off-diagonal element of the mass matrix is

$$M_{12} = \frac{\langle K^0|H_{\Delta S=2}|\bar{K}^0\rangle}{2M_{K^0}} + \frac{1}{2M_{K^0}}P\sum_n \frac{\langle K^0|H_{\Delta S=1}|n\rangle\langle n|H_{\Delta S=1}|\bar{K}^0\rangle}{M_{K^0} - E_n} + \cdots, \tag{11.72}$$

where the term with the double insertion of the $\Delta S = 1$ Hamiltonian gives rise to long-distance contributions, since the states $|n\rangle$ which can contribute are light. For instance, $|n\rangle$ can be a $\pi^+\pi^-$ state.

The off-diagonal element of the width matrix is given by the absorptive part of the integral:

$$\Gamma_{12} = \frac{1}{2M_{K^0}} \sum_n \langle K^0|H_{\Delta S=1}|n\rangle\langle n|H_{\Delta S=1}|\bar{K}^0\rangle(2\pi)\delta(E_n - M_{K^0}) \tag{11.73}$$

Now, the physical states K_L and K_S are the eigenstates of H_{ij} with eigenvalues $M_L - \frac{i}{2}\Gamma_L$ and $M_S - \frac{i}{2}\Gamma_S$, respectively. It is straightforward to express these quantities in terms of the M_{ij} and Γ_{ij}. Defining

$$\Delta M_K \equiv M_{K_L} - M_{K_S} \quad \text{and} \quad \Delta\Gamma_K \equiv \Gamma_{K_S} - \Gamma_{K_L}, \tag{11.74}$$

one obtains:

$$\frac{1+\tilde{\epsilon}}{1-\tilde{\epsilon}} = 2\frac{M_{12} - \frac{i}{2}\Gamma_{12}}{\Delta M_K + \frac{i}{2}\Delta\Gamma_K}. \tag{11.75}$$

Solving for $\tilde{\epsilon}$ gives us explicit expressions for computing the relationship of the physical eigenstates K_L and K_S to the flavor eigenstates K^0 and $\bar{K}^0$ through Eq. (11.68). It is worth noting that

$$\Delta M_K \simeq 2M_{12} \quad \text{and} \quad \Delta\Gamma_K \simeq -2\Gamma_{12}, \tag{11.76}$$

to first non-trivial order in the weak phases.

Having implicitly worked out the relation between physical and flavor eigenstates, we return to $K \to \pi\pi$ decays. What is actually measured are the amplitude ratios

$$\eta_{00} \equiv \frac{T[K_L \to \pi^0\pi^0]}{T[K_S \to \pi^0\pi^0]} \quad \text{and} \quad \eta_{+-} \equiv \frac{T[K_L \to \pi^+\pi^-]}{T[K_S \to \pi^+\pi^-]}. \tag{11.77}$$

Experimentally, $|\eta_{00}| \simeq 2 \times 10^{-3}$, and $|\eta_{00}/\eta_{+-}| \simeq 1$ (Nakamura *et al.*, 2010). These ratios are clearly CP violating since K_L does not decay into two pions if CP is conserved. The CP violating decays $K_L \to \pi\pi$ can occur in two ways. As seen in Eq. (11.67), $|K_L\rangle$ can acquire a small CP even component proportional to $\tilde{\epsilon}|K_+\rangle$ through K^0-$\bar{K}^0$ mixing. This component can then decay into two pions without violating CP. However, the CP odd component of $|K_L\rangle$, proportional to $|K_-\rangle$, can directly decay into two pions if CP is violated in the decay. These two decay modes are illustrated in

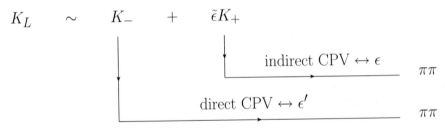

The first mode of decay is called indirect CP violation and is parametrized by a small number ϵ whose relation to $\tilde{\epsilon}$ will be given shortly. The second mode is called direct CP violation and is parametrized by an even smaller number ϵ'. These parameters are defined through

$$\epsilon = \frac{T[K_L \to (\pi\pi)_0]}{T[K_S \to (\pi\pi)_0]} \tag{11.78}$$

and

$$\epsilon' = \frac{1}{\sqrt{2}} \left(\frac{T[K_L \to (\pi\pi)_2]}{T[K_S \to (\pi\pi)_2]} - \epsilon \underbrace{\frac{T[K_S \to (\pi\pi)_2]}{T[K_S \to (\pi\pi)_0]}}_{\dot{=}\,\omega} \right). \tag{11.79}$$

A little algebra allows us to relate these CP violating parameters to the measured quantities η_{00} and η_{+-}:

$$\eta_{00} = \epsilon - 2\frac{\epsilon'}{1 - \sqrt{2}\omega} \tag{11.80}$$

$$\eta_{+-} = \epsilon + \frac{\epsilon'}{1 + \omega/\sqrt{2}}. \tag{11.81}$$

Moreover, under the assumption that $(\pi\pi)_0$ dominates the sum for Γ_{12} in Eq. (11.73), the relationship of ϵ to $\tilde{\epsilon}$ can be calculated to be

$$\epsilon \simeq \frac{\tilde{\epsilon} + i\frac{\mathrm{Im}A_0}{\mathrm{Re}A_0}}{1 + i\tilde{\epsilon}^*\frac{\mathrm{Im}A_0}{\mathrm{Re}A_0}} \simeq \tilde{\epsilon} + i\frac{\mathrm{Im}A_0}{\mathrm{Re}A_0} \tag{11.82}$$

to first non-trivial order in the small CP violating quantities $\mathrm{Im}A_0/\mathrm{Re}A_0$ and $\tilde{\epsilon}$. A bit more algebra and some simplifications allow us to express ϵ, ϵ' and ω in terms of A_0, A_2 and $\mathrm{Im}M_{12}$ (de Rafael, 1995) which, in principle, can be calculated using lattice QCD:

$$\omega \simeq \frac{\mathrm{Re}A_2}{\mathrm{Re}A_0} e^{i(\delta_2 - \delta_0)} \tag{11.83}$$

$$\epsilon \simeq e^{i\phi_\epsilon} \sin\phi_\epsilon \left\{ \frac{\mathrm{Im}M_{12}}{\Delta M_K} + \frac{\mathrm{Im}A_0}{\mathrm{Re}A_0} \right\} \tag{11.84}$$

$$\epsilon' \simeq \frac{e^{i(\delta_2 - \delta_0)}}{\sqrt{2}} \mathrm{Im}\frac{A_2}{A_0}, \tag{11.85}$$

where the phase of ϵ is approximately given by $\phi_\epsilon \simeq (2\phi_{+-} + \phi_{00})/3 \sim \pi/4$, with ϕ_{+-} and ϕ_{00} the phases of η_{+-} and η_{00}, respectively.[9] In the expression for ϵ', the imaginary part of A_2/A_0 measures the relative reality of A_2 and A_0, which is what we need for direct CP violation since the latter must arise from the interference between the two available decay channels. Note that with the approximations used here, Eq. (11.76) gave $\Delta M_K \simeq 2\mathrm{Re}M_{12}$, where $\mathrm{Re}M_{12}$ also could, in principle, be computed on the lattice.

Experimentally, the various quantities which describe the K^0-$\bar{K}^0$ system and $K \to \pi\pi$ decays are well measured (Nakamura *et al.*, 2010):

$$\Delta M_K = (3.483 \pm 0.006) \times 10^{-12}\,\text{MeV} \qquad [0.2\%] \tag{11.86}$$

$$\frac{1}{|\omega|} \simeq \left|\frac{A_0}{A_2}\right| \simeq 22.4 \qquad (\Delta I = 1/2 \text{ rule}) \tag{11.87}$$

$$|\epsilon| \simeq (2|\eta_{+-}| + |\eta_{00}|)/3 = (2.228 \pm 0.011) \times 10^{-3} \qquad [0.5\%] \tag{11.88}$$

$$\phi_\epsilon \simeq (2\phi_{+-} + \phi_{00})/3 = 43.51 \pm 0.05 \qquad [0.1\%] \tag{11.89}$$

$$\text{Re}\frac{\epsilon'}{\epsilon} \simeq \left(1 - \left|\frac{\eta_{00}}{\eta_{+-}}\right|\right) = (1.65 \pm 0.26) \times 10^{-3} \qquad [16\%]. \tag{11.90}$$

Using lattice QCD, the weak matrix element relevant for the short-distance, Standard Model contribution of $|\epsilon|$ has been calculated with a precision of less than 3% – see Lellouch (2009); Lubicz (2009); Sachrajda (2010a); Colangelo *et al.* (2010) for recent reviews. We are just beginning to provide phenomenologically relevant information for the $\Delta I = 1/2$ rule $|A_0/A_2|$ (Sachrajda, 2010a; Liu, 2010; Christ, 2010a), but $\text{Re}(\epsilon'/\epsilon)$ is still out of reach for the moment (Sachrajda, 2010a). Moreover, as already mentioned and discussed further below, ΔM_K has long-distance contributions which make its determination on the lattice difficult. However, for that also, progress has been made (Christ, 2010b).

11.3.2 K^0-$\bar{K}^0$ mixing in the Standard Model

As we have just seen, $\bar{K}^0$-K^0 mixing arises from the $\Delta S = 2$, $s\bar{d} \to \bar{s}d$ FCNC. This mixing is responsible for the K_L-K_S mass difference, ΔM_K, and indirect CP violation in $K \to \pi\pi$ decays.

In the Standard Model, it occurs at one loop through diagrams such as (Glashow *et al.*, 1970; Gaillard and Lee, 1974b)

Setting the external four-momenta to zero, the amplitude associated with this diagram is

$$-i\mathcal{M} = \left(\frac{-ig_2}{\sqrt{2}}\right)^4 \int \frac{d^4k}{(2\pi)^4} iD^W_{\mu\nu}(k) iD^W_{\rho\sigma}(k) \left(\bar{v}_{dL}\gamma_\mu iS(k)\gamma_\sigma v_{sL}\right)\left(\bar{u}_{dL}\gamma_\nu iS(k)\gamma_\rho u_{sL}\right), \tag{11.92}$$

where the u's and v's are the usual particle and antiparticle spinor wavefunctions, and we have defined the amplitude with a minus sign for convenience. The W boson propagator in the Feynman gauge is

$$D_{\mu\nu}^W(k) = \frac{-g_{\mu\nu}}{k^2 - M_W^2 + i\epsilon} \tag{11.93}$$

and the sum of the up quark propagators is

$$S(k) = \sum_{U=u,c,t} \frac{\lambda_U}{k - m_U + i\epsilon}, \tag{11.94}$$

with $\lambda_U \equiv V_{Us}V_{Ud}^*$.

The unitarity of the CKM matrix implies that $\sum_{U=u,c,t} \lambda_U = 0$. Thus, we have (dropping $i\epsilon$ for the moment)

$$S(k) = \sum_{U=c,t} \lambda_U \left(\frac{1}{k - m_U} - \frac{1}{k - m_u} \right). \tag{11.95}$$

From this we see the GIM mechanism (Glashow *et al.*, 1970) in action: if $m_u = m_c = m_t$ there would be no K^0-$\bar{K}^0$ mixing. In fact, this process was used to estimate the charm quark mass before it was actually discovered (Gaillard and Lee, 1974*b*).

After performing some Dirac algebra, we find:

$$\mathcal{M} = \frac{G_F^2 M_W^2}{2\pi^2} (\lambda_t^2 T_{tt} + \lambda_c^2 T_{cc} + 2\lambda_c \lambda_t T_{ct}) \times (\bar{v}_{dL}\gamma_\mu v_{sL})(\bar{u}_{dL}\gamma^\mu u_{sL}), \tag{11.96}$$

where we have used $G_F = g_2^2/(4\sqrt{2}M_W^2)$. Setting $m_u = 0$,

$$T_{U_1 U_2} = \frac{4i}{\pi^2 M_W^2} \int d^4k \frac{1}{k^2(1 - k^2/M_W^2)^2} \frac{m_{U_1}^2}{k^2 - m_{U_1}^2} \frac{m_{U_2}^2}{k^2 - m_{U_2}^2}. \tag{11.97}$$

With the spinor wavefunction factors appropriately replaced by quark field operators, $\mathcal{M}$ can be interpreted as an effective Hamiltonian whose matrix element between a K^0 and a $\bar{K}^0$ state yields the off-diagonal matrix element (11.72) of the mass matrix of Eq. (11.69). Of course, to obtain the full effective $\Delta S = 2$ Hamiltonian, one must include the contributions of all of the diagrams which contribute to the process (Inami and Lim, 1981).[10]

Now, the values of the CKM matrix elements as well as of the masses m_u, m_c and m_t imply that $\mathrm{Re}M_{12} \gg \mathrm{Im}M_{12}$ and that $\mathrm{Re}M_{12}$ is dominated by the cc term. Naively,

$$\frac{G_F^2 M_W^2}{4\pi^2} T_{cc} = \frac{iG_F^2}{\pi^4} \int d^4k \frac{m_c^4}{(k^2 + i\epsilon)[k^2 - m_c^2 + i\epsilon]^2} + O\left(\frac{1}{M_W^6}\right)$$

$$= \frac{G_F^2 m_c^2}{\pi^2} + O\left(\frac{1}{M_W^6}\right). \tag{11.98}$$

However, a closer look at this loop integral indicates that it is dominated by momenta in the range between 0 and m_c and we should not forget that all sorts of gluons with or without quark loops can be exchanged between the quarks in the diagram. Since these momenta include scales of $O(\Lambda_{\text{QCD}})$ or below, it is clear that α_s corrections are out of control and $\text{Re} M_{12}$ cannot be calculated in this way.

Said differently, box diagrams, such as the one of Eq. (11.91), cannot be viewed like a point interaction, but rather receive long-distance contributions from intermediate $c\bar{c}$ states. In the language of Eq. (11.72), these diagrams contribute to the second term through $(\bar{d}_L \gamma_\mu s_L)(\bar{c}_L \gamma_\mu c_L)$ effective operators in $\mathcal{H}_{\Delta S=1}$.

The situation is very different for the calculation of the CP violating parameter $\epsilon \sim \text{Im} M_{12}/\text{Re} M_{12}$.[11] Indeed, indirect CP violation in $K \to \pi\pi$ decays comes from the interference between the following types of contributions

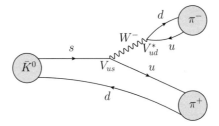

and

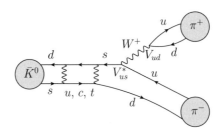

So, the relative weak phase between the two types of contributions is $\arg\{(\lambda_u^*)^2 M_{12}\}$ and thus, what we really have is

$$\epsilon \sim \frac{\text{Im}\left[(\lambda_u^*)^2 M_{12}\right]}{\text{Re}\left[(\lambda_u^*)^2 M_{12}\right]}, \tag{11.99}$$

where the $(\lambda_u^*)^2$ had been canceled in earlier expressions, using the fact that λ_u is real in our conventions.

Now, using Eq. (11.96) with the replacement of wavefunctions by operators discussed after Eq. (11.97), we have:

$$(\lambda_u^*)^2 M_{12} \sim \frac{G_F^2 M_W^2}{16\pi^2} \Big(\underbrace{(\lambda_u^* \lambda_t)^2 T_{tt}}_{I_{udts}^{(4)}} + \underbrace{(\lambda_u^* \lambda_c)^2 T_{cc}}_{I_{udcs}^{(4)}} + 2(\lambda_u^* \lambda_c)(\lambda_u^* \lambda_t) T_{ct} \Big)$$

$$\times \langle K^0 | (\bar{d}s)_{V-A}(\bar{d}s)_{V-A} | \bar{K}^0 \rangle, \tag{11.100}$$

where

$$(\bar{d}s)_{V-A} = \bar{d}\gamma_\mu(1 - \gamma_5)s \tag{11.101}$$

and where the $I^{(4)}$ are the quartic rephasing invariants defined in Eq. (11.52). As shown in Section 11.2.3, $\mathrm{Im}I^{(4)}_{udcs} = J = -\mathrm{Im}I^{(4)}_{udts}$, where J is the Jarlskog invariant. Therefore,

$$\mathrm{Im}[(\lambda_u^*)^2 M_{12}] \sim \frac{g_2^4}{8M_W^4} J \left\{ \mathrm{Re}(\lambda_u^*\lambda_t)(T_{tt} - T_{ct}) - \mathrm{Re}(\lambda_u^*\lambda_c)(T_{cc} - T_{ct}) \right\}$$

$$\times \langle K^0 | (\bar{d}s)_{V-A}(\bar{d}s)_{V-A} | \bar{K}^0 \rangle, \tag{11.102}$$

and J appears as it should in a CP violating quantity.

Moreover, the integrals in $(T_{tt} - T_{ct})$ and $(T_{cc} - T_{ct})$ are dominated by momenta between m_c and m_t. The same is true of the integrals which appear in the other diagrams (Inami and Lim, 1981) that contribute to this $\Delta S = 2$ process. That means that the QCD corrections are calculable to the extent that m_c can be considered a perturbative scale. Thus, under this assumption, and using the experimental value of ΔM_K in lieu of $\mathrm{Re}\left[(\lambda_u^*)^2 M_{12}\right]$, we can reliably calculate ϵ with the replacement

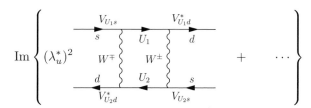

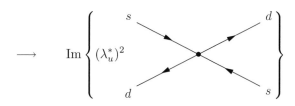

i.e. by replacing the box diagram with the local, four-quark operator of Eq. (11.102), and the appropriate short-distance QCD corrections, omitted here. As mentioned in note 9, corrections to this approximation have been examined by Buras *et al.* (2010).

11.3.3   The theory of $K^0 - \bar{K}^0$ mixing

The calculation of M_{12} of the previous section has actually been performed to leading-log (LL) (Vainshtein *et al.*, 1977*b*; Witten, 1977; Vainshtein *et al.*, 1977*a*; Gilman and Wise, 1983) and next-to-leading-log order (NLL) (Buras *et al.*, 1990) in QCD; for a review see Buchalla *et al.* (1996). The resulting NLL, $\Delta S = 2$ effective Hamiltonian is given by

$$\mathcal{H}_{\Delta S=2} = \frac{G_F^2}{16\pi^2} M_W^2 \left[\lambda_t^2 \eta_{tt} S_{tt} + \lambda_c^2 \eta_{cc} S_{cc} + 2\lambda_c \lambda_t \eta_{ct} S_{ct} \right]$$
$$\times C(\mu) \times (\bar{d}s)_{V-A} (\bar{d}s)_{V-A}, \tag{11.103}$$

where the running of matrix elements of the four-quark operator is canceled by the coefficient $C(\mu)$, the $S_{qq'}$ are the Inami–Lim functions (Inami and Lim, 1981) which correspond to the electroweak box contributions in the absence of the strong interaction, and the $\eta_{qq'}$ are short-distance QCD corrections. In Eq. (11.103) all Standard Model degrees of freedom with masses down to, and including, that of the charm quark are integrated out. Thus, in Eq. (11.103), all QCD quantities must be evaluated with *three* active quark flavors.

It is useful to consider the case of a general number of active quark flavors, N_f, and of colors, N_c, when discussing the running of this $\Delta S = 2$ operator. Indeed, the running is the same for other Standard Model, $\Delta F = 2$ operators, such as $\Delta B = 2$ and $\Delta C = 2$, except that N_f must be chosen appropriately. Moreover, working with general N_c allows one to consider the large-N_c limit, to be discussed shortly. With $a_s \equiv \alpha_s/4\pi$, the running of α_s and of $C(\mu)$ is given by:

$$\frac{d \ln a_s}{d \ln \mu^2} = -\beta(a_s) = -\beta_0 a_s - \beta_1 a_s^2 + O(a_s^3), \tag{11.104}$$

$$\frac{d \ln C}{d \ln \mu^2} = \gamma(a_s) = \gamma_0 a_s + \gamma_1 a_s^2 + O(a_s^3), \tag{11.105}$$

where $\beta(a_s)$ is the QCD β function and $\gamma(a_s)$ is the anomalous dimension of the $\Delta S = 2$ operator. It is well known that β_0, β_1, and LO anomalous dimensions are renormalization scheme independent. At two loops, we have (Nakamura *et al.*, 2010)

$$\beta_0 = \frac{11 N_c - 2 N_f}{3}, \qquad \beta_1 = \frac{34}{3} N_c^2 - \frac{10}{3} N_c N_f - 2 C_F N_f, \qquad C_F = \frac{N_c^2 - 1}{2 N_c}, \tag{11.106}$$

and in the $\overline{\text{MS}}$-NDR scheme for γ_1,

$$\gamma_0 = 3 \frac{N_c - 1}{N_c}, \qquad \gamma_1 = \frac{N_c - 1}{4 N_c} \left[-21 + \frac{57}{N_c} - \frac{19}{3} N_c + \frac{4}{3} N_f \right]. \tag{11.107}$$

It is straightforward to integrate Eq. (11.104):

$$\frac{1}{a_s(\mu)} - \beta_0 \ln \left(\frac{\mu}{\mu_0} \right)^2 + \frac{\beta_1}{\beta_0} \ln \left[\frac{a_s(\mu)}{1 + (\beta_1/\beta_0) a_s(\mu)} \right] \tag{11.108}$$

$$= \frac{1}{a_s(\mu_0)} + \frac{\beta_1}{\beta_0} \ln \left[\frac{a_s(\mu_0)}{1 + (\beta_1/\beta_0) a_s(\mu_0)} \right]$$

For reasonably small $a_s(\mu_0) \ln(\mu/\mu_0)^2$, the coupling at μ can be related to the one at μ_0 through:

$$a_s(\mu) = a_s(\mu_0) \left\{ 1 - \beta_0 a_s(\mu_0) \ln\left(\frac{\mu}{\mu_0}\right)^2 - a_s(\mu_0)^2 \ln\left(\frac{\mu}{\mu_0}\right)^2 \right. \tag{11.109}$$

$$\left. \times \left[\beta_1 - \beta_0^2 \ln\left(\frac{\mu}{\mu_0}\right)^2 \right] \right\} + O(a_s^3). \tag{11.110}$$

Alternatively, we can define Λ_{QCD} as the value of μ_0 at which $a_s(\mu_0)$ is infinite, yielding:

$$a_s(\mu) = \frac{1}{\beta_0 \ln\frac{\mu^2}{\Lambda_{\text{QCD}}^2}} \left[1 + \frac{\beta_1 \ln \ln \frac{\mu^2}{\Lambda_{\text{QCD}}^2}}{\beta_0^2 \ln \frac{\mu^2}{\Lambda_{\text{QCD}}^2}} + \cdots \right]. \tag{11.111}$$

It is also straightforward to integrate Eq. (11.105),

$$\frac{C(\mu)}{C(\mu_0)} = \exp\left\{ - \int_{a_s(\mu_0)}^{a_s(\mu)} \frac{da_s}{a_s} \frac{\gamma(a_s)}{\beta(a_s)} \right\}. \tag{11.112}$$

The coefficient $C(\mu)$ is only defined up to an integration constant. For consistency with Eq. (11.103), I consider

$$C(\mu) = [4\pi a_s(\mu)]^{-\gamma_0/\beta_0} \exp\left\{ - \int_0^{a_s(\mu)} \frac{da_s}{a_s} \left[\frac{\gamma(a_s)}{\beta(a_s)} - \frac{\gamma_0}{\beta_0} \right] \right\}. \tag{11.113}$$

Thus, at NLO

$$C(\mu) = [4\pi a_s(\mu)]^{-\gamma_0/\beta_0} \left[1 + \frac{\beta_1}{\beta_0} a_s + O(a_s^2) \right]^{\left(\frac{\gamma_0}{\beta_0} - \frac{\gamma_1}{\beta_1}\right)}. \tag{11.114}$$

For the Standard Model $\Delta F = 2$ operator relevant for K^0-$\bar{K}^0$ mixing, we use the anomalous dimension coefficients of Eq. (11.107), with $N_f = 3$ and, of course, $N_c = 3$. Thus, at NLO,

$$C(\mu) = \alpha_s(\mu)^{-2/9} \left[1 + \frac{307}{162} \frac{\alpha_s(\mu)}{4\pi} \right]. \tag{11.115}$$

Then, in Eq. (11.103) the QCD corrections $\eta_{qq'}$ are of the form

$$\eta_{qq'} \propto \frac{[1 + O(\alpha_s)]}{C(m_c)}, \tag{11.116}$$

and where $x_q \equiv (m_q/M_W)^2$. For details, please see Buchalla *et al.* (1996).

To calculate ϵ, we must compute the matrix element $\langle K^0 | (\bar{d}s)_{V-A} (\bar{d}s)_{V-A}(\mu) | \bar{K}^0 \rangle$, where the kaons are at rest. This is clearly a non-perturbative QCD quantity because the typical energies of the quarks within the kaons are on the order of $100\,\mathrm{MeV}$, a regime where perturbation theory fails and confinement effects must be taken into account. This is where lattice QCD enters.[12]

For historical reasons, and because it is very convenient in lattice computations, we define a normalized matrix element

$$B_K(\mu) \equiv \frac{\langle \bar{K}^0|(\bar{s}d)_{V-A}(\bar{s}d)_{V-A}(\mu)|K^0\rangle}{\frac{8}{3}\langle \bar{K}^0|\bar{s}\gamma_\mu\gamma_5 d|0\rangle\langle 0|\bar{s}\gamma_\mu\gamma_5 d|K^0\rangle}, \tag{11.117}$$

where we have considered the $\Delta S = -2$ matrix element to conform with convention. The benefit of this normalization on the lattice, is that the resulting B_K parameter is dimensionless. Therefore it does not suffer from ambiguities due to scale setting, something which was particularly bad in old quenched-calculation days. Moreover, the numerator and denominator are very similar, and both statistical and systematic uncertainties cancel in the ratio. Of course, the convenience of this normalization would be limited if the denominator were an unknown, non-perturbative quantity. However, the matrix elements in the denominator define the leptonic decay constant of the kaon, f_K,[13]

$$\langle 0|\bar{s}\gamma_\mu\gamma_5 d(x)|K^0(p)\rangle = if_K p_\mu e^{-ip\cdot x}, \tag{11.118}$$

which is well measured experimentally or straightforward to compute on the lattice. (In the convention used here, $f_K \simeq 156\,\mathrm{MeV}$.) Thus, once the B-parameter has been computed, the normalizing factor is a known quantity and the desired matrix element is easily obtained, from

$$\langle \bar{K}^0|\underbrace{(\bar{s}d)_{V-A}(\bar{s}d)_{V-A}(\mu)}_{O^{\mathrm{SM}}_{\Delta S=-2}}|K^0\rangle = \frac{8}{3}f_K^2 M_K^2 B_K(\mu), \tag{11.119}$$

where M_K^2 is best taken from experiment.

Historically, the denominator in Eq. (11.117) was an approximation used to estimate the matrix element of $O^{\mathrm{SM}}_{\Delta S=-2}$. It is called the vacuum saturation approximation, or VSA for short. It is obtained by inserting the vacuum in all possible ways between all possible quark-antiquark field pairs formed from the fields of the four-quark operators, using Fierz transformations if necessary to bring the fields together. For the case of interest here,

$$\langle 0|\bar{s}_b\gamma_\mu(1-\gamma_5)d^a(x)|K^0(p)\rangle = -i\frac{\delta^a_b}{N_c}f_K p_\mu e^{-ip\cdot x}, \tag{11.120}$$

where a and b are color indices and the dependence on the number of colors $N_c = 3$ is made explicit. Therefore

$$\langle \bar{K}^0|O^{\mathrm{SM}}_{\Delta S=-2}(\mu)|K^0\rangle_{VSA} = 2\Big(\langle \bar{K}^0|\bar{s}_a\gamma^\mu(1-\gamma_5)d^a|0\rangle\langle 0|\bar{s}_b\gamma_\mu(1-\gamma_5)d^b|K^0\rangle$$

$$+\langle \bar{K}^0|\bar{s}_a\gamma^\mu(1-\gamma_5)d^b|0\rangle\langle 0|\bar{s}_b\gamma_\mu(1-\gamma_5)d^a(x)|K^0\rangle\Big), \tag{11.121}$$

where repeated color indices are summed over and where the factor of 2 comes from the fact that the two factors of $(\bar{s}d)_{V-A}$ in $O^{\mathrm{SM}}_{\Delta S=-2}$ are interchangeable. Plugging (11.120) in (11.121) yields

$$\langle \bar{K}^0 | O^{\mathrm{SM}}_{\Delta S=-2}(\mu) | K^0 \rangle_{VSA} = \frac{2}{N_c^2} f_K^2 p^2 \left[\delta_a^a \delta_b^b + \delta_a^b \delta_b^a \right]$$

$$= 2 \frac{N_c + 1}{N_c} M_K^2 f_K^2. \tag{11.122}$$

This is clearly a rather crude approximation, as the LHS of Eq. (11.121) is μ dependent while the RHS is not. This approximation introduces a renormalization scale dependence which is unphysical.

A more modern approximation to the matrix element is obtained by keeping the leading term in a large-N_c expansion. The large-N_c, or 't Hooft limit ('t Hooft, 1974), is defined by taking $N_c \to \infty$ while holding $\alpha_s N_c$ fixed. By counting the number of α_s and loop factors of N_c in the various contributions to the relevant correlation functions (see below), it is straightforward to convince oneself that in the large-N_c limit, $B_K(\mu) = 3/4$. This corresponds to dropping the second term in Eq. (11.121), which is clearly suppressed by a factor of $1/N_c$ compared to the first. As in the VSA approximation $B_K(\mu)$ is μ independent, but here the μ dependence is also absent in the short-distance, Wilson coefficient, as can be seen by taking the large-N_c limit in Eqs. (11.106)–(11.113): the large-N_c approximation is a well-defined and self-consistent approximation scheme.

Before closing this section, it is worth mentioning that one can define a renormalization scheme and scale independent B-parameter, B_K^{RGI}, by multiplying $B_K(\mu)$ by $C(\mu)$ of Eq. (11.113) (with $N_c = 3$ and $N_f = 3$):

$$B_K^{\mathrm{RGI}} = C(\mu) \times B_K(\mu). \tag{11.123}$$

11.3.4 Computation of bare B_K

On the lattice, the numerator of B_K is obtained from three-point functions. Quark propagators are given by:

$$S_q[\vec{x}, t; \eta, t_s; U] = \sum_{\vec{x}_s} D^{-1}[\vec{x}, t; \vec{x}_s, t_s; m_q; U] \eta(\vec{x}_s), \tag{11.124}$$

where D is the lattice Dirac operator associated with the chosen fermion action, m_q the quark q's mass, U the gauge field configuration on which the propagator is computed, $\eta(\vec{x}_s)$ is a three dimensional source which may be a delta function or may have some spatial extent and t_s is the timeslice at which the source is placed. If only propagators from a point source at, for instance, $t_s = 0$ and $\vec{x}_s = \vec{0}$ (i.e. $\eta(\vec{x}_s) = \delta_{\vec{x}_s, \vec{0}}$), are available, then we can consider the following three-point function, in Euclidean space-time of course:

$$C_3(t_i, t_f) = \sum_{\vec{x}_i, \vec{x}_f} \langle \bar{d}\gamma_5 s(\vec{x}_f, t_f) O^{SM}_{\Delta S=-2}(0) \bar{d}\gamma_5 s(\vec{x}_i, t_i) \rangle, \tag{11.125}$$

where the argument of $O^{SM}_{\Delta S=-2}$ is its space-time position, not the renormalization scale. For $T/2 \gg -t_i \gg 1/\Delta E_K$ and $T/2 \gg t_f \gg 1/\Delta E_K$, $\bar{d}\gamma_5 s(\vec{x}_i, t_i)$ creates a K^0 at $t = t_i$, this kaon then propagates to $t = 0$ where $O^{SM}_{\Delta S=-2}(0)$ transforms it into a $\bar{K}^0$ and $\bar{d}\gamma_5 s(\vec{x}_f, t_f)$ destroys the resulting $\bar{K}^0$ at $t = t_f$. ΔE_K is the energy of the first excited state in the neutral kaon channel minus M_K. The sums over $\vec{x}_i$ and $\vec{x}_f$ put the initial and final kaons at rest. Thus, in this limit

$$C_3(t_i, t_f) \xrightarrow{T/2 \gg -t_i \gg 1/\Delta E_K, T/2 \gg t_f \gg 1/\Delta E_K} \frac{e^{-M_K(t_f - t_i)}}{4M_K^2} \langle 0|\bar{d}\gamma_5 s(0)|\bar{K}^0(\vec{0})\rangle \tag{11.126}$$

$$\times \underbrace{\langle \bar{K}^0(\vec{0})|O_{\Delta S=-2}(0)|K^0(\vec{0})\rangle}_{\text{desired mat. elt.}} \langle K^0(\vec{0})|\bar{d}\gamma_5 s(0)|0\rangle.$$

This result is obtained by inserting a complete set of hadron states between $O^{SM}_{\Delta S=-2}(0)$ and $\bar{d}\gamma_5 s(\vec{x}_i, t_i)$, between $\bar{d}\gamma_5 s(\vec{x}_f, t_f)$ and $O^{SM}_{\Delta S=-2}(0)$ and between $\bar{d}\gamma_5 s(\vec{x}_i, t_i)$ and $\bar{d}\gamma_5 s(\vec{x}_f, t_f)$, keeping the sequence of states which gives rises to the smallest exponential suppression and which has the appropriate quantum numbers to give non-vanishing matrix elements. If $T/2 \gg t_f, -t_i$ is not realized, then contributions which are not significantly exponentially suppressed compared to the one in Eq. (11.126) will have to be added.

Similarly, to obtain the matrix elements required to construct the denominator of Eq. (11.117), we can consider the following two-point function

$$C_{2,\mu}(t) = \sum_{\vec{x}} \langle \bar{d}\gamma_5 s(\vec{x}, t) \bar{s}\gamma_\mu \gamma_5 d(0) \rangle \xrightarrow{t \gg \Delta E_K} \tag{11.127}$$

$$\frac{(e^{-M_K t} - e^{-M_K(T-t)})}{2M_K} \langle 0|\bar{d}\gamma_5 s(0)|\bar{K}^0(\vec{0})\rangle \underbrace{\langle \bar{K}^0(\vec{0})|\bar{s}\gamma_5\gamma_\mu d(0)|0\rangle}_{\text{denom. mat. elt.}}, \tag{11.128}$$

where I have not assumed here that $T/2 \gg t$ to illustrate the additional contributions which arise in that case, and where I have used the properties of the correlation function under time reversal.

Then, B_K is obtained from the ratio

$$\frac{C_3(t_i, t_f)}{\sum_\mu C_{2,\mu}(t_f) C_{2,\mu}(t_i)} \xrightarrow{T/2 \gg -t_i \gg 1/\Delta E_K, T/2 \gg t_f \gg 1/\Delta E_K} B_K(a), \tag{11.129}$$

where I have reinstated $T/2 \gg t_f - t_i$ to get rid of "backward" contributions. In Eq. (11.129), the argument a of B_K is there to indicate that this is the value of B_K in the lattice regularized scheme and that it still requires renormalization.

To actually compute $C_3(t_i, t_f)$, we have to take the propagators of Eq. (11.124), contract them in the following way and average these contractions over the gauge ensemble, i.e.

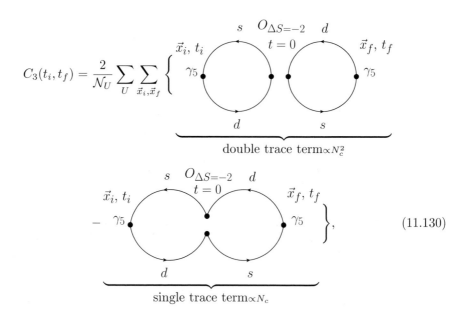

$$C_3(t_i, t_f) = \frac{2}{\mathcal{N}_U} \sum_U \sum_{\vec{x}_i, \vec{x}_f} \left\{ \right.$$

$$\left. \right\}, \qquad (11.130)$$

where $\mathcal{N}_U$ is the number of independent gauge configurations. To get the time-reversed propagator required to construct the correlation function, one usually uses the γ_5 hermiticity which most lattice Dirac operators have,

$$S_d[\vec{0}, 0; \vec{x}, t; U] = \gamma_5 S_d[\vec{x}, t; \vec{0}, 0; U]^\dagger \gamma_5. \qquad (11.131)$$

In Eq. (11.130), one clearly sees how, in the large-N_c limit, only the first contraction survives. The second contraction provides a $1/N_c$ suppressed correction. In practice, the two contractions have the same sign, and a cancellation operates in the calculation of B_K.

The method described above is actually a poor way to obtain B_K because the matrix elements of interest, $\langle \bar{K}^0(\vec{0})|O_{\Delta S=-2}(0)|K^0(\vec{0})\rangle$ and $\langle 0|\bar{\gamma}_5\gamma_\mu d(0)|K^0(\vec{0})\rangle$, are sampled at only one point on the lattice: at the origin. We would gain a factor of roughly $(LM_\pi)^3$ in statistics if we could sample them over the whole three-dimensional volume of the lattice.[14] Thus, a better way to calculate B_K is to have two zero momentum sources for the quark propagators, one at t_i and the other at t_f. One can consider, for instance, wall sources:

$$S_q[\vec{x}, t; W, t_W; U] = \sum_{\vec{x}_s} D^{-1}[\vec{x}, t; \vec{x}_s, t_W; m_q; U] \sum_{\vec{y}} \delta_{\vec{x}_s, \vec{y}}. \qquad (11.132)$$

Then, one constructs the following three-point function

$$C_3(t) = \frac{1}{N_U} \sum_U \sum_{\vec{x}} \gamma_5 \quad O_{\Delta S=-2}(\vec{x}, t) \quad \gamma_5, \qquad (11.133)$$

where $t_i < 0$ and $t_f > 0$ are chosen so that there is a range of t around $t = 0$ such that the correlation function is dominated by the propagation of a zero momentum K^0 state between t_i and t and the propagation of a zero momentum $\bar{K}^0$ state between t and t_f. The gauge field configurations are usually gauge-fixed on the walls because wall sources correspond to meson sources and sinks of the form $\sum_{\vec{x}, \vec{y}} \bar{d}_a(\vec{x}, t_W) \gamma_5 s^a(\vec{y}, t_W)$ which are clearly not gauge invariant, except for terms along the "diagonal" $\vec{x} = \vec{y}$. The gauge is usually fixed to the Coulomb gauge. However, one can also not fix the gauge, the result being that the sums over the two quark positions in the sources and sinks reduce to a single diagonal sum over the positions after the average over gauge configurations is taken.

Using the wall sources, we also construct the two two-point functions ($t_W = t_i, t_f$),

$$C_{2,\mu}(t, t_W) = \frac{1}{N_U} \sum_U \sum_{\vec{x}} \gamma_5 \quad \vec{x}, t \quad \gamma_\mu \gamma_5. \qquad (11.134)$$

Then we study the ratio of correlators

$$R(t) \equiv \frac{C_3(t)}{\sum_\mu C_{2,\mu}(t, t_f) C_{2,\mu}(t, t_i)} \qquad (11.135)$$

as a function of t. For $t_i \ll t \ll t_f$, $R(t)$ develops a plateau (see Fig. 11.5) such that

$$R(t) \xrightarrow{t_i \ll t \ll t_f} B_K(a), \qquad (11.136)$$

so that $B_K(a)$ is obtained by either averaging $R(t)$ or fitting it to a constant over the plateau region.

11.3.5   Renormalization of the Standard Model $|\Delta S| = 2$ operator

We are not done, however. As we already stated, simply inserting the lattice operator $O_{\Delta S=-2}^{\mathrm{SM}}$ in $C_3(t)$ and computing $R(t)$ yields the bare $B_K(a)$. This quantity is divergent in the continuum limit and must be renormalized. And it must be done so in a renormalization scheme which matches the one used in the perturbative calculation of the short-distance Wilson coefficients.

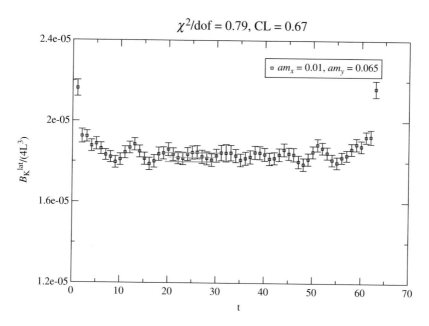

Fig. 11.5 Plateau fit to $R(t)/(4L^3)$ on the coarse $N_f = 2 + 1$, staggered, MILC, $am_l/am_h = 0.007/0.05$ ensemble. The legend shows the non-degenerate pair of domain-wall, valence quark masses making up the kaon in the three-point correlation function. The correlated χ^2/dof and confidence level of the fit to a constant (horizontal lines) are given in the title. Taken from Aubin *et al.* (2010).

It is straightforward to show that the full set of $\Delta S = -2$, $\Delta D = 2$ operators of dimension $d \leq 6$ can be written as:

$$O_1 = O_{\Delta S=-2}^{\text{SM}} = (\bar{s}d)_{V-A}(\bar{s}d)_{V-A} \quad \text{(unmix)}, \tag{11.137}$$

$$O_{2,3} = (\bar{s}d)_{S-P}(\bar{s}d)_{S-P} \quad \text{(unmix, mix)}, \tag{11.138}$$

$$O_{4,5} = (\bar{s}d)_{S-P}(\bar{s}d)_{S+P} \quad \text{(unmix, mix)}, \tag{11.139}$$

where the subscripts $(\bar{s}d)_{S-P}$ and $(\bar{s}d)_{S+P}$ are defined in analogy with $(\bar{s}d)_{V-A}$ in Eq. (11.101). In Eqs. (11.137)–(11.139) "unmix" and "mix" refer to the color indices. In the "unmix" case, the color indices of the quark-antiquark pairs within parentheses are contracted; in the "mix" case, the color index of the quark of one pair is contracted with the color index of the antiquark of the other pair, and vice versa. Because O_1 Fierz transforms into itself, the "mix" and "unmix" O_1 are the same operator.

To understand the renormalization patterns of $O_{\Delta S=-2}^{\text{SM}}$ and the other $\Delta S = -2$ operators, it is useful to consider their transformation properties under various symmetry groups. Because we will only work in massless renormalization schemes, the $SU(3)_L \times SU(3)_R$ chiral group is a symmetry which is relevant here. Under the action of this group, $(\bar{s}d)_{V-A}(\bar{s}d)_{V-A}$ transforms in the $(27, 1)$ representation, i.e. it is a 27 under $SU(3)_L$ and clearly a singlet under $SU(3)_R$, since it is composed only of left-handed fields. It is straightforward to derive the $SU(3)_L \times SU(3)_R$ representations to which the five $\Delta S = -2$ operators belong:

$$O_1 \sim (27, 1) \xrightarrow{SU(3)_V} 27 \otimes 1 = 27,$$

$$O_{2,3} \sim (6, \bar{6}) \xrightarrow{SU(3)_V} 6 \otimes \bar{6} = 27 \oplus 8 \oplus 1, \tag{11.140}$$

$$O_{4,5} \sim (8, 8) \xrightarrow{SU(3)_V} 8 \otimes 8 = 27 \oplus 10 \oplus \overline{10} \oplus 8 \oplus 8 \oplus 1.$$

In Eq. (11.140) I have also worked out the reduction of these representations to the diagonal, $V = L + R$, Eightfold Way, $SU(3)_V$ group, for reasons which will be clear shortly. Note that the "mix" versus "unmix" feature of these operators has no bearing on their flavor transformation properties as these features pertain solely to color.

From this, we see that $O_{\Delta S=-2}^{\rm SM}$ is the only $\Delta S = -2 = -\Delta D$ operator of dimension 6 or less, which transforms as $(27, 1)$. Thus, in any regularization which preserves $SU(3)_L \times SU(3)_R$ symmetry (or at least in the valence sector), $O_{\Delta S=-2}^{\rm SM}$ renormalizes multiplicatively. This includes overlap and domain wall fermions, for sufficiently large fifth dimension. Similarly, Eq. (11.140) indicates that the operator pairs (O_2, O_3) and (O_4, O_5) may mix within each pair under renormalization, but not with any of the other $\Delta S = -2 = -\Delta D$ operators.

The situation is very different for Wilson fermions. Indeed, the Wilson-Dirac operator breaks the chiral symmetry of continuum QCD explicitly, down to the vector flavor symmetry $SU(3)_V$. As Eq. (11.140) shows, $SU(3)_V$ is not sufficient to forbid $O_{\Delta S=-2}^{\rm SM}$ from mixing with the four other $\Delta S = -2 = -\Delta D$ operators, $O_{2,\cdots,3}$, under renormalization.

To push things further, we turn to parity. Since parity is preserved by Wilson fermions, we can consider separately the renormalization of the parity even and parity odd components of the operators. For B_K we are clearly interested in the parity even part:

$$\langle \bar{K}^0 | (\bar{s}d)_{V-A} (\bar{s}d)_{V-A} | K^0 \rangle = \langle \bar{K}^0 | (\bar{s}d)_V (\bar{s}d)_V + (\bar{s}d)_A (\bar{s}d)_A | K^0 \rangle. \tag{11.141}$$

So let us begin with the renormalization of the parity even part.

At this point, it is useful to invoke a discrete symmetry transformation known as CPS (Bernard *et al.*, 1985). It consists in performing a CP transformation, followed by a switching $s \leftrightarrow d$. Note that this vector flavor symmetry is only softly broken by the mass terms in the action. Therefore, violations must appear multiplied by factors of $(m_s - m_d)$. Under CPS, we have:

$$\bar{s}\gamma^\mu d \xrightarrow{CPS} -\bar{s}\gamma_\mu d \tag{11.142}$$

$$\bar{s}\gamma^\mu \gamma^5 d \xrightarrow{CPS} -\bar{s}\gamma_\mu \gamma^5 d \tag{11.143}$$

$$\bar{s}d \xrightarrow{CPS} \bar{s}d \tag{11.144}$$

$$\bar{s}\gamma^5 d \xrightarrow{CPS} -\bar{s}\gamma^5 d. \tag{11.145}$$

Thus, the parity even components of the $\Delta S = -2$ operators transform under CPS as

$$O_1^+ = (\bar{s}d)_V(\bar{s}d)_V + (\bar{s}d)_A(\bar{s}d)_A \overset{CPS}{\longrightarrow} O_1^+, \tag{11.146}$$

$$O_{2,3}^+ = (\bar{s}d)_S(\bar{s}d)_S + (\bar{s}d)_P(\bar{s}d)_P \overset{CPS}{\longrightarrow} O_{2,3}^+, \tag{11.147}$$

$$O_{4,5}^+ = (\bar{s}d)_S(\bar{s}d)_S + (\bar{s}d)_P(\bar{s}d)_P \overset{CPS}{\longrightarrow} O_{4,5}^+. \tag{11.148}$$

All of these operators are CPS eigenstates, with eigenvalue $+1$. Like $SU(3)_V$ symmetry, CPS does not forbid O_1^+ to mix with $O_{2,\dots,3}^+$, under renormalization. In fact, there is no symmetry which forbids O_1^+ to mix with the other operators and one finds, in practice, that they do mix. We have,

$$O_1^+(\mu) = Z_1^+(a,\mu)\left[O_1^+(a) + \sum_{i=2}^{5} z_{1i}(a)O_i^+(a)\right], \tag{11.149}$$

where $Z_1^+(a,\mu)$ is logarithmically divergent in the continuum limit, while the mixing factors, $z_{1i}(a)$, are finite (Testa, 1998). Using the fact that the values of B_K^{RGI} of Eq. (11.123), obtained in the continuum renormalization scheme and in the lattice regularization scheme, must be identical, one easily obtains a formal expression for the renormalization constant $Z_1^+(a,\mu)$, to all orders in perturbation theory:

$$Z_1^+(a,\mu) = \exp\left\{\int_0^{a_s(\mu)} \frac{da_s}{a_s}\frac{\gamma(a_s)}{\beta(a_s)} - \int_0^{a_s^{\mathrm{lat}}} \frac{da_s}{a_s}\frac{\gamma^{\mathrm{lat}}(a_s)}{\beta^{\mathrm{lat}}(a_s)}\right\}, \tag{11.150}$$

with $a_s^{\mathrm{lat}} = g_0^2/(4\pi)^2$, where g_0 is the bare lattice coupling. Expansions in the bare lattice coupling are notoriously poorly behaved, and one can usually improve their convergence by expressing them in terms of a renormalized continuum coupling, such as $a_s(\mu)$ at $\mu = 1/a$ for example. Applying this to Eq. (11.150), we find

$$Z_1^+(a,\mu) = \exp\left\{\int_0^{a_s(\mu)} \frac{da_s}{a_s\beta(a_s)}\gamma(a_s) - \int_0^{a_s(1/a)} \frac{da_s}{a_s\beta(a_s)}\bar{\gamma}^{\mathrm{lat}}(a_s)\right\}, \tag{11.151}$$

where $\bar{\gamma}^{\mathrm{lat}}$ is the anomalous dimension obtained by rewriting γ^{lat} in terms of the continuum coupling a_s. Obviously this change of variable would make no difference to $Z_1^+(a,\mu)$ were it to be computed to all orders in perturbation theory. However, at finite order this reordering of the expansion may improve the convergence of the series.

It is interesting to expand Eq. (11.151) to one loop – this could be done for Eq. (11.150) instead – to explicitly see the relationship between the coefficients of the expansion and the anomalous and beta function coefficients. We find:

$$Z_1^+(a,\mu) = 1 - \frac{\alpha_s}{4\pi}\left(\gamma_0 \ln(a\mu)^2 + \frac{\bar{\gamma}_1^{\mathrm{lat}} - \gamma_1}{\beta_0}\right) + O(\alpha_s^2). \tag{11.152}$$

One clearly sees that the constant $O(a_s)$ term knows about the two-loop anomalous dimension (it is a subleading log) whereas the leading log term is given by γ_0. At this same order, the mixing coefficients are finite and given by

$$z_{1i}(a) = z_{1i}^{(1)} \frac{\alpha_s}{4\pi} + O(\alpha_s^2) \qquad i = 2, \cdots, 5, \tag{11.153}$$

where the $z_{1i}^{(1)}$ are constants.

Now, the $SU(3)_L \times SU(3)_R$ properties of the operators indicate that the physical contributions in $\langle \bar{K}^0 | O_1^+(a) | K^0 \rangle$ are chirally suppressed compared to the non-physical ones: $O(p^2)$ vs. $O(1)$ in χPT counting. Thus, even though the mixing of O_1 with $O_2, \ldots, O_5$ is α_s suppressed, this suppression can easily be compensated by an $O(10$–$20)$ enhancement from the matrix element (Babich *et al.*, 2006). This means that the necessary subtractions are delicate and it is preferable to avoid calculating B_K with Wilson fermions.

It is also interesting to study the transformation properties of the parity odd components of $O_1, \cdots, O_5$ under CPS. We find

$$O_1^- = -2(\bar{s}d)_V(\bar{s}d)_A \xrightarrow{CPS} O_1^- \tag{11.154}$$

$$O_{2,3}^- = -2(\bar{s}d)_S(\bar{s}d)_P \xrightarrow{CPS} -O_{2,3}^- \tag{11.155}$$

$$O_{4,5}^- = 0. \tag{11.156}$$

Thus, CPS forbids O_1^- to mix with $O_{2,3}^-$, and $O_{4,5}^-$ vanishes anyway. We conclude that O_1^-, unlike O_1^+, renormalizes multiplicatively:

$$O_1^-(\mu) = Z_1^-(a, \mu) O_1^-(a). \tag{11.157}$$

This turns out to be useful for twisted-mass QCD (tmQCD), which is described in detail by Vladikas (2010). For instance, twisting (u, d) by an angle α and leaving the strange quark s untwisted, we have

$$[O_{VV+AA}]_{\text{QCD}} \equiv [(\bar{s}d)_V(\bar{s}d)_V + (\bar{s}d)_A(\bar{s}d)_A]_{\text{QCD}} \tag{11.158}$$

$$= \cos\alpha \, [O_{VV+AA}]_{\text{tmQCD}} - i \sin\alpha \, [O_{VA+AV}]_{\text{tmQCD}}.$$

So, picking $\alpha = \pi/2$ (i.e. maximal twist), the above equation implies the following relationship between the renormalized matrix elements:

$$\langle \bar{K}^0 | O_{VV+AA}(\mu) | K^0 \rangle_{\text{QCD}} = -i \langle \bar{K}^0 | O_{VA+AV}(\mu) | K^0 \rangle_{\text{tmQCD}}. \tag{11.159}$$

Then, since CPS symmetry is only softly broken in tmQCD, $\langle \bar{K}^0 | O_{VA+AV}(a) | K^0 \rangle_{\text{tmQCD}}$ renormalizes multiplicatively.[15] Thus, by working in tmQCD at maximal twist, one can compute $B_K(\mu)$ performing only multiplicative renormalization (please see Tassos' course notes (Vladikas, 2010) for details).

Though we discussed renormalization mostly in terms of perturbation theory, I greatly encourage you to perform this renormalization non-perturbatively, with one of the methods explained in Peter's (Weisz, 2010) or Tassos' (Vladikas, 2010) course notes. You may also want to look into the renormalization procedure put forward by Dürr *et al.* (2010a,b), which enhances the RI/MOM method of Martinelli *et al.* (1995) with non-perturbative, continuum running (see also Arthur and Boyle (2010)). Or if you choose to renormalize perturbatively, you should at least do so to two loops to ensure that you have some control over the perturbative series. Of course, one may argue that the short-distance coefficients in Eq. (11.103) are only known to NLO, and that there is no point in doing much better in the lattice to continuum matching. Moreover, there are other uncertainties in the relation (11.84) of ϵ to B_K, such as the one associated with the error on the determination of $|V_{cb}|$ or with the neglect of $1/m_c^2$ and of the $\mathrm{Im}A_0/\mathrm{Re}A_0$ corrections (see, e.g. Lellouch (2009)). However, it is admittedly a pity to perform a careful non-perturbative computation of the bare matrix elements only to introduce perturbative uncertainties through the matching procedure.

11.3.6 Final words on $K^0 - \bar{K}^0$ mixing

Given a preferably non-perturbatively renormalized B_K, it must be matched to the scheme used in the computation of the Wilson coefficients which appear in Eq. (11.103), and computed for a variety of lattice spacings and quark masses. Then you must use the methods described by, for instance, Dürr *et al.* (2008); Lellouch (2009); Dürr *et al.* (2010*a,b*) and/or in Pilar's (Hernández, 2010), Peter's (Weisz, 2010) and Maarten's (Golterman, 2010) course notes to extrapolate (preferably interpolate) to the physical values of m_{ud}, m_s and extrapolate to the continuum limit. Finally, you must perform a complete systematic error analysis, such as those performed by Dürr *et al.* (2008, 2010*c,a,b*). For a recent, comprehensive review of lattice calculations of B_K, see for instance Colangelo *et al.* (2010).

Before concluding this discussion of K^0-$\bar{K}^0$ mixing, it is worth mentioning that lattice QCD can also provide information that is relevant for this process beyond the Standard Model. Quite generically, when one adds heavy degrees of freedom to those of the Standard Model, such as in supersymmetric extensions, one finds that the full set of $\Delta S = -2$ operators of Eqs. (11.137)–(11.139) contribute to the low-energy effective Hamiltonian. Of course, the lattice can also be used to compute the matrix elements of the four additional operators, between K^0 and $\bar{K}^0$ states. This has been studied in the quenched approximation by Donini *et al.* (1999); Babich *et al.* (2006). Babich *et al.* (2006) finds ratios of non-Standard Model to Standard Model matrix elements which are roughly twice as large as those of Donini *et al.* (1999). As explained by Babich *et al.* (2006), this is most probably due to discretization errors present in Donini *et al.* (1999). This picture appears to be confirmed by the preliminary $N_f = 2$ results of Dimopoulos *et al.* (2010).

11.3.7 Phenomenology of the $\Delta I = 1/2$ rule

The goal here is to compute non-leptonic weak decay amplitudes, such as those for $K \to \pi\pi$, directly in the Standard Model, without any extraneous model assumptions, nor potentially uncontrolled approximations such as LO, $SU(3)$ χPT. This is critical for showing that QCD is indeed responsible for surprising phenomena such as the $\Delta I = 1/2$ rule, or deciding whether New Physics is hidden in the experimental measurement of direct CP violation in $K \to \pi\pi$ (i.e. of ϵ'). In the following I will concentrate on the $\Delta I = 1/2$ rule as the calculation of ϵ' is significantly more difficult. For the latter, the renormalization of the relevant matrix elements is more complicated (Dawson *et al.*, 1998). There are more matrix elements involved and there appear to be important cancellations between them (see e.g. Buchalla *et al.* (1996)). Moreover, final-state interactions, such as the ones discussed by Pallante and Pich (2001); Buras *et al.* (2000), seem to play an important role. Indeed, recent attempts to exhibit the $\Delta I = 1/2$ rule and to determine $\mathrm{Re}(\epsilon'/\epsilon)$ using soft pion theorems appear to fail due to the large chiral corrections required to translate the $K \to \pi$ and $K \to 0$ amplitudes computed on the lattice into physical $K \to \pi\pi$ amplitudes (Li and Christ, 2008; Sachrajda, 2010a; Christ, 2010a).

Experimentally, the partial widths measured in the different $K \to \pi\pi$ decay channels, together with the corresponding isospin changes between the final two-pion and initial kaon states, are Nakamura *et al.* (2010):

$$\Gamma_{+-} = \Gamma(K_S \to \pi^+\pi^-) = 5.08 \times 10^{-12}\mathrm{MeV} \qquad \Delta I = \frac{1}{2}, \frac{3}{2} \qquad (11.160)$$

$$\Gamma_{00} = \Gamma(K_S \to \pi^0\pi^0) = 2.26 \times 10^{-12}\mathrm{MeV} \qquad \Delta I = \frac{1}{2}, \frac{3}{2} \qquad (11.161)$$

$$\Gamma_{+0} = \Gamma(K^+ \to \pi^+\pi^0) = 1.10 \times 10^{-14}\mathrm{MeV} \qquad \Delta I = \frac{3}{2}. \qquad (11.162)$$

Using these results, we find

$$\frac{\Gamma_{+-}}{\Gamma_{+0}} = 463. \quad \text{and} \quad \frac{\Gamma_{00}}{\Gamma_{+0}} = 205., \qquad (11.163)$$

whereas Γ_{+-}/Γ_{+0} should be ~ 1 in the electroweak theory and in the absence of the strong interaction. Now, in terms of the amplitudes, the rates are

$$\Gamma_{+-} = \frac{\gamma}{3} \left[2|A_0|^2 + |A_2|^2 + 2\sqrt{2}\mathrm{Re}\left(A_0 A_2^* e^{i(\delta_0 - \delta_2)}\right) \right], \qquad (11.164)$$

$$\Gamma_{00} = \frac{\gamma}{3} \left[|A_0|^2 + 2|A_2|^2 - 2\sqrt{2}\mathrm{Re}\left(A_0 A_2^* e^{i(\delta_0 - \delta_2)}\right) \right], \qquad (11.165)$$

$$\Gamma_{+0} = \frac{3}{4}\gamma|A_2|^2, \qquad (11.166)$$

with $\gamma = \sqrt{M_K^2 - 4M_\pi^2}/(16\pi M_K^2)$. Considering $\Gamma_{+-} + \Gamma_{00}$ and Γ_{+0}, and taking $M_\pi = 134.8\,\mathrm{MeV}$ and $M_K = 494.2\,\mathrm{MeV}$ (i.e. isospin limit values), we find

$$|A_0| = 4.66 \times 10^{-4}\,\mathrm{MeV} \quad \text{and} \quad |A_2| = 2.08 \times 10^{-5}\,\mathrm{MeV} \qquad (11.167)$$

$$\to \frac{|A_0|}{|A_2|} = 22.4, \qquad (11.168)$$

whereas the combined chiral and large-N_c limit predicts $\sqrt{2}$ (Lellouch, 2001)! It is the huge enhancement of over 400 in the rate or 20 in the amplitude which is known as the $\Delta I = 1/2$ enhancement. It was termed a "rule" because this enhancement of

$\Delta I = 1/2$ over $\Delta I = 3/2$ transitions is also observed in other decays, such as $\Lambda \to N\pi$, $\Sigma \to N\pi$ and $\Xi \to \Lambda\pi$.

11.3.8 The $\Delta I = 1/2$ rule in the Standard Model

At leading order in the weak and strong interactions, $\Delta S = -1$, $\Delta D = 1$ transitions occur through the tree level diagram

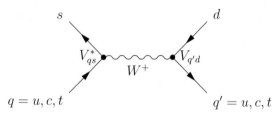

As usual, we integrate out the heavy W boson and t quark. At leading order in QCD, this yields the following four-quark, local vertex

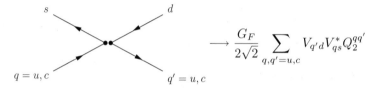

with

$$Q_2^{qq'} = (\bar{s}q)_{V-A}(\bar{q}'d)_{V-A}. \tag{11.169}$$

Now, if we include α_s corrections (Gilman and Wise, 1979), such as

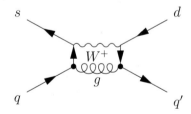

we generate a new operator:

$$Q_1^{qq'} = (\bar{s}d)_{V-A}(\bar{q}'q)_{V-A}. \tag{11.170}$$

These corrections also lead to the penguin diagram

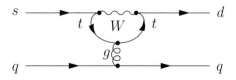

with $q = u, d, s, c, b$. However, one finds that they yield tiny contributions to the CP conserving parts of $K \to \pi\pi$ decays. Thus, we neglect these here. Moreover, we are only interested in external states with u, d and/or s quarks. So we do not need operators with a single charm leg. To simplify the operator structure, we can also use the unitarity of the CKM matrix:

$$V_{cd}V_{cs}^* = -V_{ud}V_{us}^* - V_{td}V_{ts}^*, \tag{11.171}$$

where the second term on the RHS of this equation can be neglected because it is multiply Cabibbo suppressed by a factor of $\lambda^4 \sim 0.003$ compared to the first term. Then, for CP conserving $\Delta S = -1$, $\Delta D = 1$ transitions, we have the effective Hamiltonian

$$\mathcal{H}_{\text{CPC}}^{\Delta S=-1} = \frac{G_F}{\sqrt{2}} V_{ud}V_{us}^* \sum_{i=\pm} C_i(\mu)O_i \tag{11.172}$$

with

$$O_\pm = [(\bar{s}u)_{V-A}(\bar{u}d)_{V-A} \pm (\bar{s}d)_{V-A}(\bar{u}u)_{V-A}] - [u \to c], \tag{11.173}$$

where the second term, in which u is replaced by c with an overall minus sign, implements the GIM suppression mechanism (Glashow *et al.*, 1970). It is straightforward to show that O_+ is in the $(84, 1)$ representation of the $SU(4)_L \times SU(4)_R$ chiral group, appropriate for classifying operators composed of u, d, s and c quarks for renormalization in massless schemes: it is a completely symmetric and traceless tensor in two fundamental and two conjugate $SU(4)_L$ indices. O_-, on the other hand, is in the $(20, 1)$ representation of this group (it is a completely antisymmetric and traceless tensor in two fundamental and two conjugate $SU(4)_L$ indices). Thus, O_+ and O_- do not mix under renormalization. In Eq. (11.172), the short-distance Wilson coefficients are given, at $O(\alpha_s)$, by (Gaillard and Lee, 1974a; Altarelli and Maiani, 1974; Altarelli *et al.*, 1981):

$$C_\pm(\mu) = 1 + \frac{\alpha_s}{4\pi}\left(\gamma_\pm^{(0)} \ln \frac{\mu^2}{M_W^2} + \delta_\pm\right), \tag{11.174}$$

with, in the $\overline{\text{MS}}$-NDR scheme:

$$\gamma_\pm^{(0)} = \pm 3\frac{N_c \mp 1}{N_c} \quad \text{and} \quad \delta_\pm = \pm\frac{11}{2}\frac{N_c \mp 1}{N_c}. \tag{11.175}$$

One often pushes the short-distance calculation further and integrates out the charm quark. But the GIM mechanism is very useful for reducing the divergences of relevant weak matrix element on the lattice. (Unfortunately, we will not have the time to cover this here, but I refer the interested reader to Dawson *et al.* (1998), for instance, for a discussion of this and related issues.)

A straightforward analysis of the isospin structure of the operators of Eq. (11.173) show that O_- is pure $I = \frac{1}{2}$, while O_+ has both $I = \frac{1}{2}$, $\frac{3}{2}$ components.[16] One may wonder then, whether the $\Delta I = 1/2$ enhancement comes from the running of the

Wilson coefficients $C_{\pm}(\mu)$ from the scale $\mu \sim M_W$, where the ratio $C_-(\mu)/C_+(\mu)$ is 1 plus small corrections of order $\alpha_s(M_W) \sim 0.1$, down to a scale $\mu \sim 2 \, \text{GeV}$. This would mean that the $\Delta I = 1/2$ rule could be understood as a short-distance enhancement. At leading-log order (LL) (Gaillard and Lee, 1974a; Altarelli and Maiani, 1974), using Eq. (11.112), we find

$$\frac{C_-(2 \, \text{GeV})}{C_+(2 \, \text{GeV})} = \left(\frac{\alpha_s(2 \, \text{GeV})}{\alpha_s(m_b)}\right)^{18/25} \left(\frac{\alpha_s(m_b)}{\alpha_s(M_W)}\right)^{18/23} \left(\frac{C_-(M_W)}{C_+(M_W)}\right) \sim 2.$$

So there is some short-distance enhancement, but not enough by a long shot to explain the $\Delta I = 1/2$ rule.[17] In turn, this means that most of the enhancement in

$$\left|\frac{A_0}{A_2}\right| = \frac{\langle(\pi\pi)_0|C_+O_+ + C_-O_-|K^0\rangle}{\langle(\pi\pi)_2|C_+O_+|K^0\rangle} \tag{11.176}$$

must come from long distances.

One may also wonder what role the charm quark plays in these decays. In particular, one might consider an imaginary world in which the GIM mechanism is exact, i.e. $m_c = m_u$. Does the $\Delta I = 1/2$ enhancement persist in that limit? One way to address this problem is to work in the $SU(4)$ chiral limit and determine the LECs corresponding to O_- and O_+, i.e. the $(20, 1)$ and $(84, 1)$ couplings (see comments after Eq. (11.173)) (Giusti *et al.*, 2007). Numerically, in the quenched approximation, it is found that there is an enhancement of $|A_0/A_2|$ over naive expectation (e.g. large-N_c), but that this enhancement is roughly a factor of 4 too small.

An interesting way to understand this $SU(4)$ chiral limit enhancement is to consider the three-point function contractions required to determine the matrix elements $\langle\pi^+|O_{\pm}|K^+\rangle$. It is straightforward to see that in the case of O_+, with the appropriate flavor replacements, the contractions are the same as those of Eq. (11.130), up to the overall factor of 2. On the other hand, for the pure $\Delta I = 1/2$ operator O_-, the contractions are those of Eq. (11.130), but with a plus sign between the double and single trace terms. Thus, in the large-N_c limit, the two matrix elements coincide and we find the value of $|A_0/A_2| = \sqrt{2}$ discussed after Eq. (11.168). However, for finite N_c, the single trace term contributes, and does so with opposite signs to $\langle\pi^+|O_+|K^+\rangle$ and to $\langle\pi^+|O_-|K^+\rangle$. Since both trace contractions are positive, the larger the single trace term, the larger is $\langle\pi^+|O_-|K^+\rangle$ and the smaller is $\langle\pi^+|O_+|K^+\rangle$. This creates a $\Delta I = 1/2$ enhancement in which both the numerator $|A_0|$ is enhanced and the denominator $|A_2|$ is depressed. The argument also implies an anticorrelation between B_K, the B-parameter of K^0-$\bar{K}^0$ mixing, and the $\Delta I = 1/2$ enhancement of $|A_0/A_2|$. Indeed, in the chiral limit, the smaller B_K is compared to its large-N_c value of $3/4$, the larger $|A_0/A_2|$ is compared to $\sqrt{2}$, as first noted by Pich and de Rafael (1996).

Given the argument which we made earlier, one would think that LQCD is well suited to study $K \to \pi\pi$ decays. However, there are many conceptual problems one encounters when trying to study these decays on the lattice. Some of these are:

- the renormalization of the $\Delta S = -1$ effective Hamiltonian is difficult on the lattice, even more so for the CP violating part (see e.g. Dawson *et al.* (1998));

- lattices are only a few fermi in size and the final-state hadrons cannot be separated into isolated, asymptotic states;
- only approximately evaluated Euclidean correlation functions are available.

There are also technical challenges. For instance, the study of these decays requires the calculation of four-point functions. Moreover, power divergences must be subtracted, if the charm quark is integrated out in the case of CP conserving decays, and once the W and t are integrated out in the CP violating case. Both these points make the study of $K \to \pi\pi$ decays very demanding numerically.

11.3.9 Euclidean correlation functions and the Maiani–Testa theorem

For well over a decade, it was believed that $K \to \pi\pi$ amplitudes could not be studied directly on the lattice. Indeed, these amplitudes have both real and imaginary strong-interaction contributions while, in the Euclidean, correlation functions are purely real (or imaginary). Thus, it was difficult to see how such amplitudes could be extracted from a lattice calculation, necessarily performed in the Euclidean.

Of course, the Osterwalder–Schrader theorem (Osterwalder and Schrader, 1973, 1975) guarantees that Euclidean correlation functions can be continued to Minkowski space, at least in principle. However, in practice such analytical continuations are essentially impossible with approximate Euclidean results.

This led Maiani and Testa, in 1990, to investigate what can be extracted from Euclidean correlation functions without analytical continuation (Maiani and Testa, 1990). They considered the following type of center-of-mass frame Euclidean correlation function,

$$\langle \pi(\vec{p}, t_1)\pi(-\vec{p}, t_1)\mathcal{H}_{\mathrm{eff}}^{\Delta S=-1}K^\dagger(\vec{0}, t_i)\rangle, \tag{11.177}$$

and asked the question: what information does this correlation function contain regarding physical $K \to \pi\pi$ decay amplitudes in the usual lattice, asymptotic limit, i.e. $t_1, t_2 \gg 1/M_\pi$, $-t_i \gg 1/M_K$?

To "simplify" the problem and disentangle Euclidean from other possible lattice effects, they chose to work in a large, quasi-infinite volume. This apparently innocuous assumption has rather important consequences. For one, in infinite volume, the $\pi\pi$ spectrum is continuous. This means that in the limit $-t_i \gg 1/M_K$ and $t_1, t_2 \gg 1/M_\pi$, only the ground state contribution can be picked out: there is no known numerical technique to isolate an excited state in a continuous spectrum. In turn, this implies that one can only extract information about the matrix element $\langle \pi(\vec{0})\pi(\vec{0})|\mathcal{H}_{\mathrm{eff}}^{\Delta S=-1}|K(\vec{0})\rangle$, in which all mesons are at rest: the physical decay is not directly accessible on the lattice.

This statement became known as the "Maiani–Testa theorem." It was a formalization of the general belief that $K \to \pi\pi$ decays could not be studied directly on the lattice but rather that approximations (Bernard *et al.*, 1985) or models (Ciuchini *et al.*, 1996) were needed to obtain information about physical $K \to \pi\pi$ decays from lattice calculations.

However, as is often the case with "no-go" theorems, the solution is found by questioning the underlying, apparently innocent assumptions. Here, it was the infinite volume assumption that brought in all of the difficulties while, for simulations performed in boxes with sides of a few fermi at most, it does not even approximately hold.

11.3.10 Two-pion states in finite volume

In a finite box with sides L,[18] two pions cannot be isolated into non-interacting asymptotic states. Rather, the $\pi\pi$ eigenstates are the result of a stationary scattering process. In addition, boundary conditions generically imply that the particles' momenta are quantized. For periodic boundary conditions, they come in discrete multiples of $2\pi/L$: $\vec{k} = \vec{z}(2\pi/L)$, with $\vec{z} \in \mathbb{Z}^3$. In turn, this means that the spectrum is discrete and the splitting is actually rather large for box sides of a few fermi. In such boxes, the typical spacing between momenta is $\Delta p = 2\pi/L = 1.2\,\text{GeV}/L[\text{fm}]$. This is clearly quite different from the continuous spectrum found in infinite volume.

In the free theory, in the center-of-mass frame and in the A_1^+ (i.e. cubic spin-0) sector, the energy of the nth excited state is given by:

$$W_n^{(0)} = 2\sqrt{M_\pi^2 + n\left(\frac{2\pi}{L}\right)^2},\qquad(11.178)$$

for $n \le 6$ – there is no integer three-vector with norm squared, 7, nor 8 for that matter. This spectrum is shown as the dashed lines in Fig. 11.6.

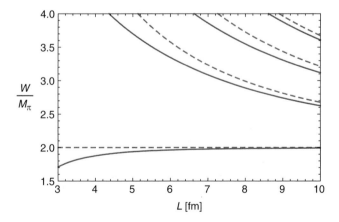

Fig. 11.6 The two-pion spectrum in a box of volume L^3 as a function of L in the $I = J = 0$ channel, under the four-pion threshold $W/M_\pi = 4$. The free spectrum is depicted by the dashed curves. The solid curves show the interacting spectrum as obtained from Eq. (11.183), using the one-loop $I = J = 0$ phase shift from Gasser and Meissner (1991) and Knecht *et al.* (1995).

In the presence of interactions, the energy W_n of the fully interacting state was worked out by Martin to all orders in relativistic quantum field theory (Lüscher, 1986, 1991) for two-pion energies below the four-pion threshold, up to corrections which fall off exponentially with the box size (i.e. up to finite-volume, vacuum polarization effects).

The ground state $n = 0$ requires special treatment. In the isospin $I = 0$ and spin $J = 0$ channel, we have (Lüscher, 1986):

$$W_0 = 2M_\pi - \frac{4\pi a_I}{M_\pi L^3} \left\{ 1 + c_1 \frac{a_I}{L} + c_2 \left(\frac{a_I}{L} \right)^2 \right\} + O(L^{-6}) \tag{11.179}$$

where

$$c_1 = -2.837297, \qquad c_2 = 6.375183, \tag{11.180}$$

where the S-wave scattering length in the appropriate channel is a_I with

$$a_I = \lim_{k \to 0} \frac{\delta_I(k)}{k}, \tag{11.181}$$

and where k is the momentum of the pions in the center-of-mass frame.

For excited states, the results are given in terms of the scattering phase, δ_I, of the relevant isospin channel. The appearance of scattering phases should not be surprising. We know from scattering theory that, under reasonable conditions, potentials can be reconstructed from these phases. Martin finds that (Lüscher, 1991)

$$W_n = 2\sqrt{M_\pi^2 + k_n^2} \qquad n = 1, 2, 3, \ldots, \tag{11.182}$$

where k_n is a solution of the quantization equation

$$n\pi - \delta_I(k_n) = \phi(q_n), \tag{11.183}$$

with $q_n \equiv k_n L/(2\pi)$ and with

$$\tan \phi(q) = -\frac{\pi^{3/2} q}{Z_{00}(1; q^2)}. \tag{11.184}$$

$\phi(q)$ is defined for $q \geq 0$. It is such that $\phi(0) = 0$ and it depends continuously on q. It is given in terms of the zeta function of the Laplacian

$$Z_{00}(s; q^2) = \frac{1}{\sqrt{4\pi}} \sum_{\vec{n} \in \mathbb{Z}^3} (\vec{n}^2 - q^2)^{-s}, \tag{11.185}$$

for Re $s > 3/2$ and by analytic continuation elsewhere. A useful integral representation for evaluating $Z_{00}(1; q^2)$ numerically is given in Section 11.4.

Solving Eq. (11.183), one generically finds

$$W_n = 2\sqrt{M_\pi^2 + n \left(\frac{2\pi}{L} \right)^2} + O\left(\frac{1}{L^3} \right), \tag{11.186}$$

where the equation actually gives the whole tower of $1/L$ corrections once the scattering phase is specified. As already stated, the equation holds for $n \leq 6$ and there is no integer three-vector whose squared norm is 7 or 8. Then, beginning at 9, there are integer vectors, $\vec{z}$, which are not related by cubic rotations but which have the same squared norm, e.g. $\vec{z}_1 = (2, 2, 1)$ and $\vec{z}_2 = (3, 0, 0)$. In a treatment where only the $O(3)$ spin-0 component of the A_1^+ representation is taken into account,[19] the states associated with such three-vectors in the free case do not feel the interaction in the interacting case either, and have $k = |\vec{z}|(2\pi/L)$. This is because they combine into states with $O(3)$ spin-4 and/or higher spins. Such states will be ignored in the following and we will consider only $n \leq 6$, which is certainly not a limitation in practice.

The full solution (taking for instance $I = J = 0$) is shown in Fig. 11.6. The first remark which can be made is that the two-pion spectrum on lattices which can be considered in the foreseeable future is far from being continuous. The second is that distortions due to interactions are quite small: the volume suppression of the corrections is effective when $L \geq 3$ fm. Finally, it is clear that by studying the energies $W_n(L)$ as a function of box size L, we can turn Eq. (11.183) around and reconstruct, at least for a few discrete momenta, the scattering phase $\delta_I(k)$.

Now, suppose that there is a single resonance R in this $\pi\pi$ channel, with mass $M_R < 4M_\pi$, i.e. under the inelastic threshold, and width Γ_R. To explain how Martin's equation works in that case, it is useful to turn to quantum mechanics. A quantum mechanics approach is actually justified because corrections suppressed exponentially in L are neglected in the derivation of the quantization formula. Such suppression factors generically correspond to tunneling phenomena. Here they are associated with features of relativistic quantum field theories which are absent in quantum mechanics: the exchange of a virtual particle around the box.

We begin by decomposing the total Hamiltonian H of the two-pion system into a free part H_0 and an interaction H_{int}:

$$H = H_0 + H_{\text{int}}. \tag{11.187}$$

Then we consider the nth free $\pi\pi$ state $(n \leq 6)$, $|n_0\rangle$, in our chosen channel (here isospin $I = 0$ or 2 and $J^P = 0^-$). It is such that $\langle n_0|n_0\rangle = 1$ and

$$H_0|n_0\rangle = W_n^{(0)}|n_0\rangle. \tag{11.188}$$

To understand what this energy becomes in the presence of interactions and of the resonance, it is useful to consider a perturbative expansion in the interaction H_{int}, though the final result is accurate to all orders. Denoting the resulting energy W_n, and $|n\rangle$, the corresponding fully-interacting eigenstate, we have, to second order in H_{int}:

$$W_n = \langle n|H|n\rangle \tag{11.189}$$

$$= W_n^{(0)} + \langle n_0|H_{\text{int}}|n_0\rangle + \sum_\alpha \frac{\langle n_0|H_{\text{int}}|\alpha\rangle\langle\alpha|H_{\text{int}}|n_0\rangle}{W_n^{(0)} - W_\alpha} + \cdots .$$

Here α runs over 2, 4, ... pion states, as well as any other state which appears in the given channel, and W_α is the corresponding energy. Factors of the form $(W_n^{(0)} - W_\alpha)$

also appear in the higher-order terms of the perturbative expansion represented by the ellipsis.

As long as L is such that $W_\alpha = M_R$ is far from the free two-pion energy, $W_n^{(0)}$, all of the terms in the perturbative series are regular and can therefore be resummed. Now, if the resonance is narrow, the coupling of the resonance to the n_0, $\pi\pi$ state will be small. In turn, this means that the matrix element $\langle \pi\pi | H_{\text{int}} | R \rangle$ is small compared to the mass and typical energies of the system.[20] Moreover, the leading correction which the resonance brings to the interacting two-pion energy, W_n, appears at second order in the expansion. Thus, it is of order $|\langle \pi\pi | H_{\text{int}} | R \rangle|^2$ and is therefore small.

When $L = L_R$ such that $W_n^{(0)}(L_R) = M_R$, the effect of the resonance is radically different. Its second and higher order contributions to W_n blow up. In such a situation, we have to resort to degenerate perturbation theory and first diagonalize H in the two-state subspace $\{|n_0\rangle, |R\rangle\}$. This means diagonalizing the 2×2 matrix

$$\begin{pmatrix} \langle n_0 | H | n_0 \rangle & \langle n_0 | H | R \rangle \\ \langle R | H | n_0 \rangle & \langle R | H | R \rangle \end{pmatrix} = \begin{pmatrix} M_R & M_n \\ M_n^* & M_R \end{pmatrix}, \tag{11.190}$$

where $M_n \equiv \langle n_0 | H | R \rangle$ is the transition amplitude between the resonance and the two-pion state $|n_0\rangle$. A straightforward diagonalization yields

$$W_n^\pm = M_R \pm |M_n|, \tag{11.191}$$

thereby lifting the degeneracy and giving rise to a typical level repulsion phenomenon. Thus, in solving Martin's formula (11.183), we would find a dependence of the two-pion energy as a function of L, which looks like what is depicted in Fig. 11.7.

11.3.11 $K \to \pi\pi$ in finite volume

What Martin and I realized over ten years ago is that to study $K \to \pi\pi$ decays in finite volume, we could treat the kaon as an infinitesimally narrow resonance in the weak interaction contribution to the scattering of the two pions (Lellouch and Lüscher, 2001). In that case, one considers the free Hamiltonian to be the QCD Hamiltonian, i.e.

$$H_0 = H_{\text{QCD}}, \tag{11.192}$$

and the perturbation, H_{int}, to be the effective weak Hamiltonian relevant for $K \to \pi\pi$ decays, i.e.

$$H_{\text{int}} = H_W = \int_{x_0=0} d^3x \, \mathcal{H}_W(x). \tag{11.193}$$

Then, since the amplitudes for $K \to \pi\pi$ decays, $T(K \to \pi\pi)$, are computed at $O(G_F)$, we can perform all of our computations to that order.

Following what was done in the preceding section for the resonance, we tune the size of the box to $L = L_K$, such that for some level n,

$$W_n(L_K) = M_K. \tag{11.194}$$

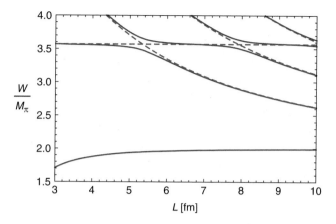

Fig. 11.7 The two-pion spectrum in a box of volume L^3 as a function of L in the $I = J = 0$ channel with an additional, fictitious resonance of mass $M_R = M_{K_S}$ and width $\Gamma_R = 10^{12} \times$ $(\Gamma_{+-} + \Gamma_{00} - (4/3)\Gamma_{+0}) \simeq 7.3\,\mathrm{MeV}$ with Γ_{+-}, Γ_{00}, and Γ_{+0} given in Eqs. (11.164)–(11.166). The dashed curves represent the two-pion spectrum interacting through δ_0 (they correspond to the solid curves in Fig. 11.6). The solid curves show the interacting spectrum as obtained from Eq. (11.183), with the contribution of the resonance to the phase shift. At the points at which the resonance (i.e. the horizontal dashed curve at $M_{K_S}/M_\pi \sim 3.57$) crosses the excited $I = J = 0$ states, one clearly sees a level repulsion effect. The effect is rather small and limited to a smallish region around the crossing point because the resonance is narrow: $\Gamma_R/M_R \simeq 1.5\%$.

In that case, the corresponding pion momentum is the momentum, k_π, which the pions would have in the physical kaon decay, i.e.

$$k_n(L_K) = k_\pi \equiv \sqrt{\frac{M_K^2}{4} - M_\pi^2}. \tag{11.195}$$

Then, the transition matrix element in the finite volume $V = L_K^3$,

$$M_n^I \equiv {}_V\langle(\pi\pi)_I n|H_W|K\rangle_V, \tag{11.196}$$

is an energy conserving matrix element. Since finite-volume corrections to a single, stable particle state are exponentially small in L and since we neglect such corrections here, $|K\rangle_V$ is identical to $|K\rangle$, up to a purely kinematic normalization factor.

However, we are not interested in the finite-volume matrix element of Eq. (11.196). What we want is the infinite-volume transition amplitude,

$$T_I \equiv \langle(\pi\pi)_I n, out|\mathcal{H}_W|K\rangle, \tag{11.197}$$

where the corresponding A_I of Eq. (11.65) can be made real in the CP conserving case.

Because it is important here, let us pause to say a few words about the normalization of states used. In finite volume we use the usual quantum mechanical normalization of states to unity. Thus, for a spinless particle of mass m and momentum $\vec{p}$:

$$_V\langle \vec{p}\,|\vec{p}\,'\rangle_V = \delta_{\vec{p},\vec{p}\,'}. \tag{11.198}$$

In infinite volume, it is the standard relativistic normalization of states, i.e.

$$\langle p|p'\rangle = 2p^0(2\pi)^3\delta^{(3)}(\vec{p}-\vec{p}\,'), \tag{11.199}$$

with $p^0 = \sqrt{m^2 + \vec{p}^2}$, which is implemented.

To obtain the relationship between these two amplitudes, we compute the shift in energy brought about by the presence of the weak interactions in two different ways, and require the two results to agree.

We begin by repeating the degenerate perturbation theory of the preceding section. We obtain

$$W_n^\pm = M_K \pm |M_n^I|, \tag{11.200}$$

where M_n^I is clearly $O(G_F)$. Then, we turn to Martin's finite-volume quantization formula (11.183). In the presence of the weak interaction, the phase shift receives a weak contribution, δ_W. Thus, in that formula, we have to perform the replacement

$$\delta_I \to \bar{\delta}_I = \delta_I + \delta_W. \tag{11.201}$$

We are interested in this phase shift at the values of momenta $k = k_n^\pm$, corresponding to the perturbed energies $W_n^\pm$ of Eq. (11.200):

$$k_n^\pm = k_n \pm \Delta k = k_n \pm \frac{W_n|M_n^I|}{4k_n} + O(G_F^2). \tag{11.202}$$

Indeed, we know that those values must come out of the quantization formula because the two methods of determining the energy shifts must give the same result. Because $W_n^\pm$ are "infinitesimally" close to M_K (in our LO counting in G_F), $\delta_W(k_n^\pm)$ is dominated by the s-channel kaon exchange depicted in Fig. 11.8. Any other contribution will be at least $O(G_F^2)$. In the s-channel, however, the factor of G_F^2 coming from the two K-$\pi\pi$ vertices is compensated by a propagator enhancement, due to the fact that we are sitting only $O(G_F)$ away from the peak of the resonance. Indeed, the scattering amplitude corresponding to Fig. 11.8 is

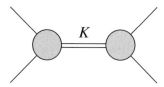

Fig. 11.8 Kaon contribution to the elastic weak scattering of two pions in the S-channel. The shaded vertices represent the $K \to \pi\pi$ transition amplitudes at $O(G_F)$.

$$S[\text{Fig. 11.8}] = -\frac{\overbrace{T_I^* T_I}^{O(G_F^2)}}{\underbrace{(W_n^\pm)^2 - M_K^2}_{O(G_F)} \underbrace{-iM_K\,\Gamma_K}_{O(G_F^2)}} + O(G_F^2) \qquad (11.203)$$

$$= \mp \frac{|A_I|^2}{2W_n^{(0)}|M_n^I|} + O(G_F^2), \qquad (11.204)$$

where we have used the fact that the vertices in Fig. 11.8 are the on-shell transition amplitudes up to higher order corrections in G_F. To translate this amplitude into a scattering phase, we use the partial wave decomposition of a $\pi\pi$ scattering amplitude S in the center-of-mass frame:[21]

$$S = 16\pi\, W \sum_{l=0}^{\infty} (2l+1) P_l(\cos\theta)\, e^{i\delta_l^I(k)} \frac{\sin\delta_l^I(k)}{k}, \qquad (11.205)$$

where $W = 2\sqrt{M_\pi^2 + k^2}$ is the energy of the pion pair. Since the amplitude of Eq. (11.203) has no angular dependence, it only leads to a zero angular momentum phase shift, which is the weak phase shift of interest. Thus, we find

$$\delta_W(k_n^\pm) = \mp \frac{k_n^\pm |A_I|^2}{32\pi (W_n^{(0)})^2 |M_n^I|} + O(G_F^2). \qquad (11.206)$$

We can now include this contribution into the total phase shift $\bar{\delta}_I$ and write down the resulting quantization equation:

$$n\pi - \delta_I(k_n \pm \Delta k) \pm \frac{k_n^\pm |A_I|^2}{32\pi (W_n^{(0)})^2 |M_n^I|} = \phi(q_n \pm \Delta q) + O(G_F^2), \qquad (11.207)$$

with $\Delta q = L\Delta k/(2\pi)$. Expanding this equation to $O(G_F)$, we finally find the relationship between the desired, infinite-volume amplitude A_I and the finite-volume matrix elements M_n^I, computed on the lattice (Lellouch and Lüscher, 2001):

$$|A_I|^2 = 8\pi \left\{ q\frac{\partial}{\partial q}\phi(q) + k\frac{\partial}{\partial k}\delta_I(k) \right\}_{k=k_n} \frac{(W_n^{(0)})^2 M_K}{k_n^3} L^6 |\mathcal{M}_n^I|^2, \qquad (11.208)$$

were we have used the definition

$$\mathcal{M}_n^I \equiv {}_V\langle(\pi\pi)_I n|\mathcal{H}_W(0)|K\rangle_V = M_n^I/L^3. \qquad (11.209)$$

In our derivation we have tuned the size of the box to L_K such that $W_n^{(0)} = M_K$ and thus, $k_n = k_\pi$ (Eqs. (11.194)–(11.195)). However, Eq. (11.208) is also valid for $W_n^{(0)} \neq M_K$, as can be seen in Section 11.3.12 and as was derived using the fact that the matching factor is related to the density of interacting two-pion states in finite volume (Lin *et al.*, 2001). An interesting discussion of this and other ways of looking at this formula is given by Testa (2005). Kim *et al.* (2005) further generalized

Eq. (11.208) to moving frames, i.e. frames in which the center of mass has a non-vanishing momentum. And Kim and Sachrajda (2010) showed how partially-twisted boundary conditions can be used to obtain the phase shift $\delta_2(k)$ and its derivative in the isospin-2 channel. With twisted boundary conditions, one allows some of the quark flavors to be periodic only up to a phase. This phase forces the flavors concerned to carry a momentum which is proportional to the phase. The boundary conditions are called partially-twisted when it is only the valence flavors which are given a twist.

The proportionality factor (11.208) is to a large extent kinematic, as it accounts for the difference in normalization of states in finite and infinite volumes, given in Eqs. (11.198)–(11.199). This can easily be seen in the absence of interactions. To reach the nth two-pion with energy $W_n^{(0)}$ and pion momentum $k_n^{(0)}$, the cube must have sides

$$L_n = \frac{2\pi}{k_n^{(0)}} \sqrt{n}. \tag{11.210}$$

Then, Eq. (11.208) assumes the form

$$|A_I|^2 = \frac{4}{\nu_n} (W_n^{(0)})^2 M_K L^3 |M_n^I|^2 \tag{11.211}$$

where

$$\nu_n \equiv \text{number of } \vec{z} \in \mathbb{Z}^3 \ni \vec{z}^2 = n. \tag{11.212}$$

The proportionality constant is precisely the relative normalization of free kaon and two-pion states in finite and infinite volume projected onto the A_1^+ and spin-0 sectors, respectively (see also Section 11.3.12). The constant is the product of the ratio of squared norms of the kaon state, $2M_K L^3$, and of the two-pion state, $2(W_n L^3)^2$,[22] times the square of the factor relating H_W and $\mathcal{H}_W(0)$, i.e. $1/L^6$, times $1/\nu_n$ since the finite-volume, A_1^+ state is obtained by summing over the ν_n pion momentum directions.

11.3.12 $K \rightarrow \pi\pi$ in finite volume: a simple relativistic quantum field theory example

To understand how the finite-volume effects predicted by Eqs. (11.183)–(11.208) show up in correlation functions similar to those one would use in numerical simulations, it is useful to consider them in the context of a relativistic field theory in which all quantities of interest can be computed analytically. Because the form of the finite-volume formulas (11.183) and (11.208) does not depend on the details of the dynamics, we choose to work in a world in which this dynamics is simplified, so as not to obscure the discussion of finite-volume effects with superfluous technical details. The calculations below were summarized by Lellouch and Lüscher (2001).

Specification of the model

We consider a theory of a single, neutral, spinless pion field $\pi(x)$, of mass M_π. In the notation of Section 11.3.11, the unperturbed Hamiltonian density is

$$\mathcal{H}_0 = \mathcal{H}_{\mathrm{kin}} + \mathcal{H}_S, \tag{11.213}$$

where $\mathcal{H}_{\mathrm{kin}}$ is the usual kinetic Hamiltonian of a scalar field and the "strong" interaction between the pions is given by

$$\mathcal{H}_S = \frac{\lambda}{4!}\pi^4. \tag{11.214}$$

We assume here that the theory is perturbative in λ and we will work to first non-trivial order in λ, i.e. $O(\lambda)$. Because the lattice calculations are performed in Euclidean space-time, we rotate this theory into the Euclidean and consider Euclidean correlation functions. Moreover, our calculations will be performed in a three-volume L^3 with periodic boundary conditions, but the time direction will be considered of infinite extent.

To make the perturbation theory completely well-defined, we introduce a Pauli–Villars cutoff Λ. At tree level the Euclidean pion propagator is then given by

$$S_\pi(x) = \int_x e^{ik\cdot x}\langle \pi(x)\pi(0)\rangle = \frac{1}{p^2 + M_\pi^2} - \frac{1}{p^2 + \Lambda^2}. \tag{11.215}$$

The cutoff should be large enough so that ghost particles cannot be produced at energies below the four-pion threshold, but in view of the universality of Eq. (11.183) and (11.208) there is no need to take Λ to infinity at the end of the calculation.

Since we are going to be computing correlation functions in the time-momentum representation, it is useful to have the pion propagator in this same representation. We have

$$S_\pi(t; \vec{k}) \equiv \int_{\vec{x}} e^{-i\vec{k}\cdot\vec{x}}\langle \pi(t, \vec{x})\pi(0)\rangle$$

$$= \frac{1}{2E_k} e^{-E_k |t|} - \frac{1}{2\mathcal{E}_k} e^{-\mathcal{E}_k |t|}, \tag{11.216}$$

where $E_k = \sqrt{\vec{k}^2 + M_\pi^2}$ and $\mathcal{E}_k = \sqrt{\vec{k}^2 + \Lambda^2}$.

As far as the kaon and its decays into two pions are concerned, the least complicated possibility is to describe it by a free hermitian field $K(x)$ with mass M_K and to take

$$\mathcal{H}_W = \frac{g}{2}K\pi^2 \tag{11.217}$$

as a weak Hamiltonian density. We will only work here to leading order in the weak coupling, also.

Determination of the phase shift

Let us first determine the phase shift, $\delta(k)$, in the model of Eq. (11.214). This is an infinite-volume, Minkowski space calculation, though at the order at which we work this fact makes very little difference. The partial wave decomposition of the invariant, scattering amplitude S, in the center-of-mass frame, is given in Eq. (11.205). At $\mathcal{O}(\lambda)$, $S = -\lambda$, and therefore

$$\delta(k) = -\frac{\lambda}{16\pi}\frac{k}{W} + O(\lambda^2), \tag{11.218}$$

where $W = 2E_k$ is the free two-pion energy.

Two-pion energies using Eq. (11.183)

In the absence of interactions, i.e. for $\delta(k) \equiv 0$, the solutions of Eq. (11.183), for $n = 1, 2, \ldots, 6$, are the free, finite-volume momentum magnitudes, $k_n^{(0)} \equiv \sqrt{n}(2\pi/L)$.[23] For weakly interacting pions, the solutions are small perturbations about these values. Thus, the rescaled momenta are

$$q_n = \frac{k_n L}{2\pi} = q_n^{(0)} + \Delta q_n = \sqrt{n} + \Delta q_n, \tag{11.219}$$

where Δq_n is the small perturbation.

To first order in Δq_n and λ

$$\tan\phi(q_n) = \frac{4\pi^2}{\nu_n}n\Delta q_n + \mathcal{O}(\Delta q_n^2), \tag{11.220}$$

and Eq. (11.183) yields

$$\Delta q_n = \lambda\frac{\nu_n}{32\pi^2\sqrt{n}}\frac{1}{W_n^{(0)}L} + \mathcal{O}(\lambda^2), \tag{11.221}$$

where $W_n^{(0)} = 2\sqrt{M_\pi^2 + (k_n^{(0)})^2}$ is the energy of two free pions with opposite momenta of magnitude $k_n^{(0)}$. Thus, the energy of the corresponding two-pion state, in the presence of interactions, is

$$W_n = W_n^{(0)}\left(1 + \lambda\frac{\nu_n}{2}\frac{1}{(W_n^{(0)}L)^3} + \mathcal{O}(\lambda^2)\right). \tag{11.222}$$

Two-pion energies from perturbation theory

In perturbation theory, the two-pion energy corresponding to pions whose momenta would have magnitude $k_n^{(0)}$, $n = 1, \ldots, 6$, in the absence of interactions, can be extracted from the $\pi\pi \to \pi\pi$ correlation function

$$C_{\pi\pi\to\pi\pi}(t) = \langle\mathcal{O}_n(t)\mathcal{O}_n(0)\rangle_{\text{conn}}, \tag{11.223}$$

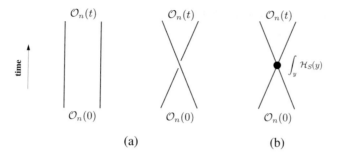

Fig. 11.9 Diagrams which contribute to $C_{\pi\pi \to \pi\pi}(t)$ at $\mathcal{O}(\lambda^0)$ (a) and $\mathcal{O}(\lambda)$ (b).

where

$$\mathcal{O}_n(t) = \frac{1}{\nu_n} \sum_{\{\vec{k}_n\}} \int_{\vec{x}_1 \vec{x}_2} e^{i\vec{k}_n \cdot (\vec{x}_2 - \vec{x}_1)} \pi(t, \vec{x}_2) \pi(t, \vec{x}_1) \qquad (11.224)$$

is an operator which has overlap with zero-momentum, cubically invariant, two-pion states and ν_n is the number of momenta $\vec{k}_n$ such that $|\vec{k}_n| = k_n^{(0)}$ (see Eq. (11.212)). The sum in Eq. (11.224) is over these momenta, all related by cubic transformations. The operator $\pi(x)$ has overlap with single pion states. In the limit of large t, the contribution of the two-pion states, $|\pi\pi\, l\rangle_V$, to the correlation function of Eq. (11.223), is

$$C_{\pi\pi \to \pi\pi}(t) \longrightarrow \sum_{l=0}^{6} |\langle 0|\mathcal{O}_n(0)|\pi\pi\, l\rangle_V|^2\, e^{-W_l t} + \cdots, \qquad (11.225)$$

where the ellipsis stands for terms which decay more rapidly. The states $|\pi\pi\, l\rangle_V$ are normalized to one.

A straightforward calculation of the diagrams of Fig. 11.9, using the propagator of Eq. (11.216), gives for the correlation function of Eq. (11.223), at $\mathcal{O}(\lambda)$,

$$C_{\pi\pi \to \pi\pi}(t)\big|_n = \frac{2}{\nu_n} \left(\frac{L^3}{W_n^{(0)}} \right)^2 e^{-W_n^{(0)} t} \left\{ 1 - \lambda \frac{\nu_n}{2L^3} \left(\frac{t}{(W_n^{(0)})^2} - \frac{1}{(W_n^{(0)})^3} \right. \right.$$

$$\left. \left. - \frac{R_\Lambda \left((k_n^{(0)})^2, (k_n^{(0)})^2 \right)}{2} \right) + \mathcal{O}(\lambda^2) \right\} \qquad (11.226)$$

$$= \frac{2}{\nu_n} \left(\frac{L^3}{W_n} \right)^2 e^{-W_n t} \left\{ 1 + \lambda \frac{\nu_n}{2L^3} \left(\frac{1}{(W_n^{(0)})^3} + \frac{R_\Lambda \left((k_n^{(0)})^2, (k_n^{(0)})^2 \right)}{2} \right) + \mathcal{O}(\lambda^2) \right\},$$

where we have only retained the contribution which decays exponentially with the rate corresponding that of the two-pion state, $|\pi\pi\, n\rangle_V$. In Eq. (11.226), the two-pion

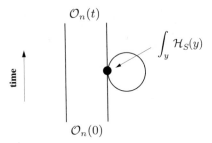

Fig. 11.10   Example of a tadpole contribution to $C_{\pi\pi\to\pi\pi}(t)$ at $\mathcal{O}(\lambda)$.

energy, W_n, is the same as that obtained from the finite-volume formula of Lüscher (1991) (see Eq. (11.222)) and the regulator contribution is given by

$$R_\Lambda\left(\vec{p}^{\,2},\vec{k}^2\right) = \frac{4(E_p + \mathcal{E}_p)}{E_p\mathcal{E}_p\left[(E_p + \mathcal{E}_p)^2 - 4E_k^2\right]} - \frac{1}{\mathcal{E}_p(E_p^2 - \mathcal{E}_p^2)}, \tag{11.227}$$

with self-explanatory notation.

Comparison of this result with Eq. (11.225) further gives the matrix element of $\mathcal{O}_n$ between the vacuum and the two-pion state, $|\pi\pi\, n\rangle_V$:

$$|\langle 0|\mathcal{O}_n(0)|\pi\pi\, n\rangle_V| = \sqrt{\frac{2}{v_n}}\left(\frac{L^3}{W_n}\right)\left\{1 + \lambda\frac{v_n}{4L^3}\left[\frac{1}{(W_n^{(0)})^3} + \frac{R_\Lambda\left((k_n^{(0)})^2,(k_n^{(0)})^2\right)}{2}\right]\right.$$

$$\left. +\mathcal{O}(\lambda^2)\right\}. \tag{11.228}$$

The observant reader will have noticed that we have not taken into account contributions from diagrams such as that of Fig. 11.10, which also appear at $\mathcal{O}(\lambda)$. As can be verified explicitly, these diagrams amount to a shift of the pion mass by terms which are independent of L, up to exponentially small corrections. Since such corrections are neglected here, these contributions will affect none of our finite-volume results, once the mass has been appropriately renormalized in infinite volume. The details of this renormalization are irrelevant here and we assume that the renormalization has been adequately performed. Furthermore, the coupling λ and the field $\pi(x)$ only get renormalized at $O(\lambda^2)$, which is beyond the order at which we are working.

Matching of finite to infinite matrix elements using Eq. (11.208)

Here we consider a weak transition between the state of a kaon at rest and a two-pion state, $|\pi\pi\, n\rangle_V$, in finite volume. The amplitude for this transition is

$$M = \int_{\vec{x}} {}_V\langle \pi\pi\, n|\mathcal{H}_W(0,\vec{x})|K\rangle_V. \tag{11.229}$$

The corresponding infinite-volume transition amplitude, T, is given by

$$T = \langle \pi(\vec{p})\pi(-\vec{p}), out | \mathcal{H}_W(0) | K(\vec{0}) \rangle. \tag{11.230}$$

Again, infinite-volume states are relativistically normalized here.

To compare with the perturbative results obtained below, we must compute the factor relating $|T|$ and $|M|$ in Eq. (11.208) to $O(\lambda)$. Using the expressions in Eqs. (11.219)–(11.221) for q_n, and after some algebra, we find

$$q_n \phi'(q_n) = \frac{4\pi^2 n^{3/2}}{\nu_n} \left\{ 1 + \frac{\lambda}{8\pi^2} \frac{1}{W_n^{(0)} L} \left[\frac{\nu_n}{n} + z_n \right] + O(\lambda^2) \right\}, \tag{11.231}$$

where z_n is the constant given by

$$z_n = \lim_{q^2 \to n} \left\{ \sqrt{4\pi} \mathcal{Z}_{00}(1; q^2) + \frac{\nu_n}{q^2 - n} \right\}. \tag{11.232}$$

Now, using the result of Eq. (11.218) for $\delta(k)$ and Eqs. (11.219)–(11.221) for k_n, we find

$$k_n \delta'(k_n) = -\frac{\lambda}{8} \frac{\sqrt{n}}{W_n L} \left[1 - n \left(\frac{4\pi}{W_n^{(0)} L} \right)^2 + O(\lambda^2) \right]. \tag{11.233}$$

Combining Eqs. (11.231)–(11.233), we find for the factor which relates $|T|$ and $|M|$ in Eq. (11.208),

$$8\pi \left\{ q \frac{\partial \phi}{\partial q} + k \frac{\partial \delta_0}{\partial k} \right\}_{k_n} \frac{M_K W_n^2}{k_n^3} \tag{11.234}$$

$$= \frac{4 M_K W_n^2 L^3}{\nu_n} \left\{ 1 + \frac{\lambda}{2} \frac{1}{W_n^{(0)} L} \left[\frac{z_n}{4\pi^2} + \frac{\nu_n}{(W_n^{(0)} L)^2} \right] + O(\lambda^2) \right\},$$

where we have used the expression for k_n obtained in Eq. (11.219) and (11.221). It should be remarked that, unless the pions interact strongly, the size of this factor is essentially determined by mismatches in the definitions of T and M and in the normalization of states in finite and infinite volume (see Eq. (11.211) and subsequent discussion).

Matching of finite to infinite matrix elements from perturbation theory

The relevant correlation function here is

$$C_{K \to \pi\pi}(t_i, t_f) = \int_{\vec{x}} \langle \mathcal{O}_n(t_f) \mathcal{H}_W(0) K(-t_i, \vec{x}) \rangle. \tag{11.235}$$

At $O(\lambda^0)$, it is given by the diagram in Fig. 11.11(a). One trivially obtains,

$$C_{K \to \pi\pi}^{(11.11a)}(t_i, t_f) \Big|_n = -g \frac{e^{-M_K t_i}}{2 M_K} \frac{e^{-W_n^{(0)} t_f}}{(W_n^{(0)})^2}. \tag{11.236}$$

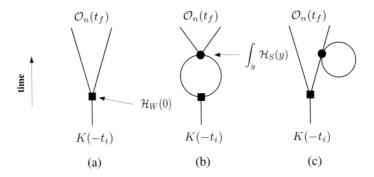

Fig. 11.11 Diagrams which contribute to $C_{K\to\pi\pi}(t_i, t_f)$ at $O(\lambda^0)$ (a) and $O(\lambda)$ (b,c).

At $O(\lambda)$, the contribution from the diagram in Fig. 11.11(b) gives, using the pion propagators defined in Eq. (11.216),

$$C_{K\to\pi\pi}^{(11.11.b)}(t_i, t_f)\Big|_n = \lambda \frac{g}{2} \frac{e^{-M_K t_i}}{2M_K} \int_{t_y} \frac{1}{L^3} \sum_{\vec{p}} \{S_\pi(t_y; \vec{p})\}^2 \times \{S_\pi(t_f - t_y; \vec{k}_n)\}^2$$

$$\longrightarrow - C_{K\to\pi\pi}^{(11.11a)}(t_i, t_f)\Big|_n \times \lambda \left\{ \frac{\nu_n}{2L^3} \left(\frac{t}{(W_n^{(0)})^2} + \frac{1}{(W_n^{(0)})^3} - \frac{R_\Lambda\left((k_n^{(0)})^2, (k_n^{(0)})^2\right)}{2} \right) \right.$$

$$\left. + \frac{1}{8} \mathcal{A}(k_n^{(0)}) \right\}, \tag{11.237}$$

where

$$\mathcal{A}(k_n^{(0)}) = \frac{1}{L^3} \sum_{\vec{p}}' \left[\frac{1}{E_p(\vec{p}^2 - (k_n^{(0)})^2)} - R_\Lambda\left(\vec{p}^2, (k_n^{(0)})^2\right) \right], \tag{11.238}$$

and where we have only kept, in the second line of Eq. (11.237), the terms in the momentum sum of the first line which fall off as $e^{-W_n^{(0)} t_f}$.

The sum in Eq. (11.238) is restricted to momenta $\vec{p}$ such that $|\vec{p}| \neq k_n^{(0)}$. To evaluate it, we use the asymptotic, large-volume expansion of Lüscher (1986). Up to terms that vanish more rapidly than any power of $1/L$, we find

$$\mathcal{A}(k_n^{(0)}) \sim I_1(k_n^{(0)}) + \frac{z_n}{2\pi^2 W_n^{(0)} L} + \frac{\nu_n}{L^3} \left[\frac{4}{(W_n^{(0)})^3} + R_\Lambda\left((k_n^{(0)})^2, (k_n^{(0)})^2\right) \right], \tag{11.239}$$

where z_n is defined in Eq. (11.232). In Eq. (11.239), $I_1(k_n^{(0)})$ is the infinite-volume contribution:

$$I_1(k_n^{(0)}) = \int_{\vec{p}} \left\{ \frac{1}{E_p} \text{Re}\left[\frac{1}{\vec{p}^2 - (k_n^{(0)})^2 - i\epsilon} \right] - R_\Lambda\left(\vec{p}^2, (k_n^{(0)})^2\right) \right\}. \tag{11.240}$$

As discussed following Eq. (11.228), tadpole diagrams such as the one of Fig. 11.11.c solely contribute to the renormalization of the mass of the corresponding leg and do not affect our finite-volume expressions.

Combining Eqs. (11.236), (11.237) and (11.239), we find, at $O(\lambda)$,

$$C_{K \to \pi\pi}(t_i, t_f; k_n^{(0)}) \longrightarrow -g \left\{ 1 - \frac{\lambda}{8} I_1(k_n^{(0)}) \right\} \frac{e^{-M_K t_i}}{2M_K} \frac{e^{-W_n t_f}}{W_n^2} \times \tag{11.241}$$

$$\times \left\{ 1 - \frac{\lambda}{8} \left(\frac{z_n}{2\pi^2 W_n^{(0)} L} - \frac{\nu_n R_\Lambda \left((k_n^{(0)})^2, (k_n^{(0)})^2 \right)}{L^3} \right) + O(\lambda^2) \right\},$$

where, again, W_n, is given by Eq. (11.222).

Putting it all together

The contributions of the two-pion states, $|\pi\pi\, l\rangle_V$, to $C_{K \to \pi\pi}(t_i, t_f)$ are, in the limit of large t_i and t_f,

$$C_{K \to \pi\pi}(t_i, t_f) \longrightarrow \sum_{l=0}^{6} e^{-M_K t_i - W_l t_f} \langle 0 | \mathcal{O}_n(0) | \pi\pi\, l \rangle_V \, {}_V \langle \pi\pi\, l | \int_{\vec{x}} \mathcal{H}_W(0, \vec{x}) | K \rangle_V$$

$$\times \, {}_V \langle K | K(0) | 0 \rangle + \cdots, \tag{11.242}$$

where the ellipsis stands for terms which decay more rapidly. In Eq. (11.242), $|K\rangle_V$ is a zero-momentum state and, again, all states are normalized to 1. With these normalizations,

$$_V \langle K | K(0) | 0 \rangle = \sqrt{\frac{1}{2M_K L^3}}. \tag{11.243}$$

Combining this matrix element with Eqs. (11.241)–(11.242) and the result of Eq. (11.228) for $|\langle 0 | \mathcal{O}_n(0) | \pi\pi\, n \rangle_V|$, we find

$$|M| = g \left\{ 1 - \frac{\lambda}{8} I_1(k_n^{(0)}) \right\} \frac{1}{2} \sqrt{\frac{\nu_n}{M_K W_n^2 L^3}} \left\{ 1 - \frac{\lambda}{4} \frac{1}{W_n^{(0)} L} \left[\frac{z_n}{4\pi^2} + \frac{\nu_n}{(W_n^{(0)} L)^2} \right] \right.$$

$$\left. + O(\lambda^2) \right\}, \tag{11.244}$$

Now, a straightforward evaluation of the infinite-volume analogs of the diagrams of Fig. 11.11 yields:

$$T = -g \left\{ 1 - \frac{\lambda}{8} \int_{\vec{p}} \left[\frac{1}{E_p (\vec{p}^2 - (k_n^{(0)})^2 - i\epsilon)} - R_\Lambda \left(\vec{p}^2, (k_n^{(0)})^2 \right) \right] \right\}. \tag{11.245}$$

Therefore,

$$\frac{|T|}{|M|} = 2\sqrt{\frac{M_K W_n^2 L^3}{\nu_n}} \left\{ 1 + \frac{\lambda}{4} \frac{1}{W_n^{(0)} L} \left[\frac{z_n}{4\pi^2} + \frac{\nu_n}{(W_n^{(0)} L)^2} \right] + O(\lambda^2) \right\}, \qquad (11.246)$$

which is in perfect agreement with the result of Eq. (11.234), predicted by the finite-volume formula of Eq. (11.208). In obtaining Eq. (11.246), we have used the fact that $|T| = |\mathrm{Re}\, T| + O(\lambda^2)$.

$K \to \pi\pi$ in finite volume: physical kaon decays

For illustration, let us suppose that the S-wave scattering phases δ_I, $I = 0, 2$, are accurately described by the one-loop formulas of chiral perturbation theory (Gasser and Meissner, 1991; Knecht *et al.*, 1995). The two-pion energy spectrum can then be calculated in the isospin I channel and in a box of size L where level $n = 1$ coincides with the kaon mass. With this input, the proportionality factor in Eq. (11.208) is easily evaluated and one ends up with (cf. Table 11.1)

$$|A_0| = 44.9 \times |M_0|, \qquad (11.247)$$

$$|A_2| = 48.7 \times |M_2|, \qquad (11.248)$$

$$|A_0/A_2| = 0.92 \times |M_0/M_2|. \qquad (11.249)$$

As these results show, the large difference between the scattering phases in the two isospin channels (about $45°$ at $k = k_\pi$) does not lead to a big variation in the proportionality factors. In fact, if we set the scattering phases to zero altogether, Eqs. (11.210)–(11.211) give $|A_I| = 47.7 \times |M_I|$ for $n = 1$, which is not far from the results quoted above. This may be surprising at first sight, since the interactions of the pions in the spin and isospin 0 state are quite strong. However, one should take into account the fact that the comparison is made for box sizes L which are greater than 5 fm. Hence, it is quite plausible that the finite-volume matrix elements already include most of the final-state interaction effects. Apart from a purely kinematic factor, only a small correction is then required to obtain the infinite-volume matrix elements, from the finite-volume ones.

The proportionality factor in Eq. (11.208) thus appears to be only weakly dependent on the final-state interactions. In particular, if the theory is to reproduce the $\Delta I = 1/2$ enhancement, the large factor has to come from the ratio of the finite-volume

Table 11.1 Calculation of the proportionality factor in Eq. (11.208) at the first level crossing.

I	L [fm]	q	$q\partial\phi/\partial q$	$k\partial\delta_I/\partial k$
0	5.34	0.89	4.70	1.12
2	6.09	1.02	6.93	−0.09

matrix elements M_I. In fact, if you carry out this calculation, you should see an 8% enhanced $\Delta I = 1/2$ enhancement!

To carry out this calculation you will have to address a couple of issues which have not yet been discussed here. The first is that, in the absence of twisted boundary conditions, at least the first excited $\pi\pi$ energy will have to be extracted from the lattice calculation. This requires cross-correlator techniques constructed from operators such as the one given in Eq. (11.224) and solving the resulting generalized eigenvalue problem (GEVP), as described by Lüscher and Wolff (1990). The second issue is that of the renormalization of the lattice matrix elements. This is an important, but fairly technical problem, which depends sensitively on the fermion discretization used. It is usually referred to as the *ultraviolet* problem, as opposed to the *infrared* problem which we dealt with here, which is associated with the continuation of the theory to Euclidean space-time and the use of a finite volume in numerical simulations. Unfortunately I will not have the time to cover the ultraviolet problem here. This problem has been studied quite extensively, and I refer you to the original literature, as well as to the lectures of Peter (Weisz, 2010) and Tassos (Vladikas, 2010) in this volume. For Wilson fermions, which explicitly break chiral symmetry, but retain a full flavor symmetry, the problem has been studied in the following series of papers: Bochicchio *et al.* (1985); Maiani *et al.* (1987); Bernard *et al.* (1988); Dawson *et al.* (1998). For twisted-mass QCD, the reference is Frezzotti and Rossi (2004). When considering domain-wall fermions with a finite fifth dimension, one should follow the renormalization set forth for Wilson fermions. However, for domain-wall fermions the required subtractions should be significantly smaller, since the chiral symmetry breaking should be significantly suppressed compared to what it is for Wilson fermions. Regarding discretizations which have the full, continuum chiral-flavor symmetry at finite lattice spacing (e.g. overlap fermions, or domain-wall fermions with a practically infinite fifth dimension), the renormalization is much simplified and will proceed as in the continuum. Finally, for staggered fermions, these issues are discussed by Sharpe *et al.* (1987); Sharpe and Patel (1994).

To conclude, if you wish to be the first particle theorist to unambiguously see the $\Delta I = 1/2$ rule in $K \to \pi\pi$ decays and determine ϵ' with controlled errors, I hope that you began working on the problem immediately after the course was given. If not, you would do better to hurry because the RBC-UKQCD collaboration is making quick progress on these problems (Christ, 2010*a*; Liu, 2010; Sachrajda, 2010*b*).

11.4   Appendix: integral representation for $Z_{00}(1; q^2)$

Here we derive an integral representation for the zeta function $Z_{00}(1; q^2)$ of Eq. (11.185), which is a meromorphic function of q^2, with poles at $q^2 = \vec{n}^2$, $\vec{n} \in \mathbb{Z}^3$. This representation is particularly effective for evaluating numerically the kinematic function $\phi(q)$ that appears in Martin's two-particle momentum quantization formula Eq. (11.183). It differs from the one given in Appendix C of Lüscher (1991).

The definition of $Z_{00}(s; q^2)$ is given by Eq. (11.185):

$$Z_{00}(s; q^2) = \frac{1}{\sqrt{4\pi}} \sum_{\vec{n} \in \mathbb{Z}^3} \left(\vec{n}^2 - q^2 \right)^{-s}.$$

We define

$$Z_{00}^{>\Lambda}(s; q^2) = \frac{1}{\sqrt{4\pi}} \sum_{\vec{n}^2 > \Lambda} (\vec{n}^2 - q^2)^{-s}, \qquad (11.250)$$

where the sum runs over all $\vec{n} \in \mathbb{Z}^3$ such that $\vec{n}^2 > \Lambda$ with $\Lambda \geq \operatorname{Re} q^2$. For $\operatorname{Re} s > 0$,

$$Z_{00}^{>\Lambda}(s; q^2) = \frac{1}{\Gamma(s)} \sum_{\vec{n}^2 > \Lambda} \int_0^\infty dt\, t^{s-1}\, e^{-t(\vec{n}^2 - q^2)} \qquad (11.251)$$

$$= \frac{1}{\Gamma(s)} \sum_{\vec{n} \in \mathbb{Z}^3} \int_0^1 dt\, t^{s-1}\, e^{-t(\vec{n}^2 - q^2)} + \Delta(s; q^2),$$

with

$$\Delta(s; q^2) = \frac{1}{\Gamma(s)} \left\{ \sum_{\vec{n}^2 > \Lambda} \int_1^\infty dt - \sum_{\vec{n}^2 \leq \Lambda} \int_0^1 dt \right\} t^{s-1} e^{-t(\vec{n}^2 - q^2)}, \qquad (11.252)$$

where the second sum runs over all $\vec{n} \in \mathbb{Z}^3$ such that $\vec{n}^2 \leq \Lambda$.

To evaluate the sum in Eq. (11.252), we use a Dirac comb:

$$\sum_{\vec{n} \in \mathbb{Z}^3} f(\vec{n}) = \sum_{\vec{n} \in \mathbb{Z}^3} \int d^3x\, f(\vec{x})\, \delta^{(3)}(\vec{x} - \vec{n})$$

$$= \sum_{\vec{n} \in \mathbb{Z}^3} \int d^3x\, f(\vec{x})\, e^{i2\pi \vec{n} \cdot \vec{x}}. \qquad (11.253)$$

With $f(\vec{n}) = e^{-t\vec{n}^2}$ and

$$\int_{-\infty}^{+\infty} dx\, e^{-tx^2 + i2\pi nx} = \sqrt{\frac{\pi}{t}}\, e^{-\frac{\pi^2 n^2}{t}}, \qquad (11.254)$$

Eq. (11.253) yields

$$\sum_{\vec{n} \in \mathbb{Z}^3} e^{-t\vec{n}^2} = \sum_{\vec{n} \in \mathbb{Z}^3} \left(\frac{\pi}{t}\right)^{3/2} e^{-\frac{\pi^2 \vec{n}^2}{t}}. \qquad (11.255)$$

Thus

$$Z_{00}^{>\Lambda}(s; q^2) = \frac{1}{\Gamma(s)} \sum_{\vec{n} \in \mathbb{Z}^3} I(s; \vec{n}) + \Delta(s; q^2), \qquad (11.256)$$

with

$$I(s; \vec{n}) = \pi^{3/2} \int_0^1 dt\, t^{s-5/2} e^{tq^2 - \frac{\pi^2 \vec{n}^2}{t}}. \qquad (11.257)$$

Viewed as a function of $\vec{n}$, $I(s; \vec{n})$ is singular only for $\vec{n} = \vec{0}$ when $\text{Re}\, s \leq 3/2$. In that case, I's integrand goes like $t^{s-1-3/2}(1 + tq^2 + O(t^2))$ when $t \to 0$. Thus, $\text{Re}\, s > 3/2$ yields the half-plane in complex s for which the expression of Eq. (11.256) gives a finite result. For such s we can write

$$I(s; \vec{0}) = \pi^{3/2} \int_0^1 dt\, t^{s-5/2} \left(e^{tq^2} - 1 \right) + \frac{\pi^{3/2}}{s - 3/2}, \qquad (11.258)$$

which is actually well defined for $\text{Re}\, s > 1/2$ and $s \neq 3/2$. Thus, for all s in the half-plane $\text{Re}\, s > 1/2$, we obtain

$$\sqrt{4\pi} Z_{00}(s; q^2) = \sum_{\vec{n}^2 \leq \Lambda} (\vec{n}^2 - q^2)^{-s} + \frac{\pi^{3/2}}{\Gamma(s)} \left\{ \frac{1}{s - 3/2} + \int_0^1 dt\, t^{s-5/2}(e^{tq^2} - 1) \right.$$

$$\left. + \sum_{\vec{n}^2 \neq 0} \int_0^1 dt\, t^{s-5/2} e^{tq^2 - \frac{\pi^2 \vec{n}^2}{t}} \right\} + \Delta(s; q^2), \qquad (11.259)$$

where the last runs over all $\vec{n} \in \mathbb{Z}^3$ such that $\vec{n}^2 \neq 0$.

Now, for the case $s = 1$ which is of interest to us here, it is straightforward to compute $\Delta(s; q^2)$:

$$\Delta(1; q^2) = \left\{ \sum_{\vec{n}^2 > \Lambda} \int_1^\infty dt - \sum_{\vec{n}^2 \leq \Lambda} \int_0^1 dt \right\} e^{-t(\vec{n}^2 - q^2)}$$

$$= \sum_{\vec{n}^2 > \Lambda} \frac{e^{-(\vec{n}^2 - q^2)}}{\vec{n}^2 - q^2} + \sum_{\vec{n}^2 \leq \Lambda} \frac{e^{-(\vec{n}^2 - q^2)} - 1}{\vec{n}^2 - q^2}. \qquad (11.260)$$

Using this result in the expression of Eq. (11.259) for $Z_{00}(s; q^2)$ with $s = 1$, we obtain:

$$Z_{00}(1; q^2) = -\pi + \frac{1}{\sqrt{4\pi}} \sum_{\vec{n} \in \mathbb{Z}^3} \frac{e^{-(\vec{n} - q^2)}}{\vec{n} - q^2} + \frac{\pi}{2} \int_0^1 \frac{dt}{t^{3/2}} (e^{tq^2} - 1)$$

$$+ \frac{\pi}{2} \sum_{\vec{n} \neq \vec{0}} \int_0^1 \frac{dt}{t^{3/2}} e^{tq^2 - \frac{\pi^2 \vec{n}}{t}}$$

$$= -\pi + \frac{1}{\sqrt{4\pi}} \sum_{m=0}^\infty \nu_m \frac{e^{-(m - q^2)}}{m - q^2} + \frac{\pi}{2} \int_0^1 \frac{dt}{t^{3/2}} (e^{tq^2} - 1)$$

$$+ \frac{\pi}{2} \sum_{m=1}^\infty \nu_m \int_0^1 \frac{dt}{t^{3/2}} e^{tq^2 - \frac{\pi^2 m}{t}}, \qquad (11.261)$$

where ν_m counts the $\vec{n} \in \mathbb{Z}^3$ such that $\vec{n}^2 = m$ (see Eq. (11.212)). Eq. (11.261) is the integral representation that we were after. Using this representation, it is straightforward to calculate $Z_{00}(1; q^2)$ numerically with good efficiency and high precision, using standard integration routines.

Acknowledgments

I am indebted to my fellow organizers for an enjoyable collaboration in preparing this Summer School and for unanimously designating me to give the traditional public lecture! A school is only as good as the students who attend it are, and I would like to thank them for their strong motivation, unrelenting questioning, and ability to put together great parties seven nights a week. I am also grateful to the other teachers for preparing excellent lectures which were profitable, not only for the students. Moreover, I wish to thank Leticia Cugliandolo for her masterful direction of the École des Houches, and to thank Brigitte Rousset and Murielle Gardette for their seamless running of the program. Finally, the help of Antonin Portelli and Alberto Ramos in preparing many of the Feynman diagrams in these notes, as well as the careful reading of the manuscript by Jérome Charles, Marc Knecht, Thorsten Kurth, Eduardo de Rafael, and Alberto Ramos, are gratefully acknowledged. This work is supported in part by EU grant MRTN-CT-2006-035482 (FLAVIAnet), CNRS grant GDR 2921 and a CNRS "formation permanente" grant.

Notes

1. CPT is research unit UMR 6207 of the CNRS and of the universities of Aix-Marseille I, Aix-Marseille II and SudToulon-Var; it is also affiliated with the FRUMAM (FR 2291).
2. Beyond fixed-order perturbation theory, the $U(1)_Y$ of hypercharge is *trivial:* the renormalized coupling constant vanishes when the cutoff of the regularized theory is taken to infinity, a notion first suggested by Wilson and Kogut (1974).
3. If one sticks to perturbation theory, the precision reached is actually limited by the fact that perturbative expansions in field theory are typically asymptotic expansions. Moreover, the triviality of the Higgs and $U(1)_Y$ sectors means that the cutoff, which we generically call Λ here, has to be kept finite. This limits the accuracy of predictions through the presence of regularization dependent corrections which are proportional to powers of E/Λ, where E is an energy typical of the process studied. In that sense, only asymptotically free theories can be fundamental theories since they are the only ones that can be used to describe phenomena at arbitrarily high energies.
4. Remember that we have separated out into $\mathcal{L}_\nu$ possible renormalizable neutrino mass terms.
5. The factor of 4 in $4/M_\pi$ is a conservative rule-of-thumb estimate which guarantees that finite-volume corrections to stable hadron masses, proportional to $e^{-M_\pi L}$, are typically below the percent level.

6. This was certainly true at the time of the school, but the situation has been moving quite fast since then. Please see Christ (2010*a*); Liu (2010); Sachrajda (2010*a*) for an update.

7. Here and in the following, we will assume that CPT is conserved. We will also work in the strong isospin symmetry limit.

8. The CP transformation of the neutral kaon states is chosen here to be $CP|K^0(p)\rangle = -|\bar{K}^0(p_P)\rangle$ and $CP|\bar{K}^0(p)\rangle = -|K^0(p_P)\rangle$, where p_P is the parity transformed four-momentum $p = (p^0, -\vec{p})$. In the following, we will ignore the momentum labels unless they play a relevant role.

9. In many phenomenological studies, ϕ_ϵ is fixed to $\pi/4$ and $\mathrm{Im}A_0/\mathrm{Re}A_0$ is neglected. However, at the levels of accuracy currently reached in the computation of the local contributions to $\mathrm{Im}M_{12}$, these approximations are becoming too crude. This has been emphasized by Buras and Guadagnoli (2008), where the implications of an estimate of $\mathrm{Im}A_0/\mathrm{Re}A_0$ and of the deviation of ϕ_ϵ from $\pi/4$ have been investigated. Moreover, as explained by Buras *et al.* (2010), if $\mathrm{Im}A_0/\mathrm{Re}A_0$, which approximates $-\mathrm{Im}\Gamma_{12}/2\mathrm{Re}\Gamma_{12}$, is included in Eq. (11.84), consistency requires that one also account for long-distance contributions to $\mathrm{Im}M_{12}$.

10. Equation (11.96) and the amplitudes associated with the other contributing diagrams must be multiplied by $1/2$ if their spinor factors are replaced by operators to yield the effective $\Delta S = 2$ Hamiltonian. Indeed, the operator $(\bar{d}_L\gamma_\mu s_L)(\bar{d}_L\gamma^\mu s_L)$ has twice as many contractions with the four external quark states as there really are. These extra contractions correspond to a doubling of the individual box diagrams.

11. For the sake of clarity, we neglect here the small contribution to ϵ from $\mathrm{Im}A_0/\mathrm{Re}A_0$ (see Eq. (11.84)).

12. Here again we choose to neglect the small contribution to ϵ from $\mathrm{Im}A_0/\mathrm{Re}A_0$. Computing it on the lattice is a whole other project which is related to the computation of ϵ'. That computation goes beyond the presentation I wish to make here.

13. Note that the K^0 does not actually decay leptonically: the $K^\pm$ do. However, in the isospin limit, the decay constant defined in Eq. (11.118) is equal to the physical decay constant f_K of the $K^\pm$.

14. This is because the longest correlation length in the system is $1/M_\pi$ so that regions separated by that distance should be reasonably decorrelated.

15. CPS violating terms are proportional to $(m_s - m_d)$, so that $O_{VV+AA}(a)$ can only mix with higher dimensional operators which are suppressed by powers of the lattice spacing and the latter can only contribute discretization errors.

16. In terms of $SU(3)_L \times SU(3)_R$ representations, O_- is pure $(8, 1)$ while O_+ contains both $(8, 1)$ and $(27, 1)$ representations.

17. If you have been reading these notes carefully, you will be quick to point out that this statement has its limitations. Indeed, beyond LLO, the running of the Wilson coefficients is scheme dependent. However, it is difficult to justify not going beyond that order at scales $\mu \sim 2\,\mathrm{GeV}$. So the statement should be be understood as applying to commonly used schemes.

18. L must be large enough so that the range of the interaction between the two pions is contained within the box.

19. Because the cube is not invariant under generic rotations, the irreducible representations of the cubic group are resolved into irreducible representations of $O(3)$. For the A_1^+ cubic representation, the relevant spin representations are spin $0, 4, \ldots$.

20. We assume here that the narrowness of the resonance is not only due to phase-space suppression.

21. Note that δ_l^I's subscript in Eq. (11.205) corresponds to angular momentum l, while the superscript is the isospin I. This notation will be used in this equation only. Elsewhere, δ's subscript will be the isospin, except for δ_W where W stands for weak.

22. The factor of 2 is required because the two pions in the final state are identical particles in the isospin limit which we consider here.

23. Here and in the following, quantities with the superscript (0) are computed at $O(\lambda^0)$, while those without a superscript are the same quantities in the presence of the "strong" interaction of Eq. (11.214).

Bibliography

Alavi-Harati, A. *et al.* (1999). *Phys. Rev. Lett.*, **83**, 22–27.

Altarelli, G., Curci, G., Martinelli, G., and Petrarca, S. (1981). *Phys. Lett.*, **B99**, 141–146.

Altarelli, G. and Maiani, L. (1974). *Phys. Lett.*, **B52**, 351–354.

Arthur, R. and Boyle, P. A. (2010). arXiv:1006.0422 [hep-lat].

Aubin, C., Laiho, J., and Van de Water, R. S. (2010). *Phys. Rev.*, **D81**, 014507.

Babich, R. *et al.* (2006). *Phys. Rev.*, **D74**, 073009.

Barbieri, R., Pomarol, A., Rattazzi, R., and Strumia, A. (2004). *Nucl. Phys.*, **B703**, 127–146.

Bernard, C. W., Draper, T., Hockney, G., and Soni, A. (1988). *Nucl. Phys. Proc. Suppl.*, **4**, 483–492.

Bernard, C. W., Draper, T., Soni, A., Politzer, H. D., and Wise, M. B. (1985). *Phys. Rev.*, **D32**, 2343–2347.

Bochicchio, M., Maiani, L., Martinelli, G., Rossi, G. C., and Testa, M. (1985). *Nucl. Phys.*, **B262**, 331–355.

Buchalla, G., Buras, A. J., and Lautenbacher, M. E. (1996). *Rev. Mod. Phys.*, **68**, 1125–1144.

Buras, A. J. *et al.* (2000). *Phys. Lett.*, **B480**, 80–86.

Buras, A. J. and Guadagnoli, D. (2008). *Phys. Rev.*, **D78**, 033005.

Buras, A. J., Guadagnoli, D., and Isidori, G. (2010). *Phys. Lett.*, **B688**, 309–313.

Buras, A. J., Jamin, M., and Weisz, P. H. (1990). *Nucl. Phys.*, **B347**, 491–536.

Buras, A. J., Lautenbacher, M. E., and Ostermaier, G. (1994). *Phys. Rev.*, **D50**, 3433–3446.

Cabibbo, N. (1963). *Phys. Rev. Lett.*, **10**, 531–533.

Charles, J. *et al.* (2005). *Eur. Phys. J.*, **C41**, 1–131. ICHEP 2010 update at `http://ckmfitter.in2p3.fr`.

Christ, N. (2010*a*). Kaon physics from lattice QCD. Talk given at *Future Directions in Lattice Gauge Theory*, 19 July–13 August, 2010, CERN, Geneva, Switzerland.

Christ, N. H. (2010*b*). arXiv:1012.6034 [hep-lat].

Christenson, J. H., Cronin, J. W., Fitch, V. L., and Turlay, R. (1964). *Phys. Rev. Lett.*, **13**, 138–140.

Ciuchini, M., Franco, E., Martinelli, G., and Silvestrini, L. (1996). *Phys. Lett.*, **B380**, 353–362.

Colangelo, G. *et al.* (2010). arXiv:1011.4408 [hep-lat].

Dawson, C. *et al.* (1998). *Nucl. Phys.*, **B514**, 313–335.

de Rafael, E. (1995). Chiral Lagrangians and kaon CP violation. Lecture given at *Theoretical Advanced Study Institute in Elementary Particle Physics (TASI 94): CP Violation and the limits of the Standard Model*, Boulder, CO, 29 May–24 June 1994. Published in Boulder TASI 1994:0015-86 (QCD161:T45:1994), `hep-ph/9502254`.

Dimopoulos, P. *et al.* (2010). arXiv:1012.3355 [hep-lat].

Donini, A., Gimenez, V., Giusti, Leonardo, and Martinelli, G. (1999). *Phys. Lett.*, **B470**, 233–242.

Dürr, S. *et al.* (2008). *Science*, **322**, 1224–1227.

Dürr, S. *et al.* (2010*a*). arXiv:1011.2403 [hep-lat].

Dürr, S. *et al.* (2010*b*). arXiv:1011.2711 [hep-lat].

Dürr, S. et al. (2010*c*). *Phys. Rev.*, **D81**, 054507.

Englert, F. and Brout, R. (1964). *Phys. Rev. Lett.*, **13**, 321–322.

Fanti, V. *et al.* (1999). *Phys. Lett.*, **B465**, 335–348.

Frezzotti, R. and Rossi, G. C. (2004). *JHEP*, **10**, 070.

Gaillard, M. K. and Lee, Benjamin W. (1974*a*). *Phys. Rev. Lett.*, **33**, 108–111.

Gaillard, M. K. and Lee, B. W. (1974*b*). *Phys. Rev.*, **D10**, 897–916.

Gasser, J. and Meissner, U. G. (1991). *Phys. Lett.*, **B258**, 219–224.

Gell-Mann, M. and Levy, M. (1960). *Nuovo Cim.*, **16**, 705–726.

Gilman, F. J. and Wise, M. B. (1979). *Phys. Rev.*, **D20**, 2392–2407.

Gilman, F. J. and Wise, M. B. (1983). *Phys. Rev.*, **D27**, 1128–1141.

Giusti, L. *et al.* (2007). *Phys. Rev. Lett.*, **98**, 082003.

Glashow, S. L., Iliopoulos, J., and Maiani, L. (1970). *Phys. Rev.*, **D2**, 1285–1292.

Golterman, M. (2010). Chapter 8 of this volume.

Hernández, P. (2010). Chapter 1 of this volume.

Higgs, P. W. (1964). *Phys. Rev. Lett.*, **13**, 508–509.

Inami, T. and Lim, C. S. (1981). *Prog. Theor. Phys.*, **65**, 297–314. Erratum-ibid. **65** (1981) 1772.

Jarlskog, C. (1985). *Phys. Rev. Lett.*, **55**, 1039–1042.

Kambor, J., Missimer, J. H., and Wyler, D. (1990). *Nucl. Phys.*, **B346**, 17–64.

Kambor, J., Missimer, J. H., and Wyler, D. (1991). *Phys. Lett.*, **B261**, 496–503.

Kaplan, D. B. (2010). Chapter 4 of this volume.

Kim, C. H. and Sachrajda, C. T. (2010). *Phys. Rev.*, **D81**, 114506.

Kim, C. H., Sachrajda, C. T., and Sharpe, S. R. (2005). *Nucl. Phys.*, **B727**, 218–243.

Knecht, M., Moussallam, B., Stern, J., and Fuchs, N. H. (1995). *Nucl. Phys.*, **B457**, 513–576.

Kobayashi, M. and Maskawa, T. (1973). *Prog. Theor. Phys.*, **49**, 652–657.

Lellouch, L. (2001). Phenomenology of nonleptonic weak decays in the $SU(4)_L \times SU(4)_R$ symmetry limit. Notes (November 2001).

Lellouch, L. (2009). *PoS*, **LATTICE2008**, 015.

Lellouch, L. and Lüscher, M. (2001). *Commun. Math. Phys.*, **219**, 31–44.

Lenz, A. *et al.* (2010). arXiv:1008.1593 [hep-ph].

Li, S. and Christ, N. H. (2008). *PoS*, **LATTICE2008**, 272.

Lin, C. J. D., Martinelli, G., Sachrajda, C. T., and Testa, M. (2001). *Nucl. Phys.*, **B619**, 467–498.

Liu, Q. (2010). Preliminary results of $\Delta I = 1/2$ and $3/2$, $K \to \pi\pi$ decay amplitudes from Lattice QCD. Talk given at *Future Directions in Lattice Gauge Theory*, 19 July–13 August, 2010, CERN, Geneva, Switzerland.

Lubicz, V. (2009). *PoS*, **LAT2009**, 013.

Lüscher, M. (1986). *Commun. Math. Phys.*, **105**, 153–188.

Lüscher, M. (1991). *Nucl. Phys.*, **B354**, 531–578.

Lüscher, M. and Wolff, U. (1990). *Nucl. Phys.*, **B339**, 222–252.

Maiani, L., Martinelli, G., Rossi, G. C., and Testa, M. (1987). *Nucl. Phys.*, **B289**, 505–534.

Maiani, L. and Testa, M. (1990). *Phys. Lett.*, **B245**, 585–590.

Martinelli, G., Pittori, C., Sachrajda, C. T., Testa, M., and Vladikas, A. (1995). *Nucl. Phys.*, **B445**, 81–108.

Nakamura, K. *et al.* (2010). *J. Phys.*, **G37**, 075021.

Osterwalder, K. and Schrader, R. (1973). *Commun. Math. Phys.*, **31**, 83–112.

Osterwalder, K. and Schrader, R. (1975). *Commun. Math. Phys.*, **42**, 281–305.

Pallante, E. and Pich, A. (2001). *Nucl. Phys.*, **B592**, 294–320.

Pich, A. and de Rafael, E. (1996). *Phys. Lett.*, **B374**, 186–192.

Sachrajda, C. (2010*a*). *PoS*, **LATTICE2010**, 018.

Sachrajda, C. T. (2010*b*). $K \to (\pi\pi)_{I=2}$ decay amplitudes. Talk at *Future Directions in Lattice Gauge Theory*, 19 July–13 August, 2010, CERN, Geneva, Switzerland.

Sharpe, S. R. and Patel, A. (1994). *Nucl. Phys.*, **B417**, 307–356.

Sharpe, S. R., Patel, A., Gupta, R., Guralnik, G., and Kilcup, G. W. (1987). *Nucl. Phys.*, **B286**, 253–292.

't Hooft, G. (1974). *Nucl. Phys.*, **B72**, 461–470.

Testa, M. (1998). *JHEP*, **04**, 002.

Testa, M. (2005). *Lect. Notes Phys.*, **663**, 177–197.

Vainshtein, A. I., Zakharov, Valentin I., Novikov, V. A., and Shifman, Mikhail A. (1977*a*). *Phys. Rev.*, **D16**, 223–230.

Vainshtein, A. I., Zakharov, Valentin I., Novikov, V. A., and Shifman, M. A. (1977*b*). *Sov. J. Nucl. Phys.*, **23**, 540–543.

Vladikas, A. (2010). Chapter 3 of this volume.

Weisz, P. (2010). Chapter 2 of this volume.

Wilson, K. G. and Kogut, J. B. (1974). *Phys. Rept.*, **12**, 75–200.

Witten, E. (1977). *Nucl. Phys.*, **B122**, 109–143.

Wolfenstein, L. (1983). *Phys. Rev. Lett.*, **51**, 1945–1947.

12

Lattice gauge theory beyond the Standard Model

Thomas APPELQUIST and Ethan T. NEIL

Department of Physics, Sloane Laboratory, Yale University New Haven,
Connecticut, 06520, USA

Overview

Lattice gauge theory has been very successful in deepening our understanding of the strong nuclear interactions. During the past two years, stimulated to some extent by the start-up of the Large Hadron Collider, interest is growing in applying lattice methods to new, strongly interacting theories that could play a role in extending the standard model. This chapter focuses on the use of lattice methods to study strongly coupled gauge theories with application to models of dynamical electroweak symmetry breaking.

Other applications of lattice methods to beyond-Standard-Model (BSM) physics are also being pursued, for example to supersymmetric theories. Since supersymmetry is inextricably connected to the Poincaré symmetry group, which is broken to a discrete subgroup on the lattice, the construction of a supersymmetric theory on the lattice is quite challenging. However, some progress has been made; recently, Catterall and collaborators (Catterall *et al.*, 2009) have demonstrated a formulation that in d space-time dimensions preserves 1 out of 2^d supersymmetries exactly on the lattice, analogous to the preservation of a subset of chiral symmetry in the staggered fermion formulation. In addition to supersymmetric extensions of the Standard Model, supersymmetric theories are often studied in the context of dualities. In particular, the AdS/CFT duality (Maldacena, 1998; Witten, 1998) provides a link between gauge theory and quantum gravity, which opens up the possibility of exploring gravity on the lattice.

Although not a direct study of physics beyond the Standard Model, precision flavor physics calculations on the lattice have become more commonplace. The dominant systematic error for many such quantities comes from QCD contributions, and it is possible that a lattice computation could reduce the error enough to reveal disagreements with Standard Model predictions, leading to important constraints on new physics. For a comprehensive review, see the chapter of Laurent Lellouch.

12.1 Introduction

12.1.1 Electroweak symmetry in the Standard Model

The Standard Model of particle physics describes all of the known fundamental forces (excluding gravity) as arising from the gauge symmetry group

$$\mathcal{G} = SU(3)_c \times SU(2)_L \times U(1)_Y. \tag{12.1}$$

Since the carriers of the weak force are massive, the symmetry group $\mathcal{G}$ must be broken into a subgroup at low energies. This is accomplished by the yet-unknown Higgs sector of the theory. In the simplest version, a complex-doublet scalar field (the Higgs field) breaks the gauge symmetry by acquiring a non-zero vacuum expectation value. The breaking pattern is

$$SU(2)_L \times U(1)_Y \to U(1)_{em}. \tag{12.2}$$

As a fundamental scalar field with a quartic self-interaction, the mass of the Higgs boson is subject to additive renormalization; if the loop momenta are allowed to run up to a very high scale, extremely delicate fine tuning of the bare Higgs mass is required to keep the renormalized mass near the electroweak scale. Also, the physical Higgs boson has yet to be detected experimentally.

The Standard Model also includes QCD, based on the gauge group $SU(3)_c$, which gives rise to the strong nuclear force. In fact, the QCD sector exhibits another well-known example of spontaneous symmetry breaking; the chiral symmetry $SU(2)_L \times SU(2)_R$ is broken into a diagonal subgroup by a non-zero vacuum expectation value for the chiral condensate, $\langle \bar{\psi}\psi \rangle \neq 0$. As the left-handed quarks are charged under the electroweak gauge group, this sector also makes a (small) contribution to electroweak symmetry breaking. However, the bulk of the breaking must come from another source; this is the unknown Higgs sector.

12.1.2 The electroweak chiral Lagrangian

Whatever the Higgs sector is, it must include the three Nambu–Goldstone boson (NGB) fields that provide the longitudinal components of the W and Z gauge bosons. If the new physics in addition to the NGB fields is heavy enough to be integrated out, it leaves an effective low-energy theory, consisting of the transverse gauge bosons, the NGB bosons, and the quarks and leptons. The gauge-boson and NGB sector of this theory is described by a non-linear chiral Lagrangian, whose lowest-dimension operators are (Appelquist and Wu, 1993)

$$\mathcal{L}_\chi = \frac{f^2}{4} \text{Tr}\left[(D^\mu U)^\dagger D_\mu U \right] - \frac{1}{2} \text{Tr}\left[F_{\mu\nu} F^{\mu\nu} \right] - \frac{1}{4} B_{\mu\nu} B^{\mu\nu}, \tag{12.3}$$

where

$$U \equiv \exp\left(\frac{2i}{f} \tau^a \pi^a \right). \tag{12.4}$$

D_μ is the gauge-covariant derivative,

$$D_\mu U = \partial_\mu U - i \frac{g}{2} \tau^a A_\mu^a U + i \frac{g'}{2} U \tau^3 B_\mu, \tag{12.5}$$

where A_μ^a and B_μ are the $SU(2)_L$ and $U(1)_Y$ gauge fields, respectively, and τ^a are the usual Pauli matrices. The kinetic terms for the gauge bosons are defined in the standard way, with

$$F_{\mu\nu}^a = \partial_\mu A_\nu^a - \partial_\nu A_\mu^a + ig[A_\mu^a, A_\nu^a], \tag{12.6}$$

$$B_{\mu\nu} = \partial_\mu B_\nu - \partial_\nu B_\mu. \tag{12.7}$$

In addition to the $SU(2)_L$ and $U(1)_Y$ gauge symmetries, the Lagrangian Eq. 12.3 obeys a global $SU(2)_R$ "custodial" symmetry.

The Lagrangian is non-linear, containing an infinite series of interaction terms between the π^a fields; we will work here to first order in π^a. Expanding Eq. 12.5 and multiplying by the conjugate,

$$|D_\mu U|^2 \approx \left| D_\mu \left(1 + \frac{2i}{f} \tau^a \pi^a \right) \right|^2$$

$$= \frac{4}{f^2} (\partial_\mu \pi^a)^2 + \frac{g^2}{4} (A_\mu^a)^2 + \frac{g'^2}{4} B_\mu^2 - \frac{2g}{f} \partial^\mu \pi^a A_\mu^a + \frac{2g'}{f} \partial^\mu \pi^3 B_\mu - \frac{gg'}{2} A_\mu^3 B^\mu$$

$$= \frac{g^2}{4} \left[(A_\mu^a)^2 - \frac{8}{fg} \partial_\mu \pi^a W^{\mu,a} + \frac{16}{f^2 g^2} (\partial_\mu \pi^a)^2 \right.$$

$$\left. + \frac{g'^2}{g^2} B_\mu^2 - \frac{2g'}{g} A_\mu^3 B^\mu + \frac{8g'}{fg} \partial_\mu \pi^3 B^\mu \right]$$

$$= \frac{g^2}{4} \left[\left(A_\mu^1 - \frac{4}{fg} \partial_\mu \pi^1 \right)^2 + \left(A_\mu^2 - \frac{4}{fg} \partial_\mu \pi^2 \right)^2 + \left(A_\mu^3 - \frac{g'}{g} B_\mu - \frac{4}{fg} \partial_\mu \pi^3 \right)^2 \right].$$

$$(12.8)$$

At this point it is manifestly obvious that if we make the standard field redefinitions

$$W_\mu^{1,2} \equiv A_\mu^{1,2} - \frac{4}{fg} \partial_\mu \pi^{1,2}$$

$$Z_\mu \equiv \frac{g}{\sqrt{g^2 + g'^2}} \left(A_\mu^3 - \frac{g'}{g} B_\mu - \frac{4}{fg} \partial_\mu \pi^3 \right)$$

$$A_\mu \equiv \frac{g}{\sqrt{g^2 + g'^2}} \left(\frac{g'}{g} A_\mu^3 + B_\mu \right) \qquad (12.9)$$

(where Z_μ and A_μ are rescaled so that their kinetic terms will have the standard normalization), then the π^a field is completely removed from the Lagrangian at this order, and the remaining terms from the interaction are simply mass terms for the shifted gauge fields. The fourth gauge degree of freedom A_μ, corresponding to the photon, remains massless as it should. The Lagrangian (again, to lowest order in π^a) is now given by

$$\mathcal{L}_\chi = \mathcal{L}_{\text{kinetic}} + \frac{1}{2} m_W^2 \left[(W_\mu^1)^2 + (W_\mu^2)^2 \right] + \frac{1}{2} m_Z^2 Z_\mu^2 + \mathcal{O}(\pi^3), \qquad (12.10)$$

with

$$m_W = \frac{1}{2} fg \qquad (12.11)$$

$$m_Z = \frac{1}{2} f \sqrt{g^2 + g'^2}. \qquad (12.12)$$

This gives us an immediate prediction for the W/Z mass ratio:

$$\frac{m_W^2}{m_Z^2} = \frac{g^2}{g^2 + g'^2} \equiv \cos^2 \theta_W \simeq 0.77, \qquad (12.13)$$

in agreement with experimental observation.

This success is a consequence of the extra, "custodial" $SU(2)_R$ symmetry built in to the scalar sector of the EW chiral Lagrangian terms so far. The minimal electroweak model with a single elementary Higgs doublet field possesses the same custodial symmetry, and therefore yields the same prediction for m_W^2/m_Z^2. QCD as described by chiral perturbation theory also gives the same prediction, due to the same custodial symmetry, but the Lagrangian above describes a distinct set of fields from the physical pions given by QCD, and at a very different energy scale $\Lambda_{\text{QCD}} \ll f_{ew}$.

Going beyond leading order in the low-energy expansion, there are several additional operators that can be added to the EW chiral Lagrangian (12.3). First, one can define some building blocks that respect the $SU(2) \times U(1)$ electroweak symmetry:

$$T \equiv U\tau^3 U^\dagger, \tag{12.14}$$

$$V_\mu \equiv (D_\mu U)U^\dagger, \tag{12.15}$$

and $B_{\mu\nu}$ and $W_{\mu\nu}$ that were defined above. There are 11 new terms that can be added that are dimension four and CP-invariant (Appelquist and Wu, 1993):

$$\mathcal{L}_1 \equiv \frac{1}{2}\alpha_1 g g' B_{\mu\nu} \text{Tr}(TW^{\mu\nu}) \qquad \mathcal{L}_2 \equiv \frac{1}{2}i\alpha_2 g' B_{\mu\nu} \text{Tr}(T[V^\mu, V^\nu])$$

$$\mathcal{L}_3 \equiv i\alpha_3 g \text{Tr}(W_{\mu\nu}[V^\mu, V^\nu]) \qquad \mathcal{L}_4 \equiv \alpha_4 [\text{Tr}(V_\mu V_\nu)]^2$$

$$\mathcal{L}_5 \equiv \alpha_5 [\text{Tr}(V_\mu V^\mu)]^2 \qquad \mathcal{L}_6 \equiv \alpha_6 \text{Tr}(V_\mu V_\nu)\text{Tr}(TV^\mu)\text{Tr}(TV^\nu)$$

$$\mathcal{L}_7 \equiv \alpha_7 \text{Tr}(V_\mu V^\mu)\text{Tr}(TV_\nu)\text{Tr}(TV^\nu) \qquad \mathcal{L}_8 \equiv \frac{1}{4}\alpha_8 g^2 [\text{Tr}(TW_{\mu\nu})]^2$$

$$\mathcal{L}_9 \equiv \frac{1}{2}i\alpha_9 g \text{Tr}(TW_{\mu\nu})\text{Tr}(T[V^\mu, V^\nu]) \qquad \mathcal{L}_{10} \equiv \frac{1}{2}\alpha_{10}[\text{Tr}(TV_\mu)\text{Tr}(TV_\nu)]^2$$

$$\mathcal{L}_{11} \equiv \alpha_{11} g \epsilon^{\mu\nu\rho\lambda} \text{Tr}(TV_\mu)\text{Tr}(V_\nu W_{\rho\lambda}). \tag{12.16}$$

In addition, there is another dimension-two operator that can be constructed:

$$\mathcal{L}_1' \equiv \frac{1}{4}\beta_1 g^2 f^2 [\text{Tr}(TV_\mu)]^2. \tag{12.17}$$

This operator does not respect the custodial $SU(2)$ symmetry, with the value of β_1 encapsulating any contributions from physics above the cutoff $4\pi f$ that break custodial symmetry.

An attractive UV completion of the EW chiral Lagrangian is *technicolor*. One adds to the Standard Model gauge group a new interaction $SU(N_{TC})$, and N_{TF} massless flavors of *technifermions*. The physics of chiral-symmetry breaking is fixed in terms of the fundamental scale Λ_{TC} of the theory. In a technicolor model, there is no fundamental Higgs particle, with the degrees of freedom that are "eaten" by the W and Z generated as composite Nambu–Goldstone bosons of the spontaneous breaking of chiral symmetry. In the simplest version, the technicolor sector is essentially a copy of the QCD sector: $N_{TC} = 3$, with two flavors of technifermions (U, D) possessing the same quantum numbers as the up and down quarks, except that color charge is exchanged for technicolor charge. In a more general model with N_{TD} technidoublets, the scale of electroweak symmetry breaking $f_{\text{EW}} \sim 246$ GeV$/\sqrt{N_{\text{TD}}}$. With a single technidoublet, by analogy to QCD (where $\Lambda_{\text{QCD}} \sim 300$ MeV and $f_\pi \sim 93$ MeV), the fundamental scale Λ_{TC} (roughly, the scale below which the technicolor gauge interaction becomes "strong") should be somewhere around 1 TeV.

12.1.3   Fermion mass generation and extended technicolor

To replace the Higgs boson in the breaking of electroweak symmetry, one also needs a replacement mechanism to give mass to the various standard model particles. One approach is to add a set of additional gauge interactions that couple the standard model and technicolor-sector fields to one another; this framework is known as *extended technicolor* (Eichten and Lane, 1980; Dimopoulos and Susskind, 1979). In the Higgs case, the masses are given by Yukawa couplings to the Higgs scalar field; here the mass terms must arise through some coupling to the technifermions.

One possible construction takes the extended technicolor gauge group to be $SU(3 + N_{\text{TC}})_{\text{ETC}}$, containing the technicolor gauge group $SU(N_{\text{TC}})$. All of the matter fields are charged under the ETC gauge group. To give rise to the generational structure observed in the Standard Model, the symmetry is then broken down at three distinct scales $\Lambda_1, \Lambda_2, \Lambda_3$:

$$SU(3 + N_{\text{TC}}) \xrightarrow{\Lambda_1} SU(2 + N_{\text{TC}}) \xrightarrow{\Lambda_2} SU(1 + N_{\text{TC}}) \xrightarrow{\Lambda_3} SU(N_{\text{TC}}). \qquad (12.18)$$

The diagram through which a Standard Model quark in generation i becomes massive is depicted in Fig. (12.1). The technifermion condensate $\langle \overline{U}U \rangle \sim \Lambda_{\text{TC}}^3$, while at momentum scales much smaller than Λ_i the ETC gauge boson propagator will contribute roughly $g_{\text{ETC}}^2/M_i^2 \sim 1/\Lambda_i^2$. In full, the mass arising from this diagram is given by (Appelquist *et al.*, 2004)

$$m_q^{(i)} = \frac{8\pi\eta\Lambda_{\text{TC}}^3}{3\Lambda_i^2}. \qquad (12.19)$$

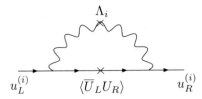

Fig. 12.1   Generation of up-type quark masses in extended technicolor. The mass scale is determined by the scale Λ_i at which the ETC gauge bosons become massive, and by the technifermion condensate $\langle \overline{U}U \rangle$.

The coefficient η represents the RG running of the technifermion condensate between Λ_{TC} and Λ_{ETC},

$$\eta \equiv \exp\left[\int \frac{d\mu}{\mu}\gamma(\mu)\right]. \qquad (12.20)$$

If the anomalous dimension $\gamma(\mu)$ of the condensate is small, then $\eta \sim 1$. The observed up-type quark masses can then be reproduced correctly by the following assignment of the ETC breaking scales, taking $\Lambda_{\text{TC}} = 400$ GeV since $N_{\text{TD}} = 4$ in this model:

$$\Lambda_1 \simeq 530 \text{ TeV}, \Lambda_2 \simeq 21 \text{ TeV}, \Lambda_3 \simeq 1.8 \text{ TeV}. \qquad (12.21)$$

In models with additional technifermion content, Λ_{TC} can be lower, reducing the above scales in turn. The scale Λ_3 associated with generation of the top-quark mass can be as low as 1 TeV, and the lack of scale separation between TC and ETC renders a simple analysis in terms of effective four-fermion interactions invalid. (It should be further noted that the large splitting between the top and bottom quark masses can lead to a large contribution to $\Delta\rho$ (Appelquist *et al.*, 1984; 1985), and that the large effective coupling to the top quark can increase the $Z \to \bar{b}b$ decay rate, allowing some models to be ruled out by existing experimental data (Chivukula *et al.*, 1992). Models that attempt to address these and other issues with the top-quark sector do exist, but we will not discuss them here.)

In order to reproduce fully observed Standard Model physics, an extended technicolor model must include not just the particle masses, but also the mixing effects encapsulated in the CKM matrix. This can be accomplished by adding mixing terms between the ETC gauge bosons (Appelquist *et al.*, 2004). However, these same interactions will also lead to the generation of flavor-changing neutral currents (FCNCs), which are tightly constrained by experiment. In particular, the contribution to the kaon long-short mass difference is given roughly by

$$(\Delta M_K)_{\text{ETC}} \simeq \frac{\text{Re}(V_{ds}^2)}{\Lambda_2^2}f_K^2 M_K < 3.5 \times 10^{-18} \text{ TeV}, \qquad (12.22)$$

which forces the ETC scale $\Lambda_2 \gtrsim 1300$ TeV (Lane, 2000), which in turn forces a quark mass that is far too low.

A way out of this contradiction is through enhancement of the technifermion condensate, which appears in the masses but not in the FCNCs. Enhancement here would come in the form of a large contribution from η, the RG running of the condensate between the TC and ETC scales. In particular, if a theory has the property that the anomalous dimension $\gamma \sim 1$ over the full range of scales to be integrated over, then

$$\eta \xrightarrow{\gamma=1} \frac{\Lambda_i}{\Lambda_{\text{TC}}}, \qquad (12.23)$$

which can provide a large enough enhancement to give the correct masses while escaping FCNC bounds.

12.1.4 The S parameter

There is one further problem that we must keep in mind, relating to precision electroweak measurements. In particular, we are concerned with the S, T, and U parameters (Peskin and Takeuchi, 1992), which describe the presence of physics beyond the standard model; by definition, the parameters are 0 in the Standard Model once the Higgs mass is fixed. Contributions to S are generally the greatest concern for technicolor theories; although the top sector in such models can lead to a large T parameter (equivalent to large $\Delta\rho$) as discussed above, a positive T is much easier to reconcile with current experimental bounds than a positive S.

S is related to the difference between vector–vector and axial–axial vacuum polarization functions $\Pi(q^2)$, evaluated in the limit $q^2 \to 0$. For a general theory described by the electroweak chiral Lagrangian, S is related simply to the parameter α_1 of Eq. (12.16).

We can write down a dispersion relation for the contribution to S of a technicolor theory in terms of the imaginary parts of the vacuum polarizations, $R(s) \equiv -12\pi \, \text{Im} \, \Pi'(s)$:

$$S = \frac{1}{3\pi} \int_0^\infty \frac{ds}{s} \left\{ [R_V(s) - R_A(s)] - \frac{1}{4} \left[1 - \left(1 - \frac{m_h^2}{s} \right)^3 \Theta(s - m_h^2) \right] \right\}. \quad (12.24)$$

In addition to R_V and R_A, this expression contains some additional terms that serve to regulate the integral in the infrared, and subtract off the Standard Model Higgs boson contribution to S at some chosen reference mass m_h.

For a technicolor model that is a scaled-up version of QCD, one can use QCD experimental input to determine the functions $R_V(s)$ and $R_A(s)$ (Peskin and Takeuchi, 1992), yielding the estimate

$$S \approx 0.32(3) \quad (12.25)$$

at reference Higgs mass $m_h = 1$ TeV. Current experimental bounds favor $S < 0$ at the same reference Higgs mass (Amsler *et al.*, 2008), so there is considerable tension here. Furthermore, naive scaling estimates and counting pseudo-NGB loop contributions both indicate that S should only increase with N_{TF} and N_{TC}, suggesting that more elaborate TC models might be in sharper conflict with experimental bounds on S.

However, it should be kept in mind that the estimate Eq. (12.25) relies on QCD phenomenology. In a more general strongly coupled theory, QCD-based estimates do not necessarily apply. In particular, S is sensitive to differences between the vector and axial spectrum of the technicolor theory. In a QCD-like theory, such differences are large, leading to a large contribution as above. If, for example, a given technicolor theory exhibits approximate parity doubling – degeneracy between the vector and axial sectors – then the contribution could be reduced (Appelquist and Sannino, 1999).

We next move on to consider the structure of more general Yang–Mills gauge theories. Our primary motivation is to search for theories that share with QCD the properties of asymptotic freedom and spontaneous chiral symmetry breaking, allowing them to form the basis for a technicolor model, and yet possess different enough dynamical properties that estimates based on QCD phenomenology are no longer valid constraints.

12.2   The conformal window and walking

12.2.1   Perturbative RG flow in Yang–Mills theories

As a strongly coupled gauge theory, QCD is far from unique. The number of colors, number of light fermion flavors, and the gauge-group representation of the fermion fields can all be varied over a wide range of choices while maintaining the crucial property of asymptotic freedom. However, the infrared physics of such theories can be strikingly different from QCD. Consider a Yang–Mills theory with local gauge symmetry group $SU(N_c)$, coupled to N_f massless Dirac fermion flavors:

$$\mathcal{L}_{YM} = -\frac{1}{4g^2} \sum_{a=1}^{N_c} F^a_{\mu\nu} F^{a,\mu\nu} + \sum_{i=1}^{N_f} \bar{\psi}_i (i\slashed{D}) \psi_i. \tag{12.26}$$

The fermions are taken to live in a representation R of the gauge group. The scale dependence of the renormalized coupling $g = g(\mu)$ is determined by the β-function, which we can expand perturbatively:

$$\beta(\alpha) \equiv \frac{\partial \alpha}{\partial(\log \mu^2)} = -\beta_0 \alpha^2 - \beta_1 \alpha^3 - \beta_2 \alpha^4 - ..., \tag{12.27}$$

with $\alpha(\mu) \equiv g(\mu)^2/4\pi$. The universal values for the first two coefficients are

$$\beta_0 = \frac{1}{4\pi} \left(\frac{11}{3} N_c - \frac{4}{3} T(R) N_f \right), \tag{12.28}$$

$$\beta_1 = \frac{1}{(4\pi)^2} \left[\frac{34}{3} N_c^2 - \left(4C_2(R) + \frac{20}{3} N_c \right) T(R) N_f \right], \tag{12.29}$$

where $T(R)$ and $C_2(R)$ are the trace normalization and quadratic Casimir invariant of the representation R, respectively. The Casimir invariants for a few commonly used representations of $SU(N)$ are shown in Table 12.1; a more exhaustive list of invariants and other group-theory factors is given in (Dietrich and Sannino, 2007). So long as $N_f/N_c < 11/(4T(R))$ so that $\beta_0 > 0$, the theory is asymptotically free. One may continue on to higher order in this expansion, at the cost of specifying a renormalization scheme; in the commonly used $\overline{\text{MS}}$ scheme, the next two coefficients are known (van Ritbergen *et al.*, 1997).

If N_f is taken sufficiently large that $\beta_1 < 0$, but not so large that $\beta_0 < 0$, then in addition to the trivial ultraviolet fixed point $\alpha = 0$, the two-loop β-function admits a second fixed-point solution,

$$\alpha_\star^{(2L)} = -\frac{\beta_0}{\beta_1}. \tag{12.30}$$

Table 12.1 Casimir invariants and dimensions of some common representations of $SU(N)$: fundamental (F), two-index symmetric (S_2), two-index antisymmetric (A_2), and adjoint (G).

Representation	$\dim(R)$	$T(R)$	$C_2(R)$
F	N	$\dfrac{1}{2}$	$\dfrac{N^2-1}{2N}$
S_2	$\dfrac{N(N+1)}{2}$	$\dfrac{N+2}{2}$	$\dfrac{(N+2)(N-1)}{N}$
A_2	$\dfrac{N(N-1)}{2}$	$\dfrac{N-2}{2}$	$\dfrac{(N-2)(N+1)}{N}$
G	N^2-1	N	N

This solution is an infrared-stable fixed point, and if $\alpha_\star^{(2L)}$ is sufficiently weak then the coupling strength will be perturbative at all scales, so that our perturbative approach is self-consistent. This condition will be satisfied if N_f is sufficiently near the value $11N_c/4T(R)$ at which asymptotic freedom is lost (Caswell, 1974; Banks and Zaks, 1982). Since confinement of color charges and spontaneous breaking of chiral symmetry, two defining properties of QCD, are strong-coupling effects, they are absent in a theory that is completely perturbative. These theories instead recover an approximate conformal symmetry in the infrared limit.

As N_f is decreased, the value of the fixed-point coupling increases, quickly causing the two-loop perturbative description to break down. Even so, it is clear that the properties of confinement and chiral symmetry breaking must be recovered at sufficiently small N_f as we approach QCD and the quenched $(N_f = 0)$ limit. Theories within the range

$$N_f^c < N_f < \frac{11N_c}{4T(R)}, \tag{12.31}$$

where N_f^c marks the transition point between conformal and confining infrared behavior, are said to lie in the "conformal window". Perturbation theory is almost certainly unreliable to describe physics in the vicinity of the infrared fixed point as N_f approaches N_f^c, so study of the transition region requires some non-perturbative approach, which in turn demands the choice of a renormalization scheme. The question of scheme dependence of any results should always be kept in mind when dealing with quantities such as the running coupling, although any physical, measurable predictions should certainly be independent of the scheme.

12.2.2 Infrared conformality and walking behavior

Theories that lie inside the conformal window, although interesting in their own right, are not generally useful in describing electroweak symmetry breaking due to the lack of chiral-symmetry breaking (although it is possible to trigger chiral-symmetry breaking even in the conformal window by including explicit fermion-mass terms, as in (Luty, 2009; Evans *et al.*, 2010).) Our focus will be on theories that lie outside the window but close to the transition, that may show interesting and novel dynamics while still breaking chiral-symmetry breaking spontaneously.

Suppose that within some scheme, there is a critical coupling α_c, which when exceeded will trigger the spontaneous breaking of chiral symmetry. Now consider a theory with a beta function (in the same scheme) such that the coupling is approaching a somewhat supercritical fixed point $\alpha_\star > \alpha_c$. As the coupling gets near $\alpha_\star$, the magnitude of the β-function approaches zero, and the running slows down. However, when α_c is exceeded, confinement and chiral-symmetry breaking set in. The fermions that were responsible for the existence of the fixed point develop masses and are screened out of the theory, causing the coupling to run as in the $N_f = 0$ theory below the generated mass scale. The situation is depicted in Fig. (12.2).

This idea is known as "walking technicolor". The "plateau" depicted in Fig. (12.2) results in a separation of scale between the UV physics (in the context of ETC models, Λ_{ETC}) where the coupling runs perturbatively and the IR scale (Λ_{TC}) at which confinement sets in. This dynamical scale separation is exactly what is needed in order to address the FCNC problem in extended technicolor. The conflict there was between trying to simultaneously match the Standard Model particle masses and suppress FCNC-generating effects, both of which are tied to the same ETC scale Λ_i. With walking, the ultraviolet-sensitive condensate can pick up a large additional contribution from the scales between Λ_{TC} and Λ_{ETC}, allowing recovery of Standard Model masses without violation of precision electroweak experimental bounds.

While walking technicolor offers a solution to the difficulties with technicolor models, the existence of such a theory is speculative. The onset of walking is a strong-

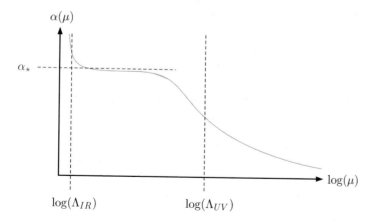

Fig. 12.2 Sketch of the running coupling evolution in a "walking" theory. The coupling (solid line) is nearly constant as it approaches the would-be fixed point at $\alpha_\star$. However, chiral symmetry breaks at $\alpha_c < \alpha_\star$, and the theory confines. Note the large separation of scales between $\Lambda_{\mathrm{IR}}(\sim \Lambda_{\mathrm{TC}})$ and $\Lambda_{\mathrm{UV}}(\sim \Lambda_{\mathrm{ETC}}.)$

coupling effect, so that perturbative methods are unlikely to be useful in locating a walking theory. The search for walking is closely linked to more general questions about the location and nature of the conformal transition at $N_f = N_f^c$.

A quasi-perturbative approach sometimes cited in the old literature is the "ladder gap equation" or "rainbow gap equation" method, which requires the use of a truncation that is not justifiable at strong coupling. In addition, the standard use of Landau gauge in this calculation raises further questions of gauge dependence. We summarize the approach here on a cautionary note.

One begins with the standard Schwinger–Dyson equation for the fermion propagator. In order to make progress, the "rainbow approximation" is then applied, which replaces the full gluon propagator and gluon–fermion vertex with their tree-level values. At weak coupling, the Feynman diagrams that are dropped in this approach are suppressed by small α, so that it can be regarded as an approximation to the full equation. However, if the coupling is sufficiently strong then the procedure becomes an unjustified truncation.

Proceeding under the "rainbow approximation", one may show that a non-zero solution for the fermion dynamical mass $\Sigma(p^2)$ exists once the coupling strength exceeds a certain value, $\alpha_C = \pi/(3C_2(R))$. If the corresponding effective potential, again constructed using the same "rainbow" truncations, is evaluated at the extremal solution for $\Sigma(p^2)$, it can be seen that this non-zero solution is energetically preferred to $\Sigma(p^2) = 0$, implying that chiral symmetry is spontaneously broken.

Because the estimated critical coupling is strong ($\alpha_c C_2(R) \sim 1$), the truncation used in order to determine it is unjustified. In spite of this, the further exercise of combining the estimated α_c with some determination of the fixed-point coupling strength $\alpha_\star(N_f)$ is often performed in order to "estimate" N_f^c. If the two-loop perturbative value of $\alpha_\star$ is used, then setting $\alpha_c = \alpha_\star(N_f^c)$ yields the values $N_f^c \sim 7.9$ at $N_c = 2$, and $N_f^c \sim 11.9$ at $N_c = 3$. There is little reason to expect these values to be robust against higher-order perturbative corrections.

12.2.3 The thermal inequality

We turn now to a more general, conjectured constraint on the structure of field theories and the value of N_f^c, known as the ACS thermal inequality (Appelquist *et al.*, 1999). This inequality connects the UV and IR degrees of freedom directly. In this way, the thermal inequality is similar in spirit to the t'Hooft anomaly-matching condition ('t Hooft *et al.*, 1980).

The basic proposal is simple: the number of relevant infrared degrees of freedom should be not larger than the number of ultraviolet degrees of freedom. In other words, the RG "blocking" transformation that takes us from UV to IR should never increase the number of degrees of freedom. Consider the thermodynamic degrees of freedom, which are encapsulated by the free energy of the theory:

$$F(T) \equiv -\frac{1}{\beta V} \log \mathcal{Z}. \tag{12.32}$$

In a free field theory with N_s scalars, N_v vector particles and N_f Dirac fermions (all massless), the free energy is given by

$$F_{\text{free}}(T) = -\frac{\pi^2 T^4}{90}\left[N_s + 2N_v + \frac{7}{8}4N_f\right].\tag{12.33}$$

More generally, one can define a function $f(T)$ that "counts" the relevant degrees of freedom:

$$f(T) \equiv \frac{90}{\pi^2 T^4}F(T).\tag{12.34}$$

For the case that the limits of $f(T)$ as $T \to 0$ and $T \to \infty$ are both finite, the proposed constraint is

$$\lim_{T\to 0} f(T) \equiv f_{\text{IR}} \leq f_{\text{UV}} \equiv \lim_{T\to\infty} f(T).\tag{12.35}$$

In the context of two-dimensional conformal field theories, Zamolodchikov's c-theorem (Zamolodchikov, 1986) makes the analogous statement rigorous: the free energy as determined by the central charge c is always reduced under RG flow from an ultraviolet to an infrared fixed point.

As an example, consider the case of an $SU(N_c)$ gauge theory with N_f fermion flavors in the fundamental representation, with N_f low enough that the theory is in the chirally broken phase. In the ultraviolet, the relevant degrees of freedom are the gluons, of which there are $(N_c^2 - 1)$, and the $N_f N_c$ fermions. Assuming that the $SU(N_f) \times SU(N_f)$ chiral symmetry is broken spontaneously to $SU(N_f)$, the relevant degrees of freedom describing the theory in the infrared will be the $N_f^2 - 1$ massless Goldstone bosons. Thus,

$$f_{\text{UV}} = 2(N_c^2 - 1) + \frac{7}{8}\cdot 4N_f N_c\tag{12.36}$$

$$f_{\text{IR}} = N_f^2 - 1\tag{12.37}$$

$$f_{\text{IR}} \leq f_{\text{UV}} \Rightarrow N_f \leq 4N_c\left(1 - \frac{16}{81N_c^2}\right)^{1/2}.\tag{12.38}$$

For $N_f \gtrsim 4N_c$, the inequality would be violated, and the chiral symmetry must therefore be unbroken. Assuming this picture is correct, one finds the restriction $N_f^c \lesssim 4N_c$, which, curiously, is nearly saturated by the gap-equation estimate in the fundamental representation noted above.

As a final observation, we note that at $N_c = 2$, the reality or pseudoreality of the group representations can lead to very different patterns of chiral-symmetry breaking from the standard case at $N_c \geq 3$ (Peskin, 1980). In particular, for N_f fermions in the pseudoreal fundamental representation, the global chiral symmetry group is now $SU(2N_f)$, and the breaking pattern preserving the largest symmetry group is $SU(2N_f) \to Sp(2N_f)$. The dimensions of the groups $SU(N)$ and $Sp(N)$ are $N^2 - 1$ and $N(N + 1)/2$, respectively, so we find that

$$f_{\text{UV}} = 2(4-1) + \frac{7}{8} \cdot 8N_f \qquad (12.39)$$

$$f_{\text{IR}} = (2N_f)^2 - 1 - \frac{1}{2}(2N_f)(2N_f + 1) \qquad (12.40)$$

$$= 2N_f^2 - N_f - 1. \qquad (12.41)$$

Then, assuming the chiral symmetry to be broken,

$$f_{\text{IR}} \le f_{\text{UV}} \Rightarrow N_f \lesssim 4.7. \qquad (12.42)$$

This estimate is in sharp contrast with the gap-equation calculation, which puts N_f^c just below 8. A lattice study in the range $5 < N_f < 8$ for the $N_c = 2$ case could thus be quite interesting.

12.3   Lattice studies of the conformal transition

12.3.1   Overview

Lattice field theory provides an ideal way to study the conformal transition, and more generally the properties of Yang–Mills theories at various N_f. Lattice simulations are truly non-perturbative, although the continuum limit must be taken carefully to recover information about continuum physics (the ability to take this limit being made possible by the asymptotic freedom of the theories in which we are interested.) Lattice simulations allow broad investigation; a large number of different observables can be computed simultaneously on a single set of gauge configurations.

Although the dynamics of the theories we are interested in are different from QCD, at the level of the Lagrangian the changes needed to investigate the conformal window are relatively minor. The addition of extra degenerate fermion flavors simply requires the insertion of multiple copies of the chosen fermion action into the simulation code. Modification of the color gauge group or fermion representation is less trivial, but the required changes should represent a small fraction of the overall size of the simulation code, with many of the algorithms and constructions used in QCD remaining applicable.

The relatively low barrier of entry to simulations of these theories can make the undertaking of such studies quite attractive for current practitioners of lattice QCD. However, as in the continuum, great care must be taken to avoid relying too heavily on intuition and experience based on working with QCD; new techniques will be needed to deal with the presence of widely separated scales. Furthermore, the expense of the simulations themselves often requires state-of-the-art computational resources, with the addition of many extra degrees of freedom increasing the cost greatly.

The following subsections will detail three different approaches to lattice study of the conformal transition that have been widely used; although most of the simulations carried out to date fall into one of these three categories, the list is not exhaustive, and new approaches are under development. In Section 12.3.2 below, we describe simulation methods that focus on the extraction of a running coupling from the lattice, with the goal of directly locating an infrared fixed point in the β-function. Section 12.3.3 discusses an alternative approach, that involves searching for a thermal phase transition that appears as a physical transition in confining theories but as a lattice artifact within the conformal window. Finally, in Section 12.3.4, we detail a different approach in which the spectrum and chiral properties of theories at various N_f are computed, with the goal of determining the N_f dependence of various observables as the conformal transition is approached from below.

12.3.2 Running coupling methods

The concept of the conformal window is usually first introduced by way of the running coupling and the β-function (as it was in this chapter). As such, a natural application of lattice simulation to investigate the transition is by the direct computation of a running coupling constant and determination of the β-function non-perturbatively.

The first step in extracting the β-function is to select a non-perturbative definition of a running coupling that can be measured on the lattice. There are a number of such choices possible, including the standard extraction of the static potential from Wilson loops, the Schrödinger functional (Lüscher *et al.*, 1992; Sint, 1994; Bode *et al.*, 2000; Appelquist *et al.*, 2008; 2009), the twisted Polyakov loop scheme (Bilgici *et al.*, 2009*b*), and constructions using ratios of Wilson loops (Bilgici *et al.*, 2009*a*; Fodor *et al.*, 2009). The Monte Carlo Renormalization Group scheme provides another approach for determining RG evolution (Hasenfratz, 2009).

Regardless of the chosen definition of the running coupling, the end goal is to map out its evolution over a large range of distance scales R. If one works at a fixed lattice spacing a, then the range of available R at which the coupling can be measured is quite small, with the computational expense quickly becoming prohibitive even in QCD. The problem is exacerbated in a theory with large N_f, where the size of the β-function is small; to go from weak to strong coupling, a change in scale of many orders of magnitude is often required. To achieve the goal, then, some way must be found to match together lattice measurements of the running coupling taken at different lattice spacings and combine them into an overall measurement of continuum evolution. A technique known as *step scaling* (Lüscher *et al.*, 1991; Caracciolo *et al.*, 1995) provides a systematic approach.

Step scaling is a recursive procedure that describes the evolution of the coupling constant $g(R)$ as the scale changes from $R \to sR$, where s is a numerical scaling factor known as the *step size*. The relation between the coupling at these two scales in the continuum is defined through the *step-scaling function*,

$$\sigma(s, g^2(R)) \equiv g^2(sR). \tag{12.43}$$

The step-scaling function σ is simply a discrete version of the usual continuum β-function, both of which describe the evolution of the coupling as a function of the coupling strength. In a lattice calculation, the step-scaling function that we extract also contains lattice artifacts in the form of a/R corrections. The lattice version of the step-scaling function is denoted by Σ, and it is related to the continuum σ by extrapolation of the lattice spacing a to zero:

$$\sigma(s, g^2(R)) = \lim_{a \to 0} \Sigma(s, g^2(R), a/R). \tag{12.44}$$

(We are neglecting finite-volume effects here, which must also be dealt with in a lattice simulation for most definitions of running coupling on a case-by-case basis.)

Generically, the implementation of step scaling begins with the choice of an initial value $u = g^2(R)$. Several ensembles at different a/R are then generated, tuning the lattice bare coupling $\beta = 2N_c/g_0^2$ so that on each ensemble one measures the chosen value of the renormalized coupling, $g^2(R) = u$. Then, one generates a second ensemble at each β, but measures the coupling at a longer scale $R \to sR$. The value of the coupling measured on the second lattice is exactly $\Sigma(s, u, a/R)$. Since $\Sigma(s, u, a/R)$ has now be computed for multiple values of a/R, one can carry out the extrapolation $a/R \to 0$ and recover the continuum value $\sigma(s, u)$. Taking $\sigma(s, u)$ to be the new starting value, the procedure is repeated, mapping $g^2(R) \to g^2(sR)... \to g^2(s^n R)$ until the coupling is sampled over a large range of R values.

There is a natural caveat on the step-scaling procedure, especially in the context of studying theories with infrared fixed points. The procedure as outlined above depends crucially on the ability to take the limit $a/R \to 0$. If $g^2(R)$ is held fixed while taking the limit $a/R \to 0$, it is important that the bare coupling $g_0^2(a/R)$, which depends on the short-distance behavior of the theory, does not become strong enough to trigger a bulk phase transition. This is satisfied automatically if the short-distance behavior is determined by asymptotic freedom, in which case $g_0^2(a/R)$ vanishes as $1/\log(R/a)$. However, in a theory with an infrared fixed point, if $g^2(R)$ is measured above the fixed point at $g_\star^2$, then $g_0^2(a/R)$ will increase as $a \to 0$, with no evidence that it remains bounded and therefore that a continuum limit exists. Even so, it is possible to extrapolate to small enough values of a/R to render lattice artifact corrections negligible, providing that $g_0^2(a/R)$ is kept small enough to avoid triggering a bulk transition into a strong-coupling phase.

Carrying out step scaling through the iterative procedure described can be quite expensive in both computational power and real time, especially in theories where many steps are required to see significant evolution in the coupling. Each tuning of β to match the initial value u may require several attempts. Furthermore, each step depends on the value of $g^2(R)$ taken from the previous step, removing the ability to parallelize the problem. A more efficient method is to measure $g^2(R)$ for a wide range of values in β and R/a, and then to generate an interpolating function. Step scaling may then be done analytically using the interpolated values. Such an interpolating function should reproduce the perturbative relation $g^2(R) = g_0^2 + \mathcal{O}(g_0^4)$ at weak coupling, but otherwise its form is not strongly constrained. One possible choice is an expansion of the inverse coupling $1/g^2(\beta, R/a)$ as a set of polynomial series in the bare coupling $g_0^2 = 2N_c/\beta$ at each R/a:

$$\frac{1}{g^2(\beta, R/a)} = \frac{\beta}{2N_c} \left[1 - \sum_{i=1}^{n} c_{i,R/a} \left(\frac{2N_c}{\beta} \right)^i \right]. \tag{12.45}$$

The order n of the polynomial is arbitrary, and can be varied as a function of R/a to achieve the optimal fit to the available data.

The remainder of this section will focus in detail on the use of one formulation of a running coupling, the Schrödinger functional (SF), which is particularly robust with respect to finite-volume effects. The SF running coupling is defined through the response of a system to variation in strength of a background chromoelectric field. It is a finite-volume method, with the coupling strength defined at the spatial box size L, so that we identify $R = L$ and can discard finite-volume corrections. Formally, the Schrödinger functional describes the quantum-mechanical evolution of some system from a given state at time $t = 0$ to another given state at time $t = T$, in a spatial box of size L with periodic boundary conditions (Lüscher *et al.*, 1992; Sint, 1994; Bode *et al.*, 2001). The temporal extent T is fixed proportional to L, so that the Euclidean box size depends only on a single parameter. The initial and final states are described as Dirichlet boundary conditions that are imposed at $t = 0$ and $t = T$, and for measurement of the coupling constant are chosen such that the minimum-action configuration is a constant chromoelectric background field of strength $O(1/L)$. This can be implemented both in the continuum (Lüscher *et al.*, 1992) and on the lattice (Bode *et al.*, 2000).

We can represent the Schrödinger functional as the path integral

$$\mathcal{Z}[W, \zeta, \overline{\zeta}; W', \zeta', \overline{\zeta}'] = \int [DAD\psi D\overline{\psi}] e^{-S_G(W, W') - S_F(W, W', \zeta, \overline{\zeta}, \zeta', \overline{\zeta}')}, \quad (12.46)$$

where A is the gauge field and ψ, $\overline{\psi}$ are the fermion fields. W and W' are the boundary values of the gauge fields, and $\zeta, \overline{\zeta}, \zeta', \overline{\zeta}'$ are the boundary values of the fermion fields at $t = 0$ and $t = T$, respectively. The fermionic boundary values are subject only to multiplicative renormalization (Sommer, 2006), and as such are generally taken to be zero in order to simplify the calculation.

As noted above, the gauge boundary fields W, W' are chosen to given a constant chromoelectric field in the bulk, whose strength is of order $1/L$ and controlled by a dimensionless parameter η (Lüscher *et al.*, 1994). The Schrödinger functional (SF) running coupling is then defined by the response of the action to variation of η:

$$\frac{k}{\overline{g}^2(L, T)} = -\frac{\partial}{\partial \eta} \log \mathcal{Z} \bigg|_{\eta=0}, \quad (12.47)$$

where (with the standard choice of gauge boundary fields for $SU(3)$), the normalization factor k is

$$k = 12 \left(\frac{L}{a}\right)^2 \left[\sin\left(\frac{2\pi a^2}{3LT}\right) + \sin\left(\frac{\pi a^2}{3LT}\right)\right]. \quad (12.48)$$

The presence of k ensures that $\overline{g}^2(L, T)$ is equal to the bare coupling g_0^2 at tree level in perturbation theory. In general, $\overline{g}^2(L, T)$ can be thought of as the response of the system to small variations in the background chromoelectric field.

For most fermion discretizations, at this point we can take $T = L$ in order to define the running coupling as a function of a single scale, $\overline{g}^2(L)$. However, if staggered fermions are used (as they are often in order to offset the cost of simulating additional fermion flavors), then an additional complication arises that can be envisioned geometrically. The staggered approach to fermion discretization can be formulated as splitting the 16 spinor degrees of freedom available up over a 2^4 hypercubic sublattice. Clearly, such a framework requires an even number of lattice sites in all directions. If all boundaries are periodic or antiperiodic, then setting $T = L$ can be done so long as L is even. However, with Dirichlet boundaries in the time direction, the site $t = T$ is no longer identified with $t = 0$, so that a total of $T/a + 1$ lattice sites must exist. In order to accommodate staggered fermions, T/a must be odd.

Thus, when using staggered fermions, the closest we can come to our desired choice of T is $T = L \pm a$. In the continuum limit, the desired relation $T = L$ is recovered. However, at finite lattice spacing $O(a)$ lattice artifacts are introduced into observables. This is especially undesirable, since staggered fermion simulations contain bulk artifacts only at $O(a^2)$ and above. Fortunately, there is a solution: simulating at both choices $T = L \pm a$ and averaging over the results has been shown to eliminate the induced $O(a)$ bulk artifact in the running coupling (Heller, 1997). We define $\bar{g}^2(L)$ through the average:

$$\frac{1}{\bar{g}^2(L)} = \frac{1}{2} \left[\frac{1}{\bar{g}^2(L, L-a)} + \frac{1}{\bar{g}^2(L, L+a)} \right]. \qquad (12.49)$$

To be precise, we can only write $\bar{g}^2(L)$ unambiguously in the continuum; on a lattice with spacing a and bare coupling g_0 given by $\beta \equiv 2N_c/g_0^2$, the coupling that we measure can be written as $\bar{g}^2(\beta, L/a)$.

As a case study of many of the above concepts and formulas in use, we now show some simulation data and results from a Schrödinger functional running coupling study of the $SU(3)$ fundamental, $N_f = 8$ and 12 theories, using staggered fermions (Appelquist *et al.*, 2009). We begin with the $N_f = 8$ theory, for which data was gathered in the range $4.55 \leq \beta \leq 192$ on lattice volumes given by $L/a = 6, 8, 12, 16$. The lower limit on β was determined to keep the lattice coupling too weak to trigger a bulk phase transition. A selection of the data, together with interpolating function fits of the form Eq. (12.45), are shown in Fig. 12.3. Note that at any fixed value of β, the coupling strength $\bar{g}^2(L)$ increases with L/a, showing no evidence of the "backwards" running that we would expect to observe in a theory with an infrared fixed point.

Although the study of Fig. 12.3 is indicative that the $N_f = 8$ theory lies outside the conformal window, it is possible for results at fixed β to be misleading; we must take the continuum limit in order to recover information about the continuum theory. We apply the step-scaling procedure detailed above in order to extract the continuum step-scaling function $\sigma(2, u)$, by extrapolation of $a/L \to 0$ with each doubling of the scale L. A selection of extracted values for the lattice step-scaling function $\Sigma(2, u, a/L)$ are shown in Fig. 12.4 for the two available steps $6 \to 12$ and $8 \to 16$. Our results for $\sigma(2, u)$ will depend on the choice of continuum extrapolation, i.e. the model function for a/L dependence of $\Sigma(2, u, a/L)$. As this is a staggered fermion study, the leading bulk lattice artifacts are expected to be of $O(a^2)$, but there are additional boundary artifacts of $O(a)$ that are only partially cancelled off by subtraction of their perturbative values. However, in this case the a/L dependence is weak, with the associated systematic error

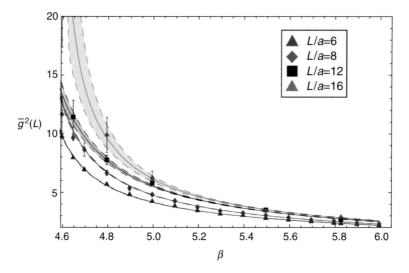

Fig. 12.3 Measured values $\bar{g}^2(L)$ versus β for $N_f = 8$. The interpolating curves shown represent the best fit to the data, using the functional form Eq. (12.45). The errors are statistical.

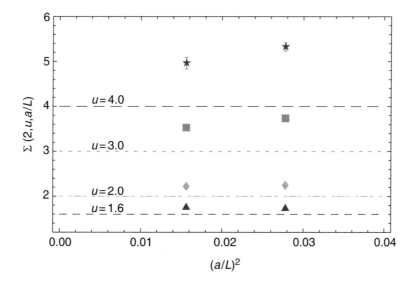

Fig. 12.4 Step-scaling function $\Sigma(2, u, a/L)$ at various u, for each of the two steps $L/a = 6 \to 12$ and $8 \to 16$ used in the $N_f = 8$ analysis. Note that $\Sigma(2, u, a/L) > u$ in each case, with the difference increasing as u increases.

dominated by the statistical errors on the points, so that a constant extrapolation (i.e. weighted average of the two points) is used to extract $\sigma(2, u)$ here.

The resulting continuum running of $\bar{g}^2(L)$ for $N_f = 8$ is shown in Fig. 12.5. L_0 is an arbitrary length scale here defined by the condition $\bar{g}^2(L) = 1.6$, anchoring the step-scaling curve at a relatively weak value. The points shown correspond to repeated doubling of the scale L relative to L_0. Derivation of statistical errors uses a bootstrap technique, and is described in detail in the reference (Appelquist *et al.*, 2009). Perturbative running at two and three loops is also shown for comparison up through $\bar{g}^2(L) \approx 10$, beyond which the accuracy of perturbation theory is expected to degrade. The coupling measured in this simulation follows the perturbative curve closely up through $\bar{g}^2(L) \approx 4$, and then begins to increase more rapidly, reaching values that exceed typical estimates of the coupling strength needed to induce spontaneous chiral-symmetry breaking (for example, the gap equation estimate of Section 12.2 above.) As there is no evidence for an infrared fixed point, or even for an inflection point in the running of $\bar{g}^2(L)$, this study supports the assertion that the $N_f = 8$ theory lies outside the conformal window.

We now move on to consider the $N_f = 12$ theory. We do not show the data and interpolating fits here, but they are available in the reference. The lattice step-scaling function $\Sigma(2, u, a/L)$ for selected values of u is shown in Fig. 12.6. Three steps are now available: $6 \rightarrow 12$, $8 \rightarrow 16$, and $10 \rightarrow 20$. As above, we choose a constant continuum extrapolation, i.e. weighted average of the three points. Again, we note the sharp contrast with the analogous plot Fig. 12.4 for $N_f = 8$; we observe that $\Sigma(2, u, a/L)$ approaches the starting coupling u as u increases.

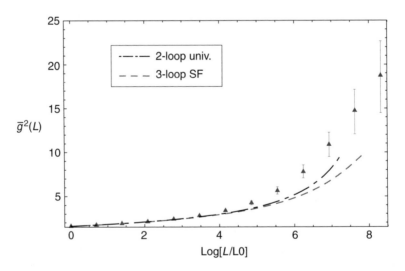

Fig. 12.5 Continuum running for $N_f = 8$. Dater points are derived by step-scaling using the constant continuum-extrapolation of Fig. 12.4. The error bars shown are purely statistical. Two-loop and three-loop perturbation theory curves are shown for comparison.

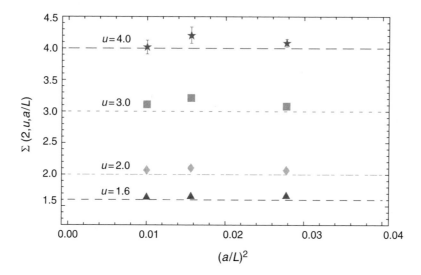

Fig. 12.6 Step-scaling function $\Sigma(2, u, a/L)$ at various u, for each of the three steps $L/a = 6 \to 12$, $8 \to 16$, $10 \to 20$ used in the $N_f = 12$ analysis. Note that $\Sigma(2, u, a/L) \to u$ as the starting coupling u approaches the fixed-point value.

Results for continuum running, again from the starting value $\bar{g}^2(L_0) = 1.6$, are shown in Fig. 12.7. Two-loop and three-loop perturbative curves are shown again for reference. The figure clearly shows the running coupling tracking towards an infrared fixed point, whose exact value lies within the statistical error band and that is consistent with the value predicted by three-loop perturbation theory. It should be noted that the error bars of Fig. 12.7 are highly correlated, with correlation approaching 100% near the fixed point, due to the use of an underlying interpolating function. This causes the error bars to approach a stable value asymptotically, even as we increase the number of steps towards infinity.

The infrared fixed point here also governs the infrared behavior of the theory for values of $\bar{g}^2(L)$ that lie above the fixed point. As discussed previously, we cannot naively apply the step-scaling procedure in this region, since we can no longer approach the ultraviolet fixed point at zero coupling strength in order to take the continuum limit. Instead, we can restrict our attention to finite but small values of a/L, small enough to keep lattice artifacts small and yet large enough so that $g_0^2(a/L)$ does not trigger a bulk phase transition for $\bar{g}^2(L)$ near (but above) the fixed point. With these caveats in mind, the step-scaling procedure can then be applied and leads to the running from above the fixed point shown in Fig. 12.7. The observation of this "backwards-running" region is crucial to distinguishing theories with true infrared fixed points from walking theories, in which the β-function may become vanishingly small before turning over and confining.

Having shown evidence for the existence of an infrared fixed point in the $N_f = 12$ theory and demonstrated its absence up to strong coupling at $N_f = 8$, we have constrained the edge of the conformal window for the case of $N_c = 3$ with fermions in

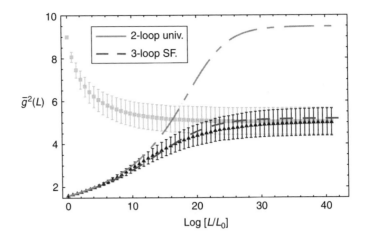

Fig. 12.7 Continuum running for $N_f = 12$. Results shown for running from below the infrared fixed point (triangles) are based on $\bar{g}^2(L_0) \equiv 1.6$. Also shown is continuum backwards running from above the fixed point (squares), based on $\bar{g}^2(L_0) \equiv 9.0$. Error bars are again purely statistical, although strongly correlated due to the underlying interpolating functions. Two-loop and three-loop perturbation theory curves are shown for comparison.

the fundamental rep, $8 < N_f^c < 12$. Similar measurements at other values of N_f can allow us to further constrain N_f^c. Furthermore, if a walking theory exists just below the transition value, a lattice measurement of the scale dependence of the coupling could directly reveal the expected plateau behavior and resulting separation between infrared and ultraviolet scales. In addition, the non-perturbative β function can be used in conjunction with additional lattice measurements to extract the anomalous dimension γ_m of the mass operator (Bursa *et al.*, 2010).

12.3.3 Thermal transition methods

For a pure Yang–Mills gauge theory on the lattice and in the strong-coupling limit, it can always be shown that the theory confines, i.e. that Wilson loops obey area-law scaling (Osterwalder and Seiler, 1978). With the addition of unimproved Wilson or staggered fermions, analysis at strong coupling and large N_c reveals behavior consistent with confinement and spontaneous breaking of chiral symmetry (Blairon *et al.*, 1981; Kluberg-Stern *et al.*, 1981; Kawamoto and Smit, 1981). If a theory with enough fermion content to place it inside the conformal window confines in a simulation performed at strong coupling, to be consistent with the lack of confinement in the continuum or weak-coupling limit the lattice theory must undergo a phase transition as a function of the bare coupling. This phase transition is purely a lattice artifact; in a continuum, asymptotically free Yang–Mills theory with an IR fixed point, the strong-coupling limit can never be reached, but on the lattice the bare coupling may be freely adjusted to any value.

The deconfining phase transition is characterized by a first-order jump in the values of the Wilson line $\langle |W_3\rangle$ and chiral condensate $\langle \overline{\psi}\psi\rangle$. Within the conformal window, it is a "bulk" transition driven entirely by physics at the lattice spacing, and therefore occurs at the same bare coupling as the lattice volume is varied. A similar transition can be observed in theories outside the conformal window, with the same behavior of the observables noted above as the bare coupling is varied. However, in a QCD-like theory this deconfining and chiral-symmetry-restoring transition is driven by infrared physics; the presence of a deconfined phase in a QCD-like theory is a finite-volume or finite-temperature effect. As such, the bare coupling associated with the transition will vary as a function of the box size.

In most lattice simulations investigating the conformal window, such as the running coupling studies detailed in Section 12.3.2 above, the phase transition presents an obstacle to be avoided; the strongly coupled physics on the confining side of the transition cannot tell us anything about the continuum theories in which we are interested. However, the difference in bare-coupling dependence between the bulk phase transition within the conformal window and the thermal transition outside it provides another possible approach to distinguishing such theories from one another through lattice simulation. If the transition can be accurately located as a function of the bare coupling g_0 on a range of lattice volumes, then the presence of variation in the transition coupling with box size will indicate the existence of a continuum thermal phase transition, placing the theory outside of the conformal window.

The choice of fermion action is important in attempting such a phase-transition study. The strong-coupling results referred to above have been demonstrated only for the simplest, unimproved fermions, and are not guaranteed to hold if an improved action is used. Furthermore, even for unimproved Wilson fermions it has been observed that with $N_f \geq 7$ in the SU(3) case, the phase diagram becomes more complicated and extrapolation to zero-current quark mass becomes impossible (Iwasaki *et al.*, 2004). In the examples to follow, we will restrict our attention to staggered fermions, which show no such signs of N_f dependence in their observed strong-coupling behavior (Damgaard *et al.*, 1997). However, with the use of staggered fermions away from the continuum limit, the possibility of taste-breaking effects reducing the effective number of fermion flavors should always be kept in mind.

For a finite-temperature simulation the temperature $T = 1/(N_t a(g_0))$, where N_t is the number of sites in the time direction and $a(g_0)$ is the lattice spacing. In a theory outside the conformal window, there exists a physical deconfining transition temperature T_c that is related to the critical bare coupling $g_{0,c}$ by $T_c = (N_t a(g_{0,c}))^{-1}$. Since T_c is fixed, $g_{0,c}$ should vanish as $N_t \to \infty$, with perturbative scaling of a describing the evolution with N_t so long as $g_{0,c}$ is sufficiently weak. The standard approach is to relate the lattice spacing to the perturbative Λ parameter; applying the two-loop perturbative β-function, one finds

$$a(g_0)\Lambda \propto (\beta_0\alpha_0)^{-\beta_1/2\beta_0^2} \exp\left(\frac{-1}{2\beta_0\alpha_0}\right), \tag{12.50}$$

with $\alpha_0 \equiv g_0^2/4\pi$. If the temporal extent is now changed from N_t to N_t', the expected scaling of the critical bare coupling is determined by the relation

$$a(g'_{0,c}) = \frac{N_t}{N'_t} a(g_{0,c}), \tag{12.51}$$

which can be used in conjunction with Eq. (12.50) above to solve for $g'_{0,c}$ given a value for $g_{0,c}$.

As an example of this technique, Deuzeman and collaborators have studied the theory with $N_f = 8$, $N_c = 3$ and the fermions in the fundamental representation, searching for a thermal phase transition (Deuzeman *et al.*, 2008). Their simulations were carried out using staggered fermions and an improved gauge action, and focused on finite-temperature simulations at $N_t = 6$ and $N_t = 12$, with multiple spatial volumes at $N_t = 6$ used in order to study finite-volume effects. As N_t is increased from 6 to 12, they observe a clear shift in the transition towards weaker bare coupling.

The precise critical coupling values determined are $\beta_c = 4.1125 \pm 0.0125$ at $N_t = 6$, and $\beta_c = 4.34 \pm 0.04$ for $N_t = 12$, where $\beta \equiv 6/g_0^2$. Applying the scaling formula Eq. (12.51) to the value $\beta_c = 4.34 \pm 0.04$ at $N_t = 12$ yields the prediction $\beta_c = 4.04 \pm 0.04$ for $N_t = 6$, which is reasonably close to the observed value of 4.1125 ± 0.0125. Observing roughly the expected perturbative scaling further supports the case that the transition seen corresponds to a thermal deconfining phase transition of the continuum $N_f = 8$ theory.

An early application of this approach to theories potentially within the conformal window was carried out by (Damgaard *et al.*, 1997), using staggered fermions and the conventional Wilson gauge action. Their focus was on the $N_c = 3$ theory with $N_f = 16$ degenerate flavors in the fundamental representation, a case that is unambiguously within the conformal window due to the presence of a very small, perturbative IR fixed point. One therefore expects to find a confining phase transition at a fixed value of the bare coupling that does not change as the extent of the lattice is varied.

This is exactly what Damgaard *et al.* observe, in the chiral condensate (as shown in Fig. 12.8) and other observables; the phase transition that they identify occurs at a fixed critical coupling $\beta_c \sim 4.1$ for all lattice volumes considered. A similar study carried out for the $N_f = 12$ theory again shows a transition that does not vary with box size (Deuzeman *et al.*, 2009), supporting the placement of $N_f = 12$ within the conformal window.

Locating a phase transition and establishing that it does not appear to change as the lattice volume is varied is not definitive evidence that a theory is inside the conformal window. The ability to resolve movement of the transition coupling is limited by computational power, which restricts both the maximum lattice volume L/a and the resolution in β. Order parameters corresponding to short-distance physics, e.g. the plaquette, will show a first-order jump for a bulk transition but not for a physical one, in principle allowing us to distinguish the two cases. However, a given theory measured with a given lattice action can easily have both transitions occur very close to one another, and the two could be easily confused. To definitively place a theory with an apparent bulk phase transition inside the conformal window, some other method must be used. For the $N_f = 16$ theory, Heller went on to measure the Schrödinger functional running coupling, demonstrating the presence of a phase in which the sign of the β-function is reversed, so that the coupling appears to run "backwards" (Heller,

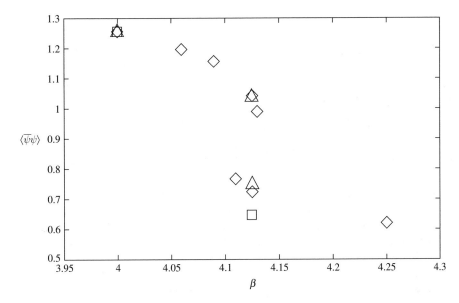

Fig. 12.8 Confining phase transition in the $N_f = 16$ theory as observed in the chiral condensate $\langle \bar{\psi}\psi \rangle$, taken from (Damgaard *et al.*, 1997). Different symbols denote different lattice sizes: $8^3 \times 16$ (diamonds), 12^4 (triangles), and $6^3 \times 16$ (squares). Error bars are within the symbols. The fixed position of the critical coupling β_c as the lattice volume is varied indicates that this is a lattice transition, rather than a continuum thermal phase transition, as expected for a theory within the conformal window.

1998). Such a phase can only exist for a theory within the conformal window, with a β-function that has a non-trivial zero.

12.3.4   Study of spectral and chiral properties

If a given theory is to be considered for a candidate technicolor model, compatibility with experimental bounds generally requires certain effects to be generated by the dynamics; namely, enhancement of the chiral condensate to avoid flavor-changing neutral current problems, and reduction of the S-parameter contribution compared to naive estimates. Aside from such practical concerns, determining the variation with N_f of observables related to chiral symmetry and confinement are important in finding the order of the conformal transition, and novel phenomena such as walking may be revealed as the transition value is approached.

In this subsection, we discuss such studies that focus on the evolution with N_f of various observables on the broken side of the conformal transition. There are several lattice groups attempting to measure correlation functions corresponding to standard QCD bound states and other quantities for theories that are ambiguously or definitely within the conformal window (Fodor *et al.*, 2009; Jin and Mawhinney, 2009; Del Debbio *et al.*, 2009); the interpretation of the lattice results here is much less clear, and we will omit such theories from our discussion here, focusing on the case where chiral symmetry is definitely broken.

In order to meaningfully compare any quantity between theories with different N_f, we must first identify a physical scale to hold fixed. In the context of technicolor theories, a natural choice is the pion decay constant F, which is identified with the scale of electroweak physics through the chiral Lagrangian as described in Section 12.1.2. However, the extraction of F from lattice simulations can be challenging. The rho meson mass m_ρ is much more easily determined, due to the lack of a chiral logarithm at next-to-leading order (NLO) in a χPT-derived fit (Leinweber *et al.*, 2001). However, in the end we are more interested in the evolution of physics with respect to F than m_ρ. In QCD the two scales are connected, $m_\rho \sim 2\pi F$, but it is not known *a priori* whether this connection will persist near the edge of the conformal window. The Sommer scale r_0 (Sommer, 1994), associated with the scale of confinement, is another possible choice with similar advantages and drawbacks to m_ρ. In the present discussion we will assume that these scales do not evolve with respect to each other, so that holding any one constant with N_f is sufficient. This assumption is supported by available data, but the choice of scale is an open question going forward.

As lattice simulations are necessarily performed at finite mass, while we are interested in the behavior of theories in the chiral limit, extrapolation of results $m \to 0$ is crucial. Chiral perturbation theory provides a consistent way to carry out this extrapolation. The familiar expressions for the Goldstone boson mass M_m, decay constant F_m and chiral condensate $\langle \overline{\psi}\psi \rangle_m$ (with the subscript denoting evaluation at finite quark mass m) are easily generalized to theories with arbitrary $N_f \geq 2$ by inclusion of known counting factors. The next-to-leading order (NLO) expressions for a theory with 3 colors are (Gasser and Leutwyler, 1987):

$$M_m^2 = \frac{2m\langle \overline{\psi}\psi \rangle}{F^2} \left\{ 1 + zm \left[\alpha_M + \frac{1}{N_f} \log(zm) \right] \right\}, \tag{12.52}$$

$$F_m = F \left\{ 1 + zm \left[\alpha_F - \frac{N_f}{2} \log(zm) \right] \right\}, \tag{12.53}$$

$$\langle \overline{\psi}\psi \rangle_m = \langle \overline{\psi}\psi \rangle \left\{ 1 + zm \left[\alpha_C - \frac{N_f^2 - 1}{N_f} \log(zm) \right] \right\}, \tag{12.54}$$

where $z = 2\langle \overline{\psi}\psi \rangle/(4\pi)^2 F^4$. These expressions have also been computed at next-to-next-to-leading order (NNLO) and for fermions in the adjoint representation and the pseudoreal representations of the 2-color theory (Bijnens and Lu, 2009).

Each of the unknown coefficients $\alpha_M, \alpha_F, \alpha_C$ also contain terms that grow linearly with N_f. α_C also contains a unique, N_f-independent "contact term" that remains even in the absence of spontaneous chiral-symmetry breaking. This contribution is linear in m, quadratically sensitive to the ultraviolet cutoff (here the lattice spacing a^{-1}.) This term dominates the chiral expansion of $\langle \overline{\psi}\psi \rangle_m$, making numerically accurate extrapolation of the condensate more difficult. Finally, the growth of the chiral log term in F_m with N_f forces the use of increasingly smaller fermion masses m as N_f is increased, in order to keep the NLO terms small enough relative to the leading order so that χPT is trustworthy.

The goal here is to search for enhancement of the condensate relative to the scale F. One way to proceed is to construct the ratio $\langle\overline{\psi}\psi\rangle_m/F_m^3$, and extrapolate directly $m \to 0$; however, as noted above the presence of the contact term can make such an extrapolation difficult to carry out precisely. By making use of the additional quantity M_m^2 and the Gell-Mann–Oakes–Renner (GMOR) relation $M_m^2 F_m^2 = 2m\langle\overline{\psi}\psi\rangle_m$, incorporated into the NLO formulas shown above, we can construct other ratios at finite m that will also extrapolate to $\langle\overline{\psi}\psi\rangle/F^3$ in the chiral limit: the other two possibilities are $M_m^2/(2mF_m)$, and $(M_m^2/2m)^{3/2}/\langle\overline{\psi}\psi\rangle_m^{1/2}$. Due to the contact term in $\langle\overline{\psi}\psi\rangle_m$, $M_m^2/(2mF_m)$ should have the mildest chiral extrapolation of the three ratios.

A lattice study of the type outlined here, investigating the evolution from $N_f = 2$ to $N_f = 6$ in the $SU(3)$ fundamental case, is described in (Appelquist *et al.*, 2010). We refer to this reference for details of the simulation and analysis, and here give only selected details. The physical scales chosen to be matched are the rho mass m_ρ and the Sommer scale r_0^{-1}; each observable was first measured in the $N_f = 6$ case, and then matched by tuning the bare lattice coupling at $N_f = 2$. The resulting chiral extrapolation of these quantities is shown in Fig. 12.9, and shows good agreement, so that the lattice cutoffs are well matched between $N_f = 2$ and $N_f = 6$.

Determination of the presence or absence of condensate enhancement is done through comparison of the quantity $\langle\overline{\psi}\psi\rangle/F^3$ between the $N_f = 6$ and $N_f = 2$ theories, by way of the equivalent ratio $M_m^2/(2mF_m)$. We can directly construct a "ratio of ratios"

$$R_m \equiv \frac{[M_m^2/2mF_m]_{N_f=6}}{[M_m^2/2mF_m]_{N_f=2}}, \tag{12.55}$$

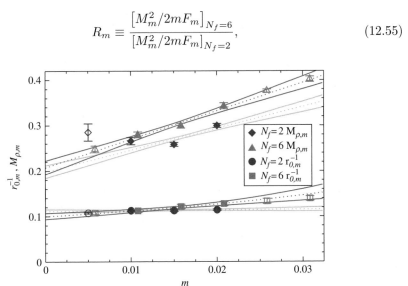

Fig. 12.9 From Appelquist *et al.* (2010). Linear chiral extrapolations of $M_{\rho,m}$ and the Sommer scale $r_{0,m}^{-1}$, in lattice units, based on the (solid) points at $m_f = 0.01 - 0.02$. Both show agreement within error between $N_f = 2$ and $N_f = 6$ in the chiral limit.

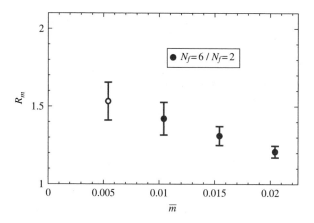

Fig. 12.10 From Appelquist *et al.* (2010). $R_m \equiv [M_m^2/2mF_m]_{6f}/[M_m^2/2mF_m]_{2f}$, versus $\overline{m} \equiv (m(2f) + m(6f))/2$, showing enhancement of $\langle \bar{\psi}\psi \rangle/F^3$ at $N_f = 6$ relative to $N_f = 2$. The open symbol at $m = 0.005$ denotes the presence of possible systematic errors.

A value of $R_m > 1$ then implies enhancement of the condensate as N_f increases. The result is shown in Fig. 12.10, and indicates that $R_m \gtrsim 1.5$ in the chiral limit, barring a downturn in R_m – an unlikely outcome, as the curvature of the NLO logarithm is naturally upwards in the chiral expansion of R_m itself. The magnitude of R_m is significant and larger than expected; an $\overline{\text{MS}}$ perturbation theory estimate of the enhancement from $N_f = 2$ to 6 by integrating the anomalous dimension of the mass operator γ_m leads to an expected increase on the order of 5–10%. Some care must be taken in comparing this value to our lattice result, since the condensate $\langle \bar{\psi}\psi \rangle$ and by extension $\langle \bar{\psi}\psi \rangle/F^3$ depends on the renormalization scheme chosen. The conversion factor $Z^{\overline{\text{MS}}}$ between the lattice-cutoff scheme with domain-wall fermions and $\overline{\text{MS}}$ is known from Ref. (Aoki *et al.*, 2003). From that reference, for this simulation the required factor to convert R_m is $Z_6^{\overline{\text{MS}}}/Z_2^{\overline{\text{MS}}} = 1.449(29)/1.227(11) = 1.18(3)$. This increases the perturbative estimate of expected enhancement to the order of $20 - 30\%$, so the observed $R_m \gtrsim 1.5$ is still significantly larger than anticipated.

A direct computation of the S-parameter is also important to the study of general Yang–Mills theories with a focus on technicolor, and is well within the reach of existing lattice techniques. We will not discuss such a calculation in detail here, but include some references for further reading on the topic (Shintani *et al.*, 2008; Boyle *et al.*, 2010).

References

Amsler, C. *et al.* (2008). Review of particle physics. *Phys. Lett.*, **B667**, 1.

Aoki, Sinya, Izubuchi, Taku, Kuramashi, Yoshinobu, and Taniguchi, Yusuke (2003). Perturbative renormalization factors in domain-wall QCD with improved gauge actions. *Phys. Rev.*, **D67**, 094502.

Appelquist, Thomas, Avakian, Adam, Babich, Ron, Brower, Richard C., Cheng, Michael, Clark, Michael A., Cohen, Saul D., Fleming, George T., Kiskis, Joseph, Neil, Ethan T., Osborn, James C., Rebbi, Claudio, Schaich, David, and Vranas, Pavlos (2010). Toward TeV conformality. *Phys. Rev. Lett.*, **104**, 071601.

Appelquist, Thomas, Bowick, Mark J., Cohler, Eugene, and Hauser, Avi I. (1984). Isospin symmetry breaking in electroweak theories. *Phys. Rev. Lett.*, **53**, 1523.

Appelquist, Thomas, Bowick, Mark J., Cohler, Eugene, and Hauser, Avi I. (1985). The breaking of isospin symmetry in theories with a dynamical Higgs mechanism. *Phys. Rev.*, **D31**, 1676.

Appelquist, Thomas, Cohen, Andrew G., and Schmaltz, Martin (1999). A new constraint on strongly coupled field theories. *Phys. Rev.*, **D60**, 045003.

Appelquist, Thomas, Fleming, George T., and Neil, Ethan T. (2008). Lattice study of the conformal window in QCD-like theories. *Phys. Rev. Lett.*, **100**, 171607.

Appelquist, Thomas, Fleming, George T., and Neil, Ethan T. (2009). Lattice study of conformal behavior in SU(3) Yang-Mills theories. *Phys. Rev.*, **D79**, 076010.

Appelquist, Thomas, Piai, Maurizio, and Shrock, Robert (2004). Fermion masses and mixing in extended technicolor models. *Phys. Rev.*, **D69**, 015002.

Appelquist, Thomas and Sannino, Francesco (1999). The physical spectrum of conformal SU(N) gauge theories. *Phys. Rev.*, **D59**, 067702.

Appelquist, Thomas and Wu, Guo-Hong (1993). The electroweak chiral Lagrangian and new precision measurements. *Phys. Rev.*, **D48**, 3235–3241.

Banks, Tom and Zaks, A. (1982). On the phase structure of vector-like gauge theories with massless fermions. *Nucl. Phys.*, **B196**, 189.

Bijnens, Johan and Lu, Jie (2009). Technicolor and other QCD-like theories at next-to-next- to-leading order. *JHEP*, **11**, 116.

Bilgici, Erek *et al.* (2009*a*). A new scheme for the running coupling constant in gauge theories using Wilson loops. *Phys. Rev.*, **D80**, 034507.

Bilgici, Erek *et al.* (2009*b*). Search for the IR fixed point in the twisted Polyakov loop scheme.

Blairon, J. M., Brout, R., Englert, F., and Greensite, J. (1981). Chiral symmetry breaking in the action formulation of lattice gauge theory. *Nucl. Phys.*, **B180**, 439.

Bode, Achim *et al.* (2001). First results on the running coupling in QCD with two massless flavors. *Phys. Lett.*, **B515**, 49–56.

Bode, Achim, Weisz, Peter, and Wolff, Ulli (2000). Two loop computation of the Schrödinger functional in lattice QCD. *Nucl. Phys.*, **B576**, 517–539. Erratum-ibid.B608:481,2001.

Boyle, Peter A., Del Debbio, Luigi, Wennekers, Jan, and Zanotti, James M. (2010). The S parameter in QCD from domain wall fermions. *Phys. Rev.*, **D81**, 014504.

Bursa, Francis, Del Debbio, Luigi, Keegan, Liam, Pica, Claudio, and Pickup, Thomas (2010). Mass anomalous dimension in SU(2) with two adjoint fermions. *Phys. Rev.*, **D81**, 014505.

Caracciolo, Sergio, Edwards, Robert G., Ferreira, Sabino Jose, Pelissetto, Andrea, and Sokal, Alan D. (1995). Finite size scaling at ζ/L much larger than 1. *Phys. Rev. Lett.*, **74**, 2969–2972.

Caswell, William E. (1974). Asymptotic behavior of nonabelian gauge theories to two loop order. *Phys. Rev. Lett.*, **33**, 244.

Catterall, Simon, Kaplan, David B., and Unsal, Mithat (2009). Exact lattice supersymmetry. *Phys. Rept.*, **484**, 71–130.

Chivukula, R. Sekhar, Selipsky, Stephen B., and Simmons, Elizabeth H. (1992). Nonoblique effects in the Z b anti-b vertex from ETC dynamics. *Phys. Rev. Lett.*, **69**, 575–577.

Damgaard, P. H., Heller, Urs M., Krasnitz, A., and Olesen, P. (1997). On lattice QCD with many flavors. *Phys. Lett.*, **B400**, 169–175.

Del Debbio, L., Lucini, B., Patella, A., Pica, C., and Rago, A. (2009). Conformal vs confining scenario in SU(2) with adjoint fermions. *Phys. Rev.*, **D80**, 074507.

Deuzeman, Albert, Lombardo, Maria Paola, and Pallante, Elisabetta (2008). The physics of eight flavours. *Phys. Lett.*, **B670**, 41–48.

Deuzeman, A., Lombardo, M. P., and Pallante, E. (2009). Evidence for a conformal phase in SU(N) gauge theories.

Dietrich, Dennis D. and Sannino, Francesco (2007). Walking in the SU(N). *Phys. Rev.*, **D75**, 085018.

Dimopoulos, Savas and Susskind, Leonard (1979). Mass without scalars. *Nucl. Phys.*, **B155**, 237–252.

Eichten, Estia and Lane, Kenneth D. (1980). Dynamical breaking of weak interaction symmetries. *Phys. Lett.*, **B90**, 125–130.

Evans, Jared A., Galloway, Jamison, Luty, Markus A., and Tacchi, Ruggero Altair (2010). Minimal conformal technicolor and precision electroweak tests. JMEP, 1010, 086

Fodor, Zoltan, Holland, Kieran, Kuti, Julius, Nogradi, Daniel, and Schroeder, Chris (2009). Nearly conformal gauge theories in finite volume. *Phys. Lett.*, **B681**, 353–361.

Gasser, J. and Leutwyler, H. (1987). Light quarks at low temperatures. *Phys. Lett.*, **B184**, 83.

Hasenfratz, Anna (2009). Investigating the critical properties of beyond-QCD theories using Monte Carlo renormalization group matching. *Phys. Rev.*, **D80**, 034505.

Heller, Urs M. (1997). The Schrödinger functional running coupling with staggered fermions. *Nucl. Phys.*, **B504**, 435–458.

Heller, Urs M. (1998). The Schroedinger functional running coupling with staggered fermions and its application to many flavor QCD. *Nucl. Phys. Proc. Suppl.*, **63**, 248–250.

Iwasaki, Y., Kanaya, K., Kaya, S., Sakai, S., and Yoshie, T. (2004). Phase structure of lattice QCD for general number of flavors. *Phys. Rev.*, **D69**, 014507.

Jin, Xiao-Yong and Mawhinney, Robert D. (2009). Lattice QCD with 8 and 12 degenerate quark flavors. *PoS*, **LAT2009**, 049.

Kawamoto, N. and Smit, J. (1981). Effective Lagrangian and dynamical symmetry breaking in strongly coupled lattice QCD. *Nucl. Phys.*, **B192**, 100.

Kluberg-Stern, H., Morel, A., Napoly, O., and Petersson, B. (1981). Spontaneous chiral symmetry breaking for a U(N) gauge theory on a Lattice. *Nucl. Phys.*, **B190**, 504.

Lane, Kenneth D. (2000). Technicolor 2000. Baston Unverscity Report No. BUHEP-00-5(arXivpreprint number hep-ph/0007304)

Leinweber, Derek Bruce, Thomas, Anthony William, Tsushima, Kazuo, and Wright, Stewart Victor (2001). Chiral behaviour of the rho meson in lattice QCD. *Phys. Rev.*, **D64**, 094502.

Lüscher, Martin, Narayanan, Rajamani, Weisz, Peter, and Wolff, Ulli (1992). The Schrödinger functional: A renormalizable probe for non-Abelian gauge theories. *Nucl. Phys.*, **B384**, 168–228.

Lüscher, Martin, Sommer, Rainer, Weisz, Peter, and Wolff, Ulli (1994). A precise determination of the running coupling in the SU(3) Yang-Mills theory. *Nucl. Phys.*, **B413**, 481–502.

Lüscher, Martin, Weisz, Peter, and Wolff, Ulli (1991). A numerical method to compute the running coupling in asymptotically free theories. *Nucl. Phys.*, **B359**, 221–243.

Luty, Markus A. (2009). Strong conformal dynamics at the LHC and on the lattice. *JHEP*, **04**, 050.

Maldacena, Juan Martin (1998). The large N limit of superconformal field theories and supergravity. *Adv. Theor. Math. Phys.*, **2**, 231–252.

Osterwalder, K. and Seiler, E. (1978). Gauge field theories on the lattice. *Ann. Phys.*, **110**, 440.

Peskin, Michael Edward (1980). The alignment of the vacuum in theories of technicolor. *Nucl. Phys.*, **B175**, 197–233.

Peskin, Michael Edward and Takeuchi, Tatsu (1992). Estimation of oblique electroweak corrections. *Phys. Rev.*, **D46**, 381–409.

Shintani, E. *et al.* (2008). S-parameter and pseudo-Nambu-Goldstone boson mass from lattice QCD. *Phys. Rev. Lett.*, **101**, 242001.

Sint, Stefan (1994). On the Schrodinger functional in QCD. *Nucl. Phys.*, **B421**, 135–158.

Sommer, R. (1994). A New way to set the energy scale in lattice gauge theories and its applications to the static force and alpha-s in SU(2) Yang-Mills theory. *Nucl. Phys.*, **B411**, 839–854.

Sommer, Rainer (2006). Non-perturbative QCD: renormalization, $o(a)$-improvement and matching to heavy quark effective theory.

't Hooft, Gerard, (ed.) *et al.* (1980). Recent developments in gauge theories. Proceedings, Nato Advanced Study Institute, Cargese, France, August 26–September 8, 1979. New York, USA: Plenum (1980) 438 P. (Nato Advanced Study Institutes Series: Series B, Physics, 59).

van Ritbergen, T., Vermaseren, J. A. M., and Larin, S. A. (1997). The four-loop beta function in quantum chromodynamics. *Phys. Lett.*, **B400**, 379–384.

Witten, Edward (1998). Anti-de Sitter space and holography. *Adv. Theor. Math. Phys.*, **2**, 253–291.

Zamolodchikov, A. B. (1986). Irreversibility of the flux of the renormalization group in a 2D field theory. *JETP Lett.*, **43**, 730–732.